BIOMASS

A GROWTH OPPORTUNITY
IN
GREEN ENERGY
AND
VALUE-ADDED PRODUCTS

Proceedings of the 4th Biomass Conference of the Americas

VOLUME 1

(In Two Volumes)

Titles of Related Interest

OVEREND and CHORNET
Making a Business from Biomass

OHTA
Energy Technology - Sources, Systems and Frontier Conversion

ELIASSON, RIEMER and WOKAUN
Greenhouse Gas Control Technologies

RIEMER, SMITH and THAMBIMUTHU
Greenhouse Gas Mitigation - Technologies for Activities Implemented Jointly

Journals of Related Interest
(Sample copy sent on request)

Applied Energy
Biomass & Bioenergy
Bioresource Technology
Energy - The International Journal
Energy Conversion and Management
Renewable Energy
Renewable and Sustainable Energy Reviews
Solar Energy

BIOMASS

**A GROWTH OPPORTUNITY
IN
GREEN ENERGY
AND
VALUE-ADDED PRODUCTS**

Proceedings of the 4th Biomass Conference of the Americas

Oakland Marriott City Center, Oakland, California, USA,
August 29-September 2 1999

Edited by

Ralph P. Overend and Esteban Chornet

Volume 1

Pergamon
An imprint of Elsevier Science

ELSEVIER SCIENCE Ltd
The Boulevard, Langford Lane
Kidlington, Oxford OX5 1GB, UK

© 1999 Elsevier Science Ltd. All rights reserved.

This work is protected under copyright by Elsevier Science, and the following terms and conditions apply to its use:

Photocopying
Single photocopies of single chapters may be made for personal use as allowed by national copyright laws. Permission of the Publisher and payment of a fee is required for all other photocopying, including multiple or systematic copying, copying for advertising or promotional purposes, resale, and all forms of document delivery. Special rates are available for educational institutions that wish to make photocopies for non-profit educational classroom use.

Permissions may be sought directly from Elsevier Science Rights & Permissions Department, PO Box 800, Oxford OX5 1DX, UK; phone: (+44) 1865 843830, fax: (+44) 1865 853333, e-mail: permissions@elsevier.co.uk. You may also contact Rights & Permissions directly through Elsevier's home page (http://www.elsevier.nl), selecting first 'Customer Support', then 'General Information', then 'Permissions Query Form'.

In the USA, users may clear permissions and make payments through the Copyright Clearance Center, Inc., 222 Rosewood Drive, Danvers, MA 01923, USA; phone: (978) 7508400, fax: (978) 7504744, and in the UK through the Copyright Licensing Agency Rapid Clearance Service (CLARCS), 90 Tottenham Court Road, London W1P 0LP, UK; phone: (+44) 171 631 5555; fax: (+44) 171 631 5500. Other countries may have a local reprographic rights agency for payments.

Derivative Works
Tables of contents may be reproduced for internal circulation, but permission of Elsevier Science is required for external resale or distribution of such material.
Permission of the Publisher is required for all other derivative works, including compilations and translations.

Electronic Storage or Usage
Permission of the Publisher is required to store or use electronically any material contained in this work, including any chapter or part of a chapter.

Except as outlined above, no part of this work may be reproduced, stored in a retrieval system or transmitted in any form or by any means, electronic, mechanical, photocopying, recording or otherwise, without prior written permission of the Publisher.
Address permissions requests to: Elsevier Science Rights & Permissions Department, at the mail, fax and e-mail addresses noted above.

Notice
No responsibility is assumed by the Publisher for any injury and/or damage to persons or property as a matter of products liability, negligence or otherwise, or from any use or operation of any methods, products, instructions or ideas contained in the material herein. Because of rapid advances in the medical sciences, in particular, independent verification of diagnoses and drug dosages should be made.

First edition 1999

Library of Congress Cataloging in Publication Data
A catalog record from the Library of Congress has been applied for.

British Library Cataloguing in Publication Data
A catalogue record from the British Library has been applied for.

ISBN: 0 08 043019 8

⊗ The paper used in this publication meets the requirements of ANSI/NISO Z39.48-1992 (Permanence of Paper).
Printed in The Netherlands.

CONTENTS

Volume 1

Preface — xli

Do We Need a Biomass Energy Policy? The California Experience — 1
Gregory Morris, Future Resources Assoc., Inc., and the Green Power Institute

BIOMASS PRODUCTION AND INTEGRATION WITH CONVERSION TECHNOLOGIES

Biomass Resources: Willows

A Profile and Analysis of Willow Growers in Sweden — 11
H. Rosenqvist, A. Roos, E. Ling and B. Hektor, Swedish Univ. of Agricultural Sciences (Uppsala, Sweden)

Soil Organic Carbon Sequestration under Two Dedicated Perennial Bioenergy Crops — 17
B. Mehdi, C. Zan, P. Girouard and R. Samson, REAP-Canada (Québec, Canada)

The Role and Process of Monitoring Willow Biomass Plantations — 25
C.A. Nowak, T.A. Volk, B. Ballard, L.P. Abrahamson, R.C. Filhart, R.F. Kopp, D. Bickelhaupt and E.H. White, State Univ. of New York (Syracuse, New York, USA)

The Influence of Site and Wastewater Sludge Fertilizer on the Growth of Two Willow Species in Southern Québec — 31
M. Labrecque and T.I. Teodorescu, Montreal Botanical Garden (East Sherbrooke, Québec, Canada)

Willow Short Rotation Coppice Profitability Assessment. A Case Study in Belgium 39
F. Goor, J.-M. Jossart and J.-F. Ledent, Laboratory of Main Crop Ecology (Louvain-la-Neuve, Belgium)

Second Rotation Willow Coppice in Upland Wales 45
F.M. Slater, R.J. Heaton, R.W. Samuel and P.F. Randerson, Cardiff Univ. (Powys, Wales)

Short Rotation Coppice of Willow and Shelterbelt Effect 47
J.M. Jossart and J.F. Ledent, Laboratory of Main Crop Ecology (Louvain-la-Neuve, Belgium)

Arborescent Willow Biomass Production in Short Rotations 55
D. Kajba, Faculty of Forestry (Zagreb, Croatia)

Biomass Resources: Regional Analysis

Potential of Short Rotation Wood Crops as a Fiber and Energy Source in the U.S. 63
M.E. Walsh, Oak Ridge National Laboratory (Oak Ridge, Tennessee, USA); P.J. Ince, USDA; Forest Service (Madison, Wisconsin, USA); De la Torre Ugarte, Univ. of Tennessee (Knoxville, Tennessee, USA); D. Adams, Oregon State Univ. (Corvallis, Oregon, USA); R. Alig, USDA Forest Service (Corvallis, Oregon, USA); J. Mills, USDA Forest Service (Portland, Oregon, USA); H. Spelter and K. Skog, USDA Forest Service (Madison, Wisconsin, USA); S.P. Slinsky and D.E. Ray, Univ. of Tennessee (Knoxville, Tennessee, USA); and R.L. Graham, Oak Ridge National Laboratory (Oak Ridge, Tennessee, USA)

Geographic Information System Modeling of Rice Straw Harvesting and Utilization in California 69
L.G. Bernheim, B.M. Jenkins, R.E. Plant and L. Yan, Univ. of California (Davis, California, USA)

Progress towards Making Willow Biomass Crops the Fuel of the Future in the Northeastern United States 75
E.H. White, State Univ. of New York (Syracuse, New York, USA); E.F. Neuhauser, Niagara Mohawk Power Corporation (Syracuse, New York, USA); L.P. Abrahamson, T.A. Volk and C.A. Nowak, State Univ. of New York (Syracuse, New York, USA); J.M. Peterson, NYSERDA

(Albany, New York, USA); *C. Demeter and C. Lindsey*, ANTARES Group, Inc. (Landover, Maryland, USA)

Lessons Learned from Existing Biomass Power Plants 79
G.A. Wiltsee, Appel Consultants, Inc. (Valencia, California, USA)

BIOCOST-Canada: A New Tool to Evaluate the Economic, Energy, and Carbon Budgets of Perennial Energy Crops 85
P. Girouard, Resource Efficient Agricultural Production (Quebec, Canada); *M.E. Walsh*, Oak Ridge National Laboratory (Oak Ridge, Tennessee, USA); *and D.A. Becker*, Science Applications International Corp. (Oak Ridge, Tennessee, USA)

Optimal Locations for Biomass Conversion Facilities in Florida: A Geographic Information System Application 91
M. Rahmani, A.W. Hodges and C.F. Kiker, Univ. of Florida (Gainesville, Florida, USA)

Lignocellulosic Feedstock Resource Assessment 99
T.E. Rooney and S.G. Haase, McNeil Technologies (Golden, Colorado, USA); *and A.E. Wiselogel*, NREL (Golden, Colorado, USA)

The Use of a GIS Model to Evaluate the Economic Potential for Biomass in Northampton County, Pennsylvania 107
D.S. Breger and H. Snyder, Lafayette College (Easton, Pennsylvania, USA)

Herbaceous Crops and Trees Can Provide Bioenergy Needs in Humid Lower South, USA 113
G.M. Prine and D.L. Rockwood, Univ. of Florida (Gainesville, Florida, USA)

Utilization of Biomass Energy Resources in Ohio: A Linear Programming Model 119
B.S. Shakya, Public Utilities Commission of Ohio (Columbus, Ohio, USA); *and D. Southgate*, Ohio State Univ. (Columbus, Ohio, USA)

Biomass Resources: Carbon Impacts

Biomass Crop Production: Benefits for Soil Quality and Carbon Sequestration 127
V.R. Tolbert, Oak Ridge National Laboratory (Oak Ridge, Tennessee, USA); *J.D. Joslin, F.C. Thornton and B.R. Bock*, Tennessee Valley

Authority (Muscle Shoals, Alabama, USA); *D.E. Pettry*, Mississippi State Univ. (Starkville, Mississippi, USA); *W. Bandaranayake, D. Tyler and A. Houston*, Univ. Tennessee (Grand Junction, Tennessee, USA); *and S. Schoenholtz*, Mississippi State Univ. (Starkville, Mississippi, USA)

Woody Biomass Production in Phytoremediation Systems 133
G.R. Alker, D.L. Rockwood, L.Q. Ma, K. Komar and A.E.S. Green, Univ. of Florida (Gainesville, Florida, USA)

Giant Reed (*Miscanthus*) as a Solid Biofuel Crop: Yield, Quality, and Energy Use Efficiency 139
I. Lewandowski, J. Clifton-Brown and U. Schmidt, Univ. of Hohenheim (Stuttgart, Germany)

Reproductive Characteristics and Breeding Improvement Potential of Switchgrass 147
C.M. Taliaferro, Oklahoma State Univ. (Stillwater, Oklahoma, USA); *K.P. Vogel*, USDA-ARS (Lincoln, Nebraska, USA); *J.H. Bouton*, Univ. of Georgia (Athens, Georgia, USA); *S.B. McLaughlin and G.A. Tuskan*, Oak Ridge National Laboratory (Oak Ridge, Tennessee, USA)

Growth, Water, and Radiation Use Efficiency of Kenaf (*Hibiscus cannabinus* L.) Cultivated in the Mediterranean Conditions 155
N. Losavio, D. Ventrella, N. Lamascese and A.V. Vonella, Istituto Sperimentale Agronomico (Bari, Italy)

Introduction of *Miscanthus*: First Experience 161
V.A. Godovikova, S.B. Nesterenko, G.V. Seliverstov and V.K. Shumny, Russian Academy of Science (Novosibirsk, Russia)

Grey Alder and Hybrid Alder as Short-Rotation Forestry Species 167
V. Uri and H. Tullus, Estonian Agricultural Univ. (Tartu, Estonia)

Potential Yield, Yield Component and Biomass Quality of Fifteen *Miscanthus* Genotypes in Southern Portugal 175
K. Tayebi, G. Basch and F. Teixeira, Universidade de Évora (Areia, Portugal)

Biomass Resources: International

Carbon Dioxide Fixation and Biomass Production Process Using Microalgae 179
S. *Hirata* and M. *Hayashitani*, Kawasaki Heavy Industries, Ltd. (Hyogo, Japan); and Y. *Ikegami*, Research Institute of Innovative Technology for the Earth (Tokyo, Japan)

Investigation of Rhizosphere Microflora to Improve Forest Tree Species for Increased Biomass Production in Afforestation and Soil Reclamation Programmes 185
A. *Saravanan*, R. *Babu* and M. *Vivekanandan*, Bharathidasan Univ. (Tamilnadu, India)

Evaluation of Decentralised Power Generation from Biomass in Competition with Alternative Technologies in Marajó Island, Brazil 187
B.R.P. *da Rocha*, Campus Univ. do Guamá (Belém Para, Brazil); C. *Monteiro*, Inst. de Engenharia de Sistemas de Computação (Porto, Portugal); E.C.L. *Pinheiro*, I.T. *da Silva*, I.M.O. *da Silva*, S.B. *Moraes*, A.O.F. *da Rocha*, Campus Univ. do Guamá (Belém Para, Brazil); V. *Miranda* and J.P. *Lopes*, Inst. De Engenharia de Sistemas de Computação (Portugal)

Bioenergy Technologies in China: Their Development and Commercialization 193
L. *Dai*, Energy Research Institute (China); G. *Wang*, Energy and Environment Institute (China); M. *Su*, Tsinghua Univ. (China); F. *Qu*, Rural Energy Office (China); and X. *Liu*, Energy Research Institute (China)

Geographic Information Systems (GIS)-Based Assessment of Rice Hull Energy Resource in the Philippines 203
S.C. *Capareda*, University of the Philippines at Los Baños (Laguna, Philippines)

Malaysia's Bioenergy Utilization Scenario 209
K.O. *Lim*, Univ. Sains Malaysia (Penang, Malaysia)

New GIS Tools for Biomass Resource Assessment in Electrical Power Generation 215
C. *Monteiro*, B.R.P. *da Rocha*, V. *Miranda* and J.P. *Lopes*, Inst. de Engenharia de Sistemas de Computadores (Porto, Portugal)

Biomass Resource Assessment for China 217
J. Li and A. Zhou, Energy Research Institute of State Planning and Development Commission (China)

Technologies for Electric Energy Production Using Biomass in Marajó Island, Brazil 233
S.B. Moraes and B.R.P. da Rocha, Campus Univ. do Guamá (Belém Para, Brazil); *C. Monteiro*, Inst. de Engenharia de Sistemas de Computação (Porto, Portugal); *I.M.O. da Silva, A.O.F. da Rocha and E.C.L. Pinheiro*, Campus Univ. do Guamá (Belém Para, Brazil); *V. Miranda and J.P. Lopes*, Inst. de Engenharia de Sistemas de Computação (Porto, Portugal)

Availability of Wood Residues for Cogeneration in Ghana 241
S.B. Atakora and A. Brew-Hammond, Kumasi Inst. of Technology and Environment (Kumasi, Ghana)

Land Availability and Productivity for Biomass Energy Plantations in Northeast Brazil 247
E.D. Larson and C. Tudan, Princeton Univ. (Princeton, New Jersey, USA); *E. Carpentieri and A.C. Leao*, Hydroelectric Company of Sao Francisco (Pernambuco, Brazil)

Influence of Laser Irradiation on Some Energy Multipurpose Crops 249
A.G. Aladjadjiyan, Higher Institute of Agriculture (Plovdiv, Bulgaria)

Crop Growth Modeling of Eucalyptus for Electricity Generation in Nicaragua 255
M. Hoogwijk and R. van den Broek, Utrecht Univ. (Utrecht, The Netherlands); *and L. Vleeshouwers*, Wageningen Agricultural Univ. (Wageningen, The Netherlands)

Feedstocks: Characterization, Drying, and Densifying

Developing New Analytical Techniques for the Characterization of Lignocellulosics Based on Multivariate Chemometric Analysis of Whole Fluorescence Spectra (AFFLUENCE) 265
E.G. Koukios and E. Billa, National Tech. Univ. of Athens (Athens, Greece)

Quality of Solid Biofuels – Database and Field Trials 273
H. Hartmann, L. Maier and T. Böhm, Munich Univ. of Technology (Freising-Weihenstephan, Germany)

Drying in a Biomass Gasification Plant for Power or Cogeneration 281
J.G. Brammer and A.V. Bridgwater, Aston Univ. (Birmingham, UK)

Forest and Agricultural Handling to Electricity Production Using Biomass
in Marajó Island, Brazil 289
I.M.O. da Silva, I.T. da Silva and B.R.P. Da Rocha, Campus Univ. de
Guamá (Belém-Pará, Brazil); *C. Monteiro, V. Miranda and J.P. Lopes*,
Inst. De Engenharia de Sistemas de Computação (Porto, Portugal); *E.C.L.
Pinheiro, S.B. Moraes and A.O.F. da Rocha*, Campus Univ. de Guamá
(Belém-Pará, Brazil)

Integration between Gas Turbine Plants and Biomasses Drying Processes 291
R. Cipollone, Univ. of L'Aquila (L'Aquila, Italy); *D. Cocco*, Univ. of
Cagliari (Cagliari, Italy); *and E. Bonfitto*, Regional Agency for the
Development of the Agricultural Services (Avezzano, Italy)

Physico-Chemical Upgrading of Agroresidues as Solid Biofuels 299
E.G. Koukios, S. Arvelakis, National Tech. Univ. of Athens (Athens,
Greece); *and B. Georgali*, Hellenic Cement Research Center (Athens,
Greece)

Extrusion Cork Powder Briquettes 305
L. Gil and C. Nascimento, INETI (Barcarena, Portugal)

Feedstocks: Harvesting, Handling and Storage

Development of Production Methods, Costs, and Use of Wood Fuels in
Finland 311
I. Nousiainen and V.-J. Aho, VTT Energy (Jyväskylä, Finland)

Supply Systems for Biomass Fuels and Their Delivered Costs 319
A. Hunter, J. Boyd, H. Palmer, J. Allen and M. Browne, Scottish
Agricultural College (Penicuik, UK)

Field Chopping as an Alternative to Baling for Harvesting and Handling
Switchgrass 325
D.I. Bransby, Agtec Development LLC (Auburn, Alabama, USA)

Harvesting and Handling of *Miscanthus giganteus, Phalaris arundanicea*,
and *Arundo donax* in Europe 327
W. Huisman, Wageningen Agricultural Univ. (Wageningen, The
Netherlands)

Equipment Performance and Economic Assessments of Harvesting and
Handling Rice Straw 335
R. Bakker-Dhaliwal, B.M. Jenkins and H. Lee, Univ. of California (Davis, California, USA)

Rapeseed as an Alternative Crop for Non-Food Uses in Portugal 343
S. Ferreira-Dias, A.C. Correia and J.V. Mazumbe, Inst. Superior de Agronomia (Lisbon, Portugal); *E.V. Lourenço*, Univ. de Évora (Évora, Portugal); *and J.E. Regato*, Escola Superior Agrária de Beja (Beja, Portugal)

Suitability of Chain-Flail Debarking Technology for the Integrated Production of Wood Fuel and Pulpwood – Technology Transfer from USA to Finland 349
V.-J. Aho and I. Nousiainen, VTT Energy (Jyväskylä, Finland)

The Recovery of Poplar Rootwood 355
R. Spinelli and R. Spinelli, CNR-Istituto per la Ricerca sul Legno (Florence, Italy)

Crop Residues as Feedstocks of Choice for New Generation Cooperative Processing Plants 363
D.L. Van Dyne and M.G. Blase, Univ. of Missouri (Columbia, Missouri, USA)

Biomass: The Carbon Connection

Contribution of Biomass toward CO_2 Reduction in Europe (EU) 371
A. Bauen and M. Kaltschmitt, King's College (London, UK)

The Net CO_2 Emissions and Energy Balances of Biomass and Coal-Fired Power Systems 379
M.K. Mann and P.L. Spath, NREL (Golden, Colorado, USA)

Producing a Life-Cycle Energy Balance for an Integrated Sweet Sorghum/Sugarcane System in the Semi-Arid Southeast Region of Zimbabwe 387
J. Woods and D.O. Hall, King's College (London, UK)

First Total Evaluation of the Ecological Comparison: Biofuels Versus Conventional Fuels 393
G. Zemanek and G.A. Reinhardt, Inst. für Energie- und Umweltforschung Heidelberg (Heidelberg, Germany)

Bioenergy and Carbon Cycling – the New Zealand Way 401
R.E.H. Sims, Massey Univ. (Palmerston North, New Zealand)

Peat and Finland, An Example of the Sustainable Use of an Indigenous
Energy Source 407
T. Nyrönen and P. Selin, Vapo Oy (Jyväskylä, Finland); *and J. Laine*,
Univ. of Helsinki (Helsinki, Finland)

Hydrogen Production and Biofixation of CO_2 with Microalgae 413
J.R. Benemann, Univ. of California (Berkeley, California, USA); *J.C.
Radway*, Univ. of Hawaii (Honolulu, Hawaii, USA); *and A. Melis*, Univ.
of California (Berkeley, California, USA)

A Sustainability Analysis of Biomass Use by Information Entropy Theory 419
M.X. Ponte, Univ. of Missouri (Columbia, Missouri, USA)

Environmental Costs of Energy from Two Biogas Plants in Denmark 421
P.S. Nielsen, Technical Univ. of Denmark (Lyngby, Denmark)

Carbon Credits – How to Measure Them for the Petroleum Refining
Industry 427
J.J. Marano and S. Rogers, Federal Energy Technology Center (Pittsburgh,
Pennsylvania, USA)

Total Costs of Electricity from Biomass in Spain 429
R. Sáez, Y. Lechón, H. Cabal and M. Varela, CIEMAT (Madrid, Spain)

Greenhouse Gas Emissions of Bioenergy Systems Compared to Fossil
Energy Systems 437
G. Jungmeier, L. Canella and J. Spitzer, Inst. of Energy Research (Graz,
Austria)

The Need for and the Benefits of a Domestic Protocol to Approve and
Accredit GHG Emission Offsets from U.S. Projects 445
C.E. Parker III, Materials Development Corp. (Carlisle, Maine, USA)

Systems Integration

EKOKRAFT™ Program – Municipality of Hedemora, Sweden 455
M.O. Wilstrand, Salix Maskiner (Hedemora, Sweden)

IEA Bioenergy Feasibility Studies 463
Y. Solantausta, VTT Energy (Espoo, Finland); D. Beckman, Zeton, Inc. (Burlington, Ontario, Canada); E. Podesser, Joanneum Research (Austria); R.P. Overend, NREL (Golden, Colorado, USA); and A. Östman, Kemiinformation (Stockholm, Sweden)

Economic Evaluation of Technical, Environmental and Institutional Barriers on Biomass Residue Collection Cost in California 471
P. Sethi and V. Tiangco, California Energy Commission (Sacramento, California, USA); Y. Lee, Lawrence Livermore National Laboratory (Livermore, California, USA); V. Dee, California Energy Commission (Sacramento, California, USA); D. Yomogida, Henwood Energy Services, Inc. (Sacramento, California, USA); and G. Simons, California Energy Commission (Sacramento, California, USA)

Long Term Perspectives for Production of Fuels from Biomass: Integrated Assessment and RD&D Priorities 477
A.P.C. Faaij and A.E. Agterberg, Utrecht Univ. (Utrecht, The Netherlands)

Opportunities for Efficient Use of Biomass in the Pulp and Paper Industry 479
K. Maunsbach, V. Martin and G. Svedberg, Royal Inst. of Technology (Stockholm, Sweden)

Energy Crops versus Wastepaper: A System Comparison of Paper Recycling and Paper Incineration on the Basis of Equal Land-Use 487
M. Hekkert, R. van den Broek and A. Faaij, Utrecht Univ. (Utrecht, The Netherlands)

Optimizing Used Recycled Pulp in Paper Production 495
M.D. Berni and S.V. Bajay, State Univ. of Campinas (Campinas, Brazil)

A Linear Programming Tool for Optimizing Bio-energy and Waste Treatment Systems 497
B. Meuleman and A. Faaij, Utrecht Univ. (Utrecht, The Netherlands)

The Concept of an FAO Integrated Energy Farm: A Strategy Towards Sustainable Production of Food and Energy 503
N. El Bassam and W. Bacher, Institute of Crop Science (Braunschweig, Germany)

Economic Evaluation of Technical, Environmental and Institutional
Barriers for Direct-combustion Biomass Power Plants in California 509
P. Sethi, V. Tiangco, V. Dee and G. Simons, California Energy Commission,
Research and Development Office (Sacramento, California, USA); *D.
Yomogida*, Henwood Energy Services, Inc. (Sacramento, California, USA);
and *Y. Lee*, Lawrence Livermore National Laboratory (Livermore,
California, USA)

BIOMASS TRANSFORMATION INTO VALUE-ADDED CHEMICALS, LIQUID FUELS, HEAT AND POWER

Value-Added Products: Lignins, Pyrolytic Oils, and Carbons

Application of the Slow Pyrolysis Eucalyptus Oil to Make PF Resins 513
J.D. Rocha, S.S. Kelley and H.L. Chum, NREL (Golden, Colorado, USA)

Wood Composite Adhesives from Softwood Bark-Derived Vacuum
Pyrolysis Oils 521
C. Roy, Pyrovac Inst., Inc. (Sainte-Foy, Québec, Canada); *L. Calvé*,
Forintek Canada Corp. (Sainte-Foy, Québec, Canada); *and X. Lu, H.
Pakdel and C. Amen-Chen*, Pyrovac Inst., Inc. (Sainte-Foy, Québec,
Canada)

Foam Concrete Using Bagasse Pyrolysis Tar 527
L.A.B. Cortez, Univ. of Campinas (Campinas, Brazil); *L.E. Brossard
Perez*, Univ. of Oriente (Santiago de Cuba, Cuba); *E. Izquierdo*, Univ.
of Oriente (Santiago de Cuba, Cuba); *E. Olivares*, Univ. of Campinas
(Campinas, Brazil); *and G. Bezzon*, State Univ. at Campinas (Campinas,
Brazil)

Plasticizers that Transform Alkylated Kraft Lignins into Thermoplastics 533
S. Sarkanen and Y. Li, Univ. of Minnesota (St. Paul, Minnesota, USA)

Production and Performance of Wood Composite Adhesives with
Air-Blown, Fluidized-Bed Pyrolysis Oil 541
D.A. Himmelblau, Biocarbons Corp. (Woburn, Massachusetts, USA); *and
G.A. Grozdits*, Louisiana Tech Univ. (Ruston, Louisiana, USA)

Preparation of Activated Carbons Using Poplar Wood and Bark as
Precursors   549
A. Bahrton, G. Horowitz, G. Cerrella, P. Bonelli, M. Cassanello and A.L. Cukierman, Univ. de Buenos Aires (Buenos Aires, Argentina)

Lignin – The Raw Material for Industry in the Future   555
P. Zuman and E. Rupp, Clarkson Univ. (Potsdam, New York, USA)

Modern Technologies of the Biomass Conversion for Chemicals, Carbon Sorbents, Energy, Heat and Hydrocarbon Fuels Production   563
V. Anikeev, Boreskove Inst. of Catalysis (Novosibirsk, Russia)

Study on Gluability of Copolymer Resins of Biomass Extracts for Laminating CCA-Treated Lumber under Room Temperature   571
C.-M. Chen and D.L. Nicholls, Univ. of Georgia (Athens, Georgia, USA)

Agronomic Evaluation of Ash Following Gasification of Five Biomass Feedstocks   577
D.I. Bransby and G.R. Mullins, Auburn Univ. (Auburn, Alabama); *and B. Bock*, Tennessee Valley Authority (Muscle Shoals, Alabama, USA)

Alkaline Bagasse Tar Solutions as Foamers in Copper Mining   583
E. Olivares, Univ. of Campinas (Campinas, Brazil); *L.E. Brossard*, Univ. of Oriente (Santiago de Cuba, Cuba); *L.A.B. Cortez*, Univ. of Campinas (Campinas, Brazil); *N. Varela*, Univ. of Oriente (Santiago de Cuba, Cuba); *and G. Bezzon*, State Univ. of Campinas (Campinas, Brazil)

Efficiency Test for Bench Unit Torrefaction and Characterization of Torrefied Biomass   589
F. Fonseca Felfli, C.A. Luengo, P. Beaton and J.A. Suarez, Univ. of Campinas (Campinas, Brazil)

Value-Added Products: Bioproducts and Fibers

Production of Levulinic Acid and Use as a Platform Chemical for Derived Products   595
D.C. Elliott, Pacific Northwest National Laboratory (Richland, Washington, USA); *S.W. Fitzpatrick*, Biofine, Inc. (Waltham, Massachusetts, USA); *J.J. Bozell*, NREL (Golden, Colorado, USA); *J.L. Jarnefeld*, NYSERDA (Albany, New York, USA); *R.J. Bilski*, Chemical Industry Services,

Inc. (West Lafayette, Indiana, USA); *L. Moens*, NREL (Golden, Colorado, USA); *J.G. Frye, Jr., Y. Wang and G.G. Neuenschwander*, Pacific Northwest National Laboratory (Richland, Washington, USA)

Decentralised Production of Biodegradable Lubricants, Renewable Fuels, Fodder, and Electricity from Oil Crops on the Production-Site, in Mobile Units   601
U.C. Knopf, Agrogen Foundation (Freiburg, Switzerland)

Agri-Pulp™ Newsprint   605
A. Wong, Arbokem, Inc. (Vancouver, British Columbia, Canada)

Polymers from the Exploitation of Biomass   613
A. Gandini, Ecole Française de Papeterie et des Industries Graphiques (St. Martin d'Hères, France)

Utilization of *Chlorella* sp. for Plastic Composite after CO_2 Fixation Using High Density Polyethylene   621
T. Otsuki and M. Yamashita, Ishikawajima-Harima Heavy Industries Co., Ltd. (Yokohama, Japan); *Z. Farao, Y. Ikegami, M. Yoshitake and H. Tsutao*, RITE (Tokyo, Japan); *H. Kabeya, R. Kitagawa and T. Hirotsu*, MITI (Takamatsu, Japan)

Valorisation of Flax-byproducts by Means of Mycelial Biomass Production in Different Industrial Applications   629
P.V. Vilppunen, O.K. Mäentausta and P.I. Kess, Univ. of Oulu (Oulu, Finland)

Utilization of *Chlorella* sp. for Plastic Composite after CO_2 Fixation Using PVC   635
T. Otsuki and M. Yamashita, Ishikawajima-Harima Heavy Industries Co., Ltd. (Yokohama, Japan); *Z. Farao, Y. Ikegami, M. Yoshitake and H. Tsutao*, RITE (Tokyo, Japan); *H. Kabeya, R. Kitagawa and T. Hirotsu*, MITI (Takamatsu, Japan)

New Green Products from Cellulosics   641
E.G. Koukios, A. Pastou, D.P. Koullas, V. Sereti, H. Stamatis and F. Kolisis, National Tech. Univ. of Athens (Athens, Greece)

Papermaking Pulp from *Hesperaloe* Species, an Arid-Zone Native Plant from Northern Mexico   649
A. Wong, Arbokem Inc. (Vancouver, British Columbia, Canada); *and S. McLaughlin*, Univ. of Arizona (Tucson, Arizona, USA)

Anaerobic Metabolism in the Marine Green Alga *Chlorococcum littorale*   655
Y. Ueno and N. Kurano, Kamaishi Laboratories (Kamaishi, Japan); *and*
S. Miyachi, Marine Biotechnology Inst. (Tokyo, Japan)

Evaluation of a Soy-Based Heavy Fuel Oil Emulsifier for Emission
Reduction and Environmental Improvement in Industries and Utilities in
Mexico   663
P.K. Lee and B.F. Szuhaij, Central Soya Co., Inc. (Fort Wayne,
Indiana, USA)

Anaerobic Processes

Anaerobic Bioconversion of Lignocellulosic Materials   667
A. Padilla, E. Marcano and D. Padilla, Univ. de Los Andes (Mérida,
Venezuela)

Construction and Operation of a Covered Lagoon Methane Recovery
System for the Cal Poly Dairy   673
D.W. Williams, California Polytechnic State Univ. (San Luis Obispo,
California, USA); M.A. Moser, Resource Conservation Mgmt. (Oakland,
California, USA); *and* G. Norris, USDA Resource Conservation Service
(Templeton, California, USA)

Langerwerf Dairy Digester Facelift: What We Found When We Took
Apart a 16-Year-Old Dairy Plug Flow Digester   681
M.A. Moser, Resource Conservation Mgmt., Inc. (Berkeley, California, USA); *and* L. Langerwerf, Langerwerf Dairy, Inc. (Durham,
California, USA)

Making Energy Recovery from Organic Wastes Work in a Deregulated
Electricity Marketplace   689
G. Simons and V. Tiangco, California Energy Commission (Sacramento,
California, USA); R. Yazdani, Yolo County Department of Public Works
(Davis, California, USA); *and* M. Kayhanian, Univ. of California (Davis,
California, USA)

Yolo County Controlled Landfill Project   691
D. Augenstein, IEM (Palo Alto, California, USA); R. Yazdani, K.
Dahl, A. Mansoubi and R. Moore, Yolo County Department of Public
Works (Davis, California, USA); *and* J. Pacey, Emcon (San Mateo,
California, USA)

Contents

Wastewater Treatment for a Biomass-to-Ethanol Process: System Design
and Cost Estimates — 699
K.L. Kadam and R.J. Wooley, NREL (Golden, Colorado, USA); *F.M. Ferraro and R.E. Voiles*, Merrick & Co. (Aurora, Colorado, USA); *J.J. Ruocco and F.T. Varani*, Phoenix Bio-Systems, Inc. (Lakewood, Colorado, USA); *and V.L. Putsche* (Denver, Colorado, USA)

Prefeasibility Study for Establishing a Centralized Anaerobic Digester in
Adams County, Pennsylvania — 707
P. Lusk, Resource Development Associates (Washington, D.C., USA); *and R. Mattocks*, Environomics (Riverdale, New York, USA)

Pretreatment of Domestic Sewage by the Modification of an Existing
IMHOFF Tank — 715
S. Di Berardino, S. Antunes and M. Bergs, INETI-ITE (Lisbon, Portugal); *and A. Alegria*, Serviços Municipalizados de Sintra (Sintra, Portugal)

Biogas Production from Wastes in Portugal: Present Situation and
Perspectives — 723
S. Di Berardino, INETI (Lisbon, Portugal)

Computer-Aided Design Model for Anaerobic-Phased-Solids Digester
System — 729
Z. Zhang, R. Zhang, Univ. of California (Davis, California, USA); *and V. Tiangco*, California Energy Commission (Sacramento, California, USA)

Accounting for the Fate of Manure Nutrients and Solids Processed in
Mesophilic Anaerobic Digesters — 735
R.P. Mattocks, ENVIRONOMICS (Riverdale, New York, USA); *and M. Moser*, RCM Inc. (Berkeley, California, USA)

AgSTAR Charter Farm Program: Experience with Five Floating Lagoon
Covers — 743
K.F. Roos, U.S. Environmental Protection Agency (Washington, D.C., USA); *M.A. Moser*, Resource Conservation Mgmt., Inc. (Berkeley, California, USA); *and A.G. Martin*, ICF, Inc. (Sherman Oaks, California, USA)

Methane from Manure: An Energy-Saving Solution for Iowa Pork
Producers — 751
L. Vannoy, Iowa Department of Natural Resources (Des Moines, Iowa, USA)

Anaerobic Digestion of Brewery Wastewater for Pollution Control and Energy 759
D.W. Williams, California Polytechnic State Univ. (San Luis Obispo, California, USA); *D. Schleef*, SLO Brewing Co. (Paso Robles, California, USA); *and A. Schuler*, California Polytechnic State Univ. (San Luis Obispo, California, USA)

Biogas Slurry Utilization in Ghana 765
C. Asser, Ministry of Mines and Energy (Accra, Ghana)

Feasibility of Using Rice Straw for Biogas Energy Production in California 773
R. Zhang, Univ. of California (Davis, California, USA); *and J. Turnbull*, Peninsula Energy Partners (Los Altos, California)

Biofuels: Thermochemical Conversion and Biodiesel

Unified Approach to Next Generation Biofuel – Fischer-Tropsch Fuel Blends 781
G.J. Suppes and M.L. Burkhart, Univ. of Kansas (Lawrence, Kansas, USA)

The HTU® Process for Biomass Liquefaction: R&D Strategy and Potential Business Development 789
J.E. Naber and F. Goudriaan, Biofuel B.V. (Heemskerk, The Netherlands); *S. van der Wal*, Stork Engineers & Contractors (Amsterdam, The Netherlands); *J.A. Zeevalkink*, TNO-MEP (Apeldoorn, The Netherlands); *and B. van de Beld*, Biomass Technology Group (Enschede, The Netherlands)

Transesterification of Rapeseed Oils in Supercritical Methanol to Biodiesel Fuels 797
S. Saka and K. Dadan, Kyoto Univ. (Kyoto, Japan)

Production of Methanol and Hydrogen from Biomass via Advanced Conversion Concepts 803
A. Faaij and C. Hamelinck, Utrecht Univ. (Utrecht, The Netherlands); *and E. Larson and T. Kreutz*, Princeton Univ. (Princeton, New Jersey, USA)

Technical Performance of Vegetable Oil Methyl Esters with a High Iodine Number 805
H. Prankl, M. Wörgetter and J. Rathbauer, Federal Inst. of Agricultural Engineering (Wieselburg, Austria)

Lignin Conversion to High-Octane Fuel Additives 811
J. Shabtai, W. Zmierczak and S. Kadangode, Univ. of Utah (Salt Lake City, Utah, USA); *E. Chornet and D.K. Johnson*, NREL (Golden, Colorado, USA)

Performance Advantages of Cetane Improvers Produced from Soybean Oil 819
G.J. Suppes, T.T. Tshung, M.H. Mason and J.A. Heppert, Univ. of Kansas (Lawrence, Kansas, USA)

Fluidized Bed Catalytic Steam Reforming of Pyrolysis Oil for Production of Hydrogen 827
S. Czernik, R. French, C. Feik and E. Chornet, NREL (Golden, Colorado, USA)

Methanol Production from Biomass Using the Hynol Process 833
P. Sethi and S. Chaudhry, California Energy Commission (Sacramento, California, USA); *and S. Unnasch*, Arcadis Geraghty & Miller (Mountain View, California, USA)

A Green Approach for the Production of Bio-Cetane Enhancer for Diesel Fuels 837
A. Wong, Arbokem, Inc. (Vancouver, British Columbia, Canada); *and E. Hogan*, Natural Resources Canada (Ottawa, Ontario, Canada)

Biomass Conversion to Fischer-Tropsch Liquids: Preliminary Energy Balances 843
E.D. Larson and H. Jin, Princeton Univ. (Princeton, New Jersey, USA)

A Preliminary Assessment of Biomass Conversion to Fischer-Tropsch Cooking Fuels for Rural China 855
E.D. Larson and H. Jin, Princeton University (Princeton, New Jersey, USA)

Biofuels: Ethanol and Biotechnology

Production of Low Cost Sugars from Biomass: Progress, Opportunities, and Challenges 867
C.E. Wyman, BC International (Dedham, Massachusetts, USA)

Status of Biomass Conversion to Ethanol and Opportunities for Future Cost Improvements 873
D. Glassner, NREL (Golden, Colorado, USA)

Collins Pine/BCI Biomass to Ethanol Project 875
M.A. Yancey, NREL (Golden, Colorado, USA); N.D. Hinman, BC
International (Englewood, Colorado, USA); J.J. Sheehan, Plumas
Corp (Quincy, California, USA); and V.M. Tiangco, California Energy
Commission (Sacramento, California, USA)

Technological and Ecological Aspects of Ethanol Production from Wood 881
Y.I. Kholkin, V.L. Makarov, V.V. Viglazov and V.A. Elkin, St. Petersburg
Forest Technical Academy (St. Petersburg, Russia); and H.D. Mettee,
Youngstown State University (Youngstown, Ohio, USA)

Simultaneous Saccharification and Co-fermentation of Peracetic Acid
Pretreated Sugar Cane Bagasse 887
L.C. Teixeira, Fundação Centro Tecnológico de Minas Gerais (Belo
Horizonte, Brazil); J.C. Linden and H.A. Schroeder, Colorado State Univ.
(Fort Collins, Colorado, USA)

Transgenic Fungal-Based Conversion of Waste Starch to Industrial
Enzymes 895
J. Gao, B.S. Hooker, R.S. Skeen and D.B. Anderson, Pacific Northwest
National Laboratory (Richland, Washington, USA)

Simulation of Low-Power Agitation Systems for Large-Scale Biomass
Conversion Reactors 903
S.P. Svihla, C.K. Svihla and T.R. Hanley, Univ. of Louisville (Louisville,
Kentucky, USA)

Production of Ethanol, Protein Concentrate, and Technical Fibers from
Clover/Grass 911
S. Grass, G. Hansen, M. Sieber and P.H. Müller, Biomass and Bioenergy
(Dübendorf, Switzerland)

Bark-Rich Residual Biomass as Feedstock for Ethanol and Co-products 915
J.M. Garro and P. Jollez, Kemestrie, Inc. (Sherbrooke, Quebec, Canada);
D. Cameron and R. Benson, Tembec, Inc. (Sherbrooke, Quebec, Canada);
W. Cruickshank, NRCan (Canada); G.B.B. Le, Ministere des Ressources
Naturalles (Sherbrooke, Quebec, Canada); Q. Nguyen and E. Chornet,
NREL (Golden, Colorado, USA)

Performance of Diesel-Ethanol Fueled Compression Ignition Engine 923
F. Guo and B. Zhang, China National Rice Research Inst. (Hangzhou,
China)

Steam Pretreatment Conditions to Optimize the Hemicellulose Sugar
Recovery and Fermentation of Softwood-Derived Feedstocks 929
Y. Cai, J. Robinson, S.M. Shevchenko, D.J. Gregg and J. N. Saddler, Univ.
of British Columbia (Vancouver, British Columbia, Canada)

Volume 2

Gasification: Gas Conditioning

Biomass Gasification with Air in Fluidized Bed: Hot Gas Cleanup and
Upgrading with Steam-Reforming Catalysts of Big Size 933
J. Corella and M.A. Caballero, Univ. Complutense (Madrid, Spain); *M.P.
Aznar and J. Gil*, Univ. of Saragossa (Saragossa, Spain)

An Experimental Investigation of Alkali Removal from Biomass Producer
Gas Using a Fixed Bed of Solid Sorbent 939
*S.Q. Turn, C.M. Kinoshita, D.M. Ishimura, T.T. Hiraki, J. Zhou and S.M.
Masutani*, Univ. of Hawaii (Honolulu, Hawaii, USA)

CO_2 Gasification of Pine Sawdust Using a Coprecipitated Ni-Al Catalyst 947
L. Garcia, M.L. Salvador, R. Bilbao and J. Arauzo, Univ. of Zaragoza
(Zaragoza, Spain)

An Integrated Modular Hot Gas Conditioning Technology 953
N. Abatzoglou, D. Bangala and E. Chornet, Kemestrie, Inc. (Sherbrooke,
Quebec, Canada)

Determination of Vapor Phase Alkali Content during Biomass Gasification 961
J. Smeenk, R.C. Brown and D. Eckels, Iowa State Univ. (Ames, Iowa, USA)

Kinetics of CO_2 Gasification of Black Liquors from Alkaline Pulping of
Straw 969
G. Gea, M.B. Murillo and J. Arauzo, Univ. of Zaragoza (Zaragosa, Spain)

Catalytic Gasification of Biomass 977
S. Rapagná, J. Nader and P.U. Foscolo, Univ. di L'Aquila (L'Aquila, Italy)

A Full-Flow Catalytic Reactor at Pilot Scale for Hot Gas Cleanup in
Biomass Gasification with Air   979
M.A. Caballero and M.P. Aznar, Univ. Complutense (Madrid, Spain); *J.
Corella*, Univ. of Saragossa (Saragossa, Spain); *J. Gil*, Univ. Complutense
(Madrid, Spain); *and J.A. Martin*, Univ. of Saragossa (Saragossa, Spain)

Conversion of Various Biomass-Derived Chars to Hydrogen/High Btu
Gas by Gasification with Steam   985
N.N. Bakhshi and A.K. Dalai, Univ. of Saskatchewan (Saskatchewan,
Canada); *and S.T. Srinivas*, Pennsylvania State Univ. (University Park,
Pennsylvania, USA)

Effects of Injecting Steam on Catalytic Reforming of Gasified Biomass   991
J. Zhou, D.M. Ishimura and C.M. Kinoshita, Univ. of Hawaii at Manoa
(Honolulu, Hawaii, USA)

Gasification: Co-Feeding and Modeling

Superficial Velocity – The Key to Downdraft Gasification   1001
T.B. Reed, The Biomass Energy Foundation (Golden, Colorado, USA);
R. Walt, Community Power Corp. (Aurora, Colorado, USA); *S. Ellis*,
Colorado School of Mines (Golden, Colorado, USA); *A. Das*, Original
Sources (Boulder, Colorado, USA); *and S. Deutsch*, NREL (Golden,
Colorado, USA)

Conversion of Biomass and Biomass-Coal Mixtures: Gasification, Hot
Gas Cleaning and Gas Turbine Combustion   1009
W. de Jong, J. Andries, P.D.J. Hoppesteyn and Ö Ünal, Delft Univ. of
Technology (Delft, The Netherlands)

Biomass Co-Gasification with Coal: The Process Benefit Due to Positive
Synergistic Effects   1017
G. Chen, Q. Yu, C. Brage, C. Rosén and K. Sjöström, Royal Inst. of
Technology (Stockholm, Sweden)

Dynamic Modelling of the Two-Stage Gasification Process   1025
B. Gøbel, Technical Univ. of Denmark (Lyngby, Denmark); *J.D. Bentzen*,
COWI (Lyngby, Denmark); *U. Henriksen and N. Houbak*, Technical Univ.
of Denmark (Lyngby, Denmark)

Synergistic Effect on Tar Formation in Co-Gasification of Energy Crops
and Coal 1033
Q. Yu, C. Brage, G. Chen and K. Sjöström, Royal Inst. of Technology
(Stockholm, Sweden)

Optimisation of Gasification Experimental Conditions of Mixtures of
Biomass with Plastic Wastes 1041
F. Pinto, I. Gulyurtlu, C. Franco and I. Cabrita, INETI-ITE-DTC (Lisbon,
Portugal)

Modeling a Fluidized Bed Gasifier of Biomass in Stationary State with
In-Bed Use of Dolomite: Abrasion, Erosion and Carry Over of the
Dolomite 1049
J. Corella, Univ. Complutense (Madrid, Spain); *M.P. Aznar, J. Gil and
M.A. Caballero*, Univ. of Saragossa (Saragossa, Spain); *J.A. Martín*, Univ.
Complutense (Madrid, Spain)

Use of Bio-Nut shells for Gasification 1051
M. Dogru, Univ. of Newcastle (Newcastle, UK); *A.A. Malik*, Univ. of
Northumbria at Newcastle (Newcastle, UK); *C.R. Howarth*, Univ. of
Newcastle (Newcastle, UK); *and H. Olgun*, Blacksea Technical Univ.
(Trabzon, Turkey)

Gasification: Projects

Commercial Demonstration of the Battelle/FERCO Biomass Gasification
Process: Startup and Initial Operating Experience 1061
M.A. Paisley, Battelle (Columbus, Ohio); *M.C. Farris and J. Black*,
FERCO (Norcross, Georgia, USA); *J.M. Irving*, Burlington Electric
Department (Burlington, Vermont, USA); *and R.P. Overend*, NREL
(Golden, Colorado, USA)

Camp Lejeune Energy from Wood (CLEW) Project 1067
J.G. Cleland, Research Triangle Inst. (Research Triangle Park, North
Carolina, USA); *and C.R. Purvis*, U.S. Environmental Protection Agency
(Research Triangle Park, North Carolina, USA)

Biomass Gasification with a Difference 1075
D. Wiles, R. Sunter, B. Ramsay, J. Neufeld, G. Neufeld, H. Burke, Malahat
Systems Corp. (Victoria, British Columbia, Canada)

The Realisation of a Biomass-Fueled IGCC Plant in Italy 1079
H.J. de Lange, Bioelettrica S.p.A. (Pisa, Italy); *and P. Barbucci*, ENEL Ricerca (Pisa, Italy)

Installation, Operation and Economics of Biomass Gasification System in Indonesia 1087
P. DeLaquil, III *and F.S. Fische*, BG Technologies L.L.C. (Washington, D.C., USA)

The "Turbo" Wood-Gas Stove 1093
T.B. Reed, The Biomass Energy Foundation (Golden, Colorado, USA); *and R. Walt*, Community Power Corp. (Aurora, Colorado, USA)

Energy Losses Due to Elutriation in Fluidized Bed Gasification of Sugarcane Bagasse 1099
D.L. Larson, Univ. of Arizona (Tucson, Arizona, USA); *E. Olivares and L.A.B. Cortez*, University of Campinas (Campinas, Brazil); *L.E. Brossard*, Univ. of Oriente (Santiago de Cuba, Cuba); *and G. Bezzon*, Univ. of Campinas (Campinas, Brazil)

Fluidized Bed Air Gasification of Sugar Cane Bagasse. Bed Temperature and Air Factor Influence 1105
E. Esperanza, Univ. of las Villas (Santa Clara, Cuba); *J. Arauzo and G. Gea*, Univ. of Zaragossa (Zaragossa, Spain)

Bagasse Gaseous Fuel: An Environmentally Safe Replacement for Natural Gas 1113
S. Peres, Univ. de Pernambuco (Recife, Brazil)

Development of a Novel Reverse-Flow Slagging Gasifier for Small-Scale Cogeneration Applications 1119
J.G. Brammer, Aston Univ. (Birmingham, UK); *L. van de Beld*, Biomass Technology Group (Enschede, Netherlands); *A.V. Bridgwater*, Aston Univ. (Birmingham, UK); *and D. Assink*, Biomass Technology Group (Enschede, Netherlands)

Potential Assessment of BIG-GT Technology in Cuban Sugar Cane Mills 1127
F. Ponce, Univ. of Las Villas (Santa Clara, Cuba); *and A. Walter*, State Univ. of Campinas (Campinas, Brazil)

Pyrolysis: Kinetics

Modeling of Flash Pyrolysis of a Single Wood Particle 1137
A.M.C. Janse, R.W.J. Westerhout and W. Prins, Twente Univ. (Enschede, The Netherlands)

Tar Production from Biomass Pyrolysis in a Fluidized Bed Reactor: A Novel Turbulent Multiphase Flow Formulation 1145
J. Bellan and D. Lathouwers, California Institute of Technology (Pasadena, California)

Pyrolysis of Biomass – Aerosol Generation: Properties, Applications, and Significance for Process Engineers 1153
J. Piskorz, P. Majerski and D. Radlein, Resource Transforms International (Waterloo, Ontario, Canada)

Products Formed under Pressurized Pyrolysis 1161
P. Girard, S. Numazawa, S. Mouras and A. Napoli, Cirad-Forêt (Montpellier Cedex, France)

Release of Potassium and Chlorine during Straw Pyrolysis 1169
P.A. Jensen, Technical Univ. of Denmark (Lyngby, Denmark); *B. Sander*, ELSAMPROJEKT A/S (Fredericia, Denmark); *and K. Dam-Johansen*, Technical Univ. of Denmark (Lyngby, Denmark)

Pyrolysis and Combustion Behaviour of Wood: Temperature Profiles and Solid Conversion 1177
J. Ceamanos, R. Bilbao, M.E. Aldea, M. Betran and J.F. Mastral, Univ of Zaragoza (Zaragoza, Spain)

Biomass Thermal Decomposition in the Pyrolysis Process 1185
K.M. Abdullayev, National Academy of Sciences (Baku, Azerbaijan)

Pyrolysis of Olive Stones in Different Reactors 1193
J. Ruiz, M.J. Sanz, J. Gómez and J. Arauzo, Univ. of Zaragoza (Zaragoza, Spain)

Design of Biomass Fast Pyrolysis Systems: Heat Transfer to the Biomass Particle 1199
G.V.C. Peacocke, Conversion and Resource Evaluation, Ltd. (Birmingham, UK)

Pyrolysis of an Agricultural By-Product: A Characterization Study   1201
A.L. Cukierman, P.A. Della Rocca, P.R. Bonelli and E.G. Cerrella, Univ. de Buenos Aires (Buenos Aires, Argentina)

Biomass Pyrolysis of Organic Waste in a Rarefied Layer   1209
M. Gubinsky, Y. Shishko, R. Cheifetz and A. Usenko, State Metallurgica' Academy of Ukraine (Dniepropetrovsk, Ukraine)

Pyrolysis: Systems and Oil Properties

Fast Pyrolysis Technology   1217
A.V. Bridgwater, Aston Univ. (Birmingham, UK); *S. Czernik,* NREL (Golden, Colorado, USA); *D. Meier,* BFH-Inst. for Wood Chemistry (Hamburg, Germany); *and J. Piskorz,* RTI Ltd. (Waterloo, Ontario, Canada)

BioTherm^{TM}: A System for Continuous Quality, Fast Pyrolysis BioOil   1225
K.W. Morris, DynaMotive Technologies Corp. (Vancouver, British Columbia, Canada); *J. Piskorz and P. Majerski,* RTI (Vancouver, British Columbia, Canada)

The Pyrocycling^{TM} Process: New Developments   1227
C. Roy, Inst. Pyrovac Inc. (Sainte-Foy, Québec, Canada)

Analysis, Characterization, and Test Methods of Fast Pyrolysis Liquids   1229
A. Oasmaa, VTT Energy (VTT, Finland); *and D. Meier,* Inst. for Wood Chemistry and Chemical Technology of Wood (Hamburg, Germany)

A Review of Physical and Chemical Methods of Upgrading Biomass-Derived Fast Pyrolysis Liquids   1235
S. Czernik, NREL (Golden, Colorado, USA); *R. Maggi,* Univ. Catholique de Louvain (Louvain-la-Neuve, Belgium); *and G.V.C. Peacocke,* Conversion and Resource Evaluation Ltd. (Birmingham, UK)

Investigation of Flame Characteristics and Emissions of Pyrolysis Oil in a Modified Flame Tunnel   1241
P. Wickboldt, R. Strenziok and U. Hansen, Univ. of Rostock (Rostock, Germany)

Fuel Oil Quality of Biomass Pyrolysis Oils   1247
A. Oasmaa, VTT Energy (Espoo, Finland); *and S. Czernik,* NREL (Golden, Colorado, USA)

Fast Pyrolysis of Biomass in a Circulating Fluidized Bed Reactor 1253
I.P.H. Boukis, Center for Renewable Energy Sources (Pikermi, Greece); and *A.V. Bridgwater*, Aston Univ. (Birmingham, UK)

Catalytic Pyrolysis for Improved Liquid Fuel Quality 1255
E.H. Salter, Aston Univ. (Birmingham, UK); *P. Wulzinger*, Inst. for Wood Chemistry (Hamburg, Germany); *A.V. Bridgwater*, Aston Univ. (Birmingham, UK); *and D. Meier*, Inst. for Wood Chemistry (Hamburg, Germany)

Implementation of Biomass Fast Pyrolysis in Highly Competitive Markets 1263
M. Lauer, Inst. of Energy Research (Graz, Austria)

Pyrolysis Environment, Health and Safety Issues Output from the PyNe Network 1269
P. Girard and S. Mouras, Cirad Forêt (Montpellier Cedex, France)

Biomass Power: Cofiring

Cofiring Biomass with Coal 1277
L. Baxter and A. Robinson, Sandia National Laboratories (Livermore, California, USA)

Gas Cofiring for Performance Improvement and Emission Reduction in Biomass-Fired Boilers 1285
H.B. Mason and L.R. Waterland, Arcadis Geraghty & Miller (Mountain View, California, USA); *S.A. Drennan*, Coen Co., Inc. (Burlingame, California, USA); *I.S. Chan*, Gas Research Institute (Chicago, Illinois, USA); *V. Tiangco*, California Energy Commission (Sacramento, California, USA); *C. Knight*, Burney Mountain Power (Burney, California, USA); *and R.J. Auzenne*, Fairhaven Power Co. (Eureka, California, USA)

Biomass Cofiring in Coal-Fired Boilers: Test Programs and Results 1287
D.A. Tillman, Foster Wheeler Development Corp. (Clinton, New Jersey, USA); *S. Plasynski*, Federal Energy Technology Center (Pittsburgh, Pennsylvania, USA); *and E. Hughes*, Electric Power Research Institute (Palo Alto, California, USA)

Experiences on Biomass Cofiring in Finland 1293
S. Helynen, VTT Energy (Jyväskylä, Finland)

Cofiring of Coal and Straw 1299
P. Overgaard, MIDTKRAFT Energy Company (Aarhus, Denmark); *B. Sander*, ELSAMPROJEKT A/S (Fredericia, Denmark); *and N.O. Knudsen*, Nordjyllandsvaerket (Vodskov, Denmark)

Southern Company Evaluation of Switchgrass 1307
P.V. Bush, Southern Research Inst. (Birmingham, Alabama, USA); *D.M. Boylan*, Southern Company Services (Birmingham, Alabama, USA); *and D.I. Bransby*, Auburn Univ. (Auburn, Alabama, USA)

Cofiring Biofuel in a PC Boiler Using Direct Injection of Wood Waste 1309
D.A. Tillman, Foster Wheeler Development Corp. (Clinton, New Jersey, USA); *and J.R. Battista*, GPU Genco (Johnstown, Pennsylvania, USA)

Mixing Effects in the Application of SNCR to Biomass Combustion 1315
M. Oliva, M.U. Alzueta, A. Millera, J.C. Ibáñez and R. Bilbao, Univ. of Zaragoza (Zaragoza, Spain)

Feasibility of Cofiring (Biomass + Natural Gas) Power Systems 1321
A. Walter and M.R. Souza, State Univ. of Campinas (Campinas, Brazil); *R.P. Overend*, NREL (Golden, Colorado, USA)

NO_x Emissions Reduction: An Application of Biomass Reburning Technology 1329
J.J. Sweterlitsch and R.C. Brown, Iowa State Univ. (Ames, Iowa, USA)

Hybrid Combustion-Gasification Concept for Environmentally Safe Utilization of Biomass Energy 1335
V.M. Zamansky, Energy and Environmental Research Corp. (Irvine, California, USA); *B.M. Jenkins*, Univ. of California (Davis, California, USA); *T.R. Miles*, T.R. Miles Technical Consultants, Inc. (Portland, Oregon, USA); *and V.M. Tiangco*, California Energy Commission (Sacramento, California, USA)

Multiple Fuel Oven for Untreated Biofuels and Solid Waste 1341
P.H. Heyerdahl, Agricultural Univ. of Norway (Aas, Norway)

Cofiring Biomass with Coal Utilizing Water-Cooled Vibrating Grate Technology 1343
T.A. Giaier and M.A. Eleniewski, Detroit Stoker Co. (Detroit, Michigan, USA)

Cofiring Multiple Opportunity Fuels for Cost-Effective Biomass
Utilization 1349
D.A. Tillman, Foster Wheeler Development Corp. (Clinton, New
Jersey, USA); and P.J. Hus, Northern Indiana Public Service Co.
(Merrillville, Indiana, USA)

Biomass Power: Fouling and Inorganics Management

Combustion of Leached Rice Straw for Power Generation 1357
B.M. Jenkins, R.B. Williams, R.R. Bakker, S. Blunk and D.E. Yomogida,
Univ. Of California (Davis, California, USA); W. Carlson and J. Duffy,
Wheelabrator-Shasta Energy Co. (Anderson, California, USA); R. Bates,
Woodland Biomass Power, Ltd. (Woodland, California, USA); K. Stucki,
Wadham Energy Limited Partnership (Williams, California, USA); and V.
Tiangco, California Energy Commission (Sacramento, California, USA)

Ash-Induced Operational Difficulties in Fluidised Bed Firing of Biofuels
and Waste 1365
A.L. Hallgren and K. Engvall, TPS Termiska Processer AB (Nyköping,
Sweden); and B.-J. Skrifvars, Åbo Akademi Univ. (Åbo, Finland)

Use of Biomass Derived Gases in a Catalytic Combustor for a 3 MW_e
Gas Turbine 1371
E. Lebas and G. Martin, Inst. Français du Pétrole (Vernaison Cedex,
France)

Formation, Composition and Particle Size Distribution of Fly-Ashes from
Biomass Combustion Plants 1377
I. Obernberger, J. Dahl and T. Brunner, Technical Univ. Graz (Graz,
Austria)

Combustion Properties of Lignin Residue from Lignocellulose Fermentation 1385
S.L. Blunk and B.M. Jenkins, Univ. of California (Davis, California, USA);
and K.L. Kadam, NREL (Golden, Colorado, USA)

Investigation of Superheater Fouling in Biomass Boilers with Furnace
Exit Gas Temperature Control 1393
R.B. Williams, B.M. Jenkins and R.R. Bakker, Univ. of California
(Davis, California, USA); and L.L. Baxter, Sandia National Laboratories
(Livermore, California, USA)

Use of Combustion Turbines to Produce Electricity from Sugarcane Bagasse   1401
R.F. Tamaro and C. Echeverria, Tazcogen Development Co. (Moraga, California, USA)

Model Performance of a Biomass-Fueled Power Station with Variable Furnace Exit Gas Temperature for Fouling Control   1409
D.E. Yomogida, Henwood Energy Services, Inc. (Sacramento, California, USA); *B.M. Jenkins and B.R. Hartsough*, Univ. of California (Davis, California, USA); *V.M. Tiangco*, California Energy Commission (Sacramento, California, USA)

NO_x Reduction by Primary Measures on a Traveling-Grate Furnace for Biomass Fuels and Waste Wood   1417
A. Weissinger and I. Obernberger, Technical Univ. Graz (Graz, Austria)

Fluidized Bed Combustion of Leached Rice Straw   1425
R.R. Bakker, B.M. Jenkins, R.B. Williams and D. Pfaff, Univ. of California (Davis, California, USA)

Influence of Jet Angles on the Performance of a Boiler Furnace for Suspension Burning of Biomass: A Numerical Study   1433
R.F. Mut, P. Beaton and J. Martin, Univ. de Oriente (Santiago de Cuba, Cuba)

Effects of Management Factors on Energy Content and Slagging Potential of Switchgrass   1435
D.I. Bransby, Auburn Univ. (Auburn, Alabama, USA); *P. Vann Bush*, Southern Research Inst. (Birmingham, Alabama, USA); *and D.M. Boylan*, Southern Company Services (Birmingham, Alabama, USA)

Biomass Power: Electricity and Heat Systems

The Flex-Microturbine™ for Biomass Gases – A Progress Report   1439
E. Prabhu, Reflective Energies (Mission Viejo, California, USA); *V. Tiangco*, California Energy Commission (Sacramento, California, USA)

Micro-Scale Biomass Power   1445
L. Bowman and N.W. Lane, Sunpower, Inc. (Athens, Ohio, USA)

Biomass Energy Use in Small-Scale Commercial Operations   1453
K.M. Sachs, Carbon Cycle Co. (Woodland, California, USA)

Adaptation and Evaluation of a Rice Hull Gasifier in the Philippines 1457
R.E. Aldas and E.U. Bautista, Philippine Rice Research Inst. (Nueva Ecija, Philippines); and V.M. Tiangco, California Energy Commission (Sacramento, California, USA)

Commercialization of Innovative Biomass-Fired Furnaces for Heating Poultry Houses 1465
J. Wimberly, Foundation for Organic Resources Mgmt. (Fayetteville, Arkansas, USA)

The Use of Exhausted Olive Husks as Fuel in the Calabrian Bread-Baking Industry 1467
G. Nicoletti, Univ. of Calabria (Arcavacata di Rende, Italy)

An Advanced Modular Fluid-Bed Combustor Concept for Biomass Utilization 1475
D.D. Schmidt, J.H. Pavlish and M.D. Mann, Energy & Environmental Research Center (Grand Forks, North Dakota, USA); and M.F. Robb, King Coal Furnace Corporation (Bismarck, North Dakota, USA)

Sulphur Balances for Biofuel Combustion Systems 1481
S. Houmøller and A. Evald, Teknik Energy & Environment (Søborg, Denmark)

Feasibility Studies on Cogeneration from Industrial Wood-Processing Residues in Ghana 1487
A. Brew-Hammond and S.B. Atakora, Kumasi Inst. of Technology and Environment (Kumasi, Ghana)

Feasibility Study of a California Landfill Sited 10 MW_e Waste-to-Energy Project Driven by Landfill Cutbacks and Electricity Deregulation 1495
M.D. Lefcort and F.G. Ghahremani, HEITE Associates (Irvine, California, USA)

Assessment and Dissemination of a Pyrolysis Gasifier Stove in Ghana – Project Description 1501
P.S. Nielsen, Technical Univ. of Denmark (Lyngby, Denmark); and T. Kuuyuor, Ministry of Interior (Accra, Ghana)

Straw Combustion in Grate Furnaces 1507
R.P. van der Lans, L.T. Pedersen, A. Jensen, P. Glarborg and K. Dam-Johansen, Technical Univ. of Denmark (Lyngby, Denmark)

Novel Approach for the Analysis of Heat Transfer in Bagasse-Fired
Furnaces 1509
A.L. Brito and P.A. Beaton, Univ. of Oriente (Santiago de Cuba, Cuba);
J. Ballester and C. Dopazo, Univ. of Zaragoza (Zaragoza, Spain)

Retrofitting and Optimisation of Bagasse Boilers 1517
P.A. Beaton and A.L. Brito, Univ. of Oriente (Santiago de Cuba, Cuba);
J. Ballester and C. Dopazo, Univ. of Zaragoza (Zaragoza, Spain)

BIOMASS AND BIOENERGY POLICY – PUBLIC ISSUES AND PRIVATE INITIATIVES

Green Electricity

Wood Energy and European Trade Patterns: Why Sweden Is the No. 1
Biofuel Importer in Europe 1527
B. Hillring and J. Vinterbäck, Swedish Univ. of Agricultural Sciences
(Uppsala, Sweden)

Fuel Cells: A Solution for Pollution 1533
P.S. Patel, Energy Research Corp (Danbury, Connecticut, USA); *V. Tiangco*, California Energy Commission (Sacramento, California, USA);
and *K. Craig*, NREL (Golden, Colorado, USA)

Including Biomass in Wisconsin's Public Benefits Pilot 1535
D.B. Wichert and A.F. De Pillis, Wisconsin Department of Administration
(Madison, Wisconsin)

Power Generation from Poultry Litter Biomass 1541
T.A. Giaier and R.S. Morrow, Detroit Stoker Co. (Detroit, Michigan, USA)

Green Power Marketing in Retail Competition: An Early Assessment 1547
R. Wiser, Lawrence Berkeley National Laboratory (Berkeley, California, USA); *K. Porter and J. Fang*, NREL (Washington, DC, USA)

Landfill Fuel Cell Application: A Concept to Utilize All Outputs through
Integration with a Greenhouse 1555
T.O. Manning, D.R. Specca and H.W. Janes, New Jersey EcoComplex
(New Brunswick, New Jersey, USA)

Sugarcane Bagasse and Black Liquor Substituting for Fossil Fuels in
Power Generation in Brazil as Responses to Global Climate Change   1557
M.D. Berni and S.V. Bajay, State Univ. of Campinas (Campinas, Brazil)

Opportunities for Biomass in the APX Green Power Market™   1559
J.C. Pepper, Automated Power Exchange, Inc. (Cupertino, California, USA)

The Impacts of Assembly Bill 1890 on California's Biomass Energy
Industry   1561
G. Simons, V. Tiangco, P. Kulkarni and M. Masri, California Energy
Commission (Sacramento, California, USA)

Landfill Leachate Management Using Short Rotation Forestry Plantations   1563
D. Riddell-Black, WRC plc (Marlow, UK)

A Technical-Environmental Comparison of Some Agricultural Biomasses   1565
G. Nicoletti, F. Anile and C. Marandola, Univ. of Calabria (Arcavacata di
Rende, Italy)

The Effect of Wisconsin's 50 MW Renewables Mandate on Biomass
Power Plant Development   1573
A.F. DePillis and D.B. Wichert, Wisconsin Energy Bureau (Madison,
Wisconsin)

Policy and Programs

DOE Activities to Support Opportunities and Benefits of Biomass
Cofiring in Coal-Fired Utility and Industrial Boilers   1577
M.C. Freeman, P.M. Goldberg and S.I. Plasynski, Federal Energy
Technology Center (Pittsburgh, Pennsylvania, USA); R. Costello, U.S.
Department of Energy (Washington, DC, USA)

Biomass Policies and Biogas Utilisation in Denmark   1579
P.S. Nielsen and K. Salomonsen, Technical Univ. of Denmark (Lyngby,
Denmark); and J.B. Holm-Nielsen, South Denmark University (Esbjerg,
Denmark)

Biomass Programme of the Bavarian State Government   1587
T. Weber and L. Wanner, Bavarian Ministry of Food, Agriculture and
Forestry (Munich, Germany)

Biomass Energy Research, Development and Demonstration Needs and
Directions in California under a Deregulated Electricity Marketplace 1593
G. Simons, V. Tiangco and P. Kulkarni, California Energy Commission
(Sacramento, California, USA)

The Role of the Australian Biomass Taskforce in Bioenergy Development 1595
S.M. Schuck, Australian Biomass Taskforce (Killara, Australia)

Biomass Fuels, Technical Interventions, and Environmental Impacts in
the Indian Himalaya 1601
R. Prasad and A. Sharma, TERI (New Delhi, India)

Biomass Cofiring R&D in the FETC Combustion and Environmental
Research Facility 1607
M.C. Freeman, P.M. Goldberg, R.A. James and M.P. Mathur, Federal
Energy Technology Center (Pittsburgh, Pennsylvania, USA); *and G.F.
Walbert*, Parsons Infrastructure & Technology Group, Inc.

The Advanced Technology Program's Support of High-Risk R&D
Involving Biomass 1609
R.L. Bloksberg-Fireovid, National Inst. of Standards and Technology
(Gaithersburg, Maryland, USA)

Asia/Pacific Rim Renewable Energy Market Assessments by the State of
Hawaii 1611
D.M. Ishimura, C.M. Kinoshita and S.Q. Turn, Univ. of Hawaii at Manoa
(Honolulu, Hawaii, USA); *M.H. Kaya, J. Tantlinger and J.P. Dorian*, State
of Hawaii Department of Business, Economic Development and Tourism
(Honolulu, Hawaii, USA)

A Market-Based Development Strategy for Biomass and Bioenergy in
China 1619
Z. Zhang, A. Zhou, Z. Lv, B. Zeng and W. Luo, Energy Research Institute
of State Planning and Development Commission (China); Rural Energy
Office, Shandong Province (China); Sanxing Energy and Environmental
Engineering Corp (Shanghai, China); Rural Energy Office (Zhejiang,
China)

Development of Green Energy Market in the Netherlands and the
Perspectives of Biomass 1629
K.W. Kwant and C. van Leenders, Netherlands Agency for Energy and
the Environment (Utrecht, Netherlands)

Strategies and Assessments

IEA Bioenergy, Liquid Biofuels Activity: Results of the Past Three-Year Period   1633
M. Wörgetter, Federal Inst. of Agricultural Engineering (Wieselburg, Austria)

Biomass Feedstocks: Integration of Research and Development for Multiple Products and Multiple Sponsors   1639
L. Wright and J. Cushman, Oak Ridge National Laboratory (Oak Ridge, Tennessee, USA); *S. Sprague and J. Kaminsky*, U.S. Department of Energy (Washington, DC, USA)

PyNe – The Pyrolysis Network   1647
A.V. Bridgwater, N. Ahrendt and C.L. Humphreys, Aston Univ. (Birmingham, UK)

Review Report of the Regional Biomass Energy Program Technical and State Grant Projects   1655
P.D. Lusk, Resource Development Associates (Washington, DC, USA)

Changing Paradigm: Converting California's Waste Biomass and Waste Sugars to Value-Added Products   1663
S. Shoemaker, Univ. of California (Davis, California, USA)

Using a Direct Public Offering as a Means to Fund Your Company   1665
D.A. Johnson and D.A. Johnson, Pinnacle Technology, Inc. (Lawrence, Kansas, USA); *B.S. Davis*, Corporate Counsel Group LLP (Kansas City, Missouri, USA); *and W. Eckinger*, Henderson, Warren & Eckinger PC, PA (Overland Park, Kansas, USA)

A Brief History of Energy Biomass in Brazil   1673
M.A. dos Santos, Ed. Tecnologia (Rio de Janeiro, Brazil)

Effects of the California Energy Commission's Renewable Technology Program on California's Biomass Industry   1679
T.A. Gonçalves and M. Masri, California Energy Commission (Sacramento, California, USA)

Proposals for the Improvement of Biomass Participation in the Brazilian Energy Matrix: The "Declaration of Recife"   1685
S.T. Coelho, J.R. Moreira, I.A. Campos and A.C. Oliveira, Univ. of São Paulo (São Paulo, Brazil)

Wood as a Primal Fuel for Rural Areas of Estonia — 1691
P. Muiste, Estonian Agricultural Univ. (Tartu, Estonia); and Ü Kask, Tallinn Technical Univ. (Tallinn, Estonia)

Resource Potential and Opportunities for Biogas Energy Projects in California under a Deregulated Electricity Marketplace — 1695
G. Simons, V. Tiangco and J. Young, California Energy Commission (Sacramento, California, USA); and M. Moser, Resource Conservation Management (Berkeley, California, USA)

Toward a Biomass-Intensive Sustainable Energy Strategy for Indiana — 1697
J. Eflin, Ball State Univ. (Muncie, Indiana, USA)

Minnesota Wood Energy Scale-Up Project: A Progress Report 5 Years after Establishment — 1703
M. Downing, Oak Ridge National Laboratory (Oak Ridge, Tennessee, USA); D. Langseth, Champion International (Alexandria, Minnesota, USA); T. Lundblad, WesMin Resource Conservation & Development (Alexandria, Minnesota, USA); R. Pierce, Champion International (Alexandria, Minnesota, USA); and R. Stoffel, Minnesota Department of Natural Resources-Forestry (Alexandria, Minnesota, USA)

Accelerating the Commercialization of Biomass Energy Generation within New York State — 1711
G.J. Proakis, New York State Technology Enterprise Corp. (Rome, New York, USA); J.J. Vasselli, Syracuse Research Corp. (North Syracuse, New York, USA); E. Neuhauser, Niagara Mohawk Power Corp. (Syracuse, New York, USA); and T.A. Volk, State Univ. of New York (Syracuse, New York, USA)

The German Strategy to Increase Renewable Resources for Energy and Industry — 1717
P. Kornell and E. Langer, C.A.R.M.E.N. (Rimpar, Germany)

The Ethanol Fuel Race — 1725
S. Vyas, McNeil Technologies, Inc. (Springfield, Virginia, USA)

Biomass and Bioenergy: Public Issues

Community Outreach and Education: Key Components of the Salix Consortium's Willow Biomass Project　1733
T.A. Volk, State Univ. of New York (Syracuse, New York, USA); *S. Edick*, South Central New York Resource Conservation and Development Project, Inc. (Norwich, New York, USA); *S. Brown*, Cornell Cooperative Extension (Ithaca, New York, USA); *and M. Downing*, Oak Ridge National Laboratory (Oak Ridge, Tennessee, USA)

Biofuels in the Industry: Reduction of CO_2 Emissions from the Industrial Sector　1739
S. Houmøller, J. Cramer, B.H. Christensen and L. Jørgensen, Teknik Energy & Environment (Søborg, Denmark)

The Value of Survey Information on the Potential to Grow Switchgrass for Energy in Alabama　1747
D.I. Bransby, Auburn Univ. (Auburn, Alabama, USA)

Energy Crop Farmer Education and Closed-Loop Biomass Power Plant Development in Wisconsin　1753
A.F. De Pillis and D.B. Wichert, State of Wisconsin Department of Administration (Madison, Wisconsin)

Oklee Tree Project　1759
E. Wene, Agricultural Utilization Research Inst. (Crookston, Minnesota, USA); *and W. Johnson*, University of Minnesota (Crookston, Minnesota, USA)

Hybrid Poplar on Conservation Reserve Program Land: Farm Producer Information and Education　1765
M. Downing, Oak Ridge National Laboratory (Oak Ridge, Tennessee, USA); *R. Pierce*, Champion International (Alexandria, Minnesota, USA); *and R. Stoffel*, Minnesota Department of Natural Resources-Forestry (Alexandria, Minnesota, USA)

Switchgrass for Energy in Southern Iowa: Developing and Maintaining Producer Involvement　1771
J.T. Cooper and M.L. Braster, Chariton Valley Resource Conservation and Development, Inc. (Centerville, Iowa, USA)

Barriers to Implementation of Waste-to-Energy (WTE) Technologies in Brazil 1777
S.S.P. Mercedes, I.L. Sauer and S.T. Coelho, Univ. of São Paulo (São Paulo, Brazil)

Author Index I1

PREFACE

Conference proceedings are unique barometers that indicate the current issues and progress of a given scientific field. Breakthroughs in technology, as well as the political, economic, business and social context in which they are developed, are captured in them. Very often, many of the contributions represent work in progress. Indeed, some of the papers in this volume are only abstracts because the lead time for the production of these proceedings meant that final papers about work in progress were unavailable for inclusion at press time.

The Conference of the Americas series is held biennially in the 'odd' years in opposition to those of the European Union, which holds its major meetings in the 'even' years. Since biomass and bioenergy are global issues, it is no surprise that each year there are contributions from countries all over the world. Contributions from South America have been increasing ever since conference inception in 1993, and it is particularly pleasing to note the extensive contributions to this volume from Brazil.

Between the Third Biomass Conference of the Americas and this one, two conferences of the parties (COP) to the United Nations Framework Convention on Climate Change (UNFCCC) occurred. In December of 1997 in Kyoto, Japan, COP3 established a list of mainly industrialized countries committed to a range of greenhouse gas (GHG) reductions by 2008 to 2010. These countries included all 29 countries belonging to the Organization for Economic Co-operation and Development (OECD). The Inter-governmental Panel on Climate Change is investigating how biomass, bioenergy, and the biosphere fit into the question of GHG mitigation. A significant theme of these proceedings is the role of biomass and bioenergy in carbon management. One of the tools with which this issue is increasingly addressed is life-cycle analysis (LCA). Although it originated in the assessment of sustainability—a theme that came into prominence with Agenda 21 of the United Nations Conference on Environmental Development in Rio de Janeiro in 1992—it has become so pervasive that it no longer merits a topic area of its own in this conference series.

The major part of the contributions in these proceedings fill in the intersections of two vectors: the applications of biomass and bioenergy, and the production chain from biomass to bioenergy. Applications of biomass can be seen in six major categories:
- Daily Living - cooking and space heating at the household level
- Community needs
- Industry captive uses of bioenergy
- Materials and chemicals uses in conjunction with energy production or substitution
- Environmental services
- Secondary energy forms such as electricity and biofuels that carry biomass derived energy from the rural and forestry sectors to the increasingly urban populations of the world

The production chain addresses biomass resources, feedstocks, and their conversion to heat, biofuels, and value-added products.

If the contents of these proceedings are scrutinized, it is immediately clear that of the 25 technical areas, almost one-quarter represent a third dimension in our taxonomy. Linking the two-dimensional surface of the production chain and the applications to their realization are systems integration needs; green electricity; technology and resource assessments; implementation strategies; and municipal, state or province, and country-level policies and programs.

A slogan of our era is the need to 'think globally but act locally.' A feature of the Conference of the Americas series is that it is hosted by different cities and states in the Americas. The first two conferences were held in the Northeastern and the Northwestern United States and the third conference was in Canada. Each of these conferences highlighted the special issues that challenge or facilitate the adoption of biomass and bioenergy at local levels in each region. These proceedings are from the fourth conference, held in Oakland, California, a State with a population of over 30 million. California encompasses a wide variety of biomass, from an almost Mediterranean climate to extreme deserts. Some of the world's highest concentrations of high technology industries and one of the world's most productive and high-value agricultural sectors fit within its borders. Biomass and bioenergy has featured in the economic growth of California from the beginning, and California has been a leader in the development of biomass power. About 15% of the U.S.A.'s grid-connected biomass power generation has been put into place in California since the early 1980s.

California is now at the forefront of innovation in several areas that will become widespread in the United States in the next decade. It leads in the deregulation of its electricity sector. The need for renewable energy contributions must be maintained under circumstances where the State had previously used subsidy mechanisms (such as the Public Utility Regulatory Policies Act of 1978 [PURPA]) to create a large biomass and wind power sector. This need has been addressed in several ways that are detailed in these proceedings. Environmental applications of biomass and bioenergy are being addressed not only in technologies such as anaerobic digestion, which is a major section of the proceedings, but also through environmental and ecosystem policies that integrate biomass and bioenergy into the larger scheme of sustainable development. California is already adopting laws to reduce the burning of agricultural residues such as rice straw and seeking forest management strategies that will minimize forest fire degradation to maintain forest health and protect watersheds. In this context, biomass and bioenergy are integral to the implementation of the policies and strategies being used to address these issues.

The theme of these proceedings is that biomass is a growth opportunity in green energy and value-added products. The papers detail how the modernization of the biomass energy sector, which started with the energy crises of the 1970s and 1980s, is now starting to mature. Novel processes such as fast pyrolysis, which did not even exist before 1970, are now reaching the market place. Biomass has moved from being merely a proposed energy source to being an advanced and modern one. Many elements of the biomass and bioenergy production chain that were proposed a decade ago are now in these pages as on-going demonstrations. Demonstrations of short-rotation wood crops as fiber and energy sources are reaching large scales; technology demonstrators of both electricity and liquid fuels production are also widespread and are detailed here. Major biomass and bioenergy and value-added-materials implementation programs are underway or about to start in many areas of world. The experiences of those programs currently underway are recorded in these proceedings, and the analysis of these experiences is a major contribution to be found in these volumes.

On behalf of all those who have contributed to these proceedings, we hope that the results and ideas in these pages represent not just the current health and pulse rate of a new industry, but that they can be used as guides through the myriad challenges of moving from the 20th to the 21st century.

Acknowledgments

The proceedings of the Fourth Biomass Conference of the Americas would not have been compiled without the commitment, intensive effort, creativity and patience of Dee Scheaffer in working with the authors, her assistants (Dimi Currey, Rick Minnotte, David Wright, Nancy Wells, and Lynda Wentworth), NREL's word processing team (Irene Medina, Patricia Haefele, and Judy Hulstrom), and the production team at Elsevier Science. The editors also thank Mary-Margaret Coates for her able editorial assistance.

The sponsorship of the U. S. Department of Energy, Natural Resources Canada, and the California Energy Commission for these proceedings is gratefully acknowledged.

Ralph P. Overend and Esteban Chornet
Golden, Colorado, U.S.A., and Sherbrooke, Québec, Canada.
June 1999

DO WE NEED A BIOMASS ENERGY POLICY? THE CALIFORNIA EXPERIENCE

Gregory Morris

Future Resources Assoc., Inc., and the Green Power Institute, 2039 Shattuck Ave., Suite 402, Berkeley, CA 94704

California is the world leader in the production of electricity from biomass. Biomass energy facilities have become a crucial part of the state's solid waste disposal infrastructure, providing a beneficial-use disposal option for a wide variety of the state's forestry, agricultural, and urban biomass residues. In the peak year of 1990, the biomass energy industry converted more than ten million tons of the state's solid waste into two percent of its electricity supply. However, those heady days are a thing of the past. Deregulation of the electric utility industry has led to the closure of thirty percent of the industry since 1994, and the low energy prices expected in the competitive energy market threaten to bring more closures and cutbacks. The future will be bleak unless policies are developed to reward biomass power generation for the waste-disposal services that, until now, have been provided free of charge.

1. THE CALIFORNIA BIOMASS ENERGY INDUSTRY

Two oil-price shocks during the 1970s led to the enactment of a variety of programs and incentives designed to stimulate the development of renewable energy sources. California led the country in the development of all kinds of renewables, including biomass. The biomass facilities built in California were designed to burn agricultural, forestry, and urban biomass wastes. By the early 1990s, almost 750 MW of biomass generating capacity were operating in the state, and the future appeared to be bright. Changing circumstances in world energy markets, however, culminated in the initiation of the electric utility deregulation process in California in 1994. One immediate and unintended consequence was a spurt of contract buyouts and facility shutdowns that caused biomass power production in the state to shrink by about thirty percent during a two year period. Transition support for biomass during the deregulation process has stabilized the industry for now, but there is great uncertainty ahead.

1.1. The 1980s: incentives for development

The federal PURPA law of 1978 established the framework for sales to utility companies of electricity produced by private generating sources. This was augmented by a series of tax credits and other incentives for investments in biomass and renewable energy technologies. At the same time, California, which was particularly dependent on oil, enacted a series of parallel policies to promote the development of renewables in the state. For a variety of reasons, California provided fertile ground for renewable energy development.

When the biomass energy development cycle was initiated in the early 1980s, there were less than 50 MW of biomass generating capacity in California. Ten years later, more than 750 MW of generating capacity was in operation, and an additional 100 MW of facilities were in advanced stages of construction. Sixty-one biomass energy facilities were eventually built. The rapid growth in biomass generating capacity and fuel demand peaked during 1990, when 230 MW of new capacity were placed into service. Figure 1 shows a map of the biomass generating facilities in California.

Annual biomass fuel consumption during the 1980s increased from less than 0.85 million tons (0.5 mil. bdt) at the beginning of the decade to more than ten million tons (5.5 mil. bdt) by the end. This rapid growth in demand put considerable pressure on the supply of biomass fuels. By 1990, average biomass fuel prices in the state had more than doubled to more than $40 per bdt. Many observers began to refer to the situation as a fuel crisis, but fuel prices were simply responding to increased demand. Figure 2 shows the supply curve for biomass fuels in California.

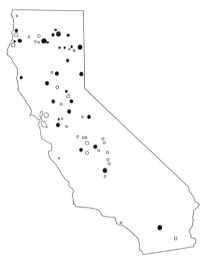

Figure 1. California Biomass Energy Facilities, 1999. Closed: operating. Open: shut down or dismantled.

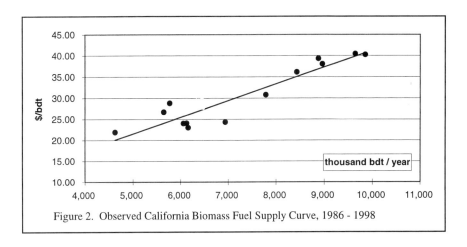

Figure 2. Observed California Biomass Fuel Supply Curve, 1986 - 1998

1.2. The 1990s: incentives for contraction

The California biomass energy industry operated at a fairly steady level during the early 1990s. The high fuel prices that predominated during this period led to the shut down of some of the least profitable facilities, while the last of the facilities from the standard-offer cohort entered operations. The situation changed dramatically beginning in April, 1994, when the California Public Utilities Commission (CPUC) published its initial proposal to restructure the electric utility industry. The proposal envisioned a market in which energy sources competed on the basis of price alone, without regard to environmental or social impacts. This caused a flurry of activity on the part of the regulated utility companies to shed their most costly supply sources. During the next two years 16 biomass facilities (200 MW) agreed to contract buyouts and halted operations. Three million tons per year of biomass re-entered the state's solid waste stream.

After three years of intense debate, California passed comprehensive electric utility restructuring legislation in late 1997. AB 1890 provided funding to allow renewable generating sources to make the transition to the competitive marketplace, and it recognized biomass generation explicitly for its provision of valuable environmental services. Cal/EPA was directed to report on measures to shift some of the costs of biomass power generation away from electric ratepayers and onto the beneficiaries of the waste disposal services, in order to allow biomass power generation to be able to compete in the deregulated market.

The industry stabilized again by the end of 1995, after the initial flurry of contract buyouts that followed the initial restructuring proposal. The surviving industry consisted of 550 MW of operating biomass capacity, providing for the annual disposal of six million tons (3.5 mil. bdt) of the state's solid waste. However, this stability is imperiled by the convergence of several factors that threaten to shut down some of the currently operating facilities, and reduce operations for many of the ones that manage to remain in operation. Figure 3 shows the history of biomass fuel use in California since 1980. The figure also shows projections of

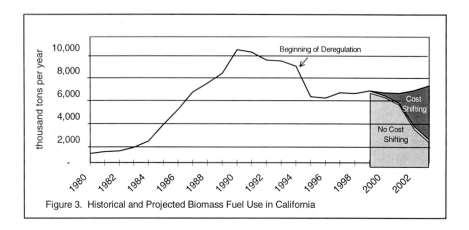

Figure 3. Historical and Projected Biomass Fuel Use in California

future biomass use under two possible scenarios, one in which cost-shifting measures are enacted, and one in which no cost-shifting measures are enacted.

2. THE CALIFORNIA BIOMASS INDUSTRY AT THE CROSSROAD

Today the California biomass energy industry finds itself caught in an untenable position. On the one hand, the state has come to depend on the ancillary services provided by biomass energy production. On the other hand, biomass power is costly to produce, and the basic premise of electric utility deregulation is to lower the cost of electricity for consumers. The future commercial viability of many California biomass power generators is very much in doubt. Only an active public policy that is focused on balancing the costs and benefits of biomass energy production can insure that the future direction taken by the California biomass energy industry is the optimal one for society.

2.1. The first and second laws of biomass dynamics

The future of biomass energy production will have to conform with two immutable laws of nature, which I call the two laws of biomass dynamics. Any successful policy to deal with biomass matters will have to conform to the strictures of these laws. The short form of the first and second laws can be expressed as follows:

1. Biomass energy production provides valuable ancillary services.
2. Biomass energy production is expensive.

Like all renewable energy sources, biomass energy displaces fossil fuel use and its associated environmental impacts. However, biomass energy is unique in providing additional environmental benefits to society by providing an environmentally preferred disposal option for a portion of society's solid waste stream. Biomass fuels would otherwise end up being disposed of by burial in landfills, by open burning in agricultural and forested regions, or as an accumulation of excess biomass in the state's forests. Landfill burial of wood wastes leads to leachates and greenhouse gas emissions, and it accelerates landfill capacity depletion. California counties are under state mandate to divert fifty percent of its solid waste disposal by the year 2000, and energy production represents a legitimate, socially beneficial diversion option. Open burning of agricultural and forest residues are major, and at times highly visible, sources of air pollution in some regions of the state. Controlled combustion in energy generating facilities reduces emissions of most pollutants by at least two orders of magnitude compared with open burning, in addition to displacing the burning and emissions of fossil fuels in conventional power plants. Overstocking of biomass in the state's forests increases wildfire hazards, depresses watershed productivity, and degrades overall forestry health and productivity. California's ability to comply with its various environmental obligations depends, in no small part, on the waste-disposal contributions provided by the biomass energy industry.

On the other side of the ledger, biomass energy production has costs that are inherently high compared with those of other electric generation options. This is an inevitable consequence of two characteristics of biomass energy systems. First, biomass fuels are low-density materials, which means that they are expensive to gather, process, transport, and handle. Second, biomass generating facilities tend to be smaller than conventional generating sources, which means that biomass power generation is not able to realize the kinds of economies of scale that conventional generating sources enjoy. Table 1 shows the cost of biomass power generation in California today (Morris, 1998). Modern gas-fired facilities can generate power for costs in the range of 2.5 - 3.5 ¢/kWh.

Table 1
Cost of biomass power generation in California.

	¢/kWh
Capital	2.1
Fuel	2.2
Non-fuel O&M	2.3
Total	6.6

2.2. Striking a balance

Biomass power production clearly provides major social and environmental benefits to California. However, biomass energy costs more to produce than other sources available in the marketplace. In the era of high conventional energy prices that spawned the modern biomass energy industry, the price of electricity was high enough to pay for all of the ancillary benefits. In the modern world of low-cost fossil fuels, however, biomass energy production is not the low-cost electric-generation alternative, and electricity sales cannot support all of the ancillary benefits.

The inevitable consequences of the first and second laws of biomass dynamics leave society with two options: (1) Let the competitive electricity market function solely on the basis of economic factors, unfettered by social and environmental considerations, and lose many of the environmental services provided by biomass energy production. (2) Perform some kind of public-policy balancing act that rewards biomass energy for its social and environmental benefits, so that the reward, combined with the compensation available from electricity sales, is sufficient to make biomass power production a viable enterprise. Only the second option has the potential to preserve the industry and promote its future growth.

The modern biomass energy industry in California was built as a result of a series of incentives that were enacted in response to the energy crises of the 1970s. During the policy debates of that era, biomass advocates argued that if incentives were created for biomass energy systems, society would benefit by receiving cleaner and more secure domestic energy. Now, twenty years later, oil supplies appear to be plentiful, and biomass advocates find themselves arguing that if effective biomass policies are not once again enacted, the environmental services that the state has come to depend on will disappear, with distinctly negative environmental and social consequences.

3. THE COMPETITIVE ELECTRICITY MARKET

California's electric utility restructuring legislation contains language encouraging the development of policies to shift some of the costs of biomass energy production away from the electric ratepayer and onto the beneficiaries of the environmental services. A number of cost-shifting policies have been proposed, but little progress has been made in implementing them. Table 2 lists some of the most promising of the proposed cost-shifting measures under consideration.

Table 2
Cost-shifting policies for biomass energy.

General support measures
    Biomass production tax credit
    Minimum content requirement
    Surcharge-supported production credit

Targeted measures
    Expanded credit for wood diverted from landfill disposal
    Agricultural fuels tax credit
    Agricultural burning permit program
    Allocation of state and federal forest service funds for fire prevention treatments
    Water-sales surcharge for watershed improvement treatments
    Grants and low-interest loans for fuel production equipment

AB 1890 provides for the payment of decreasing transition subsidies to biomass power producers during the four-year implementation period for electric utility restructuring, which ends after 2001. The legislation envisions the phase-in of cost-shifting measures that will substitute for the subsidy payments and provide for the long-term competitiveness of biomass power generation. So far, the development of effective cost-shifting measures for biomass power generation in California has proceeded very slowly, and no measures are yet in place.

An optimal cost-shifting program for biomass power production would consist of a collection of measures, which can be grouped into two categories. First, an overall support program for biomass power production is needed, in order to recognize the fact that all production of energy from biomass provides valuable benefits to society. Modification of the federal biomass tax credit (IRS § 45) would be one way to accomplish this, and would be effective nation-wide. Second, targeted cost shifting policies should be designed to promote the use of designated types of biomass residues to solve specific environmental problems associated with the disposal of those residues. Together, these programs could insure a bright future for biomass power generation in California.

REFERENCE

Morris, G. (1998). The Implications of Deregulation for Biomass and Renewable Energy in California. Report No. NREL/SR-570-24851. Published by the National Renewable Energy Laboratory (Golden, Colo.)

Biomass Production and Integration With Conversion Technologies

Biomass Resources: Willows

A PROFILE AND ANALYSIS OF WILLOW GROWERS IN SWEDEN

H. Rosenqvist, A. Roos, E. Ling, and B, Hektor

Department of Forest Management and Products, Swedish University of Agricultural Sciences, P. O. Box 7054, S-750 07 Uppsala, Sweden

Willow plantations on Swedish farmland increased considerably between 1991 and 1996. The main driving forces behind this development were (1) the 1991 introduction of an agricultural deregulation policy in Sweden which created lower grain prices and simultaneously introduced compensation for set-aside land and subsidies for willow plantations on surplus arable land, (2) higher taxes on fossil fuels, and (3) the existence of a biofuel market in Sweden based on forest fuels. This paper presents a statistical study of 1,158 willow growers in southern and central-eastern Sweden. The resulting profile of growers will help policy makers and agents in the bioenergy business design information campaigns and marketing strategies. Willow growers are described according to geographical distribution, willow parcel sizes, farm sizes, and farm types. They are compared with the population of farmers who are not growing willow. Willow growers are more often between 50-65 years of age, and they have larger farms than non-willow growers. They are less often focused on animal and milk production and more often on cereal and food crop production than are other farmers.

1. INTRODUCTION

During the past two decades Swedish authorities and energy policy makers have regarded willow as an important alternative crop that could produce woodfuel on Swedish farmland. The expansion of full-scale willow plantations for energy use, however, took place only in the early 1990s, for three main reasons. First, subsidies for set-aside land and subsidies for planting willows were generous between 1991 and 1996, amounting to 1200 ECU/ha, plus 480 ECU/ha for fencing. The subsidies were one part of the transition program of the new, deregulated agricultural policy set in 1990, which sought to remove 500,000 ha from price-regulated food production into other uses (Anon., 1990). Second, biofuels became more competitive as environmental and energy taxes on fossil fuels increased considerably in 1991. Third, a major biofuel market for the district heating sector was already established and growing in Sweden in the early 1990s. These three factors contributed to a dynamic expansion of willow cultivation: the area planted with willows increased from almost nil to 17,000 ha between 1991 and 1996 (Larsson and Rosenqvist, 1996).

Few studies to date contain detailed analyses of how an increase in biofuel crop production can develop. One key question is: where, and on what type of farm, is short-rotation coppice production likely to take place? Several studies from the United States use basic biological and economic factors to model biomass fuel production using a geographic information system (GIS) (Downing and Graham, 1996; Graham et al., 1996; Noon and Daly, 1996). This empirical study of willow growing in Sweden in the 1990s provides empirical information about the effects of incentives for increasing bioenergy feedstock production. Moreover, the profile of the farm types and willow-growing farmers can show where and on what type of farm energy crop production is most likely to take place.

This paper describes the willow growers in southern and central Sweden in terms of the age of the farmer and the geographical location, size, and type of farm enterprise.

2. METHOD AND DATA

The investigation concerns farm enterprises in southern and central Sweden that were growing willows in 1995. Data is obtained from the 1995 Farm Register (FR) compiled by Statistics Sweden (Anon., 1995). The sample from the large FR includes information about all 1,158 farms that grow willow. For comparison, a sample of 535 non-willow growers was also obtained from the same register. This sample was stratified according to farm size class, frequency, and known farm size distribution of the willow growers. Weights were assigned according to sample possibilities.

3. RESULTS OF DATA ANALYSIS

There were 1,158 recorded willow growers in southern and central Sweden in 1995, growing about 13,300 ha of willows. In relative terms, 1.6% of the farmers in the region grew willow on 0.54% of the total area of farmland. Willow was most likely to be grown in counties with average grain yield and least likely in counties with lower-than-average grain yield.

The average willow area on a willow farm is 11.5 ha. About 74% of the willow area is composed of plots larger than 10 ha, owned and managed by 34% of the willow growers. Eight farms planted 100 ha or more to willow; the largest willow plantation covered 143 ha. In relative terms, average willow farmers use 10.5% of their arable land for willows.

Sixty three percent of the farmers growing willows are between 40 and 65 years old. Their mean age, 52.8 years, is slightly lower than that of the comparison group, 55.5 years.

The mean arable land area of the willow-growing farms is 109 ha, compared to 33 ha for farms without willow. More than 77% of the willow-covered area lies on farms with more than 30 ha of arable land. The combined effect of farm size and age class on the likelihood that willows are grown is shown in Table 1 below.

Table 1
Percentage of willow growers per area class and age of farmers.

Area, ha	Owner type/Age class				
	Institutional and Ltd.	0–30 yrs	30–50 yrs	50–65 yrs	65-above yrs
2-10	0.72	0	0.37	0.25	0.19
10-50	1.19	0.53	0.94	1.44	0.69
50-100	3.53	0.30	2.80	4.35	1.93
100	10.20	1.14	9.71	9.27	7.61

Farmers on large estates in the 30–50 and 50–65 year age classes are relatively likely to grow willows, while the percentage decreases for smaller farms and for both younger and older farmers. Of farms larger than 100 ha, as many as 10% grow willow.

Table 2 shows the percentage of different sub-groups among willow growers and others. The last column in the table presents the change in percentage units of probability for willow production if the farm belongs to a given sub-category.

Willow growers and non-willow growers own about the same amount of forest land and pasture land. Tenancy is related only modestly to willow growing, but leasing out is positively associated with willow production. Most food crops show positive correlations with use of land for short-rotation willow coppice. Milk production, cattle, pasture, and fodder production, on the other hand, all are negatively related to willow growing.

4. DISCUSSION

The conditions conductive to willow cultivation around Lake Mälaren in eastern Sweden may be analysed in terms of the production capacity of the land. The grain yield around Lake Mälaren is somewhere in the middle of the studied counties. Profitability assessments (Rosenqvist, 1997) show that willow is less competitive on high-yield grain land. Land with lower-than-average grain yield, on the other hand, usually produces fodder or cattle. Therefore, land with average grain productivity, as represented by the counties around Lake Mälaren, seems to offer more suitable conditions for willow growing. The area around Lake Mälaren also has a relatively large share of set-aside land, which indicates a high potential for land for willow plantation. A high percentage of set-aside land is an indicator of willow plantation in itself.

Willows are generally produced on larger farms, which might be better able to assess and diversify risks connected with new crops. Wålstedt et al. (1993) have noted, for example, that recognition of the need to compensate for risks increases with the size of the farm enterprise. Leasing out of farmland may also indicate a labour-extensive management style; willow production requires only moderate labour inputs after the plantation year.

Table 2

Willow growers and non-willow growers on various types of farm.

Type of Farm	Willow Growers, %	Non-Willow Growers, &	Probability of Growing Willow, Change in %
Land class			
Forest	76	72	+0.07
Pasture	56	57	-0.03
Leasing			
Tenancy of farmland	53	51	+0.07
Leasing out	15	7	+1.52
Land use			
Winter wheat	49	19	+4.12
Spring wheat	19	5	+1.81
Barley	66	49	+0.52
Oats	52	43	+0.34
Ray	15	6	+2.0
Potatoes	14	11	+0.53
Sugar beets	12	7	+1.07
Fodder production	82	61	-0.41
Oil crops	40	13	+3.3
Animals			
Milk cows	7	20	-1.01
Cattle	29	48	-0.64
No milk or cattle	71	51	+0.59

Willow-producing farms less commonly include pasture or fodder, cattle, or milk production than do other farms. Perhaps farms with animals need most of the land for animal fodder, whereas food crops are positively correlated with willow, being produced on the farm. Larger farms more often produce cereal and crops, whereas small farms often produce milk and animals.

This investigation identified several differences between average Swedish farms and willow farms. Willows are generally grown on large farms and less commonly on farms with an animal or milk orientation. The effects of the Swedish incentive program for willow production have important implications. It is important for policy makers and agents in the bioenergy business to have sound knowledge of likely willow growers to effectively design information campaigns and marketing strategies.

REFERENCES

Anon. (1990). Regeringens proposition 1989/90:146 om livsmedelspolitiken. Nordstedts, Stockholm.

Anon. (1995). Information om Lantbruksregistret 1995 (Information about the Farm Register). Statistics Sweden. Örebro (In Swedish).

Downing, M. and R.L. Graham (1996). The potential supply and cost of biomass energy crops in the Tennessee Valley Authority Region. Biomass and Bioenergy, 11, pp. 283–303.

Graham, R.L., W. Liu., H.I. Jager, B.C. English, C.E. Noon, and M.J. Daly (1996). A regional-scale GIS-based modeling system for evaluating the potential costs and supplies of biomass from biomass crops. Proceedings of Bioenergy '96 - The Seventh National Bioenergy Conference: Nashville, Tennessee, pp. 444–450.

Larsson, S. and H. Rosenqvist (1996). Willow production in Sweden—politics, cropping development and economy. Paper presented at the European Energy Crops Conference 30 September–1 October 1996, Enschede, The Netherlands. Organised by BTG Biomass Technology Group B.V.

Noon, C.E. and M.J. Daly (1996). GIS-based biomass resource assessment with BRAVO. Biomass and Bioenergy, 10, pp. 101–109.

Rosenqvist, H. (1997). Salixodling—kalkylmetoder och lönsamhet. Department of Forest-Industry-Market Studies. Swedish University of Agricultural Sciences, Dissertation 24. Uppsala. (In Swedish)

Wålstedt, K., B-A. Bengtsson, H. Rosenqvist, and I.B. Carracedo (1992). Lantbrukets beslutfattande under osäkerhet. Påverkansfaktorer och lantbrukares reaktioner på förändringar i omvärlden. Dept. of Economics. Swedish University of Agricultural Sciences, Rapport 50. Uppsala. (In Swedish)

SOIL ORGANIC CARBON SEQUESTRATION UNDER TWO DEDICATED PERENNIAL BIOENERGY CROPS

B. Mehdi, C. Zan, P. Girouard, and R. Samson

Resource Efficient Agricultural Production (REAP)-Canada
Box 125, Glenaladale House, Ste. Anne de Bellevue, QC., H9X 3V9 Canada

Certain dedicated bioenergy crops, such as SRF willow and switchgrass, may have greater soil carbon storage potentials than conventional row crops such as corn, owing to their perennial nature and greater root biomass. The increased carbon sequestration with these crops may reduce CO_2 GHG emissions because the soil acts as a carbon sink. In 1993, two sites in southwestern Quebec were established to assess the potential of two dedicated perennial bioenergy crops, switchgrass and SRF willow. The potential for soil carbon sequestration of these crops was assessed in the fall of 1996 and 1998 at a 0-60 cm depth. In 1996 the soil organic carbon values showed switchgrass and willow at the Ecomuseum site to have substantially higher amounts of soil organic carbon than corn. In 1998, at 0-15 cm, willow and switchgrass had higher soil organic carbon than corn (35, 33, and 27 Mg ha^{-1}, respectively), but not from 15-45 cm depth. At 45-60 cm depth, willow had the highest soil organic carbon compared to switchgrass and corn (18, 12, and 10 Mg ha^{-1}, respectively). At the Seedfarm site, which had inherently lower soil fertility, the perennial crops in 1996 and 1998 did not differ significantly in soil organic carbon content from the corn. From 1996 to 1998 a decline in soil organic carbon was measured (-15 Mg ha^{-1} at the Ecomuseum, and –9 Mg ha^{-1} at the Seedfarm), which is most likely a result of the low rates of fertilization in the perennial crops. Perennial bioenergy crops, such as SRF willow and switchgrass, have the potential to increase soil carbon levels compared to corn. However, further information is required on the impact of fertilization practices on productivity and the impact on maintaining or increasing soil carbon levels in these crops.

1. INTRODUCTION

Agricultural soils have long been recognised by soil scientists as potential carbon reserves. However, temperate agricultural soils are not a large source or sink of carbon under current agricultural practices (Cole et al., 1997). Key strategies to increasing the carbon sequestration potential of a soil include increasing the time under which the soil is vegetated; reducing, or eliminating, soil tillage; boosting primary production and the return of organic matter to the soil; increasing the soil fertility; and increasing the use of perennial grasses and legumes. However, the greatest agricultural potential for mitigating CO_2 lies in increasing the amount and variety of plant biomass used directly for energy production as a substitute for fossil fuel energy (Cole et al., 1997).

2. MATERIALS AND METHODS

2.1 Experimental layout

The two experimental sites (Seedfarm and Ecomuseum) were located at the Emile A. Lods Agronomy Research Centre of McGill University, in southwestern Quebec (42°25'N lat., 75°56'W long.). Both sites consisted of undulating landscapes, with soils of moderate moisture holding capacities. Aboveground and soil sampling was restricted to areas of Chicot sandy loam exclusively. Prior to the experimental layout, the Seedfarm site was under a corn–alfalfa rotation for 3 years. The Ecomuseum site was under continuous corn for over four years and in addition to mineral fertilizer received regular applications of approximately 20 Mg ha^{-1} yr^{-1} of dairy manure.

In 1993, the sites were divided into four blocks with three treatments in each block as a randomized complete block design. The three treatments consisted of switchgrass (*Panicum virgatum* L.), short rotation forestry (SRF) willow (*Salix alba* sp.) and corn (*Zea mays* L.). In 1993, the SRF willows cuttings were planted at a rate of 11 000 ha^{-1}, in 0.92 m wide rows. The trees were coppiced in January 1996; they were fertilized with 77 kg N ha^{-1} in 1996 and with 125 kg N ha^{-1} in 1998. Switchgrass was planted in May 1993 and fertilized with 45 kg N ha^{-1} in 1994, 1995, 1996, as well as with 75 kg N ha^{-1} in 1997 at the Seedfarm. At the Ecomuseum, the switchgrass was fertilized with 30 kg N ha^{-1} and with 45 kg N ha^{-1} in June 1995, 1996, and 1997. In 1998, both sites received 50 kg N ha^{-1}. Each spring the overwintered switchgrass was harvested at a 15 cm height at both sites, and baled. In 1996, the corn at the Seedfarm was fertilized with 205 kg N ha^{-1}, and with 161 kg N ha^{-1} at the Ecomuseum. In 1997, the Seedfarm received 162 kg N ha^{-1} of fertilizer, and the Ecomuseum received 176 kg N ha^{-1}. In 1998, the corn received 148 kg N ha^{-1} at both sites.

2.2 Soil sampling

At six locations (in 1996, and four locations in 1998) in each plot, a 15 cm × 6 cm diameter cylindrical metal soil core was manually driven into the soil. The procedure was repeated at 15 cm depth increments unto a depth of 60 cm. Soil samples were mechanically crushed and sieved to 2 mm to obtain a uniform subsample, then further ground with a mortar and pestle to pass through a 100 mesh sieve.

2.3 Organic carbon determination

In 1996, the soil samples were analysed for organic carbon content by using the modified Mebius procedure (Yeomans and Bremner, 1988). Carbon storage under different management systems was calculated on an equivalent mass basis (Mg ha^{-1}) (Ellert and Bettany, 1995). This was the average soil mass across all systems from 0-60 cm, which was 7090 Mg ha^{-1} in 1996, and 6910 Mg ha^{-1} in 1998.

In 1998 the organic carbon was measured using a LECO Carbon Analyser 1400 C. Due to the different methodologies used in the two sampling years, 60 randomly chosen 1996 soil samples (equally representing each cropping system and depths) were re-analysed using the LECO analyser. From these results a regression equation ($r^2 = 0.99$) was used to adjust the 1996 organic carbon values to enable comparisons to be made between years.

2.4 Statistical analysis

Data was analysed as a randomized complete block design, with four blocks, the three crops as treatments, and the two sites as main factors. Statistical analyses were performed using SAS, GLM procedure. Data obtained from root biomass and soil carbon were analysed as repeated measure analyses of variance, using depth as a repeated factor. Significant ANOVA means were compared using the Student Newman Keuls (SNK) multiple range test.

3. RESULTS AND DISCUSSION

3.1 1996 root biomass yields

The total switchgrass root production was greater at the Seedfarm site (8068 kg ha^{-1}) than at the Ecomuseum site (5965 kg ha^{-1}). The willow fine roots, root crown and total roots were not significantly different between the two sites. In contrast, the fine root yields for corn were significantly higher at the Seedfarm (522 kg ha^{-1}) than at the Ecomuseum (337 kg ha^{-1}) (Figure 1), as measured by Zan (1998).

3.2 Soil organic carbon

In 1996, willow plots at the Ecomuseum had significantly higher amounts of soil organic carbon (138 Mg ha^{-1}), followed by switchgrass at the Ecomuseum (116 Mg ha^{-1}) which had significantly higher organic carbon compared to the remaining systems. The remaining systems were not found to differ in terms of organic carbon contents from 0-60 cm. Analysis were also conducted on individual depths, however, no significant differences were detected between any of the systems in any of the soil layers (Zan, 1998).

Table 1
1998 Soil Organic Carbon values for individual depths and 0-60 cm, for each site.

	Soil Organic Carbon (Mg ha^{-1})				
	0-15 cm	15-30 cm	30-45 cm	45-60 cm	*0-60 cm
Ecomuseum					
Willow	35 a	33 a	16 a	18 a	98 a
Swg	33 a	27 a	15 a	12 b	90 a
Corn	27 b	27 a	14 a	10 b	83 a
Seedfarm					
Willow	33 a	27 a	14 a	19 a	88 a
Swg	29 a	22 a	10 a	13 a	77 a
Corn	26 a	25 a	15 a	13 a	84 a

Means with different letters within each site are not significantly different at $p = 0.05$ according to the SNK test. Only the 0-60 cm column was calculated on an equivalent mass basis.

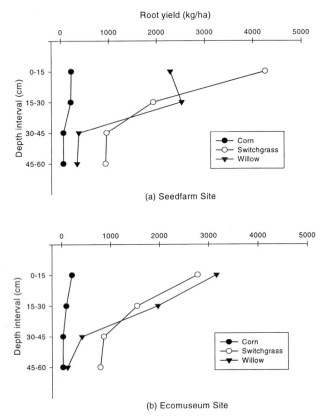

Figure 1. Mean root yields of corn, switchgrass, and willow, at the Seedfarm and Ecomuseum sites, in 1996, at various depth intervals. Only includes roots sampled with soil cores (no crowns) (Zan, 1998).

In 1998 however, there were differences between the individual depths at the Ecomuseum only (Table 1). Willow and switchgrass had higher soil organic carbon than corn at 0-15 cm, which may be due to the incorporation of surface residue in the case of corn. At 45-60 cm, willow had substantially more organic carbon than the other systems, possibly due to the greater root biomass. However, overall from 0-60 cm, there was no detectable difference between any of the crops, at either site.

Although an overall decrease in soil organic carbon was observed in 1998, by comparing individual cropping systems (Figure 2), willow at the Ecomuseum had significantly higher soil organic carbon than switchgrass at the Seedfarm. This indicates that willows may have greater potential to increase soil organic carbon levels, relative to switchgrass and corn.

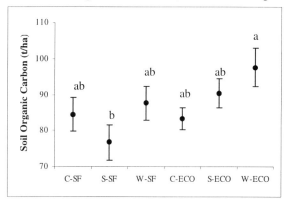

Figure 2. Mean 1998 soil organic carbon in willow, switchgrass and corn plots, in an equivalent soil mass basis (Mg ha^{-1}). Error bars represent standard errors of means. Means with different letters are not significantly different at $p=0.05$ according to the SNK test. C=corn; S=switchgrass, W=willow; SF=Seedfarm; Eco=Ecomuseum.

In both 1996 and 1998, the Ecomuseum was found to have a significantly higher amount of organic soil carbon than at the Seedfarm (Table 2). However, the mean level of organic carbon in 1998 decreased from 1996 at both sites. From 0-60 cm, average organic carbon loss at the Ecomuseum was 15 Mg C ha^{-1} yr^{-1}, while at the Seedfarm, the average loss since 1996 was 9 Mg C ha^{-1} yr^{-1}.

Table 2
Mean soil organic carbon from 0-60 cm in each cropping system.

Site	Crop	Organic Carbon (Mg ha^{-1})	
		1996	1998
Ecomuseum	Corn	105	83 (-11)*
	Switchgrass	119	90 (-14)
	Willow	137	98 (-20)
Mean			91 a (-15)
Seedfarm	Corn	108	85 (-12)
	Switchgrass	93	77 (-8)
	Willow	103	88 (-8)
Mean		101 b	83 b (-9)

*Values in brackets represent the amount of annual incremental organic C increase since the last sampling period (Mg C ha^{-1} yr^{-1})

The decrease in soil organic carbon from 1996 to 1998 was not anticipated. The conversion from annually tilled systems (prior to planting switchgrass and willow) to an undisturbed system, coupled with the addition of litterfall (approximately 2.0 Mg ha^{-1} in switchgrass plots and 1.8 Mg ha^{-1} in the willow), and the turnover of root carbon were expected to markedly increase the soil organic carbon. The decline in soil organic carbon in corn may be explained by soil erosion, as well as by the moldboard plowing, which increases the decomposition rate of organic matter.

The continuous removal of switchgrass biomass each spring may have decreased the microbial population owing to the lack of organic input, thereby slowly reducing the microbial population, as their source of nutrients became depleted. Campbell et al. (1997) found that soil organic matter decreased from 1959 to 1996 at 0-15 cm depth when wheat straw was removed every two out of three years. Several studies point to the importance that fertilization levels have on soil organic carbon sequestration (Campbell, 1997; Mahli et al. 1997). A study by Mahli et al. (1997) found that the total C increased with increasing N fertilizer application (rates varied from 0-336 kg ha^{-1}), by up to 64 g kg^{-1} at the 112 kg N ha^{-1} rate (from 50.33 g kg^{-1} in the control) over 27 years. Hay was grown on those plots with average dry matter yields of 1.17 Mg ha^{-1} on the no-N fertilized plots to 5.45 Mg ha^{-1} on the 336 kg N ha^{-1} plots. In our study, switchgrass 1998 aboveground biomass yields averaged 12.7 Mg ha^{-1} at the Seedfarm, and 12.9 Mg ha^{-1} at the Ecomuseum, with fertilization rates of 50 kg N ha^{-1}.

The relatively low fertilization rates used were chosen to minimize the amount of switchgrass lodging, pest outbreaks in the willows, and input costs. Optimal rates of fertilization for economical biomass production may not be optimal for soil organic carbon storage. This study should not discredit previous work conducted on carbon sequestration with perennial crops. Rather, it should be cautioned that relatively high yielding biomass crops are efficient nutrient feeders and have significant nutrient demands.

REFERENCES

Campbell, C.A., F. Selles, G.P. Lafond, B.G. McConkey, and D. Hahn (1997). Effect of crop management on C and N in long-term crop rotations after adopting no-tillage management: Comparison of soil sampling strategies, Can. J. Soil Sci., 78, pp. 1–12.

Cole, C.V., J. Duxbury, J. Freney, O. Heinemeyer, K. Minami, A. Mosier, K. Paustian, N. Rosenberg, N. Sampson, D. Sauerbeck, and Q. Zhao (1997). Global estimates of potential mitigation of greenhouse gas emissions by agriculture, Nutrient Cycling Agroecosyst, 49, pp. 221–228.

Ellert, B.H. and J.R. Bettany (1995). Calculation of organic matter and nutrients stored in soils under contrasting management regimes, Can. J. Soil Sci., 75, pp. 529–538.

Malhi, S.S., M. Nyborg, J.T. Harapiak, K. Heier, and N.A. Flore (1997). Increasing organic C and N in soil under bromegrass with long-term N fertilization, Nut. Cycl. Agroecosyst, 49, pp. 225–260.

Yeomans, J.C. and J.M. Bremner (1998). A rapid and precise method for routine determination of organic carbon in soil, Commun. Soil Sci. Plant Anal., 19, pp. 1467–1476.

Zan, C. (1998). Carbon storage in two biomass energy plantations, M.Sc thesis, McGill University, 108 pp.

THE ROLE AND PROCESS OF MONITORING WILLOW BIOMASS PLANTATIONS

C.A. Nowak, T.A. Volk, B. Ballard, L.P. Abrahamson, R.C. Filhart, R.F. Kopp, D. Bickelhaupt, and E.H. White

State University of New York, College of Environmental Science and Forestry, One Forestry Drive, Syracuse, New York 13210

Short rotation woody crop programs are growing in prominence in the United States. Programs have progressed from field experiments to commercial-scale operations across the country. Production systems have generally been developed using small-scale field experiments. Commercial-scale plantations are often established and maintained in ways unknown to the production system. Monitoring is critical to learn about the effects of scale-up. An adaptive management approach is useful, wherein each planting is treated as an experimental unit, management activities are viewed as treatments, and monitoring is used to quantify treatment effects, the results of which are reported to the manager to improve the system.

After a decade of research on short rotation willow biomass crops, efforts are underway to expand the area of plantations to move the system towards commercialization. The State University of New York College of Environmental Science and Forestry (SUNY-ESF), in conjunction with the Salix Consortium, is in the second year of this effort, as part of the Department of Energy's Biomass Power for Rural Development program. SUNY-ESF has been developing an adaptive management / monitoring program in support of scale-up operations, as described in this paper.

1. THE SHORT-ROTATION WOODY CROPS PROGRAM AT SUNY-ESF

Short-rotation woody crops (SRWCs) research has been conducted at the State University of New York College of Environmental Science and Forestry (SUNY-ESF) for nearly two decades, first with hybrid poplar (White et al., 1989, 1991), now with willow (Abrahamson et al., 1990; White et al. 1992; Abrahamson et al., 1998). Program development has featured small-scale field experiments, including trials for genetic selection, clone-site relations, fertilization, spacing/cutting cycle, irrigation, and herbicides for site preparation and release. Many of these trials, particularly early on, were conducted in partnership with the Ontario Ministry of Natural Resources and the University of Toronto. A production system was developed from research results with an aim at culturing high yields of wood fiber (Kopp et al., 1997). Experiences with willow in Europe, particularly Sweden (Ledin and Willebrand, 1996), were integrated into the system.

The development of the production system resulted in many successes. Survival and early growth of the crop was kept at high rates. Wood production has consistently been over 20 o.d.t. ha^{-1} yr^{-1} with fertilization and irrigation. Fertilization regimes have been developed with both organic and inorganic sources. The need for crop protection from pests has been minimal; when needed, standard methods have been effective. Whole-tree harvesting in the dormant season has lead to reliable regeneration of the next stand through coppicing. Sustainability is being demonstrated through long-term maintenance of crop health and productivity.

The success of the research program and growing interest in closed loop biomass have promoted the expansion of the system. To date, 50 hectares of large field plantings have been established. Additional plantings of at least 200 hectares are planned for the next few years. Most of the larger scale plantings are being established on sites of varied quality and land use history, and in environments different from those used to develop the system. Another challenge is that machinery used in many aspects of the commercial scale system is different from small-scale research plantings. Most operations at the research scale were done with hand labor. The large scale of the operations requires more time to plant, cutback, and harvest—all three phases have been executed at times of year different from those of the original system. For example, research plots are usually planted in early May, commercial plots are planted from May to July. A chief question of SUNY-ESF's willow biomass program is, how do scale-up efforts affect the survival, growth, and yield of the crop? This paper reviews the role of research and monitoring in SRWC programs as a demonstration of an ongoing attempt by SUNY-ESF to address this question. This paper describes research and monitoring in the context of adaptive management and some lessons learned.

2. THE ROLE OF RESEARCH AND MONITORING IN THE WILLOW BIOMASS PROGRAM

2.1. What is research?

Studies that attempt to improve our understanding of how the components of the production system work are considered research. Components of the production system can be categorized as crop establishment, tending, and harvesting. Research produces information that is used to describe cause–effect pathways, thresholds for effects, and mechanisms that describe how the crop interacts with the environment and management activities. Controlled greenhouse and field experiments are the common tools of research.

2.2. What is monitoring?

Studies that provide data on the status of the system as a whole are considered monitoring. Monitoring can also be used to improve our understanding of how the system works, although this is not its primary purpose (Cairns et al., 1979).

2.3. Why is it important to make a distinction between research and monitoring?

A distinction between research and monitoring formalizes the importance of both and promotes full and separate lines of organizational support. Without separation, efforts in each are not made clear. Research, monitoring, or both are likely to be marginalized in the scale-up effort.

2.4. Why is it important to draw a relationship between research and monitoring?

Monitoring and research have the same goal: to learn about the system. What differs between the two is the scale and control of effects measured. At the research scale, complete aspects of individual components of the system can be studied in controlled settings. At the monitoring scale, the system is studied as it is being applied in the field under less controlled situations. In order to meet the challenges of scale and control, a different approach to study is needed with monitoring: adaptive management.

2.5. What is adaptive management?

Adaptive management is a formal approach where operations are taken to manage the plantations intentionally to enhance the manager's learning about the system. It is "management with a built in learning process" (Baskerville, 1985). As further described by Baskerville (1985, p. 171), with adaptive management, the *design of management goals, the design of actions, and the measurement of progress are carried out in a manner that allows the manager to learn about the system from his management of it. As the manager learns about the systems, he is able to redesign (i.e., adapt) his management approach to be more efficient. Because the adaptive process forces the recognition of error, and therefore facilitates learning, it is a particularly good approach to use in initiating management in systems for which the dynamics structure is not well known, and where it is important to avoid irreversible error.* Both are the case in the willow biomass program at SUNY-ESF.

3. THE PROCESS OF MONITORING

Given an adaptive management framework, the process of monitoring is shown by answers to a series of questions: what, where, when and how to do it (Cairns et al., 1979; Draggan et al., 1987).

3.1. What to monitor?

Key factors of the production system—those that directly affect outputs—are usually chosen as variables for monitoring. These may include such variables as weed competition, crop survival, pest activity, and wood yield. All of these are being monitored in SUNY-ESF's program.

3.2. Where and when to monitor?

In conventional monitoring programs, choice of where and when is an exercise in sampling randomly chosen plantations, and subsamples are taken within plantations across space and time. In the case of SUNY-ESF's program, all plantations established so far are monitored because the scale is relatively small and information from each plantation is beneficial. Measurements are timed to coincide with critical points in the ecological dynamics and management of plantations. For example, foliage sampling (as a diagnostic of the site's capacity to provide nutrients) is done late in the growing season. At this time, foliar element levels have stabilized in a manner that accurately reflect levels of soil nutrient availability.

3.3. How to monitor?

The most fundamental step in monitoring plantation situations is developing a complete and unambiguous statement of objectives and related specifications, and then adhering to these rigorously in the design of the sampling plan and in the actual execution of the work (Cairns et al., 1979). Standardization of operating procedures (SOPs) for monitoring is critical. Included in SOPs are such elements as a statement of purpose, specification of the sampling universe, timing of sampling, precision of estimates desired, confidence levels, cost criteria, and limitations on efforts (Cairns et al., 1979). These elements are included in SOPs for sampling crop survival, site nutrient supply, pest incidence, soil condition, and wood yield as part of the SUNY-ESF SRWC program.

4. LESSONS LEARNED / FUTURE CHALLENGES

Some lessons learned in the ongoing development of monitoring and adaptive management for SUNY-ESF's short rotation willow program are as follows:
- schedule a role and a funding base for monitoring from the inception of the scale-up efforts,
- elevate the status of monitoring to that of research in terms of attention to details of accuracy and precision of collected data (see a statistician early),
- keep monitoring separate from research, yet recognize and utilize their interdependence,
- look to learn from the monitoring effort using an adaptive management model, emphasize the use of SOPs, and
- dedicate staff to the process of monitoring, from field visits to report writing.

Given that the SUNY-ESF willow program is in the early stages of scale-up towards commercialization, much remains to be learned from and within the process of monitoring.

Monitoring will be a recurring need for the foreseeable future. As the area planted increases and the distribution of plantations broadens regionally, it will not be possible to monitor every plantation with great detail. Some aspects of monitoring should be done with every plantation (e.g., insect and disease incidence); others not (e.g., survivorship). Initial efforts are underway to evaluate whether precision agricultural remote sensing techniques can be used with monitoring. As willow biomass farming becomes a commercial enterprise, monitoring will shift from a goal of learning about the system to one of judging the status of each plantation. This shift may not occur for another decade. In the interim, plantations established as part of the scale-up need to be treated as experiments, each as an opportunity to learn more about large scale production of willow biomass crops.

REFERENCES

Abrahamson, L.P., D.J. Robison, T.A. Volk, E.H. White, E.F. Neuhauser, W.H. Benjamin, and J.M. Peterson (1998). Sustainability and environmental issues associated with willow bioenergy development in New York (U.S.A.), Biomass and Bioenergy, 15, pp. 17–22.

Abrahamson, L.P., E.H. White, C.A. Nowak, and R.F. Kopp (1990). Yield potential of willow in New York State: Evidence from two years research in an ultrashort-rotation system, Proceedings, Energy From Biomass and Wastes XIII Conference, New Orleans, Louisiana. Institute of Gas Technology, Chicago, pp. 261–274.

Baskerville, G. (1985). Adaptive management: Wood availability and habitat availability, Forestry Chronicle, 61, pp. 171–175.

Cairns, Jr., J., G.P. Patil, and W.E. Waters (eds.) (1979). Environmental biomonitoring, assessment, prediction, and management–certain case studies and related quantitative issues, International Co-operative Publishing House, Fairfield, Maryland.

Draggan, S., J.J. Cohrssen, and R.E. Morrison (eds.) (1987). Environmental monitoring, assessment, and management: The agenda for long-term research and development, Praeger Publishers, New York.

Kopp, R.F., L.P. Abrahamson, E.H. White, T.A. Volk, and J.M. Peterson (1997). Willow bioenergy producer's handbook, New York State Energy Research and Development Authority, Albany, New York.

Ledin, S. and E. Willebrand (eds.) (1996). Handbook on how to grow short rotation forests, Department of Short Rotation Forestry, Swedish University of Agricultural Sciences, Uppsala, Sweden.

White, E.H., L.P. Abrahamson, R.F. Kopp, C.A. Nowak, and J. Sah (1992). Integrated woody biomass systems in eastern North America, Proceedings, Problems and Perspectives of Forest Biomass Energy, ed. by C.P. Mitchell, et al., Swedish Univ. Agric. Sci., Dept. Ecol. and Environ. Res., Sect. Short Rotation For, Rpt. 48, pp. 68–75.

White, E.H., L.P. Abrahamson, C.A. Maynard, R.D. Briggs, D.J. Robison, and C.A. Nowak (1989). Biotechnology and genetic selection of fast-growing hardwoods. Published by the New York State Energy Research and Development Authority, Albany, New York, Energy Authority Report 89–11.

White, E.H., L.P. Abrahamson, C.A. Nowak, M.L. Edmonds, and D.J. Robison (1991). Cultivation of fast-growing hardwoods, New York State Energy Research and Development Authority, Albany, New York, Energy Authority Report 91–10.

THE INFLUENCE OF SITE AND WASTEWATER SLUDGE FERTILIZER ON THE GROWTH OF TWO WILLOW SPECIES IN SOUTHERN QUEBEC

Michel Labrecque and Traian I. Teodorescu

Institut de recherche en biologie végétale, Montreal botanical Garden, 4101 Sherbrooke East, Montreal, Quebec, Canada, H1X 2B2

Salix discolor Mühl. and *Salix viminalis* L. were planted under short-rotation intensive culture (SRIC) on three unirrigated abandoned farmland sites with different drainage conditions, one well-drained and the other two poorly drained. One dose of dried and granulated sludge equivalent to 150 (T1) kg of available N/ha^{-1} was applied to some plots in the spring of the second season while others were left unfertilized (T0). The aims of the experiment were i) to investigate plant response (growth and productivity) to plantation site conditions and sludge application; and ii) study nutrient status by foliar analysis. During three seasons, growth in height and aboveground biomass were greater for *S. viminalis* than for *S. discolor* on all sites. *S. viminalis* planted on a poorly drained site had the highest biomass yield (45.28 t/ha^{-1}). Both species showed best height and diameter growth on poorly drained sites. For both species, best performances were obtained on wastewater sludge fertilized plots. Comparative foliar analysis suggested that sandy soil (S1) and low foliar nitrogen concentration and content were important limiting factors in the performance of the two species. Soil nitrate concentration increased as a result of sludge application. Heavy metal accumulation from sludge does not represent a risk to the environment. It was concluded that *S. viminalis* had the best productivity on clay sites, and that a moderate dose of dried and pelleted sludge (150 kg of available N/ha^{-1}) may be a good fertilizer during the establishment of willows in SRIC, and may reduce nitrate leaching.

1. INTRODUCTION

Productivity-oriented agriculture as practiced in North America, including Quebec, causes thousands of hectares of less fertile farmland to be abandoned. This land is characterized by heavy (clay) or light (sandy) soils, and it could be reconverted for the production of woody biomass of carefully chosen species using adapted culture techniques (Labrecque et al., 1993).

Experiments conducted in the south of Quebec for the past ten years with two willow species (*S. viminalis* and *S. discolor*) on small areas produced yields between 15 and 20 tons of dry-matter/ha/yr (Labrecque et al., 1993, 1997). Fertilizing the willows with wastewater sludge both significantly increased biomass productivity on the treated plots and contributed to recycling waste (Hytönen, 1994; Labrecque et al., 1995, 1997). A pilot study of the establishment of large culture areas has recently been conducted in the Upper Saint-Laurent region in southern Quebec, Canada. The objectives of the present study were to compare the growth, performance, and nutrient status of willow species cultivated on extensive sites with different characteristics which are considered marginal for agriculture, and to assess the effect of fertilization with wastewater sludge on yields after the first rotation cycle (three years after establishment).

2. METHODS

2.1. Study area

Experiments were conducted in the Upper Saint-Laurent region (45°05' N, 74°20' W) 90 km southwest of Montreal. Meteorological data were collected at a permanent station of the Environment ministry of Quebec. Mean total precipitation during the growth season (from May to September) calculated for 21 years (1961 to 1990) for the Upper Saint-Laurent region was 427 mm. Rainfall was lower than normal by 86 mm and 71 mm respectively during the 1995 (341 mm) and 1997 (356 mm) seasons. In 1996 (497 mm), values were 70 mm above normal. So, the 1995 season started with a prolonged draught (May and June) which was certainly unfavorable to the rooting of cuttings. During the 1996 and 1997 seasons, precipitation was generally more abundant (except in August 1997). Temperatures registered for the first three seasons were favorable to the growth of willows, with the exception of the beginning of 1995, and toward the end of the 1996 season, which were slightly warmer than normal.

2.2. Site characteristics

Marginal sites abandoned by traditional agriculture for several years were used. Of three one-hectare sites, S1 had a sandy loam, well-drained soil and S2 and S3 had clay loam, poorly drained soil.

2.3. Species and experimental design

Two willow species were chosen: *S. viminalis* L., a species introduced from Europe, and *S. discolor* Mühl. a species indigenous to Quebec.

Three large (10,000 m² each) experimental plots were set up in the spring of 1995. Each design comprised six blocks in which the two species and two fertilization treatments were randomized: a control treatment (T0) and a treatment corresponding to the application of 12.5 t of granulated sludge (equivalent to 150 kg ha^{-1} available N). The sludge dose was calculated on the basis of the theoretically available N content for the first season following the application (available N = inorganic N + 0.30 organic N) determined per kg of dry-matter of sludge.

Willows were planted in May 1995 with a mechanical type planter (two rows) commonly used in vegetable growing. The plantation design comprised several plots 3 m apart from each other. Each plot had 6 rows 1.5 m apart. With an interval of 0.30 m between plants on the same row, a density of 20,000 plants ha^{-1} could be obtained. In all, 60,000 willow cuttings (50% *S. discolor* and 50% *S. viminalis*) were planted. Eight lysimeter per site (24 in all) were installed at a depth of 1 m in plots planted with *S. discolor* and *S. viminalis* (fertilized and unfertilized).

Plots were fertilized in the spring of 1996 (beginning of the second growth season) with a dose of 12.5 t ha^{-1} of dried sludge (< 10% water). In order to conform to established standards for the valorization of sludge in Quebec, total contents of some elements and metals were tested.

2.4. Measurement and sampling

Foliar element concentration and content were determined from a sample of green leaves taken in August 1997. On each of the sites, forty-eight samples consisting of ten leaves were collected above the tenth leaf of the main stem of two randomly chosen plants for each species, treatment and block. On each of the three sites, 144 randomly chosen plants (six plants for each species, treatment and block) were sampled at the end of November 1997. The diameter at the base of the main stem, the height of the main stem from soil surface to apex, and the number of stems produced by each cutting were measured. In order to estimate the productivity of the willows, the same plants were coppiced and weighed in the field with a spring scale. All chemical analyses (foliar tissue, soil, and sludge) were conducted by Agri-Direct, a private laboratory associated with the Quebec ministry of Agriculture.

3. RESULTS AND DISCUSSION

3.1. Growth and productivity

At the end of November 1997, *S. viminalis* was significantly taller than *S. discolor* in all plots and on all plantation sites, while diameter growth was similar for both species (Table 1).

For both species height and diameter were greater on clay sites (S2 and S3), compared to plants grown on the sandy site (S1). However, stems of *S. viminalis* grown on S2 fertilized plots were the tallest. Fertilization at the beginning of the second growth season increased the growth in diameter of both species on all experimental sites. On the sandy site, sludge application did not affect stem height, while on the clay sites it affected only the height of *S. viminalis* stems. After the first culture cycle, *S. discolor* had developed a significantly greater number of stems than *S. viminalis*. However, this number does not seem to depend on the plantation site and the fertilization treatment.

Biomass yield of the two species on the sandy site was significantly lower than on other sites (Table 1). For both species, greatest productivity was obtained on the fertilized clay plots of site 2. In contrast, the lowest productivity was observed on the unfertilized plots of site 1 planted with *S. discolor*. Fertilization increased the growth of *S. viminalis* in every situation. The application of a sludge dose equivalent to 150 kg of available N ha^{-1} caused significant increases in yield: 63% on site 1, 42% on site 2, and 55% on site 3. *S. discolor* also tend to react positively to sludge application, but variations in yield are not significant.

Table 1

Comparison of growth and yield of *S. discolor* (Sd) and *S. viminalis* (Sv) planted on three sites fertilized (T1) and control (T0) at the end of the first rotation

Parameters	Site 1				Site 2				Site 3			
	Sd		Sv		Sd		Sv		Sd		Sv	
	T1	T0	T1	T0	T1	T0	T1	T0	T1	T0	T1	T0
Height (m)	2.01	2.14	3.38	3.17	2.95	2.66	4.33*	3.72	2.84*	2.6	3.76*	3.59
Site effect (p<0.05)	S1<S3=S2 (Sd); S1<S3=S2 (SvT1); S1<S3=S2 (SvT0)											
Species effect (p<0.05)	Sd<Sv (S1, S2, S3)											
Diameter (mm)	28.3*	27.1	28.6*	24	35.2*	30.1	34.6*	30.5	30.9*	30.2	32.2*	29.4
Site effect (p<0.05)	S1<S2=S3 (Sd, Sv)											
Species effect (p<0.05)	Sd=Sv (S1, S2, S3)											
Number of sprouts	2.89	3.36	2.83	2.53	3.44	4.01	2.61	2.56	2.78	3.08	2.94	2.64
Site effect (p<0.05)	S1=S2=S3 (Sd, Sv)											
Species effect (p<0.05)	Sd>Sv (S1, S2, S3)											
Dry biomass (t/ha)	15.8	14.6	24.9*	15.2	30.7	28.5	45.3*	31.9	18.6	15.9	32.7*	21.1
Site effect (p<0.05)	S1=S3<S2 (Sd, Sv)											
Species effect (p<0.05)	Sd<Sv (S1, S2, S3)											

Note: Means within lines followed by an asterisk indicate a significant effect of fertilization treatment at p<0.05.

Growth and productivity performances observed at the end of the first growth cycle demonstrate that clay sites are more conducive to the development of willows. Indeed, once they are rooted, the willows benefit from much better fertility conditions than on the sandy site. After three years *S. viminalis* had the best growth and productivity performances on all plots and all plantation sites. Growth of *S. discolor* was reduced by a late frost in the spring of 1996 that destroyed part of the branches on outside rows more exposed to winds. The dead stems were quickly replaced by new shoots that greatly increased the total number of stems. However, too many stems impede maintenance and harvesting. In previous trials conducted on small areas that were protected from the wind, *S. discolor* yields were often higher than those of *S. viminalis* during the first years of growth (Labrecque et al., 1993, 1997). Planted on larger culture areas that are exposed to winds, *S. discolor* is sensitive to frost, which reduces its growth and productivity.

Fertilization with dried and pelleted sludge at the beginning of the second season increased productivity especially for *S. viminalis*. The increase in yield due to fertilization was greater on the sandy site, which is poorer than the other two. The timing of the sludge application (beginning of the second season), and the dose of fertilizer applied (150 kg/ ha available N) seem favorable to the growth of willows in the south of Quebec. This choice was mostly based on previous results (Labrecque et al., 1997, 1998) and is confirmed by other authors (Alriksson et al., 1997).

Table 2
Comparison of nutrient status (concentration and content) in leaves of *S. discolor* (SD) and *S. viminalis* (Sv) fertilized (T1) and control (T0) planted on three sites (S1, S2, S3) and sampled at the end of August 1997

Nutrient	*S. discolor*						*S. viminalis*					
	S1		S2		S3		S1		S2		S3	
	T1	T0	T1	T0	T1	T0	T1	T0	T1	T0	T1	T0
	Concentration (% dry mass)											
N	2.56*	2.34	3.06	2.94	2.83*	2.72	2.73*	2.39	3.81	3.37	3.32	3.20

Site effect (p<0.05): S1<S3<S2 (SdT1, SvT1); S1<S2=S3 (SdT0, SvT0)
Species effect (p<0.05): Sd=Sv (S1); Sd<Sv (S2, S3)

P	0.19	0.21	0.22	0.22	0.20	0.21	0.22	0.27	0.22	0.19	0.24	0.19

Site effect (p<0.05): S1=S2=S3 (Sd, Sv)
Species effect (p<0.05): Sd=Sv (S1, S2, S3)

K	1.50	1.29	1.21	1.56	1.46	1.66	1.80	1.79	1.76	1.65	.868*	1.74

Site effect (p<0.05): S1=S2=S3 (Sd, Sv)
Species effect (p<0.05): Sd<Sv (S1T0, S2T1, S3T1); Sd=Sv (S1T1, S2T0, S3T0)

Ca	1.29*	1.47	1.25*	1.17	1.35*	1.19	0.89*	0.82	1.3*	1.07	1.19*	1.10

Site effect (p<0.05): S1=S2=S3 (Sd); S1<S2=S3 (Sd, Sv)
Species effect (p<0.05): Sd<Sv (S1); Sd=Sv (S2, S3)

Mg	0.28	0.30	0.47*	0.39	0.45*	0.40	0.28	0.27	0.51	0.48	0.49	0.54

Site effect (p<0.05): S1<S2=S3 (Sd, Sv)
Species effect (p<0.05): Sd<Sv (S2T0, S3T0); Sd=Sv (S1, S2T1, S3T1)

	Content (mg/leave)											
N	3.25*	2.75	5.43*	5.04	4.99*	4.35	3.38*	2.67	7.31*	5.75	5.8*	6.08

Site effect (p<0.05): S1=S2=S3 (Sd); S1<S2=S3 (Sv)
Species effect (p<0.05): Sd=Sv (S1); Sd<Sv (S2, S3)

P	0.32	0.37	0.40	0.37	0.35	0.33	0.28	0.31	0.43	0.32	0.42	0.37

Site effect (p<0.05): S1=S2=S3 (Sd, Sv)
Species effect (p<0.05): Sd=Sv (S1, S2, S3)

K	2.64	2.36	2.16	2.59	2.62	2.62	2.13	2.11	3.38	2.77	3.26	3.32

Site effect (p<0.05): S1=S2=S3 (Sd); S1<S2=S3 (Sv)
Species effect (p<0.05): Sd=Sv (S1); Sd<Sv (S2, S3)

Ca	2.22	2.61	2.20	2.02	2.35	1.89	1.12	0.94	2.49	1.84	2.04	2.10

Site effect (p<0.05): S1=S2=S3 (Sd); S1<S2=S3 (Sv)
Species effect (p<0.05): Sd>Sv (S1); Sd=Sv (S2, S3)

Mg	0.48	0.54	0.83*	0.65	0.79*	0.63	0.36	0.30	0.98	0.81	0.86	1.03

Site effect (p<0.05): S1=S2=S3 (Sd,T1); S1<S2=S3 (SdT0, Sv)
Species effect (p<0.05): Sd=Sv (S1, S2, S3T1;); Sd<Sv (S3T0)

Note: Means within lines followed by an asterisk indicate a significant effect of fertilization treatment at $p<0.05$.

3.2. Foliar nutrient status

The content and concentration of the various elements which were dosed in samples from both species are shown in Table 2.

In general, values for N content and concentration are lower on the sandy site. Differences were also observed according to species, such as higher foliar N content and concentration in *S. viminalis* than in *S. discolor* leaves on the clay sites. However, these were comparable on the sandy site. The application of wastewater sludge significantly increased the foliar N content of the species while the percentage of N increased only in the leaves of plants from site 1 (both species) and site 3 (for *S. discolor*). Kopinga and Burg (1995) found that foliar element concentrations that are considered normal for willows are 2.3 to 3% N; 0.17 to 0.21% P; 0.85 to 1.9% K; and 0.17 to 0.30% Mg. According to these authors, optimal values should be higher than the highest normal value. Compared to these limits, our values for foliar N concentration can be considered normal (for *S. discolor*) and optimal (for *S. viminalis*) only on the clay sites. Foliar P concentrations seem closer to optimal values while those for K are normal everywhere and very close to deficit values. Values for foliar Mg are optimal on clay sites and normal on the sandy site. Soil analyses of the sandy site show that it is deficient in Ca (slightly acid soil, pH 5.6) and has low values for K and Mg, which decrease progressively with depth. The sludge is poor in K, and could not satisfy the needs of willows under short rotation intensive culture SRIC (Labrecque et al., 1998). A potassium supplement from chemical fertilizers can become necessary on certain soil types. Both species produced the greatest quantity of above-ground biomass with a foliar N concentration of 3.81% for *S. viminalis* and 3.06% for *S. discolor* (obtained on fertilized plots of clay site 2). Our results therefore suggest that to obtain optimal growth during the establishment of willows under SRIC requires a foliar N concentration above 3.5% for *S. viminalis* and at least 3% for *S. discolor*.

3.3. Element and metal leaching

Mean concentrations of the various elements and metals dosed in the soil solution from fertilized and unfertilized were compared with the limits recommended in Canada for drinking water. The data shown here demonstrate that the effect of fertilization with wastewater sludge on the contamination of groundwater was greater during the year it was applied. Thus, in 1996, analyses reveal higher nitrate concentrations in the leachate of fertilized plots. In all plots of site 2, and in plots of site 3 planted with *S. discolor*, nitrate quantities exceed the acceptable limits established. However, on the sandy site, nitrates were lower than these limits.

During the second season after the sludge application, nitrate concentrations decrease markedly in the leachate on all sites. On all the plots planted with *S. viminalis*, the quantity of nitrates leached into groundwater is extremely low. In contrast, there are greater quantities in the leachate from plots planted with *S. discolor* and, in certain cases, acceptable limits were exceeded.

During the last two years Fe exceeded acceptable limits on all plots (fertilized and unfertilized) of all plantation sites. It seems that this element was already present in the soil before the sludge added an substantial quantity (109 kg/ha). However, values for environmentally hazardous metals (Cd, Cr, Cu, Mn, Ni, Zn) dosed in the leachate never exceeded the norms for drinking water.

4. CONCLUSION

For both species, the best stem biomass yield was obtained on the fertilized clay sites. *S. viminalis* was the more productive species, and important quantities of dry biomass

(42.28 t ha-1 after the first production cycle) were harvested from the fertilized plots of site 2. Thus, poorly drained clay sites (willow's natural habitat) are greatly recommended for SRIC plantations. On the other hand, sandy sites enriched with organic fertilizers also have some potential for willow culture if fertilization after each harvest cycle can be contemplated.

Dried and granulated wastewater sludge is a good fertilizer for willows under SRIC. The chosen dose and timing of the application generally satisfy the nutritional needs of the willows, reduce environmental risks, and could improve the profitability of this biomass production system. Pilot trials conducted on larger areas that were not as well protected against the winds as the smaller experimental surfaces allowed us to discover *S. discolor*'s sensitivity to the cold, a characteristic that hindered its growth and productivity.

The plantation configuration developed by our research team is different from that used in other countries such as Sweden, but it seems well-adapted to the specific conditions of Quebec. It allowed us to obtain optimal plant density, and to perform mechanical work (e.g., weeding, phytosanitary treatments) with equipment normally used on any farm.

REFERENCES

Alriksson, B., S. Ledin, and P. Seeger (1997). Effect of nitrogen fertilization on growth in a *Salix viminalis* stand using a response surface experimental design, Scand. J. For. Res., 12, pp. 321–327.

Hytönen, J. (1994). Effect of fertilizer application rate on nutrient status and biomass production in short-rotation plantations of willows on cut-away peatland areas, Suo 45, 3, pp. 65–77.

Kopinga, J. and J. van den Burg (1995). Using soil and foliar analysis to diagnose the nutritional status of urban trees, Journal of Arboriculture, 21, 1, pp. 17–24.

Labrecque, M., T.I. Teodorescu, A. Cogliastro, and S. Daigle (1993). Growth patterns and biomass productivity of two *Salix* species grown under short-rotation, intensive-culture in south western Quebec, Biomass & Bioenergy, 4, 6, pp. 419–425.

Labrecque, M., T.I. wastewater sludge on growth and heavy metal bioaccumulation of two *Salix* species, Plant and Soil, 171, pp. 303–316.

Labrecque, M., T.I. Teodorescu, and S. Daigle (1997). Biomass productivity and wood energy of *Salix* species after two years growth in SRIC fertilized with wastewater sludge, Biomass and Bioenergy, 12, 6, pp. 409–417.

Labrecque, M., T.I. Teodorescu, and S. Daigle (1998). Early performance and nutrition of two willow species in short-rotation intensive culture fertilized with wastewater sludge and impact on soil characteristics, Can. J. For. Res., 28, pp. 1621–1635.

WILLOW SHORT-ROTATION COPPICE PROFITABILITY ASSESSMENT. A CASE STUDY IN BELGIUM

F. Goor, J.-M. Jossart, and J.-F. Ledent

Laboratoire d'Ecologie des Grandes Cultures, Faculté des Sciences Agronomiques, UCL, Place Croix du Sud 2 bte 11, 1348 Louvain-la-Neuve, Belgique

This paper deals with the economics of a willow short rotation coppice (SRC) production and valorisation pathway: a case of small scale electricity production by gasification in Belgium. A detailed specific economic model (ECOP), run with the key values of a reference scenario, calculated a cost of SRC delivered to the conversion plant of 92,5 EURO odt^{-1}, and a global project net present value of 18,000 EURO ha^{-1} (project duration: 28 years). The model was also used to identify the most relevant parameters determining the project profitability and their sensitivity, and it considered the impact of the harvest option.

1. THE SRC PRODUCTION AND VALORISATION PATHWAY

Short-rotation coppice is a way of growing fast-growing tree species on agricultural land with a relatively short rotation length. In coppice plantations, shoots regenerate from the stumps after the main stems are cut down. The rotation length depends on the final use of the wood (such as heat or electricity production, paper, sticks). Various fast-growing species can be grown as coppice, but the most frequently used species are willow and poplar.[1]

In 1995, the Walloon Region and Electrabel (the Belgian electricity manufacturer) decided to collaborate on a four-year pilot project to generate peak electricity using conventional SRC gasification.[2] Low power gasifiers were tested which can be fed with the wood production from 10-12 ha of SRC plantations. From the agricultural viewpoint, the interesting aspect of this pathway is that it allows a farmer (or a group of farmers) to control and to be responsible for all the operations (SRC cultivation, wood storage and crushing, electricity production) and then to produce a finished good (the electricity), which can have a higher added-value than the raw wood. Moreover, the farmer is then less dependent on intermediaries who might keep a large part of the wood valorisation return. On the other hand, farmers are not very familiar with woody perennial crops or with the valorisation of a finished good on the open market. A sound economic study is therefore necessary to evaluate the potential annual cash-flows as well as to compare the profitability of SRC with these of annual crops, on a common basis. Consequently, a detailed economic model (ECOP, economic assessment of coppice plantations) was developed from existing studies[3] (ETSU, pers. com.), which covers all the stages from the plantation of SRC to the sale of electricity.

2. THE SRC ECONOMIC MODEL: BASES AND CALCULATION METHODS

The model is driven by the hours of annual electricity production and the power of the conversion unit (kW). The electricity production (kWh) can be converted into wood quantity to supply using the wood's inferior calorific power. Finally, the required crop surface is calculated from the wood supply needed and from the curve of SRC yield with time.

To compare SRC, a perennial crop, with traditional annual crops requires a breakdown of crop management costs and that income and expenditures occurring with time be discounted. We here use the net present value as the economic indicator, which consists in bringing back to the reference time "t" the cash flows relative to various periods, on the basis of a reference interest rate "r" (zero risk investment). The model also provides a non-discounted cost-benefit analysis (allowing the world to extract the labour costs of the farmer and his family) and evaluates the intermediate costs of production—storage, transport and crushing.[4] Finally, the model calculates a constant equivalent annuity, corresponding to the annual revenue the farmer would receive if depositing (without risk, with an interest rate equal to the discount rate) all the discounted project incomes along the plantation lifetime.[5]

The model user may access various parameters to define precisely the scenario to study: rotation length, pruning or not, spreading of the plantation, subsidies, etc. To be realistic, the model also proposes a contractor price to limit artificially some machinery costs (mainly the specific plantation and harvest tools) when the plantation is small. In our simulation hypotheses (see point 3), the "subcontracting limit" is fixed at 25 ha. The harvest is probably the most important operation, determining the subsequent storage or crushing operations. Four harvest options have been included in the model: manual harvest, harvest combined with crushing (using a silage harvester with an adapted head [circular saws]), harvest in stems with direct bundling and harvest in stems without bundling. Harvest with an adapted silage harvester produces wet chips which are directly loaded and transported close to the gasifier, and stored in a barn to dry off (a fan may be used to limit fermentation losses). In the case of harvest in stems, the stems are stored in piles (covered or not) on the field edge and dry off by natural ventilation. Electricity is produced in a conversion unit (gasifier) in a farm building. The selling price of the electricity is fixed by conventions[6] and depends among others on the quantity produced and on the reliability of the installation.

3. THE SRC ECONOMIC MODEL: REFERENCE CASE AND SENSITIVITY ANALYSIS

As the SRC-Gazel pathway does not yet work in practice, a lot of uncertainties remain which may affect the results of the economic assessment. Therefore, we have calculated our economic indicators in a reference case for which all the parameters have been given average classic values. After that, the relevance of this scenario has been discussed by studying its sensitivity to the variation of a key parameter: the harvest option. The reference case assumptions are summarised in Table 1, the cash-flow breakdown is presented in Figure 1 and the net income evolution against time in Figure 2.

Table 1
Reference case data for the parameters of greatest interest

Interest reference rate :	5%
Rotation length :	3 years (plantation and harvest are spread over 3 years)
Harvest :	adapted maize harvester (chips)
SRC plantation yield :	12 odt* ha^{-1} year^{-1}
Storage :	in a barn, with ventilation
Generator power :	300 kW
Annual production :	350 hours (~ peak hours)
Area of the SRC plantation :	8 ha

odt = oven dry tons. Average annual value after the plantation establishment phase

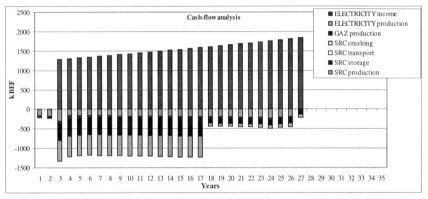

Figure 1. Cash-flow breakdown for the reference case (for all the 8 ha). The crushing cost = 0 for our reference case (harvest and crushing are integrated).

In our reference case, the total cost of the wood delivered to the gasifier is 3700 BEF (92.5 EURO) oven dry ton (odt)$^{-1}$, and it can be subdivided into production cost (2864 BEF [71.6 EURO] odt^{-1}), storage cost (722 BEF [18.1 EURO] odt^{-1}), and transport cost (114 BEF [2.9 EURO] odt^{-1}). The electricity production by the conversion unit costs ~7 BEF (0.18 EURO) kWh^{-1}. Without considering the storage, the "production + transport" cost of SRC chips is estimated at ~2864+114 = 2978 BEF (74.5 EURO) odt^{-1}. This value is close to the results presented in the literature.[7] Indeed, if we conform to the hypothesis of Ledin, except for the cost of machinery, we obtain a "production + transport" cost of ~2250 BEF (56.3 EURO) odt^{-1}. According to Ledin, if the farmer sells his SRC 2600 BEF (65 EURO) odt^{-1}, he realises a benefit of 3700 BEF (92.5 EURO) ha^{-1} or 300 BEF (7.5 EURO) odt^{-1} (based on an average yield of 12 odt ha^{-1} year^{-1}). Finally, the net spending for the SRC "production + transport" amounts to 2550 BEF (63.8 EURO) odt^{-1}. This value is relatively close to the 2978 BEF (74.5 EURO) calculated with our model.

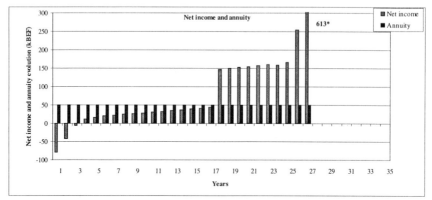

Figure 2. Net income evolution against time and annuity*: The two first and two last incomes are higher because they correspond to a lower surface (plantation and harvest are spread over three years).

According to our hypotheses, the NPV of the global project is positive (~720 000 BEF [18,000 EURO] ha^{-1}), which corresponds to an annuity (year by year breakdown of the NPV on the total project duration) of ~50 000 BEF (1250 EURO) ha^{-1}. Moreover, this value includes the farmer's salary (~2000 BEF [50 EURO] ha^{-1} year^{-1} during 28 years). This salary, considered as a cost for the global project, is in fact a revenue for the farmer and can be added to the annuity.

The SRC production cost is more or less constant (~2864 BEF [71.6 EURO] odt^{-1}). By analogy, we can suppose that this cost will also remain constant with other plantation or harvest hypothesis. The storage and transport costs are not very high.

As already said, the harvest option is a key parameter that strongly affects the SRC production cost and also determines the subsequent storage operations. The influence of the harvest option on the global project profitability (NPV) and on the cost of SRC wood delivered to the gasifier are presented in the figures 3a and 3b.

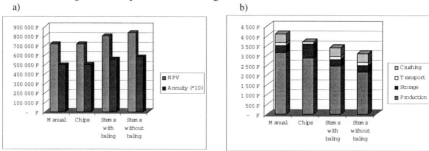

Figure 3. Influence of the harvest option on a) the project NPV, and b) the price of SRC delivered to the gasifier.

From the NPV analysis, and not considering the manual harvest which requires many costly laborers (and is difficult to use on an extended plantation), the harvest in stems is the most interesting. The difference between the options with and without bundling comes from the price of the harvester, three times higher in the first case. Figure 3b confirms the advantage of harvest in stems. The harvest in chips allows the integration of the harvest and crushing, but it also faces high storage costs (a barn with a cooling fan). Transport of wet chips is also costly. The dimensions and uniformity of the chips produced by the different harvesting methods must be considered, because they influence the good course and the energy performance of the conversion unit. However, we don't yet have enough information to add this parameter to our model.

Figure 4 presents the harvest costs against the SRC area harvested every year, and it shows that this area determines the choice of machinery: the Rodster will be preferred to the Bundler[1] for little plantation areas. Whatever the chosen machine, the investment is surely not profitable for a single farmer's limited plantation area, and the contractor's option is therefore considered. The machinery costs corresponding to the minimum area (25 ha in our hypotheses) can therefore be considered as a maximum.

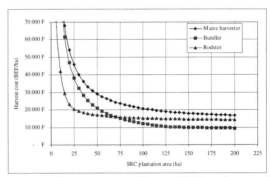

Figure 4: Evolution of the harvest costs against the SRC area harvested every year

4. DISCUSSION AND CONCLUSION

Our study has shown the effect of the method of harvest on the project's profitability. Similarly, our model highlighted that this value is very sensitive to the reference interest rate, to the subsidy for the specific material, to the SRC plantation yield and to the generator power, all other parameters remaining constant. The rotation length has only little influence.

The costs of depreciation and financing the SRC-specific tools and conversion unit, as well as the way the price of electricity is fixed, have a very large influence on the project's profitability. Once the pathway is developed, we can expect favourable scaling effects and a costs decrease due to the technological developments. And, finally, the most important

---

[1] The Rodster and the Bundler are both developed in Sweden by Salix Maskiner AB.

parameter to consider when planting SRC seems to be the insurance of a reasonable market price for the electricity in the next years.[7]

To conclude, the NPV itself is not the only criterion to be considered by the farmer, but this value must be compared with other potential uses of his soil, taking into account reasonable yields for other potential crops. Generally, the traditional crops are compared using the "labour income" parameter, defined as the difference between all the receipts and all the (direct and indirect) costs related to this crop. Its average value was 27,384 BEF (684.6 EURO) ha^{-1} in the Walloon region (Belgium) in 1996/97.[8] This value may be compared with the annuity from the SRC-Gazel pathway, which is 50,023 BEF (1250.6 EURO), enhanced by the farmer's salary, which is 2174 BEF (54.5 EURO), resulting in a global value of 52,197 BEF (1304.9 EURO) ha^{-1}. The difference between SRC and other crops is significant; nevertheless, in the case of the SRC-Gazel pathway, a higher NPV may be required to compensate the high risk, the many uncertainties of this kind of innovative project and the negative incomes during the early years of the project.

REFERENCES

1. Jossart, J.M. (1994). Le taillis à courtes rotations : alternative agricole, Laboratoire d'Ecologie des Grandes Cultures, Université Catholique de Louvain, Louvain-la-Neuve, Belgium, 78 pp.
2. SRC-Gazel (1996). TtCR-Gazel : un projet pilote grandeur nature, brochure de présentation du projet TtCR-Gazel, Louvain-la-Neuve, Belgium, 10 pp.
3. Rosenqvist, H. (1997). Calculation method and economy in Salix production, in Biomass for the Energy and the Environment, Proceedings of the 9th European Bioenergy Conference, Vol. 3, pp. 1967–1972.
4. Meekers, E., J.-C. Jacquemin (1997). Projet TtCR-Gazel – Mode d'emploi du Modèle de Simulation Economique, Faculté des sciences économiques, sociales et de gestion, Faculté Universitaire Notre-Dame de la Paix, Namur, Belgium, 36 pp.
5. François, I. (1995). Le boisement des terres agricoles et les cultures pluriannuelles sur jachères: étude de la rentabilité économique de 5 cas, thesis, ICHEC, Bruxelles, Belgium, 134 pp.
6. CCEG (Comité de Contrôle de l'Electricité et du Gaz) (1998). Note de secrétariat et projet de recommandation datés du 29 juin, Bruxelles, Belgique, pp. 14–20.
7. Ledin, S. (1996). Willow wood properties, production and economy," Biomass and Bioenergy, 11, pp. 75–83.
8. MRW (Ministère de la Région Wallonne) (1998). Evolution de l'économie agricole et horticole de la Région wallonne, rapport final, Unité d'Economie Rurale, Faculté universitaire des Sciences Agronomiques, Gembloux, Belgique, 249 pp.

SECOND ROTATION WILLOW COPPICE IN UPLAND WALES

F.M. Slater, R.J. Heaton, R.W. Samuel, and P.F. Randerson

Cardiff University, Wales, Cardiff University Field Centre, Newbridge-on-Wye, LD1 6NB, U.K.

Throughout this decade we have been conducting research into the production of short rotation willow coppice at altitudes over 250 m in mid-Wales. The purpose of the research is to determine whether willow is a viable biomass crop in upland situations, in order to help diversify agriculture in an area dependent upon sheep farming. A second reason for the work is to examine the use of willow as a filtration crop for on-farm use. Both aspects represent pioneer work, particularly in upland situations. Several new but lower-altitude trials are underway in Wales, but few have reached their first harvest, and none except our own have been harvested at their second rotation.

At our first rotation harvest, one cultivar, "dasyclados" clearly outperformed all the rest. With no further fertilization, at second harvest – seven years after planting – "dasyclados" was clearly outperformed by at least two other cultivars. Our work demonstrates the importance of longer-term results in assessing the results of short rotation willow coppice trials.

SHORT ROTATION COPPICE OF WILLOW AND SHELTERBELT EFFECT

J.-M. Jossart and J.-F. Ledent

Laboratory of Main Crop Ecology (ECOP), Place Croix du Sud, 2 bte 11, 1348 Louvain-la-Neuve, Belgium

For environmental reasons short rotation coppice (willow grown at high density and harvested every three years) could be grown in strips or windbreaks, but this leads to complex interactions with neighbouring food crops. Trials have been conducted in Belgium since 1994. Depending on the crops and the height of the windbreak, food crops were negatively or positively affected by windbreaks within 3 to 6 h (distances equivalent to 3 to 6 times the height of the windbreak), but the effect is rather small at a level of the entire field.

Regarding microclimate, willow windbreaks increase temperature leeward and reduce wind velocity up to 20 h. Temperature and wind effects close to the windbreak seem to indicate that turbulences are occurring.

1. INTRODUCTION

In the framework of biomass development in the Walloon region (southern Belgium) several R&D projects studied the implementation of short rotation coppice (SRC) for electricity and heat production. SRC is considered here as a willow plantation at high density (more than 10 000 plants per hectare). Harvests occur every three years, for a total duration of the crop of about 25 years. Within the pilot project SRC-Gazel, the whole chain of SRC production on agricultural land and energy production in small scale conversion units can be managed by farmers.

The plantation design is an important issue because it affects the relationship between agriculture and environment. Several regional programmes were set up to plant new hedges and maintain existing ones for landscape and biodiversity reasons. Nevertheless currently promoted hedges are often disregarded by farmers because of psychological resistance, high maintenance cost, and technical problems. In this context SRC planted in strips could be considered as a real crop by farmers and as a hedge by environmentalists. It is a kind of compromise in which maintenance is not a problem for the farmers (because it will be part of the regular harvesting operations), and it will not be as good as environmentalists dream of (only one species too homogeneous).

The European project "Evaluation of combined food and energy systems for more efficient land use and environmentally benign sustainable production" (CT96-1449) studied the effects of willow strip plantations, or windbreaks, on microclimate as well as on neighbouring crops.

2. MATERIAL AND METHODS

Four strips of 6 m width were planted in 1994 and 1997 according to general recommendations for short rotation coppice plantation (double rows 75 – 125 cm, 18000 cuttings.ha^{-1}). Figure 1 shows the design of one trial, with 52 - 76 m distances between strips. The height h of the windbreaks varied between 1.3 and 4 m.

Food crops will be grown inbetween according to the normal rotation (barley, sugar-beet, wheat). A catch crop might be planted as well (mustard).

For four years microclimatic parameters such as wind speed and direction, air and soil temperature, relative humidity, irradiance were recorded at several distances from the windbreaks. Figure 1 shows the experimental design for 1998. Probes were situated at about 20 cm above the canopy of the food crop (adapted along the year, except for wind direction and rainfall, 2 and 1 m). Data were automatically recorded at 30-minute intervals.

Measurements were carried out on neighbouring crops like sugar-beet, maize, chicory, triticale and wheat for development stage, final yield and quality, at different distances from the windbreaks.

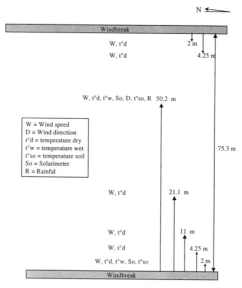

Figure 1. Design of the trial.

3. RESULTS AND DISCUSSION

3.1. Windbreak effect on neighbouring food crops

Change in yield according to the distance to the windbreak varies according to the crop considered and the height h of the windbreak. Figure 2 represents firstly trials with a positive impact (global crop yield increased within a few h of the windbreak) and secondly those with a negative impact. In nearly all cases (except for maize in 1995) there is a sharp yield decrease of about 30-40% at 1 h (a distance equivalent to 1 times the height of the windbreak). In the positive cases the yield increase at 3-5 h compensates this effect. In the negative cases, yield seems to be stabilised beyond 4 h.

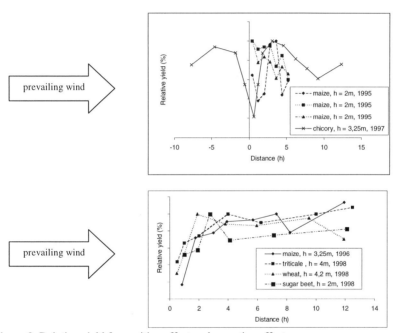

Figure 2. Relative yield for positive effect and negative effect.

From this yield behaviour it is possible to calculate the effect of the windbreak on global yield (Table 1). For each trial an average value was calculated for the most distant measurement points and relative values were determined for the measurement points near the windbreak. The global impact compares these relative values to the average.

Yields of triticale, wheat and sugar-beet seems to be badly affected while maize and chicory are favoured by the shelterbelt. Maize had both a positive and negative effect, probably explained by the height of the windbreak. A higher windbreak increased the competition nearby.

Table 1
Calculation of windbreak effect on yield

Crop, name of trial, year	h (m)	Yield difference
Triticale, Triandra, 1998	4	- 16.0% on 3h
Wheat, Ferme, 1998	4.2	- 13.1% on 3 h
Maize, Ferme, 1996	3.25	- 12.7% on 6 h
Sugar beet, CFE, 1998	2	- 2.4% on 4 h
Maize, Ferme, 1995	2	0 %
Chicory, Décharge, 1997	3.25	+ 7.4% on 6 h
Maize, Triandra, 1995	2	+ 8.5% on 3 h
Maize, Décharge, 1995	2	+ 11.6% on 3 h

However such differences in yield are not important at the field level because it is likely that the food crop will be cultivated far beyond the distance in which we have measured the windbreak effect.

Regarding quality, the statistical analysis of the dry matter content evolution versus distance (Figure 3) shows that when there is an effect (maize 1996, wheat 1998), the dry matter content decreases at about 1 h and there is no effect beyond 3 h.

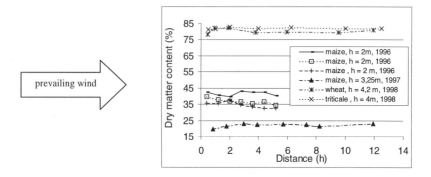

Figure 3. Windbreak effect on dry matter content of maize, triticale and wheat.

3.2. Microclimatic effects

Only effects on temperature and wind speed will be presented here. To study the impact of the windbreak, 20 typical days were chosen, 10 days of western wind and 10 days of eastern wind.

For 10 days of western wind, there was no significant effect of the distance (Figure 4) during entire days but the effects were significant and opposite during diurnal and night periods. During daylight, temperature typically increased leeward close to the windbreak in comparison with longer distances (because wind speed reduction tends to maintain the natural temperature gradient which decreases with altitude during daylight. As a way air mixture between air layers is less important, preventing cold air to lower the temperature leeward). This effect is more important at lower wind speed.

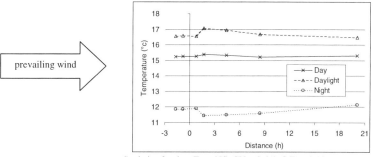

Statistics for day: Ftest NS, CV = 8,4%, S.E. = 0,08, LSD = 0,17
Statistics for daylight: Ftest HS, CV = 12,8%, S.E. = 0,19, LSD = 0,38
Statistics for night: Ftest HS, CV = 12,3%, S.E. = 0,15, LSD = 0,31

Figure 4. Temperature by western wind *vs.* distance (average of 10 days).

The relative difference between temperatures at 1 h and 2 h of the windbreak is interesting. During daylight and night, the maximal temperature is sometimes found at 1 h, sometimes at 2 h. This can be explained by the wind speed (Figure 5). If wind speed increases during daylight, the relative temperature at 1 h will increase in comparison with temperature at 2 h, becoming higher at around 2.6 m.sec^{-1}. At night the situation is just the opposite.

These observations seem to indicate that turbulence is playing a role. Stronger winds produce turbulence at longer distances from the windbreak, mixing the air, destroying the natural temperature gradient and therefore decreasing temperature during the day and increasing it at night.

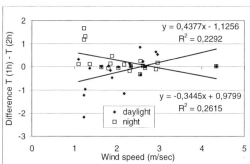

Figure 5. Difference between temperature at 1 h and at 2 h [T (1h)-T (2h)] during daylight and night versus wind speed (20 days).

From 20 May to 14 October the average temperature was significantly higher within about 8 meters leeward from the windbreak. A sum of temperature calculated over days and diurnal periods on a base 6°C (starting temperature for willows) confirms the higher temperature during daylight (4.7% higher than reference).

The profile of wind speed (Figure 6) on 20 days shows that wind velocity decreased by 30-40% at 0.8 h and returned to normal at 20 h leeward. At 0.8 h windward, wind speed was reduced by 5-20% and the effect was limited to about 3 h.

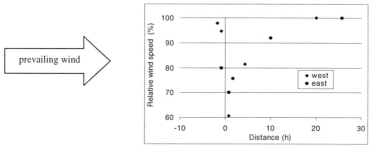

Figure 6. Wind speed profile by western and eastern wind (20 days)

Relative variation of wind speed between 0.8 and 4.4 h depended on wind speed. Figure 7 shows that when wind ranges from 1.2 to 4.3 m.sec^{-1}, relative wind speed between 0.8 and 4.4 h can be opposite. In addition, the difference between speed at mid-distance (about 4 h) and speed close to windbreak (about 1 h) becomes higher when wind increases.

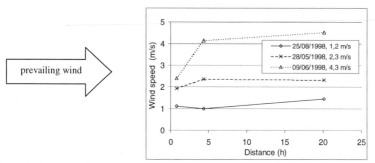

Figure 7. Relation of wind speed versus distance for different wind speeds.

This analysis confirms the conclusion for temperature. A higher wind speed appears to create turbulence at longer distances from windbreak, with a direct impact on temperature.

ACKNOWLEDGEMENT

This work was supported financially by the Commission of the European Communities, Agriculture and Fisheries (FAIR) specific RTD programme, CT96-1449, "Evaluation of combined food and energy systems for more efficient land use and environmentally benign sustainable production". The content of this paper does not represent the views of the Commission or its services and in no way anticipates the Commission's future policy in this area.

ARBORESCENT WILLOW BIOMASS PRODUCTION IN SHORT ROTATIONS

D. Kajba

Faculty of Forestry, Svetosimunska 25, 10000 Zagreb, Croatia

Clonal tests of the arborescent willow biomass production in short rotation have determined that genotypical clones differ in production of dry matter per hectare. At the age of 4 to 5 years the trispecies willow hybrid (*S. alba* × *S. fragilis* × *S. caprea*) produced considerably more dry matter than did tested white willow clones (*Salix alba*). The proportion of biomass above the ground increased with age, and the most productive trispecies hybrid had the most favourable relation between the underground and aboveground plant parts. The influence of clone and site governs production, and in addition the existence of a clone × spacing interaction has been determined.

1. INTRODUCTION

Today, pioneer soft broadleaved tree plantations have been established that use mixed cultures and special-purpose biomass production in short rotations. They are used to meet the requirements of log production in the mechanical wood processing and the wood-pulp industries and as energy forests. Increasing wood consumption puts greater pressure on natural stands, but by establishing intensive fast-growing broadleaved tree plantations on marginal forest and agricultural lands a global shortage of wood mass can be diminished considerably while conserving the natural ecosystem.

Bioenergy plantations and biomass production are in line with world trends, and they better utilize renewable energy sources without producing additional CO_2 (abundant in fossil fuels). Biomass is produced by photosynthesis; it forms oxygen and uses atmospheric CO_2, and it has advantages in relation to other energy sources.

Arborescent willows, as compared with other species of broadleaved and coniferous tree species, are the most suitable for biomass production in short rotations because of their very abundant growth in the first years. They produce in very short rotations an amount of dry matter such as other species need much more time to produce. Because they can be propagated by cuttings the maximal genetic gain can be obtained, and their strong sprouting power enables many vegetative generations at minimal production costs. Arborescent willows grow well even under dense canopy, thus growing many plants per surface unit. Nowadays in Croatia, a large number of selected and recognized willow clones are available that can be grown on lands unfavourable for other forest tree species.

This paper discusses the relationship of dry matter production (biomass production), in an energy plantation of selected arborescent willow clones, to the stand, the clone, and the spacing.

2. MATERIALS AND METHODS

In the Cakovec Forest Office area, near Podturen, a clonal test of arborescent willows was established in spring 1993 to try biomass production in short rotations. The experiment surface was 0.3 ha, and it was established on an alluvial soil, fluvisol (Komlenovic & Krstinic, 1994). The clonal test was established as a factorial test in four repetitions, with subtreatments of three different spacings (1.2 × 1.2 m, 1.2 × 0.8 m, 1.2 × 0.4 m). The experiment used seven clones, of which six belonged to the species *Salix alba* and one to the trispecies hybrid of white, fragile and sallow willows (*Salix × savensis* Trinajstic et Krstinic hybr. nov.). Before the beginning of the second vegetation the seedlings were cut, and later the excess shoots were removed, so that for each clone and each spacing, in all repetitions, one or three shoots per tree stump were left.

At the age of 2 to 3 years, the weight of seedlings (both fresh and dry) and nutrient concentrations per biomass components were determined (Komlenovic et al., 1996). Moreover, for seed material aged 1 to 2, 2 to 3, and 4 to 5 years, the total height and the diameter above the ground level were measured.

In autumn 1997, at the age of 4 to 5 years, the average seedling per clone in fresh condition, at a spacing of 1.2 × 1.2 m, were measured. The seedling weights for other spacings were determined from the regression relations of heights, diameters and weights of the analyzed seedlings. The kilograms of dry matter per hectare (kg/ha) was determined according to the actual number of shoots (sprouts) for the given spacings.

The statistical subtreatment relations in the factorial test, which refer to the influence of the clone, the spacing, and one or three shoots per tree stump, as well as to their interaction, were calculated using variant analysis (Statistica Programme 4.5).

3. RESULTS AND DISCUSSION

At 1 to 2 years, all clones contained about the same amount of fresh matter, but more important is the large share of underground biomass, which in some clones exceeded 45% of the total amount. By 2 to 3 years, the hybrid clone V221 contained several times the biomass in other tested clones. The share of underground biomass was considerably reduced, although it still exceeded 36% in some clones. The most productive V221 clone maintained its positive characteristics with the smallest underground share (21.6%) and the largest share of tree trunk wood (78.4 %). The ability to develop a more abundant root system is a genetic specificity of each clone, and in this hybrid clone this ability is very likely due to the genes of the sallow willow, which cannot be routinely propagated from cuttings.

Earlier research showed that, because of their higher demands for nutrients, the white and fragile willow hybrids grew well only on favourable sites, so good results with these hybrids can be obtained only on highly productive soils. On extremely poor soils, pure white willow clones, which consume fewer nutrients, especially nitrogen, are well adaped and survive (Komlenovic & Krstinic, 1969, 1987; Krstinic et al., 1989). Thus, good results may be expected from this trispecies hybrid (V221) on favourable sites. This clone is characterized by high calcium and zinc concentrations in leaves and by low contents of nitrogen and other nutrients, especially sulphur (Komlenovic et al., 1996). Of the studied genotypes, the V221 clone also used nitrogen the most efficiently to produce tree trunk dry matter.

At 4 to 5 years this clone again had the highest content of fresh matter in the seedling aboveground (8950.0 g), and it had the best ratio of aboveground to underground mass. At that age, the share of the underground part ranged between 17.3 and 26.0%.

Measured heights, diameters above the ground, and estimated dry matter/ha of seedling aboveground part at the age of 4 to 5 years are shown in Table 1.

Each clone was tested at three spacings, and in each spacing the subtreatments were separated (one or three shoots per tree stump). Survival (the number of sprouts) was used to determine the dry matter amount in the aboveground part/ha.

In countries that use selected willow clones for biomass production on rotations of 3 to 5 years, production ranges between 36 and 60 tons of dry matter/ha. After felling in the second production cycle, production may reach 12 to 15 tons of dry matter/ha, if the soil is provided with sufficient quantities of nutrients, chiefly nitrogen. The production can be maintained in vegetative cycles for 20 years (Sennerby-Forsse, 1986; Zsuffa et al., 1993; Christersson et al., 1993).

The hybrid clone V221, even at 4 to 5 years, has the best production (Table 1). At a spacing of 1.2 × 0.4 m (in total 41 660 sprouts/ha), its production was 130 tons of dry matter/ha. Although the nitrogen content of this plot was not good, this production is much higher than the biomass production in the selected willow clones at the same age anywhere else in the world. Next, research should determine the specific wood weight and other characteristics of the V221 clone that are essential for dry matter production.

The possibilities for biomass production with the selected arborescent willow clones in Croatia are optimal with respect to the stand production potential and selected genotype assortment. Biomass production can be improved considerably by fertilization, a standard practice abroad that can be avoided by selecting clones that thrive on poor soil.

Variant analysis was applied to total height, the diameter above the ground, and dry matter/ha.

The clone and the spacing influenced all three measured properties in a statistically very significant way; and they dominated total biomass production. Moreover, for all three properties a statistically significant clone × spacing interaction was determined.

Table 1
Biomass production of arborescent willows in short rotations - age 4–5 years

No.	Clone sign	Botanical Name	Spacing (m)	No. of sprouts (no./ha)	Height (cm)					Diameter above the ground level (mm)					Dry matter		
					$\bar{x}$	Range of variability	s	C.V. %		$\bar{x}$	Range of variability	s	C.V. %	Above the ground (g)	kg/ha
1.	V 158	Salix alba	1.2 × 1.2	4874	530	400-650	85.6	16.1		42	30-55	9.7	23.1	1145	5582
			1.2 × 1.2	14999	515	350-600	85.1	16.5		42	25-50	9.1	21.7	1112	16687
			1.2 × 0.8	7875	550	400-650	74.5	13.5		43	30-60	11.5	26.7	1172	9235
			1.2 × 0.8	25313	560	400-650	70.0	12.5		32	10-50	14.6	45.6	872	22091
			1.2 × 0.4	14625	570	400-650	75.3	13.2		45	30-60	12.2	27.1	1227	17949
			1.2 × 0.4	49611	575	550-600	26.3	4.6		27	15-40	10.6	39.3	736	36514
2.	107/65/6	Salix alba	1.2 × 1.2	5874	574	400-700	97.9	17.0		52	45-60	6.7	12.9	2263	13292
			1.2 × 1.2	17625	660	600-800	61.5	9.3		39	15-65	17.3	44.4	1697	29913
			1.2 × 0.8	8626	553	450-650	70.6	12.8		50	35-60	9.3	18.6	2176	18769
			1.2 × 0.8	24188	575	400-700	9.2	16.0		36	20-55	12.7	35.3	1566	37894
			1.2 × 0.4	18374	570	400-750	115.9	20.3		49	40-60	8.21	16.7	2132	39181
			1.2 × 0.4	45562	560	400-650	99.4	17.7		41	20-50	10.5	25.6	1784	81283
3.	V 221	Salix alba X S. fragilis X S. caprea	1.2 × 1.2	4749	720	650-750	34.9	4.8		59	50-70	8.9	15.1	4191	19902
			1.2 × 1.2	14999	735	700-800	33.7	4.6		56	25-80	15.7	28.0	3977	59663
			1.2 × 0.8	8063	720	700-750	25.8	3.6		64	55-75	8.2	12.8	4546	36654
			1.2 × 0.8	24188	770	750-800	25.8	3.3		48	20-75	16.3	33.9	3409	82469
			1.2 × 0.4	15750	610	550-650	31.6	5.2		43	20-50	10.3	23.9	3054	48106
			1.2 × 0.4	41660	700	650-800	47.1	6.7		44	20-70	15.6	35.4	3125	130187
4.	MB 15	Salix alba X Salix alba var. Vitellina	1.2 × 1.2	5250	695	600-800	5.2	8.4		46	25-60	12.1	26.3	2025	10631
			1.2 × 1.2	16500	610	400-700	71.3	11.7		32	20-50	11.8	36.9	1409	23243
			1.2 × 0.8	8813	306	200-400	76.9	25.1		23	10-30	7.6	33.0	1012	8923
			1.2 × 0.8	24188	320	200-450	78.9	24.6		21	10-35	10.2	48.6	924	22360
			1.2 × 0.4	16875	390	300-500	65.8	16.9		28	20-35	5.7	20.3	1232	20800
			1.2 × 0.4	43537	560	400-700	96.6	17.2		29	15-45	11.4	39.9	1276	55553

Table 1 (continued)

5.	79/64/2	*Salix alba*	1.2 x 1.2	5375	560	450-650	61.5	11.0	50	45-55	3.5	7.0	2328	12515
			1.2 x 1.2	18000	570	450-650	71.5	12.5	30	20-50	11.1	37.0	1397	25147
			1.2 x 0.8	8813	565	400-700	91.4	16.2	37	30-45	5.7	15.4	1723	15185
			1.2 x 0.8	25313	580	500-700	71.5	12.3	31	15-55	13.7	44.2	1443	36542
			1.2 x 0.4	15375	605	500-700	72.6	12.0	36	25-50	9.6	26.7	1676	25775
			1.2 x 0.4	46574	695	600-800	64.3	10.6	31	10-45	12.9	41.6	1483	69069
6.	V 160	*Salix alba*	1.2 x 1.2	4874	520	400-650	71.5	13.7	43	30-55	9.1	21.1	1075	5239
			1.2 x 1.2	14999	575	500-700	67.7	67.7	38	20-60	13.9	36.6	950	14249
			1.2 x 0.8	8063	670	600-700	34.9	5.2	61	55-70	5.5	9.0	1525	12296
			1.2 x 0.8	24188	610	450-700	69.9	11.4	38	10-55	14.9	39.2	985	23825
			1.2 x 0.4	15375	665	600-750	53.0	7.9	55	45-65	7.9	14.4	1375	21140
			1.2 x 0.4	42524	690	600-800	51.6	7.5	53	35-85	21.4	40.4	1325	56344
7.	V 0240	*Salix alba* var. *calva* X *Salix alba*	1.2 x 1.2	5874	605	450-800	114.1	6.8	42	35-50	5.7	14.8	1545	9076
			1.2 x 1.2	16124	790	650-900	81.0	8.8	43	15-90	23.9	39.7	1229	19819
			1.2 x 0.8	9188	815	650-900	94.4	12.4	55	40-65	10.6	25.7	1896	17424
			1.2 x 0.8	27563	695	600-900	11.2	5.0	45	20-85	22.9	35.1	1159	31943
			1.2 x 0.4	16349	815	650-900	94.4	7.8	55	40-65	10.6	10.8	1229	20196
			1.2 x 0.4	44550	755	600-850	724	12.5	54	35-75	17.5	37.5	1123	50029

59

4. CONCLUSIONS

1. In the clonal test for biomass production in short rotations, a genotypical differentiation of clones with respect to dry matter production/ha has been determined.
2. For the trispecies willow hybrid (*S. alba* x *S. fragilis* x *S. caprea*) aged 4 to 5 years, the estimated dry matter/ha was 130 tons, a value much higher than shown by the tested white willow clones (*Salix alba*).
3. The share of the biomass aboveground increased with age, and the most productive trispecies hybrid had the best relation between the underground and overground plant parts.
4. The clone and the canopy influence on the production is dominant, and the existence of clone × spacing interaction was determined as well.

REFERENCES

Christersson, L., L. Sennerby-Forsse and L. Zsuffa (1993). The role and significance of woody biomass plantations in Swedish agriculture, The Forestry Chronicle, 69, pp. 687–693.

Komlenovic, N. and A. Krstinic (1987). Genotipske razlike izmeu nekih klonova stablastih vrba s obzirom na stanje ishrane, Topola, 133-134, pp. 29–40.

Komlenovic, N., A. Krstinic, and D. Kajba (1996). Mogucnosti proizvodnje biomase stablastih vrba u kratkim ophodnjama u Hrvatskoj, in Unapredenje proizvodnje biomase šumskih ekosustava, ed. by B. Mayer, Šumarski fakultet Sveučilišta u Zagrebu i Šumarski institut, Jastrebarsko.

Komlenovic, N., A. Krstinic, and D. Kajba (1996). Selection of arborescent willow clones suitable for biomass production in Croatia. Proceedings, 1996 20th Session of the International Poplar Commission. Budapest, pp. 297–308.

Krstinic, A., N. Komlenovic, and M. Vidakovic (1989). Selection of white willow clones (*Salix alba* L.) suitable for growing mixed plantations with black alder (*Alnus glutinosa* (L.) Gaertn.). Annales forestales, 15, pp 17–36.

Sennerby-Forsse, L. (1986). Handbook for Energy Forestry. Swedish University of Agricultural Sciences.

Zsuffa, L., L. Sennerby-Forsse, H. Weisgerber and R.B. Hall (1993). Strategies for clonal forestry with poplars, aspens and willows, in Clonal Forestry II, ed. by M. R. Ahuja and W. J. Libby, Springer-Verlag.

Biomass Production and Integration With Conversion Technologies

Biomass Resources: Regional Analysis

POTENTIAL OF SHORT ROTATION WOOD CROPS AS A FIBER AND ENERGY SOURCE IN THE U.S.

M. E. Walsh,[a] P. J. Ince,[b] D. De La Torre Ugarte,[c] D. Adams,[d] R. Alig,[e] J. Mills,[f] H. Spelter,[b] K. Skog,[b] S. P. Slinksy,[c] D. E. Ray,[c] and R. L. Graham[a]

[a]Oak Ridge National Laboratory, P.O. Box 2008, Oak Ridge, Tennessee 37831-6205
[b]U.S.D.A. Forest Service, Forest Products Laboratory, Madison, Wisconsin 53705
[c]Agricultural Policy Analysis Center, University of Tennessee, Knoxville, Tennessee 37901-1071
[d]Oregon State University, College of Forestry, Corvallis, Oregon, 97331
[e]U.S.D.A. Forest Service, Pacific Northwest Station, Corvallis, Oregon 97331
[f]U.S.D.A. Forest Service, Pacific Northwest Research Station, Portland, Oregon 97208-3890

The use of short rotation wood crops (SRWC) as dedicated energy sources is constrained by their high price. An alternative approach is to produce co-products from SRWC, including higher-value fibers (i.e., paper and pulp, veneers, engineered wood products, etc.) and lower-value energy. The U.S. Departments of Energy and Agriculture are working together to explore the economic potential of this approach. An agricultural sector model (POLYSYS) has been modified to include switchgrass, hybrid poplar, and willow, and it is currently being used to evaluate the price, quantity, and location of energy crop production and the potential effects (prices, quantities, and net returns) of large scale energy crop production on traditional crops. This modeling is being extended to evaluate the economic competitiveness of using SRWC for fiber and energy by linking the modified POLYSYS model with U.S. Forest Service models (NAPAP—paper and pulp; TAMM/ATLAS—lumber supply and wood panels market). The linked models will estimate the potential SRWC fiber demand, SRWC supply, equilibrium SRWC fiber quantities and prices, and selected effects on traditional forest management activities. Additionally, SRWC residue quantities potentially available for energy use will be estimated.

1. INTRODUCTION

Short rotation wood crops (SRWC) such as hybrid poplar have been developed as a feedstock for electricity, liquid fuel, and chemicals. The high cost of producing these crops limits the opportunities to use them as a dedicated energy feedstock. The production of multiple products from a single feedstock offers a potential opportunity to reduce the cost of using at least a portion of the feedstock for energy. For hybrid poplars, high-value fiber uses combined with energy uses offer a multiple use opportunity. Hybrid poplar can be used in a variety of fiber uses (e.g., paper and pulp, oriented strandboard, other composite materials) and the higher prices that can be paid for poplar in these uses provide an economic opportunity to begin production of hybrid poplar. Process residuals (bark, tops, limbs) can potentially be available for energy use. The fiber industry in the U.S. has begun showing interest in producing SRWC. In 1996, approximately 47,000 hectares (117,000 acres) of SRWC involving at least 25 companies were in research or commercial production in the U.S. (Wright, 1996). The U.S. Department of Energy (DOE) and the U.S.D.A. Forest Service (FS) have joined together to more thoroughly analyze the potential for hybrid poplar to become a significant fiber source in the U.S. during the next century.

2. ANALYTICAL APPROACH

Forest Service fiber demand and supply models (NAPAP, TAMM/ATLAS) are being linked with POLYSYS, an agricultural sector model that has been modified to include dedicated energy crops such as hybrid poplar. The linked models will be used to calculate the equilibrium quantities and price of hybrid poplar and evaluate the effects that this production will have on the agricultural and forestry sectors. Agricultural effects will include the geographic location of production, effects on traditional agricultural crop quantities and prices and net returns to producers. Forest sector effects will include implications for equilibrium fiber price and quantities and the intensity of harvest in public and private forests.

2.1. POLYSYS model description

POLYSYS was developed by, and is maintained by, the University of Tennessee Agricultural Policy Analysis Center (Ray et al., 1997). It is used by the USDA Economic Research Service to conduct policy analysis. POLYSYS contains data describing major agricultural crops (corn, wheat, soybeans, cotton, rice, grain sorghum, barley, oats, alfalfa, other hay), livestock (swine, beef, dairy, poultry, sheep), and food, feed, industrial, and export demand for agricultural products. POLYSYS is aggregated into 305 Agricultural Statistical Districts which can be combined to provide local, state, regional, and national level information. POLYSYS allocates land to each crop based on a comparison of its relative profitability (i.e., price × yield − production costs). The introduction of energy crops into agriculture results in the shifting of acres from traditional crop production, thus altering the prices and quantities of these crops. These changes are estimated using feedback mechanisms and repeated iterations until equilibrium prices and quantities are achieved. The model is anchored to the USDA baseline (which is being extended from the year 2007 to the year 2025).

In a joint project between the U.S. Department of Agriculture (USDA) and DOE, POLYSYS has been modified to include dedicated energy crops (switchgrass, hybrid poplar, willow). Hectares deemed suitable for energy crop production are currently limited to the hectares that are classified as cropland and planted to the major crops in the eastern half of the United States (i.e., land east of central North Dakota, South Dakota, Kansas, Nebraska, Oklahoma, and Texas) and the Pacific Northwest (104 million hectares, 257 million acres). Hectares that are idle (either in CRP or for other reasons) have not yet been incorporated into the model. Because energy crops require multi-year rotations, POLYSYS utilizes a net present value approach to compare the relative profitability of energy crops and traditional agricultural crops. Additionally, the model incorporates into the decision function, the expected price changes of traditional crops resulting from large-scale energy crop production (Walsh et al., 1998, de la Torre Ugarte et al., 1998). Conventional and energy crop production costs were estimated using the ABS budget system (POLYSYS budget generator) and BIOCOST (ORNL energy crop production cost model) and energy crop yields were obtained from the ORECCL database (Slinsky et al., 1996; Walsh and Becker, 1996; Graham et al., 1996). A workshop, attended by energy crop experts from DOE and USDA, reviewed the yield and energy crop management practices used in the analysis.

2.2. NAPAP model description

The North American Pulp and Paper (NAPAP) Model is an economic model of the U.S. and Canadian pulp and paper sector. It was developed jointly by the U.S.D.A. Forest Service, Forestry Canada, and the University of Wisconsin. NAPAP analyzes the domestic and export demand for paper and paperboard products in North America and is used by the Forest Service to conduct the Resource Planning Act Assessments (Ince, 1998).

NAPAP uses a price endogenous linear programming system that combines regional information on paper and pulp supply, demand, manufacturing processes, and transportation costs into a spatial sector model. It maximizes consumer and producer surplus throughout the sector. In its static phase, NAPAP computes a multi-region, multi-commodity equilibrium quantity and price at a given point in time. In its dynamic phase, it predicts the evolution of the spatial equilibrium over time. Shifts in supply and demand and capacity changes are determined endogenously in the model (Zhang et al., 1996). NAPAP includes two demand regions (domestic and export markets) and five U.S. supply regions. It includes virgin hardwood and softwood pulpwood and residuals, recycled materials, and the major classes of paper (e.g., newsprint, free sheets, sanitary), paperboard, and other paper materials (e.g., construction paper). It incorporates a variety of manufacturing processes to reflect existing technologies and to allow incorporation of new technologies into the analysis. Paper and paperboard demand is modeled as a function of price, gross national product, and population while supply is modeled as a function of price, regional timber inventory, sawtimber stumpage price, and discount rates (Zhang et al., 1996). NAPAP is run in conjunction with the TAMM/ATLAS model.

2.3. TAMM/ATLAS model description

The TAMM/ATLAS model was developed by the U.S.D.A. Forest Service Pacific Northwest Research Station. TAMM/ATLAS (Timber Assessment Market Model/Aggregate TimberLand Analysis System) is a spatial model of the solid wood and timber inventory elements of the U.S. forest sector. It estimates annual projections of volume and prices in solid wood products and sawtimber stumpage markets and estimates total timber harvest and inventory by geographic regions for periods of up to 50 years. TAMM evaluates demand for softwood lumber, softwood plywood, oriented strandboard/waferboard, and engineered wood products for 18 wood product uses (e.g., use in home construction, home maintenance, nonresidential construction, etc.) and allows for price-based substitution among lumber, plywood, oriented strandboard, and other non-wood materials. Net revenues (total consumer payments minus producer variable costs and transport costs) are maximized (Adams and Haynes, 1997).

ATLAS projects timber growth and wood supply from forests (Mills and Kincaid, 1992). Harvest is a function of prices, inventory, interest rates, and income from non-forest sources for non-industrial owners. ATLAS provides regional estimates of privately owned timberland (acres) and growing stock yields by age class, forest type, site class, and management intensity class. The ATLAS model incorporates exogenous projections of land area changes, such as shifts among major land uses and conversion of natural forest types to plantations, that are provided by land use and land cover models and also estimates future rates of increase in private forest management intensity (such as increases in harvest frequency and shifts in forest management practices) (Alig and Wear, 1992).

3. RESULTS

Results from the linked POLYSYS, NAPAP, and TAMM/ATLAS models are not yet available at the time this paper is being written. However, preliminary analysis using the models separately has been conducted. The POLYSYS model has been used to evaluate equilibrium quantities and locations of energy crops at selected prices. The three energy crops compete with each other as well as with conventional agricultural crops for land with the land allocated to the most profitable crops. Assuming the same price (on an energy content basis) for switchgrass, hybrid poplar, and willows, the model indicates that switchgrass is likely to be the dominant energy crop produced with significantly fewer acres allocated to short rotation wood crops (Walsh et al., 1998). POLYSYS analysis with substantially higher SRWC prices relative to switchgrass has not been completed.

At a farmgate price of $38.50/dry Mg ($35.00/dry ton), an estimated 3.91 million hectares (9.66 million acres) of switchgrass could be produced at a profit at least as great as for traditional agricultural crops produced on the same land. Total production is about 45 million dry Mg annually (50 million dry tons). At an equivalent energy price ($42.23/dry Mg and $38.39/dry ton for poplar; $40.73/dry Mg and $37.03/dry ton for willow), 24,000 hectares (60,000 acres) of poplar and 32,000 hectares (80,000 acres) of willow could be grown resulting in an additional annual production of 627,000 dry Mg (690,000 dry tons) of biomass. Net farm returns increased by an estimated $878 million. At a switchgrass farmgate price of $55.00/dry Mg ($50.00/dry ton), an estimated 7 million hectares (17.3 million acres) of switchgrass can be grown resulting in an annual production of 79 million dry Mg (87 million dry tons) of biomass. At an equivalent energy price ($60.32/dry Mg and $54.84/dry ton for poplar; $58.19/dry Mg and $52.90/dry ton for willow), 32,000 hectares (80,000 acres) of poplar and 77,000 hectares (190,000 acres) of willow could be grown resulting in an additional annual production of 1.2 million dry Mg (1.3 million dry tons) of biomass. Net farm returns increased by an estimated $3.5 billion dollars.

The analysis indicates that switchgrass can be produced in substantial quantities at lowest price in the Lake States, North Plains, South Plains, and Southeast. Parts of the Midwest and Northeast enter the solution as switchgrass price increases. Short rotation wood crop production occurs primarily in Tennessee, Louisiana, Minnesota, and Oregon for poplar, and the Northeast and Lake States for willow. The regions of the country where energy crop production is lowest also corresponds to regions with substantial forestry activities.

Using production costs and yields similar to those used in the POLYSYS analysis, preliminary analysis using the NAPAP model indicates that short rotation wood crops become economically feasible on a substantial scale of production beginning in the year 2010. By the year 2050, the analysis indicates that SRWC could supply up to 20 percent of the total pulpwood supply and would substitute for harvest of hardwood pulpwood on forest lands (Ince, 1998).

Preliminary analysis suggests that a significant opportunity exists for SRWC to be produced for fiber uses. Process residuals (limbs, tops, bark) could be used for energy and could potentially be available in significant quantities at substantially lower prices than if produced as a dedicated energy crop. Linkage of the POLYSYS model with the NAPAP and TAMM/ATLAS models will provide substantial insights as to the extent of this potential.

REFERENCES

Adams, D.M. and R.W. Haynes (1997). Long-term projections of the U.S. forest sector: The structure of the timber assessment market model, models needed to assist in the development of a national fiber supply strategy for the 21st Century: Report of a workshop, Resources for the Future, Washington, D.C., pp. 39–57.

Alig, R. and D. Wear (1992). Changes in private timberland: statistics and projections for 1952 to 2040, Journal of Forestry, 90, pp. 31–36.

de La Torre Ugarte, D.G., S.P. Slinsky, and D.E. Ray (1998). The economic impacts of biomass crop production on the U.S. agriculture sector, Working paper, Agricultural Policy Analysis Center, The University of Tennessee, Knoxville, Tennessee.

Graham, R.L., L.J. Allison, and D.A. Becker (1996). ORRECL-Oak Ridge Energy Crop County Level Database. Proceedings, BIOENERGY '96—The Seventh National Bioenergy Conference, Nashville, Tennessee, Southeastern Regional Biomass Energy Program, pp. 522–529.

Ince, P.J. (1998). A long-range outlook for U.S. paper and paperboard demand, technology and fiber supply-demand equilibria. Proceedings, 1998 Society of American Foresters National Convention, Society of American Foresters, Bethesda, Maryland (in press).

Mills, J. and J. Kincaid (1992). The Aggregate TimberLand Assessment System—ATLAS: A comprehensive timber projection model. USDA Forest Service, Pacific Northwest Research Station, Portland, Oregon, PNW-GTR-281.

Ray, D.E., D.G. de La Torre Ugarte, M.R. Dicks, and K.H. Tiller (1997). The POLYSYS modeling framework: A documentation, Working paper, Agricultural Policy Analysis Center, The University of Tennessee, Knoxville, Tennessee.

Slinsky, S.P., D.E. Ray, and D.G. de La Torre Ugarte (1996). The APAC Budgeting System: A User's Manual, Unpublished manuscript, Agricultural Policy Analysis Center, The University of Tennessee, Knoxville, Tennessee.

Walsh, M.E. and D.A. Becker (1996). BIOCOST: A software program to estimate the cost of producing bioenergy crops. Proceedings, BIOENERGY '96—The Seventh National Bioenergy Conference, Nashville, Tennessee, Southeastern Regional Biomass Energy Program, pp. 480–486.

Walsh, M.E., D.G. de La Torre Ugarte, S. Slinsky, R.L. Graham, H. Shapouri, and D. Ray (1998). Economic Analysis of Energy Crop Production in the U.S.—Location, Quantities, Price and Impacts on Traditional Agricultural Crops. Proceedings, BIOENERGY '98: Expanding Bioenergy Partnerships Conference, Madison, Wisconsin, Great Lakes Regional Biomass Energy Program, pp. 1302–1311.

Wright, L.L. and S. Berg (1996). Research and commercialization of short-rotation woody crops for energy, fiber, and wood products. Proceedings, BIOENERGY '96—The Seventh National Bioenergy Conference, Nashville, Tennessee, Southeastern Regional Biomass Energy Program, pp. 508–514.

Zhang, D., J. Buongiorno, and P.J. Ince (1996). A recursive linear programming analysis of the future of the pulp and paper industry in the United States: Changes in supplies and demands, and the effects of recycling, Annals of Operations Research, 68, pp. 109–139.

GEOGRAPHIC INFORMATION SYSTEM MODELING OF RICE STRAW HARVESTING AND UTILIZATION IN CALIFORNIA

L.G. Bernheim,[a] B.M. Jenkins,[a] R.E. Plant,[b] and L. Yan[a]

[a]Department of Biological and Agricultural Engineering, University of California, Davis, California, USA 95616
[b]Agronomy Department, University of California, Davis, California, USA 95616

We developed a geographic information system (GIS) model for rice in California's Sacramento Valley to evaluate regional effects of straw harvest and use, and to optimize commercial site locations. This GIS work is part of a larger project investigating straw harvesting and handling methods, potential machine improvements, and agronomic and environmental effects of large-scale straw utilization to replace open burning and soil incorporation. The GIS model includes resource and infrastructure information that allows network modeling of handling and transportation to identify least-cost storage, processing, and transformation sites. A variety of map coverages are incorporated including Public Lands Survey (section-township-range polygons) data, road system, rail system, wetlands, cities and urban areas, and soils coverages. Sensitivity analyses are performed on resource distribution and site locations to evaluate cost effects on potential facilities using straw. Traffic effects such as heavily trafficked road segments, infrastructural limitations, potential hazards, and emission loads were also assessed. Preliminary results of the network models using actual field data coupled with sensitivity analyses are presented.

1. DATA COLLECTION

The geographic information system is created in ESRI's[*] ARC/INFO[+], with several types of data included. Rice acreage information was combined with a variety of ARC/INFO map coverages to create the network model and for analysis. Figure 1 gives basic resource distribution information.

1.1. Rice acreage data
Exact locations of rice fields within each county must be known, in order to study the costs of moving rice straw to processing and storage sites. The database of pesticide

---

[*] Environmental Systems Research Institute, Inc., Redlands, CA 92373 USA
[+]ARC/INFO, ARCEDIT, and ARC/INFO NETWORK are trademarks of environmental Systems Research Institute, Inc.

application permits from each County Agricultural Commissioner's Office was the best source for this information. For the 1997 growing season, text files of acreage identified by township, range and section were obtained from each of the ten counties in the study area, Butte, Colusa, Glenn, Placer, Sacramento, San Joaquin, Sutter, Tehama, Yolo, and Yuba. Once combined into a single file, a record exists for each unique section of approximately one mile by one mile specifying the acreage of rice in that section. Sections are identified by meridian ("Mt. Diablo" for all these Sacramento Valley counties), township code (for example, "02W"), range code (for example, "15N") and section number (an integer 1 – 36).

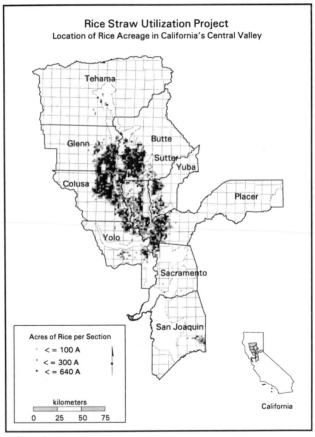

Figure 1

1.2. Maps

Polygon map coverages giving county and section boundaries, and a line coverage of the road systems are needed to create the model. All three of these maps come from the Teale Data Center of the State of California. Additional maps used for further analysis include urban areas, elevation, soil types, railroads, powerlines, and others.

The Teale Data Center's Public Lands Survey coverage (PLS) provides the section boundaries in the ten county area. Not all sections are included on this map because some land grant and wetland areas were never mapped.

2. BUILDING THE NETWORK MODEL

Acreage data were combined with the section map, and then an ARC/INFO NETWORK with nodes for rice acreage and potential processing and storage sites was created.

2.1. Combining rice acreage data with maps

Rice acreage data are joined to the section map polygons, since each is identified by unique section code in the same format. The features of ARC/INFO's INFO system are used throughout this project for combining and adding map features. Of the 1,400 sections for which rice acreage data are known, 269 do not have map polygons in the PLS map because they fall into the unmapped areas. To use the map to represent all the rice acreage, these unmapped sections were added.

Because the exact location of each rice field within a section is not given in this data set, the centroid point within each section is assumed to have the acreage associated with it. For unmapped sections, the approximate center of the section was manually entered using ARCEDIT. This resulted in a point coverage with 1,400 points, each with an associated acreage amount. Straw tonnage was computed from these acreage amounts.

2.2. Creating the NETWORK

An ARC/INFO NETWORK is a line coverage with demand points and centers points. To model the transportation of rice straw, the "demand points" represent the resource locations, i.e., fields aggregated at the section level. The "centers" are the potential processing sites to which this straw will be transported.

Here, the line coverage is the road coverage, restricted to roads of USGS classification 1 through 4. These are roads designated primary, secondary, thoroughfare and residential. The points with rice acreage associated, described in the previous section, were added as nodes to this line coverage and connected to the existing roads by short, straight line segments. As these added segments do not represent actual roads, they are designated pseudoroads. They are merely a way to connect the section centroid points to the actual road network. The process of adding these nodes, and constructing the road segments were accomplished in a series of ARC/INFO commands stored in macro files.

A set of potential processing or storage facilities for rice straw was identified on the map by the road coverage point closest to the site. In some cases, ARCEDIT was used to add a node (point) to the road coverage. These are stored in a "centers file."

Several values must be assigned to complete the NETWORK including the cost per ton-mile of transporting rice straw by road class and the tons of rice straw per acre of rice.

3. SITE SELECTION

Because not all the potential centers will actually be constructed as rice straw storage or processing sites, we identified a sub-set of the potential sites to which straw might be delivered most cost effectively. Of the fifteen potential sites, the model can be used to select the sites most cost effective from a transportation standpoint.

3.1. Assigning the variables
For any given site selection run of the model, several variables need to be assigned:
- the number of sites to be selected, for example three of the fifteen potential sites;
- the transportation cost per ton-mile by USGS road class;
- the number of tons of rice straw per acre of rice; for example 2.0 tons per acre.

3.2. Using "LocateAllocate" to select the sites
Once the variables are assigned, and some NETWORK set-up commands run, the NETWORK command "LocateAllocate" selects the sites based on a least-cost optimization. Results from the selection of three of the fifteen sites are shown in Figure 2. On this map, the underlying road network is not shown, but in fact was used in determining transportation cost.

4. EFFECT ON ROADS

The model can be used to study effects on roads from several perspectives. The term "trial" is used here to indicate a model run with a given set of road transportation costs and tons of straw per acre. Average transportation cost per ton of straw from several trials may be compared to gain an understanding of the effects of the variation of trial parameters, as well as an understanding of the effect of various road types. For a potential processing site, a detailed mapping may be made of the effects on roads for a given site capacity.

4.1. Site-specific analysis
By setting the trial parameters, and the capacity of the processing center, the model provides detailed results on cost per ton-mile to deliver straw by USGS road type, as well as overall cost. The specific road segments used to deliver the straw to the center can be mapped by effect of tonnage.

4.2. Example
Figure 3 shows the road impact mapping for a potential center in Sacramento County. The darkest road segments are those which are most heavily affected by rice straw transportation. Parameters used in running the trial are shown on the map.

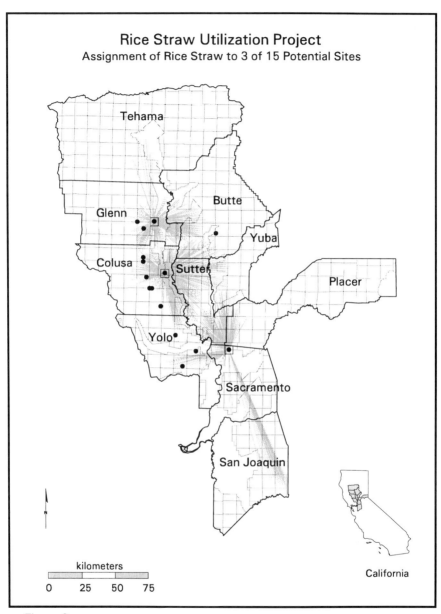

Figure 2

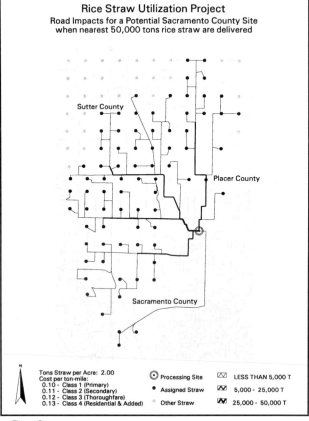

Figure 3

5. FURTHER DEVELOPMENT

The rice acreage data for the 1998 growing season will be added to the model, providing an opportunity to compare the data for two years, as well as to create the analyses with another data set. Output maps and data tables will be made available over the internet to registered users, with the possibility of some user interaction with the model.

PROGRESS TOWARDS MAKING WILLOW BIOMASS CROPS THE FUEL OF THE FUTURE IN THE NORTHEASTERN UNITED STATES

E.H. White,[a] E.F. Neuhauser,[b] L.P. Abrahamson,[a] T.A. Volk,[a] C.A. Nowak,[a] J.M. Peterson,[c] E. Gray,[d] C. Demeter,[d] and C. Lindsey[d]

[a]State University of New York College of Environmental Science and Forestry, Syracuse, New York
[b]Niagara Mohawk Power Corporation, Syracuse, New York
[c]New York State Energy Research and Development Authority, Albany, New York
[d]ANTARES Group, Inc., Landover, Maryland

The Salix Consortium is a dynamic partnership of over 25 groups and organizations representing research institutions, farming and environmental groups, government and industry (Table 1). The Consortium's goal is to facilitate the development of willow biomass crops as a locally produced source of renewable energy by simultaneously optimizing production systems, developing producer interest and participation, and expanding markets.

Considerable progress has been made over the past year towards this goal. The State University of New York College of Environmental Science and Forestry (SUNY-ESF) continues to develop and implement a strong applied research program based upon 20 years of effort that underpins the commercialization of willow biomass crops. Research focuses on both the production system (i.e., nitrogen fertilizer rate studies, insect and disease work, genetics and breeding) and environmental benefits associated with willow biomass crops (i.e., changes in soil carbon under willow biomass crops, avian and soil microarthropod diversity). Results to date have been translated into initial recommendations for scale-up activities.

Over 60 landowners have expressed interest in participating in the willow biomass program. These individuals represent over 1500 hectares of potential willow production. Leases on over 150 hectares of private land have been secured, of which about 60 hectares of willow biomass crops were planted in central and western New York in 1998. An additional 100 hectares were prepared in the fall of 1998 for planting in the spring of 1999. A Step Planter has been ordered for the program, which should increase planting rates by 200–300%.

Table 1
SUNY-ESF Biomass-Bioenergy Program and Salix Consortium Members and Associates

PRINCIPAL MEMBERS

Burlington Electric Department (Vermont)
Cornell University: Cooperative Extension, Dept. of Agricultural Engineering, Laboratory of Ornithology
Electric Power Research Institute (California)
GPU GENCO (Pennsylvania)
New York State Electric and Gas Corporation (NYSEG)
New York State Energy Research and Development Authority (NYSERDA)
Niagara Mohawk Power Corporation (NMPC)
Ontario Hydro Technologies
South-Central New York Resource Conservation and Development
State University of New York College of Environmental Science and Forestry
University of Toronto
US Department of Agriculture:
    Cooperative State Research, Education and Extension Service (CSREES)
    Forest Service (FS)
    Natural Resources Conservation Service (NRCS)
US Department of Energy:
    Oak Ridge National Laboratory (ORNL)
    National Renewable Energy Laboratory (NREL)

ASSOCIATE PARTNERS

Coalition of Northeast Governors
Gas Research Institute (Illinois)
National Audubon Society
New York Center for Forestry Research and Development at SUNY-ESF
New York Farm Bureau
New York Gas Group
New York State Association of Environmental Management Councils
New York State Department of Agriculture and Markets
New York State Department of Environmental Conservation
New York State Rural Development Council
New York Wine and Grape Foundation
Miner Agricultural Research Institute
Research on Energy and Materials Conservation Program at SUNY-ESF
State University of New York at Plattsburgh
Sullivan Trail Resource Conservation and Development

At the market end of the system, Niagara Mohawk Power Corporation (NMPC) will complete co-firing retrofits at one of the Dunkirk station's 100 MW boilers. Co-firing will commence in 1999, including trials with willow biomass. The staff's attitude at the power plant has changed from one of initial skepticism and reluctance to one of enthusiasm and active participation. This has been an important component in making the plant retrofits successful. Plans are also underway to test willow biomass as a feedstock for the gasifier at the McNeil station in Burlington, Vermont.

The following are essential findings of these long-term research efforts:

The production system for short-rotation intensive culture willow includes willow clones planted in a double-row configuration at approximately 15,300 trees per hectare on a 3-4 year coppice harvest cycle which consistently yields 10-15 dry tons per hectare per year. Yield improvement to 20 dry tons per hectare per year is projected for the post-2005 period.

The successful adoption and modification of established European technology, including commercially available equipment, has permitted rapid adoption of agriculturally based mechanical planting and harvesting with significant cost reductions in producing the biomass.

Open land with suitable soils are available within acceptable distances of the generating facilities to provide near-term co-firing capability of up to 400MW.

The most promising near-term biopower business scenario involves independent growers, a willow biomass planting–harvesting–processing cooperative, and a co-firing utility market. Post-2005 business expansion includes markets for new generation capacity based on biomass-fired integrated gasification power systems, as well as production of liquid, gaseous fuels and high-value chemical products from cellulosic conversion.

Grower revenues for the second generation enterprise include land rents and a 6% internal rate of return. After a five-year demonstration phase is completed and operations are scaled up to the 1200 hectare level of production, the Cooperative is projected to earn a 10% internal rate of return on investments in farm and processing machinery. Purchasing power producers are expected to be able to buy blended biomass fuels at prices competitive with coal while earning pollution abatement credits (SO_x, NO_x and CO_2 reductions).

Bottom-line, full-cost accounting price of willow biomass delivered to a co-firing facility without subsidies is $2.00 per MMBtu by 2001.

Willow biomass crop establishment and scale-up could reach 16,000 to 25,000 hectares in central and western New York by 2010.

Natural forest harvests and wood processing industry residues in the study areas could provide large quantities of wood for an integrated biomass resource.

Predicted emissions for a co-fired PC unit show a reduction of SO_x by 9% at 10% co-firing. Displacement of coal with biomass, a CO_2 neutral fuel, equates to a 9.8% reduction in greenhouse gas emissions. Depending on firing conditions, a reduction in NO_x is also predicted.

Environmental issues of biomass energy crops are similar, but of lesser magnitude, than those of traditional annual cropping agriculture. Willow biomass crops are expected to diversify farm landscapes and reduce fertilizer and other chemical inputs in farm production, as well as to use farm waste streams as significant sources of organic nitrogen and phosphorus for the crop.

The major utilities serve the rural areas of New York. Rural economic development benefits include the creation of 287 jobs and over $9.1 million annually in income by 2006.

Both utilities and growers can forge a long-term business relationship that offers fuel diversity, fuel cost competitiveness and environmental benefits for the utility partners while reinvigorating central and western New York business in the agriculture sector. Growers can bring idle land and land being farmed at a loss back into profitable production while reducing environmental impacts associated with more traditional row crops. The Consortium is gearing up to put in place the growers contracts and the acreage necessary to take the first steps to prove and develop a major new business opportunity for rural New York.

The Salix Consortium is a diverse group developing willow as a locally produced source of renewable fuel. The electric generating partners within the Consortium, led by NMPC, remain the primary end-use market. As consumers, the power generators provide the economies of scale necessary to make a willow business feasible. Therefore, recent steps taken by New York to deregulate and restructure its electric utilities are very important to the Consortium. Asset sales and a shift in corporate focus away from R&D present the Consortium with new challenges, but they also provide new opportunities.

Power companies operating in a deregulating market will focus on generating electricity at the lowest possible incremental cost, while complying with ever tightening environmental limits. Some power companies are already shifting R&D funding towards value-added services, expanding markets and establishing new business areas. Some of these new businesses include non-traditional investments in areas such as telecommunications.

The challenge for the Salix Consortium will be to demonstrate that willow energy crops can compete as a fuel in an environment where emphasis is placed on obtaining the lowest production cost. The key to accomplishing this will be translating all of the benefits of willow energy crops to a power generator's bottom line. For example, demonstrating a delivered fuel cost of $2.00/MMBtu for willow would be a major step forward for energy crop development. However, on average, that price is still $0.57/MMBtu more expensive than coal under long-term contracts in New York State. To compete, the Consortium must convince end-users that tax incentives, emission credits and other environmental benefits effectively reduce willow's cost as a fuel.

While this will remain a challenge throughout the demonstration, the power companies involved in the Consortium provide an excellent test bed for new ideas. NMPC, NYSEG and GPU have already felt the effects of deregulation and restructuring. Over the next five years, the Consortium will have a unique opportunity to interact with these organizations which are living under new rules and changing markets. The collaboration will ensure that the Consortium maintains a realistic perspective and the opportunity to address end-user concerns as they arise.

Technological progress and research on cellulosic conversion of willow biomass to high-value chemicals will be helpful in addressing barriers to successful commercialization utilizing these wood-based renewable resources. However, science alone will not overcome all of the barriers without strong federal and state visions providing supportive policies and regulations that enhance rather than hinder renewable biomass as a viable market competitor to a barrel of oil or a ton of coal.

LESSONS LEARNED FROM EXISTING BIOMASS POWER PLANTS

G.A. Wiltsee

Appel Consultants, Inc., 23905 Plaza Gavilan, Valencia, California 91355

Information was collected from 20 biomass power plants about each plant's history and outlook, design, fuels, operating experience, environmental performance, economics, and lessons learned. The lessons learned by these successful projects relate to such diverse subjects as fuel sources and acquisition methods; fuel handling and preparation; design for fuel flexibility; location; reliability and dependability; joint venture business arrangements; cofiring with fossil fuels; and benefits created.

1. INTRODUCTION

This paper summarizes information on 20 biomass power plants—18 in the United States, one in Canada, and one in Finland—from a report sponsored by the National Renewable Energy Laboratory, the Western Regional Biomass Energy Program, and the Electric Power Research Institute.[1] Table 1 lists the 20 plants in the order in which they began operating or started using biomass fuel on a commercial basis. The online dates of the plants span about 18 years, from December 1979 to January 1998. The nominal sizes of the plants range from 10 MW to 79.5 MW.

1.1. Electricity generation and fuel consumption

Table 2 lists the plants in order of electricity generation, in gigawatt-hours/year (GWh/yr). For some of the plants, the generation numbers are actual statistics from a recent year; for plants that did not provide these statistics, the generation rates were estimated. The same is true for the annual capacity factors (CF, %) and net plant heat rates (Btu/kWh).

1.2. Capacity factors

Annual capacity factors range from 19% to 106%. Some of the plants with low capacity factors (e.g., Multitrade and McNeil) are peaking units. The plants with very high capacity factors have special circumstances. Shasta and Colmac were still under the first ten years of California Standard Offer contracts when the data were obtained. Williams Lake is able to operate at up to 15% over its rated capacity, and frequently is able to sell extra power.

Table 1
Biomass power plants listed chronologically

Plant	Online	Fuels	Boiler(s)	MWe
Bay Front	Dec-79	Mill, TDF, coal	2 modified coal stokers	30
Kettle Falls	Dec-83	Mill	1 traveling grate stoker	46
McNeil	Jun-84	Forest, mill, urban	1 traveling grate stoker	50
Shasta	Dec-87	Mill, forest, ag,	3 traveling grate stoker	49.9
El Nido (closed)	Oct-88	Ag, forest, mill,	1 bubbling FBC	10
Madera (closed)	Jul-89	Ag, forest, mill,	1 bubbling FBC	25
Stratton	Nov-89	Mill, forest	1 traveling grate stoker	45
Chowchilla II (closed)	Feb-90	Ag, forest, mill	1 bubbling FBC	10
Tracy	Dec-90	Ag, urban	1 water-cooled vib grate	18.5
Tacoma (cofiring)	Aug-91	Wood, RDF, coal	2 bubbling FBCs	12
Colmac	Feb-92	Urban, ag, coke	2 CFB boilers	49
Grayling	Aug-92	Mill, forest	1 traveling grate stoker	36.17
Williams Lake	Apr-93	Mill	1 water-cooled vib grate	60
Multitrade	Jun-94	Mill	3 fixed grate stokers	79.5
Ridge	Aug-94	Urban, tires, LFG	1 traveling grate stoker	40
Greenidge (cofiring)	Oct-94	Manufacturing	1 tangentially-fired PC	10.8*
Camas (cogen)	Dec-95	Mill	1 water-cooled vib grate	38-48
Snohomish (cogen)	Aug-96	Mill, urban	1 sloping grate	43
Okeelanta (cogen)	Jan-97	Bagasse, urban,	3 water-cooled vib grate	74
Lahti (cofiring, cogen)	Jan-98	Urban, RDF	1 CFB gasifier + PC	25**

*108 total net MW, 10% from wood and 90% from coal.
**167 total net MW, 15% from biofuels and 85% from coal.

1.3. Heat rates

The Williams Lake plant has the largest single boiler (60 MW) and the lowest heat rate (11,700 Btu/kWh) of any 100% biomass-fired power plant. Biomass-cofired coal plants such as Greenidge can achieve slightly lower heat rates (11,000 Btu/kWh on the cofired biomass portion, compared to 9818 on coal alone). The least efficient plants have heat rates of about 20,000 Btu/kWh. Typical values are about 14,000 Btu/kWh (24.4% efficiency, HHV).

1.4. Cogeneration

The four cogeneration plants use the latest technology, in traditional niches for biomass power: two at pulp and paper mills (Snohomish and Camas), one at a sugar mill (Okeelanta), and one at a municipal district heating plant (Lahti). Table 2 shows statistics just for the solid biomass input. At the pulp and paper mills, waste liquor recovery boilers produce large fractions of the total steam. At Lahti, coal and natural gas produce a majority of the energy; wood wastes and RDF are gasified to provide low-Btu gas to the boiler. Okeelanta burns bagasse about six months of the year; it burns urban and other wood wastes at other times.

Table 2
Plant electricity generation and biomass fuel consumption estimates

Plant	Location	MWe	GWh/yr	CF, %	Btu/kWh	Tons/yr*
Williams Lake	British Columbia	60.0	558	106	11,700	768,000
Okeelanta (cogen)	Florida	74.0	454	70	13,000	694,000
Shasta	California	49.9	418	96	17,200	846,000
Colmac	California	49.0	393	90	12,400	573,000
Stratton	Maine	45.0	353	90	13,500	561,000
Kettle Falls	Washington	46.0	327	82	14,100	542,000
Snohomish (cogen)	Washington	39.0	205	60	17,000	410,000
Ridge	Florida	40.0	200	57	16,000	376,000
Grayling	Michigan	36.0	200	63	13,600	320,000
Bay Front	Wisconsin	30.0	164	62	13,000	251,000
McNeil	Vermont	50.0	155	35	14,000	255,000
Lahti (cogen)	Finland	25.0	153	70	14,000	252,000
Multitrade	Virginia	79.5	133	19	14,000	219,000
Madera	California	25.0	131	60	20,000	308,000
Tracy	California	18.5	130	80	14,000	214,000
Camas (cogen)	Washington	17.0	97	65	17,000	194,000
Tacoma	Washington	40.0	94	27	20,000	221,000
Greenidge	New York	10.8	76	80	11,000	98,000
Chowchilla II	California	10.0	53	60	20,000	125,000
El Nido	California	10.0	53	60	20,000	125,000

*Tons/year are calculated, assuming 4250 Btu/lb.

1.5. Fuel costs

The costs of mill wastes and urban wood wastes can range from about $0 to about $1.40/million Btu (MBtu), depending on the distance from the fuel source to the power plant. Getting to zero fuel cost depends on locating a power plant in an urban area next to a wood waste processor, or next to a large sawmill or group of sawmills. Plants that have come close to zero fuel cost are Williams Lake, which is located very close to five large sawmills, and Ridge, which accepts raw urban wood wastes and whole tires, and also burns landfill gas.

Agricultural residues (primarily orchard tree removals) can be processed into fuel and delivered to nearby biomass power plants for about $1.00/MBtu. Only if open burning of residues is prohibited will it be possible to transfer some of this cost to the orchard owners.

Forest residues are more costly ($2.40-3.50/MBtu), owing to the high costs of gathering the material in remote and difficult terrain. There are strong arguments for government programs to bear the costs of forest management and (in the West) fire prevention. Only if such programs are created will forest residues be cost-competitive in the future.

2. LESSONS LEARNED

Experience gained from these projects can lead future project developers in the direction of an improved biomass power industry. While each plant is unique, many of the lessons are similar and tend to be related to a few key subjects, as summarized below.

2.1. Fuel sources and acquisition methods

The highest priority at almost every biomass power plant is to obtain the lowest-cost fuels possible. This involves tradeoffs in fuel quality, affects the design and operation of the system, and frequently is limited by permit requirements. At Bay Front, the conversion from coal and oil to biomass and other waste fuels kept an old generating station operating. At McNeil, long-term fuel contracts required by lenders created some costly problems. The plant now runs more economically by buying wood fuel under short-term contracts.

Maintaining fuel supply in the midst of a declining regional timber industry has been the single biggest challenge for the Shasta plant. From an initial list of permitted fuels that included only mill waste and logging residues, Shasta added agricultural residues, fiber farm residues, land and road clearing wood wastes, tree trimmings and yard wastes, and natural gas.

Chowchilla II, El Nido, and Madera used low-cost agricultural wastes such as grape pomace, green waste, onion and garlic skins and bedding materials. The most difficult to burn agricultural residues were mixed in small percentages with better fuels, primarily wood.

Ridge, Tacoma, and Tracy show that urban wood waste can be an inexpensive fuel if the plant is located close to the urban area. Tacoma found that focusing on fuel cost (¢/kWh) rather than fuels that provide highest efficiency (Btu/kWh) saved the plant $600,000/year. Tacoma and Ridge are specifically designed as urban-waste recycling facilities.

2.2. Fuel handling and preparation

The one area of a biomass power plant that nearly always causes problems is the fuel yard and fuel feed system. Most plants spend a great deal of time and money solving problems such as fuel pile odors and heating, excessive equipment wear, fuel hangups and bottlenecks, tramp metal separation, and wide fluctuations in fuel moisture to the boiler—or making changes in the fuel yard to respond to market opportunities. Stratton spent $1.8 million to improve the operation of the fuel yard. Tacoma personnel stress the need to take extra care at the beginning of a project with design of the fuel feed system. The one area of the Williams Lake plant that was modified after startup was the fuel handling system.

After adding a debarker, high speed V-drum chipper, chip screen and overhead bins, the Shasta plant offered to custom chip logs, keeping the 35% of the log not suitable for chips. Shasta marketed the program to large forest landowners in California.

The Greenidge Station has found that the technology for preparing biomass fuel for cofiring in a pulverized coal boiler needs further economic evaluation, research and development. The fuel product needs to processed by equipment that produces a chip.

2.3. Design for fuel flexibility

Many biomass plants change fuels over the years, as new opportunities arise or old fuel sources dry up. These changes are often not predictable. The best strategy is to have a plant design and permits that allow as much fuel flexibility as possible.

Bay Front was a coal-fired stoker plant that converted to wood firing and cofiring in 1979. The plant now operates in either 100% coal or 100% wood firing mode.

In 1989, the ability to burn natural gas was added to McNeil Station. Summer pricing for Canadian gas was more attractive than wood prices at that time. Gas prices rose in the mid-1990s, and McNeil burned almost no natural gas in 1997 and 1998.

At the Shasta plant, a large hammer mill was added to the fuel processing system to allow the use of a broader range of fuels. This reduced fuel costs by allowing the plant to process "opportunity fuels" such as railroad ties, brush, and prunings.

The Tacoma plant was constrained by a limited fuel supply and permit, and it developed more options to use opportunity fuels—such as waste oil, asphalt shingles, petroleum coke.

Colmac found it worthwhile to modify its permit to allow the use of petroleum coke. At times, waste fossil fuels can be more economical than biomass.

2.4. Location

As realtors say, "Location, location, location!" Biomass residues and wastes are local fuels with very low energy densities compared to fossil fuels. Transport costs become very high beyond about 20 miles and usually are prohibitive beyond 100 or 200 miles. The ability to have the waste generators deliver the fuel to the plant site at their own expense requires a location very close to the sources of waste. The Colmac plant shows that urban wood waste can be an expensive fuel (~$1.50/MBtu) if the plant is located far outside the urban area.

The primary lesson learned from McNeil is that siting the plant in a residential neighborhood of a small city caused a number of problems and extra expenses: a permit requirement to use trains for fuel supply, high taxes, high labor rates, local political involvement, and neighborhood complaints about odors and noise.

2.5. Reliability and dependability

Several of the plant managers with the best long-term operating records stressed the necessity for placing a high value on reliability and dependability. This is true both during plant design and equipment selection, and during operation. Other than planned outages, the Kettle Falls plant has an availability factor of about 98% over a continuous 16-year period. The plant superintendent has high praise for the people on the staff. The plant is always exceptionally clean and neat. The Shasta general manager advises: "Always place a high value on reliability and dependability, for these will allow you to be considered a 'player' and thus a participant in the development of special programs with the utility." At Williams Lake, which has an outstanding performance record, the chief engineer stresses that it is essential to stay on top of maintenance programs at all times.

2.6. Joint venture business arrangements

The most successful projects have developed formal or informal partnerships with their key customers and suppliers. The relationship with the utility company that buys the power is usually the most important. Cogeneration plants by definition must have close relationships with their steam users. Sometimes there are a few large fuel suppliers (such as sawmills) with whom special relationships are crucial.

Grayling is operated as a cycling plant, running at about a 70-80% capacity factor during peak demand periods, and at about a 40-50% capacity factor during off-peak periods. The McNeil, Multitrade, and Ridge plants are other examples of cycling plants.

At Camas, the utility-financed turbine and generator provides the mill with an additional source of cash flow, without changing the steam system within the mill in any significant way. The utility has added about 50 MW of reliable generating capacity to its system for a relatively small investment, and it has strengthened its relationship with a major customer.

2.7. Cofiring with fossil fuels

The primary issue addressed in most retrofit biomass cofiring projects is how to feed the fuel (and in what form to feed it) to the existing fossil power plant. Other issues include effects on boiler operations, plant capacity, emissions, and ash quality. Bay Front was able to use standard wood sizing and feeding equipment because its coal-fired boilers were stokers. The bubbling fluidized bed combustors at Tacoma are capable of firing 0-100% wood, 0-50% coal, and 0-50% RDF (permit limitation). Greenidge demonstrated that a separate fuel feed system can effectively feed wood wastes to a pulverized coal (PC) unit. The Lahti cofiring project uses a circulating fluidized bed gasifier to convert wood wastes and RDF to low-Btu gas that is burned in a PC boiler.

2.8. Benefits

The 20 biomass projects provide many concrete illustrations of environmental and economic benefits. The Kettle Falls, Williams Lake, and Multitrade plants provide air quality benefits in rural settings where sawmills used to pollute the air with teepee burners. The Ridge, Tacoma, and Lahti plants burn urban waste fuels cleanly. The Okeelanta, Tracy, and San Joaquin plants burn agricultural residues cleanly, which formerly were burned with no emission controls. The Shasta, McNeil, and Grayling plants serve the forest management operations in their areas, by cleanly burning unmerchantable wood, brush, and limbs.

REFERENCES

1. Wiltsee, G.A. (1999). Lessons learned from existing biomass power plants, published by National Renewable Energy Laboratory, Golden, Colorado; Western Regional Biomass Energy Program, Lincoln, Nebraska; and Electric Power Research Institute, Palo Alto, California.

BIOCOST-CANADA: A NEW TOOL TO EVALUATE THE ECONOMIC, ENERGY, AND CARBON BUDGETS OF PERENNIAL ENERGY CROPS

P. Girouard,[a] M.E. Walsh,[b] and D.A. Becker[c]

[a]Resource Efficient Agricultural Production (REAP) - Canada, Box 125, Ste. Anne de Bellevue, Quebec, Canada H9X 3V9
[b]Oak Ridge National Laboratory, Bioenergy Feedstock Development Program, P.O. Box 2008, Oak Ridge, Tennessee 37831-6205
[c]Science Applications International Corporation, 800 Oak Ridge Turnpike, Oak Ridge, Tennessee 37831

The ability of biofuels to generate positive energy balances and to reduce greenhouse gas emissions will strongly influence the role they will play in future energy portfolios. In a collaborative effort, REAP-Canada and Oak Ridge National Laboratory (ORNL) developed a software program to evaluate the economics, energy and carbon (C) balances associated with producing two perennial energy crops under eastern Canadian soil and climatic conditions. Energy and C budgets generated by the software include fossil fuel energy and C used during cropping activities and the manufacture of farm inputs. A key feature of the software is that the user is able to include a provision for soil organic carbon (SOC) sequestration resulting from crop production. Thus, the impact of modifying rates of organic C sequestration on the net C balance of each bioenergy supply system can be evaluated. Methods and defaults used to develop the software and results from recent simulations are presented.

1. INTRODUCTION

When growing perennial crops to produce energy and mitigate greenhouse gas emissions (GHG), the renewable nature and economic viability of the fuel produced are important concerns. In a joint effort between REAP-Canada and ORNL, a model evaluating economic costs and estimating energy and C balances associated with the production of switchgrass (*Panicum virgatum* L.) and short rotation forestry (SRF) (willows *Salix* sp.), under eastern Canadian agricultural conditions, was developed in 1998. Based on ORNL's BIOCOST software, first released in 1996, BIOCOST-Canada version 1.0 is a user-friendly, EXCEL-based software featuring pop-up windows allowing users to change selected parameters.

2. METHODOLOGY

2.1. Economic model

The economic models used in BIOCOST-Canada are based on the modeling effort performed by both REAP-Canada and ORNL. The economic models being developed by REAP-Canada since 1992 were updated with the latest information available and integrated into the existing models of BIOCOST. Approaches, methodologies and assumptions used by both sets of models were analysed and modified (when necessary) to develop a model that best reflects the conditions of eastern Canada.

BIOCOST-Canada's default values are set to estimate the full economic costs of growing switchgrass and SRF willows in eastern Canada. Estimates for western Canada can be generated by changing default values. Costs computed include variable cash expenses (seeds, chemicals, fertilizers, fuel and repairs), fixed cash costs (overhead, taxes, interest payments) and the cost of owned resources (producer's own labour, equipment depreciation, land rental and the opportunity cost of capital investments). The approach used is consistent with the methods used by the USDA to estimate the cost of producing field crops (Walsh, 1996). Full economic costs estimates are higher than when only variable cash expenses are used, the approach farmers traditionally use for making year-to-year planting decisions. BIOCOST-Canada provides the flexibility of computing costs on either basis, including other subsets. Full economic cost estimates are most useful for policy analysis and in evaluating the long term survival and expansion potential of a farm operation.

BIOCOST-Canada can be tailored to individual needs by changing input prices and quantities, fixed and owned resource costs, yields, and management operations. Production cost outputs are presented annually while per oven dry Megagram (odMg) costs are estimated as the net present value over the entire rotation. Costs are expressed in Canadian dollars ($). For more details on the assumptions used by the economic model, see Walsh (1996) and Girouard et al. (1999).

2.1.1. SRF willow default economic values

SRF willows are grown for a 20-year period with five harvests in a four-year harvest cycle. The crop is planted as 30-cm long cuttings and the default price per cutting and for planting ranges from 0.04-0.20$ and 0.01-0.20$, respectively. Establishment begins by spraying the site the previous fall with a broad-spectrum herbicide (Roundup®), followed by plowing. The site is harrowed twice and planted in the spring. A combination of pre-emergent (Simazine®) and post-emergent (Assure®) herbicides is sprayed for weed control during the growing season in addition to rotary hoeing and inter-row cultivation. No weed control is required in the second season. Chemical control thereafter is assumed to be necessary only in the year following harvest. Fertilizer doses are as outlined in Samson et al. (1995).

Willow harvest is by custom operation, using a Claas-Jaguar harvester that chips the trees as it harvests and blows them into a trailing wagon. Equipment productivity (h/ha) is adjusted for yield and is based on data collected in harvesting trials in the United Kingdom (Deboys, 1996). Additional business costs incurred by a custom operation, such as transporting equipment to the harvest site, storing equipment not in use, labor costs for mechanics and supervisory personnel, general overhead costs not already covered, a profit margin, and income taxes are included in the estimated harvesting costs. The default transportation cost of the chips to a conversion facility is estimated to be $11.47/odMg.

Mean annual biomass growth increments estimated to be realistic at the present time in eastern Canada range between 7-11 odMg/ha/yr. Productivity is lower during establishment years, therefore annual growth in the first and second years is assumed to be 15 and 65%, respectively, of the mean annual increment obtained over a 4-year cycle of an established crop. In the third and fourth years, annual growth is assumed to be 100%, resulting in the mean annual increment over the 1st cycle to be 70% of what is achieved in subsequent cycles.

Annual production costs are totaled and discounted to provide the present value of these costs. The present value of each production year is summed to obtain the total present value of costs incurred during the life span of the crop. Annual harvestable biomass yields are also discounted to the present value and summed to estimate the total present yield value. Average production cost per odMg is computed by dividing total discounted costs by total discounted yield.

2.1.2. Switchgrass default economic values

The model assumes switchgrass is grown for a 10-year period. In the fall, prior to planting, fields are sprayed with a broad-spectrum herbicide (Roundup®) and plowed. In spring, fields are harrowed twice and planted with a cereal grain drill equipped with a forage seed box. Broadleaf weed control in the establishment year is achieved by spraying Laddock® herbicide. No weed control is assumed to be necessary for the following nine years.

Fertilizer requirements range from 50 to 75 kg of N/ha, and 10 kg P/ha for a spring harvest. No fertilization is required in the establishment year. For a fall harvest, fertilizer requirements are estimated at 70 kg N/ha, 15 kg P/ha, and 30 kg K/ha. Crop harvest is accomplished with a self-propelled haybine and a large square baler. Large square bales are stored on-farm, then delivered to a conversion plant located within a 60-km radius. No provision for storage costs is allowed in the budget. Off-farm transportation costs are estimated to be $11.47/odMg.

Current annual harvestable yields once the crop is fully established vary between 6-10 odMg/ha for a spring harvest and 8-13 odMg/ha for a fall harvest. Since switchgrass fields usually become fully established in the third growing season, the model assumes that productivity during the establishment year is 35% of the full yield potential, 80% in the second growing season and attaining full productivity in the third growing season.

2.2. Energy model

A main concern of growing crops for energy production relates to the amount of fossil fuel energy required in the production process. BIOCOST-Canada's energy model is integrated into the economic and C emission models. Fossil fuel energy used in manufacturing farm inputs, diesel fuel energy used for cropping activities, and energy in oil and lubricants are taken into consideration.

2.2.1. SRF willow default energy values

The energy used to produce willow cuttings is estimated in Girouard et al. (1999). Annual doses of herbicide and fertilizer used in the economic model are automatically integrated into the energy model. Values from the literature determine the energy associated with the use of one elemental unit of N, P, and K fertilizers and one unit of active ingredient of herbicide.

Diesel fuel energy required (liters/ha) for each pre-harvest and harvest operation is estimated using a formula relating engine horsepower rating to fuel consumption. The values are then multiplied by the energy content of 1 litre of diesel fuel (0.03868 gigajoule [GJ]). For off-farm transport, fuel consumption per kilometer (km) traveled is estimated from a trucking cost study (see Girouard et al., 1999). Energy contained in oil and lubricants is assumed to be 15% of the diesel fuel energy required for each operation. Finally, the energy output:input ratio of the biomass produced is estimated using the higher heating value (HHV) of willow biomass (19.5 GJ/odMg).

2.2.2. Switchgrass default energy values

The switchgrass energy model is similar to the SRF model, with two exceptions. First, the amount of energy needed for seed production is 10% of the energy required for growing the crop during the first growing season, and second, switchgrass HHV is estimated to be 17.5 GJ/odMg.

2.3. Carbon model

BIOCOST-Canada's C model is integrated into the economic and energy models. Carbon released during the manufacture of farm inputs, from burning diesel fuel during cropping activities, and from the use of oil and lubricants, is accounted for in the budget.

2.3.1. SRF willows default carbon values

To determine C emissions resulting from the use of one elemental unit of N, P, and K fertilizers, the quantity of each type of fossil fuel required to manufacture each fertilizer is retrieved from the energy model and multiplied by the C released during combustion of 1 GJ of each fuel (Girouard et al., 1999). The same procedure is used to estimate C emitted from herbicide use. Carbon emissions from diesel fuel used for cropping activities are computed by multiplying fuel amount in GJ/ha required for a specific operation by 19.28 kg C/GJ. Emissions related to oil and lubricants are 15% of the estimates for diesel fuel for each farming activity. Carbon emissions during the production of cuttings use the estimate from Girouard et al. (1999).

Total estimated fossil fuel C emissions are then analyzed in terms of their intensity compared with the quantity of biomass energy produced. The analysis involves three levels, in each of which total kg of fossil fuel C emitted is divided by total number of GJ of harvestable biomass energy produced, a ratio known as the C requirement. The first level analysis (C1) estimates the C requirement of the crop with no provision for soil or biomass C changes resulting from introducing the crop to a piece of land. The second level analysis (C2) introduces the concept of SOC accumulation in the computation of the C requirement. The third level analysis (C3) includes changes in biomass carbon stocks in addition to changes in SOC. The complete details of the carbon requirement analysis are presented in Girouard et al. (1999).

2.3.2. Switchgrass default carbon values

Carbon emission calculations follow the same methodology discussed for SRF willows, except for emissions related to seed production which are assumed to be equivalent to 10% of the carbon emitted in the establishment year.

A summary of the information taken into consideration or generated by BIOCOST-Canada is presented in Table 1.

Table 1
Summary output information generated by BIOCOST-Canada

1a: Economic ($/ha)	1b: Energy (GJ/ha)	1c: Carbon (kg C/ha)
Variable cash costs	Diesel - on farm & transportation	Diesel - on farm & transportation
Fixed cash costs	Fertilizers	Fertilizers
Owned resources	Herbicides	Herbicides
	Seeds/Cuttings	Seeds/Cuttings
Total Economic Cost	Total Energy Required	Total Carbon Input Required

2. Energy Ratio: (Total harvestable biomass energy [GJ/ha])/(Total energy used [GJ/ha])

3. Carbon Requirement Estimates:
    C1: $\underline{\text{Fossil fuel carbon required (kg C/ha)}}$
        Total harvestable biomass energy (GJ/ha)
    C2: $\underline{\text{Fossil fuel carbon required - SOC accumulation}}$
        Total harvestable biomass energy
    C3: $\underline{\text{Fossil fuel carbon required - SOC accumulation - Change in biomass carbon stocks}}$
        Total harvestable biomass energy

3. SIMULATION RESULTS

The simulations performed on SRF willows and switchgrass (Table 2) indicate that SRF willows are more expensive to grow than switchgrass but would have a slightly higher energy balance and lower carbon requirement (C1). All three scenarios investigated have largely positive energy balances along with carbon requirements (C1) in the order 1 kg C GJ^{-1} of biomass energy produced, delivered to a conversion facility. Even taking into consideration carbon emissions resulting from conversion and distribution activities, the resulting biofuels will still most probably emit much less carbon than do fossil fuels, as gasoline for instance releases 18.54 kg C GJ^{-1}.

In the event that soil organic carbon can be increased while growing these crops, these crops could be become a carbon sink as shown by the negative carbon requirement values

(C2) in Table 2. Assuming that both switchgrass and SRF willows are established on a land previously cropped with corn, a preliminary analysis suggests that the carbon requirement (C3) of the bioenergy crops could be further improved. A complete analysis of these simulations can be found in Girouard et al. (1999).

Table 2
Results summary of simulations performed using BIOCOST-Canada

	SRF Willows		Switchgrass			
			Spring Harvest		Fall Harvest	
Yield (odMg/ha/yr)	7	11	6	10	8	13
Total Variable Cost ($/odMg)	74	55	35	30	34	31
Total Economic Cost ($/odMg)	108	76	81	61	71	57
Energy Ratio	18	24	16	19	17	18
Carbon Requirement Estimates						
C1 (kg C/GJ)	0.99	0.74	1.05	0.89	1.04	0.97
C2^a (kg C/GJ)	-2.90	-1.74	-4.15	-2.22	-2.85	-1.43
C3$^{a,\,b}$ (kg C/GJ)	-7.28	-4.53	-6.54	-3.65	-4.64	-2.53

^{a}assumes a SOC accumulation rate of 500 kg C ha^{-1} yr^{-1}.
^{b}see Girouard et al. (1999) for more details.

REFERENCES

Deboys, R. (1996). Harvesting and Comminution of Short Rotation Coppice. Department of Trade and Industry, Energy Technology Support Unit. Harwell, England. United Kingdom, ETSU/B/W2/00262/REP.

Girouard, P., R. Samson, C. Zan, and B. Mehdi (1999). Economics and Carbon Offset Potential of Biomass Fuels. Final Report. Contract awarded through Environment Canada's Environmental Program and funded by the Federal Panel on Energy Research and Development (PERD). Contract 23341-6-2010/001/SQ.

Samson, R., P. Girouard, J. Omielan, Y. Chen, and J. Quinn (1995). Technology Evaluation and Development of Short Rotation Forestry for Energy Production. Final Report. Report completed under the Federal Panel on Energy Research and Development (PERD) for Natural Resources Canada and Agriculture Canada. Contract 23440-2-94933/01SQ.

Walsh, M. and D. Becker (1996). Biocost: a Software Program to Estimate the Cost of Producing Bioenergy Crops. Proceedings, Bioenergy '96—The Seventh National Bioenergy Conference, pp. 480–486.

OPTIMAL LOCATIONS FOR BIOMASS CONVERSION FACILITIES IN FLORIDA: A GEOGRAPHIC INFORMATION SYSTEM APPLICATION

M. Rahmani, A.W. Hodges, and C.F. Kiker

Food and Resource Economics Department, IFAS, University of Florida, PO Box 110240, Gainesville, Florida 32611-0240, U.S.A. Ph. (352)392-1881 Ext. 315, Fax (352)392-9898

Biomass used for energy production must be transported to processing facilities. Geographic information systems (GIS) are a powerful tool for site-specific analysis of natural resources. In this study GIS was used to create maps showing the spatial relationships of potential biomass production and conversion sites. A geographically referenced database of biomass resources in Florida was developed, the potential statewide aggregate supply of each biomass resource type was estimated, and sites for potential biomass crops were identified using a network transport cost minimization algorithm. Data used for this analysis included (1) an area type (polygon) geographic coverage of the Florida Land Use/Land Cover Classification System (FLULCCS) database for peninsular Florida; (2) a polygon geographic coverage of the State Soil Geographic Data Base (Statsgo) for peninsular Florida; (3) a geographic line coverage of primary United States highways; (4) a point geographic coverage of populated places in Florida. Results show that peninsular Florida can potentially produce over 13 million metric tons (Mg) of biomass feedstocks per year from forest, cropland, and reclaimed land-use types. The possibility of biomass production for energy in Florida exists for the entire study area, but there are areas with very high concentrations of biomass resources. Considering land availability, soil productivity, and transportation costs, the best locations for new biomass conversion facilities in Florida were identified and mapped.

1. INTRODUCTION

Biomass resources are typically widely dispersed across the landscape, either as dedicated biomass crops, forest or agricultural residues, or urban yard wastes. The low energy density of biomass materials means that transportation costs are high, given the large volumes of biomass required to supply commercial energy facilities. For example, a 69 MW thermo-electric power plant fired by biomass feedstocks having a typical energy value of 6,000 BTU per pound would require a supply of 685,000 dry tons/yr. Such a power plant would require nearly 34,000 acres of crops having an annual dry matter yield of 44 Mg/ha (20 tons/acre). Assuming that this level of production was available from 10 percent of the land base in a given region, a circular area of 20.8 km (13 miles) radius would be required to supply these energy needs.

Utilization of biomass for energy production, additionally, poses important logistical problems for transportation to processing facilities. In a feasibility study of a biomass energy system for the phosphate mine region of central Florida, transportation costs were estimated at $3.3 to $13.2 per dry Mg ($3 to $12 per dry ton), depending upon the type of biomass material and transport method (Rahmani et al., 1997). There was, in this case, a large land base potentially available for dedicated biomass crop production, and the average haul distance was only 6 miles. In other areas where the biomass resource is less concentrated, transportation costs would be higher–and possibly prohibitive.

The specific purpose of this study was to identify the optimal locations for new biomass conversion facilities in Florida. Following Liu et al. (1992) and Noon et al. (1996), a geographic information system was used to geographically display maps showing the location of biomass growing sites and to spatially analyze sites in relation to potential processing plant sites. The biomass resources evaluated included pine forests, mixed hardwoods and invasive, non-native trees.

2. DATA SOURCES

The study area for this project was peninsular Florida (specifically, Jefferson County and east). Data from the following sources were obtained and used for this study.

2.1. Land-use

An area type (polygon) geographic coverage of the Florida Land-Use/Land-Cover Classification System (FLULCCS, State of Florida, 1985) database for peninsular Florida contains all data available for Florida. Data were compiled by Florida Water Management Districts and by the University of Florida Geoplan Center, Department of Urban and Regional Planning; they were provided in universal transverse mercator (UTM) coordinates. Land-use types evaluated included natural forests, plantation forests, cropland, pasture, reclaimed land, and industrial sites.

2.2. Soils

A polygon geographic coverage of the State Soil Geographic Data Base (Statsgo) for peninsular Florida was used to estimate potential woody biomass production for different soil types. The attribute table included fields for potential woody biomass productivity (m^3/ha/yr), forest site indices, land-use suitability codes, and codes for restrictions on equipment or seasonal accessibility. This data was obtained from Natural Resource Conservation Service (SCS, 1993), as compiled by Florida Water Management Districts and the Geoplan Center. The 117 distinct soil types have an average productivity of 8.36 m^3/ha/yr (range from 3–11 m^3/ha/yr), occupy 70,265 parcels, and cover a total area of 2,456,762 ha–more than 5.5 million acres).

2.3. Populated places

A point geographic coverage of 650 populated places in Florida was used to evaluate potential sites for biomass processing plants. The attribute table included population and socioeconomic characteristics from the 1990 U.S. Census (Caliper Corporation, Boston, Massachusetts).

2.4. Highways

A geographic line coverage of primary U.S. highways digitized from official U. S. highway maps was used to evaluate transport of biomass. This database included 3,533 separate highway segments that were joined into a transportation network. The attribute table included speed, transit time and number of lanes in each direction for each highway segment (Caliper Corp.).

2.5. State and county boundaries

A geographic polygon coverage of the Florida counties was used to summarize information at the county level. State boundaries were digitized at high resolution (Caliper Corp.).

3. ANALYSES

Two geographic information system (GIS) software packages, Arc-Info (ESRI, Redlands, CA) and TransCAD (Caliper Corporation) were used to analyze land-use and transportation data and to produce maps. Database software such as Statistica (StatSoft, Tulsa, OK) and Lotus 123 were also used to analyze and tabulate the results.

Geographic data files for land-use and soils were entered and set up for preliminary processing in Arc-Info. Land uses of interest included agricultural croplands, pasture, forests, reclaimed lands, power generation, and other industrial uses. The minimum economic size of land parcels used for biomass production was assumed to be 5 ha (12.5 acres); smaller parcels were eliminated. Approximately 70,000 parcels were identified as potential biomass production sites. The soil data layer was overlaid on the land-use data, and the soil type having the largest area in each land-use parcel was assigned as an attribute to that parcel.

To minimize subsequent computational processing and storage requirements, the area type (polygon) coverage was converted to point coverage, and the location of each parcel was represented by its geographic centroid. Biomass productivity values were assigned to each land-use parcel based on the values for potential woody biomass productivity listed for each soil type in the Statsgo database. The biomass productivity values ranged from 3.0 to 10.65 m^3/ha/yr. The potential production of biomass from each parcel was calculated as $B = P \times A \times C$, where B is the annual biomass production (Mg/yr), P is biomass productivity (m^3/ha/yr), A is the parcel area (ha), and C is a conversion constant (0.68 Mg/m^3) representing the typical air-dry (20% moisture) density of woods in the southern United States (Koch, 1985). Land-use parcels of at least 10 ha were further analyzed, reducing the total number of parcels to 35,291.

The potential supply of biomass to each existing biomass-using facility and each populated place was determined within a 48 km (30 miles) radius using the "aggregate" procedure in TransCAD. The transportation cost from each potential biomass production site to each biomass-using facility or populated place was calculated using the TransCAD cost matrix procedure (Bertsekas and Tseng, 1988; Ahvja et al, 1993). The transport cost was based on travel time and was optimized for the shortest path through the highway network. The distance from each site to the nearest highway network link was calculated as the shortest possible straight line.

The best location for new biomass-using facilities was determined with the TransCAD facility location procedure to minimize the average cost of transportation (Dresner, 1995).

4. RESULTS

4.1. Potential biomass production by land-use type.

More than 70,000 parcels of forest, agricultural, reclaimed, or industrial land with a total area of 2.5 million ha (6.17 million acres) exist in the study area. Land-use and cover classification defined these parcels being used as crop and pasture land, natural or developed forest, reclaimed land, power plants, waste treatment facilities, food processing, other processing facilities, and transportation terminals (Table 1). The crop and pasture land-use type with an area of 877,000 ha has the highest potential biomass production, more than five million Mg/yr. Potential biomass production for all types of land-use in peninsular Florida exceeds 13 million Mg/yr.

Table 1
Potential biomass producing sites in Florida by land use type

Land-use types	Number of parcels larger than 5 ha	Area (1000 ha)	Potential biomass production (1000 Mg/yr)
Crop & pasture	15,841	877	5,021
Pine forest	17,142	522	2,931
Hardwood plantations	8,332	462	2,796
Mixed hardwoods, dead trees	15,499	278	1,628
Oak, exotic forest	7,584	195	1,024
Reclaimed land	667	29	167
Industrial, other	5,200	94	NA
All types	70,265	2,457	13,567

4.2. Potential biomass production by county

By overlaying the county data layer on the land-use data, the potential biomass production by county was summarized. Marion, Volusia, Osceola, Polk, Hillsborough, and Putnam counties in central Florida, Alachua County in north-central, and Nassau County in north-east Florida had the highest potential land use area and potential biomass production, all more than 400,000 Mg/yr. Marion County had by far the greatest total at 1.1 million Mg/yr.

4.3. Optimal locations for potential biomass conversion facilities

Considering availability of land, soil productivity, and transportation cost, the optimal locations for hypothetical new biomass conversion facilities requiring one million Mg biomass feedstock per year were determined. The best nine locations in peninsular Florida are Greenville, Bellair-Meadowbrook Terrace, Branford, Ocala, Sanford, Dade City, Vero Beach, Arcadia, and Immoklee (Figure 1).

5. CONCLUSIONS

Considering land availability and the cost of transportation, central- and north-central Florida are the best areas in peninsular Florida for development of a biomass-to-energy system. In certain locations the cost of transportation can be minimized within a 30 mile radius for new biomass conversion facilities. The 10 locations with the highest potential biomass production per year within 48 km (30 miles) radius are mostly in three north-central Florida counties: Alachua, Marion and Putnam.

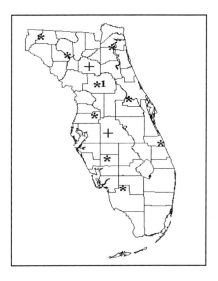

Figure 1. Optimal locations for one, two, or nine new biomass conversion facilities in peninsular Florida.

1 One new facility (Ocala)
+ Two new facilities (Bartow, Gainesville)
* Nine new facilities (Greenville, Branford, Bellair-Meadowbrook Terrace, Ocala, Sanford, Dade City, Verona Beach, Arcadia, Immoklee)

Several caveats should be expressed about these results. The land identified as having biomass or the potential of growing biomass generally is being used for other economic activities at present. Some of the forest lands are national or state forests with multiple uses. Much forest land is owned by timber and paper companies, and the biomass production is captively used in these industries. Other lands are in some form of commercial agriculture. For these lands to be dedicated to producing biomass for energy, this new use would need to outbid the existing uses.

Environmental effects are other concerns. Forested lands provide a range of environmental services, and these would have to be considered. In other cases harvesting biomass for energy could provide an additional environmental service: if invasive, non-native plants could be harvested for an economic use, this income could help defray the cost of managing these lands. Additionally, there may be potential for biomass production to help reclaim lands that have been mined and environmentally disturbed, again helping to defray costs. Last, the analysis presumed that present roadways could handle the increase in use caused by biomass transport. The highways and roads in the areas identified are already used. The increased cost caused by increased usage would have to be determined.

6. ACKNOWLEDGMENTS

This study received financial support from the Center for Biomass Programs, University of Florida. The cooperation of L.G. Arvanitis and J.A. Saarinen of the GIS Laboratory, the University of Florida School of Forest Resources and Conservation, the Florida Water Management Districts and the University of Florida Geoplan Center is greatly appreciated.

REFERENCES

Ahvja, R.K., T.L. Magnanti, and J.B. Orlin (1993). Network Flows: Theory, Algorithms, and Applications. Prentice Hall, Englewood Cliffs, New Jersey.

Bertsekas, D.P. and P. Tseng (1988). The relax codes for linear minimum network flow problems, Annals of Operation Research, 13, pp. 125-190.

Caliper Corporation (1992). TransCAD, Transportation GIS Software, Newton, Massachusetts.

Dresner, Z., ed. (1995). Facility Location: A Survey of Applications and Methods. Springer Series in Operations Research, Springer-Verlag, New York.

Easterly, J.L. and J.L. Reinertsen (1996). Southeast biomass facility information system, Proceeding, the Seventh National Bioenergy Conference, Nashville, Tennessee, September 15-20, pp. 905-910.

Environmental System Research Institute Inc. (1997). Arc/Info, GIS Software, Redlands, California.

Koch, P. (1985). Utilization of Hardwoods Growing on Southern Pine Sites, Southern Forest Experiment Station, Agriculture Handbook No. 605, U.S. Department of Agriculture, Forest Service.

Liu, Wei, V.D. Phillips, and D. Singh (1992). A spatial model for the economic evaluation of biomass production systems, Biomass and Bioenergy, 3, 5, pp. 345–356.

Noon, C.E., M.J. Daly, R.L. Graham, and F.B. Zahn (1996). Transportation and site location analysis for regional integrated biomass assessment (RIBA), Proceedings, the Seventh National Bioenergy Conference, Nashville, Tennessee, September 15–20, pp. 487–493.

Rahmani, M., A.W. Hodges, J.A. Stricker, and C. F. Kiker (1997). Economic analysis of biomass crop production in Florida, Proceedings, the Third Biomass Conference of the

Americas, Making a Business from Biomass in Energy, Environment, Chemicals, Fiber, and Materials, Montreal, Canada, August 23–29, pp. 91–99.

Soil Conservation Service (1993). State Soil Geographic Data Base (Statsgo) Data Users Guide, SCS Misc. Pub. 1492.

State of Florida, Department of Transportation, State Topographic Bureau (1985). Florida Land Use Cover and Forms Classification System, Thematic Mapping Section, Procedure No. 550-010-001-A.

LIGNOCELLULOSIC FEEDSTOCK RESOURCE ASSESSMENT

T.E. Rooney,[a] S.G. Haase,[a] and A.E. Wiselogel[b]

[a]McNeil Technologies, Inc., 350 Indiana Street Suite 800, Golden, Colorado 80401
[b]National Renewable Energy Laboratory, MS 1634, 1617 Cole Boulevard, Golden, Colorado 80401

In 1997, the National Renewable Energy Laboratory (NREL) undertook a national biomass resource assessment to address a lack of basic resource information for scientists, industry, and policy-makers. The objective of this resource assessment was to describe the state-level distribution, quantity and market value of lignocellulosic feedstocks for ethanol production in the United States. Lignocellulosic feedstocks, derived from plant materials, are composed primarily of cellulose, hemicellulose, and lignin. Cellulose and hemicellulose are polymers of simple sugars that can be chemically fermented to produce ethanol. Lignin plays a role in binding cellulose and hemicellulose together in plant cell walls. This study focused on lignocellulosic by-products of agriculture, food processing, forest products industry, and consumers.

The United States generates 306 million metric tons (dry weight) of lignocellulosic materials annually that could be used to manufacture ethanol. The bulk of this potential feedstock supply is made up of agricultural residues left following crop harvesting, such as corn stover. Corn stover makes up 70 percent of the total biomass resource. If agricultural residues are used for ethanol production, the price paid for agricultural residues will have to meet or exceed their current soil nutrient and animal fodder values. Food processing residues such as corn gluten feed and meal, distillers' dried grains, and spent brewer's grains are used as animal feed additives, which makes them high in cost in relation to other feedstocks. Forest products residues are mostly used for fuel, pulp, animal bedding, or mulch. The unutilized portion of these residues may be available for ethanol production. Recycled paper, sugarcane bagasse, rice straw, paper sludge, and urban tree residue show potential for use as ethanol feedstocks due to their low cost.

1. INTRODUCTION

The United States (U.S.) Department of Energy's (DOE) Office of Fuels Development (OFD) and the NREL Biofuels Program (BFP) are developing the technology to convert lignocellulosic resources into ethanol for use as a liquid transportation fuel. The OFD and BFP developed a multi-year technical plan for the commercialization of biomass-to-ethanol technology. While preparing this plan, OFD and BFP found that there was a lack of basic resource information on many potential ethanol feedstocks. NREL commissioned and funded this resource assessment to address this information gap. This paper summarizes the results of this work effort, which was completed in January 1998. The work effort provided information on a state and national basis on the quantity and cost of potential biomass feedstocks for ethanol production in the United States. This study focused on the production of residues from agriculture, food processing, forest products industry, and consumers.

The data in this report are suitable for targeting states as potential candidates for biomass-based ethanol manufacturing facilities. Decisions as to whether a cellulose-to-ethanol facility should be constructed require detailed, local feasibility studies.

2. DATA SOURCES AND ANALYTICAL APPROACH

This section describes the data and analytical approach used to estimate the quantities and prices for each residue. The feedstocks are divided into four categories: agricultural residues, forest products industry residues, food processing residues, and recycled paper.

2.1. Agricultural residues

Agricultural residues consist of stalks, leaves, hulls, and other debris left over after crop harvest. Agricultural crop residues examined include corn stover, hay-alfalfa, small grain straw (wheat, rye, barley, oats, and sorghum) and rice straw. Residue quantity estimates are based on United States Department of Agriculture National Agricultural Statistics Service (USDA NASS) crop production data (USDA NASS, June 1997) and NREL conversion factors for residue yield per unit of crop production (Tyson, 1991). Residue market prices estimates include residue collection costs and opportunity costs of equipment and nutrients. This method of price estimation follows prior efforts by Walsh and Becker (1996). The University of Minnesota Department of Applied Economics and Iowa State University publications provided detailed equipment operating specifications (Lazarus, 1997; Ayres and Williams, 1983) used to estimate residue collection costs.

2.2. Forest products industry residues

Forest products industry residues examined in this study included primary and secondary mill residues, recycled primary paper pulp sludge, and urban tree residues (UTR). The approach used to develop price and quantity information for each resource is described below.

Mill residues consist of woody slabs, edgings, cants, ends, bark, sawdust, and planer shavings associated with milling. Primary mill residues are from firms that use whole logs for a portion of their raw material. Secondary mills use only raw materials that have had some level of primary processing. Primary mill residue quantities and prices are from the DOE Energy Information Administration (EIA) Renewable Fuels module of the National Energy Modeling System (NEMS) (Decision Analysis Corporation of Virginia, July 1995).

Secondary mill residue quantities are estimates based on the number of secondary processing establishments (Jamski, 1997) and the amount of residue generated by secondary mills (Prosek et al., 1994). Prosek et al. (1994) residue estimates account for differing mill conversion efficiencies by breaking down residue generation for mills in different employment size classes. Smaller mills frequently have lower overall quantities of residue. Prices for secondary mill residues are also taken from EIA data (Decision Analysis Corporation of Virginia, July 1995). Mill residue quantities include 50 percent of the total resource. This percentage represents the proportion of the total resource for which price can persuade producers to test alternative outlet markets for residues (Decision Analysis Corporation of Virginia, July 1995).

Pulp sludge is a byproduct of processing wood chips and recovered paper for paper pulp production. The resource assessment focuses on primary pulp sludge from waste paper deinking mills. Approximately 270 dry kilograms of sludge are generated per metric ton of deink pulp (NCASI, September 1992). Deink pulp quantities are taken from a national pulp and paper industry database (Dyers, 1997). Chartwell Information Publishers, Inc. (Thompson, October 1997) landfill tipping fees are used as a proxy for recycled paper pulp sludge prices.

UTR include tree limbs, tops, brush, leaves, stumps, and grass clippings from the urban forestry and landscaping industries. Quantity data are from a national survey by NEOS Corporation in 1993 (NEOS Corporation et al., 1994). Landfill tipping fees (Thompson, October 1997) are used as a proxy for UTR prices, since most UTR is disposed of in landfills or given away (NEOS Corporation et al., 1994).

2.3. Food processing residues

The food processing residues examined in this study are corn gluten feed (CGF) and meal (CGM) from wet corn milling, distiller's dried grains (DDG) from dry corn milling, spent grains from breweries, and sugarcane bagasse. CGF, CGM, DDG, and spent brewer's grain quantities are derived from ethanol production capacity. For example, it is assumed that each bushel of corn yields approximately 9.9 liters of ethanol and 8.6 kilograms of DDG for distilleries (Schaffer, October 1997). Multiplying plant capacities by these approximate yields provides estimates of DDG production. CGM, CGF, and spent brewer's grain production are estimated in a similar manner. The Corn Refiners Association (CRA) published 1996 national wet corn milling industry shipment figures for all corn products including both CGM and CGF (CRA, 1997). The Renewable Fuels Association (RFA) and Information Resources, Inc. (IRI) provided dry corn milling plant production capacities (RFA, October 1997; Schaffer, October 1997). The Institute for Brewing Studies provides brewery production capacities on a state level for the United States (Institute for Brewing Studies, 1996). Brewing capacities are converted to spent grain production using a conversion factor of 0.003 dry metric tons/barrel of beer produced (Tyson, 1991). Prices for CGF, CGM, and spent brewers' grains are based on Department of Agriculture regional livestock feed reports (J. Van Dyke, U.S. Department of Agriculture, Agricultural Marketing Service, personal communication, July 1997). Sugarcane bagasse quantities and prices are estimated in the same manner as for agricultural residues; namely, a constant residue yield per unit of harvested crop was assumed.

2.4. Recycled paper

Newsprint represents approximately 10.4 percent of recovered MSW by weight in the United States, while mixed office paper accounts for 5.4 percent of MSW (Franklin Associates, Ltd., June 1997). State-level data on generated and recovered MSW are taken from *Biocycle* (Goldstein and Glenn, May 1997). Estimates of recovered newsprint and mixed office paper are the product of multiplying state-level recovered MSW quantities by the national MSW newsprint and office paper composition percentages. Market paper prices are based on average reported monthly prices for July 1996 through June 1997 (Advanstar, July 1996 - June 1997).

3. RESULTS

For the resources included in the study, the United States generates more than 305 million metric tons of materials annually that could be used for ethanol production. Table 1 shows that some materials are available at low or negative costs, while the high costs of others precludes their use for ethanol production.

Table 1
Summary of 1996 feedstock quantity and price (ranked by price).

	Quantity		Price
	Thousand dry metric tons	Percent (%) of total	U.S. $/dry metric ton
Recycled primary paper pulp sludge	3,053	1.0	(35.43)
Urban tree residue	34,503	11.3	(30.18)
Mixed office paper	4,415	1.4	2.29
Sugarcane bagasse	645	0.2	6.42
Newsprint	10,211	3.3	16.37
Softwood secondary mill residue	553	0.2	22.03
Hardwood secondary mill residue	401	0.1	22.11
Hardwood primary mill residue	890	0.3	24.55
Softwood primary mill residue	1,141	0.4	25.03
Rice Straw	2,424	0.8	26.18
Corn stover	212,377	69.6	27.79
Hay-alfalfa	25,971	8.5	71.73
Corn gluten feed	5,150	1.7	104.75
Spent brewers' grains	967	0.3	143.92
Distillers' dried grains	1,635	0.5	145.12
Corn gluten meal	1,018	0.3	227.68

	Quantity		Price
	Thousand dry metric tons	Percent (%) of total	U.S. $/dry metric ton
Small grain straw	-	0.0	NA
Total/*Average*	305,353	100.0	*50.02*

Note: NA signifies not applicable: small grain straw removal was not considered because of soil conservation considerations.

Materials that are currently being landfilled represent a significant opportunity for ethanol producers, particularly if the ethanol facility is able to charge a fee for accepting the material.

The largest potential source of biomass for ethanol production is agricultural crop residues. Currently, the primary competing use for field crop residues is animal fodder. Forest products residues from primary and secondary mills are sold as pulp chips, fuel, mulch, compost, animal bedding, and other products. They may be available at a competitive price for use as an ethanol feedstock. A proportion is also used in the manufacture of engineered wood products such as wood pellets and composite board. Paper availability is based on current recovery rates, which are slightly more than 44 percent in the United States (American Forest and Paper Association, 1996). CGF, CGM, spent brewer's grains, and DDG are fully used for value-added products. CGF, CGM, and DDG are used almost exclusively for animal feed supplements. Sugarcane bagasse and rice straw is used for boiler fuel, ethanol, and in building products. Recycled paper pulp sludge and UTR either are used in low-value applications or are disposed of by landfilling.

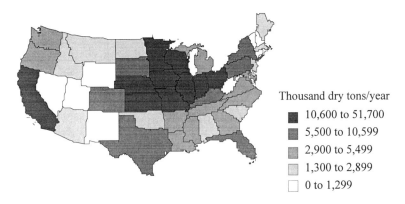

Figure 1. 1996 lignocellulosic feedstock distribution in the United States; Hawaii (not shown) produced 600 thousand dry metric tons of crop residues. Alaska was not included in the study.

Regions of the United States that are highly populated (e.g., California and New York) and areas with high agricultural productivity (e.g., the Great Plains, the Mid-West, and the Lake States) have the highest biomass potential (Figure 1).

4. CONCLUSIONS

Recycled primary paper pulp sludge and UTR are appropriate candidates for further research due to their negative cost. Limited quantities of sugarcane bagasse and rice straw are currently used as feedstocks for ethanol production. Mixed office paper, newsprint, and corn stover (listed in order of increasing cost) costs were sufficiently low in value to warrant attention as viable future candidates for ethanol production technologies. If the value of corn stover as animal fodder to the farmer does not exceed its price as an ethanol feedstock, this feedstock could play a large role when the application of biomass-based ethanol technologies is more widespread. Small grain straw is currently produced in smaller quantities than corn stover. In many cases, its removal could impact soil nutrient levels and affect crop productivity. The high value of primary forest products residues for fuel and fiber made it unlikely that they will play a part in ethanol production efforts without significant changes in either pulp markets or the price of ethanol. Depending on the transportation costs secondary forest products industry residues may be viable ethanol feedstocks. Under current economic conditions, spent brewers' grains, CGF, CGM, DDG, and hay-alfalfa straw, are not likely to be viable ethanol feedstocks because they have high value as ruminant and poultry feeds.

REFERENCES

Advanstar Communications (July 1996 through June 1997). Transacted paper stock prices. Official Board Markets 72 [28] through 73 [24].

American Forest and Paper Association (1996). Recovered Paper Statistical Highlights. Published by the American Forest and Paper Association, Washington, D.C.

Corn Refiners Association, Inc. (CRA) (1997). 1997 Corn Annual. Published by the Corn Refiners Association, Washington, D.C.

Decision Analysis Corporation of Virginia (July 1995). Data Documentation for the Biomass Cost- Supply Schedule. Published by the Department of Energy's Energy Information Administration, Washington, D.C.

Dyers, H. (1997). 1997 Lockwood Post's Directory of the Pulp, Paper, and Allied Trades. Published by Miller Freeman Publications (New York)

Franklin Associates, Ltd. (June 1997). Characterization of Municipal Solid Waste in the United States. Published by the U.S. Environmental Protection Agency, Washington, D.C.

Goldstein, N. and J. Glenn (May 1997). The state of garbage in America, BioCycle Journal of Composting and Recycling. 38, pp. 60–75.

Institute for Brewing Studies (1996). Brewers Resource Directory. Published by the Institute of Brewing Studies, Boulder, Colorado.

Jamski, J. (1997). Unpublished report. U.S. Department of Commerce, Bureau of the Census, Suitland, Maryland.

Lazarus, W. (1997). Farm Machinery Economic Costs for 1997. Published by the University of Minnesota Department of Applied Economics, St. Paul, Minnesota.

National Council of the Paper Industry for Air and Stream Improvement, Inc. (NCASI) (September 1992). Solid Waste Management and Disposal Practices in the U.S. Paper Industry. Published by NCASI, New York.

NEOS Corporation, International Society of Arboriculture Research Trust, Allegheny Power Service Corporation, and the National Arborist Foundation (September 1994). Urban Tree Residues: Results of the First National Inventory; Final Report. Published by the International Society of Arboriculture Research Trust, Champaign, Illinois.

Prosek, C., J. Edmonds, P. Peterson, and P. Vieth (1994). Minnesota Wood Waste Studies: One Man's Waste is Another Man's Gold. Published by the Minnesota Department of Natural Resources, Division of Forestry, St. Paul, Minnesota.

Renewable Fuels Association (RFA) (October 21, 1997). About the producers. Published by the Renewable Fuels Association, Washington, D.C.

Schaffer, S. (October 1997). Unpublished report. Produced by Information Resources, Inc., Washington, D.C.

Thompson, J. (October 1997). Unpublished report. Produced by Chartwell Information Publishers. (Alexandria, Virginia)

Tyson, K.S. (1991). Resource Assessment of Waste Feedstocks for Energy Use in the Western Regional Biomass Energy Area. Published by the Western Regional Biomass Energy Program, Golden, Colorado.

U.S. Department of Agriculture National Agricultural Statistics Service (USDA NASS) (June 1997). Agricultural Statistics 1997. Published by the Government Printing Office, Washington, D.C.

Walsh, M. and D. Becker (1996). Biocost Documentation. Published by Oak Ridge National Laboratory, Oak Ridge, Tennessee.

THE USE OF A GIS MODEL TO EVALUATE THE ECONOMIC POTENTIAL FOR BIOMASS IN NORTHAMPTON COUNTY, PENNSYLVANIA

D. S. Breger and H. Snyder

Department of Civil and Environmental Engineering, Bachelor of Arts in Engineering Program, Lafayette College, Easton, Pennsylvania 18042, U.S.A.

This paper describes the development and use of a geographical information system model to evaluate the technical and economic potential for biomass energy (particularly willows) in a county of Pennsylvania. The model uses GIS coverages of land use, soil type, and riparian zones to evaluate the applicability and cost of biomass production and to generate a supply curve for a biomass economy. The model can be extended to consider energy end-use facilities and transportation costs to analyze the willingness-to-pay for biomass fuels by large energy users. The GIS model is designed to produce a county-level supply-and-demand curve for biomass energy, and the potential for market activity. The spatial distributions of supply-and-demand economics are valuable to target efforts to initiate biomass activities.

1. INTRODUCTION

A transition towards the greater use of renewable energy resources can be justified on economic and environmental grounds; and many government and non-governmental studies conclude that biomass energy will be a predominant renewable energy source in the near future. The recent international agreement in Kyoto on reducing greenhouse gas emissions establishes the importance and commitment governments throughout the world will now need to place on this energy transition. While most biomass energy consumed today comes from residual sources, meeting substantial amount of our nation's demand will require the production of biomass energy crops—that is, crops grown specifically as an energy resource. Substantial research, development, and commercialization activities are currently underway in biomass energy both in the U.S. and internationally.

Willows, or *Salix sp.* are among the most promising biomass crops to produce wood fuels for thermal or electric energy (as opposed to liquid fuels for transportation). The Salix Consortium is a partnership of over 30 industrial, government agency, outreach/technology transfer, farming research and academic institutions committed to making wood biomass crops for energy a viable enterprise. The Consortium is engaged in a commercialization effort in New York State (White et al., 1997) and is extending its effort other states, including Pennsylvania, where a first willow clone site trial was established in Easton in spring 1998.

Analysis indicates the potential of willow biomass crops to improve the economic prosperity of both the farmers and energy users. To stimulate economic activity in this area, innovative and coordinated efforts are needed to establish local economic conditions, disseminate information, and develop and implement a plan to reduce risk and uncertainty.

Previous work using GIS for biomass economic assessment has been consulted. The GIS modeling at Oak Ridge National Laboratory (Graham et al., 1997) is applied to a larger geographical area to assess regional or state level biomass potential supply. Their model incorporates soil quality, climate, land use and road network information, with transportation, economic and environmental models to predict the "farmgate price" of supplying biomass from energy crops and where regional energy crops can be grown and transported to specific locations.

The work by Goor et al. (1998) in Belgium provided an important basis for the model reported in this paper. This work presents a similar approach using GIS to evaluate county level biomass supply costs. Our model has structured this approach to the available coverages and extends the analysis be producing a biomass supply curve, and proposes to additionally evaluate the regional biomass demand to determine an equilibrium level of biomass activity. GIS data and other information have been acquired from the Northampton County Mapping office. The model has been developed using ESRI's ArcView GIS software.

2. APPROACH AND APPLICATION OF MODEL

A purpose of this study was to develop the GIS model in a generic manner so that other small region and economic planning offices can easily adapt the model for their own use. We obtained the local GIS database from the Northampton County mapping office, which is typical for county or regional planning needs. From the large amount of data provided, the following coverages were extracted: land use (residential, industrial, commercial, governmental, agricultural, vacant, etc.); parcel size, soil type and productivity (based on corn crops); woodlands, and watersheds. The model of the biomass supply potential applies a grid (one-acre pixels) to the geographic region, and determines for each cell a cost of biomass production using three steps.

The first step was to determine the applicability of each cell to be used for biomass crop production. Using the Land Use coverage, a cell was given a value of one if it was vacant or rural land, and a value of zero otherwise (residential, urban lands). The Woodland coverage was not functioning, but would be used to assign a zero value to avoid the consideration of undisturbed wooded regions for biomass crops. This step led to the Biomass supply applicability map shown in Figure 1.

The second step was to generate a map that quantifies the expected biomass yield of each cell. This calculation used the Soil Productivity rating in the county data and the Applicability value from step one. Equation 1 scales the biomass yield to the soil productivity and was applied to each cell to create the biomass yield map shown in Figure 2.

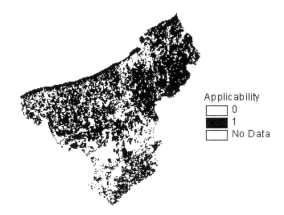

Figure 1. Biomass supply applicability map.

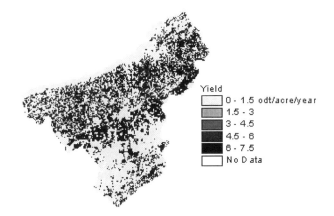

Figure 2. Biomass yield map.

$$Yield = Max\ Yield \times Applicability \times \left(\frac{Productivity\ Rating\ of\ Soil}{Max.\ Productivity}\right) \qquad (1)$$

where *Yield* is in oven dried tons (odt) per acre per year, *Max. Yield* is the current maximum yield expected from willow biomass plantations (7.5 odt/acre/year), *Productivity Rating of Soil* is the actual rating of the cell given in the county data (bushels/acre), and the *Max. Productivity* is the maximum rating in the data (130 bushels/acre).

The final step is to generate cell values for the cost ($/odt) of growing biomass. This calculation assumes a nominal lifecycle cost per acre of biomass production, and adjusts this cost as a function of *Parcel Size* to account for the economy of scale of larger plantations and the economic inefficiencies associated with dedicating small parcels to biomass production. Dividing the cost by the biomass yields produces the cost of growing biomass for each cell. The biomass cost equation is given in Equation 2.

$$Cost = \frac{Nominal\ Cost}{Yield} \times \left(\frac{1}{a\ (Parcel\ Size)^{\alpha}} + 1 \right) \qquad (2)$$

The calculations were made with parameter values of a = 0.08, and α = 1. The *Nominal Cost* parameter was set to $95/acre/year which was derived from the lifecycle yearly cost analysis of willow plantations in New York [EPRI, 1995]. The resulting biomass production cost map is shown in Figure 3.

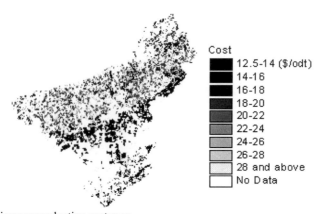

Figure 3. Biomass production cost map.

This biomass production cost does not adequately represent the value needed to develop a biomass supply curve. As is clearly discussed in the ORNL model (Graham et al., 1997), the supply will be influenced by the production cost and by the opportunity cost of the farmer to forgo the profit of conventional farming. On the larger geographical scale employed by ORNL, they were able to assume profitability trends based on county level data. Within a county, and with farm specific data not typically available, this approach is not feasible and requires further consideration.

The minimum production costs determined and shown on the map are on the order of $12-15/odt. The price of willow wood fuel to energy end-users would need to include transportation costs and profit to farmers comparable to profits on conventional crops. Information from the local agricultural extension service indicates normal profits on corn to be roughly $100/acre. Including this profit margin on top of the biomass production cost per acre would lead to a biomass price (before transportation) of approximately $26/odt. This price is comparable to (wet) wood chip supplies currently available in the region.

Using the biomass production costs, a supply cost curve can be generated from the GIS data. This is done by sorting the (one-acre) cells by their production costs and plotting the costs versus the cumulative acreage. This plot is shown in Figure 4 and indicates the acreage within the county that can produce biomass at (or below) a specified cost level.

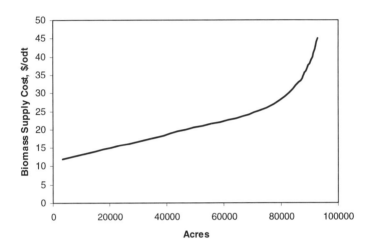

Figure 4. Biomass supply cost curve.

3. FURTHER STEPS

As noted earlier, Woodlands coverage should be used to eliminate wooded areas from being applicable for energy crop production. For watershed and sensitive riparian zones subject to agricultural runoff, the GIS model could prescribe a riparian buffer zone and consider a policy option under which biomass production cost is reduced by a government incentive.

The GIS approach can also be used to model the demand side of the biomass economy by considering energy end-users and their willingness-to-pay for biomass fuel. This demand price would be based on the end users current fuel and price, the age of the energy conversion equipment, and the annualized cost of biomass conversion equipment. Again, availability of site specific data and privacy of end-user information may present a barrier to this approach.

REFERENCES

White, E.H. et al. (1997). Creating a climate for commercializing willow biomass for bio-energy: The development strategy of the New York-based Salix Consortium. Proceedings, Third Biomass Conference of the Americas, Elsevier Science, Montreal, p. 1403.

EPRI (Electric Power Research Institute) (1995). Economic Development Through Biomass System Integration – Final Report, TR-105205, Project 3407-20 (Palo Alto, CA).

Goor, F., A. Alami, X. Dubuisson, and J.-M. Jossart (1998). Interest of a geographical information system to assess the potential for willow short rotation coppice at a low scale level, Procedings, Biomass for Energy and Industry, 10th European Conference and Technology Exhibition, C.A.R.M.E.N., Wurzburg, Germany, pp. 853–856.

Graham, R.L., B.C. English, C.E. Noon, H.I. Jager, and M.J. Daly (1997). Predicting switchgrass farmgate and delivered costs: An 11 state analysis, Proceedings, Third Biomass Conference of the Americas, Elsevier Science, Montreal, pp. 121–129.

HERBACEOUS CROPS AND TREES CAN PROVIDE BIOENERGY NEEDS IN HUMID LOWER SOUTH, USA

G.M. Prine[a] and D.L. Rockwood[b]

[a]Agronomy Department, University of Florida, P.O. Box 110500, Gainesville, Florida 32611-0500, USA
[b]School of Forest Resources and Conservation, University of Florida, P.O. Box 110410, Gainesville, Florida 32611-0410

The Humid Lower South (HLS) can be a leading producer of bioenergy. It has adequate land area, long warm growing season, high rainfall and a subtropical climate which allows production of a number of tropical perennial grasses, the tropical tree legume leucaena, and short rotation woody trees. The vegetatively propagated tall grasses, sugarcane and energy cane (*Saccarhum* sp.), elephantgrass (*Pennisetum purpureum*) and erianthus (*Erianthus arundenaceum*), all have high linear growth rates of 17 to 27 g m^{-2} d^{-1} for 140 to 196 days or longer. Dry matter yields of 20 to 60 Mg ha^{-1} yr^{-1} are possible depending on crop, location, season, soils, management and climate. The seed propagated switch grass (*Panicum virgatum*) offers sustainable dry biomass production of 15 to 22 Mg ha^{-1} yr^{-1}. Tall castorbean (*Ricinus communis*) has produced two-year-old dry stem yield of 65 Mg ha^{-1}. Leucaena (*Leucaena* sp.) has an annual or multiple season woody stem production with potential dry matter annual yields of 19 to 31 Mg ha^{-1}. Besides these herbaceous crops the vast natural forests can be supplemented by short rotation woody crops having potential yields of 15-35 dry Mg ha^{-1} yr^{-1}. The northern portion of HLS can grow short rotation cottonwood (*Populus deltoides*) and slash pine (*Pinus elliottii*) on many soils. *Eucalyptus* species can be grown in peninsular Florida. Culture of the bioenergy crops is enhanced by application of sewage sludge and waste waters. Industries can take advantage of various mixtures of these available high-biomass-yielding crops for energy use.

1. INTRODUCTION

The Humid Lower South (HLS) is a prime region for growing bioenergy crops. The HLS covers all of the state of Florida, the southern third of Georgia, Alabama, Mississippi, Louisiana, and East Texas, the coastal quarter of South Carolina and a narrow band near the ocean in North Carolina. It has a climate with high rainfall and long warm season growing seasons and mild winters. This climate is preferred by many annual and perennial plants. Land not farmed or urban development is forests of pine and hardwood.

The bioenergy industries in HLS will get their biomass (bioenergy) needs from multiple sources: (1) perennial intermediate and tall grasses and other herbaceous plants harvested annually, (2) short rotation wood crops (SWRC) harvested every 4 to 10 years, (3) established forests of pine and hardwood, (4) crop residues of yards, field and forest, and (5) urban and industrial wastes. We will discuss sources (1) and (2) in this paper as these will, in most cases, need to be newly planted from farmland and renovated forest land. Prine and Rockwood discussed these two bioenergy sources from HLS in an earlier paper.[1] Prine et al.[2] reported on tallgrass and leucaena as energy crops for HLS.

2. TALLGRASSES

Many temperate plants grow in HLS but the highest biomass yields have come from tropical plants that take advantage of the long warm growing season. The tall grasses having C_4 metabolism are elephantgrass, erianthus, energy cane and sugarcane. They grow the entire warm season (more than 200 growing days). The tall grasses have linear growth rates up to 27 g m^2 da^{-1} for 140 to 196 days and sometimes longer.[3] The biomass production possible for various linear growth rate periods at a 25 g m^2 d^{-1} growth rate is shown in Table 1. An elephantgrass or energycane growing at this rate from May through October (180 d) has a potential biomass yield of 45 Mg ha^{-1} (20 tons/acre). Such warm growth periods are quite normal for the HLS, and even longer growth periods are available in peninsular Florida.

Table 1
The estimated biomass and energy equivalent production of tall grasses over time based on a linear growth rate of 25 g m^{-2} d^{-1}

	Oven dry biomass		Oil equivalent		
				Barrels*	
Growth period (da)	Mg ha^{-1}	Tons/A	Mg ha^{-1}	ha^{-1}	Acre
100	25.0	11.6	10.2	67.0	27.1
120	30.0	13.4	12.2	80.3	32.5
140	35.0	15.6	14.3	93.9	38.0
160	40.0	17.9	16.3	107.2	43.4
180	45.0	20.1	18.4	120.6	48.8
200	50.0	22.3	20.4	134.0	54.2
365	91.3	40.7	37.2	245.0	99.0

* 159 liter (42-gallon) barrel[1] where 2.45 g oven dry biomass = 1 g No. 2 diesel oil.

Plants don't always reach their growth potential owing to droughts or inadequate drainage, poor nutrition owing to poor soils or inadequate fertilization and liming, climatic catastrophes such as wind and hail storms and tornados, plant pests of all types, and sparse stands. Dry biomass yields of 20 to 60 Mg ha^{-1} have been measured in the HLS. The tall grasses are resistant to most stresses and recover quickly following stress and continue their linear growth rate. The tall grasses are good bioenergy plants because they are stout enough to hold up all the biomass produced over a season without serious lodging and the green top growth maintains a complete and effective canopy until killed by frost. They are efficient in using nutrients, which translocate out of old leaves to new leaves as plants grow. Because of high biomass yields, a very low percentage of a nutrient in biomass can add up to large amounts of nutrient per hectare. For example, a 0.5% N content of a 45.0 Mg ha^{-1} biomass yield is 225 kg ha^{-1}. To have a sustainable production, the needed nutrients will need to be applied annually from fertilizer, animal manures, composts and sewage sludge or effluents if the top growth is removed from each year. The chemical composition of tall grasses was reported by Prine et al.[4]

Tall grasses are planted vegetatively, which is usually expensive, because they are all susceptible to freeze damage which kills or reduces the stand in the next growing season. Only adapted cold-tolerant plants should be planted because a planting can be killed or thinned by a cold winter. Adapted tall grasses should survive five or six years before replanting is needed.

3. SWITCHGRASS

Switchgrass is an intermediate height grass in HLS but would be called a tall grass in other parts of the U.S. Switchgrass yields (15 to 22 Mg ha^{-1} yr^{-1}) are usually lower than tall grasses but are not as affected by winter kill. Switchgrass can be sustained for longer periods with fewer inputs than tallgrasses. Switchgrass is propagated from seed but establishing the initial stand is some times difficult. Bransby and Sladden[5] reported Alamo switchgrass yields of 25.8 Mg ha^{-1} yr^{-1} in Alabama.

4. CASTORBEAN

Tall growing castorbean forms a tree 9 to 12 m tall in the tropics. Where top growth is not killed annually but is damaged by frost, plants grow 5 to 9 m tall. Tall castorbean planted in April grew to 6.7 m in height and had dry matter stem yields of 40 Mg ha^{-1} at Gainesville, Florida in 1997. Unharvested plants surviving winter frost grew only 5.3 m tall and had a two-season stem yield of 65.4 Mg ha^{-1} in 1998. Castorbean needs further study as an energy crop.

5. LEUCAENA

Leucaena (*L.* sp., mainly *leucocephala*) is a tropical legume shrub or tree that takes two years to establish before annual or multiple year biomass harvests can be made. Several year old harvests of leucaena are possible where freezes do not kill the stems in the winter. Table 2 shows the yields of leucaena harvested annually and after four years of growth at Gainesville, Florida. Freeze-killed stems usually will stand for one season, allowing two years growth of old and new stems to be harvested in the winter of the second year where top growth is killed annually. In colder upper portions of HLS leucaena would be harvested annually or biannually and in warmer portions of Florida peninsular every three to five years and managed as a SRWC. Leucaena tends to be a sustainable crop because as a legume it can fix its needed nitrogen and usually the leaf portion with its high nutrient content falls to the ground after harvest or freezes and recycles for next season's crop. Leucaena is not adapted to poorly drained or highly acidic soils.

Table 2
The oven dry stem yields of selected leucaena accessions harvested annually and every four years at Gainesville, Florida

	Stem yields ($Mg\ ha^{-1}$)			
	Harvested annually		Harvested once	
No. of accessions	4 year total	Annual	4 year total	Annual
12	125.6	31.4	-	-
10 (unadjusted)	-	-	135.6	33.9
(Adjusted to space)	-	-	76.9	19.2

Based on Cunilio and Prine, 1992[6] and 1995.[7]

6. CITY WASTE

Cities often have sewage effluent and sludge and composts which are useful in the sustainable growth of bioenergy crops. PI 300086 elephantgrass had annual yields of 52 $Mg\ ha^{-1}$ when grown under sewage effluent at Leesburg, Florida for two years.[1] Elephantgrass and energycane grew well on droughty sand at Tallahassee, Florida irrigated with sewage effluent.[8] Bioenergy crops are a good way to turn city waste products into useful energy.

7. MINED LANDS

In peninsular Florida, mined out phosphate mines cover more than a quarter million acres with over 100,000 acres in clay settling ponds. Recovered mined out land would be a good site for energy crops. Yields of tall grasses have been especially high (up to 60 Mg ha^{-1} yr^{-1}) on settling pond clays. Segrest et al.[9] and Stricker et al.[10] have discussed the bioenergy potential of energy crops in central Florida phosphatic mining lands.

8. SHORT ROTATION WOOD CROPS

Intensive culture of SRWCS in the HLS can be practiced in many ways, including the application or use of wastewater or materials. Eastern cottonwood, a woody crop potentially plantable on good sites throughout the whole HLS, can even be grown on typically dry sandhills if sewage effluent is applied.[11] However, its productivity was up to 30% greater in the first growing season when sewage effluent was augmented with mulch and/or compost (Table 3). Castorbean, a perennial in the absence of freeze, has also been responsive to these amendments

Table 3
Height (H in m) and survival (S in %) responses of castorbean (CB), cottonwood (CW), cypress (CY), *Eucalyptus amplifolia* (EA), *E. camaldulensis* (EC), *E. grandis* (EG), and *Leucaena leucocephala* (LL) to sewage effluent only (E), sewage effluent + mulch (E + M), sewage effluent + compost (E + C), and sewage effluent + compost + mulch (E + C + M) cultures near Winter Garden, Florida

Species	Age (mos)	Culture								Species mean	
		E		E + M		E + C		E + C + M			
		H	S	H	S	H	S	H	S	H	S
CB	8	2.31	89	2.82	86	2.47	87	2.97	95	2.63	88
CW	9	2.51	39	2.92	71	2.94	85	3.26	87	2.94	68
CY	9	0.73	31	0.76	45	1.61	30	0.87	18	0.93	32
EA	8	1.07	33	1.54	69	1.38	43	1.93	65	1.53	52
EC	6	0.34	19	0.84	31	1.20	46	1.02	83	0.92	41
EG	5	-	-	-	-	1.01	71	1.39	100	1.24	85
LL	7	1.79	58	1.54	69	1.69	50	1.84	70	1.71	62
Culture mean		1.81	48	2.00	65	2.07	59	2.26	70	2.04	60

to effluent. With its rapid early growth from seed leading to quick dominance of other vegetation, castorbean may be especially useful in warm peninsular Florida as an easy-to-establish energy crop that can out-compete aggressive vegetation, such as cogongrass, that frequently is difficult to control in the initial months following planting of other energy crops. Some *Eucalyptus* and *Leucaena* species may also respond favorably to compost incorporated into sandy soils receiving sewage effluent. Should these productivity enhancements continue through harvest, woody crop yields could exceed 35 dry Mg ha^{-1} yr^{-1} in Florida.

9. CONCLUSIONS

The availability of land, a multitude of high yielding bioenergy crops and vast natural forests makes HLS a likely area for developing bioenergy industries.

REFERENCES

1. Prine, G.M. and D.L. Rockwood (1998). Proc. Bioenergy 98 Conf., 2, p. 1192.
2. Prine, G.M., et al. (1997). Proc. 3rd Biomass Conf. of Americas, 1, p. 227.
3. Woodard, K.R. and G.M. Prine (1993). Crop Science, 32, p. 818.
4. Prine, G.M., et al. (1995). Proc. 2nd Biomass Conf. of the Americas, p. 278.
5. Bransby, D.L. and S.E. Sladden (1995). Proc. 2nd Biomass Conf. of Americas, p. 261.
6. Cunilio, T.V. and G.M. Prine (1992). Proc. Soil and Crop Sci. Soc. Fla., 51, p. 120.
7. Cunilio, T.V. and G.M. Prine (1995). Proc. Soil and Crop Sci. Soc. Fla., 54, p. 44.
8. Prine, G.M. and W.V. McConnell (1996). Proc. Bioenergy 96 Conf., p. 770.
9. Segrest, S.A., et al. (1998). Proc. 10th European Biomass for Energy and Industrial Conf., p. 1472.
10. Stricker, J.A., et al. (1995). Proc. 2nd Biomass Conf. of Americas, p. 1608.
11. Rockwood, D.L., et al. (1996). Proc. Bioenergy 96 Conf., 1, p. 254.

May 30, 1999

UTILIZATION OF BIOMASS ENERGY RESOURCES IN OHIO: A LINEAR PROGRAMMING MODEL

Bibhakar S. Shakya[a] and Douglas Southgate[b]

[a]Ohio Biomass Energy Program, Public Utilities Commission of Ohio, 180 East Broad St., Columbus, Ohio 43215, USA
[b]Department of Agricultural, Environmental, and Development Economics, Ohio State University, 2120 Fyffe Road, Columbus, Ohio 43210, USA

Biomass energy has a promising future because it is a cleaner source of energy and there is a growing concern regarding global greenhouse gas emissions. Prospective regulatory changes in the electric industry will open the market for competition. Consumers will be able to choose alternative sources of energy, such as biomass, and commercial energy generated by using biomass resources can be marketed as "green power."

Energy from biomass minimizes global warming since zero net carbon dioxide is emitted, which means the amount of carbon dioxide emitted is equal to the amount absorbed from the atmosphere during the biomass growth phase. In addition, other environmental benefits of biomass energy include mitigating pressure to landfills, reducing soil erosion and improving water quality and wildlife habitats.

This study reviews the current and potential use of biomass energy resources in Ohio. A Linear Programming (LP) model was developed to identify potential sites for biomass power plants based on availability of forest and industrial wood residues. This LP model can be used to examine various policy issues having to do with the use of industrial wood and forest residues in Ohio, and this model can easily be adopted in other states.

1. INTRODUCTION

Most states in the U.S. are currently considering regulatory changes that will have a major effect on the electricity market. This move toward increased competition in the electric utility industry parallels what has happened in the U.S. markets for natural gas and long-distance telecommunication services. Similar regulatory changes in Ohio could make the market more competitive and provide incentives to utilize biomass resources for electricity generation (DOE, 1992; Douglas, 1994; Hyman, 1994; NRRI, 1995; US Congress, 1989).

Biomass energy has a promising future since it is a cleaner source of energy and there is a growing concern regarding greenhouse gas emissions at the global level. Consumers will be attracted to "green power" generated by biomass owing to its various environmental benefits which include minimizing global warming effects because zero net carbon dioxide is emitted. This means the amount of carbon dioxide emitted is equal to the amount absorbed from the atmosphere during the biomass growth phase. Energy from biomass produces no or low sulfur emissions, helping to mitigate acid rain. It also reduces pressure to create new landfills and extends the life of existing landfills. And energy crops require less fertilization and herbicides and provide much more vegetative cover throughout the year, providing protection against soil erosion and deterioration of watersheds as well as improved wildlife habitats.

Ohio has a substantial area covered by forests, which ensures abundant supplies of woody biomass, a renewable resource that can be used for producing electricity and other forms of energy. The wood manufacturing industry also generates a large volume of industrial residues; this sector, which is important to Ohio's economy, comprises more than 300 sawmills and more than 1,000 secondary wood manufacturing companies (Heiligmann et. al., 1993; ODNR, 1993; Shakya, 1995). There are also significant prospects in Ohio for energy crops, such as trees and grasses, which will expand the supply of biomass feedstock for electricity generation. A biomass energy crop can be a profitable alternative, provide an additional source of income for farmers, and can be grown on currently underutilized or marginal agricultural land.

The generation of commercial energy from biomass would increase the demand for biomass from forests as well as industrial wood residues tremendously. It would also open the market for energy crops in Ohio. However, owing to the bulky nature of biomass, it is important to have the biomass power plant sites close to the source of biomass supply to minimize transportation costs. This study develops an LP model to identify both demand and supply sites for biomass resources in Ohio based on the current availability of forest and industrial wood residues.

2. LINEAR PROGRAMMING METHODOLOGY

Linear programming (LP) is the application of the economic concept of constrained optimization to real-world problems and issues. For example, an LP can be set up to identify a mix of inputs and outputs that maximizes a firm's profits subject to regulatory restrictions or limited availability of some given set of resources. The LP model developed for this study is a least cost transportation model to meet the demand for biomass resources for a given set of cogeneration facilities.

2.1. Objective function and activities

The objective function (C), which is to be minimized, equals the sum of the cost of transporting industrial wood and forest residues among various regions around Ohio as well as the aggregate cost of landfills for industrial wood residues. The objective function, supply and demand activities of the model, are given below:

$$\text{Total Cost} = \sum_i \sum_j CT * D_{ij} ((CW * X_{ij}) + (CF * Z_{ij})) + \sum_i CL * L_i \qquad (1)$$

It is subject to the following:,

Supply: (a) Wood residues $\quad \sum_j X_{ij} + L_i = X_i$ (2)

(b) Forest residues $\quad \sum_j Z_{ij} <= Z_i$ (3)

Demand: $\quad \sum_i X_{ij} + \sum_i Z_{ij} = Y_j$ (4)

where:
- CF = Cost of collecting a ton of forest residues
- CL = Cost of land filling a ton of wood residues
- CT = Cost of transporting a ton of residues one mile
- CW = Cost of wood residues per ton
- D_{ij} = Miles between center point of regions i and j
- L_i = Tons of wood residues landfilling in a region i
- X_i = Tons of wood residues available in a region i
- Y_j = Tons of residues demanded in a region j
- Z_i = Tons of forest residues available in a region i
- X_{ij} = Tons of wood residues shipped from regions i to j
- Z_{ij} = Tons of forest residues shipped from regions i to j

The transportation cost of forest and wood residues is based on the rate of $1.75/ 25-ton load/ mile, which is equivalent to $0.07/ton/mile. It is assumed in the model that the excess supply of industrial wood residues is landfilled. The landfill cost is currently at $20/ton, which includes transportation cost and a landfill fee. And Ohio is divided into 50 regions based on the location of wood manufacturing companies, its timber resources, and metropolitan areas.

2.2. Supply specification

Using data from a survey of wood manufacturers for wood residues, carried out by this report's coauthor (Shakya, 1995), residue the amount of in each of the 50 regions was estimated. Based on the field interviews and the survey report on wood residues, the cost of wood residues is estimated to be $8.00/ton (Shakya, 1995). Estimates of forest residues are based on data obtained from Forest Statistics for Ohio, 1991 (NFES, 1993). If not collected, forest residues are left on the site. The collection cost for forest residues is estimated to be $15.00/ton. The supply constraint for forest residues on the above equation describes the supply of forest residues from a region "i" to the demand regions is less than or equal to the availability of forest residues in a region "i."

2.3. Demand specification

There are 26 boiler facilities around the state which are located in 19 regions. Most of these boiler facilities are located in the northwest and the northeast regions of Ohio, while the southeast and the southwest regions have only a few biomass power plants. These 19 regions demand wood and forest residues based on their respective boiler capacity.
The average annual woody biomass demand from a biomass power plant is 250,000 tons. The demand constraint on the above equation means that the demand for residues in region "j" must

be equal to the sum of the supply of wood residues from regions "i" to "j," plus the sum of the supply of forest residues from regions "i" to "j."

3. RESULTS AND DISCUSSION

This LP model is used to identify the best transportation schedule for any type of use of the residues to a given location, for example, building a reconstituted furniture manufacturer in a certain region. The potential location for establishing a new company can also be identified, based on the availability of residues within a specific mile radius. Any firm or company which uses wood residues as an input could take advantage this LP model to identify ideal locations closer to the residues source points.

Similarly, establishment of a wood manufacturer in a certain region would add to the supply of wood residues in that region. The change in transportation or landfilling costs can be incorporated in the model by modifying their corresponding values. The concise representation and flexibility of the least-cost transportation LP model are designed for its larger application to address the issues regarding the use of wood and forest residues in Ohio.

The LP model designed for this study was solved using GAMS software to present and analyze the information generated by the program. Given that the per-unit transport and landfilling costs are $0.07/ton and $20/ton respectively, and given the residues demands and mill waste availability, the model identifies a least-cost pattern for utilizing and disposing of wood and forest residues. In addition, shadow prices obtained by running the model identify the regions in Ohio where biomass resources are relatively scarce. The latter information is particularly useful in making rational economic decisions regarding the possible locations of cogeneration plants.

In Ohio, there are twenty-six known wood manufacturing companies that have wood-burning boilers to generate both heat and electricity (Shakya, 1995). The LP results show that the wood residue resource from wood industries around the state exceeds the demand of existing wood burning boilers by a wide margin; therefore, interregional shipments are limited. Because of the high collection cost of forest residues and the excess supply of wood residues, it is cheaper to use industrial wood residues than forest residues. However, the data on forest residues provides the biomass base in each region for future use, if needed.

The other uses of wood residues can be added in this LP model to analyze the transportation schedule and the residue availability scenarios under the new demand. Similarly, sensitivity analysis can be carried out to gauge the impact of per-unit transportation costs, landfill expenses, and other variables.

4. CONCLUSION

Presently, forest residues are little used because of the excess supply of wood residues from wood manufacturing companies. If biomass power plants become viable for commercial energy production in Ohio, the utilization of forest residues will take place in the future. In other states like Vermont, Michigan, Wisconsin, Washington, and Oregon, utilities also operate wood-fired power plants which range from 10 to 50 MW capacity (Swezey et al., 1994). Most of these states also have energy crops to produce biomass feedstock. Ohio can also grow such crops to supply biomass for generating commercial electricity in the future. This LP model can be used to identify both demand and supply points for biomass resources in Ohio.

ACKNOWLEDGMENTS

This paper was prepared for the Ohio Biomass Energy Program, Public Utilities Commission of Ohio (PUCO) and was sponsored by the Great Lakes Regional Biomass Program (GLBRP) of the U.S. Department of Energy. However, this paper does not necessarily reflect the views of PUCO or GLRBP. We would like to thank Anne Goodge, Carl Tucker and Michael Long for their comments and suggestions. Special thanks to all the wood manufacturing companies for providing valuable information for this study.

REFERENCES

Department of Energy (1992). Electricity from Biomass, Solar Thermal and Biomass Power Division, DOE/CH10093-152, April 1992.

Douglas, J. (1994). "Buying and selling power in the age of competition." EPRI Journal, 19, pp. 6–13.

Heiligmann, R.B., T.R. Weidensaul, and R.L. Romig (1993). Forestry Data Summary. OSU Extension/School of Natural Resources.

Hyman, L.S. America's Electric Utilities: Past, Present, and Future. Public Utilities Report, Inc. August 1994.

National Regulatory Research Institute (NRRI) (1995). Missions, strategies, and implementation steps for State Public Utility Commissions in the year 2000: Proceedings of the NARUC/NRRI Commissioners Summit, Columbus, Ohio: NRRI, Ohio State University. NRRI 95-08, May 1995.

Northeastern Forest Experiment Station (NFES) (1993). Forest Statistics for Ohio, 1991. USDA, Forest Service, Resource Bulletin NE-128. March 1993.

Ohio Department of Natural Resources (ODNR) (1993). Wood Industry Directory - 1993. ODNR, Division of Forestry.

Shakya, Bibhakar S. (1995). Survey of Wood Manufacturers for Wood Residues in Ohio, Ohio Biomass Energy Program, PUCO, Columbus, Ohio..

Swezey, B.G., K. L. Porter, and J. S. Feher (1994). The Potential Impact of Externalities Considerations on the Market for Biomass Power Technologies. National Renewable Energy Laboratory (NREL), NREL/TP-462-5789, February 1994.

U.S. Congress, Office of Technology Assessment (1989). Electric Power Wheeling and Dealing: Technological Considerations for Increasing Competition. Washington, DC: US Government Printing Office. OTE-E-409. May 1989.

Williams, S. and K. Porter (1989). Power Plays: Profiles of America's Independent Renewable Electricity Developers. Washington, D.C.: Investor Responsibility Research Center. 1989.

Biomass Production and Integration With Conversion Technologies

Biomass Resources: Carbon Impacts

BIOMASS CROP PRODUCTION: BENEFITS FOR SOIL QUALITY AND CARBON SEQUESTRATION[*]

V.R. Tolbert,[a] J. D. Joslin,[b] F.C. Thornton,[b] B.R. Bock,[b] D.E. Pettry,[c] W. Bandaranayake,[d] D. Tyler,[d] A. Houston,[d] and S. Schoenholtz[c]

[a] Bioenergy Feedstock Development Program, Oak Ridge National Laboratory, Oak Ridge, Tennessee
[b] Tennessee Valley Authority, Muscle Shoals, Alabama
[c] Mississippi State University, Starkville, Mississippi
[d] Ames Plantation, University of Tennessee, Grand Junction, Tennessee

Research at three locations in the southeastern U.S. is quantifying changes in soil quality and soil carbon storage that occur during production of biomass crops compared with row crops. After three growing seasons, soil quality improved and soil carbon storage increased on plots planted to cottonwood, sycamore, sweetgum with a cover crop, switchgrass, and no-till corn. For tree crops, sequestered belowground carbon was found mainly in stumps and large roots. At the Tennessee site, the coarse woody organic matter storage below ground was 1.3 Mg ha^{-1} yr^{-1}, of which 79% was stumps and large roots and 21% fine roots. Switchgrass at the Alabama site also stored considerable carbon belowground as coarse roots. Most of this carbon storage occurred mainly in the upper 30 cm, although coarse roots were found to depths of greater than 60 cm. Biomass crops contributed to improvements in soil physical quality as well as increasing belowground carbon sequestration. The distribution and extent of carbon sequestration depends on the growth characteristics and age of the individual biomass crop species. Time and increasing crop maturity will determine the potential of these biomass crops to significantly contribute to the overall national goal of increasing carbon sequestration and reducing greenhouse gas emissions.

1. INTRODUCTION

Conversion from agricultural crops to biomass crop production has been projected to improve soil physical quality and soil carbon storage (Smith 1995, Grigal and Berguson, 1998). The greatest gains in soil carbon storage are expected to occur on lands that were previously in agriculture or barren (Smith, 1995). Hansen (1993) concluded that after a

---

[*] Research sponsored by the Biofuels Systems Division, U.S. Department of Energy, under contract DE-AC05-96OR22464 with Lockheed Martin Energy Research Corporation.

markedly increased during the 6-12 year rotation, primarily in the upper 30 cm. Hansen hypothesized that the initial carbon losses occurred from the clean-tilled surface soil and as a result of mineralization during the first several years of planting establishment. Makeschin (1994) found that soil carbon storage nearly doubled after three years under hybrid poplar compared with adjacent arable fields in Germany.

Johnson (1993) identified the need to determine the effects of management practices and soil properties on soil carbon storage in intensive production systems. The goal of this research was to determine changes in soil physical properties and belowground carbon sequestration during the early years of biomass crop production on sites converted from agricultural crops to short-rotation woody crops and switchgrass. A further objective of this study was to determine the contribution of cover crops to site quality and belowground carbon sequestration.

2. METHODS

On three sites in the Tennessee Valley, typical of areas having identified biomass-crop potential, replicated 0.4- to 0.6-ha plots were enclosed within berms, planted to the appropriate agricultural crops or biomass crops, and instrumented for long-term monitoring to quantify differences in runoff and water quality. Sweetgum (*Liquidambar styraciflua*) with and without a cover crop between rows and switchgrass (*Panicum virgatum*) were established at a site in northern Alabama for comparison with no-till corn. Eastern cottonwood (*Populus deltoides*) was established in western Mississippi for comparison with cotton, and American sycamore (*Platanus occidentalis*) was established in western Tennessee for comparison with no-till corn. At the Tennessee site, existing 12-year-old sycamore were included as a comparison. The Alabama site is on the Decatur silt loam soil, the Mississippi site, a Bosket silt loam soil, and the Tennessee site, a Memphis-Loring silt-loam intergrade. Soil physical characteristics of bulk density, penetration resistance, infiltration, and aggregate stability were measured on all plots at each site in 1995 at the time of establishment and again at the end of the third growing season (1997) to determine changes in soil quality over time. More detailed site descriptions, methods, and results of water quality monitoring can be found in Tolbert et al. (1998), Joslin and Schoenholtz (1998), and Thornton et al. (1998).

Belowground biomass and soil carbon were determined for each tree crop at the end of the third growing season. For each tree-crop treatment, six stumps and those roots extracted with the stumps were removed, weighed, dried, and reweighed to determine organic matter content. Replicate root cores were taken at 0-1.25 cm, 1.25-7.5 cm, and 7.5-15 cm depths to determine root biomass for both tree and switchgrass crops. Soil carbon storage under land converted to switchgrass was determined to 30-cm depths. Additional estimates for switchgrass were made based on root distributions to greater depths for comparison with samples from CRP lands.

3. RESULTS AND DISCUSSION

Comparisons of aggregate stability from the 3-year old cottonwood and continuous cotton plots at the Mississippi site showed that the percent stability was twice as great under the tree crop after both 8 and 32 minutes of immersion and oscillation. At the Tennessee site, aggregate stability was significantly greater ($P > 0.05$) under 12-year-old sycamore than under the 3-year old sycamore and no-till corn. The aggregate stability did not greatly improve under the sycamore plots at the 0-15 cm depth during the first four years of the tree-crop rotation; large improvements, however, were observed under the no-till corn plots. These differences at these shallow depths could be attributed to the more extensive root system of the no-till corn with a winter cover crop as compared with a less fibrous root system and less even root distribution in the sycamore.

Both penetration resistance and bulk density decreased under cottonwood compared with cotton at the Mississippi site. After the first growing season, the soil traffic pan, which initially lay at 15 to 30 cm, was not evident under the cottonwood but persisted under the cotton rotation. The penetration resistance dropped by half (from 3.0 to 1.5 MPa) in the upper 10-20 cm. At the Tennessee site, bulk density values for the profile (from surface to 1.5 m depth) showed that significant changes occurred within the upper-30-cm depth for 12-year-old sycamore plantings compared with no-till corn and the 3-year-old sycamore. At a depth of 0-3 cm, the bulk density under the young sycamore showed a significant decline from year 1 to year 3. Bulk density under the 3-year-old sycamore at this depth was 1.17 Mg ha^{-1}, which was comparable with the 12-year-old sycamore at the same depth. Bulk density under no-till corn at the same depth was significantly greater (1.34 Mg ha^{-1}).

At the Tennessee site, the greatest organic carbon increases occurred within 25 cm of the stem center, largely due to the contribution of coarse organic matter, particularly stumps (Figure 1). At the Mississippi site, the total mass of soil carbon (excluding stumps and coarse organic matter) under cottonwood was 15 Mg ha^{-1}. With stump and coarse roots included, the belowground biomass was approximately 18 Mg ha^{-1}. During the first growing season, the cottonwood that was established from cuttings did not develop a tap root but formed a dense horizontal root system that extended across the 3.6-m distance between rows at depths of 10-25 cm. Approximately 60% of the soil organic carbon under the cottonwood occurred in the upper 30 cm (Figure 2); small roots extended to depths of ≥ 1 m. This greater distribution of cottonwood roots in the upper 30 cm is consistent with distributions found to this depth by Tuskan (personal communication) for cottonwood (70%), sweetgum (71%), and sycamore (59%).

Sycamore and no-till corn established on former traditional agricultural lands at the Tennessee site increased soil carbon in the upper 15 cm by 27% and 34%, respectively, by the end of the third growing season. The increase in soil carbon storage under the sycamore was approximately 1.3 Mg ha^{-1} yr^{-1}. This number is similar to that found by Hansen (1993) for soil carbon accretion rates under hybrid poplar in the north-central states. Soil carbon under cottonwood at the Mississippi site increased by 19% by the end of the third growing season; there was no change in soil carbon for the cotton plots (Figure 3). The increase in soil carbon on the plots converted from traditional agricultural crops to no cultivation is consistent with the projections made by Smith (1995) that reforestation of agricultural lands could increase carbon sequestration in soils.

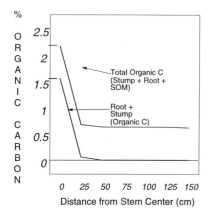
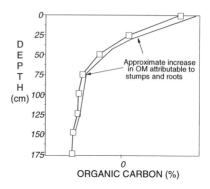

Figure 1. Lateral distribution of soil organic C within the upper 30 cm under sycamore at the Tennessee site.

Figure 2. Distribution of roots through the soil profile. The solid line represents the additional contribution of larger roots and coarse organic matter not passing through a 2-mm sieve.

There was a significant increase ($p \leq 0.02$) in soil carbon under switchgrass, sweetgum with a cover crop, and no-till corn during a three year period at the Alabama site. By contrast soil carbon in the plots of sweetgum without a cover crop actually decreased by 6% in the same period (Figure 4). The response of the sweetgum without a cover crop treatment is consistent with Hansen's (1993) results which showed decreases in soil carbon during the first few years of tree crop establishment when clean cultivation was practiced to minimize weed competition. Soil carbon has been found to significantly increase under switchgrass across a wide range of sites from Texas through Virginia when compared with fallow plots and initial samples (McLaughlin and Walsh, 1998).

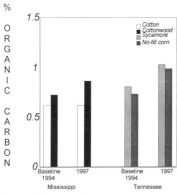

Figure 3. Comparison of baseline soil C and soil C at the end of the third growing season at the Mississippi and Tennessee sites for the 0-15 cm depth increment.

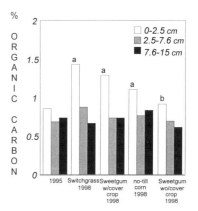

Figure 4. Change in % organic C by depth increments at Alabama A&M after 3 years compared with baseline samples.

4. SUMMARY

Conversion of traditional agricultural lands to production of short-rotation woody and herbaceous crops as energy feedstocks offers considerable potential to sequester carbon in the belowground components of these crops. Crop rotations of 5 to 20 years for these perennial crops offer potential long-term storage of soil carbon below ground. Data from the three sites show the value of soil cover provided by switchgrass, sweetgum with a cover crop, no-till corn, sycamore, and cottonwood. The development of extensive perennial rooting systems and litter layers appear to be major factors accounting for increasing carbon sequestration. Questions still to be considered are how the vertical distribution of soil carbon within the soil column changes with time and the duration of the belowground carbon storage with conversion of sites back to agricultural crop production.

REFERENCES

Grigal, D.F. and W.E. Berguson (1998). Soil carbon changes associated with short-rotation systems, Biomass and Bioenergy, 14, pp. 371–378.

Hansen, E. A. (1993). Soil carbon sequestration beneath hybrid poplar plantations in the north central United States, Biomass and Bioenergy, 5, pp. 431–436.

Johnson, D.W. (1993). Carbon in forest soils – research needs, New Zealand J of Forestry Sci., 23, pp. 354–366.

Joslin, J.D. and S.H. Schoenholtz (1998). Measuring the environmental effects of converting cropland to short-rotation woody crops: A research approach, Biomass and Bioenergy, 13, pp. 301–311.

Makeschin, F. (1994). Effects of energy forestry on soils, Biomass and Bioenergy, 6, pp. 63–79.

McLaughlin, S.B. and M.E. Walsh (1998). Evaluating environmental consequences of producing herbaceous crops for bioenergy, Biomass and Bioenergy, 14, pp. 317–324.

Smith, C. T. (1995). Environmental consequences of intensive harvesting, Biomass and Bioenergy, 9, pp. 161–179.

Thornton, F.C., J.D. Joslin, B.R. Bock, A. Houston, T.H. Green, S. Schoenholtz, D. Pettry, and D.D. Tyler (1998). Environmental effects of growing woody crops on agricultural land: First year effects on erosion and water quality, Biomass and Bioenergy, 15, pp. 57–69.

Tolbert, V.R., F.C. Thornton, J.D. Joslin, B.R. Bock, W.E. Bandaranayake, D.D. Tyler, D. Pettry, T.H. Green, R. Malik, A.E. Houston, S. Schoenholtz, M. Shires, L. Bingham, and J. Dewey (1998). Soil and water quality aspects of herbaceous and woody energy crop production: Lessons from research-scale comparisons with agricultural crops. Proceedings, BioEnergy'98: Expanding BioEnergy Partnerships, Madison, Wisconsin, pp. 1272–1281.

WOODY BIOMASS PRODUCTION IN PHYTOREMEDIATION SYSTEMS

G. R. Alker,[a] D. L. Rockwood,[a] L. Q. Ma,[b] K. Komar,[b] and A. E. S. Green[c]

[a]School of Forest Resources and Conservation,
[b]Department of Soil and Water Sciences, and
[c]Department of Mechanical Engineering
All of the University of Florida, Gainesville, Florida, USA 32611

Woody biomass production for energy and other uses can be increased while concurrently removing contaminants from soil and water. An interdisciplinary study to (1) identify the most effective tree genotypes for contaminant uptake, (2) quantify their phytoremediation potential, (3) develop guidelines for establishing and managing phytoremediation systems using these genotypes, and (4) explore environmentally benign uses of the woody biomass produced in these systems has assessed growth and phytoremediation potential of castorbean (CB, *Ricinus communis*), cottonwood (CW, *Populus deltoides*), cypress (CY, *Taxodium distichum*), *Eucalyptus amplifolia* (EA), *E. camaldulensis* (EC), *E. grandis* (EG), *Leucaena leucocephala* (LL), and several other tree species for a range of applications and field sites.

1. INTRODUCTION

Phytoremediation is the use of plant-based systems to remediate contaminated soils, sediments, and water and can take place in many ways. For phytostabilisation, plants can be used to stabilise a contaminated soil by decreasing wind and soil erosion while decreasing water infiltration and leaching of the contaminant. Phytoextraction removes the contaminant from the rhizosphere and accumulates it in the plant's biomass for removal from the system when the plant is harvested and treated. Other methods include rhizofiltration, the removal of contaminants from water, and rhizodegradation, the symbiotic relationship between plants and microorganisms that results in the breakdown of soil contaminants. Compared to most traditional remediation and engineering techniques, it is a relatively new technology that addresses a wide variety of surficial contaminants.

Hyperaccumulating plants take up, translocate, and tolerate levels (1-5% is considered hyperaccumulative) of certain heavy metals that would normally be toxic.[1] Although these plants are taxonomically widespread, hyperaccumulation is relatively rare.

Although phytoremediation is a low-tech and low-cost technique, there are limitations and constraints associated with this process. Perhaps the most important is that the contaminated site must sustain plant growth. Another limitation is that plants can accumulate only contaminants that are accessible in the rhizosphere.[2] The type of accumulation mechanism in the plant, the soil type, and the origin of the contaminant can also limit the effectiveness of phytoremediation. The greatest limitation can be the amount of time that is required for the plants to accumulate the contaminant and to reduce the soil contaminant to acceptable levels.[3]

The distribution of heavy metals among the different plant parts (leaves, stems, roots) is of great importance in the evaluation of the capacity of trees to decontaminate soils fertilized with sludge.[4] Metals in stems and roots of trees could be important for immobilization of these elements for long periods of time. Nitrogen and P uptake from nutrient rich wastewaters by fast growing tree species has been studied by a number of authors.[5-7]

2. DENDROREMEDIATION RESEARCH IN FLORIDA

Research to determine the feasibility of using trees to address some of Florida's environmental concerns is currently underway at the University of Florida. Many of the state's soils have been polluted by agricultural and industrial processes such as CCA wood treatment, which uses salts of copper, chromium, and arsenic, and the dipping of cattle in a mixture of arsenic, toxaphene, and DDT. With both of these processes, contamination of the surrounding soil from leaks, spills, and deliberate dumping was possible. Additionally, sewage effluent can contain high levels of N and P.

In 1997-98, sewage effluent disposal, cattle dip, and wood treatment sites were selected for investigation (Table 1) with three objectives: (1) To determine the levels of nutrients or metals in the contaminated sites and to identify and analyze the existing plant species that effectively remove, accumulate, or tolerate these contaminants; (2) To determine the effectiveness of selected tree species to accumulate nutrients or metals, identify the most effective genotypes within species for nutrient or metal uptake, and quantify their remediation potential; and (3) To improve the ability of these trees to accumulate contaminants by using chelating agents, fertilizers, amendments, pH, and microbial inoculations.

The dendroremedial potential of 21 tree species: EA, EC, EG, CY, LL, CB, CW, and 14 other primarily very fast growing and widely grown species are being assessed. The *Eucalyptus* species are highly productive (height growth up to 6 m yr^{-1}), resprout after harvest, are heavy consumers of nutrients and water, and have excellent potential for efficiently extracting toxic chemicals or heavy metals from soil or water in sites contaminated with a variety of hazardous wastes.[8] EA is suitable for the frost-frequent subtropics, while EC and EG are suited to the warmer subtropics. No single *Eucalyptus* species is the most productive in all regions of Florida nor most suitable for all dendroremediation applications.

CW is relatively easy to establish and is fast growing. Its high transpiration rate and wide-spreading root system make it ideal to intercept, absorb, degrade, or detoxify contaminants while reducing soil erosion. These trees are amenable to coppicing and short-rotation harvest

Table 1
Field studies estimating woody biomass production in dendroremediation systems

Dendroremediation Application	Species	Florida Location	Description
Agriculture	EA, EC, EG.	Belle Glade	Irrigation runoff pond
Stormwater	EA, EC, EG	Tampa	Retention pond
Effluent spray	CB, CW, CY, EA, EC, EG, LL, Pine	Winter Garden	Water Conserv II sewage effluent application
Cu, Cr and As contaminated soil	CW, CY, EA.	Archer	Former CCA wood treatment facility

which maintains root vigor.[9] Furthermore, widespread root systems of this species are more effective than confined root systems in uptake of nutrients.[10]

Where possible, the most productive genotypes of the candidate species have been included in the studies. Cuttings of the most productive CW clones[11] were secured from clone banks in Mississippi and Florida. Superior seedlots of EA, EG, and CY were selected from ongoing tree improvement programs. Plantlets of three EC clones were obtained. CB seed from specific collections in central and south Florida was arranged, as was seed from three varieties of LL tested in Florida.

2.1. Field studies—Winter Garden

A sewage effluent application study (Water Conserv II), located on sandhills west of Orlando and totaling 2.8 ha, was initiated in March 1998. The experimental area utilises a split-split-block design with three replications to compare four silvicultural options (no compost or mulch; compost only; mulch only; and compost + mulch) for enhancing the uptake of sewage effluent applied at the rate of 100 l tree^{-1} d^{-1} to EA, EC, EG, CY, LL, CB, and CW.

Measurements taken at the end of the first growing season (Table 2) suggest significant culture and genetic effects in dendroremediating sewage effluent. Sewage effluent alone (E) resulted in the slowest growth and lowest survival. With compost incorporated into the sandy soil (E + C), survival and height increased by about 10% and 0.2 m, respectively. When mulching with hay was done in addition to the effluent (E + M), overall survival increased to more than 65%, but tree height was comparable to E + C. With both compost and mulch combined with effluent (E + C + M), survival and height were highest at 70% and 2.3 m, respectively.

Overall differences among species and differential species responses to cultural treatments were detected that are likely to impact sewage effluent dendroremediation effectiveness (Table 2). CB started vigorously from seed and suppressed vigorous weed competition in all four cultures, averaging at least 86% survival. CB's greatest height of nearly 3 m was achieved with the E + C + M culture. CW was not as effective in suppressing weed competition (especially in the E-only culture), tended to have the tallest trees, and reached heights averaging 3.3 m after nine months under the E + C + M culture. EA also tolerated competition poorly but with the

E + C + M culture was 1.9 m tall after eight months. EG, established as large seedlings in two cultures with C, grew best when mulched. LL survived poorly without weed control and had only 70% survival but 1.8 m height in the E + C + M culture. Under the cultural conditions and propagation practices followed in 1998, E + C + M culture of CB, CW, EA, and perhaps EG would be recommended for most effective dendroremediation, if tree size and survival are the main criteria.

Table 2
Height (H in m) and survival (S in %) responses of CB, CW, EA, EG, and LL to sewage effluent only (E), sewage effluent + mulch (E + M), sewage effluent + compost (E + C), and sewage effluent + compost + mulch (E + C + M) cultures near Winter Garden, FL

Species	Age (mos)	Culture								Species Mean	
		E		E + M		E + C		E + C + M			
		H	S	H	S	H	S	H	S	H	S
CB	8	2.31	89	2.82	86	2.47	87	2.97	95	2.63	88
CW	9	2.51	39	2.92	71	2.94	85	3.26	87	2.94	68
EA	8	1.07	33	1.54	69	1.38	43	1.93	65	1.53	52
EG	5	-	-	-	-	1.01	71	1.39	100	1.24	85
LL	7	1.79	58	1.54	69	1.69	50	1.84	70	1.71	62
Culture Mean		1.81	48	2.00	65	2.07	59	2.26	70	2.04	60

EG has shown remediation potential for wastewater applications in Florida. Annual transpiration rates of up to 165 cm of water and N and P uptakes of 170 and 330 kg ha^{-1} yr^{-1} respectively have been recorded.[12]

2.2. Field studies—Archer

Preliminary soil characterization and species screening were conducted at a former CCA wood treatment site in Archer, Florida, which covers approximately 0.4 ha including the former treating facility. Four of 18 subunits had high concentrations of metals. Triplicate soil samples were randomly collected in two of these four subunits at depths of 20 and 100 cm, mixed at the site, and preserved in polyethylene bags. In the lab, they were dried at 70°C for three days. Soil samples were prepared for analysis according to EPA method 3051 and analysed for As, Cu, Cr, Fe, Ca, Zn and K using atomic absorption spectrophotometry.

The two plots, which had high soil concentrations of As, Cr, and Cu (530, 120, and 240 mg kg^{-1}, and 330, 530, and 180 mg kg^{-1}, respectively) in 1992, had lower but still high concentrations in 1998. The total arsenic levels of 156-184 mg kg^{-1} were well above the natural background levels of 0.1-6.1 mg kg^{-1} for Florida soils. Many As-contaminated sites in Florida, such as former cattle dipping ponds, still have levels exceeding 600 mg kg^{-1}.

Six tree and five herbaceous species growing on the Archer site were collected with roots and preserved in polyethylene bags. In the lab, the plant material was washed with tap water to remove soil, dust and contaminants on the surface of the plant, preserved in brown paper covers and dried at 60°C for three days. The whole plant without separation into stems, leaves and roots was ground in Laboratory Mill model-4 and analyzed for metal concentrations using AAS.

For Cu and Cr, all species had concentrations similar to the normal ranges of corn (5-25 and 0.05-1.0 mg kg^{-1}, respectively). For As (Table 3), all trees and herbaceous species except one had similar concentrations and accumulation ratios. Brake fern (*Pteris vittata*) had As concentrations as high as 4,980 mg kg^{-1} and an accumulation ratio > 20, qualifying it as a hyperaccumulator of As. On subsequent As analysis of the leaf, stem, and root components of *P. vittata*, statistically significant (P < 0.05) differences were detected. Leaf As concentrations were 94% higher than root concentrations and 103% higher than stem concentrations.

Table 3
Elemental concentrations (mg kg^{-1}) for As, Cu, and Cr, and accumulation ratio for As, in the highest accumulating tree and annual in a former CCA wood treatment site

Plant	Arsenic		Copper	Chromium
	Concentration	Ratio	Concentration	Concentration
Boxelder	8.9 - 9.5	0.059	10.1 - 14.2	0.9 - 1.9
Brake Fern	3280 - 4980	23.69	15 - 28	1.7 - 2.3

3. FUTURE RESEARCH

A greenhouse study is underway to determine the effectiveness of high biomass-producing trees in the removal of heavy metals from contaminated soil. Concurrent laboratory experiments will determine methods to increase the bioavailability of metals for removal by trees. Chelating agents, fertilizers, amendments, pH, and microbial inoculations will be examined as techniques for improving the ability of these trees to accumulate contaminants. Promising trees and methods will then be combined in subsequent greenhouse and field studies. Additional genotypes (tested progenies or clones) of the best species will be included to expand the genotypes that could be used in future studies.

The Water Conserv II study will be monitored through 1999 to estimate the species that best take up N and P and use high amounts of effluent. These species will then be planted in plots as large as one hectare employing the cultural treatment that maximizes growth so that operational scale costs and productivity can be documented for short rotation cropping of trees that combine effluent remediation with commercial production of fuelwood, mulchwood, or even more conventional timber products. Soil and water monitoring of the original and new plots is also planned to determine the fate of nutrients and water added by the effluent and compost.

Gasification procedures for concentrating contaminants contained in trees will be initiated in late 1999. Gasification properties and yields for biomass components of promising genotypes will be determined, and solid residues will be analyzed to evaluate if indirectly heated gasifiers offer a benign method of distilling off and utilizing organics while retaining metals or other contaminants in residual char.

4. ACKNOWLEDGMENTS

The research reported here was supported in part by a Section 3011 Hazardous Waste Management State Program grant from the U.S. Environmental Protection Agency through a contract with the Hazardous Waste Management Section of the Florida Department of Environmental Protection. Additional support has been provided by Woodard & Curran on behalf of the City of Orlando.

REFERENCES

1. Baker, A.J.M. and R.R. Brooks (1989). Biorecovery, 1, pp. 81–126.
2. Bolton, H. Jr. and Y.A. Gorgy (1995). An overview of the bioremediation of inorganic contaminants, in Bioremediation of Inorganics, ed. by R.E. Hinchee, J.L. Means and D.R. Burris, Battelle Press, Columbus, Ohio.
3. Kelley, R.J. and T.F. Guerin (1995). Feasibility of using hyperaccumulating plants to bioremediate metal contaminated soil, in Bioremediation of Inorganics, ed. by R.E. Hinchee, J.L. Means and D.R. Burris. Battelle Press, Columbus, Ohio.
4. Labrecque, M., T.I. Teodorescu, and S. Daigle (1995). Effect of wastewater sludge on growth and heavy metal bioaccumulation of two Salix species, Plant Soil, 171, pp. 303–316.
5. Stewart, H.T.L., P. Hopman, and D.W. Flint (1990). Nutrient accumulation in trees and soil following irrigation with municipal effluent in Australia. Env. Pollution, 63, pp. 155–177.
6. Hasselgren, K. (1984). Municipal wastewater reuse and treatment in energy cultivation in Proc. Water Reuse Symposium 3, 26–31 August 1984, San Diego, California, Amer. Water Works Assoc. Research Foundation, Denver, Colorado, 1, pp. 414–427.
7. Alker, G., D. Riddell-Black, S. Smith, D. Butler, and A. Butler (1998). Nitrogen removal from a nutrient rich wastewater by Salix grown in soil-less system, 10th European conference and technology exhibition. Biomass for energy and industry. Proceedings of the International conference, Würzberg, Germany.
8. Rockwood, D.L. (1997). Eucalyptus—Pulpwood, Mulch, or Energywood?, Florida Cooperative Extension Service Circular 1194, 6 p.
9. Dix, M.E., N.B. Klopfenstein, J.W. Zhang, S.W. Workman, and M. Skim (1997). Potential use of Populus for phytoremediation of environmental pollution in riparian zones, in Micropropagation, Genetic Engineering and Molecular Biology of Populus, ed. by N.B. Klopfenstein et al., Gen. Tech. Rep. RM-GTR.297. Fort Collins, Colorado: U.S. Dept. of Agriculture, Forest Service, Rocky Mountain Research Station, pp. 206–211.
10. Licht, L.A. (1992). Salicaceae family trees in sustainable agroecosystems, For. Chron., 68, pp. 214–217.
11. Rockwood, D.L., S.M. Pisano, and W.V. McConnell (1996). Superior cottonwood and Eucalyptus clones for biomass production in waste bioremediation systems. Proc. Bioenergy 96, 7th National Bioenergy Conf., September 15-20, 1996, Nashville, Tennessee, pp. 254–261.
12. Pisano, S.M. and D.L. Rockwood (1997). Stormwater phytoremediation potential of Eucalyptus. Proceedings 5th Stormwater Research Conference, Nov 5-7. Tampa, Florida.

GIANT REED (*MISCANTHUS*) AS A SOLID BIOFUEL CROP: YIELD, QUALITY, AND ENERGY USE EFFICIENCY

I. Lewandowski,[a] J. Clifton-Brown,[a] and U. Schmidt[b]

[a]Institute for Crop Production and Grassland Research (340), University of Hohenheim, 70599 Stuttgart, Germany
[b]Institute of Soil Science (310), University of Hohenheim, 70599 Stuttgart, Germany

Annual yields from field trials in Germany with Giant Reed (*Miscanthus x giganteus*) of 25.8 t dry matter or 410 963 MJ ha^{-1} have been achieved. *Miscanthus* is a perennial crop requiring low inputs. It has an energy output:input ratio which ranges between 16 and 51. Annual fertilization with 50 kg N ha^{-1} is sufficient to maintain high yields. For combustion purposes the biomass quality is often improved by delaying harvest.

1. HISTORY AND PROPERTIES OF GIANT REED

Giant Reed (*Miscanthus*) is a perennial C_4 grass with origins in East Asia. In Europe field trials started in 1987 to examine the growth and yield potential of this crop. For these experiments *Miscanthus x giganteus*, a triploid hybrid, was used. Interest in *Miscanthus* arises from a number of promising characteristics:
- It has C_4-photosynthesis and consequently a high yield potential.
- The productive life of the perennial crop is estimated to be 20 years. This allows low-input crop production.
- *Miscanthus* is generally lodging resistant, allowing harvests in late winter that are essential to reduce the moisture content.
- The translocation of mineral nutrients from the shoots to underground rhizomes occurs in autumn and results in efficient use of elements such as nitrogen, since it is recycled in the plant. Translocation also reduces nitrogen contents in the harvested biomass.
- The biomass has a number of uses including production of cellulose for paper, building material such as pressed boards and as solid biofuel.

However, to date the productivity of *Miscanthus* is limited by the following problems:
- The management system for the production of *Miscanthus* as a biomass crop and the technical equipment for harvesting have not yet been fully developed.
- *Miscanthus x giganteus* is triploid and can not form fertile seeds. Micropropagation is costly and makes the crop costly to establish.

The purposes of this study were to test the performance of giant reed with respect to its biofuel characteristics: energy yield, energy use efficiency and chemical composition.

2. MATERIALS AND METHODS

2.1. Field trial and quality analysis

In all experiments the genotype *Miscanthus x giganteus* was used. The experimental sites are described in Table 1. The water supply levels are determined according to reference 1.

Table 1
Description of experimental sites

Location	Average Temperature (°C)			Level of soil water supply			Soil	Texture	N_t soil concentration
	1994	1995	1996	1994	1995	1996			
Durmersheim (DUR) (Upper Rhine Valley)	10.5	9.6	n.a.	4.5	5	4	Haplic luvisol	loamy sand	0.07 %
Ihinger Hof (IHO) (near Stuttgart)	9.0	8.2	8.0	7	7	7	Haplic luvisol	loamy clay	0.19 %
Gutenzell (GUT) (Upper Swabia)	8.3	8.9	7.8	n.a.	6.5	6.5	Gleysol	loamy sand	0.73 %

In DUR and IHO identical trials were planted in a split plot design in May 1992. The trial in GUT was planted in block design 1993. All trials were run with four replications. The plant density in all trials was two plants m^{-2}. The plots were irrigated in the year of planting. During the first two years manual weed control was employed. From the third year on weeds were not controlled. To test the effect of N-fertilization on yield and quality of *Miscanthus* in DUR and IHO the N-fertilizer was given annually at rates of 0, 50 and 100 kg $ha^{-1}a^{-1}$ when new shoots emerged. In GUT the content of mineralized N in the soil was very high. For this reason no mineral N was applied. All trials were fertilized with 22 kg P and 166 kg K $ha^{-1}a^{-1}$.

The trials were harvested twice in December and February by cutting manually one m^2 per plot at a height of 5 cm above soil level. the chemical analyses of the samples were performed according to the method described in ADDIN.[2]

2.2. Energy yield and energy balance

The energy yield in GJ ha^{-1} was derived by multiplying the yield in t fresh matter ha^{-1} with the heating value in GJ t^{-1} fresh matter. The energy balance was calculated according to the method of [3]. This method considers the energy consumption for production, transport and application of consumables (for example, fertilizer) and the fuel used by machines for soil cultivation, harvest, and transport, but it does not include the energy used in the production of the machines. Data from these plot trials were extrapolated to a field size of 5 ha. Those manual procedures in the plot trial (e.g., fertilization) were assumed to be carried out by machine. Life of the stand was assumed to be 15 yr.

3. RESULTS

3.1. Dry matter (DM) and energy yield

The yield harvested in February differed markedly between the three locations tested (Table 2). The highest average yield was harvested in IHO. This location is characterized by good water availability mainly from a nearby lake. DUR is a warmer location but the sandy soil has a very low water-holding capacity and drought occurs in summer. GUT is a location with good water supply but it is cooler than the other locations. It is important to note that the crop reaches maturity in the third growing season. In 1994, the crop at GUT was only in its second growing season, explaining the rather low yields compared to the other sites which were then in their third growing season.

Table 2
Dry matter (DM) and energy yield, averages of all treatments at the three locations at February harvest

	DUR		IHO		GUT	
	t DM ha^{-1}	MJ ha^{-1}	t DM ha^{-1}	MJ ha^{-1}	T DM ha^{-1}	MJ ha^{-1}
1994	12.3	221 136	25.8	410 963	15.32	261 497
1995	12.8	220 120	20.4	319 122	21.9	347 658
1996	9.8	167 363	23.8	362 428	22.8	376 269

Applying N fertilizer did not increase yields. The delayed harvest decreased yield owing to leaf losses and the breaking off of shoot tips. Losses recorded from December 1994 to February 1995 reached up to 26% of total dry matter in GUT (Figure 1). A further delay between February 1995 and March 1995 led to minor losses of up to 9% at IHO.

3.2. Biomass quality

Biomass quality decreases with increasing contents of water, ash, N, K, Cl and S. These substances can complicate the power plant operation or cause environmentally harmful emissions such as NO_x, SO_2, HCl, and dioxins.[2] The major tool for influencing the quality of *Miscanthus* biomass is harvest time. Delayed harvest generally leads to a significant decrease in moisture, ash, K and Cl contents (see Figure 1). Minerals like Cl and K are leached by precipitation during the winter and leaves that are high in ash and N[2] are lost owing to death and detachment.

N fertilization did not influence the biomass quality in IHO. At this location N availability was high, owing to high soil N mineralization rates. In DUR soil N-contents were lower than at the other locations, which led to lower N contents in the biomass (Figure 1). In some years the contents of water, ash, N, and K increased with increasing N fertilization. In 1995 the N concentration increased from 0.15 to 0.28% in the dry matter (DM) when 100 kg N ha^{-1} were applied. In addition, an increase in concentration of K from 0.59 to 0.88% and of ash from

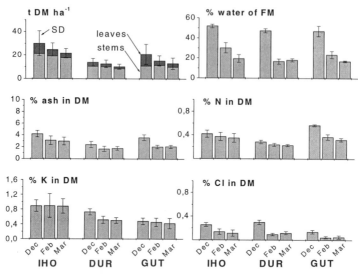

Figure 1. Effect of delayed harvest on yield, water, ash, N, K, and Cl content of *Miscanthus* biomass at three locations (DM = dry matter; FM = fresh matter).

1.9 to 2.4% in the DM were recorded. The concentration of Cl was not affected by N fertilization.

Sulphur concentration in *Miscanthus* biomass did not vary much between locations and years (Figure 2). Both S and Cl concentrations had low levels with a mean of 0.09 and 0.14% in the DM. High variation was observed in the concentration of ash, N, and K (Figure 2).

3.3. Energy balance

Output: input relations of 16 to 51 were obtained under different climatic and local conditions. Grass harvested in February produced an energy yield of 160,486 to 506,044 MJ $ha^{-1}a^{-1}$. The calculated energy consumption for annual *Miscanthus* production ranged from 7,730 to 16,075 MJ $ha^{-1}a^{-1}$. The propagation of plantlets and the provision of N fertilizer required the most energy. At an optimal production intensity, (using 50 kg N ha^{-1} and micropropagated plants at two plants m^{-2}) total energy consumption was calculated to be 9,729 MJ $ha^{-1}a^{-1}$. Under these conditions 22% went into the provision and application of N fertilizer, 37% into the propagation and planting of plantlets and 9% into the diesel fuel consumption of machines (see Figure 3).

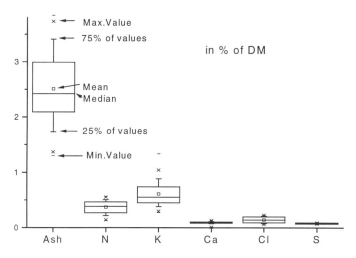

Figure 2. Variation of biomass quality of *Miscanthus* harvested in February at three locations IHO, DUR, GUT, and for the harvests of the years 1994 to 1996.

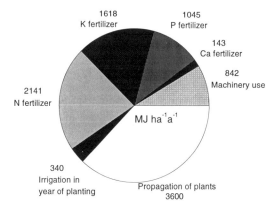

Figure 3. Distribution of energy consumption for the production of *Miscanthus*. Total energy requirement is calculated to be 9 729 MJ ha^{-1} a^{-1} when plants are planted at 2 plants m^{-2} and the N fertilization is 50 kg ha^{-1}a^{-1}.

4. DISCUSSION

Miscanthus has several characteristics that make it a suitable biomass plant. *Miscanthus* can recycle nutrients by translocation from the shoots into the underground rhizomes in late summer and autumny.[4,5] For this reason N concentration of the harvested biomass decreases until harvest. Excessive N fertilizer application does not increase yield but decreases the biomass quality. From the experiments presented here it can be concluded that an application of 50 kg N ha^{-1} is sufficient. *Miscanthus* has strongly lignified stems which are resistant to lodging, which makes delayed harvest in early spring possible. Delayed harvest allows time for leaching of unwanted substances like Cl, K, ash and for drying of the biomass.[2] As a perennial crop, *Miscanthus* needs to be planted only every 15 to 20 years. Weed control is necessary only in the first two years and there is no need for pest control at present. These characteristics make *Miscanthus* an attractive biomass crop. In Germany water supply is a yield-limiting factor despite the higher water use efficiency of C_4 crops than C_3 crops.[6]

The N fertilizer is the main energetic input in biomass production after establishment. Owing to its low N fertilizer demand (in comparison to standard C_3 crops) the energetic balance of *Miscanthus* compared to other crops is favorable. In addition, the low N fertilizer demand and the long period of soil cover in *Miscanthus* mean that the nitrogen pollution risk from *Miscanthus* fields is low. Under south German conditions no winter losses following the first growing season were recorded. The methods used in the plots for establishment of the plants (planting, irrigation, weed control) can not be transferred to large scale biomass production because they are too expensive although all these problems are tractable.

5. CONCLUSIONS

Miscanthus has a very promising future as a solid fuel biomass crop and as an industrial raw material for processes requiring fiber. Further research has to be done to find cheaper ways to establish the crop.

REFERENCES

1. Müller, U. (1997). Auswertungsmethoden im Bodenschutz: Dokumentation zur Methodenbank des Niedersächsischen Bodeninformationssystems NIBIS, Kettler Verlag, Bönen
2. Lewandowski, I. and A. Kicherer (1997). Combustion quality of biomass: practical relevance and experiments to modify the biomass quality of *Miscanthus x giganteus,* European Journal of Agronomy, 6, pp. 163–177.
3. Kaltschmitt, M. and G.A. Reinhardt (1997). Nachwachsende Energieträger. Grundlagen, Verfahren, ökologische Bilanzierung, ed. by M. Kaltschmitt, G.A. Reinhardt, Vieweg, Braunschweig, and Wiesbaden.

4. Christian, D.G., P.R. Poulton, A.B. Riche, and N.E. Yates (1997). The recovery of ^{15}N-labelled fertilizer applied to *Miscanthus x giganteus,* Biomass and Bioenergy, 12, pp. 21–24.
5. Himken, M., J. Lammel, D. Neukirchen, U. Czypionka-Krause, and H-W. Olfs (1997). Cultivation of *Miscanthus* under West European conditions: Seasonal changes in dry matter production, nutrient uptake and remobilization, Plant and Soil, 189, pp. 117–126.
6. Sinclair, T.R. and T. Horie (1989). Leaf nitrogen, photosynthesis, and crop radiation use efficiency: A review, Crop. Science, 29, pp. 90–98.

REPRODUCTIVE CHARACTERISTICS AND BREEDING IMPROVEMENT POTENTIAL OF SWITCHGRASS

C. M. Taliaferro,[a] K. P. Vogel,[b] J. H. Bouton,[c] S. B. McLaughlin,[d] and G. A. Tuskan[d]

[a]Plant & Soil Sciences Department, Oklahoma State University, Stillwater, Oklahoma 74078, U.S.A.
[b]USDA-ARS, 344 Keim Hall, University of Nebraska, P.O. Box 830973, Lincoln, Nebraska 68583, U.S.A.
[c]Crop & Soil Sciences Department, University of Georgia, Athens, Georgia 30602, U.S.A.
[d]Oak Ridge National Laboratory, P.O. Box 2008, Oak Ridge, Tennessee 37831, U.S.A.

The genetic improvement of switchgrass, *Panicum virgatum* L., is an important component of the Department of Energy's (DOE) Biofuels Feedstock Development Program (BFDP) effort to develop the species as a biofuels crop. This paper reports research in elucidating reproductive characteristics and improvement potential of the species. Recent research confirms switchgrass to be highly outcrossed, highly self-incompatible, and comprised mainly of tetraploid (2n=4x=36) and octaploid (2n=8x=72) cytotypes. Hybridization potential is very low between plants of different ploidy and relatively high between plants of the same ploidy, regardless of ecotype. Breeding results based on phenotypic vs. genotypic recurrent selection procedures are discussed.

1. INTRODUCTION

Switchgrass, *Panicum virgatum* L., is a warm-season (C_4) perennial species indigenous to the North American tall-grass prairie.[23] Presently, it serves as the DOE-BFDP model species for developing herbaceous biofuels crops.[14] An important component of the development effort is the breeding improvement of switchgrass to provide adapted cultivars with enhanced biomass yield potential for varied climatic and edaphic conditions. BFDP sponsored switchgrass breeding is now focused at Lincoln, Nebraska; Stillwater, Oklahoma; and Athens, Georgia with some additional effort in Tennessee and Wisconsin. This paper reports research in determining basic reproductive characteristics of switchgrass, characterizing genetic relationships among native switchgrass ecotypes, and identifying breeding procedures that most effectively and efficiently increase biomass production.

2. ECOLOGICAL DIVERSITY

Switchgrass is separated into two major ecotypes, lowland and upland, broadly classified on the basis of morphology and habitat preference. The lowland ecotypes are coarse stemmed, robust, tall growing and best adapted to alluvial soils. The upland ecotypes tend to be shorter in stature, finer-stemmed, and more tolerant of drier conditions than the lowland ecotypes.[18] Many sub-ecotypic forms within each of the two major ecotypes have resulted from natural selection for adaptation to specific environments differing in climatic and edaphic characteristics.[16]

3. REPRODUCTIVE CHARACTERISTICS

3.1. Chromosome numbers

Reported switchgrass chromosome numbers range from $2n=2x=18$ to $2n=12x=108$.[17] Native switchgrass populations may contain plants of different ploidy (7,15,17). Recent studies employed flow cytometry to measure DNA concentration within switchgrass nuclei and used this information to estimate ploidy level (9,10,11). Hopkins et al. (9) found 25 tetraploid ($2n=4x=36$), 2 hexaploid ($2n=6x=54$), and 14 octaploid ($2n=8x=72$) types among 41 populations. Mean nuclear DNA contents for the tetraploid and octaploid populations were 3.0 and 5.2 pg, respectively. The two hexaploid plants had 3.9 and 4.2 pg DNA nucleus^{-1}. Hultquist et al.[10,11] classified 87 switchgrass populations into tetraploid and octaploid types with mean nuclear DNA contents of approximately 3.0 and 6.0 pg, respectively. Collectively, these studies indicate that the tetraploid and octaploid chromosome numbers predominate in switchgrass. All lowland ecotypes in the Hopkins et al.[9] and Hultquist et al.[10,11] studies, with one possible exception, were tetraploid. Tetraploid, hexaploid, and octaploid levels were found among upland types.

3.2. Cross- and self-fertility

Early research indicated switchgrass to be highly cross-pollinated,[12] with outcrossing enforced by strong genetic self-incompatibility.[19,21] However, information has only recently become available on the extent to which hybridization can occur between switchgrass plants of different ploidy or ecotype. In controlled crossing studies at Stillwater, Oklahoma, the hybridization success between plants of different ploidy was very low whether the plants were of the same (0.65%) or different (0.21%) ecotype (22). Only seven hybrid plants (6x) were obtained from several thousand attempted hybridizations of 8x by 4x parent plants, and are the first such documented hybrids. Hybridization success among plants of the same ploidy level averaged 11.65% (range 0-63.8%) and was not affected by ecotype. Controlled crosses between Kanlow (4x lowland ecotype) and Summer (4x upland ecotype) have been made at relatively high frequency (up to 20%) at Lincoln, Nebraska (13). At both Stillwater and Lincoln, F_1 plants from upland x lowland crosses (same ploidy) were normal, demonstrating that interfertility exists between these cytotypes. Self-fertilization studies at both locations resulted in very low seed set presumably due to strong self-incompatibility.

4. BREEDING IMPROVEMENT

The switchgrass breeding programs seek to develop regionally adapted varieties with enhanced biomass yield potential. Associated research is geared to germplasm enhancement and preservation and improvement of breeding methodologies including application of biotechnology in breeding. The breeding methodologies being used focus on population improvement and the development of synthetic or F_1 hybrid cultivars.

4.1. Population improvement

Recurrent selection is a powerful procedure for increasing the frequency of genes conditioning quantitative traits in plant populations.[20] Its effectiveness is contingent on adequate additive genetic variation in the breeding population and the ability to accurately identify superior genotypes. Heritability estimates for switchgrass biomass production range from 0.25 to 0.78[4,21] suggesting the presence of adequate heritable variation. A goal of the breeding programs is to identify the screening techniques that most efficiently and effectively assess the breeding value of switchgrass plants.

The breeding programs are testing recurrent breeding strategies employing either phenotypic or genotypic selection. Phenotypic and genotypic selection refer to selection of plants based respectively on their own performance or the performance of their progenies. Recurrent restricted phenotypic selection (RRPS) is a "fast track" breeding procedure proposed and used by Burton[3] to annually complete one selection cycle in bahiagrass, *Paspalum notatum*. Application of RRPS for increased biomass yield in switchgrass in Oklahoma[22] and in Nebraska[8] failed to produce substantive positive results. Several major problems were encountered with the 'fast track' RRSP that mitigated against its effectiveness with switchgrass. Attempted intercrossing of selected plants by using detached flowering culms or by growing plants in the greenhouse during winter gave erratic results. Seed quantity and quality was frequently poor, though there was large plant to plant variation in seed yield and quality. Mature seed from plants intercrossed in the greenhouse was not available until April, delaying field plantings until mid-summer. A major concern was the degree of correlation between establishment year and subsequent performance of individual plants. Rank correlations between individual plant yields in yr1 and yr2, though positive and significant, were not as high as correlations based on half sib progeny performance.[22]

The Georgia switchgrass breeding program, initiated in 1996, is practicing phenotypic selection with progeny testing in lowland populations using a 'honeycomb' design.[5] The unique attributes of the moving circles of the honeycomb should minimize environmental and competition effects on individual plant biomass expression and is suitable for selection among as well as within progenies (families). Though it extends the time required to complete a breeding cycle, progeny testing appears to be the best indicator of the breeding value of switchgrass plants, and it is now a common practice of the programs. Several narrow genetic-base synthetic varieties were developed from the broader genetic-base RRPS populations in the Oklahoma program. The parent plants in most of the synthetics were chosen on the basis of half-sib progeny performance. Preliminary data from field performance trials indicate some of these to have yield advantages over standard cultivars (Table 1).

Table 1
Dry biomass yields of switchgrass experimental synthetic and standard lowland cultivars

Cultivar	Test (Yields in Mg ha^{-1})			
	96-1[1]	96-3[2]	96-5[3]	97-1[4]
		Mg ha^{-1}		
SL 94-1[5]	18.60	15.13	5.89	12.96
NL 94-1[5]	16.45	13.01	5.49	10.24
Alamo (SL)	16.72	13.56	5.59	9.07
Kanlow (NL)	14.90	11.42	4.41	7.17
LSD (0.05)	2.31	1.95	0.82	3.54
CV (%)	9.39	11.08	15.63	22.42

[1] 97-98 mean, Chickasha, OK; [2] 97-98 mean, Perkins, OK; [3] 96-98 mean, Stephenville, TX
[4] 98 mean, Chickasha, OK; [5] SL=southern lowland ecotype, NL=northern lowland ecotype

4.2. Effect of yield environment on breeding response

An important question is whether switchgrass varieties developed under high yield environments will perform well when grown in low yield (stress) environments. Research is underway in Oklahoma to answer this question. Identical sets (clonal) of half-sib progeny families are being performance tested under high and low yield environments. The results should ultimately indicate whether the yield environment under which they were developed affects the performance stability of varieties.

4.3. Hybrid switchgrass varieties

Research is underway to determine the feasibility of producing F_1 hybrid switchgrass varieties. The potential benefit of F_1 hybrid varieties is increased biomass yield due to favorable dominance and epistatic gene interactions. These types of gene interactions are frequently responsible for heterosis, or "hybrid vigor." The presence of heterotic response is indicated in first harvest year yields of three F_1 hybrid populations in comparison to their respective parents in a replicated field test

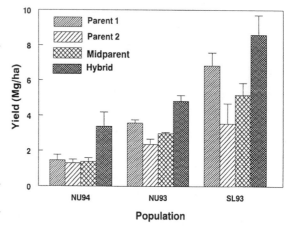

Figure 1. Biomass yields of Clonal Parents and F_1 hybrids of three switchgrass crosses.

at Stillwater, Oklahoma (Figure 1). The NU93, NU94, and SL93 F_1 hybrid populations had biomass yields 63, 44, and 66% greater than the respective mid-parent yields. First harvest year data from Oklahoma trials show other F_1 hybrid populations to have greater biomass production than standard varieties (data not shown). Production of hybrid switchgrass varieties is contingent on the parental plants being highly self-incompatible and the ability to clonally propagate the two parents in seed production fields. We believe most switchgrass plants meet the first requirement. Laboratory switchgrass propagation methods perfected at the University of Tennessee as part of the BFDP program are expected to satisfy the second.[1,2] Switchgrass hybrids are also being evaluated in the Nebraska breeding program. Seed production research at Nebraska indicates that it will probably be feasible to produce F_1 hybrid seed from space-transplanted seed production fields with plants spaced 1.1 m apart.

5. FUTURE OUTLOOK

The BFDP sponsored switchgrass breeding–genetics research during the past few years has substantially broadened our knowledge in these areas. Cytogenetic and breeding characteristics of the species are now more clearly defined. Refinements in procedures for characterizing the breeding value of individual plants and plant populations have been made. Higher biomass production potential of some initial products of the breeding programs are indicated by preliminary performance data. Molecular genetic technology has provided useful information on switchgrass genetic relationships and is expected to become increasingly important in its development as a biofuels crop. Hultquist et al.[10,11] reported a chloroplast DNA (cpDNA) restriction fragment length polymorphism (RFLP) that distinguished upland and lowland cytotypes. Gunter et al.[6] used random amplified polymorphic DNA (RAPD) polymorphisms and cluster analysis to assess the genetic relatedness of 14 switchgrass cultivars. Currently, expanded research is being considered in the areas of genetic mapping of switchgrass and the identification of genetic markers to assist in marker-aided selection. Advances in each of these areas could greatly facilitate the breeding improvement of switchgrass. We believe that biomass yield gains on the order of 5% or greater per breeding cycle are not unrealistic as refined techniques and new technology are put into practice.

REFERENCES

1. Alexandrova, K.S., P.D. Denchev, and B.V. Conger (1966). Micropropagation of switchgrass by node culture, Crop Sci., 36, pp. 1709–1711.
2. Alexandrova, K.S., P.D. Denchev, and B.V. Conger (1966). In vitro development of inflorescences from switchgrass nodal segments, Crop Sci., 36, pp. 175–178.
3. Burton, G.W. Improved recurrent restricted phenotypic selection increases bahiagrass forage yields, Crop Sci., 22, pp. 1058–1061.
4. Eberhardt, S.A. and L.C. Newell (1959). Variation in domestic collections of switchgrass, *Panicum virgatum* L., Agron. J., 51, pp. 613–616.
5. Fasoulas, A.C. and V.A. Fasoula (1995). Honeycomb selection designs, in Plant Breeding Reviews, ed. by Jules Janick, 13, pp. 87–139.

6. Gunter, L.E., G.A. Tuskan, and S.D. Wullschleger (1996). Diversity among populations of switchgrass based on RAPD markers. Crop Sci., 36, pp. 1017–1022.
7. Henry, D.S. and T.H. Taylor (1989). Registration of KY 1625 switchgrass germplasm,. Crop Sci., 29, p. 1096.
8. Hopkins, A.A., K.P. Vogel, and K.J. Moore (1993). Predicted and realized gains from selection for in vitro dry matter digestibility and forage yield in switchgrass, Crop Sci., 33, pp. 253–258.
9. Hopkins, A.A., C.M. Taliaferro, C.D. Murphy, and D. Christian (1996). Chromosome number and nuclear DNA content of several switchgrass populations, Crop Sci., 36, pp. 1192–1195.
10. Hultquist, S.J., K.P. Vogel, D.J. Lee, K. Arumuganathan, and S. Kaeppler (1996). Chloroplast DNA and nuclear DNA content variations among cultivars of switchgrass, *Panicum virgatum* L., Crop Sci., 36, pp. 1049–1052.
11. Hultquist, S.J., K.P. Vogel, D.J. Lee, K. Arumuganathan, and S. Kaeppler (1997). DNA content and chloroplast DNA polymorphisms among switchgrasses from remnant midwestern prairies, Crop Sci., 37, pp. 595–598.
12. Jones, M.D. and J.G. Brown (1951). Pollination cycles of some grasses in Oklahoma. Agron. J., 43, pp. 218–222.
13. Martinez, J.M. and K.P. Vogel (1997). Hybridization between upland and lowland types of switchgrass, Abstracts 50th Annual Meeting Society for Range Management, p. 14.
14. McLaughlin, S.B. (1993). New switchgrass biofuels research program for the southeast, in Proc. 1992 Annual Automotive Technol. Dev. Contractors Coordinating Meeting. 2-5 Nov. 1992. Dearborn, Michigan, pp. 111–115.
15. McMillan, C. and J. Weiler (1959). Cytogeography of *Panicum virgatum* in central North America. Am. J. Bot., 46, pp. 590–593.
16. Moser, L.E. and K.P. Vogel (1995). Switchgrass, big bluestem, and indiangrass, in Forages, ed. by R.F. Barnes, D.A. Miller, C.J. Nelson, Vol. 1, An introduction to grassland agriculture. Iowa St. Univ. Press, Ames, Iowa, pp. 409–420.
17. Nielsen, E.L. (1944). Analysis of variation in *Panicum virgatum*, J. Agric. Res., 69, pp 327–353.
18. Porter, C.L. Jr. (1966). An analysis of variation between upland and lowland switchgrass, *Panicum virgatum* L., in central Oklahoma, Ecology, 47, pp. 980–982.
19. Reeves, D.L. (1963). Hybridization in *Panicum virgatum* L., M.S. thesis, Kansas State University.
20. Sprague, G.F. (1982). The adequacy of current plant breeding procedures, evaluation techniques, and germplasm resources relative to future food, feed, and fiber needs. In Report of the 1982 Plant Breeding Research Forum, 11-13 Aug., Des Moines, Iowa, Pioneer Hi-Bred Internatl., Des Moines, Iowa, pp. 205–221.
21. Talbert, L.E., D.H. Timothy, J.C. Burns, J.O. Rawlings, and R.H. Moll (1983). Estimates of genetic parameters in switchgrass, Crop Sci., 23, pp. 725–728.
22. Taliaferro, C.M. and A.A. Hopkins (1996). Breeding characteristics and improvement potential of switchgrass, Proc. 3rd Liquid Fuel Conf.: Liquid Fuels and Industrial Products from Renewable Resources, Nashville, Tennessee, Sept. 15-17, pp. 2–9.

23. Weaver, J.E. (1968). Prairie plants and their environment. Univ. Nebraska Press, Lincoln, Nebraska, 276 pp.

GROWTH, WATER AND RADIATION USE EFFICIENCY OF KENAF (*HIBISCUS CANNABINUS* L.) CULTIVATED IN THE MEDITERRANEAN CONDITIONS [*]

N. Losavio, D. Ventrella, N. Lamascese, and A.V. Vonella

Istituto Sperimentale Agronomico - MiPA
Via C. Ulpiani, 5 - 70125 Bari (Italy)

The research reported in this paper determined the relationships between growth and radiative regime of kenaf cultivated with enough water irrigation to replace all of the calculated maximum evapotranspiration.

The trial was carried out during the 1996 season at Metaponto in southern Italy (lat. 40°24' N; long. 16° 48' E; altitude of 8 m a.s.l.) on clayey soil.

Cultivar "Tainung 2" was sown at 0.5 m spacing between rows and thinned to 40 plants m^{-2}. Irrigation water was applied by a localized irrigation method.

During the biological cycle, the following measurements were taken: solar radiation (both incident and transmitted to the soil), photosynthetically active radiation (both incident (PARi) and transmitted to the soil (PARt), leaf area index (LAI), and dry matter accumulation. These measurements were used to determine three parameters: extinction coefficient (k), radiation use efficiency, and water use efficiency.

The value of k was 0.35, while the values of radiation use efficiency and water use efficiency were 2.65 g MJ^{-1} and 2.91 kg m^{-3}, respectively.

1. INTRODUCTION

Industrialized countries are increasingly interested in non-food crops as economic and ecological factors have converged.

Kenaf is an herbaceous annual fiber crop that is being studied as a potential source of paper pulp. Kenaf has been reported to be three to five times as productive, per unit area, as pulpwood trees, and it produces a pulp of quality equal or superior to that of many wood pulps. Paper produced from kenaf pulp retains ink well and its high tensile strength is ideal for high-speed presses (Francois et al., 1992).

---

[*] The work is to be equally attributed to the four authors and it falls within the ad hoc project "PRisCA".

The whole kenaf plant can be utilized. The stems contain two distinct fibers: the long fibers of the cortex produces high-quality paper; the shorter wood fibers makes paper of lower quality. The plant tops, when ground, are highly digestible and can be used as source of roughage and protein for cattle and sheep (Hays, 1989).

The economic potential of this crop depends on the gradually diminishing supply of hardwoods and softwoods in the world and the increasing per capita consumption of paper and paperboard materials.

Many have studied the agronomic potential of kenaf in Australia and in the U.S.A. (Campbell and White, 1982; Bhangoo et al., 1986). Now that kenaf has been introduced into the Mediterranean countries (as a potential new crop for southern Italy), research is needed to examine the adaptability of kenaf in this environment. This work determined the relationship between growth and radiative regime of kenaf cultivated with enough water to replace all of the calculated maximum evapotranspiration

2. MATERIAL AND METHODS

The research was carried out during 1996 at Metaponto in southern Italy (40° 24' N; 16° 48' E; 8 m a.s.l.) on a silty-clay soil (a Typic Epiaquerts). Kenaf was grown on a large, well-irrigated area (0.2 ha) that replaced all of the calculated maximum evapotranspiration.

The evaporation rate from Class A pan and meteorological data were recorded hourly by an automated data-logger located in the experimental area, in order to calculate the daily maximum evapotranspiration of the crop (Doorenbos and Pruitt, 1977). The crop was fertilized with 100 kg P_2O_5 ha^{-1} as superphosphate and 60 kg N ha^{-1} as ammonium sulphate before seeding and after, as top-dressing, with 40 kg N ha^{-1} as urea.

Cultivar "Tainung 2" was sown on 28 May at 0.5 m spacing between the rows and thinned to 40 plants m^{-2}. Irrigation water was applied by a localized irrigation method.

During the biological cycle the following measurements were taken: solar radiation, continuously monitored (both incident (Rgi) and transmitted to the soil (Rgt)) and, every two weeks, photosynthetically active radiation (both incident (PARi) and transmitted to the soil (PARt)), leaf area index (LAI), and dry matter accumulation.

These measurements were used in order to determine the extinction coefficient (k), radiation use efficiency and water use efficiency.

Three sample areas of 1 m^2 (2 by 0.5 m) were harvested on 11 October from the center of plot. Plant density, crop height, green, and above ground dry matter were determined.

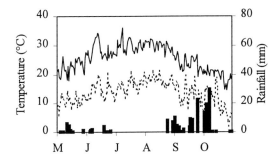

Figure 1. Daily air temperature (maximum: continuous lines; minimum: dashed lines) and rainfall (bars) recorded during the kenaf crop cycle.

3. RESULTS

The temperature and rainfall pattern during the kenaf crop cycle is reported in Figure 1. Rainfall was very low until the second ten days of August, followed by higher intensity and frequency at the end of August and September. Minimum and maximum temperatures were close to the long-term averages.

Water use, dry matter production, and the principal biometric parameters of the kenaf are reported in Table 1. The water requirements of the kenaf, cultivated with enough water to replace all of the calculated maximum evapotranspiration, were definitely high (5,234 m^3 ha^{-1}). The total and stem dry matter were equal to 15.2 t ha^{-1} and 10.8 t ha^{-1}, respectively.

The LAI and the accumulation of dry matter trends are reported in the Figure 2. Until the 37th day after emergence the values of LAI and dry matter accumulated were very low; subsequently they increased almost exponentially. The maximum value of LAI was observed 109 days after emergence, whereas the maximum accumulation of dry matter was reached only 123 days after emergence.

Table 1
Water use, dry matter production, and the principal biometric parameters of kenaf cultivated in southern Italy.

Water use (mm)	523.4
Water use efficiency (kg m^{-3})	2.9
Total dry matter (t ha^{-1})	15.2
Stem dry matter (t ha^{-1})	10.8
Basal stem diameter (mm)	14.4
Plant height (cm)	206.0

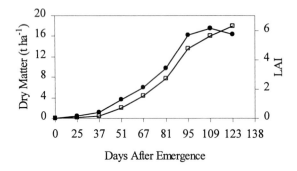

Figure 2. Trends of LAI (-•-•-•-) and dry matter accumulation (-□-□-□-).

The differing trends of such parameters could be explained by the different redistribution of dry matter between leaves and stems. From the 109th day after emergence, in fact, the biomass translocation towards the leaves decreased and instead increased towards the stems.

The climatic efficiency of the site (ε = PARi/Rgi) was equal to 0.48, a value similar to that obtained by other authors in the Mediterranean region.

The extinction coefficient (k) has been calculated by means of relationship between measured values of the transmittance (PARt/PARi) and LAI, according to the following equation (Monsi and Saeki, 1953): PARt/PARi = exp (-k · LAI).

The extinction coefficient (k) was equal to 0.35 (Figure 3), a value similar to that obtained by Rana et al. (1997) in research carried out at the same site, but lower than that obtained by Perniola et al. (1997) in research carried out in the Mediterranean environment.

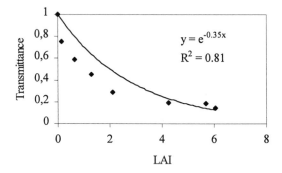

Figure 3. Transmittance as a function of LAI.

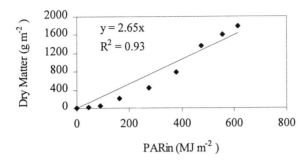

Figure 4. Dry matter accumulation as a function of the PARin.

Figure 4 shows the relationship between intercepted photosynthetically active radiation (PARin = PARi - PARt) and dry matter accumulation up to maximum value of LAI. The slope of the regression is the radiation use efficiency, 2.65 g MJ^{-1} (R^2 = 0.93). That value is similar to the value found by Perniola et al. (1997) in the Mediterranean environment and higher than those obtained in others C3 species.

The calculated radiative parameters can be used to define the radiation energetic yield of a crop (Monteith, 1977) cultivated in a certain area and in a determined period according to the following relationship: E = c · RUE · ε · A, where RUE is radiation use efficiency.

Assuming the climatic efficiency of the site (ε), the interception efficiency (A = PARin/PARi) and the energetic equivalent of the dry matter accumulated (c) equal to 0.48, 0.54 and 0.012 MJ g^{-1} respectively, the calculated radiation energetic yield of the kenaf, up to the maximum value of LAI, was equal to 0.0081. This value is equal to that obtained by Perniola et al. (1997) on the same crop in the Mediterranean environment.

4. CONCLUSIONS

This research examined interception efficiency, conversion efficiency, and radiation energetic yield of a kenaf crop cultivated with enough water to replace all of the calculated maximum evapotranspiration. The results of this work allow the following conclusions:
- the water use efficiency was equal to 2.91 kg m^{-3};
- the interception efficiency was equal to 0.54;
- the radiation use efficiency was equal to 2.65 g MJ^{-1};
- the radiation energetic yield was equal to 0.0081.

This study has confirmed the high adaptability of this crop under the Mediterranean conditions and well-watered treatment. The efficiency of water use and of photosynthetically active radiation for dry matter production was comparable with that of other highly efficient species such as corn and sweet sorghum.

REFERENCES

Bhangoo, M.S., H.S. Tehrani, and J. Henderson (1986). Effect of planting date, nitrogen levels, row spacing, and plant population on kenaf performance in the San Joaquin Valley, California. Agronomy Journal, 78, pp. 600–604.

Campbell, T.A. and G.A. White (1982). Population density and planting date effects on kenaf performances. Agronomy Journal, 74, pp. 74–77.

Doorenbos, J. and W.O. Pruitt (1977). Guidelines for predicting crop water requirements. Irrigation and Drainage, 24 (revised), FAO, Rome.

Francois, L.E., T.J. Donovan and E.V. Maas (1992). Yield, vegetative growth, and fiber length of kenaf grown on saline soil. Agronomy Journal, 84, pp. 592–598.

Hays, S.M. (1989). Kenaf tops equal high-quality hay. USDA-ARS, Agricultural Research, 37, pp. 18.

Monsi, M. and K. Saeky (1953). Uber den Lichtfaktor in den Pflanzengeseilschaften und seine Bedentung fur die Stofproduktion. Japanese Journal of Botany, 14, pp. 22.

Monteith, J.L. (1977). Climate and the efficiency of crop production in Britain. Phil. Trans. R. Soc. London, 281, pp. 277–294.

Perniola, M., E. Tarantino and F. Quaglietta Chiarandà (1997). Intercettazione, efficienza d'uso e rendimento energetico della radiazione in colture di sorgo zuccherino (*Sorghum bicolor* L. Moench var. *saccharatum*) e kenaf (*Hibiscus cannabinus* L.) in condizioni di diversa disponibilità idrica. Rivista di Agronomia, 31, pp. 590–599.

Rana, G., N. Losavio, N. Lamascese and A.V. Vonella (1997). Efficienza di conversione della radiazione solare in biomassa per il kenaf (*Hibiscus cannabinus* L.) allevato in ambiente meridionale. Rivista di Agronomia, 31, pp. 611–615.

INTRODUCTION OF *MISCANTHUS*: FIRST EXPERIENCE

V.A. Godovikova, S.B. Nesterenko, G.V. Seliverstov, and V.K. Shumny

Institute of Cytology and Genetics Siberian Department of Russian, Academy of Science, Novosibirsk, Russia

In 1992, light of an international program of research and introduction of new species of plants as source of a raw material and phytobiomass, we began to investigate endemic species of *Miscanthus*. Preliminary estimates of growth intensity, biomass accumulation, specification of the root system, and common response to soil and climate conditions open a wide ecological range of *Miscanthus* applications.

Five years of experiments show that it is possible to cultivate these perennial tall grasses in most parts of West Siberia. The absence of seeds and vegetative propagation by rhizomes require additional work to plan material production. Cell and tissue culture methods and micro clone propagation were used.

We now have collected ecotypes of three endemic species and plant regenerantes of *Miscanthus* which can be used as germplasm for breeding selection of new genotypes better adapted to local conditions.

1. INTRODUCTION

Traditionally an ornamental plant, *Miscanthus* is now one of the most promising energy plants for wide introduction as an alternative wood renewable resource for cellulose. Its potential uses are wide and interesting (Figure 1).

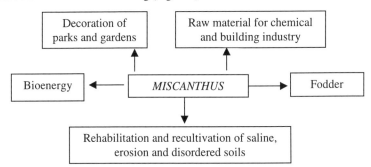

Figure 1. Schema of potential application of *Miscanthus*.

Conflicts between environmental conditions required mankind and environmental alteration in the name of science and technical progress have led to questions about the future of civilization as a whole. Considering an ecosystem as the sum of biotic and abiotic components changed and evolved together, specialists speak about the necessity of "stable ecological development."

According to data of the Russian State Soil Committee, of 185 Mha of agricultural area 53.6 Mha is eroded soil, 40 Mha is saline lands, and 5 Mha is polluted by radionuclides. Deserts are expanding in the southern regions. The total area of disturbed and contaminated soils in 1995 was 1.3 Mha (Chadisov, 1998). Of 177.8 thousand km^2 of common area in the Novosibirsk region 4.18 is eroded soil and 13.46 is dangerously eroded soil.

Soil erosion and industrial and farm wastes increasingly pollute habitat. In Russia alone 1159 stock-raising complexes and 718 poultry farms annually produce more than 160 Mt of organic wastes, which as a rule are not utilized. However, economic calculations demonstrate that the cost of using all farm waste is smaller than the damage to habitat by pollution (Gridnev et al., 1998). New governmental, ecological and economic approaches are needed to protect soil.

Together with the Novosibirsk Eco Committee of enviromental protection and natural resources we have prepared two special programs that use *Miscanthus*.

1. Covering and stabilizing sand dunes in the dry bottom of Chanu Lake.
2. Biological purification and recultivation of soils polluted by farm wastes.

Some fundamental questions must be solved before the real introduction of this nontraditional use of our agriculture crop:
- an estimate of the plant's ability to acclimatize to environmental conditions of West Siberia,
- selection on the basis of modern biotechnology and genetics methods some adaptive clones with high productivity and good tolerance similar to wild varieties in native conditions,
- develop methods of micro-clone propagation at a commercial scale.

2. DESCRIPTION OF MATERIAL

Miscanthus (or silver grass) is a tall perennial grass in the family Poaceae, subfamily Pooideae, tribe Andropogoneae. Taxonomically it is a complex genus because of wide interspecies polymorphism that contains about 14 species widespread in South-east Asia, America and the savannah of Africa (Greef and Deuter, 1993). There are three endemic species of genus *Miscanthus* in far eastern Russia on Sakhalin and Kurilsky islands, where savannah-like compositions are formed by their grouping with species of *Arundinella Radii.*, *Mulenbergia Schreb.* and *Quercus mongoli a L.* (Tsvelev, 1987). There are *M. purpurascens Anderss*, *M. sinensis Anderss*, *M. sacchariflorus Maxim.* (Voroshiliv, 1982).

We have used plant material gathered during two Far East expeditions (in 1993 and 1994). We have original plant material of all endemic species and subspecies as well as a clone of *M. "giganteus"* kindly given by Ursula Dathe (IN VITROTEC, Germany).

All genotypes were first planted in greenhouses, but in 1994 the first field experimental plantation was founded. This collection of *Miscanthus* represents unique bioproductive, decorative, cold- and salt-tolerant forms, and distinct varieties of the plant regenerantes.

3. RESULTS AND DISCUSSION

Summary data of morphogenic analysis of three species of *Miscanthus* are presented in Table 1.

Table 1
Characterization of main parameters of *Miscanthus* species

Genotype	*M. sinensis*	*M. purpurascens*	*M. sacchariflorus*
Type of development	bush	bush	creeper
Length, cm	200	100	150
Diameter of straw, mm	6	5	3
Number of shoot	718	56	150
Number of nodes	9	8	8
Leaf max, cm	72.5	64.2	60.2
Leaf flag, cm	34.2	27.6	37.5
Dry weight of shoot, g	13.4	10.1	5.5
Pollen fertility, %	65.05	85.22	94.25
Flowering	+/-	+	+

Data of the annual plant length alternation among some genotypes of *Miscanthus* presented in Figure 2 demonstrate length variation between species as well as some dependence on warm temperatures.

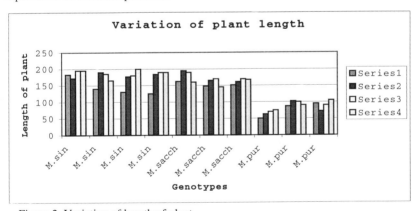

Figure 2. Variation of length of plant.

In spite of a short summer growing season in the Novosibirsk region, all plants attained their species specific biological productivity similar to the plants in natural conditions of the Far East where length of plants varied from 0.8 to 2.5 m. However, the flowering stage of *M. sinensis* and *M. giganteus* began only at the end of August, when night temperatures fell sharply and first frosts were observed.

Although *Miscanthus* has a high adaptive potential, we did not receive seeds in the greenhouse or the in field for some genotypes. Reasons for this phenomenon are as follows:
- environmental conditions of West Siberia. It's well known that flowering—one of the most important stage of plant development—is very sensitive to the photo period and temperature (Summerfield et al., 1994);
- high level of self-incompatibility of *Miscanthus* (Hirayoshi et al., 1955);
- hybrid origin of *Miscanthus*.

According to the doctrine of descent the most modern species of grasses have a hybrid origin. Evolutionary stabilization of these species occurs by amphiploidy—doubling of the chromotype, which is expressed in a high number of chromosomes, excession of genetic information, increasing of variability, and a decrease in or absence of fertility. As a rule, lowering of fertility is compensated for by creeping roots or rhizomes and vegetative propagation. Sometimes perennial hybrids are sterile only in the first flowering and they can restore some fertility in following years.

Miscanthus is well described by this schema. Recent phylogenetic research has shown the presence of DNA introgression between the large species of genus *Andropogoneae* and *Saccharinae* (Aljanabi et al., 1994).

Cytologic analyses of pollen fertility of *miscanthus* in Novosibirsk region show variation in fertility ranging from 65 to 95%. Abnormal pollen grains with five or more nuclei were observed between grains with normal structure.

Seeds gathered in 1994 and in 1997 in wild populations have numerous morphological disturbances in embryo structure—as a rule an abortive embryo or its complete absence and thereby a very low germinating ability. For various reasons all plants were sterile.

To overcome plant sterility, cell and tissue culture methods of propagation were used. There are two directions of investigations (Figure 3, 4).
- cultivation of cell and tissue to raise interspecies biodiversity and somaclone variation;
- working out the methods of micro clone propagation of the best clones for commercial production of plant material.

Modification cell and tissue culture methods were applied (Godovikova, 1995; Gavel et al., 1990; Nielson et al., 1993). We used apical and nodular meristem, immature inflorescences to initiate callusogenesis, and a morphogenic process. Plant regenerantes were received for all types of genotypes, although the process was both complicated and expensive, and it was characterized by a high level of callus necrosis and a short period of morphogenic ability. Embriogenic potential was about 15-20 plants on explant.

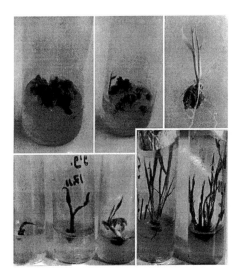

Figure 3. Cell and tissue culture of *Miscanthus sinensis*; top left—callusogenesis from apical meristem; top middle—induction of morphogenic process; top right—rooting of shoot; bottom left and right—in vitro tilling.

Figure 4. Vegetative propagation by pieces of creeper roots of *Miscanthus sacchariflorus*; top left and right, bottom left—different stages of vegetative propagation; bottom right—total field view of three endemic species of *Miscanthus*. August 1998.

As for micro clone propagation, tilling of the explant was begun in the first month of in vitro cultivation and continued during one year for all species. At our conditions the coefficient of propagation was 4 on tiller in a month:

$$N = 4^m \tag{1}$$

where N = common number of tillers, m = number of months. Under laboratory conditions during the period of in vitro cultivation from September to April 1998, we can receive, according to formula (1), about 15,000–17,000 tillers on one explant. It is substantially more than the level of vegetative propagation in natural conditions.

4. CONCLUSIONS

At present we have shown the adaptation of *Miscanthus* in West Siberia (Novosibirsk region). Using *Miscanthus* to rehabilitate and recultivate degraded soil is more promising. We have used the complex of in vitro methods for high biodiversity and received new adaptive genotypes. The results of our studies could contribute to their introduction into agriculture.

REFERENCES

Aljanabi, S.M. et al. (1994). TAG, 88, p. 933.
Chadisov, L. (1998). International Agricultural Journal, 3, p. 6.
Gavel, N.J., C.D. Robacker, and W.L. Corley (1990). Hort. Science, 25, p. 1291.
Godovikova, V.A., E.A. Moiseeva, and V.K. Shumny (1995). Proc. 2nd Biomass Conference of the Americas, Portland, USA, p. 343.
Greef, J.M. and M. Deuter (1993). Angew. Bot. 67, p. 87.
Gridnev, P.I., T.N. Kolesnikova, and T.T. Gridneva (1998). Proc. of Russian Agriculture Academy of Science, 6, p. 32.
Hirayoshi, I.K., Nishikava, and R. Kato (1955). Jap. J. Breeding, 5, p. 199.
Lafferty, J. and T. Lelley (1994). Plant Breeding, 113, p. 246.
Nielson, J.M., K. Brandtand, and J. Hansen (1993). Plant Cell, Tissue and Organ Culture, 35, p. 173.
Summerfield, R.J. et al. (1994). Euphytica, 96, p. 246.
Tsvelov, N.N. (1987). Systema slakov and its evolution, Leningrad.
Voroshilov, V.N. (1982). The Far East Plant Definition. Moskow. Ru.

GREY ALDER AND HYBRID ALDER AS SHORT-ROTATION FORESTRY SPECIES

V. Uri and H. Tullus

Institute of Silviculture, Faculty of Forestry, Estonian Agricultural University, Kreutzwaldi 5, Tartu, 51014 Estonia

This study investigated grey alder and hybrid alder plantations as short-rotation forests on former agricultural land. Biomass production was evaluated during four years after the establishment of plantations. To develop rational methods of afforestation of abandoned agricultural land with alders, experimental plantations were established where different planting material was used. Preliminary results show that biomass production of the grey alder stand exceeds that of the hybrid alder stand.

1. INTRODUCTION

During the last decade the Estonian economy has undergone drastic changes. First, agricultural production has decreased, as a result of which approximately 300,000 – 400,000 hectares of agricultural land were abandoned. Second, the price of fossil fuels has risen, which has stimulated research into the wide-scale use of wood and peat in energetics and heating management.

In turn, problems in establishing and using energy forests have come to the fore. Generally, energy forests represent plantations of fast-growing short-rotation tree species, the aim of which is to obtain the highest possible biomass production during a short time (Šlapokas, 1991; Hytönen et al., 1995). In the northern countries nearest to Estonia (Finland and Sweden), the use of willows, birches, and grey alder in energy forestry is best studied both on mineral and peat soils (Granhall, 1982; Šlapokas, 1991; Saarsalmi, 1995; Hytönen, et al., 1995; Rytter, 1996), while willows have generally proved the most fast-growing species (Saarsalmi, 1995). In Estonia, the first energy plantations of willows were founded in 1993-1995 in co-operation with the Swedish University of Agricultural Sciences (Koppel, 1995). Although grey alder generally has a smaller production capacity than some willow species and clones, it has several advantages that make it a good prospect as a short-rotation forestry species.

The natural distribution of grey alder is wide, forming 4.2% of the whole area under stands and 11.1% of the area under private forests. Grey alder is the fastest-growing local tree species, averaging growth of 7.57 m^3 ha^{-1} yr^{-1}. Therefore, natural grey alder stands have been studied as potential sources of energy in Estonia (Tullus et al., 1995; Tullus et al., 1996; Tullus and Uri, 1998). It has many advantages: it grows rapidly, is symbiotically N_2-fixing by the actinomycete *Frankia*, and has only a few pests and diseases. Decomposition of alder litter improves the soil. After cutting, a new alder generation emerges by coppicing from the root system, and thus artificial reforestation is not needed. Grey alder seedlings withstand direct sunlight and frost. Young grey alder grows faster than other tree species; however, growth slows earlier and maturity is reached at an age of 30-50 years. It can be harvested with ordinary equipment (e.g., power-saw). In Estonia, grey alder has been cultivated only recently. The first experimental man-made stand was planted on former agricultural land in the spring of 1995.

In Estonia as well as in the other Baltic countries hybrid alder (*Alnus incana* x *Alnus glutinosa*) is rather rare in nature. Considering literature data, its growth in natural conditions can be more rapid than that of grey alder or black alder (Pirag, 1962). Hybrid alders are as yet not well studied in Estonia. However, literature data suggest that they may be promising for short-rotation forestry (Granhall, 1982).

The aim of the present paper is to analyse the first results of the use of grey alder and hybrid alder as short-rotation forestry species on abandoned agricultural lands. The effect of planting stock of different origins on their survival and further production was studied as well. Dynamics and production of biomass of grey alder and hybrid alder were estimated during the first years after planting.

2. MATERIAL AND METHODS

2.1. Plantations

The experimental areas are located in Põlva county in south-eastern Estonia; they were established on abandoned agricultural land in the spring of 1995, two years after the area was last farmed. The soil was not prepared before planting. Since grey alder has not been cultivated in Estonia, a nursery-grown planting stock was not available. Three different types of transplants of natural origin aged 1-2 years were used. Among them, 62% were naturally regenerated bare-root seedlings, 33% were of root-suckering origin with a soil ball and 5% were bare-root plants of root-suckering origin. The total area of the experimental field is 0.08 hectares. A planting arrangement of 0.7 x 1.0 m was employed. Because of the high density of natural grey alder stands as well as the energy forestry experience of the Nordic countries, where primary density of 20,000 – 40,000 per ha is used, the crop was established with high primary density (Saarsalmi, 1995). In addition, 120 stem cuttings were planted with the aim to find out the propagation possibilities of grey alder by this method. The soil in grey alder as well as in hybrid alder plantations was classified as planosol, according to the FAO classification.

The experimental area of hybrid alder was set up at the same site in 1996. Biennial seedlings from the nursery were used. As hybrid alder grows higher than grey alder, a planting arrangement of 1.0 x 1.5 m was employed, which has approximately half the density as that used for the grey alder crop. The total area of the experimental plantation of hybrid alder was 0.2 ha. In either group, no weed control, fertilization or other operation was performed.

2.2. Biomass and productivity estimation

The above-ground biomass of crops was determined each year late in the summer when leaf mass was greatest. Dimension-analysis techniques (Borman and Gordon, 1984) were used to estimate the above-ground biomass and productivity of alder plantations. At both test sites the height and the diameter were measured (root-collar diameter in 1995 and 1996, and breast-height diameter in 1997 and 1998). Using a random procedure, based on height or diameter distribution, seven model trees per plantation were felled, except in the first year after planting when one medium model tree from the crop of grey alder and three model trees from the crop of hybrid alder were taken for determination of biomass. Model trees were divided into fractions: leaves, primary growth of branches, secondary growth of branches and dead branches. The stem was divided into sections according to annual increment. Subsamples were taken for estimating bark and wood proportions and their dry weight percentages. Since the size of plants in the first year after planting was small, the biomass of the above-ground part of the crop was estimated through the medium tree.

To estimate the biomass of the above-ground part of the plantation during the following years a regression equation was used:

$$y = ax^b \qquad (1)$$

Table 1
The parameters of the regression equation (1) used in dimension analysis to estimate biomass (g); level of significance $p < 0,001$ in all cases, D—diameter (cm), H—height (m), r^2—coefficient of determination, s.e.e. —standard error of estimate

Year	x	a	B	r^2	s.e.e.
		Grey alder			
1996	DH	3.059	1.406	0.991	0.12
1997	DH	3.791	1.252	0.960	0.17
1998	DH	2.503	1.655	0.996	0.06
		Hybrid alder			
1997	D^2H	3.331	0.881	0.976	0.20
1998	D^2H	4.838	0.633	0.906	0.34

The parameters of the regression equation are presented in Table 1. Annual production of stemwood, bark and branches was calculated as the difference between the masses of respective fractions for the studied year and for the previous year.

3. RESULTS

3.1. Cultivation and survival

In the autumn of the same year in which the plantation was established, the diameter of root collar, plant height and the increment of the main shoot of the previous year were measured for all plants. After the first year of growth, a check of the distributions of the characteristics showed that the distributions did not correspond to a normal distribution. Distributions were checked for normality by the χ^2–test (at significance level $p < 0.05$). The lack of a normal distribution in the case of any measured characteristics can be accounted for by the fact that plants had been gathered from different sites, and the consequent high variability was not levelled during the first vegetation period. When the difference from a normal distribution was significant, the nonparametric Kruskal-Wallis analysis of variance was used.

The difference between the plants of seed origin and the plants of sprout origin, as well as the difference between transplants with a ball and bare-root transplants of sprout origin were checked. It was found that by the autumn of the first year of growth there was an insignificant difference in the root collar diameter between the plants of sprout origin and the plants of seed origin ($p > 0.05$). At the same time, the height of plants of seed origin and their height increment were significantly larger compared with the respective parameters of plants of sprout origin ($p < 0.05$). Comparison of the plants of sprout origin revealed that diameter and height did not differ significantly between bare-root plants and plants with a soil ball, whereas they did differ significantly in increment.

During the following three years plants of different origin were compared on the basis of average biomass (Figure 1). Plants of generative origin had a significantly larger height and higher productivity than plants of root suckering origin. Comparison of plants of sprout origin showed that the height of plants with a soil ball was significantly larger.

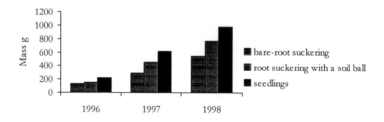

Figure 1. The average biomass of plants of different origin.

The growth of herbaceous plants after planting was extremely intensive. Despite the exceedingly intensive growth of herbaceous plants, the average survival percentage of the grey alder plantation after the first growing season was high, 93.5%, and later growth and development were good. Propagation of grey alder by stem cuttings failed, and survival after the first growing season was nil. Plants of seed origin displayed the best growth during the first years after planting, and this tendency continued in all four years of study.

By the end of the first year after planting, survival as a planting average was 94.3% in the experimental plantation of hybrid alder. The average height of seedlings (± standard deviation) was 87.7 ± 36.5 cm, root collar diameter 0.9 ± 0.29 cm and height increment 42.4 ± 24.4 cm.

3.2. Biomass production

In the autumn of the first year of planting, biomass in the above-ground part of the grey alder plantation was 370 kg DM ha^{-1}. Literature provides scanty data on such young crops. According to Rytter the mass of sprouts from root suckering can reach 1 t DM ha^{-1} in the first year after cutting (Rytter, 1996). The biomass and production of grey alder and hybrid alder crops, estimated in different years, are presented in Table 2. Comparison of biomass dynamics in the crops of grey alder and hybrid alder shows that in the first years after planting grey alder appears considerably more productive, although this difference can disappear in older stands; in this case the production of hybrid alder can exceed that of grey alder.

Because the preliminary density of crops in this experiment was different, then comparison of average biomass was based on one tree. After the first vegetation period the biomass of the above-ground part of one grey alder is 0.025 kg and 0.024 kg in the hybrid alder. In the autumn of the third year, these measurements are 0.58 kg and 0.43 kg, respectively. Consequently, the crop of grey alder is much more productive than the crop of hybrid alder. However, growth conditions are more favorable for hybrid alder because the trees grow more sparsely and there are more nutrient resources and light per tree. When comparing the biomasses of the above-ground part of grey alder with similar data provided by the literature, Saarsalmi reports that the biomass production of the above-ground part of a 5-year-old crop of *Alnus incana* (without leaves) can amount to 5.8-6.1 t DM ha^{-1} yr^{-1} depending of fertilisation (Saarsalmi, 1995). In experimental crops of grey alder, which were fertilised and irrigated, the dry mass of the current increment of the above-ground biomass, was 8 t DM $ha^{-1}yr^{-1}$ for a four-year-old crop and approximately up to 12 t DM $ha^{-1}yr^{-1}$ for a five-year-old crop (Rytter, 1996).

Table 2
Biomass and production of grey alder and hybrid alder plantation on former agricultural land

	1995		1996		1997		1998	
	Biomass $t.ha^{-1}$	Annual increment $t.ha^{-1}.y^{-1}$	Biomass $t.ha^{-1}$	Annual increment $t.ha^{-1}.y^{-1}$	Biomass $t.ha^{-1}$	Annual increment $t.ha^{-1}.y^{-1}$	Biomass $t.ha^{-1}$	Annual increment $t.ha^{-1}.y^{-1}$
Grey alder	0.37	0.28	2.68	2.45	7.61	5.80	12.27	6.74
Hybrid alder	-	-	0.16	0.14	0.94	0.86	2.72	2.20

4. CONCLUSIONS

1. Cultivation of grey alder is simple; the competitiveness of planted plants is high and plants are extremely viable, even without tending. Thus the expenses of establishment and tending are low.
2. Growth of planting material of generative origin is more vigorous than growth of plants of vegetative origin. Since this difference does not disappear even with ageing of the stand, planting material of generative origin should be used for the establishment of grey alder crops.
3. Use of stem cuttings for propagation of grey alder did not yield positive results.
4. The biomass production of grey alder on former agricultural lands is high; in the fourth year of planting the total biomass of the above-ground part is 12.3 t DM ha^{-1} and annual current increment 6.7 t DM ha^{-1} yr^{-1}.
5. Compared with grey alder, the production of the above-ground part of hybrid alder is lower; at young age the production of grey alder exceeds that of hybrid alder in similar growth conditions.

5. ACKNOWLEDGEMENTS

This study was supported by the Estonian Science Foundation grant No 3529.

REFERENCES

Bormann, B.T. and J.C. Gordon (1984). Stand density effects in young red alder plantations: productivity, photosynthate partitioning and nitrogen fixation, Ecology, 2, pp. 394–402.

Granhall, U. (1982). Use of *Alnus* in energy forest production. Report, 1982 The Second National Symposium on Biological Nitrogen Fixation, Helsinki, pp. 273–285.

Hytönen, J., A. Saarsalmi, and P. Rossi (1995). Biomass production and nutrion uptake of short-rotation plantations, Silva Fennica, 29, pp. 117–139.

Koppel, A. (1996). Above-ground productivity in Estonian energy forest plantations. Report, 1995 Energy Forestry an Vegetation Filters, Uppsala, pp. 81–86.

Pirag, D.M. (1962). The growth and structure of timber on hybrid alder in Latvia. Academic dissertation, Latvian Agricultural Academy, Elgava, Latvia. (In Russian).

Rytter, L. (1996). The potential of grey alder plantation forestry. Report, 1995 Energy Forestry and Vegetation Filters, Uppsala, pp. 89–94.

Saarsalmi, A. (1995). Nutrition of deciduous tree species grown in short rotation stands. Academic dissertation, University of Joensuu, Finland.

Šlapokas, T. (1991). Influence of litter quality and fertilization on microbial nitrogen transformations in short-rotation forests. Doctoral dissertation, Swedish University of Agricultural Sciences, Uppsala, Sweden.

Tullus, H., V. Uri, and K. Keedus (1995). Grey alder as an energy resource on abandoned agricultural lands. Proceedings, 1995 Intl. Conference Land on Use Changes and Nature Conservation in Central and Eastern Europe, Palanga, pp. 55–56.

Tullus, H., Ü. Mander, K. Lõhmus, K. Keedus, and V. Uri (1996). Sustainable Forests Management in Estonia. Planning and Implementing Forest Operations to Achieve Sustainable Forests. USDA Forest Service North Central Forest Experiment Station General Technical Report NC 186, pp. 99–101.

Tullus, H. and V. Uri (1998). Grey alder *(Alnus incana)* as energy forests in Estonia. Proceedings, 1998 Intl. Conference Biomass for Energy and Industry, Würtzburg, pp. 919–921.

POTENTIAL YIELD, YIELD COMPONENT AND BIOMASS QUALITY OF FIFTEEN *MISCANTHUS* GENOTYPES IN SOUTHERN PORTUGAL

K. Tayebi, G. Basch, F. Teixeira

Universidade de Évora, Rua Sto. Izidro, Lote 62, 2750 Cascais - Areia; Portugal

A total of 15 different *Miscanthus* genotypes (*M. Sinensis* [5], *M. Sinensis* Hybrid [5], *M. x Giganteus* [4] and *Sacchariflorus* [1]) were grown and tested under Mediterranean climate conditions. There were three repetitions of each genotype, each consisting of 49 plants per plot at two plants/m^2. All plots were irrigated during the vegetative period. The trial was carried out in 1997 and 1998, and each year two planting times, spring and autumn, were tried. Plants were harvested in autumn and spring. The parameters measured were number of shoots, height of canopy and panicle, flowering time (beginning and end of flowering), sprouting time (beginning and end of sprouting) number of panicles per plant, senescence, yield, overwintering rate and seed germination. The *M. x Giganteus* group (N°s 1 to 4) had the highest yield in both years. It was observed that yields were influenced, among other factors, by the flowering time (early, mid-early, late and very late). In all genotypes production was lower in spring than in autumn.

In both spring and autumn harvests, the moisture content was extremely high (more than 50%), which was due mostly to the existence of green leaves in autumn and rains in spring. Ash content was about 30% higher in the autumn harvest than in the spring harvest.

Biomass Production and Integration With Conversion Technologies

Biomass Resources: International

CARBON DIOXIDE FIXATION AND BIOMASS PRODUCTION PROCESS USING MICROALGAE

S. Hirata,[a] M. Hayashitani[a] and Y. Ikegami[b]

[a]Akashi Technical Institute, Kawasaki Heavy Industries, Ltd., 1-1 Kawasaki-cho, Akashi, Hyogo, Japan 673-8666
[b]Project Center for CO_2 Fixation and Utilization, Research Institute of Innovative Technology for the Earth; 2-8-11 Nishi-shinbashi, Minato-ku, Tokyo, Japan 105-0003

For the fixation of CO_2 in exhaust gas from a thermal power plant and the production of microalgal biomass using photosynthetic reaction, some laboratory-scale experiments and theoretical studies were conducted. Culture conditions and growth rates of the cells were correlated by using several strains of microalgae that captured CO_2 at high rates. The procedure for predicting cell growth from the structure and operational conditions of a photobioreactor were established. Some equipment required for the photobioreactor system was investigated and the operational cost of the system was roughly examined. It was found that this process for designing a photobioreactor not only reduced CO_2 exhausted from a thermal power plant, but also helped identify areas to produce biomass according to the locations of the plant and the required biomass.

1. INTRODUCTION

Global warming is due to the rapid increase of the concentration of CO_2 in the atmosphere. Much effort has been made to reduce the concentration of CO_2 everywhere in the world. In Japan, several government and industrial programs strive to control emissions of CO_2. As a part of this effort, a project on biological CO_2 fixation and utilization sought to develop a technology that used photosynthetic microorganisms to fix and reuse CO_2 in the exhaust gas released by a thermal power plant.

The conceptual scheme of the biological CO_2 fixation and utilization process is shown in Figure 1. Some of the very large amount of CO_2 emitted from a LNG thermal power plant is effectively fixed by microalgal cells using sunlight. To achieve this, microalgal stains that captured CO_2 at high rates and contained high-value materials were necessary. Microalgae were screened from natural environments and bred through genetic engineering methods. In this process, a photobioreactor system that used light energy at a high efficiency were also indispensable. Therefore, microalgal cultivating system equipped with solar light collecting devices were also studied and developed. This study established how to predict cell growth from the structure and operational conditions of a photobioreactor, and it evaluated the performance of a sunlight collecting and inner-irradiated type photobioreactor by experimental and theoretical calculations.

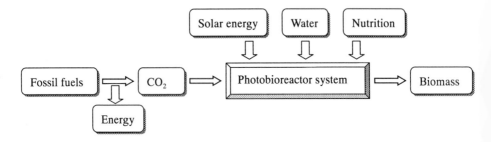

Figure 1. Concept of the biological CO_2 fixation and utilization process.

2. CHARACTERISTICS OF MICROALGAE

Three freshwater microalgal strains screened from nature in Japan were used in this study. *Chlorella* sp. UK001 was a unicellular green alga with a mean diameter of 2 to 4 µm. *Synechocystis aquatilis* SI-2 was a unicellular bluegreen alga with a mean diameter of 2 µm. *Botryococcus braunii* SI-30 was a colony-forming green alga; the colony varied in diameter from 20 µm to 1 mm according to the culture condition. It was well known that this algal strain produced large amounts of hydrocarbons in the cells (Casadevall et al., 1985). The optimum culture conditions of these microalgae are summarized in Table 1.

Table 1
Optimum culture conditions of microalgae used in this study

Strains	*Chlorella* sp. UK001	*Synechocystis aquatilis* SI-2	*Botryococcus braunii* SI-30
Temperature [°C]	35	40	25
pH [–]	6.0	7.5	5.0
CO_2 concentration [%]	10	2-5	5-10

CO_2 concentration was the mole ratio of CO_2 in the gaseous mixture supplied to the medium.

In photoautotrophic cultures of microalgae, CO_2 is the sole carbon source and light is the sole energy source. Therefore, the cell growth is significantly influenced by CO_2 concentration and light intensity irradiated to the cells. If sufficient carbonation and mixing were carried out, the CO_2 concentration in the medium was nearly uniform; however, light intensity was proportional to cell concentration and light path length. For the purpose of predicting the cell growth in a photobioreactor, the light transmittance in the cell suspension and the correlation between the light intensity and the growth rate of the cells had to be clarified.

Regarding the light transmittance $I(L)/I_0$ and light path length L at a given concentration X of the cell suspension, the following linear relationship was obtained according to Lambert-Beer's law.

$$\ln(I(L)/I_0) = -\alpha \times L \tag{1}$$

where α is an effective absorption coefficient of light. For three microalgal strains, the values of α were examined and plotted against X on a logarithmic scale as shown in Figure 2. Consequently the following empirical equation for light absorption in the cell suspension was also shown as solid lines in Figure 2.

$$\alpha = a\, X^b \tag{2}$$

where a = 157 and b = 0.715 for *Chlorella*, a = 145 and b = 0.771 for *Synechocystis*, and a = 30.1 and b = 0.902 for *Botryococcus*, respectively.

The light intensity at a distance of L from the irradiation face I(L), could be calculated by Eqs. (1) and (2) against prescribed values of I_0 and X, on the basis of the following assumptions: (a) incident light rays strikes uniformly over the entire irradiation face; (b) reflection and scattering of light rays by bubbles can be neglected; (c) cells are homogeneously suspended in the medium.

Under optimum conditions except light intensity, three algal strains were batch cultured at various light intensities. Figure 3 represents the relationships between the light intensity and the specific growth rate μ of the cells. The experimental data were fitted to an following empirical equation (Hirata et al., 1996).

$$\mu = \frac{1}{X} \cdot \frac{dX}{dt} = \mu_{max} \cdot \frac{1 - I/I_m}{1 + K_s/I} - m \tag{3}$$

where μ_{max} = 0.33, K_s =121, I_m = 8530, and m = 0.0124 for *Chlorella*, μ_{max} = 0.571, K_s =274, I_m = 5368, and m = 0.0176 for *Synechocystis*, μ_{max} = 0.142, K_s =191, I_m = 1444, and m = 0.0127 for *Botryococcus*, respectively. The specific growth rate μ(L) corresponding to light intensity I(L) was employed and the average specific growth rate in a photobioreactor was defined as follows.

$$\overline{\mu} = \iiint \mu(L)dL \Big/ \iiint dL \tag{4}$$

The growth rate and the productivity of the cells could be calculated using Eqs. 1–4 under various lighting conditions in photobioreactors. The details were described elsewhere (Hirata et al., 1996; 1997).

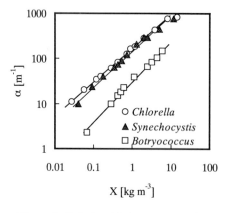 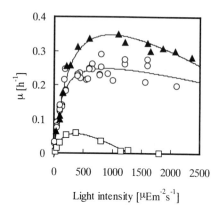

Figure 2. Relationship between X and α of three algal strains.

Figure 3. Relationship between light intensity and specific growth rate of the cells.

3. PHOTOBIOREACTOR SYSTEM FOR CO_2 FIXATION AND BIOMASS PRODUCTION

In order to fix CO_2 and produce high-value materials, a sunlight collecting and inner-irradiated type photobioreactor was developed. Figure 4 is schematic drawing of the photobioreactor. A sunlight-collecting device automatically tracked the sun and the sunlight was collected by a concave mirror. The light was transmitted via optical fibers to illumination plates equipped in the cultivation tank, and it was radiated to microalgal cells. Illumination plates made of acrylic resin were specially designed so that light intensities at the surfaces of the illumination plates were almost uniform. The light collecting and transmitting efficiency was estimated as 50%. For carbonation and mixing, a gaseous mixture containing CO_2 was injected as minute bubbles (as small as possible) from the bottom of the cultivation tank. Closed and semi-axenic cultures of microalgae could be grown using the photobioreactor system and the culture was scarcely affected by atmospheric temperature, humidity and wind. In this system, however, diffuse solar radiation could not be used and the cost of constructing the system was high.

Using a photobioreactor of 0.216 m^3 working volume, repeated batch cultures of *Chlorella* cells were grown. In these experiments, xenon lamps were used as a light source and light was supplied by optical fibers and illumination plates. Light intensity was set at 50 $\mu Em^{-2}s^{-1}$ at the surface of the illumination plates. The experimental results are shown in Figure 5. The experimental data were in fair agreement with calculated results using the procedure described above. The batch and continuous cultures of three algal strains grown at laboratory-scale (data not shown) confirmed that the calculation procedure was sufficient to predict the cell growth with accuracy under various conditions.

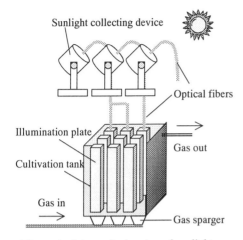
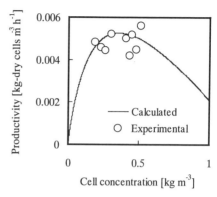

Figure 4. Schematic drawing of sunlight collecting and inner-irradiated type

Figure 5. Results of repeated batch cultures of *Chlorella* cells.

4. ESTIMATE OF CO_2 FIXATION AND BIOMASS PRODUCTION

Assuming that a light-collecting and inner-irradiated photobioreactor system was sited next to a LNG thermal power plant in Japan, the amount of fixed CO_2 and the productivity of biomass on a fine day were calculated considering light intensity change during the day. The ratio of the light collecting area to the illuminating area was fixed at 1 : 6. The calculated results are shown in Figure 6. In this system, diffuse solar radiation could not be used so that the rate of photosynthesis was very low on cloudy or rainy days. To make this system practical, other photobioreactors that receive the sun light directly must be combined with diffuse solar radiation. We are now conducting a feasibility study for the hybridization of the above photobioreactors, and we estimate that more than 50 g m^{-2} day^{-1} of CO_2 fixation rate will be available. This value is considered to be about ten times that of forests in the temperate region. If a CO_2 fixation plant could be built at an existing power plant or in the surrounding area covering about 1 km^2 (by using for example, the roof of buildings for turbine generators), about 50 ton CO_2 could be fixed per day. That amount corresponds to only 1% of the total emission of a 1-GW LNG power plant operated under the 40% load on average through a year.

The value of 1% seems to be very tiny from the viewpoint of CO_2 reduction, but we need to pay attention to the biomass production of about 35 tons per day, which corresponds to about 70% weight of CO_2, and the potential applications of it to useful substances. A large amount of fossil resources has been consumed in order to obtain the above useful materials. If we could decrease the consumption of fossil resources or forests, the reduction of CO_2 emissions is expected to be several times that of practical CO_2 fixation by the plant developed in this project. The problem is how much energy is necessary for the construction of the CO_2 fixation plant and for the operations of the process. We are now quantitatively estimating the balance of the fixed and the emitted CO_2 by the total system and have a prospect that the initially emitted CO_2 could be paid backed in ten years.

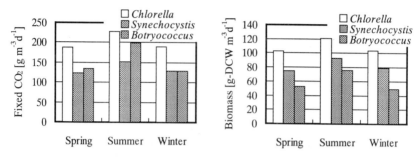

Figure 6. Calculated results of CO_2 fixation and biomass production.

5. ACKNOWLEDGMENT

This work was supported by a grant from the New Energy and Industrial Technology Development Organization (NEDO), Tokyo, Japan.

REFERENCES

1. Casadevall, E., D. Dif, C. Largeau, C. Gudin, D. Chaumont, and O. Desanti (1985). Studies on batch and continuous cultures of *Botryococcus braunii*, Biotechnol. Bioeng., 27, pp. 286–295.
2. Hirata, S., M. Taya and S. Tone (1996). Characterization of *Chlorella* cells in batch and continuous operations under a photoautotrophic condition, J. Chem. Eng. Japan, 29, pp. 953–959
3. Hirata, S., J. Hata, M. Taya and S. Tone (1997). Evaluation of *Spirulina* growth considering light intensity distribution in photoautotrophic batch culture, J. Chem. Eng. Japan, 30, pp. 355–359

INVESTIGATION OF RHIZOSPHERE MICROFLORA TO IMPROVE FOREST TREE SPECIES FOR INCREASED BIOMASS PRODUCTION IN AFFORESTATION AND SOIL RECLAMATION PROGRAMMES

A. Saravanan, R. Suresh Babu, and M. Vivekanandan

Department of Biotechnology, Bharathidasan University, Tiruchirappalli-620 024, Tamilnadu, India.

Recently rhizospere microflora have been given utmost importance in afforestation programmes. The enrichment of rhizosphere microflora is necessary for healthy growth and faster establishment of saplings in the soil. The rhizosphere zone is considered to be dynamic and enriched with nutrients providing a suitable abode for microbes supporting the growth of plants. Therefore, there is a need to understand the rhizosphere microflora of each species in the native soil (Saxena and Tilak, 1994). The present investigation promotes understanding of the rhizosphere microflora of fast-growing forest tree species, *Tectona grandis*, *Acacia nelotica* and *Casuarina equisetifolia* to improve them further in their native soil.

A significant rhizosphere microbial population (Azospirillum, Azotobacter, Rhizobia and VAMF) was observed in all the plants. VAMF colonization was highest in *Tectona grandis*. VAMF spores were also identified as *Glomous mosseae* and *Gigaspora spp*. The percentage of VAMF infection was higher in well-established trees of *Tectona grandis* (91%). There was no significant variation in rhizosphere microbial population except Azotobacter in different plant species investigated. No significant difference in pH could be observed between the rhizosphere and rhizoplane soils.

The rhizosphere microflora was always rich in beneficial microbes and contributed to faster establishment of saplings and richness of plant growth. The rhizobia occurring in the rhizosphere regions might be asymbiotic or serve as plant growth promoting bacteria (PGPB) by providing cytokinins and auxins.

Futher work is in progress in screening out rhizosphere microflora in all the species growing in problematic and non-problematic soils as well as to find out the effects of seasonal variations in the diversity of rhizosphere microflora. It is hoped that the present study will be useful in restoring forest cover that is fast deteriorating for want of proper application of microbial innocula either singly or in combination.

EVALUATION OF DECENTRALIZED POWER GENERATION FROM BIOMASS IN COMPETITION WITH ALTERNATIVE TECHNOLOGIES IN MARAJÓ ISLAND, BRAZIL

B.R.P. da Rocha,[a] C. Monteiro,[d] E.C.L. Pinheiro,[a] I.T. da Silva,[a] I.M.O. da Silva,[c] S.B. Moraes,[b] A.O.F. da Rocha,[a] V. Miranda,[d] J.P. Lopes[d]

[a]Electrical Engineering Department– Campus Universitário do Guamá – CP 8619 66075-900 Belém Pará
[b]Mechanical Eng. Department- DM/CT/UFPA
[c]Dep. of Meteorology –DM/CG/UFPA
[d]Instituto de Engenharia de Sistemas de Computação - Praça da República,93 Porto Portugal

An Integrated Methodology for use in Amazon Region was developed by MEAPA project (Integrated Methodology for Renewable Energy in Pará State, Brazil).

The use of biomass for production of electrical energy is part of the study for use of renewable energy in Marajó Island, a pilot area for the project, situated between latitudes 02° 30' S and 01°00' N; longitudes 47° 30' W and 52° 00' W, with an area of 49,606 km^2 and a population of 333,063 inhabitant.

Levelized electric cost for the electrical energy produced by different technologies using biomass was compared with those obtained for other sources such as wind, solar, and diesel. The methodology uses a geographic information system to compare several parameters for each technology and obtain maps for the different costs. Carbon emission for each technology using biomass and other sources can also be compared as well as creation of new jobs and environmental impact.

The software developed in the project can be used also to make local studies with different parameters from the ones used during the development of the project. This can help governmental agencies plan energy supplies for underdeveloped equatorial areas.

1. INTRODUCTION

Marajó is a group of fluvial islands in the Amazon estuary in the eastern part of the vast Brazilian Amazon. Its isolation in the river-inundated areas and its large area (49,606 km^2 area) and distances from continental land characterize Marajó. Preservation of local fauna and flora is very important. Sustainable development is urgent for the region and energy is the key word for that development.

Electrical energy generation for the interior of the Marajó Island is in an economic-technological impasse; demand is small, distant from production centers and of difficult access. Electric power companies, in the past State companies and now private enterprises, assist the medium cities in the Island with great expenditure of resources, financial deficit and low quality energy.

Marajó Island today needs to develop its interior and generate employment and financial resources to improve the quality of its inhabitants' life. The use of renewable energy in the decentralized production of energy can reverse today's energy picture for these communities and generate both direct and indirect employments.

Strong environmental and social benefits on the one hand, and the absence of technological alternatives for electric energy generation in the other, led to the "MEAPA Project—Integrated Methodology for Renewable Energy on Pará State, Brazil," which introduced energy and environmental planning in Marajó Island.

The objectives of MEAPA demanded the creation of a structure of information, tools, methods, institutions and people to study the integration of renewable energy in the State of Pará. More detailed information will allow the integration of politics and promotion of the best energy policies. These studies used a geographical information analysis (GIS) and produced results in an easily comprehensible form (maps). One of the largest concerns of the project was to involve and motivate a vast group of institutions and people who would be affected by renewable energy, a guarantee to a good use of the results of the project.

2. METHODOLOGY

Field surveys gathered information on perceptions of the energy problem. The mayor of the municipal district was interviewed, as were some of the population and entities that contribute to the economic sustainability in the area. Other project members collected meteorological data, confirmed locations, evaluated costs and difficulties of transport and access, and they observed the technical capacity of installation, operation and maintenance in the area.

The project was developed using a Geographic Information System (GIS) with ArcView software. Information about resources (wind, solar, biomass) and costs (such as transportation of people and equipment, and collection costs) were fed into the project data base. Other parameters such as system efficiency, CO_2 emission, creation of direct and indirect jobs were also introduced in the data base.

Data collected by the team at the Federal University of Pará (UFPa) were supplied by institutions and people in diverse fields such as meteorology, statistics, cartography, and energy.

Meteorological data included all available sources and measurements of the last 30 years. The data were statistically analyzed by the Department of Meteorology of the University of Pará, which obtained a group of information for each meteorological station on global radiation and wind speeds.

Biomass resources and possible methods of generating electricity were identified, as were sources of residues (such as sawmills, forest, agricultural residues, cattle-raising residues). Also evaluated were the energy value of the residues, the costs of concentration of residues, and transformation efficiency. To evaluate the costs of several electrification solutions we obtained a representative database of economic and technical parameters for the several technologies.

The software MEAPA developed during the project possesses a methodology based on GIS—systems of geographical information, to elaborate and accompany the plans for integrating renewable energy. This software will be used to develop a methodology that in a clearer way shows the best system of renewable energy for each place of Marajó Island, the electrification cost for the system of energy, and an electrification plan for each area of the Island.

3. RESULTS

The mapping of energy resources consists of bringing to the geographical information platform the resources at each point of the area under study.

Wind resources mapping is a quite elaborated methodology with capacity to specify data by each sector on the map, if these data are available. The model developed is applicable in areas in which topography is not an important factor, as in Marajó. Figure 1 presents wind resources. They are quite good in the NE zone of the island (classified as 7 on a scale from 0 to 10). However, in some months (February, March and April) winds are so scarce as to prevent wind systems from being completely autonomous. In the NW zone wind resources are low (3 on the scale from 0 to 10), and in some places are insufficient for energy use.

Figure 2 presents annual global radiation resources in Marajo Island. The values were obtained from interpolation values of the global radiation. Solar resources in Marajó Island can be classified as medium, in the world scale of resources (4 in a scale from 0 to 6). Values are slightly superior in the east zone of the island. Low radiation values are observed in rainy months (February, March and April).

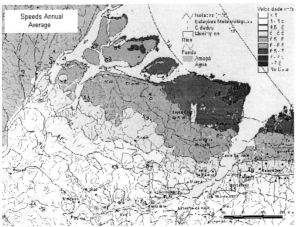

Figure 1. Map of annual average wind speed at 30 meters.

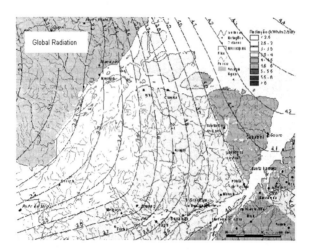

Figure 2. Map of the annual average for the global radiation (kWh/m^2/dia).

Biomass resources were calculated from vegetation maps and residues. A user may also specify a group and characteristics for these resources; they then obtain a map of energy resources, costs for its use, and the geographical characteristics of the exploration. The maps of biomass resources are conditioned by the definition of the concentration points and philosophy of biomass collection, points where the biomass will be consumed, around which are mapped areas explored to feed a certain consumption. Statistical analyses of biomass abundance were classified according to municipal districts. The map of medium values for municipal district is presented in Figure 3.

In Marajó Island transportation costs play a fundamental part in the viability of certain electrification solutions. To map these costs a software GIS tool was developed to calculate the several costs of transport in the area. Figure 4 shows a map of the minimum cost to arrive with merchandise transported to each one of the places of the study area. These transport alternatives are to calculate the viability of the systems, and they include the installation cost (system transport and technicians), operation cost (transport of fuel or biomass), and maintenance (transport of parts and technicians). The cost map depends on definitions of the parameters of consumption alternatives, the parameters of the system, economic parameters and the parameters of resources and transport costs.

Calculations showed that a hybrid diesel/wind system or a single wind system with a back-up diesel was a good solution for the NE zone of the island, which is more isolated, making the cost of transporting fuel very expensive. The main cities of NE can integrate a wind component into the current diesel systems to resolve its energy problems. To confirm the viability of these systems, for each city meteorological data should be collected and complementary studies should simulate the systems in stationary and dynamic regimes within the existing electric grids.

Biomass is also a good solution for small cities on the Island. However, when consumption increases larger areas must be searched for biomass, thus increasing the cost of the produced energy.

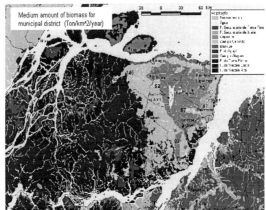

Figure 3. Map of medium resources of biomass classified by municipal district.

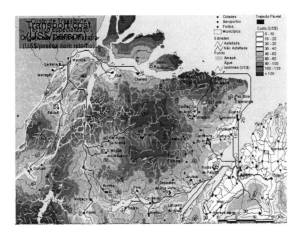

Figure 4. Map of transport costs.

4. CONCLUSIONS

The experience developed in this projects leads us to conclude that before defining an energy policy it is necessary to identify the problems and the available solutions using studies and energy plans. When the energy plans include renewable energy it is essential to map the energy resources and the viability of its use. The maps of energy resources and the tools for future studies that MEAPA supplies ensure their use by the Agencies of Energy in its management plans.

During the project we concluded that transport costs are a major cost component for Marajó Island. Because the transport methodology can attribute specific costs to each geographical element, we suggested an exhaustive evaluation of these costs by attributing specific costs to each transport road (highway, river or vegetation zone) and to each type of merchandise transported.

The methodology of the MEAPA software is to help define the areas in which each one of the systems (wind, solar, biomass and diesel) is viable. With this tool it will be possible to evaluate investment costs and consequently, to evaluate the degree of acceptability by the population. With the help of these simulations, then State institutions can use other parameters in the model, corresponding to the incentives, and to observe the alterations in the results. Electric power companies can use the methodology to do comparative studies among isolated solutions of renewable energy and conventional solutions, such as evaluating the integration potentiality in its system of production.

The results of the Project will disclose to potential buyers and users of small systems of renewable energy which system will be the best for a given place and level of consumption.

REFERENCES

1. INESC – Porto (1999). Relatório do Projeto MEAPA – Metodologias Integradas para o Mapeamento de Energias Alternativas no Estado do Pará, Janeiro.
2. Monteiro, C.D.M. (1996). Integração de Energias Renováveis na Produção Descentralizada de Electricidade Utilizando SIG, Faculdade de Engenharia da Universidade do Porto, Departamento de Engenharia Electrotécnica e de Computadores.
3. McGowin, C.R. (1996). Biomass for Electric Power in the 21st Century. Biomass and Bioenergy, 10, pp. 69–70.
4. McGowin, C.R. and G.A. Wiltsee (1996). Strategic Analysis of Biomass and Waste Fuels for Electric Power Generation. Biomass and Bioenergy, 10, pp. 167–175.

BIOENERGY TECHNOLOGIES IN CHINA: THEIR DEVELOPMENT AND COMMERCIALIZATION

Dai Lin,[a] Wang Gehua,[b] Su Mingshan,[c] Qu Feng,[d] Liu Xiaofeng[a]

[a]Energy Research Institute, State Development Planning Commission, P.R.China
[b]Energy and Environment Institute, China Rural Engineering Design Academy
[c]Tsinghua University, [d]Rural Energy Office, Si Chuan Province
[d]Energy Research Institute, State Department Planning Commission, P.R. China

By the end of 1996, the output of straw and stalks reached to 120 Mtce, and firewood 80 Mtce. An efficient cook stove for saving coal, firewood, and crop straw had been disseminated to 170 million households. To provide clean and high-efficiency energy to farmers, the Chinese government has paid great attention to the development and utilization of biomass energy. A series of bioenergy research projects and demonstration projects have been successively implemented with each five year plan. Since 1980 the Governmnent has launched RD&TD to further improve biomass and bioenergy technologies. Notable achievements include the large and medium husbandry farm biogas engineering technology, straw and stalk gasification technology for central gas supply, and refuse landfill power generation. Anaerobic digestion has been introduced at several scales with over six million rural household biogas pools producing 1.6 Gm^3/year of a biogas with about half the heating value of natural gas. About 600 large- and medium-scale biogas projects have been constructed (including biogas projects treating industrial organic waste), providing cooking fuel to 84,000 families.

1. STATUS OF BIOMASS ENERGY CONVERSION TECHNOLOGY DEVELOPMENT

The utilization of biomass energy can be roughly classified as the technology of direct burning, physical conversion technology, biological conversion technology, liquefaction technology and conversion technology for solid waste. The following three technologies have been selected since they have shown an obvious significance to rural energy improvement.

1.1. Anaerobic digestion technology for large- and medium-scale animal farms
Biogas projects on animal farms refer to the complete set of engineering equipment for the anaerobic digestion of animal excrement for the production of biogas and reducing environmental pollution. In our country biogas projects are classified into three groups according to the size (see Table 1.1.1). Generally there are five distinct process steps or stages that are required. (1) Pretreatment—including an acid treatment tank, pH adjustment tank, humidity

adjustment and solid-liquid separation. These are important in guaranteeing stable operation of the biogas project. (2) The anaerobic digester—digesters such as Anaerobic Filter (AF) and Upflow Anaerobic Sludge Blanket (UASB) have been widely used and developed into high performance systems. (3) Biogas handling—the collection, storage, transportation and distribution system, including such equipment as the gas-liquid separator, purification and desulfurization, gas storage, gas transportation and biogas combustion, which help to guarantee high-grade fuel to users and its high efficiency utilization. (4) Liquid residue post-treatment equipment—fermented residue precipitation pond, aerobic and anaerobic treatment facilities, and residue discharge facilities, that are essential in meeting the environmental regulations for water discharge to the environment. (5) Solid residue or sludge treatment system—including equipment for solid residue drying after fermentation, solid-liquid separation and the production of grain fertilizer and feedstuffs, which improves the economics of the whole project and helps to realize integrated resource utilization. Table 1.1.2 shows the common technical features of some digestion devices.

Table 1.1.1
Classification of biogas projects

Scale	Volumetric capacity of digester (M^3)	Total volumetric capacity (M^3)	Daily biogas production (M^3/d)
Large scale	> 500	> 1000	> 1000
Medium scale	50-500	50-1000	50-1000
Small scale	< 50	< 50	< 50

Table 1.1.2
Features of some common digestion devices

	Normal digester	Reactor for upper flow sludge bed (UASB)	Anaerobic filter (AF)	Reactor for anaerobic contact
Basic operating principle			Effluent	Effluent
Maximum loading (CODkg/m^3.d)	2£-3	10£-20	5£-15	4£-12
COD reduction rate (CODmg/m^3.d) under maximum loading	70£-90	90	90	80£-90
Minimum density for influent (CODmg/L)	5000	1000£-1500	1000	3000
Volumetric gas output Normal temperature (m^3/m^3.d) Mid-temperature (m^3/m^3.d)	> 0.3 0.6-0.8	0.6-0.8 1.0-2.0	£- £-	> 0.3 ~ 0.7
Power consumption	Common	Small	Small	Common
Operation & control	Comparatively easy	Comparatively difficult	Easy	Comparatively easy
Blockage	None	None	Possible	None
Land occupied	Comparatively large	Comparatively small	Comparatively small	Comparatively large
Stand striking loading	Comparatively low	Comparatively high	Common	High

As to the goal for the biogas projects in different phases, the raw materials and the digestion technologies, the following table (Table 1.1.3) is a summary:

Table 1.1.3
Historical periods of the development of anaerobic digestion

	1960s-1980s	1980s-1990s	1990s
Object	Energy	Energy + environmental protection	Energy & Environmental Protection & Economic Comprehensive Utilization
Types of the materials	Excrement + crop stalks	Excrement	Same as in the Left
Technology for digestion	Batch process	Upper fluid, anaerobic filter, anaerobic contact	Same as in the left & separation of solid and liquid
Temperature for digestion	Normal temperature	Normal temperature mid temperature	Same as in the Left
Rate of production $(m^3/m^3 \cdot d)$	0.2	0.5 (normal temperature)	> 1.0 (mid-temperature)
Equipment for aftertreatment	None	Have	Have
Biogas purification	None	Have	Have
Equipment for biogas storage	None	Have	Have

1.2. Straw gasification and central gas supply system

Biomass gasification is a thermochemical process technique that can convert solid biomass materials to convenient, clean and combustible gas through pyrolytic reactions. After many years' research abroad, a lot of experience both in theory and practice have been accumulated, and much has been achieved. As for gasifiers, there are mainly three types: fixed-bed reactors (including up-draft and down-draft gasifier), fluidized-bed gasifiers and airflow (whirlwind) beds. The status of application for biomass gasification technologies is illustrated in Table 1.2.1.

Table 1.2.1
Present status of application for biomass gasification technologies in China

Type	Diameter of Gasifier (mm)	Gasification Intensity (kg/m².h)	Power (MJ/h)	Usage	Research Unit
Up-draft	1100	240	2.9	Thermal supply for production	Guangzhou Institute of Energy Conservation, CAS
	1000	180	1.6	Thermal supply for boiler	Nanjing Institute of Forestry and Chemistry, Chinese Academy of Forestry Sciences
Down-draft	400	200	300	Tea-drying	Chinese Academy of Agricultural Mechanization Sciences
	600	200	660	Timber-drying	Same as above
	900	200	1490	Boiler gas-supply	Same as above
	700	200	900	Living gas	Research Institute of Energy, Shandong Academy of Sciences
Layer-type	2000	150	160kW	Power-generation	Ministry of Commerce and Trade
Down-draft	1200	150	60kW	Power-generation	Food Bureau of Jiangsu Province
	200	398	2-5kW	Power-generation	Guangzhou Institute of Energy Conversion, CAS
Cycling fluidized-bed	400	2000	4.2	Gas-supply for boiler	Same as above

Biomass pyrolysis gasification and centralized gas-supply system is a crop stalk-based biomass energy conversion and cooking gas-supply system in a unit of natural village, which has already been developed for gas supply by various research institutes in China. The most popular straw gasification and central gas system is Model XFF type, which includes crop stalk-based down-draft fixed-bed gasifier, gas cleaning system, and a blower suitable for a distance of 1 km. Its technical features are using down-draft fixed-bed gasifier, operating at a slightly sub-atmospheric pressure, top-open and consecutive feed. It was developed by Energy Research Institute of Shandong Academy of Science. The whole system consists of three parts: a biomass gasification station, fuel gas distributing system and indoor appliances for households. The system is compact in structure, simple and practical. The framework for technical system is illustrated in Figure 1.

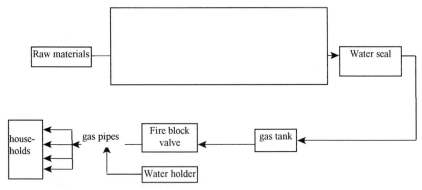

1.3. Landfill gas for power generation

Landfills help prevent the secondary pollution of a natural dump. Pollution of surface water and underground water can be prevented by controlling seepage in advance, and inflammable and explosive gas emitted from refuse can be collected by pipeline and used for power generation or for raw materials as chemicals. After it is conveyed to the landfill plant, refuse is spread as a layer 30-50 cm thick; then it is compacted and covered by a soil layer 20-30 cm thick. The refuse layer and soil layer together constitute a landfill unit. The refuse conveyed to the landfill plant every day will be compacted to be a landfill unit. A landfill layer contains a series of landfill units. A whole landfill plant contains one or more landfill layers. Whenever the landfill thickness reaches the designed height, a soil layer of 90-120 cm will cover the top layer and be compacted, and then the whole plant will be completed.

In order to solve the second pollution problem and save land resource, some technical measures have been taken and received good results as follows.
1) Raising the height of landfill plant will be useful to save land and decompose refuse to produce biogas.
2) New materials and technologies have been adopted to strengthen the impervious barrier, and a sewage disposal system is also installed to prevent pollution of water resource.
3) Refuse leaching solution and biogas collecting pipeline are built to increase the biogas collecting efficiency. Now the main uses of biogas are fuel, domestic use after it is cleaned, and power generation. The final usage is determined by the situation of the region where the landfill plant is located. The percentage of biogas used for power generation is 80% in America and 50% in Europe.

4) Compacting the refuse prevents air from entering the refuse and avoids the aerobic reaction.

Now in China, only a few units have studied municipal refuse disposal technology. For example, Chengdu Creature Institute of Chinese Academy of Science completed research on biogas production from anaerobic digestion of refuse during 1985-1986, research on the systematic disposal of municipal organic refuse during 1987-1990, and the pilot-scale test of 160 M^3 had also completed. A refuse power generation project in Hangzhou is a pilot project with a designed capacity of 6,000,000 m^3, which had filled 1.4 million ton of refuse by the end of 1995. Its square measure is 160,000 m^2, designed in Canada, to use collected biogas for power generation. The installed capacity is 1,520 kW and annual operation time is 95% of total time. The annual power output is 12.7 GWh.

2. ASSESSEMENT AND PROSPECT OF COMMERCIALIZATION OF SELECTED TECHNOLOGIES

2.1. The main considerations of technology commercialization

Technology commercialization, which means that technology will be sold or transferred in the market as a product must have four attributes:
- technology characterization including maturity, stability, reliability, safety and lifecycle considerations;
- economic feasibility;
- social acceptability;
- performance and warranties.

Based on the four considerations above, the commercialization of a technology can be illustrated at different stages: (1) R & D; (2) pilot and demonstration; (3) pre-commercial stage, which requires some incentives for its application; and (4) fully commercial, in which stage the technology is well developed for a market-oriented base.

2.2. Assessment and prospect of selected biomass energy conversion technologies

The considerations of technology commercialization metioned above can be summarized in Table 2.2.1. The prospect of commercializtion of three selected technologies can be analyzed based on Table 2.2.1.

- Large-medium biogas projects have already met the basic criteria for commercialization and can be viewed as pre-commercial. Because it increases environmental quality, the biogas technology will have bright prospect in the future. Key improvements that would increase the commercialization potential include increasing the gas yield rate, developing a commercially viable post-treatment system, and improved equipment quality and construction practices. By

Table 2.2.1
The main biomass energy conversion technologies in China

	Anaerobic digestion in husbandry farm	Biomass gasification for gas supply fed by straw	Landfill- biogas - power generation
Technical Characterization	Multi-functions on energy production, environment sounded, and comprehensive disposal Sufficient Gas output Considerable effect on pollution reduction Economic benefit from comprehensive utilization Increased reliability of main equipment	Simply applicable technology Gas Quality basically meet the requirement for residential use Lower heat value gas Specific designed and manufactured gas stove Tar problem remained	Research on landfill technology Some pilot landfill farms No specific biogas generator manufacture, using the remolded diesel engine Good environmental effectiveness
Feasibility	FIRR*: 4.53%-17.40%	FIRR: ~ 6%	FIRR: ~14%-15%
Social acceptability	Good	Good, but need the subside from government	Good, but need to reduce the investment
Performance and Warranties	Stable and good	Stable, but not meeting the State standard of gas supply for urban residential	
Comment on commercial prospective	Well developed Technology Economic feasibility depend on the extent of comprehensive utilization	Need more work on technology improvement Gas price is the main factor for its feasibility Still at demonstration stage	Need technology instruction Attractive economic feasible Still at demonstration stage

* FIRR: refers to the Financial Rate of Return, which is the index in the measurement of financial feasibility, and the discount rate when annual present value of net cash flow accumulated to zero in the project lifetime.

reducing investment costs, strengthening the technology guarantee and service system, and increased use of standards and regulations it will be possible to create favorable conditions for the complete commercialization of biogas projects.
- In the past, the efficient stoves had been developed, since farmers used crop straw and stalks as their cooking fuel. Economic growth has now allowed many farmers to use high grade fuel such as LPG. In many rural areas in China, more and more biomass gasification systems have been developed for absorbing the surplus of straw and stalks and solving the environmental pollution caused by directly burning straw in the field. Straw gasification and gas supply technology is one of the fastest biomass energy utilization technologies in recent years, because it is suitable in Chinese rural areas and thus has a broad base for development. However, this technology is not economically mature, it needs more practical experience, and some problems related to safety and stable operation still need to be resolved. R&D, experiments, and demonstrations should be further strengthened to perfect the technology, and using market mechanisms for construction and operation should be summarized and discussed to promote the technology into market as soon as possible.
- The technology of landfill gas for power generation is still in the primary period of development and there is still a long way to go before commercialization, but the technology is developing strongly and the prospective market is large.

REFERENCES

1. Biomass Energy Conversion Technologies In China (1998). Development And Evaluation, MOA/DOE Project Expert Team, 1998, China Environmental Science Press.
2. Design for Market-oriented Development Strategy of Bioenergy Technologies in China (1999). MOA/DOE Project Expert Team, 1999, China Environmental Science Press

Note: This paper describes results of a joint research project, Evaluation of Commercialization of Biomass Energy Conversion Technologies And Their Market Oriented Development Strategy, between the Ministry of Agriculture, P.R. China, and DOE of the U.S.A.

GEOGRAPHIC INFORMATION SYSTEMS (GIS)-BASED ASSESSMENT OF RICE HULL ENERGY RESOURCE IN THE PHILIPPINES

Sergio C. Capareda

Associate Professor, Agricultural Machinery Division, Institute of Agricultural Engineering, College of Engineering and Agro-Industrial Technology, University of the Philippines at Los Baños, College, Laguna, Philippines 4031

Geographic Information System (GIS) was used as a tool in encoding available rice hull production data in the Philippines to determine the exact location of rice hull resources, the potential electrical power that could be generated given a certain radius of coverage, and the trend in yearly production.

Data from the Bureau of Agricultural Statistics (1997), the National Statistics Office (1994), and the National Food Authority (1997) were used to calculate and validate available rice hull resources. Maps from the National Mapping and Resource Information Administration (1997) were digitized and the available rice hull resource was estimated for each municipality in the top producing regions. In 1996, an estimated 2.3 million metric tons of rice hull produced over a wide geographic range with an energy equivalent of about 970 MW. Central Luzon, Western Visayas and the Cagayan Valley are the regions with the largest rice hull resource. There is sufficient rice hull resource in the Philippines to merit the development of technologies for thermal energy production.

1. INTRODUCTION

Because rice is the staple crop of the Philippines, rice hull is an abundant waste. Rice hull is only minimally used at present. Disposal of rice hull is a problem as many municipalities have banned open dumping and burning. The potential for converting this waste into energy and power meshes well with a government program to provide electricity to many towns without power. More than 2.3 million metric tons of rice hull are produced each year. The country's rice production is still inadequate for the population, and thus the volume of rice hull produced is expected to increase in the near future with the expected increase in rice production. At present, yield gaps vary from 5.6-8.2 tonnes/hectare during the dry season and 4.0-5.7 t/ha during the wet season (Lansigan et al., 1996). The country maintains 3.1 million ha of land planted to rice.

2. OBJECTIVES

The main goal of this study is to estimate the rice hull resource of the country for possible power generation and waste utilization. The specific objectives include the following:
- to assess the magnitude of the country's rice hull resource by municipality, province, and region;
- to document the rice hull resource in a GIS format and determine the exact location and concentration of rice hull areas; and
- to estimate the available power that could be generated, at what locations, and at a given radius of coverage.

3. METHODOLOGY

3.1. Source of primary data
Data from the Bureau of Agricultural Statistics (BAS) were the main source of information for the rice hull resource. Each year the BAS reports the annual rice production. By using a conservative estimate of 20% rice hull waste (Kaupp, 1984), the amount of rice hull produced can be calculated. This amount was validated by data from the National Statistics Office (NSO), which conducts regular surveys on rice acreage and productivity per municipality. The magnitude of rice hull wastes was further verified by the rice hull output from rice mills in the country as reported by the National Food Authority (NFA).

3.2. Validation and encoding
Validated rice hull production data were encoded in Microsoft-Excel. Maps from the National Mapping and Resource Information Administration (NAMRIA) were digitized using AUTOCAD^{TM} (Release 14). The production data and exact geographic location were imported in IDRICI (Eastman, 1992) on a GIS format.

3.3. Maps of resource and distribution
Color-coded maps were developed for the top producing regions in the country showing the power potential per municipality. A sample output is presented in this work.

4. RESULTS AND DISCUSSION

4.1. Resource generation and power potential study
Table 1 presents the breakdown of rice hull resource in the top three producing regions: Region III, VI and II. The highest provincial production represents only 7.92% of the country's total output. The rice hull output from these three regions represents nearly half of the total output of the country. While there are provinces with high production, the widespread distribution of the rice hull waste is evident. This shows the potential to develop large-scale power plants in some selected regions (i.e., Nueva Ecija, Isabela and Iloilo) but more important, the potential for small-scale power plants far exceeds that for large scale

ones. The maximum power potential from rice hull in the country was calculated to be 970 MW. The maximum output of the top three regions was estimated at 413 MW.

Table 1
Rice hull resource in the top two producing regions in the country[1]

Region/Provinces		Rice hull	% in Country	% in Region	Maximum Power
Region III		(MT/yr)			(MW)
1. *Nueva Ecija*		178,837	7.92	47.36	77
2. Tarlac		64,684	2.87	17.13	28
3. Bulacan		50,259	2.23	13.31	22
4. Pampanga		48,356	2.14	12.81	21
5. Bataan		22,003	0.98	5.83	9
6. Zambales		13,478	0.60	3.57	6
	Sub-total	377,617	16.73	100.00	162
Region VI					
1. *Iloilo*		121,785	5.40	41.30	52
2. Negros Occidental		59,585	2.64	20.21	26
3. Capiz		47,771	2.12	16.20	21
4. Antique		36,490	1.62	12.37	16
5. Aklan		22,131	0.98	7.50	10
6. Guimaras		7,123	0.32	2.42	3
	Sub-total	294,886	13.07	100.00	127
Region II Cagayan Valley					
1. Cagayan		75,091	3.33	26.07	32
2. *Isabela*		174,110	7.72	60.45	75
3. Nueva Viscaya		29,330	1.30	10.18	13
4. Quirino		9,489	0.42	3.29	4
	Sub-total	288,020	12.76	100.00	124
Philippines		2,256,714	42.56		970

[1]Source: Bureau of Agricultural Statistics, 1997.

4.2. Rice hull production from major rice millers in Region III

Table 2 lists the top 10 rice millers in Region III, their capacity and the maximum theoretical power each mill could derive from rice hull. The highest capacity mill can generate 6.5 MW of power at 50% efficiency. During peak season, these mills operate 24 hours per day.

4.3. GIS map of the largest rice hull producing region

Figure 1 shows an example of a GIS map resulting from the resource generation and power potential study for the top-producing region (Region III). Other important data, such as the total number of rice mills and power potential were layered onto the GIS map. Only a few provinces have a power capacity exceeding 5 MW. Majority of the provinces have power potential of less than 1 MW. Thus, for the installation of a large-scale power plant, the

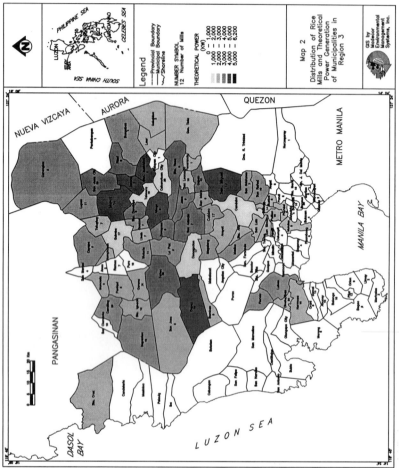

Figure 1. Example of a GIS map of a region in the Philippines with the largest rice hull production.

combined output of several provinces is required. This could pose a problem in transporting the raw material.

Table 2
Top ten rice millers in Region III (Central Luzon)[2]

Name and Location	Mill Type	Capacity (bags/hr)	Capacity (kg/hr)	Theoretical Power (MW)
1. Purina Phils., Inc. Pulilan, Bulacan	Grinder	690	3,450	13
2. Eleuterio Violago San Jose City, Nueva Ecija	Cono	194	970	3.7
1. Masagana Feeds Turo, Bucaue, Bulacan	Grinder	150	750	2.8
2. Valentina Crisostomo Gapan, Nueva Ecija	Rubber Roll	141	705	2.7
3. Ysaac, Tanjutco Guiguinto, Bulacan	Cono	136.65	683	2.6
4. Mariposa Rice Mill Mc Arthur Highway, Tarlac	Cono	110	550	2.1
5. Eduardo Manalastas Gapan, Nueva Ecija	Rubber Roll	106	530	2.0
6. Aurelio Coronel Munoz, Nueva Ecija	Rubber Roll	98.25	491	1.9
7. Victoria Cereal Center, Inc. Victoria, Tarlac	Cono	90	450	1.7
8. Jose Renato Lim Ligtasan, Tarlac	Cono	90	450	1.7

[2]Source: NFA, 1997

5. CONCLUSIONS

Important findings of the study include the following:
1. Large-scale power plants fueled by rice hull could be installed in some regions of the country, but the widespread distribution of the rice hull in most regions favors the installation of small-scale and decentralized power plants.
2. The GIS-based format makes it easier to assess the resource and power generation in the area. For a certain site of installation, the magnitude of the available resource could easily be assessed given a certain radius of coverage.
3. The market potential for power generation from rice hull is very attractive considering the current problem of disposal of these wastes and the lack of electricity in many regions of the country.

The data presented here call for the need to develop technologies suited to utilize rice hull as a valuable resource.

REFERENCES

1. Bureau of Agricultural Statistics (1997). Agricultural Production Data for 1996, Quezon City, Philippines.
2. National Statistics Office (1994). Census of Agriculture for 1991, Manila, Philippines
3. National Food Administration (1997). Data on Rice Mills by Type, Quezon City, Philippines.
4. National Mapping and Resource Information Administration (1997). Map of Region III (Central Luzon), Fort Bonifacio, Metro Manila.
5. Lansigan, F.P., B.A.M. Bouman, and P.K. Aggarwal (1996). Yield gaps in selected rice producing areas in the Philippines, in SARP Research proceedings: Towards Integration of Simulation Models in Rice Research, ed. by P.K. Aggarwal, International Rice Research Institute, College, Laguna, Philippines, pp. 11–19.
6. Kaupp, Albrecht (1984). Gasification of rice hulls: Theory and practice, Deutsche Gesellschaft fur Technische Zusammenarbeit (GTZ), GmbH, Eschborn, Germany.
7. Eastmen, J. Ronald (1992). IDRICI Users Manual, Clarke University Graduate School of Geography, Worcester, Massachusetts, 01610, U.S.A.

MALAYSIA'S BIOENERGY UTILISATION SCENARIO*

K.O. Lim

School of Physics, Universiti Sains Malaysia, 11800 Penang, Malaysia

Of late, Malaysia's economy has emphasized manufacturing activities more than agricultural activities. Even so nearly 6 million ha of a total land area of about 33 million ha are planted with crops. The major crops cultivated include oil palm, rubber, paddy, coconut, cocoa and some sugarcane. In addition, forests are being logged. Both practices generate large quantities of biowastes. Presently Malaysia consumes roughly 340 million boe of energy per year. Of this amount 14% is contributed by biomass. However, of the total amount of biowastes generated, roughly 24.5% are used for energy purposes while the rest are wasted. They are either left to rot in the fields or burnt as a means of disposal. If all of the unused biomass were to be harnessed for use as energy, then the contribution of biomass to the nation's energy consumption can be raised to about 59%, a figure that is indeed attractive and therefore should be given serious attention.

1. INTRODUCTION

Even though manufacturing activities are now given greater emphasis as opposed to agricultural activities, nearly 6 million of Malaysia's total land area of 33 million ha are still planted with crops. The major crops cultivated include oil palm (2.567 million ha), rubber (1.714 million ha), paddy (0.639 million ha), coconut (0.249 million ha), cocoa (0.235 million ha) and some sugarcane.[1] In addition, logging of forested areas is also being carried out. Large quantities of biowastes are generated. This paper reports estimates of the amount generated as well as the current status of use of these biowastes.

2. RESULTS OF THE STUDY

The major plantation crops (oil palms, rubber, paddy, coconut and cocoa) constitute more than 90% of the total area that are cultivated with crops. In this report sugarcane residues,

---

* This study was supported by an IRPA grant (No.190/9605/2802) from the Ministry of Science, Technology and Environment, Government of Malaysia.

2.1. Oil palm cultivation

In 1996, it was estimated that 8.04 million tonnes of crude palm oil and 1.13 million tonnes of palm kernel oil were produced from an area of 2.57 million hectares, of which 2.326 million ha have mature trees.[1] In Malaysia, the oil palm tree is cultivated for its oil.

To harvest the ripe fresh fruit bunches (FFB) palm fronds may have to be cut. This pruning process generates large quantities of biomass. Pruning also provides an easier assessment of fruit ripeness and easier access for pollination. The cut fronds are presently left to rot on the plantation grounds. The process of extracting oil from the FFB results in the production of both solid and liquid wastes. The solid wastes are in the form of empty fruit bunches (EFB), fibres from the pericarp and mesocarp of the fruits and shells from the nuts. The liquid wastes produced called palm oil mill effluents (POME) is the result of the extraction process which requires large quantities of water. Palm trees are replanted after 25 to 30 years. The replanting activity also generates large quantities of biomass, most of which are presently either left to rot or burnt on the plantation sites.

We estimated that the dry lignocellulose biomass yield (pruned fronds, EFB, fruit fibres, shells and replanting wastes) from oil palm plantations is 20.336 dry tonnes per ha per year with an energy content of 62.45 boe per ha per year.[2] The extraction of oil from FFB requires electricity and steam. Currently the energy requirements of all palm oil mills in Malaysia are met by the use of fibres and shells where cogeneration is widely practised. In some mills there is an excess of shells which are then sold to others that further refine the crude palm oil.[3] As such of the 62.45 boe that are available per ha per year, roughly 16.01 boe are currently being used as a source of energy.[2] The EFB are normally incinerated.

From the data provided by palm oil mills, we estimated that in 1996 the total amount of POME produced was about 28.46 million m^3. This amount, if digested to produce biogas, will result in an energy production of 2.95×10^6 boe. The data that we gathered from reports is that the current total amount of biogas utilised is about 2.16×10^4 boe per year. This is less than 1% of the 2.95×10^6 boe that were available in 1996.

2.2. Rubber cultivation

Wastes generated by the rubber industry comes from three principal sources. The first is biomass generated in the fields. This includes fallen branches and twigs, shed leaves and rubber seeds. Presently most of these are left to rot on the plantation grounds though some branches and twigs are collected for use as domestic fuel. This practice provides nutrients to the fields and as such one should perhaps not attempt to gather them for other purposes. Furthermore the shed leaves and seeds are scattered all over the plantation grounds, thus making collection uneconomical.

The second source of wastes consists of effluents that result from latex processing. The energy potential of these effluents, if converted to biogas is some 210,000 boe per annum.[3] However this potential remains untapped.

The third source of biowastes is rubberwood, where large quantities become available during replanting activities. Lim et al.[2] estimated that the amount should average 3.47 million dry tonnes per year with an energy content of about 11.12 million boe. Hong and Sim[4] reported that roughly a third of the total supply of rubber wood is still not being used. These are usually discarded and burnt on the plantation grounds. Of those consumed Lew[5] reported that about 67% are being used for energy purposes. This would imply that annually Malaysia uses 1.55 million dry tonnes of rubber wood as fuel with an energy content of 4.967 million boe. The remaining 0.76 million dry tonnes per annum are used for the production of other value-added products.

2.3. Paddy cultivation

In 1996, it was estimated that 2.128 million tonnes of paddy were produced from 639,000 ha of land.[1] Roughly 23% of paddy is husk. Thus the amount of husk produced is 489,440 tonnes with an energy content of about 1.025×10^6 boe. The ratio of straw production to grain yield under local cultivation practices is about 1.2.[3] Thus 1.064 million tonnes of straw were produced in 1996 and the energy potentially available from straw is about 2.541×10^6 boe for 1996. Thus the energy potential on a per ha basis of the above ground wastes from paddy cultivation is about 5.58 boe per year. Attempts have been made to use rice husks for energy but so far with little success. Husks have also been processed into value-added products such as fertilisers, while straw is used for mushroom cultivation. Even so most of the husks and straw are still disposed of by burning or being left to rot in the fields as well as in the vicinity of the rice mills.

2.4. Coconut cultivation

Wastes from coconut cultivation can be classified into three categories, namely fronds and debris that are shed throughout the year, wastes generated by the processing and consumption of the fruits and wastes generated during replanting. Replanting, however, is not significant now. Lim[3] estimated that the dry weight of shed fronds and debris is about 0.583 million tonnes whose energy potential is about 1.742 million boe. Though some fronds and debris are left in the fields, a large proportion is used as domestic fuel by villagers and some are used for the manufacture of articles such as brooms. Lim and Rugayah[6] estimated that in 1995 about 59% of poor rural households used mainly coconut fronds as fuel. Thus a total of about 0.528 million tonnes of coconut fronds were used. This works out to be some 90% of the total amount of fronds generated.

From the data of Lim[3] it is estimated that in 1995, 0.747 million tonnes of dry husks were generated while the corresponding figure for dry shells was 0.374 million tonnes. Hence the amount of energy potentially available from these two sources of wastes is 1.994 million boe and 1.122 million boe respectively. It is estimated that 70% of shells produced are used as fuel in copra kiln dryers.[3] An unknown amount is also used for charcoal and coconut shell flour production and as kindling material. Husks, on the other hand, are used as kindling material, as filling material for seats and mattresses, as material for the manufacture of brushes, as mulch or are just left in the fields to rot. The exact amount used for each of the above is difficult to estimate but a substantial amount of husks is used as mulch or is just left to rot in the fields. Since most coconuts are planted by smallholders the potential for large scale gathering of the unutilised shells and husks for further energy use may not be economical.

2.5. Cocoa cultivation

Due to a decline in planted area, from 452,000 ha in 1991 to 235,000 ha in 1996,[1] we can assume that all the present hectarage are of mature trees which traditionally require pruning. Pruning is done to keep mature trees at about 3-4 m so as to enable easy access for spraying and fruit harvesting. Lim[3] reported that pruning generates about 25.2 tonnes of dry organic matter per ha per year with an energy content of about 71.7 boe. Thus the total amount of pruning biomass generated in 1996 was about 5.92×10^6 tonnes whose energy content is roughly 16.8 million boe. Currently pruning wastes are left to rot in the fields. Lim[3] estimated that the average yield of dry cocoa fruit husks is about 150 kg per ha per year which has an energy content of about 0.36 boe per ha per year. As for the pruning wastes and husks generated in 1996, about 17 million boe were potentially available for use. However their potential is currently not being harnessed.

On replanting, Lim[7] estimated that the dry organic matter generated from one tree is about 48 kg. Thus the amount of biomass generated at replanting is about 57.6 tonnes per ha; the energy content of which is about 163.8 boe per ha. Since cocoa trees are normally replanted after about 25 years, the energy potentially available from replanting wastes works out to be about 6.55 boe per ha per year. This potential is similarly not being harnessed.

2.6. Sugarcane cultivation

The areas under sugarcane cultivation has stagnated (18,600 ha in 1976; 25,300 ha in 1980 and 18,000 ha in 1997[1]). Malaysia currently produces 100,000 tonnes of sugar per year but the country requires 700,000 tonnes. One tonne of sugar is produced from 10 tonnes of cane and 30% by weight of the cane end up as bagasse. Thus the total amount of bagasse produced in the country is 300,000 tonnes per year and the energy potentially available from the bagasse is 0.421×10^6 boe per annum. Presently all of the bagasse is used as boiler fuels in the mills operating in the country.

Lim[8] reported the ratio of dry weight of leaves and cane tops to the dry weight of canes is 0.685. As such on an annual basis the energy potential available from the leaves and cane tops is 0.298×10^6 boe. The current practice is to burn the leaves and cane tops in the fields before the canes are harvested. This practice not only is harmful to the environment, it also wastes potential valuable biofuels.

2.7. Logging and timber industries—biomass generation and utilisation

From studies at logging sites, it is estimated that on the average about 15% of a felled tree, i.e., its top branches and leaves, are not removed to be processed into timber but are left as wastes on the logging sites. For each cubic meter of sawlogs produced the amount of unwanted branches and leaves that end up as logging residues is be about $0.176 m^3$. The process of tree felling inevitably results in the destruction of smaller trees that happen to be in the path of the falling tree. This destruction however is minimised wherever possible. From logging site studies, it is estimated that the amount of unwanted trees that get in the way of a falling tree is comparable to the amount of unwanted branches and leaves of the felled tree. As such for one cubic meter of sawlogs produced the amount of biomass residues generated is roughly $0.35m^3$. For 1996 it was estimated that Malaysia produced 31 million m^3 of sawlogs.[1] As such the amount of logging residues generated works out to be about 6.239 million tonnes of dry biomass with an energy content of some 19.06 million boe. All this biomass is currently left in the forest to rot.

The second source of biowastes generated by the timber industry are wastes produced when logs are processed either into sawn timber or into plywood. If not exported these are then used locally for downstream activities such as in the construction and furniture industries. The wastes generated by these downstream activities are scattered and most would end up as construction and industrial wastes. As such we will not attempt to estimate the biowastes generated by these downstream activities. However from studies done at sawmills and plywood mills the following annual quantities of processing wastes are estimated: 0.181, 0.226. 1.222 million dry tonnes of bark, sawdust and fuelwood respectively and their respective energy potential is 0.553, 0.691 and 3.733 million boe.

In most mills, barks if not already removed at loging sites, are either burnt or incinerated, or allowed to decompose within the grounds of the mills. Though in some mills sawdusts are also combusted to produce energy and in a few situations briquetted to produce charcoal, the fate of most sawdust is similar to that of barks. From the above estimates, if these two sources of wastes are harnessed for energy purposes an amount of about 1 million boe per annum is available.

3. DISCUSSION AND CONCLUSION

Though there are many other crops cultivated in Malaysia such as tapioca, pepper, pineapple, groundnuts, tea, tobacco, fruits and vegetables, the hectarages involved are small and therefore the amount of biowastes generated from these crops would be small and moreover scattered. As such the total amount of biomass outline above can be considered to be fairly close to the total amount of biowastes produced in the country. From the above data we conclude that of the total 199.3 million boe currently generated per annum, about 24.5% are already utilised for energy purposes and roughly 75.5% are still unutilised and therefore wasted. 69.7% of the biomass that are currently wasted or unused are contributed by the oil palm industry, 12.7% by logging residues and 11.2% by cocoa tree prunings. Serious attempts should be made to harness their energy potential, although in the years to come the contribution from cocoa tree prunings would decrease as more and more cocoa plantations are converted to other crops and as pruning practices are reduced as some are of the opinion that pruning hurts the trees.

Currently Malaysia consumes roughly 290 million boe per annum of commercial energy which does not include biomass.[9] If the latter were included, the total energy consumption for the country is some 340 million boe per annum. This means that about 14% of the country's energy consumption is contributed by biomass. If the biomass that are currently wasted can be harnessed for energy, the contribution of biomass can be raised to around 59% of the country's total energy consumption. This is not a small figure and therefore efforts ought to be directed towards realising this possibility. By so doing, the nation's fossil fuel reserves can be conserved.

REFERENCES

1. Economic Reports (1991/92 - 1996/97). Published by the Ministry of Finance, Government of Malaysia.
2. Lim, K.O., Z. Zainal, A.Q. Ghulam, A. Mohd. Zulkifly (1999). Energy Productivity of Some Plantation Crops in Malaysia and the Status of Bioenergy Utilisation (submitted for publication).
3. Lim, K.O. (1986). The energy potential and current utilisation of agriculture and logging wastes in Malaysia, Renewable Energy Review Journal, 8, pp. 57–75.
4. Hong, L.T. and H.C. Sim (1994). Products from rubberwood—An overview, in Rubberwood: Processing and Utilisation, ed. by Hong Lay Thong and Sim Heah-Choh, Malayan Forest Records No. 39, The Forest Research Institute of Malaysia Press, Malaysia.
5. Lew, W.H. (1992). A study on the rubberwood industry in Malaysia, UNCTAD/GATT Report. Published by the International Trade Centre.
6. Lim, K.O. and Rugayah Durani (1993). Utilisation of Biomass Fuel in Rural Malaysia. (Unpublished).
7. Lim, K.O. (1986). The future energy potential of replanting wastes in malaysia, Renewable Energy Review Journal, 8, pp. 76–85.
8. Lim, K.O. (1981). Energy from agriculture wastes—it is a worthwhile consideration in Malaysia, The Planter, 57, pp. 182–187.
9. Lim, K.O. (1996). Our energy situation—more positive actions needed, The Malaysian Technologist, 1, pp. 17–27.

NEW GIS TOOLS FOR BIOMASS RESOURCE ASSESSMENT IN ELECTRICAL POWER GENERATION

C. Monteiro, B.R.P. da Rocha, V. Miranda, J.P. Lopes

Instituto de Engenharia e Sistemas de Computadores – INES/Porto, Praca da Republica, 93, 4050 Porto, Portugal, cdm@bart.inescn.pt

This abstract describes the methodology used on MEAPA project (Integrated methodology for renewable energy integration on Pará, Brazil). The methodology was applied on Marajó Island on the Amazon River. One of the project modules is biomass resource assessment. We use a geographical database, including vegetation coverage and land use, to evaluate the cost to collect and transport biomass residues.

Using GIS we estimate, based on geographical data and on other parameters specified by the user, the geographical availability of biomass resources. In a second step we apply a geographical zone calculation to compute the mean cost of collection and transportation around each location. The methodology is generalized for several kind of residues specified by the user. For a specific biomass residue the user specifies the availability, collection cost and transport cost on each land use and vegetation class.

As results we construct maps of biomass resource availability and the mean costs to concentrate the resource in each potential generation site. These maps are used on other GIS modules that include the technology parameters to evaluate the levelized electric cost for a specific biomass system.

BIOMASS RESOURCE ASSESSMENT FOR CHINA

Li Jingjing, Zhou Aiming

Energy Research Institute of State Planning and Development Commission, China

This paper calculated and assessed the biomass resource availability in China, especially straw and stalk, domestic animal excreta and municipal solid waste. The assessment showed that biomass energy will be a rich and sustainable resource in China, important for developing the social economy and improving the environment in future.

1. BACKGROUND

China hosts the largest rural population in the world. At the end of 1995, 860 million people lived in the rural areas. Recently, with the rapid development of township and village enterprises (TVEs) and increasing farmers' income, energy demand in rural areas has grown steadily. At present, 61% of the rural household energy consumption depends on traditional biomass energy resources; the energy consumption in the rapidly expanding TVEs is a quarter of the total commercial energy consumption. Moreover, China has 23% of the world's population and 7% of the world's cultivated lands. A great deal of energy is still needed to realize agricultural modernization. Therefore, energy issues in rural areas in China are important, and China needs to formulate a sustainable energy development strategy in the rural areas that is suitable for the national condition. This development strategy should coordinate the population, economy, society, environment and resources, and meet the needs not only of people today, but also the needs of the next generations.

Biomass resources in China mainly are the residue from agriculture production and forestry industry (for example, crop straw and stalk burned as fuel; fuelwood), animal excreta from medium-large scale husbandry farm, and gas resources from municipal solid waste (MSW) landfills. In this transition from rural and agricultural society to urban and industrial society the rural population is still 80% of the total national population. Straw and stalks are widely distributed and available in all of the regions.

2. RESOURCE AVAILABILITY ASSESSMENT OF STRAW AND STALK

Over China's vast land, there are lots of agricultural residues. In 1995, the crops straw and stalk output in China reached about 600 Mt. With a collection rate of 75%, the available amount of straw and stalk is 450 Mt (180 million tce) and the available amount of rice husk is about 50 Mt (20 million tce). Much is presently used for cooking and heating in rural households; other uses include forage, industrial raw material for paper production and organic fertilizer. Most is used at low efficiency; for example, domestic cooking stoves have only 10-20% conversion efficiency. The remainder of the straw is either dumped or burnt in the field.

Rapidly developing rural economics and farmer incomes drive a rapid increase in the use of commercial energy (LPG, coal and even electricity) for rural residential purposes. Straw consumption is declining rapidly, and the share of straw and stalk left in field or burned directly will increase. Already the residual straw and stalk amounts in some regions are over 60% of the production, which not only damages the environment but also waste resources. Table 1 shows the main crop straw and stalk outputs in 1995 and Figure 1 illustrated the historical main crop yields.

Table 1
The main crop straw and stalk outputs in 1995

	Output (10^6 ton)	Ratio of grain to grass	Straw and stalk (10^6 ton)	Coefficient equal to tce	Standard coal (10^6 tce)
Rice	185.23	1:0.623	115.40	0.429	49.51
Wheat	102.21	1:1.366	139.62	0.50	69.81
Corn	111.99	1:2.0	223.98	0.529	118.49
Other miscellaneous	16.69	1:1	16.69	0.05	8.35
Soybeans	17.88	1:1.5	26.81	0.543	14.56
Tubers	32.62	1:0.5	16.31	0.486	7.93
Oil bearing	22.50	1:2.0	45.01	0.529	23.81
Cotton	4.77	1:3.0	14.30	0.543	7.77
Sugarcane	65.42	1:0.1(leaf)	6.54	0.441	2.88
TOTAL			604.66		303.09

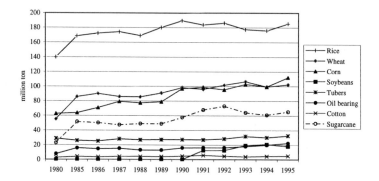

Figure 1. Trends in the production of major crops output.

Table 2
The distribution and availability of straw and stalk in 1995 (Mt)

		Total Straw and Stalk Outputs	Fertilizer and Collection loss	As Forage	As Raw material of Paper	As Energy
	National	604.664	90.7	144.99	13.879	355.094
	Shandong	71.544	10.732	15.787	0.715	44.31
	Jiangsu	36.043	5.406	1.262	0.391	28.984
East	Anhui	30.47	4.571	8.934	0.303	16.663
China	Zhejiang	11.227	1.684	0.641	0.336	8.566
	Jiangxi	13.592	2.039	4.897	0.35	6.306
	Fujian	7.199	1.08	1.625	0.723	3.771
	Shanghai	2.349	0.352	0.088	0.018	1.891
	Henan	57.056	8.558	15.969	2.496	30.034
South	Hubei	26.572	3.986	5.216	0.46	16.91
	Hunan	21.294	3.194	5.484	0.752	11.864
	Guangdong	14.603	2.19	6.006	0.782	5.625
China	Guangxi	15.809	2.371	10.154	0.239	3.045
	Hainan	2.319	0.348	1.866	0.01	0.095
Northeast	Helongjiang	38.799	5.82	6.516	0.468	25.996
	Jiling	34.268	5.14	4.895	0.613	23.62
China	Liaoning	20.859	3.129	3.842	0.623	13.264
	Hebei	44.682	6.702	7.379	1.185	29.415
North	Inner Mongolia*	17.533	2.63	-	0.153	14.89
	Shanxi	13.933	2.09	3.206	0.704	7.932
China	Beijing	4.843	0.726	0.18	0.01	3.928
	Tianjing	3.576	0.536	0.283	0.101	2.656
West South	Sichuang	45.31	6.797	14.215	1.069	23.23
	Yunnan	15.413	2.312	10.014	0.176	2.912
	Guizhou	11.787	1.768	8.271	0.05	1.699
China	Tibet*	1.06	0.159	-	-	0.901
	Xinjiang*	15.19	2.279	-	0.132	12.78
West North	Shannxi	13.896	2.084	3.543	0.858	7.412
	Ninxia*	8.096	1.214	4.717	0.068	2.097
China	Gansu*	3.24	0.486	-	0.09	2.664
	Qinghai*	2.102	0.315	-	0.004	1.783

Note: Provinces with * are the pasturing area.

The top ten provinces for straw and stalk resource availability in 1995 are listed as Table 3. The resource base in these ten provinces represents more than 70% of the national resource. Therefore, these areas should develop uses for straw and stalk in the future.

The total straw and stalk output in 2000 will be 648 Mt according to the projected grain output in 2000. Deducting the projected use for paper production, 21 Mt, and by assuming that the percentage of the straw and stalk for forage is nearly equal to that in 1995, i.e., about 178 Mt, and allowing 97.2 Mt for fertilizer usage and collection losses, the residual amount available for energy purposes is 351.9 Mt. This quantity is listed for each region listed in Table 4, and the top ten provinces in terms of resource availability are shown in Table 5.

Table 3
The top ten provinces of the straw and stalk resource availability in 1995

Region	Order	Availability for Energy (Mt)
Shandong	1	44.31
Henan	2	30.034
Hebei	3	29.415
Jiangsu	4	28.984
Helongjiang	5	25.996
Sichuang	6	23.62
Jilin	7	23.23
Anhui	8	16.91
Hubei	9	16.663
Inner Mongolia	10	14.89
TOTAL		254.052

Table 4
Straw and stalk resource distribution and its availability in 2000 (Mt)

Region	Province	Total Straw and Stalk Output	Fertilizer and Collection loss	As Forage	As Paper Raw Material	As Energy
	National	647.923	97.188	177.447	21.000	352.288
East China	Shandong	76.662	11.499	20.011	1.082	44.070
	Jiangsu	38.622	5.793	1.600	0.592	30.637
	Anhui	32.650	4.898	11.324	0.459	15.970
	Zhejiang	12.030	1.805	0.812	0.508	8.905
	Jiangxi	14.564	2.185	6.207	0.529	5.643
	Fujian	7.714	1.157	2.060	1.094	3.403
	Shanghai	2.517	0.378	0.111	0.027	2.000
South China	Henan	61.138	9.171	20.242	3.776	27.949
	Hubei	28.473	4.271	6.612	0.696	16.894
	Hunan	22.818	3.423	6.951	1.138	11.306
	Guangdong	15.647	2.347	7.614	1.183	4.504
	Guangxi	16.940	2.541	12.871	0.361	1.167
	Hainan	2.485	0.373	2.006	0.016	0.091
North East	Heilongjiang	41.575	6.236	8.259	0.708	26.372
	Jilin	36.720	5.508	6.205	0.928	24.079
	Liaoning	22.351	3.353	4.870	0.943	13.185
North China	Hebei	47.878	7.182	9.354	1.793	29.549
	Inner Mongolia*	18.788	2.818	-	0.231	15.739
	Shanxi	14.929	2.239	4.064	1.066	7.560
	Beijing	5.190	0.779	0.228	0.015	4.169
	Tianjing	3.831	0.575	0.358	0.153	2.745
South west	Sichuang	48.551	7.283	18.019	1.618	21.632
	Yunnan	16.516	2.477	12.693	0.266	1.080
	Guizhou	12.631	1.895	10.485	0.075	0.176
	Tibet*	1.136	0.170	-	-	0.966
North west	Xinjiang*	16.277	2.442	-	0.200	13.635
	Shannxi	14.890	2.234	4.491	1.298	6.869
	Ninxia*	3.472	0.521	-	0.137	2.814
	Gansu*	8.675	1.301	-	0.103	1.292
	Qinghai*	2.252	0.338	-	0.006	1.908

Table 5
Top ten provinces of the straw and stalk resource availability in 2000 (Mt)

Province	As Energy	Order
Shandong	44.07	1
Henan	30.637	2
Hebei	29.549	3
Jiangsu	27.949	4
Heilongjiang	26.372	5
Jilin	24.079	6
Sichuang	21.632	7
Hubei	16.894	8
Anhui	15.97	9
Inner Mongolia	15.739	10

A similar calculation using the projections for 2010 shows a total straw and stalk resource of about 726 Mt, when the grain output is forecast to be 560 Mt. It has the following withdrawals: 28 Mts straw and stalk for paper making, 213 Mts forage or the raw material of forage processing, and 108.9 Mts as fertilizer and collection loss, leaving the amount for energy purposes about the same as in the year 2000, 376.1 Mt.

Preliminary survey work carried out in Zhejiang, Shandong and Sichuan Province on straw and stalk utilization and the methods of collection is noted in Table 6. The collection radius and cost are closely related to the density of straw resources distribution and labor price in the local regions, resulting in significant differences in the collection radius and cost among the three provinces. The average collecting radius in Zhejiang Province is 2 km with a collection cost of 0.14-0.20 Yuan/kg; in Shandong, the radius is 0.7-3.4 km with the average cost of 0.12-0.22 Yuan/kg; in Sichuan, there are three collecting radii, which are function of the terrain, i.e., 8 km on the plain, 5 km on in low hills, and only 3 km in the mountains. The cost of collection ranged between 0.15 Yuan/kg and 0.20 Yuan/kg.

By constructing demand forecasts from the "bottom-up," it has been possible to estimate the quantities of crop straws and stalks in China and carry out an analysis of their availability for energy purposes. The following was concluded

- The quantity of straw resource which was available and could have been used for energy in 1995 was 355.1 Mt after making allowances for that straw used as forage; raw material for paper manufacture; as fertilizer, that returned to the field and straw collection losses. The actual straw consumption is estimated to have been 415 Mt including 190 Mt for fuel, 14 Mt for paper mill ,145 Mt for forage as well as 91 Mt for collection loss.
- By 2000, the quantity of straw forecast to be available for energy use 352.28 Mt, nearly the same as in 1995.
- By 2010, the total grain output is forecast to be 560 Mt with a total straw and stalk production of 726 Mt in China, excluding the straw for paper making, forage, fertilization, returning to the field and collecting loss, the total quantity of straw can be used as energy will be 376.1 Mt.
- Since straw and stalks are produced as a by-product of the food and feed production system, they are likely to be a sustainable source of biomass for energy. The amount available is

Table 6
Comparison of straw and stalk use, and collection costs in Zhejiang, Shandong and Sichuan Provinces in 1995

	Unit	Zhejiang	Shandong	Sichuan
1. Percentage of the every Straw Use				
a. Raw material for paper	%	3.0	1	2.4
b. Forage	%	5.7	22	31.4
c. Fertilizer and Collection loss	%	15	15	15
d. Available for energy	%	76.3	62	51.2
In which, burned as fuel	%	23.3	34	41.2
left in field	%	53	28	10
Total	%	100	100	100
2. Collecting radius	km	2	0.7-3.4	8, 5, 3*
3. Collecting cost				
Rice straw	Yuan/kg	0.15	0.2	0.15-0.20
Wheat straw	Yuan/kg	0.15	0.2	0.15-0.20
Cotton stalk	Yuan/kg	0.2	---	---
Rape stalk	Yuan/kg	0.14	---	---
Jute stalk	Yuan/kg	0.2	---	---
Corn stalk	Yuan/kg	---	0.12	---

Note: There are 3 types of collecting radius of straw and stalk in Sichuan Province, in plain area, less than 8 km; in mountainous area, less than 3 km; in hilly ground, less than 5 km.

forecast out to 2010 as being essentially the same as that of today in the range 350 to 370 Mt or approximately 170 Mtce. If this were used for electricity production, and process and space heating in conjunction with the provision of gaseous fuels for cooking and daily living, there is a potential for 120 GW of power generation producing 450 TWh at reasonable efficiencies of conversion.
- It can be seen from the ratio of straw for different use that the ratio of straw used as fuel has relation with the level of economic development. In the regions with rapid development of economy, the ratio of straw used as fuel is low and the ratio of straw refused is high. For example, in Zhejiang Province, 30% of the straw is used as fuel (the lowest among the three provinces surveyed), while 60% of it is refused, which is a relatively high ratio.
- The collecting radius and cost depend on the density of straw resources distribution and labor price in local region. In the region with high level of economy, the collecting radius is relatively small and the collecting cost is high. A study on economic efficiency should do a more detailed survey, combined with the project progress.

3. RESOURCE AVAILABILITY ASSESSMENT OF ANIMAL EXCRETA FROM MEDIUM-LARGE SCALE HUSBANDRY FARM

In the past decade, the Chinese government has also paid much attention to the development of agriculture products that are important to people. The "Vegetable Basket Project" was launched, which led China's domestic animal farm to intensification, large size and modernization.

Many species of livestock and poultry are kept. There are two methods of feeding. One is conventional feeding, which is mainly suitable for small-sized farms and families, or for specific animals, such as sheep, horses and ducks, whose excrement is scattered in grasslands and pools. The other way is concentrate feeding, demonstrated by large and medium-sized livestock and poultry farms. Cattle (including milk cow and beef cattle), pig (including pork and boar), chicken (including hen and chicken as food) are generally reared in pens so that the excrement can be easily collected. Here we analyze only the keeping of cattle, pig and chicken on large scale.

Table 7
General information of raising livestock and poultry (1985-1995)

Year	Cattle	Slaughtered Fattened Hogs	Commercial Chicken
	million	million	million
1985	68.82	238.75	523.69
1986	91.67	257.22	655.87
1987	94.65	161.77	787.78
1988	97.95	275.70	942.90
1989	100.75	290.23	1077.91
1990	102.88	309.91	1228.80
1991	104.59	329.87	1610.20
1992	107.84	351.70	2001.86
1993	113.16	378.24	2393.16
1994	132.22	401.03	2805.05
1995	132.06	480.51	3057.51

Source : Statistical Yearbook of China

Table 8
Growth rate of cattle, pig and chicken

	Annual Average Growth Rate %		
	1985-1994	1985-1989	1990-1994
Cattle	4.7	3.8	6.5
Pig	6.5	5.0	8.0
Chicken	20.5	19.8	22.9

Table 9
Distribution trend and forms of keeping

	Area	Method of keeping
Cattle	Southwest China-Sichuan, Guizhou, Yunnan	natural feeding, work cattle on the whole
	East China -Anhui, Hunan, Guangdong	natural feeding and pens feeding is 50% receptively
	North China -Shandong, Hebei, Henan	mainly pen feeding, most of which are milk cows and work cattle
Pig	North China -Hebei, Liaoning, Shandong	Large medium sized piggery
	East China -Jiangsu, Jiangxi, Hunan	Dominated by medium & small-sized piggery
	Southwest China -Sichuan, Yunnan, Guangxi	Mainly by household
Chicken	North China -Tianjin, Beijing, Liaoning, Shandong, Hebei	Mainly by larger & medium type, most of medium & small type
	East China -Shanghai, Guangdong	mainly by larger & medium type

Table 10
Distribution of state-owned large and medium sized cattle farms in China

Area	Number of state-owned farms	Number of milk cow farms	Number of beef cattle farms
Beijing	16	34	2
Tianjin	14	23	4
Hebei	32	11	5
Shanxi	31	13	6
Inner Mongolia	138	46	46
Liaoning	141	10	15
Jilin	163	8	10
Heilongjiang	120	18	3
Shanghai	27	86	12
Jiangsu	26	8	1
Zhejiang	65	11	1
Anhui	24	7	0
Fujian	100	16	0
Jiangxi	131	7	0
Shandong	17	15	2
Henan	94	10	3
Hubei	48	9	0
Hunan	81	10	3
Guangdong	136	24	3
Guangxi	49	8	1
Sichuan	134	87	30
Guizhou	43	7	2
Yunnan	88	9	0
Tibet	6	2	3
Shaanxi	18	7	1
Guansu	18	2	0
Qinghai	16	2	9
Ningxia	15	2	0
Xinjiang	287	53	20
TOTAL	2078	535	185

Middle China contains about 10% of all cows in large and medium milk-cow farms. Farms are located in the suburbs of big cities (not including Inner Mongolia). The scale is not big either. Both milk cows and beef cattle are mostly kept free in rural areas such as Inner Mongolia and Shanxi.

In east China, large- and medium-sized cattle farms are concentrated, especially along the coast. A farm typically has more than 500 head; the milk cows are mostly imported. These good-strain cows are kept in modern farms and have high production of milk. There are farms of more than 1000 cows in Beijing, Tianjin, Shanghai and Guangzhou and other big cities. The state farm system is planning to build 40-45 modernized farms with over a thousand milk cow.

The distribution of beef cattle farms is similar to that of milk cow. The numbers of cattle and of farms are less than that of milk cows. The scale is relatively small. Beef cattle in North China are kept mainly free range.

Table 11
Distribution of large and medium sized pig farms in China

Area	Total	Annual Production		
		500-1000 head	1000--5000 head	over 5000 head
Beijing	240	77	115	48
Tianjin	140	53	63	24
Hebei	81	33	37	11
Shanxi	10	7	2	1
Inner Mongolia	8	6	2	0
Liaoning	125	96	25	4
Jilin	65	30	25	10
Heilongjiang	79	36	29	14
Shanghai	390	154	158	78
Jiangsu	153	54	69	30
Zhejiang	79	33	31	15
Anhui	28	14	9	5
Fujian	45	16	21	8
Jiangxi	42	36	5	1
Shandong	167	103	36	28
Henan	80	34	29	17
Hubei	25	9	12	4
Hunan	44	15	20	9
Guangdong	135	30	68	37
Guangxi	19	79	9	3
Hainan	2	20	0	0
Sichuan	51	2318	18	10
Guizhou	9	6	3	0
Yunnan	12	7	4	1
Tibet	0	0	0	0
Shaanxi	7	5	2	0
Guansu	5	3	1	1
Qinghai	1	1	0	0
Ningxia	1	1	0	0
Xinjiang	2	1	1	0
TOTAL	2045	892	794	359

Table 12
Distribution of large and medium sized hen farms in China

Area	Total	50-100 thousands	100-150 thousands	over 150 thousands
Beijing	58	36	12	10
Tianjin	29	17	8	4
Hebei	12	8	2	2
Shanxi	9	5	2	2
Inner Mongolia	10	8	2	0
Liaoning	27	11	8	8
Jilin	6	2	3	1
Heilongjiang	40	34	4	2
Shanghai	48	32	8	8
Jiangsu	16	11	4	1
Zhejiang	10	7	2	1
Anhui	9	9	0	0
Fujian	5	5	0	0
Jiangxi	4	3	0	1
Shandong	31	23	5	3
Henan	23	13	7	3
Hubei	9	7	1	1
Hunan	6	5	1	0
Guangdong	10	4	3	3
Guangxi	2	2	0	0
Hainan	0	0	0	0
Sichuan	28	19	7	2
Guizhou	2	1	1	0
Yunnan	4	2	2	0
Tibet	0	0	0	0
Shaanxi	5	2	2	1
Guansu	4	2	2	0
Qinghai	0	0	0	0
Ningxia	1	1	0	0
Xinjiang	5	2	3	0
TOTAL	413	271	89	53

China's main domestic animals, cattle, pig and chicken, were considered in the analysis of availability of domestic animals excrement. The excrement resources can estimated about 850 Mt physical quantity and more 78 million tce in 1995 according to the animals kind, weight and amount of excrement by animal during 24 hours. See Table 13.

According to statistics in 1992, large- and medium-sized farms hold 9.1% of all pigs in China, 43.4% of milk cows, and 15% of chicken. In the latest three years, pigs on the large-and medium-sizes farms now make up 10% of total number; milk cows, 40%; chicken, 20%. For excrement resources on large and medium-sized farms, see Table 14. Because the excrement on large- and medium-sized farms is generally flushed with water, the mixture of excrement and water is several times larger than net excrement.

Table 13
Amount of excrement of cattle, pig and chicken and developable resource in 1995

	Cattle	Pig	Chicken	Hen
1. Weight (kg)	500	50	1.5	1.5
2. Circle of feeding (day)	365	150	60	365
3. Amount of excrement (kg/head.day)	20	4.0	0.1	0.1
4. Amount of excrement (ton/head.year)	7.3	0.6	0.006	0.0365
5. Number of animal (million/ year)	132.22	481.03	1200	1800
6. Weight of excrement (Mt/year)	964.00	288.00	7.00	66.00
7. Coefficient of excrement collection	0.6	0.9	0.2	0.2
8. Amount of developable excrement(Mt/year)	578.00	259.00	1.00	13.00
Coefficient of conversion	0.47	0.43	0.64	0.64
Dry excrement content (%)	18.00	20.00	80.00	80.00
Amount for standard (million tce)	48.90	22.30	0.51	6.66

Table 14
Available amount of mixture of excrement and water on large and medium sized farms in 1995

	Cattle	Pig	Chicken	Hen
1. Number (million)	1.73	48.00	240.0	360.0
2. Weight of mixture of excrement and water per day (kg/per animal)	50	12	0.3	0.3
3. Total weight of mixture per day (thousand tons)	86.5	576	72	108
4. Circle of feeding (days)	365	150	60	365
5. Total annual weight of mixture (Mts)	31.57	86.40	4.32	39.42
6. Coefficient of conversion	0.47	0.43	0.64	0.64
7. Dry excrement content (%)	8	7	27	27
8. Million tce	1.187	2.843	0.747	6.812

Note: As for the amount of cattle's excrement, only milk cows are included, because beef cattle are kept in pen for only a very short time every day, and thus little excrement is collected.

4. RESOURCE AVAILABILITY ASSESSMENT OF MUNICIPAL SOLID WASTE RESOURCE

The progress of social development and human activities has to extort from nature continuously. But, the earth's resources are not always inexhaustible and available. Modern civilization has produced a life style of high consumption. High consumption leads to lots of solid wastes, which in turn has led to serious environmental pollution. People are finally beginning to consider mines, oil, coal and forests as valuable and non-renewable resources. For example, based on the explored reserves of oil in the world and the growth rate of oil consumption, it is predicted that oil exploitation will continue only about 50 more years in industrialized countries. Oil can be considered as the blood that keeps the countries going.

Meanwhile, more and more solid waste will be produced that is more difficult to process. High-polymer organic substances in plastic waste won't decompose in two hundred years. The pollutants in the landfills will be stable after about several decades.

Waste is now considered a usable resource. In order to protect the environment for ourselves and for our offspring, people began to seek renewable resources from wastes. Especially since the western energy crisis in the 1970s, the use of urban waste utilization for chemicals and energy has been emphasized in developed countries. Advanced technologies for classification and collection of wastes, waste incineration for power generation, recycle of landfill gases and so on are actively promoted and developed.

As people's living standard improves, the volume of urban-living wastes grows. The disposal of urban waste becomes an issue that influences urban economic development and people's living standard. In 1995, there were 640 cities in which the quantity of urban waste cleared and transported reached 107.50 Mts, while in the same year the industrial solid waste was 645 Mts. Municipal residential solid waste formed 14.3% of total solid waste disposed of in 1995; however, the ratio was comparatively low in 1991 (11.5%). The disposal ratio of wastes is less than one-third at present, and the ratio of harmless disposal and utilization as resources of wastes is lower. The phenomenon of waste surrounding a city has appeared in many large cities such as Beijing, Shenyang, Shanghai, Xi'an and so on. Wastes not only take land and damage scenery, but they also spread diseases and influence environmental sanitation and residents' health. Therefore, the disposal of wastes without causing harm, generating less of it, and using it as a resource are an important environmental goals in cities of China.

Table 15
Resources of municipal residential solid waste

Year	1980	1985	1989	1990	1991	1992	1993	1994	1995
Residential disposal cleared (M ton)	31.32	44.77	62.91	67.67	76.36	82.62	87.91	99.81	107.5
Industrial solid waste (M ton)			525.9	571.7	578.0	588.0	618.8	617.1	617.0	644.8
Night soil disposal (M ton)	16.43	17.31	26.03	23.85	27.64	30.02	31.68	31.60	30.71

Source: Statistical Yearbook of China 1996

Table 16
Urban residential refuse cleared per capita in 100 cities in 1995

City group(10,000)	>200	100-200	50-100	20-50	<20
Cleared refuse per capita (ton)	0.46	0.420	0.568	0.703	0.830
in north China	0.620	0.432	0.669	0.739	0.915
in south China	0.353	0.383	0.485	0.667	0.704
High percentage of population using gas for household use	85.3	77.0	57.9	52.0	53.2
in north China	89.6	81.8		42.8	66.1
in south China	81.2	72.6		61.2	42.9

Source: Statistical Yearbook of China 1996

At present, the measures of disposal and resource utilization of refuse in China are mainly sanitary landfill, compost and incineration. In most cities, surface dumping, simple landfill, sanitary landfill and mechanical compost are popularly used, and refuse incineration is also partly used in some regions.

4.1. Application of landfill

For a long term, natural dump, natural landfill in pits and land level up are the commonly adopted measures in refuse disposal in cities in China. However, in recent years, improved construction of landfill yards and a series of advanced landfill yards had been completed in succession. Many cities, such as Hangzhou, Guangzhou, Suzhou, Beijing, Chengdu and Baotou, had set up comparatively perfect sanitary landfill yards in accordance with their respective situations. Meanwhile, the technology of collection and reutilization of gas from landfill yards has also progressed a lot. At present, sanitary landfill is still the major type of refuse disposal in China.

4.2. Application of fertilization

Refuse fertilization has been developed rapidly in China with relatively high research and application level. Dual fermentation has been applied since the 1980s. The process adopted obligated blow and aerobic fermentation to shorten the primary fermentation cycle, consummate piling equipment and promote industrialization of refuse fertilization. At present, domestic-designed mechanical process lines for refuse fertilization based on Chinese conditions have been set up in several cities, such as Wuxi, Changzhou, Tianjin, Mianyang, Beijing, and Wuhan. Some simply equipped refuse fertilization plants are also in operation.

4.3. Application of incineration

Research on incineration of solid refuse started in the middle of the 1980s in China. Incineration of refuse is currently the most effective method of harmless decrement disposal and use of solid refuse. At present, only several refuse incineration plants have been built in Shenzhen, Leshan, Xuzhou cities to demonstrate refuse-incineration power generation and now run in good condition. However, these plants are small and have only a small daily disposal capacity. Comparatively large scale refuse incineration plants are planned for some large cities such as Beijing, Shenyang and Guangdong.

Table 17
Disposal of municipal residential solid refuse in China (in 1995)

Type	Quantity (ton/day)	Ratio (%)
Surface dumping & simple landfill	232520	78.95
Sanitary landfill	51073	17.34
High temperature compost	7095	2.41
Incineration	2000	0.68
Others(*)	1832	0.62
TOTAL	294520	100

*including refuse-based construction materials and comprehensive utilization of refuse etc.

Using urban living wastes as resources helps realize goals of processing wastes, is a sustainable use of resources, and constructs a new industry that protects environmental resources. The departments, enterprises and institutions related to the collection, transport, disposal and utilization as resources and comprehensive management of urban wastes comprise the whole system, which is concerned with many fields and an integration of industrial technologies, social sciences and urban comprehensive management. It is an emblem of physical and mental civilization of the whole society, and it helps to secure sustainable development of society and economy. Thus, it can help build up urban environment without waste pollution and form a benign cycle of resources.

The quantity of municipal residential waste is related to many factors. Generally, as income and consumption increase, consumption and living style will change, and the quantity and composition of municipal residential refuse will change also.

Beijing can be used as an example in China. Beijing has a suitable geographic location, comparatively high economic level, and a resident living style that may become a model for the future development of cities in China. Table 18 and Figure 2 show the relation between the GDP and residential refuse per capita in Beijing from 1990 to 1995.

Table 18
GDP per capita and residential refuse per capita in Beijing (1990-1995)

Year	1990	1991	1992	1993	1994	1995
GDP per capita (RMByuan, constant price in 1995)	8700	9653	10530	11366	11592	13073
Residential refuse per capita (kg/year)	597	598	603	613	621	631

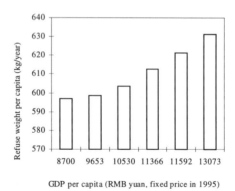

Figure 2. Relation between GDP per capita and residential refuse per capita in Beijing (1990-1995).

Table 19
Forecast of refuse resource and its distribution in China (Mt)

Region	2000	2010	2020
North China	24.9	37.0	50.4
North-east China	28.3	45.1	63.8
East China	50.4	78.1	108
Middle China	28.4	45.3	64.1
South China	15.0	24.0	33.9
South-west China	22.6	36.0	51.0
North-west China	14.8	23.7	33.0
TOTAL	184	289	405

Based on the above assumptions, the resource of refuse in China in the coming 20-30 years can be predicted (see Table 19). It can be seen that the quantity of municipal residential refuse in China will increase rapidly. The result of this scenario is a little more than the result of prediction of refuse made by the major environmental protection departments of China (see Table 20).

At present, the quantity of municipal residential refuse is increasing at the rate of 10% annually. According to the forecast of environment protection departments in China which is based on the growth rate of refuse in recent years, the quantity of refuse will reach 151 Mt and 230 Mt, and the rate of harmless disposal will be 60% and 90% by 2000 and 2010 respectively.

Table 20
Forecast of municipal residential solid refuse in China

Year	2000	2010
Refuse cleared (Mt)	151	203
Harmless disposal rate (%)	55-60	85-90
Harmless disposal quantity (1,000 tons per day)	31	54

It will be a common and great task facing most cities in China in the near future how to dispose the more and more residential refuse and make full use of these resources. Demand for refuse disposal technology and market will be substantial. Disposal of municipal refuse is a large project and should be disposal according to the local conditions.

5. CONCLUSION

Through the above calculation and analysis, it is easy to conclude that there is an abundant biomass energy resource in China. With the coming of the 21st century, people will confront the double pressures of developing economy and protecting the environment. Thus, in China, with its rich available biomass resource, it is important to produce and consume energy by modern technology that exploits biomass. So doing will help to establish a sustainable resource system, promote the development of the social economy, and improve the environment.

6. ACKNOWLEDGMENTS

Support for this work was provided by USDOE/ China MOA Expert Team.

TECHNOLOGIES FOR ELECTRIC ENERGY PRODUCTION USING BIOMASS IN MARAJÓ ISLAND, BRAZIL

S.B. Moraes,[a] B.R.P. da Rocha,[b] C. Monteiro,[d] I.M.O. da Silva,[c] A.O.F. da Rocha,[b] E.C.L. Pinheiro,[b] V. Miranda,[d] J.P. Lopes[d]

[a]Mechanical Engineering Department – Campus Universitário do Guamá – CP 8619 66075-900 Belém Pará – DM/CT/UFPA
[b]Electrical Eng. Department- DEE/CT/UFPA
[c]Department of Meteorology-DM/CG/UFPA
[d]Instituto de Engenharia de Sistemas de Computação - Praça da República, 93 Porto Portugal

In tropical regions with dispersed population that have an average low density, the cost of producing electrical energy by traditional means usually is high, mainly because of the cost of transportation fuel and maintenance of the system. Many of these areas, like the Amazon region in Brazil, are rich in biomass resources.

In order to use biomass for electrical energy production in Marajó Island, several technologies were investigated. Specific consumption, cost of the raw materials, transportation costs and technologic costs were obtained for gasification, vegetable oils in internal combustion engines, direct burning of agricultural residues, firewood and vegetable coal (charcoal), biogas produced from animal feces and some vegetables. These costs were determined for different vegetation present on the Marajo Island, Amazon Region, Brazil.

The equivalent information was also obtained for traditional electric generators using conventional diesel motors. The emission of pollutants was also determined for the different technologies.

A geographic information systems was used to map the costs for different technologies for electrical energy generation using biomass for the entire region under study.

1. INTRODUCTION

The economic base of rural sectors in Latin America is characterized by small rural family producers, who engage in primary productive activities. These producers have little financing and little or no possibility for diversifying their production. Pará consumes just 45% of the electrical energy produced in the State and exports the rest to the south of Brazil even though many areas in the State have no electricity supply. This problem is still larger in Marajó Island, owing to its great expanse and long distances from either transmission lines or power plants (mainly hydroelectric). The high cost of electrical energy (or its total nonexistence) hinders enormously the development of rural activities, blocking the diversification of production and improvement of product quality and its final value. Appropriate technologies are not available for these small producers.

Producing abundant energy at a competitive price is a decisive factor in promoting regional sustainable development.

2. BIOMASS RESOURCES

Biomass is any original material of biological matter: wood, agricultural, forest and cattle-raising residues, vegetable oils, ethanol, and gases resulting from anaerobic decomposition. Besides the production of electricity, biomass can be used in the kitchen and for heating. In this work only the production of electricity is considered.

In a general way the biomass resources in Marajó Island are geographically distributed in two different ways. Forest resources and dispersed agricultural production exist for the whole area through different homogeneous types of vegetation. In addition, at specific points residues from certain activities accumulate, as for example residues from wood industries, or other industrial or urban residues.

Several ways to produce of electricity using biomass can be considered. Combustion, carbonization, fermentation, gasification and bio-digestion are the most important technological uses of biomass in Brazil. These methods are at several degrees of development.

2.1. Combustion
The most practical and easy way to use biomass is in a thermal process that decomposes the chosen material, transforming it into energy and leaving ash as solid residue. Combustion uses three categories of equipment: fixed bed, suspension and fluidized bed.

2.2. Carbonization
Decomposition of the biomass at high temperatures produces a solid composed almost exclusively of pure carbon (coal), after the volatization of gases and liquids during the burning process. In Brazil, wood is especially used in brick ovens for the production of vegetable coal. Vapors are lost, and gases may or may not be condensed. Practically all Brazilian production of vegetable coal is destined to metallurgy, and just a small fraction goes to other applications. An improved carbonization process would use the volatiles recovered from the burning process, whose energy value cannot be ignored.

2.3. Fermentation
Basically, in this decomposition process several enzymes interact with solutions of sugars and produce preferably saccharine, starches and cellulose. Ethanol or ethyl alcohol is the main fuel produced through the fermentation. Its use in Brazil is growing, thanks to its high octane rating, the absence of burn residues, and its liquid state, which makes it easy to transport, store, and distribute.

2.4. Gasification

Quite old, it is a thermal process that decomposes the biomass (or the vegetable coal), transforming it from a solid to a gaseous fuel. The liberation of gas produces hydrogen, tar, light hydrocarbon, carbon monoxide and dioxide, nitrogen and water vapor. The resulting gas can also be used as raw material in another processes, as in the synthesis of methanol. The energy content of its components—hydrogen, carbon monoxide, hydrocarbons and tar—are very good. Today gasification is an area of dramatic technological development for use of wood and forest residues. In Brazil, CESP-IPT (of São Paulo) is developing electrothermic gasification to produce synthesis gas needed to make methanol. In the same way, ELETROBRAS-Light steers a technological program to introduce gasification units in the Amazon area, seeking to replace diesel oil by the generation of electric energy by motors that can work with poor gas.

2.5. Bio-digestion

A biological anaerobic process transforms biomass into methane, carbon dioxide and stabilized residues. A complex group of enzymatic reactions results in a gaseous mixture (the biogas), that is about 60% methane and 40% carbon dioxide; stabilized solids, composed of vegetable proteins and humus, which can be used as animal food or fertilizer; and nutrients dissolved in the water that can aid the growth of algae or fertilize plantations.

The use of biodigestors in the rural sector is increasing; they are used to process vegetable and animal residues. Bio-digestion also renders industrial wastes such as vinhoto, sewer mud, residues of slaughterhouses, butcher shops and similar industries and urban waste.

The best process for converting biomass to profitable energy depends on the source to be processed and the desired final product—vapor production, electricity, mechanical energy, automotive fuel, process heat or environmental heating.

3. METHODOLOGY

A geographic information system based in ArcInfo was used to determine the best technology, considering all parameters involved in the energy production process.

Biomass costs in a given place depend on the relationship between the amount of biomass and the distance to site of consumption. For example, if a power plant consumes 2 GWh/year, 10 km^2 around the site will produce enough biomass. On the other hand, if a power plant needs 5 GWh/year, a 30 km^2 area will need to be harvested, and that certainly will correspond to more expensive biomass. For the same technology type the relationship between biomass demand and cost of the biomass depends on the place where the biomass will be used.

The costs considered in this methodology (the cost of concentrating biomass at the consumption site) are the collection cost and the transport cost. The collection cost is the price of the available biomass where it originates. This cost depends on the biomass type: charcoal or firewood, for example, may be more costly whereas some residue may be free for hauling it away. Transport cost also depends on the biomass type, on the available transport infrastructures (highways, rivers), and the accessibility of the site (what terrain must be crossed). Unlike the collection cost, transport cost increases markedly with the distance to the site of consumption.

For energy production both the amount of biomass and its the energetic value are equally important for the definition of the exploration area. For example, the energy value of wood shavings is just 30% of the energy value of vegetable coal—but it takes 3 tons of firewood to produce a ton of coal. The energy value of a single geographical element allows to the methodology to calculate the biomass resources separately, for each biomass type, or including several biomass types together. Figure 1 shows the costs calculated for different places in Marajó Island.

Figure 2 presents the costs involved in the generation of electric power from biomass; enough biomass to produce 2500 kW costs 3 $/ton in Afuá, Breves, and Ponta de Pedras. In Soure and Cachoeira these costs rise to 10 $/ton. For Chaves and Anajás have costs slightly higher, 13 $/ton. Finally, in Tapera only 500 kW can be produced, at a biomass cost of 15 $/ton.

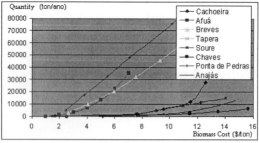

Figure 1. Relationship between the amount of biomass and its cost.

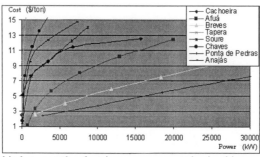

Figure 2. Relationship between the electric power generated using biomass and its cost.

In another step the amount, energy and average cost of biomass are calculated inside of each one of these areas. The tolerable maximum cost as well as the interval among the several levels will be specified by the user. Figure 3 evaluates biomass around two sites.

Based on these calculations of available biomass energy inside of each area, a generating system's efficiency parameters and characteristics of the consumption, then the size (in kW) of the powerplant that can be supported on the Island was calculated (Figure 4).

The largest biomass resources are available in the west of the island. Figure 5 shows four places located in the zone of larger vegetation presenting curves with higher inclinations (Afuá, Breves, Ponta de Pedras and Anajás). The places with smaller density of energy are Fazenda, Tapera and Cachoeira do Arari. Soure and Chaves appear with intermediate values owing to its location in places with grass fields mixed with other forest types.

Figure 3. Amount of biomass in the areas corresponding to the several cost levels.

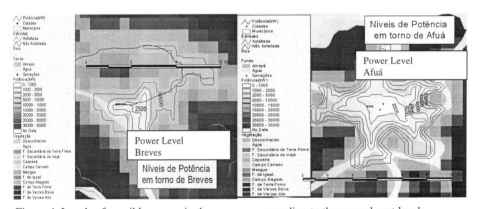

Figure 4. Levels of possible power in the areas corresponding to the several cost levels.

These graphs and maps show clearly how it is possible to evaluate different places for biomass use for electric energy production. In Marajó Island no consumption center needs more than 4 MW and the average power needed by a typical city is approximately 1 MW.

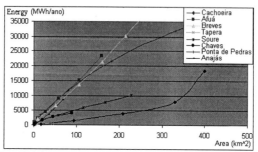

Figure 5. Density of available energy.

4. CONCLUSION

Considering the diversity of biomass in Marajo Island and the technologies that can use biomass for electric energy production, we can conclude that the use of the biomass in isolated areas of Marajó Island can be an excellent way to promote sustainable development of small places in the area. This option can help improve the quality of life of small rural farmers and of the communities of the Island of Marajó. This option favors the small farmer's development and the search for more rational alternatives.

Costs are much lower in areas that allow easier transport. The costs of electric energy using biomass are low close to small cities, although transport difficulty quickly increases these costs as power generation increases. Biomass can be an important option in the local generation of energy. Participation of inhabitants of the Island in preservation activities can also generate direct and indirect employment.

The development of programs that use biomass to generate energy in Marajó Island can collaborate with a politics of sustainable development in the area. The integration of the population of the Island, with centers of local research and State of Pará Universities, and with public and deprived actors of the agriculture-industrial and energy sectors, will also be fundamental to the success of the biomass power programs.

REFERENCES

1. Silva, I.T. and B.R.P. Rocha (1998). Possibilidade do Uso de Biomassa como Fonte de Energia Alternativa para a Ilha do Marajó – Resultados preliminares, Belém, Maio.
2. Silva, I.T. and B.R.P. Rocha (1998). Manejo adequado de Espécies Florestais para Fonte de Energia Alternativa para a Ilha do Marajó, no Estado do Pará, Belém, Julho.
3. Silva, I.T. and B.R.P. Rocha (1998). Estimativa da Biomassa como Fonte de Energia Alternativa para a Ilha do Marajó, Belém, Junho.

AVAILABILITY OF WOOD RESIDUES FOR COGENERATION IN GHANA

S. B. Atakora and A. Brew-Hammond

Kumasi Institute of Technology and Environment (KITE), P.O. Box 6534, Kumasi, Ghana

Cogeneration plants producing both electricity and process heat operate at total efficiencies as high as 80% compared with conventional power plants, which operate at 35 – 55%. Cogeneration from wood residues is also a way to mitigate unwanted climate change by cutting greenhouse gas emissions while generating positive economic returns.

In Ghana, residues from wood processing industries contribute to environmental pollution while power supply is inadequate and erratic. Cogeneration from wood residues is therefore being considered as an option for alleviating some of the pertinent environmental problems as well as contributing to the security and reliability of Ghana's power supply.

This paper reviews a number of studies on availability of wood residues for cogeneration in Ghana. The paper indicates that there is enough wood residue in the country to produce close to 200 MW of electric power alone and it argues that pursuing the cogeneration option could indeed help to make a significant contribution towards resolving Ghana's power supply problems.

1. INTRODUCTION

The wood processing industry in Ghana consists for the most part of primary producers such as sawmills, plymills, veneer plants and combinations of such plants. A small amount of secondary manufacturing produces small items such as mouldings, broom handles, parquet flooring and knockdown furniture components for furniture producers. The primary product from the sawmill industry is lumber of various grades. Many of the active mills in Ghana are straight sawmills performing log breakdown and producing sawn timber for domestic consumption and export. A smaller percentage of mills produces rotary veneer or plywood (principally for the domestic market) or sliced veneers (primarily for export). The plywood and veneer mills are nearly always sited next to a sawmill and if not physically located in the same complex, then common ownership and management exist to ensure the input log supply. The most recent estimates place the number of wood-processing firms in Ghana at 134 (110 sawmills and 24 plywood and veneer mills) for 1996.

Many wood processing facilities use wood waste fired boilers to generate steam and process heat. Most of these firms are combination mills processing sawn timber, plywood and veneer, and they generally consume about 50 to 60 percent of the wood residues.

The production of wood industry residues is primarily concentrated in a few major wood-processing centres. According to a Joint UNDP/WORLD BANK Sawmill Utilisation study in 1988, 66% of the total residue production is concentrated within an 8 km radius in the Kumasi (capital town in the Ashanti region) area, and 23% is distributed among three other centres: Takoradi (9%), Mim (9%) and Akim-Oda (5%). Close to a half of Ghana's wood-processing firms could be said to be located in the Kumasi area alone. In 1988, an estimated 342,000 tonnes of residue was released from the concentration of wood-processing firms in Ghana. Over a million (1,000,000) tonnes of residue was left to rot in the forest.

2. TYPES AND QUALITY OF RESIDUES

2.1. Forest residues

The exploitation of a closed forest zone of 31,860 square miles is controlled by the Forestry Act, which empowers the Forestry Division to ensure adequate supply. Forest operations yield quite an amount of residue, which normally is not used. Such residues become available when the trees are felled and are approximately equal to the volume of the round wood extracted and transported to the wood-processing firms. Thus, butt and top logs, branchwood and non-sawing material left in the forest from commercial logging amounts to some 1.1 million tonnes annually or 860,000 tonnes/year green weight.

Forest residues in general find no use or very limited use, although some projects are now trying. For example the Forestry Department is experimenting with mobile metal kilns in the Subin River basin to ensure efficient on-site carbonisation of the forest residues. However, the cost of kilns is high, the kilns are difficult to move from logging site to logging site, and forest fires are a hazard.

2.2. Wood processing residues

Sawmilling has continued to expand rapidly and sawn timber exports have also continued to increase. The primary product from the sawmill industry is lumber at various grades. The other products are considered as waste in the Ghanaian lumber industry. These products include bark, wood off-cuts, slabbings, edgings, sawdust (from sawmills) and shavings (from the moulding mills).

These products or industrial wood residues can be classified into two main categories:
- Solids (bark, slabbings, edgings, offcuts, veneer waste and cores)
- Fines (sawdust, planes, shavings and sander dust)

According to a UNDP/WORLD BANK Energy Sector Management Assistance Project on Sawmill Residue Utilisation in 1988, solids account for 79% whilst sawdust accounts for 21% of residues produced. More than half of the industrial wood residue is used off-site. Of the residue used off-site, about 28% is used directly as firewood for food preparation and about 70% converted to charcoal in earth mound kilns. The bark, wood off-cuts, slabbings and edgings are considered as a free fuel and in most cases have an opportunity cost and are sold as domestic or industrial fuel or as construction material.

In contrast, sawdust and shavings generated as a by-product in considerable amounts are usually considered a waste material and pose a disposal problem and can constitute fire hazards. They often have a negative value, because resources are employed to burn the material on site or transport it for disposal elsewhere. The common method of disposal by burning causes a great deal of pollution and is a menace to the general public.

Only in few cases are some of these secondary products used and these are mainly off-cuts for the boiler to produce hot water or steam for the drying kiln. At a few plants sawdust is used to generate steam for a veneer mill. In 1996, sawdust was used for generation of electricity (cogeneration in fact) by only one firm—Glisksten West Africa, Ltd., (GWA) in Sefwi Wiawso—although three other firms were reported to have advanced preparations for cogeneration.

However, the bulk of the wood-processing residues is looked upon as waste and treated as such. From the Forest Products Inspection Bureau's statistics, the output from sawmills and veneer mills in the Kumasi area has been calculated [R. Twumasi, Timber Export Development Board]. The study puts one distribution of the different products from the wood-processing industry as:

LUMBER (45%):
Sawn rough lumber, export grade and veneer................30%
Local grade lumber 15%

MILL RESIDUE (55%):
Sawdust 15%
Off - cuts 20%
Slabs and edging ribs 20%

TOTAL LOG INPUT: 100%

Sawdust, off-cuts, and slabs and edging ribs are the main targets for energy production; and they represent approximately 55% of the total log input volume.

2.3. Wood residue characteristics

Wood residue has a mass composition of 50% carbon, 43% oxygen and 6% hydrogen. The calorific value of wood residue as fuel is largely affected by its moisture content. Moist residues should as much as possible be pre-heated with any available waste heat to give optimum results. Energy content of fuel wood is largely determined by its heating value. The nominal characteristics of some average samples of wood processing residue are shown in Table 1. Table 1 shows that the energy of the residues (determined by the HHV), the ash content, moisture content, and bulk density (weight per unit volume) are all favourable for cogeneration.

Table 1
Nominal characteristics of wood processing industry residues

Oven dry density	503 kg/m^3
Green moisture content (wet basis)	36%
Green density	786 kg/m^3
Volatiles	81.3%
Ash content	1.6%
Higher heating value (oven dry) (HHV)	20 MJ/kg
Lower heating value (LHV)[a]	13.1 MJ/kg

Source: Sawmill Residue Utilisation Study, 1987
[a]LHV is at 36% moisture content wet basis

3. AVAILABILTY OF WOOD RESIDUES

3.1. Quantities produced

Off-cuts are the most abundantly produced residues at 399,600 m^3 SWE, representing 66.6% of the estimated 1994 total residual production of 599,400m^3 SWE. Sawdust at 26.9% and slabs and edgings at 6.5% follow in terms of abundance. Table 2 gives sawmill residues production (m^3 SWE) for 1990-1994.

Table 2
Sawmill residue production (m^3 SWE)[a]

Year	Slabs and edgings	Off-cuts	Sawdust	Total
1990	38,220	393,120	158,340	589,680
1991	35,490	365,040	147,030	547,560
1992	39,935	410,760	165,445	616,140
1993	41,510	426,960	171,970	640,440
1994	38,850	399,600	160,950	599,400

Source: Forestry Commission, 1995
[a]Solid weight equivalent

3.2. Reliability of supply

Supply of residues for on-site use would be assured, as the use of residues is controlled by the firms themselves. The supply of residues for off-site uses could be assured by contractual arrangement with producers. Effects of the wood product industry trends on the reliability of residue volumes and characteristics can be summarised as follows.

Greater processing of logs into export lumber. Increases in the volume of logs converted into sawn products would increase the volumes of slabs and edgings, offcuts and sawdust that are direct by-products of the log breakdown and cutting operations.

Higher product recovery. Mill improvements increasing recovery tend to reduce the amount of residues. Examples are reduced kerf diminishes sawdust volumes; improved edging practices reduce edging volumes; greater sawing accuracy increases the number of boards produced; accurate charging reduces round-up losses; improved clipping reduces veneer clips.

Increased secondary manufacture. Secondary manufacture may increase some residues and diminish others. If small cuttings are recovered from edgings and off-cuts, these residues would reduce while sawdust and shavings increase. If additional mouldings are produced edgings may reduce and sawdust and shavings increase. As wood pieces for secondary manufacture must be kiln dried, the shift is towards drier and more finely divided fuels and away from larger section residues.

The effects of these trends would be most evident in the larger operations where capital is available to be invested in upgrading and converting plants. The smaller mills would continue to produce residues as they do today. The rate of log throughput will continue to hold the greatest influence on the production levels of wood industry residues. Other effects are to some extent counterbalancing, and their overall net effect on the suitability and reliability of wood processing residues as a fuel is not expected to be significant on an industry-wide basis.

3.3. Present and projected surplus

Currently over 90% of the sawdust generated is not used (Forest Commission, 1995) and incurs a disposal cost. With the exception of a minor amount of sawdust for briquetting, no other sawdust is used for off-site energy purposes.

Sawdust is presently in abundant surplus in Ghana, accounting in 1994 for some 160,950 m^3 SWE (Source: Forestry Commission). It is anticipated that there will be future supply of wood residues if a good strategy is employed in ensuring this. Under a 'business as usual' scenario, the composition and distribution of surplus residues is not expected to change drastically and sawdust will continue to be the main unutilised residue.

It is, however, envisaged that with a creation of demand for the much larger amounts of logging residues left to rot in the forests, logging companies would retrieve and cart such products to points of sale. Again considering the high potential for dedicated biomass plantations in a country like Ghana, large commercial plantations or many small family businesses could be set up that cultivate fast-growing wood species for final sale to a cogenerating facility in Kumasi.

4. CONCLUSION AND RECOMMENDATIONS

Electricity supply in Ghana in recent times has been inadequate and rather erratic. Therefore, the need for a more reliable and secured power supply system cannot be overemphasised.

It has been established from this resource assessment study that there are large amounts of residues (or wastes) produced by the many wood processing firms in Ghana. Annual loss of revenue in wood processing firms as a result of power failure are usually very high and this waste could be put to some good use to redeem the enormous loss of revenue. Coupled with this is the environmental problem associated with wood residue, especially sawdust disposal. Cogeneration from wood residues would thus not only improve the reliability of electricity supply but also serve as a climate change mitigation option that can cut greenhouse gas emissions while generating positive economic returns.

With present technologies allowing for efficiencies of as high as 80%, cogeneration plants present a real alternative to conventional power plants. Full feasibility studies will be required to make investment decision.

Considering the concentration of wood-processing firms and thus availability of wood residues in Kumasi, the ideal location for a plant using wood residues as raw materials should be in the heart of the wood-processing centre in this city. This would ensure the adequacy and reliability of supply as well as big reductions in transport cost and a large market for the proposed project.

REFERENCES

1. World Bank (1988). Sawmill Residue Utilisation Study.
2. Hellem, S. and J.A. Sagoe (1998). The energy potential in wood waste from sawmill and moulding mill industry in Kumasi, April.
3. Hagan, E.B. (1997). Prefeasibility report on the proposed Letus Power Plant, Kaase Kumasi, October.

LAND AVAILABILITY AND PRODUCTIVITY FOR BIOMASS ENERGY PLANTATIONS IN NORTHEAST BRAZIL

Eric D. Larson,[a] Constantin Tudan,[a] Eduardo Carpentieri,[b] and Alexandre Carneiro Leao[b]

[a]Center for Energy and Environmental Studies, Princeton University, Princeton, New Jersey, USA
[b]Alternative Energy Division, Hydroelectric Company of Sao Francisco, Recife, Pernambuco, Brazil

The Hydroelectric Company of Sao Francisco (CHESF), the electric utility serving the Northeast region of Brazil, has been planning since the mid-1980s to introduce biomass-based power generation to augment generating capacity needs; after the year 2000, hydroelectric resources in the region will be fully utilized. The Northeast comprises nine states covering 150 million hectares, or 18% of Brazil's area and 10% of South America. CHESF concluded on the basis of extensive surveys in the 1980s that some 50 million hectares of the Northeast region were suitable for establishment of biomass energy plantations and that the land was potentially available without conflict with present or foreseeable future agricultural needs. The corresponding estimated potential production of biomass energy was some 12.6 EJ, or more than four times the thermal-energy equivalent of all hydroelectricity consumed in Brazil today. Using geographic information system tools, geo-referenced biogeophysical and socio-economic data that have become available since the CHESF surveys were completed, and a simple biogeophysical model of potential biomass productivity, we reassess land availability and productivity in Northeast Brazil for biomass plantations and make comparisons with the earlier CHESF work.

INFLUENCE OF LASER IRRADIATION ON SOME ENERGY MULTI-PURPOSE CROPS

A.G.Aladjadjiyan

Higher Institute of Agriculture, 12, Mendeleev St., 4000 Plovdiv, Bulgaria

Biomass comes mainly from forests and agriculture, from industry residues and municipal wastes. Biomass production by forests and agriculture is significant because it helps to reduce greenhouse gas emissions and because crops are used in many human activities. Of special interest is the use of biomass as a substitute for fossil fuels.

The south part of Bulgaria, especially the region of its main town Plovdiv, is characterized by trees of the family Fabaceae, to which the species *Caragana frutex arborescence, Cytisus Laburnum anagiroides,* and *Robinia pseudoacacia* belong. These trees show intensive growth and good sprouting ability, high adaptability to environmental conditions, and resistance to many diseases and insects.

The influence of laser irradiation on some physical properties of species *Caragana frutex arborescence, Cytisus Laburnum anagiroides,* and *Robinia pseudoacacia* was investigated to establish a fast method of controlling the plants' quality.

1. INTRODUCTION

Nowadays it is impossible to imagine mankind's development without energy. Energy consumption per capita has become a measure of progress in certain countries. However, the intensive exploitation of fossil fuels' has worsened our environmental status. The goal of sustainable development led to a search for alternative energy sources, which include renewables and especially biomass.

Energy crops have substantial advantages in terms of energy balance, reducing CO_2 emissions, protecting groundwater, decreasing erosion, and protecting biodiversity and landscapes (El Bassam, 1998).

Typical of south Bulgaria's flora are the species *Caragana frutex arborescence, Cytisus Laburnum anagiroides,* and *Robinia pseudoacacia*. They are characterized by intensive growth and good sprouting ability and can be used to retard erosion. They do not need special care because of their high adaptability to environmental conditions. These ornamental trees and bushes are also of interest for their medicinal and perfumery properties.

Physical properties of biological objects are related to their vitality and adaptivity. These properties correlate with growth conditions, and their measurement could be used for environmental pollution control.

2. MATERIALS AND METHODS

Electroconductivity and optical extinction of samples from species *Caragana frutex arborescence, Cytisus Laburnum anagiroides* and *Robinia pseudoacacia* were measured with the aim to investigate the influence of laser irradiation.

Samples of *Caragana frutex arborescence, Cytisus Laburnum anagiroides,* and *Robinia pseudoacacia,* grown in natural conditions on the hills of the town of Plovdiv were prepared by soaking 2 g of seeds in 20 ml distilled water. Samples were irradiated with a He-Ne laser, wave length 632,8 nm. Irradiation time was 2 min; irradiation intensity 100 W/m. Ten samples of each species have been prepared. Five of each group were irradiated.

Electroconductivity of samples was measured with a standard universal device LCR (E7 –11). Water extract was been placed in a special holder for electrical measurements; the holder consisted of a glass cylinder with two mobile Pt electrodes submerged in the extract and situated at a distance of 3 cm one from the other.

Optical properties have been measured by SPECOL – 11 spectrophotometer. Water extract was poured into one of the cuvettes and distilled water into the other. Absorption spectra of the samples were measured at wavelengths from 400 to 760 nm.

3. EXPERIMENTAL RESULTS AND DISCUSSION

3.1. Electroconductivity

Electroconductivity σ of investigated samples was found by the equation

$$\sigma = C/R, \tag{1}$$

where R is the resistivity of the sample, measured by LCR (E7-11) and C is the holder capacity, determined by the formula

$$C = l/S, \tag{2}$$

S being the surface in m^2 of holder' electrodes and l is the distance in m between them.

The average values of σ for irradiated and non-irradiated samples of above-mentioned species are shown in Table 1. They change from 0.01 to 01 $\Omega^{-1} m^{-1}$.

The electroconductivity of irradiated samples rose significantly compared to non-irradiated ones (see Figure 1). It appears that laser irradiation increased the permeability of the seed coat. As a consequence, diffusion through the seed coat increased, the ions' concentration in the water extract increased, and σ rose also (Aladjadjiyan, 1995). The maximum effect of laser irradiation was observed in *Cytisus Laburnum anagiroides* – its electroconductivity increased approximately 5 times; that of *Robinia pseudoacacia* increased less than 2 times.

Table 1
Electroconductivity of samples $(\Omega^{-1} m^{-1})$

Species	Irradiated	Non-irradiated
Caragana frutex arborescence	0.0555±0.005	0.0225±0.0006
Cytisus Laburnum anagiroides	0.0655±0.009	0.0135±0.0007
Robinia pseudoacacia	0.0857±0.003	0.0512±0.0011

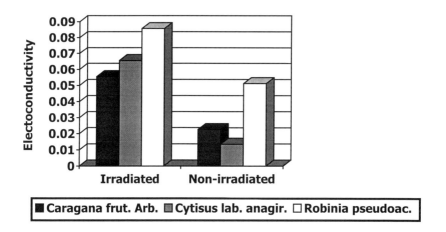

Figure 1. Influence of laser irradiation on the electroconductivity of investigated species.

3.2. Absorption spectra

Absorption spectra of irradiated and non-irradiated samples of *Caragana frutex arborescence*, *Cytisus Laburnum anagiroides*, and *Robinia pseudoacacia* are shown on Figures 2, 3, and 4 respectively.

Extinction E is defined by the equation

$$E = \ln (I/I_0), \tag{3}$$

where I_0 is the intensity of laser emission before, and I is the intensity after passing through the sample. E was measured as a function of wavelength.

Values of E changed between 0.02 and 0.5 in. the three plant species. Generally, extinction in all investigated samples decreased from the violet to the red side of spectra, in agreement with previous results obtained in some vegetable seeds (Aladjadjiyan, 1997).

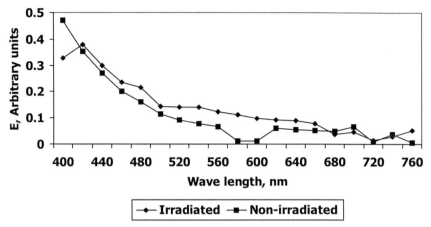

Figure 2. Absorption spectra of irradiated and non-irradiated samples of *Caragana frutex arborescence*.

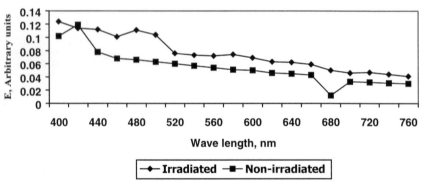

Figure 3. Absorption spectra of irradiated and non-irradiated samples of *Cytisus Laburnum anagiroides*.

Comparison of the spectra of irradiated and non-irradiated samples of different plant species suggests that laser irradiation increased optical extinction by about 10 % for the samples of *Robinia pseudoacacia*, 20 % for *Cytisus Laburnum anagiroides*, and between 2 and 30 % for *Caragana frutex arborescence*. In the first and third cases in first approximation the exponential law can be used to describe the dependence of extinction on λ. The second case is closer to linear dependence.

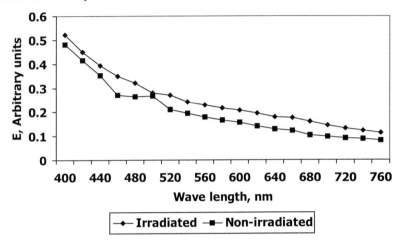

Figure 4. Absorption spectra of irradiated and non-irradiated samples of *Robinia pseudoacacia*.

The theory of optical absorption maintains that the dependence of E vs. λ can not be expressed analytically. It is a characteristic of the absorbing substance.

Nevertheless, if one assumes a linear dependence of E vs. λ, curve slopes for the three investigated plants can be defined by the equation

$$tg\alpha = dE/d\lambda. \qquad (4)$$

This assumption better fits the range of λ greater than 500 nm. Even in this range curve slopes differ markedly from one other – in the case of *Robinia pseudoacacia* the slope is three times as large as that of *Cytisus Laburnum anagiroides* (see Figure 5).

A poorly expressed peculiarity can be seen in the range 480 –520 nm for all three plants, most distinctly in *Caragana frutex arborescence*.

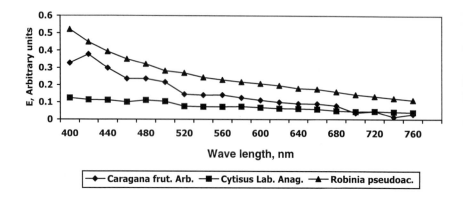

Figure 5. Comparison of irradiated-samples spectra of three plants.

4. CONCLUSION

Electrical and optical properties of *Robinia pseudoacacia*, *Cytisus Laburnum anagiroides* and *Caragana frutex arborescence* change significantly after He-Ne laser irradiation (wavelength 632,8 nm).

REFERENCES

Aladjadjiyan, A., D. Khadziatanasov, and V.Georgiev (1995). Seed coat transparency of maize, soybean and bean, Bulgarian Journal of Agricultural Sciences, 1, pp. 23–26.

Aladjadjiyan, A. (1997). Optical and electrical properties of some vegetable seeds. Proceedings of the First Balkan Symposium on Vegetables and Potatoes. V.2 (*Acta Horticulturae No.462*), pp. 445–451.

El Bassam, N. (1998). Energy plant species, ed. by James & James, London.

CROP GROWTH MODELLING OF EUCALYPTUS FOR ELECTRICITY GENERATION IN NICARAGUA

Monique Hoogwijk,[a] Richard van den Broek,[a] and Leo Vleeshouwers[b]

[a]Utrecht University, Department of Science, Technology and Society, Padualaan 14, 3584 CH Utrecht, The Netherlands, e-mail: r.vandenbroek@chem.uu.nl
[b]Wageningen Agricultural University, Department of Theoretical Production Ecology, P.O. Box 430, 6700 AK Wageningen, The Netherlands

In Nicaragua two sugarmills have extended their power production outside the sugarcane-crushing season by using eucalyptus from dedicated energy plantations. This study estimated the eucalyptus yield for power generation at the plantations of one of these sugarmills. The actual yield was estimated by extrapolating the harvest results from the eucalyptus plantation on the other sugarmill by using well-established photosynthetic principles and a simple waterbalance. The model makes a good distinction between the growth in the different months; for five months a year growth is mainly limited by radiation. The actual yield of the plantations over the total lifetime was estimated to be almost 10 $ton_{0\%}$/ha/yr. Unfortunately, the model is not validated because of the lack of data. Therefore general conclusions about its reliability could not be drawn.

1. INTRODUCTION

In Nicaragua, two sugarmills (San Antonio and Victoria de Julio) sell power to the national grid both during and outside the sugarcane crushing season. During the sugarcane crushing season the power is generated by bagasse; off-season fuel is eucalyptus from dedicated energy plantations.[1] The San Antonio sugarmill has a total area of plantations of 3000 ha and is expected to have a lifetime of 26 years. The total area of the Victoria de Julio sugarmill is 4000 ha with a lifetime of 21 years.[1] The yield per hectare is a key factor in studies of the possibility of biomass crops as an energy source.[2] No data on harvest results or expected yields are available for the plantations of the sugarmill San Antonio. The sugarmill Victoria de Julio, however, has already harvested. Although the plantations differ in climate and soil conditions, these data can be used to estimate the expected yield at San Antonio, considering the differences in radiation, precipitation, soil characteristics and rotation length. This study estimated the actual yield at San Antonio.

The potential, water-limited and actual yield of these plantations are estimated by using a crop-growth model, together with the harvest results of Victoria de Julio.

2. THE CROP-GROWTH MODEL

The potential growth is calculated in this model by using well-established photosynthetic principles. For the estimation of the water-limited yield a simplified water balance was used. The actual yield was estimated by using an extrapolation from the harvest results of the plantations of Victoria de Julio.

First, we calculate the potential yield of both sugarmills on the basis of Equations 1a and 1b.

$$G_p(d) = \varepsilon \kappa(d) * \varphi_{pa}(d) \tag{1a}$$

$$G_p(t) = \sum_{d=(t-1)*365}^{t*365} G_p(d) \tag{1b}$$

$G_p(d) =$	daily potential yield	(ton$_{0\%}$/ha/day)
$G_p(t) =$	average annual potential yield in year t	(ton$_{0\%}$/ha/yr)
$d =$	time	(day)
$t =$	time	(yr)
$\varepsilon =$	radiation utilisation coefficient	(ton$_{0\%}$/MJ)
$\kappa(d) =$	the daily reduction factor for the ability of the radiation absorption	
$\varphi_{pa}(d) =$	the absorbed photosynthetically active radiation [a,1,2]	(MJ/ha/day)

On the basis of this potential yield, the water limited yield was calculated for the plantations of both sugarmills. The water limited yield is represented in Equations 2a and 2b.

$$G_{wat\,lim}(d) = G_p(d) * f_\theta(d) \tag{2a}$$

$$G_{wat\,lim}(t) = \sum_{d=(t-1)*365}^{t*365} (G_p(d) * f_\theta(d)) \tag{2b}$$

$G_{watlim}(d) =$	estimated water-limited yield	(ton$_{0\%}$/ha/day)
$G_{watlim}(t) =$	estimated water-limited yield in year t	(ton$_{0\%}$/ha/yr)
$f_\theta(d) =$	reduction factor for the daily water limitation	

For the reduction factor for the daily water limitation we introduce the critical point (CR), defined as the amount of water in the soil that lies exactly between the wilting point

---

[a] It is assumed that 80% of the total photosynthetically active radiation (PAR) is absorbed and that PAR is 50% of the total amount of radiation.

and the field capacity. The growth is assumed to be reduced if the total amount of water in the soil is lower than this critical point.[5] The reduction factor for the water limitation, $f_\theta(d)$, is calculated with the use of Equation 3.

$$f_\theta(d) = 0 \qquad \text{if } \theta(d) \leq WP \qquad (3)$$

$$f_\theta(d) = \frac{\theta(d) - WP}{CR - WP} \qquad \text{if } WP < \theta(d) < CR$$

$$f_\theta(d) = 1 \qquad \text{if } \theta(d) \geq CR$$

$f_\theta(d) =$	daily water limitation reduction factor	
$WP =$	Wilting Point	(mm)
$CR =$	Critical Point	(mm)
$\theta(d)$	soil water content at the end of day d	(mm)

The soil water content, $\theta(d)$, is based on a waterbalance (Equation 4). For the use of the water balance it is assumed that water can enter the soil only by precipitation (the plantations are not irrigated) and leave the soil by drainage and transpiration.

$$\theta(d) = \theta(d-1) + P(d) - \frac{G_{wat\,lim}(d-1)}{WUE} \qquad (4)$$

$\theta(d-1)$	soil water content at the end of day (d-1)	(mm)
$P(d)$	precipitation at day d	(mm)
$G_{watlim}(d-1)$	water limited yield at day d-1	$(g_{0\%}/m^2)$
WUE	Water Use Efficiency	$(g_{0\%}/kg)$

The actual yield is extrapolated from the harvest results of Victoria de Julio by using the reality reduction factor (RRF), as defined in Equation (5).

$$G_{a,SA}(t) = h_{sa} * RRF(t) * G_{wat\,lim,SA}(t) \qquad (5)$$

$G_{a,SA}(t) =$	actual annual yield for the plantations of San Antonio	$(ton_{0\%}/ha/yr)$
$RRF(t) =$	reality reduction factor per rotation	
$h_{sa} =$	harvest index for the plantations of San Antonio	

The RRF is determined by the ratio between the observed actual harvest of Victoria de Julio and calculated water limited yield of Victoria de Julio. The RRF depends on the rotation time of the plantations and is defined in Equation 6.[b]

---

[b]The plantations at Victoria de Julio are first harvested after five years and each subsequent harvest is every four years. The plantations of San Antonio are first harvested after six years, and each subsequent harvest is after five years.[9,10]

$$RRF(t) = \frac{G_{a,VdJ(observed)}(t)}{h_{VdJ} * G_{wat\,lim,VdJ(calculated)}(t)} \tag{6}$$

$G_{a.VdJ\,(observed)}(t) =$ actual observed harvest results at the
plantations of Victoria de Julio (ton$_{0\%}$/ha/yr)
$h_{VdJ} =$ harvest index for the plantations of Victoria de Julio

The RRF for the first rotation at San Antonio is based on the observed actual yield of the first rotation of Victoria de Julio. The RRF for the subsequent rotations is based on the harvest results of the second rotation at Victoria de Julio. For a detailed description of the model we refer to Hoogwijk.[3]

3. THE INPUT DATA

The main input data of the model are shown in Table 1.

Table 1
The main input data.

Parameter	(mean) Value	Unit	Parameter	Value	Unit
φ_{pa}^4 (SA)	5.6[4]	MJ/m^2/day	Soil profundity	100 [10]	cm
φ_{pa}^4 (VdJ)	6.0[7]	MJ/m^2/day	κ (d) SA	0.4 for (0<d (after planting) <730) [10] 0.7 for (731<d (after planting)<2190) 0.5 for (0<d (after each harvest)<365) 1 for (366<d (after each harvest)<1825	-
ε	2.2 [4]	g$_{0\%}$/MJ	κ (d) VdJ	0.4 for (0<d (after planting) <730) [9] 0.7 for (731<d (after planting)<1825) 0.5 for (0<d (after each harvest)<365) 1 for (366<d (after each harvest)<1460)	
P(d) (SA)	5.4[7]	mm	WUE	2.7[5]	g/kg
P(d) (VdJ)	3.0 [7]	mm	Harvest results VdJ	7.1 (for 1st rotation) 10.6 (for 2nd rotation)[6]	ton $_{0\%}$/ha/yr
h, harvest index	0.85[7]	-	RRF(t)^c	0.50 for 1≤t≤6 0.51 for 6<t≤26	

4. RESULTS

The monthly results of the model on the water-limited yield vary substantially (Fig. 1). About five months per year the growth is mainly determined by the radiation (potential

---

^{c}This is not, strictly speaking, an input parameter, but an intermediate result.

growth, from June until November). In the other months, the growth is water-limited, and precipitation and soil characteristics of the plantations then determine growth.[6]

Figure 2 shows the potential, water-limited and actual yield of Victoria de Julio and San Antonio over their lifetime. The water-limited yield of Victoria de Julio appears to be higher than that of San Antonio in spite of the higher precipitation at San Antonio. This may be explained by the distribution of the precipitation; excessive precipitation during the rainy season drains off and cannot be used by the crop. Also, the water-limited yield depends strongly on the potential yield, which is higher for Victoria de Julio. Because of the extra rotation of the plantations of San Antonio the average water-limited yield over its total lifetime is slightly higher.

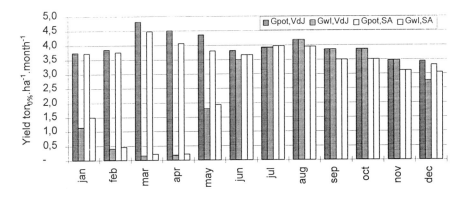

Figure 1. The monthly potential and water-limited yield distribution (with κ=1) of the plantations of the sugarmills San Antonio and Victoria de Julio in Nicaragua over one year.

5. DISCUSSION AND EVALUATION OF THE MODEL

The study shows that the crop growth model may be a useful instrument to estimate eucalyptus yields in energy plantations when relatively few data are available on weather, soil nutrients and management. However, the model makes some assumptions.

- Soil depth was assumed to be the same for both plantations. It is doubtful if the soils at San Antonio are much deeper than 200 cm, whereas the soils of Victoria de Julio may be less than 100 cm.[10] Greater soil depth at San Antonio can lead to a 12% underestimation of the yield at San Antonio.[6]
- Actual yield estimates are most strongly influenced by the RRF. This RRF assumed that the chemical and physical characteristics of the soils and the management at both

plantations are the same, except for the parameters already included. The nutrient concentration in the soil and the management at the plantations can be seen as the most important factors to influence on the RRF.
- There were no data available to validate the model used in this study. Therefore it is not possible at this moment to draw general conclusions on the uncertainty in the estimation compared to reality. We tried to validate the model using the first harvest results

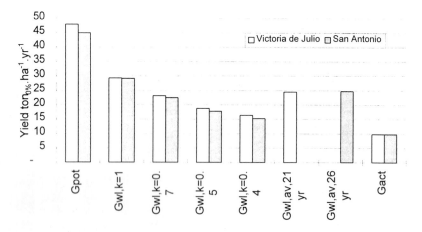

Figure 2. The mean estimated potential, water limited and actual yield of the plantations of both the sugarmills San Antonio and Victoria de Julio in Nicaragua. (wl. = water limited; av. = average). The potential yield (G_{pot}) is presented for κ=1 and the waterlimited yield for κ = 0.4, κ = 0.5, κ = 0.7, κ = 1 and as an average value over the plantation lifetime of the two sugarmills. For the actual yield (G_{act}) only the average value over the plantation lifetime is shown.

6. CONCLUSIONS

This model can be useful to gain insight into the factors that influence crop growth. It does not need many input parameters and can easily yield rapid estimations. The model has not yet been validated because data is lacking. However, because assumptions are based on observed actual yields, a large discrepancy is not expected. The resulting actual yield of the plantations of San Antonio for their total lifetime was estimated at almost 10 ton$_{0\%}$/ha/yr.

7. ACKNOWLEDGEMENTS

We would kindly like to acknowledge the people in Nicaragua for their help during the fieldwork in Nicaragua: Ricardo Coronel of the sugarmill Victoria de Julio, Mayro Antón of the Los Maribios Project and Serafín Filomeno Alves-Milho, María Engracia Detrinidad and Rogerio de Miranda of the NGO Proleña. Above all we would like to thank Pedro Silva of the sugarmill San Antonio, for his help and useful information.

REFERENCES

1. Broek, R. van den and A. van Wijk (1998). Heat and power from eucalyptus and bagasse in Nicaragua: Part A: Description of existing initiatives, Proceedings, 1998 10th European Conference and Technology Exhibition on Biomass for Energy and Industry, ed. by Kopetz et al., Wurzburg, pp. 1724–1727.
2. Broek, R. van den and A. van Wijk (1996). Methodologies for environmental, micro- and macro-economic evaluation of bioenergy systems, Paper presented at the European Conference on Biomass for Energy, Environment, Agriculture and Industry, Copenhagen.
3. Lövenstein, H., E.A., Lantiga, R. Rabbinge and H. van Keulen (1995). Principles of Production Ecology, Wageningen
4. Landsberg, J.J. and F.J. Hingston (1996). Evaluating a simple radiation/dry matter conversion model using data from Eucalyptus globulus plantations in Western Australia, Tree Physiology, 16, pp. 801–808.
5. Kropff, M.J. and H.H. van Laar (1993). Modelling Crop-Weed Interactions. Wallingford, CAB Internatioinal.
6. Hoogwijk, M. (1998). Crop growth modelling of Eucalyptus plantations in Nicaragua. Report 98076, Department of Science, Technology and Society, Utrecht University, The Netherlands, Utrecht, pp. 44.
7. Instituto Nicaragüense De Estudios Territoriales (INETER) (1998). Meteorological database, Managua.
8. Schneider, L., A. Kinzig and L. Solórzano (1996). Biomass project in NE Brazil: Identification of potential sites for biomass-energy production. Internal report, Centre for Energy and Environmental Studies-Princeton University.
9. Coronel, C. (1998). Personal communication, Tipitapa.
10. Silva de la Maza, P. (1998). Personal communication, Chichigalpa.

Biomass Production and Integration With Conversion Technologies

Feedstocks: Characterization, Drying, and Densifying

DEVELOPING NEW ANALYTICAL TECHNIQUES FOR THE CHARACTERIZATION OF LIGNOCELLULOSICS BASED ON MULTIVARIATE CHEMOMETRIC ANALYSIS OF WHOLE FLUORESCENCE SPECTRA (AFFLUENCE)

Emmanuel G. Koukios and Evaggeli Billa

Bioresource Technology Unit, Dept. Chemical Engineering, National Technical University of Athens, Zografou Campus, GR - 15700 Athens, Greece, tel: (301) 7723287; fax: (301) 7723163, e-mail: billa@chemeng.ntua.gr

The objective of this paper is to present specific selected examples of a new analytical exploitation of the fluorescence of lignocellulosics through chemometrics. In particular, information from whole fluorescence emission spectra has been subjected to principal components analysis (PCA) and the results demonstrated the potential of fluorescence spectroscopy for the characterization of paper pulps from non-wood fiber sources. Moreover, the presence of specific correlations of technical interest among the fluorescence data and the chemical properties of the samples (composition, K number, other characteristics) are investigated by partial least square (PLS) regression. Examples of the application of new techniques in the field of paper pulp production and characterization will be presented in order to illustrate the potential of the new approach of multivariate chemometric analysis of whole fluorescence spectra, named AFFLUENCE, and its distinct advantages as a process and product analytical method.

1. INTRODUCTION

Fluorescence spectroscopy has been widely used as an analytical tool in a variety of chemical and biochemical investigations: it is a cheap, rapid, sensitive, specific and non destructive method.[1] Although lignocellulosics exhibit autofluorescence, the application of fluorescence spectroscopy in this field has not been very fruitful. Interpretation of fluorescence spectra from lignocellulosics is difficult because of their physical heterogeneity and chemical complexity.[2,3] In fact, the use of excitation and emission spectra has so far been used with the aim of finding a pair of excitation-emission wavelengths where the analyte of interest is the only component giving rise to the recorded signals. This approach is questionable as far as lignocellulosics are concerned, as they contain many fluorophores.[4]

The chemometrics approach proposed by our work is based on the application of appropriate mathematical and statistical methods in order to extract reliable and relevant information from whole fluorescence spectra.[5]

In this paper, we present selected examples from recent applications of this new approach.[6,7] Fluorescence spectral data through PCA can reveal information about the origin of the fibrous raw material; they are correlated through PLS (partial least squares) models with physical and chemical properties of lignocellulosics, e.g., the yields of various pulping processes, the Kappa number and the lignin content of the corresponding paper pulps.

2. MATERIALS AND METHODS

2.1. Chemometric methods

Principal Component Analysis (PCA). Principal Component Analysis finds the main variation in a multidimensional data set by creating new linear combinations of the raw data.[5,8] In matrix form we have: **X = TP'**
X: the analyzed data matrix with dimensions **s x w,**
T: the score matrix with dimensions **s x min (s, w)**, and
P: the loading matrix with dimensions **s x min (s, w)**.

Only a significant number of principal components (PCs), f, equal to the chemical rank of the **X** matrix is relevant in describing the information in **X**. This leads to the following decomposition of **X**. **X = T$_f$P$_f$ + E**, where T$_f$ is the score matrix with dimensions **s** × **f**, P$_f$ is the loading matrix with dimensions **w** × f, and **E** is a residual matrix with the same dimensions as **X**.

Partial Least Square Regression (PLS). PLS is used in order to make a model correlating **X** and **y**, where **X** contains the fluorescence spectra and **y** (s × 1) is a vector containing the property of interest. The model performance is validated by cross validation due to our small data set. Multivariate calibration models were built correlating the fluorescence emission spectra and the physical as well as chemical properties of the samples. PCA and PLS regressions were performed with the UNSCRAMBLER program (version 6.0, Camo AS, Norway). The predictive performances of the PLS models are assessed by the RMSEP (root mean square error of prediction) and Pearson's correlation coefficient R.

2.2. Recording of fluorescence spectra

Solid lignocellulosic samples were ground in order to obtain an homogeneous surface. Fluorescence spectra were recorded on a Perkin Elmer LS 50B Luminescence Spectrometer connected with a PC. Emission spectra were recorded at excitation wavelengths of 450, 400, 350 and 280 nm whereas the emission is measured in the region of 275 to 650nm with 0.5nm intervals (in total 751 data points); for this purpose an OBEY program was written. Excitation and emission monochromator slit widths were 3 nm and a mirror, absorbing the 99% of the emitting radiation, was used. The measurement starts from the largest and finishes with the lowest excitation wavelength in order to minimize photodecomposition of the sample.

Spectral data were converted to ASCII files by a program furnished by Perkin Elmer (FL Data Manager, version 3.50). For the chemometrics analysis the ASCII files were introduced to the UNSCRAMBLER program (version 5.5).[8]

3. RESULTS AND DISCUSSION

The potential of the new approach for the evaluation of the fluorescence spectroscopic data will be illustrated by two example cases.

3.1. Analysis of wheat, sweet and fiber sorghum organosolv pulps

The raw fluorescence emission spectra of solid wheat straw and sorghum pulps, corresponding to four excitation wavelengths, were put sequentially creating a matrix with 3004 variables. Each of the 70 spectra is the average of two measurements. The spectral shapes of the different samples are similar with common high peaks corresponding to the Rayleigh scatter (Figure 1). Rayleigh scatter, which makes a large contribution to the emission spectrum at the emission wavelength corresponding to the actual excitation wavelength, was removed from the spectral emission region; so, from the 3004 initial wavelength-variables, we considered only 1354 (Figure 2). It was noted that the fluorescence intensity of sweet sorghum pulps is higher than that of fiber sorghum and wheat straw. Moreover, we observed a shift of the emission maxima according to the treatment, whether acid or alkali.

In order to see whether our samples could be classified according to their fluorescence fingerprint, a PCA model of 70 samples (matrix X: 70 × 1354) was built using test set validation. The variance explained by the first four principal components (PC) resulting from the principal component analysis (PCA) was 100%. The score plot of PC4 versus PC1, explaining 80% of the total variance, is shown in Figure 3 and demonstrates the presence of three clusters according to the plant origin, wheat straw (W), fiber sorghum (F) and sweet sorghum (S).

The score plot of PC3 versus PC2, explaining 20% of the total variance, is depicted in Figure 4. We observe the presence of six clusters according to the treatment (alkali or acid) and the plant origin: i) acid-treated sweet sorghum samples (SO); ii) alkali-treated sweet sorghum samples (SA); iii) acid-treated fiber sorghum (FO); iv) alkali-treated fiber sorghum (FA); v) acid-treated wheat straw pulps (WO); vi) alkali-treated wheat straw pulps (WA). Fiber sorghum and wheat straw clusters are very close with some misclassifications in the case of alkali- treated fiber sorghum samples (FA), indicating similarities in their structure.

A PLS model for the Kappa number of all the 70 samples was built. The Kappa number ranged from 41.6 to 87 for the fiber sorghum pulps, from 28.5 to 57.4 for the sweet sorghum and from 49.2 to 65.7 for the wheat straw organosolv pulps. The correlation coefficient (R = 0.88) was good and was considerably improved (R = 0.93) when five samples indicated by the model as outliers were taken out (Figure 5). Considering each raw material separately, the correlation coefficient for the Kappa number was higher than 0.90 (Figure 6).

Figure 1: Whole emission spectra as introduced to the UNSCRAMBLER program. Emission variables 1-751 correspond to λ_{ex} = 450nm, variable 752-1502 correspond to λ_{ex} = 400nm, variables 1502-2253 correspond to λ_{ex} = 350nm, variables 2253-3004 correspond to λ_{ex} = 280nm.

Figure 2: Concatenated emission spectra after the removal of Rayleigh peaks. Variables 1-210: λ_{ex} = 450nm, emission 470-575nm; Variables 211-461: λ_{ex} = 400nm, emission 435-575nm; Variables 462-892: λ_{ex} = 350nm, emission 375-575nm; Variables 893-1354: λ_{ex} = 280nm, emission 305-535nm.

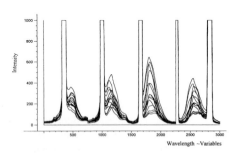
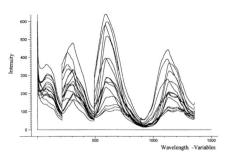

Figure 3: PCA score plot of principal component 1 (PC1) versus PC4 of the 70 paper pulp samples explaining 79% and 1% of the total variance respectively. **W**: wheat straw; **S**: sweet sorghum; **F**: fiber sorghum.

Figure 4: PCA score plot PC 2 versus PC3 of the 70 paper pulp samples explaining 12% and 8% of the total variance respectively. W:wheat straw; S: sweet sorghum; F: fiber sorghum; **O**: acid treatment; **A**: alkali treatment.

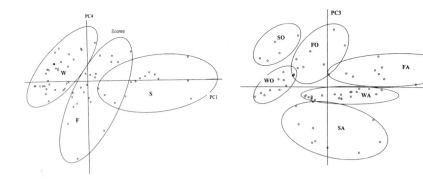

The lignin content ranged from 8.5 to 11.7% for the fiber sorghum pulps, from 6 to 12.4% for the sweet sorghum and from 7.4 to 15.8% for the wheat straw organosolv pulps. The PLS models for the Klason lignin content gave good correlation coefficients, slightly lower than the ones for the PLS models referring to the Kappa Numbers. In particular, when 65 samples were taken into account; the R value was 0.82 and became 0.94 when only the fiber and sweet sorghum samples were considered.

Figure 5: PLS model of the predicted versus measured values for the Kappa number of the 65 samples. R = 0.93 for the fifth PC.

Figure 6: PLS model of the predicted versus measured values for the Kappa number of the fiber sorghum samples. R = 0.96 for the fourth PC.

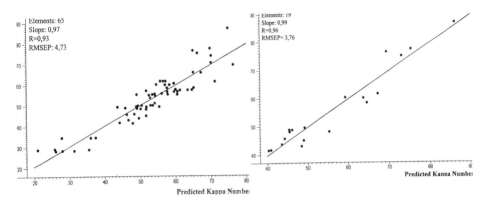

3.2. Study of juvenile and adult eucalyptus trees

In this case we looked for correlations between fluorescence data and the physical–chemical properties and the pulping data of adult and juvenile eucalyptus trees. This information is important to determine the suitable age for harvesting as well as to select criteria for the improvement of the species.[7]

Raw fluorescence emission spectra of ground adult and juvenile eucalyptus wood were obtained as described in the previous case. The four recorded emission spectra corresponding to the four excitation wavelengths were put one after the other and introduced as rows in an X matrix whose dimensions for the 20 samples is 20 × 3004. Rayleigh scattering was removed from the spectral emission region, giving rise to a 20 × 1115 matrix.

Principal component analysis of the 20 samples and their duplicate measurements gives the score plot depicted in Figure 7. The variation explained by the first three principal components (PC) is 99%. The three-dimensional representation of these PCs shows a clear clustering of juvenile wood samples (N) as well as adult wood samples (A). This indicates that juvenile and adult wood samples can be discriminated by their fluorescence spectra.

Moreover, we can remark that the duplicate measurements are close enough implying little effect of the heterogeneity of the samples' texture.

Furthermore, in Figure 8 is depicted the PLS model for the yield (%) of the kraft pulps issued from the different clones. The value of the correlation coefficient for the yield is R = 0.86. In the case of the Kappa number the R = 0.68. According to these data, adult trees are characterised by higher pulping yields and lower Kappa numbers indicating that they are more suitable for the paper pulp production.

Figure 7: A PCA 3-dimensional score plot of adult (A) and juvenile (N) eucalyptus clones.

Figure 8: PLS modeling. Predicted versus measured values for the yield of the kraft pulping of the different eucalyptus wood (third PC).

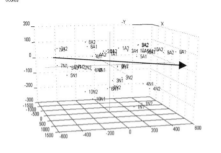
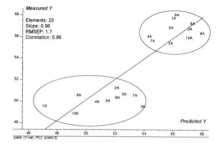

4. CONCLUSIONS

The results indicated that the fluorescence emission spectra of solid paper pulps through principal component analysis can give information about the origin of the pulp samples as well as the kind of treatment. Moreover, the existence of a good correlation between the fluorescence data and the Kappa number as well as the Klason lignin contents in the pulps from different raw materials was shown. These findings are promising indications of the potential of fluorescence through chemometrics as process analytical method applied to lignocellulosics. With further studies on various sample series obtained during a pulping process, the chemometrics interpretation of solid pulp fluorescence may become a control tool for monitoring the pulping operation.

REFERENCES

1. Munck, L. (1989). Practical experiences in the development of fluorescence analyses in an applied food research laboratory, in Fluorescence Analysis in Foods, ed. by Lars Munck, Longman Scientific & Technical, UK. pp.1–32.
2. Olmstead J.A. and D.G. Gray (1993). Fluorescence emission from mechanical pulp sheets J. Photochem. Photobiol. A: Chem., 73, pp. 59–65.
3. Beyer M., D. Steger, and K. Fischer (1993). The luminescence of lignin-containing pulps - A comparison with the fluorescence of model compounds in several media J. Photochem. Photobiol. A: Chem, 76, pp. 217–224.
4. Castellan, A., H. Choudhury, R.S. Davidson, and S. Grelier (1994). Comparative study of stone-ground wood pulp and native wood. 2. Comparison of the fluorescence of stone-ground wood pulp and native wood. J. Photochem. Photobiol. A: Chem., 81, pp. 117–122.
5. Noergaard, L. (1995). A multivariate chemometric approach to fluorescence spectroscopy, Talanta, 42, pp. 1305–1324.
6. Billa, E., E. Koutsoula, and E. Koukios (1999). Fluorescence analysis of paper pulps. Bioresource Technol., 67, pp. 25–33.
7. Billa, E., A. Pastou, E.G. Koukios, J. Romero, and B. Monties. Multivariate Chemometric analysis of the fluorescence spectra of juvenile and mature eucalyptus wood submitted.
8. Esbensen, K., S. Schonkopf, T. Midtgaard (1994). Multivariate Analysis in Practise, Camo AS, Norway.

QUALITY OF SOLID BIOFUELS - DATABASE AND FIELD TRIALS

H. Hartmann, L. Maier and T. Böhm

Munich University of Technology, Research Center of Agricultural Engineering,
Voettinger Strasse 36, D-85354 Freising-Weihenstephan, Germany

Quality aspects of solid biofuels were investigated in a new database. Most parameters varied greatly, particularly when annually harvested biomass was considered. For planning purposes the frequency distributions should be used rather than mean values. The quality of some crops may be changed by modified agricultural practices. Rainfall shortly after cutting can deplete chlorine and potassium in grass by 60 to 80%.

1. PROBLEM DEFINITION AND APPROACH

Quality aspects of solid biofuels must increasingly be known to meet technical and environmental requirements. However, the chemical composition of a relevant biomass is uncertain, particularly when new biofuel resources are regarded. This research set up a database and the conducted of field trials in order to better understand the factors that influence biofuel characteristics and to devise measures to fulfill a specific quality demand. The results of this work should aid common standardization activities of many European countries.

2. DATABASE ACTIVITIES

2.1. Database structure and data acquisition

An SQL-database was designed to hold a large range of information on each fuel sample studied. In order to precisely specify a given biomass type and origin, results of analyses, the analysis method, the data source, geographical location, special treatments and other criteria were recorded (Figure 1). An initial series of 1,200 data sets was entered until February 1999. These data were either generated during research projects or were found in other primary sources in Germany and Austria. Additionally, 200 relevant institutions or companies were surveyed and the most reliable data were selected and applied.

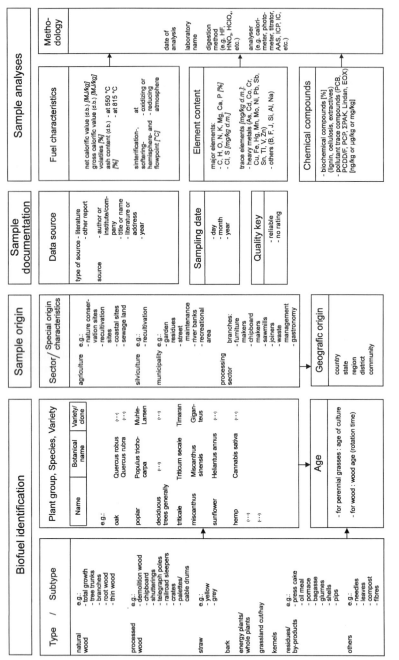

Figure 1. Structure of database for solid biofuel characteristics (simplified).

2.1. Selected database evaluation results

The database was queried for various fuel types and species. When the number of data in a data set was higher than 10, outliers were eliminated by cutting off the extreme upper and lower 10 percent of the full data range. Thus, the results should better describe the situation to be expected in practice. Nevertheless, for some parameters the variation within the remaining data is still very large, as demonstrated for the chlorine content in Figure 2. This large variation particularly applies to annually harvested crops, while wood fuels show comparatively little scatter in most parameters.

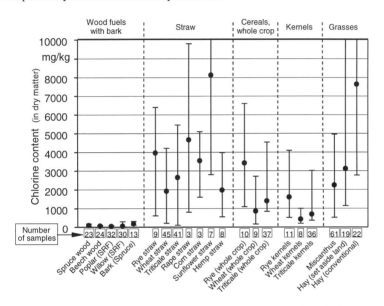

Figure 2. Mean values and range of chlorine contents in selected biofuels.

The large range in chlorine content raises questions about the reliability of using mean values. Particularly for annually harvested agricultural crops, whose quality characteristics are likely to be affected by environmental or agricultural influences, a more precise description of the fuel is desirable. Therefore the database query function was coupled with an evaluation routine for a frequency distribution analysis. These calculations were made for cereal straw (Figure 3). The results show that for a few parameters, such as the net calorific value or the nitrogen content, there is a relatively high probability of a yet unanalyzed sample lying near the mean values (here: 17.17 MJ/kg and 0.48%, respectively). However, for many other parameters, such as the chlorine content or the ash softening temperature, predictions are highly uncertain (mean values: 2,550 mg/kg and 962°C, respectively). This uncertainty is due to the rather flat and often skewed shape of the frequency distribution plot. Here, questions are raised about the kind and effect of the quality influencing factors, which are discussed in Section 3.

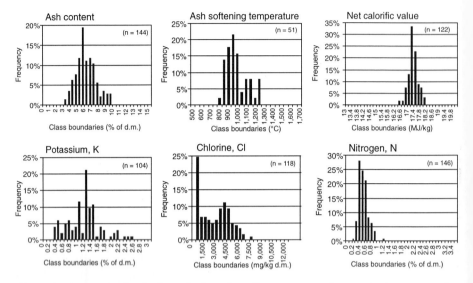

Figure 3. Frequency distribution of selected quality parameters from a database query for similar cereal straw types (wheat, rye and triticale) (Data here without outlier elimination).

3. FIELD TRIALS

The relatively large parameter variations of annually harvested crops may reflect various influencing factors during their production phase. As implied by Figure 4, these factors are manifold. The leaching effects of natural rainfall were here investigated in field trials. The results for the experiments with grass from set-aside land are given in Figure 5. The crop was spread out and left on the field after cutting. It was found that potassium and particularly chlorine were highly mobile and they therefore were depleted rapidly. The depletion actually began before the grass was cut, during a progressive dying off of the crop, as indicated by an increase of the dry matter content (8 to 13 percent-points until cutting date). As the growing or retention time increases, the texture of the died-off crop becomes increasingly porous, thus allowing any precipitation to drain away soluble elements into the ground.

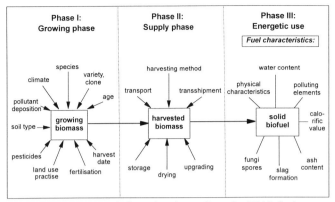

Figure 4. Factors influencing the quality of solid biofuels.

However, other elements such as nitrogen and sulfur (and magnesium and calcium, not shown in Figure 5) are largely unaffected by leaching. Smaller and shorter rainfall events are believed to be less effective than longer rainy periods, due to the interception rate of the exposed crop. This conclusion can also be drawn from the data in Figure 5. The smaller rainfall events of 1, 3 and 7 mm following the cutting did not affect element concentration. The depletion to a final level of only 23 and 39% compared to the mowing date can therefore be attributed to only two rainfall events of 37 and 27 mm.

Leaching effects on the total ash content were not observed. Rather an increase was recorded shortly before the cutting date (Figure 5, center). This increase may be attributed to secondary pollution during the mowing process (disc cutter) and by dust emissions through wind drift. These factors can obviously override any possible leaching effects.

Ash content variations usually affect the calorific value, but effects are here hardly measurable and the results for this parameter largely remained constant throughout all field trials. Ash softening behavior, however, was generally improved by an increased field retention time, which is indicated by the marked increases of the sinterification, softening, hemisphere and melting temperatures (Figure 5, bottom). Parallel field trials with wheat straw produced comparable results.

4. CONCLUSIONS

The quality of solid biofuels varies within more or less wide ranges, depending on the individual species, the parameter in question, and the history of the given material. For many characteristics therefore the simple consideration of mean values is unsuitable for planning purposes. Mean values are not suitable for those elements that are highly mobile within the plants (e.g., Cl and K) and for the ash softening behavior, which is affected by changes in the

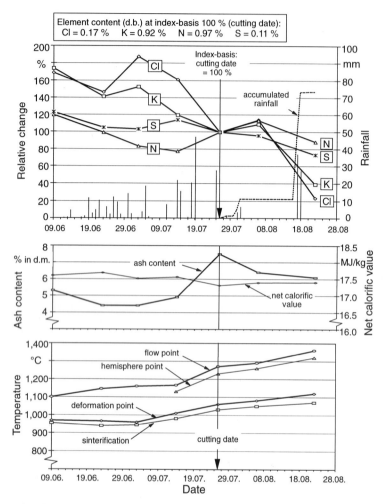

Figure 5. Effect of harvesting date and field retention time (leaching) on selected quality parameters in grass from set aside land.

ratio of the relevant elements towards each other. Database evaluations can here be helpful for a more reliable description of a given fuel resource.

Both database evaluations and field tests also imply that there is a large range of possible measures for quality manipulations. However, due to the complex interactions between the influencing factors, quantitative predictions for their magnitude are yet highly uncertain. Additionally, most of the possible corrective measures increase costs. These restrictions should be considered in the ongoing standardization activities.

5. ACKNOWLEDGEMENT

The research is funded by the Bavarian State Agency for Environmental Protection in Munich.

DRYING IN A BIOMASS GASIFICATION PLANT FOR POWER OR COGENERATION

J.G. Brammer and A.V. Bridgwater

Bio-Energy Research Group, Aston University, Birmingham B4 7ET, United Kingdom

Many biomass feedstocks for gasification-based power and cogeneration plants have a typical moisture content of 50% (wet basis - w/b) on delivery to the plant. For most gasifier types, it is essential to dry the biomass prior to gasification to obtain a satisfactory producer gas, and broadly a drier feed improves the quality of the producer gas (although moisture may be necessary for temperature control). Drying is a process requiring energy and capital equipment, however, and there is a trade-off between the cost involved and the benefit obtained. This is particularly acute in cogeneration plants where surplus heat which might otherwise be used for drying is in demand for the heat load. In this paper, issues relating to integrated drying are discussed, and a number of suitable technologies are described. The paper is based on a review carried out as part of a European Community JOULE project on slagging gasification of biomass (JOR3-CT97-0130), with the purpose of identifying suitable drying technologies for such a process (Brammer, 1998).

1. INTRODUCTION

This paper considers the drying of virgin biomass in the form of loose particulate solids, for use in a biomass gasification plant for power or cogeneration. The biomass will have been derived either from agricultural or forestry residues, or from dedicated plantations of short rotation coppice or herbaceous crops, usually comminuted to the required size.

Such material will usually have a moisture content on delivery to the gasification plant in the range 30-60% w/b, depending on type, location, time of harvest and period of storage after harvest. Most biomass gasification processes, however, require rather lower moisture contents, and it is often either necessary or desirable to dry the feed to a moisture content of 20% w/b or less. This requires an evaporative process.

Forced evaporative drying requires large amounts of energy as well as the provision of often expensive equipment. The impact of the drying operation on overall plant efficiency can however be reduced by integration, making use of surplus energy streams within the process.

Many types of dryer or drying process are encountered in industry, reflecting the wide range of drying tasks. For many materials including biomass gasifier feedstocks, a number of different types may be suited to the task, and the final choice is made after careful consideration of operational and economic factors specific to the application. In the present study, only commercially available types of dryer are considered. The reader is referred to the large number of texts in the literature covering drying theory and practice (e.g., Mujumdar, 1995).

2. THE DRYING PROCESS

2.1. Material physical characteristics

Particle size requirements are dictated largely by the gasification process. A throated downdraft gasifier requires particle sizes of 20-80 mm, whereas an updraft gasifier can accept sizes of 10-200 mm and a fluid bed gasifier typically requires a mean particle size of < 10 mm.

The material will usually have a bulk density in the range 50-300 kgm^{-3}, depending on type and moisture content. Usually the bulk material will have only moderate flow properties, but will readily permit through-circulation of the drying medium. Generation of significant amounts of additional dust within the drying process is not normally a problem, as most such materials are not friable under all but the most strongly agitated of processes.

Biomass materials of this type are hygroscopic, and equilibrium moisture content during drying can vary from 25% w/b or more if the drying medium is close to saturation, down to below 1% w/b where the medium is exceptionally dry. Under ambient air conditions typical of European locations the equilibrium moisture content of woody biomass will fall in the range 15-25% w/b. Clearly if much lower moisture contents are required, there are consequences for post-dryer handling.

2.2. Sources of heat for drying

Evaporative drying processes require heat exchange, by convection or conduction. Possible sources of heat within a biomass gasification plant which may be utilised for drying include:
- hot gasifier product gas to be cooled
- hot engine or gas turbine exhaust
- dedicated combustion of additional or by-product biomass, or diverted product gas
- hot air from the air cooled condenser in a steam or combined cycle plant
- hot water from engine cooling or condenser in a steam or combined cycle plant.

Some of these sources may only be used indirectly, via a secondary medium (air or steam). In the case of cogeneration plants in which these sources of heat are in demand to raise steam for external use, their use for drying may be in direct competition.

2.3. Gaseous emissions

Gaseous emissions from biomass drying arise either from vaporisation of volatile components in the biomass, or from thermal degradation of the biomass. Condensable volatile components create the greatest problem, giving rise to so-called "blue haze," a discoloration of the exhaust plume that can represent an odour and visual nuisance as well as a potential safety hazard. These components tend to be released at material temperatures

> 100°C. Thermal degradation products are released at high wood temperatures (> 200°C for short drying times, rather less for longer drying times), and the rate of loss accelerates rapidly as temperature is increased further. Such degradation at the drying stage represents an energy loss to the overall process.

Clean-up equipment for the exhaust gas stream may be necessary, depending on local legislative restrictions which vary greatly with location. Blue haze is composed largely of sub-micron aerosols, and these are notoriously difficult to remove with conventional gas cleaning techniques such as wet scrubbers. As a general rule, therefore, material temperatures in open systems should be kept below 100°C for all processes where possible. Further information on emissions from the drying of biomass materials is given by Wastney (1994).

2.4. Fire or explosion risk

A dryer fire or explosion can arise from ignition of a dust cloud if substantial amounts of fines are present, or from ignition of combustible gases released from the drying material.

Both causes of ignition require the presence of sufficient oxygen and either a sufficiently high temperature or some other source of ignition. Under conditions found in most dryers, the risk of fire or explosion becomes significant if the drying medium has an O_2 concentration over ~10% (vol.). Under these circumstances high-drying-medium inlet temperatures should be avoided. Problems have been encountered with wood chips in rotary dryers at inlet temperatures of well below 300°C, and a probable safe working limit is in the region of 250°C. If a low-oxygen environment can be guaranteed, much higher inlet temperatures may be used provided material temperatures do not become excessive (see previous section); however, prevention of accidental air in-leakage can be difficult and expensive.

3. DRYER TYPES FOR BIOMASS GASIFIER FEEDSTOCKS

3.1. Perforated floor bin dryer

The perforated floor bin dryer (Figure 1) is a batch through-circulation method, used primarily for drying grain. Although continuous methods are usually preferred in biomass gasification plants, batch methods can sometimes be attractive because of their low capital cost. The dryer would be suited to small plants of perhaps 1 MW_e or less.

The basic system comprises a bin with a perforated floor through which hot air is made to flow. Drying performance is strongly influenced by drying medium humidity at inlet, and exhaust gases are normally too humid to be used directly.

The wet material forms a fixed bed above the perforated floor. A relatively shallow bed depth of 0.4-0.6 m has been recommended by Nellist (1997). The batch time is typically a few hours. It is possible to automate the material loading and unloading to varying degrees, greater degrees of automation requiring greater capital outlay, particularly for larger systems. Low velocities and temperatures usually make clean-up of the gaseous exhaust unnecessary.

A drawback with fixed bed through-circulation methods is that they invariably produce a large vertical gradient in moisture content of the dried bed. The dried material therefore needs to be thoroughly mixed before use, and possibly allowed to equilibrate for a period in a buffer store. The lower levels of the bed almost always reach the inlet temperature of the drying medium, and this effectively limits the latter to 100°C or less (see Section 2.3).

Perforated floor bin dryers are not sold commercially specifically for biomass gasifier feedstocks, and they are likely to require modification, particularly to the perforated floor.

3.2. Band conveyor dryer

Continuous through-circulation dryers are most often used for materials that require gentle handling; however, they have a number of features that make them attractive for biomass gasifier feedstocks. The most widely used and the most appropriate for such feedstocks is the band conveyor dryer (Figure 2), in which the drying medium is blown through a thin static layer of material on a horizontally moving permeable band. At least one commercially available system (made by Saxlund) has been designed specifically for fuel-wood chips.

Gas flow may be upward or downward. In the single-stage single-pass design (shown in Figure 2), a continuous band runs the whole length of the dryer. In multi-stage single-pass designs, a number of bands are arranged in series horizontally, and in multi-pass designs a number of bands are installed one above the other, each discharging onto the band below.

The drying medium is usually either air or combustion products, and it is moved through the dryer by fans, either with or against the movement of the material. The dryer is usually divided up into zones through which the drying medium progressively passes. Each zone will have either a steam heater or a gas or oil burner (air systems) or a port for the progressive admixture of hot exhaust gas. Residence time and, in most designs, temperature can be closely controlled. Temperature is limited to around 350°C because of the problems of lubricating the conveyor, chain and roller drives. Because of the relatively shallow depth of material on the band (usually in the range 2-15 cm) the uniformity of drying is very good.

Entrainment of fines with this type of dryer should be low due to the low velocities and static material bed. A bag filter may still however be needed, depending on local regulations.

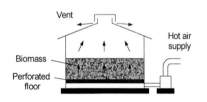
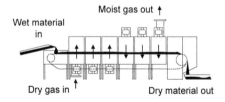

Figure 1. Perforated floor bin dryer.   Figure 2. Band conveyor dryer.

3.3. Rotary cascade dryer

The rotary cascade dryer (Figure 3) is very widely used in industry for a wide range of materials. It is by far the most common type found in the limited number of existing large-scale biomass gasification projects, as well as in large wood-chip combustion plant.

The dryer consists of a cylindrical shell, inclined at a small degree to the horizontal and rotating at 1-10 r.p.m. The shell diameter can range from < 1m to > 6m depending on throughput. Material is loaded into the shell at the upper end, passes along the shell and exits at the lower end. The drying medium (either heated air or combustion products) may flow either way, although parallel-flow is usually necessary for heat-sensitive materials including biomass gasifier feedstocks. On the inside of the shell are a number of longitudinal flights which lift the material and cascade it in a uniform curtain through the passing gases.

Overall heat transfer is less effective than for through-circulation systems, and significantly larger volumes of gas are required. Inlet temperatures of up to 1000°C have been used with very wet materials in parallel flow configurations (as the drying medium loses temperature rapidly during the initial drying stages), although temperatures above about 600°C require either expensive alloy steels or refractory lining. However, because of the risk of fires, inlet temperatures when drying wood chips using exhaust gas are often kept to below 250°C (see Section 2.5). Sealing problems leading to air in-leakage have been a weakness in many installations trying to maintain a low-oxygen environment.

Many systems are offered with heating equipment at inlet, often in the form of a gas burner. In at least one system for drying wood chips, fines separated in the exhaust cyclone are used to fire a small combustor to provide additional hot gases for the dryer. The unavoidable carry-over of entrained material and the relatively large gas volume flow rates make sizeable cyclones and bag filters usually essential.

3.4. Fluid bed steam dryer

A Danish company, Niro, has recently developed a unique design of pressurised steam fluid bed dryer (Figure 4) specifically for moist fibrous particulate materials. The system was designed primarily for the brewing, food and sugar processing industries, but at least two systems are in operation for the drying of wood chips.

Recycled moisture evaporated from the feed forms the low-pressure steam drying and fluidising medium. The bed of material is contained in 16 cells, arranged around a central high-pressure superheated steam heat exchanger. After leaving the cells the low-pressure steam passes through a cyclone for dust separation, after which the excess steam from evaporation is discharged, and the remainder passes down through the heat exchanger to be heated indirectly to about 200°C by the high-pressure steam (max. 25 bar g, 250°C), before returning to the bed distributor plates. The continuously discharged flow of evaporated steam is at about 2.8 bar g and 150°C, and may therefore be used as process steam elsewhere.

Niro claim a very efficient process because of the utilisation of the recovered steam, although a suitable external use must exist. In a biomass gasification plant this would only be the case with cogeneration applications. Niro also claim excellent environmental performance as a result of the system being fully closed with no gaseous emissions to atmosphere.

The minimum size of dryer offered has a capacity at 25 bar steam pressure of 3 tonnes per hour water evaporated, and the largest 40 tonnes per hour. Niro have suggested 5 tonnes per hour as a minimum for a viable installation drying wood chips. Capital cost is high, and the dryer is likely only to be suited to relatively large scale plant (over ~10 MW_e).

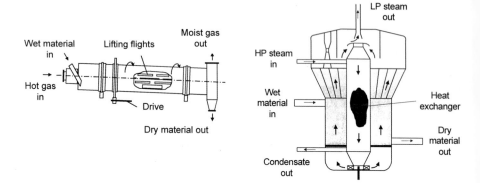

Figure 3. Rotary cascade dryer.   Figure 4. Fluid bed steam dryer (Niro A/S).

3.5. Pneumatic conveying steam dryer

Pneumatic conveying dryers achieve rapid drying with short residence times by fully entraining the material in a high velocity gas flow, usually air. They are normally used for drying very small particles. However, a closed-loop pneumatic conveying dryer has been developed in Sweden for drying biomass at sizes of up to 50 mm, using only indirectly heated steam from the liberated moisture as the conveying and drying medium. The technology, which was developed primarily for the wood products industry, is now owned and marketed as the "Exergy" steam dryer by Stork Engineering (Figure 5).

The design philosophy of the Exergy steam dryer has much in common with the Niro steam dryer, including a high degree of energy recovery and zero gaseous emissions. The drying section consists of a sequence of vertically orientated shell-and-tube heat exchangers, through which the material is conveyed and the liberated steam superheated. The heat is usually supplied by high-pressure condensing steam. The dried material is extracted in a cyclone, and the near-saturated steam continues around the loop to a steam extraction point where the excess is continuously bled off at between 2 and 6 bar, available for external use. The remainder of the steam continues to a first stage of superheater before reaching the material inlet point and re-entering the drying section. Typical residence time of material in the dryer is 10-30 seconds, and good uniformity in final moisture content is achieved.

Installations are typically large, in excess of 10 tonnes/hour evaporation (~10 MW_e). Attractions are the recovery of low-pressure steam and zero gaseous emissions. For a biomass gasification plant however, recovered steam could be used only in cogeneration applications.

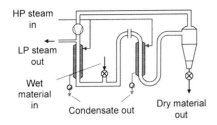

Figure 5. "Exergy" steam dryer (Stork Eng.)

4. CONCLUSIONS

An integrated drying operation is usually necessary in a biomass gasification plant. At small scales (below ~1 MW$_e$) costs are likely to dictate either a batch perforated-floor technology using heated air, or a simple band conveyor using exhaust gas. At intermediate scales (below ~10 MW$_e$), the rotary cascade dryer will probably continue to dominate, with band conveyor designs a possible alternative. At larger scales (over ~10 MW$_e$) in steam cogeneration applications, the use of steam dryers (fluid bed or pneumatic) may offer efficiency advantages.

5. ACKNOWLEDGEMENT

The financial support of the European Commission JOULE Programme is acknowledged.

REFERENCES

Brammer, J.G. and A.V. Bridgwater (1998). Drying Technologies for an Integrated Gasification Bio-energy System, EC JOULE Programme Report (Project JOR3-CT97-0130).
Keey, R.B. (1991). Drying of Loose and Particulate Materials, Hemisphere Press.
Wastney, S.C. (1994). Emissions from Wood and Biomass Drying: a Literature Review, New Zealand Forestry Research Institute Report.
Nellist, M.E. (1997). Storage and Drying of Short Rotation Coppice, ETSU Report (UK).

FOREST AND AGRICULTURAL HANDLING TO ELECTRICITY PRODUCTION USING BIOMASS IN MARAJÓ ISLAND, BRAZIL

I.M.0. da Silva,[c] I.T. da Silva,[a] B.R.P. da Rocha,[a] C. Monteiro,[d] E.C.L. Pinheiro,[a]
S.B. Moraes,[b] A.0.F. da Rocha,[a] V. Miranda,[d] J.P. Lopes[d]

[a]Departamento de Engenharia Elétrica
[b]Departamento de Engenharia Mecânica
[c]Departamento de Meteorologia
[a,b,c] all of the Federal University of Pari, Campus Universitário do Guamá – CP 1611, 66075-900 Belém – Pará-Brazil
[d]Instituto de Engenharia de Sistemas de Computação – INESC – Portugal

In the Marajó Island, Northern Brazil, many vegetable species are suitable to be used as an alternative energy source. When wood is harvested, the volume of waste can achieve values of about 9 m^3/ha. This waste in turn can yield 8.8 m^3/ha, mostly in the area of handling, should the forestry remnant be used to generate electrical energy. A planned agricultural exploitation allows the utilization of these residues to produce energy. With adequate handling, there is less waste, enabling a better use of the forestry potential and a better agricultural handling for the use of this energy source.

The loss during the removal of wood will be taken into account together with species without commercial values, and also the selection of the areas of agricultural crops. These residues can be utilized for production of energy, improving the quality of life and the sustainable development of the region, as well as keeping an ecological balance and preserving forest.

INTEGRATION BETWEEN GAS TURBINE PLANTS AND BIOMASSES DRYING PROCESSES

Roberto Cipollone,[a] Daniele Cocco,[b] Emanuele Bonfitto[c]

[a]Department of Energetics, University of L'Aquila, Italy
[b]Department of Mechanical Engineering, University of Cagliari, Italy
[c]Regional Agency for the Development of the Agricultural Services, Avezzano (AQ), Italy

In order to reduce the transport and storage cost of biomass used for energy, we propose a new way to manage chipped biomass for industrial energy uses. Raw biomass is dried near its source through a network of drier units fed with the exhaust thermal energy of small gas turbines. The gas turbine-drier units of the network produce the dry biomass feedstock for a central biomass power plant.

The drying capabilities of these gas turbine-drier units have been evaluated as a function of the main operating characteristics of the gas turbine plant, including the steam injection in the combustion chamber and the production of thermal energy. The analysis of the performances of both the biomass combustion power plant and the overall biomass network gives interesting energy results owing to the exhaust energy recovery in the gas turbine-drier units.

1. INTRODUCTION

The recent international agreements concerning the reduction of the green-house gas emissions gives to biomass a new strategic role in the energy production market. In the next decade, for example, the contribution of biomass to the primary energy supply in the EC would increase from the actual 45 Mtep to 135 Mtep, with an installed electrical power of about 20 GWe.[1,2]

Owing to its high moisture content, biomass is not easily combustible both for technological and economical reasons, so that drying is required before any thermal use. However, drying biomass, in a power plant through different drying systems, lowers plant efficiencies.[3-5] Moreover, the seasonal nature of the biomass production requires storage of chipped biomasses in protected confined areas. Storage favours a natural drying process that reduces the moisture content, but that also reduces biomass' combustible fraction (by as much as 30%), owing to unavoidable natural fermentation. Volatile organic compounds thus pollute the environment while energy (and financial) value are lost.[6]

Thermal drying of biomass before the transportation, storage and combustion can be a suitable option that reduces the transport and storage costs of the biomass and allows it to be used more effectively. In fact, a natural drying process, with 20–30% losses of dry matter and

30–40% final moisture content, can lead to an energy loss of 5–20% and a mass reduction of 45–65%. On the contrary, a thermal drying process (without dry matter losses and 10–20% final moisture) can allow for an energy increase of about 20–25% and a mass reduction of about 50–60%.

In this paper a direct biomass drier fed with exhaust thermal energy leaving a small gas turbine power plant has been studied. This simple integration requires only conventional equipment and shows really interesting perspectives, particularly in the case of a combined heat and power production. In particular, the paper analyses the performances of a network composed of some gas turbine-drier units and a central power plant which burns the dried biomass produced by the network.

2. GAS TURBINE-DRIER UNIT

Each gas turbine-drier unit of the network includes a gas turbine power plant, a biomass rotating drier and, in case, a heat recovery steam generator, HRSG (Figure 1). The gas turbine operates in simple cycle or, if the HRSG is included in the plant, with steam injection; in the latter case, part of the steam produced by the HRSG can be used to fulfill thermal requirements such as district heating or industrial needs. The exhaust thermal energy leaving the gas turbine (or HRSG) is recovered by drying the raw biomass to the desired moisture content.

A plant configuration based on a gas turbine with steam injection is more complex and expensive, but it can allow for a more flexible and effective operation, from an energetic and economic point of view, than a plant based on the simple cycle gas turbine, particularly in the presence of a thermal power demand. A suitable distribution of the energy recovery between the HRSG and the biomass drier, as well as the variation of the mass flow of the injection steam, can allow for a great flexibility in the production of electrical and thermal power and dried biomass, according to their daily or seasonal variations.

The gas turbine-drier units are based on small gas turbine power plants, which are widely used for combined heat and electrical power production. Moreover, owing to their low capital cost, high specific work, high reliability and high steam-to-air mass flow ratios (up to 0.20), these small gas turbines are particularly suitable for operating with steam injection. The gas turbine here considered operates with a net efficiency of 25.6%, a specific work of about 200 kJ/kg

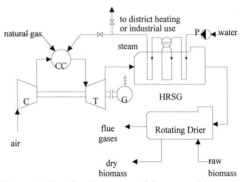

Figure 1. Functional scheme of the gas turbine-drier.

and an exhaust temperature of about 530°C; in the case of steam injection (400°C and steam/air ratio of 0.20), the net efficiency is 37.5%, the specific work 390 kJ/kg and the exhaust temperature 555°C.

The biomass drying section is based on a direct rotating drier directly fed by the hot flue gases leaving the gas turbine (or HRSG). The excess water of the biomass is vaporized through the direct heat transfer between the hot gases and the wet biomass; the produced steam is directly removed by the hot gases leaving the drier.[7] The performance of the gas turbine-drier units have been evaluated with reference to the gas turbine parameters typical of a power range between 1–3 MW and to the biomass composition reported in Table 1.

Table 1
Biomass composition

Biomass Composition (dry basis)		
C	[%]	49.0
H	[%]	6.5
N	[%]	2.0
O	[%]	40.0
S	[%]	-
Ash	[%]	2.5
LHV	[MJ/kg]	16.75

Figure 2 shows the biomass drying capabilities of a direct drier integrated with a simple cycle gas turbine plant. If the mass of raw biomass fed to the drier (60% moisture content) increases, then the temperature of the gas leaving the drier shows a remarkable decrease. Because the temperature of the dried biomass is always equal to about 70°C, increasing the mass flow of raw biomass decreases the temperature difference at the drier outlet (ΔT_U). The value of ΔT_U strongly influences the size and the effective operation of the drier: a direct rotating drier operates most economically for values of ΔT_U equal to about 40–50°C. For a given value of ΔT_U, the specific mass rate of raw biomass increases with the final moisture content; for a value of ΔT_U equal to 50°C, if the final moisture content of the dried biomass increases from 10% to 40%, the specific mass rate of raw biomass increases from 0.30 to 0.45. Figure 3 shows the biomass-drying capabilities of a direct drier integrated with a steam injected gas turbine and a HRSG; the latter produces just the steam (400°C) required by the gas turbine. For a given value of the steam/air ratio μ, owing to the thermal energy recovered in the HRSG, the temperature of the flue gas fed to the drier is lower than the temperature of the gas leaving the gas turbine, with lower drying capabilities than those previously discussed (Figure 2). In particular, the flue gas temperature at the HRSG outlet decreases from 425°C to 175°C when the steam ratio μ increases from 0.05 to 0.20. For a value of ΔT_U still equal to 50°C and for a final moisture content of 10%, for example, the specific mass rate of the raw biomass which can be dried is about 0.23 for $\mu = 0.05$ and 0.04 for $\mu = 0.20$.

In the presence of thermal power requirements near the biomass drying plant, the plant configurations based a steam-injected gas turbine can allow for a more flexible and effective operation, from an energetic and economic point of view, in comparison with those based on simple cycle gas turbine plants. In fact, the thermal power production often requires a daily or seasonal modulation, and biomass production varies seasonally. Figure 4a-b shows, for plant configurations based on a gas turbine with ($\mu > 0$) and without steam injection ($\mu = 0$), the correlation between the electrical power produced and the raw biomass fed to the drier

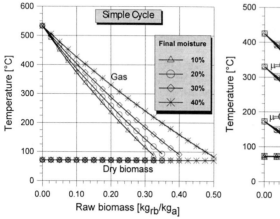 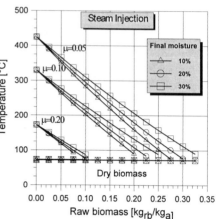

Figure 2. Temperature of gas and biomass at the drier outlet in function of the specific mass rate of the raw biomass and the final moisture content for a simple cycle gas turbine plant.

Figure 3. Temperature of gas and biomass at the drier outlet in function of the specific mass rate of the raw biomass and the final moisture content for a steam injected gas turbine plant.

(Figure 4a) and the thermal power produced for external uses (Figure 4b). Figure 4a refers to a thermal energy recovery exclusively carried out into the drier (from 60% to 10% moisture content, with a value of ΔT_U equal to 50°C); the latter is fed by the hot flue gases leaving the gas turbine ($\mu = 0$) or the HRSG ($\mu > 0$). On the contrary, Figure 4b refers to a thermal energy recovery, without any biomass drying, exclusively carried out in the HRSG, which produces the steam injected in the gas turbine and that required by the external needs. In all cases, the temperature of the steam is 400°C and the minimum flue gas temperature is 125°C.

With reference to a simple cycle gas turbine ($\mu = 0$) characterized by a net electrical power of 1 MW (point P), the exhaust energy recovery exclusively carried out through the biomass drier allows for drying about 1.5 kg/s of raw biomass; on the other hand, the exhaust energy recovery exclusively carried out in the HRSG allows for producing about 2.3 MW of thermal power. The utilization of a steam injected gas turbine plant ($\mu > 0$) leads to a higher electrical power production and a corresponding lower biomass drying or thermal production. The line P-Q shows the correlation resulting from the variation of the steam/air ratio from $\mu = 0$ (simple cycle) up to $\mu = 0.2$. The electrical power produced increases from 1 MW to about 1.8 MW, the raw biomass dried decreases from 1.5 to about 0.2 kg/s and the thermal power produced decreases from 2.3 MW to about 0.3 MW.

3. BIOMASS COMBUSTION POWER PLANT

Biomass production, both residual or from short-rotation forestry (SRF), is seasonal, so that the continuous use of biomass requires the storage of remarkable amounts of chipped biomass. Moreover, the biomass grown for industrial energy requires very wide extents of wooded land (the annual biomass production from SRF is about 12–15 t/ha of dry matter, with a LHV of about 16–17 MJ/kg, also on the dry basis), so that transportation becomes a major component in the overall cost of the biomass.[8-10]

It is more economical to dry the raw biomass near its source through a gas turbine-drier unit. However, it is not profitable to integrate each gas turbine-drier unit with a thermal power plant (based on combustion or gasification processes) fed with biomass, owing to the high specific costs of the small power plants. It is more effective to use a network of small gas turbine-drier units tied to a commercial power plant that uses all the biomass dried by these units. Decentralizing the drying process reduces the transportation and storage components of the biomass supply cost, which is the key factor for the economical feasibility of a biomass plant.

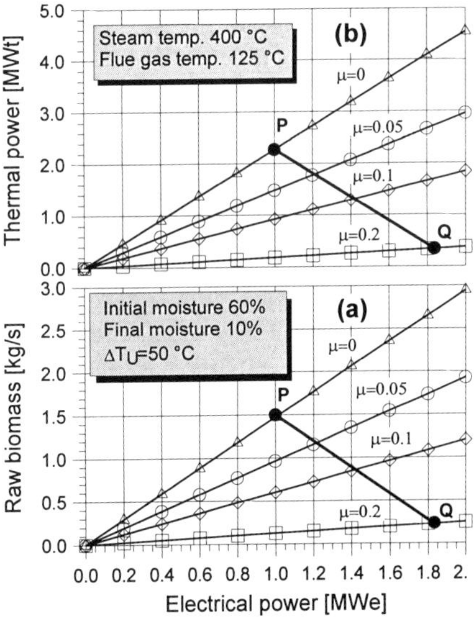

The central power plant here considered is based on a conventional steam cycle and can include a gas turbine, with or without steam injection, whose exhaust thermal energy is recovered in a biomass drier. The dry biomass produced by this drier and those dried by the network are burned in the steam boiler, which produces the superheated steam for the injection in the gas turbine and for feeding the steam turbine. The flue gas cleaning has been accounted for by means of a SCR process for NOx removal and cyclones and bag fabric filter for particles removal.[11]

Figure 4. Raw biomass dried and thermal power produced in function of the electrical power and the steam/air mass ratio.

Table 2 reports the main energy performances of a 50-MW thermal power plant fed by dry biomass (with 10% moisture content), without any thermal power production and for three different cases:

CASE A. the power plant is exclusively fed with the biomass dried by the network of gas turbine-drier units and does not include an own gas turbine-drier unit;

CASE B. the power plant includes a simple cycle gas turbine plant integrated with a biomass drier which produces a portion of the dry biomass fed to the steam boiler;

CASE C. the power plant includes a steam injected gas turbine plant integrated with a biomass drier which produces a portion of the dry biomass fed to the steam boiler. The superheated steam injected in the gas turbine is produced by the steam boiler.

The superheated steam pressure and temperature produced by the steam boiler have been assumed equal to 60 bar and 450°C, respectively, whereas the steam for the deareator has been extracted from the steam turbine at 4 bar. The excess air in the combustion process has been set to 30% and the minimum flue gas temperature has been assumed equal to 180°C. If the biomass power plant includes the internal drier (Cases B and C), the value of ΔT_U has been set to 50°C; moreover, in the case of steam injection, the value of μ has been assumed equal to 0.2. Finally, the net efficiency of the plant has been evaluated through the ratio between the electrical power produced and the overall primary energy fed to the plant (the dry biomass from the network, the raw biomass for the internal drier, if present, and the gas turbine fuel).

Table 2
Main performances of the biomass power plant

		Case A	Case B	Case C
Dry biomass fed to	[kg/s]	3.34	3.34	3.34
the steam boiler	[MW]	50.00	50.00	50.00
Dry biomass from	[kg/s]	-	0.56	0.76
the internal drier	[MW]	-	8.35	11.32
Dry biomass from	[kg/s]	3.34	2.78	2.58
the network	[MW]	50.00	41.65	38.68
Raw biomass	[kg/s]	-	1.25	1.70
(60% moisture)	[MW]	-	6.49	8.81
Gas turbine fuel	[MW]	-	3.39	4.49
Superheated steam	[kg/s]	15.58	15.58	15.58
Injection steam	"	-	-	0.79
Gas turbine power	[MW]	-	0.93	1.68
Steam turbine power	"	15.32	15.32	14.46
Overall power	"	15.32	16.25	16.14
Net efficiency	[%]	30.63	31.54	31.05

As shown in Table 2, the overall electrical power varies from 15.3 MW to 16.3 MW, with only a minor contribution of the gas turbine plant (6–10%), if present. The integration of a gas turbine-drier unit allows for higher net efficiencies owing to the energy recovery carried out in the biomass drier. Moreover, the utilization of a steam injected gas turbine determines a decrease of the net efficiency, due to the unrecoverable heat of evaporation of the injection steam, discharged with the flue gases; on the other hand, the operation with steam injection allows for a higher biomass drying. A combined thermal and electrical power production, here not considered, can allow for a higher overall efficiency (thermal + electrical) but also a lower electrical efficiency.

Table 3 synthesizes the main performances of the overall biomass drying and combustion system, composed by the thermal power plant and several simple cycle gas turbine-drier units. In particular, Table 3 refers to 7000 hours of annual operation of the biomass thermal power plant, without any thermal power production and with reference to the same three cases of Table 2. All the gas turbine-drier units of the network operate in simple cycle, with a net efficiency of 25.6%.

Starting from the same amount of raw biomass, for the three cases here considered the fraction of biomass dried by the gas turbine-drier units decrease from 100% to 77% of the overall biomass utilized by the power plant; the biomass drying process results in a reduction of about 55% of its mass and an increase of about 30% of its energy content.

Owing to the recovery of the gas turbine exhaust energy, for all cases A-C the overall efficiency of the network (35.4%) is higher than those of the simple cycle gas turbine (25.6%) and the biomass combustion power plant (31.54%). Finally, as already discussed, the utilization of a steam injected gas turbine determines a lower efficiency and a higher internal biomass drying.

4. CONCLUDING REMARKS

In this paper, new way to manage the chipped biomass for industrial energy uses has been proposed. The raw biomass is dried near its source through a network of small gas turbine-drier units, where the exhaust thermal energy leaving the gas turbine is fed to a rotating drier. The gas turbine-drier units produce the dry biomass for feeding a central combustion power plant.

With respect to a conventional management of the chipped biomass, a raw biomass drying process carried out near the locality of production though of a gas turbine-drier unit can allow for a more economical utilization of the biomass. In fact, the transport cost can be reduced because the mass of the dried biomass is much lower than that of the raw biomass, as much as 50–60%; the storage cost can also be reduced owing to the avoided losses of the combustible fraction, even of about 20–30%, which occurs during a natural drying process.

Moreover, this unconventional biomass management also gives interesting energy performances due to the exhaust energy recovery at the gas turbine outlet. In fact, the network

NOTATION

C	Compressor
CC	Combustion chamber
HRSG	Heat Recovery Steam Generator
LHV	Lower Heating Value
P	Pump
SRF	Short Rotation Forestry
T	Turbine
ΔT_U	Temperature difference at the drier outlet
μ	Steam/air mass ratio

Table 3
Overall energy performances of the network of gas turbine-drier units and the biomass combustion

		Case A	Case B	Case C
Raw biomass from the network	[t/y]	187900	156400	145060
Raw biomass from the drier	"	-	31500	42800
Raw biomass utilized [1]	"	187900	187900	187900
Dry biomass to steam boiler [2]	"	84200	84200	84200
Fuel to the network	"	10190	8480	7870
Fuel to biomass power plant	"	-	1710	2260
Fuel to the overall system [3]	"	10190	10190	10130
Electrical energy from the network	[GWh/y]	38.8	32.3	30.0
Electrical energy from the plant	"	107.3	113.8	113.0
Electrical energy produced	"	146.1	146.1	143.0
Overall efficiency	[%]	35.4	35.4	34.7

[1] LHV equal to 5.2 MJ/kg [2]; LHV equal to 15 MJ/kg; [3] LHV equal to 50 MJ/kg

shows an overall efficiency of about 35.4%, which is higher than the electrical efficiency of the simple cycle gas turbine power plant (25.6%) and the electrical efficiency of the biomass combustion power plant (31.54%) here considered.

REFERENCES

1. Libro Verde ENEA (1998). Fonti rinnovabili di energia, Conferenza Energia e Ambiente, Napoli, 4-5 Giugno.
2. COM(97) 599 def., 22.11 (1997). Energia per il futuro: le fonti energetiche rinnovabili, Libro bianco per una strategia e un piano d'azione della Comunità.
3. Hughes, W.E.M. and E.D. Larson (1998). Effect of fuel moisture content on biomass-IGCC performance, Journal of Engineering for Gas Turbines and Power, 120, pp. 455–459.
4. Hulkkonen, S., M. Raiko, and M. Aijala (1991). High efficiency power plant processes for moist fuels, IGTI Vol. 6, ASME COGEN-TURBO.
5. Hulkkonen, S., M. Aijala, and J. Holappa (1993). Integration of a fuel dryer to a gas turbine process, IGTI Vol. 8, ASME COGEN-TURBO.
6. Riva, G., J. Calzoni, C. Fabri (1997). Biomassa legnosa per finalità energetiche: analisi tecnico-economica delle problematiche inerenti lo stoccaggio e l'essiccazione, 52° Congresso Nazionale ATI, Cernobbio (Como), September 14-17.
7. Bonfitto, E., R. Cipollone, and D. Cocco (1998). Impianti di turbina a gas integrati con processi di essiccazione e combustione di fanghi di depurazione da reflui civili e agro-industriali, 53° Congresso Nazionale ATI, Florence, September 15-18.
8. Canella, M., P. Zappelli, and A. Robertiello (1997). Coltivazione ed impiego di piante energetiche: uno studio integrato, 52° Congresso Nazionale ATI, Cernobbio (Como), September 14-17.
9. De Vita, L. (1997). Produzione di energia elettrica da biomasse: alcune valutazioni di carattere economico, 8° Congresso Sistemi Energetici Complessi, Bologna, June 14-15.
10. Larson, E.D. and C.I. Marrison (1997). Economic scales for first generation biomass-gasifier/gas turbine combined cycles fueled from energy plantations, Jour. of Eng. for Gas Turbines and Power, vol. 119, April.
11. Bonfitto, E. and R. Cipollone (1997). Energy opportunities from an integrated management of sewage municipal or agroindustrial sludges and refuse derived fuels, I° National Congress Valorizzazione e riciclaggio dei residui industriali, L'Aquila, 7-10 luglio.

PHYSICO-CHEMICAL UPGRADING OF AGRORESIDUES AS SOLID BIOFUELS

E.G. Koukios,[a] S. Arvelakis,[a] and B. Georgali[b]

[a]Bioresource Technology Unit Department of Chemical Engineering, National Technical University of Athens, Zografou Campus, Athens, GR-15700, Greece
[b]Hellenic Cement Research Center (HCRC), Lucovrisi, Athens, GR-14123, Greece

Two types of agroresidues, wheat straw and olive kernels, which are considered to be of premium importance for electricity production through combustion or gasification, were studied with respect to upgrading them by use of the pretreatment techniques of fractionation, leaching and both combined. Both the pretreated and the untreated samples were analyzed in order to determine the effect of the pretreatment on the ash and the ash chemistry of the pretreated samples compared with the untreated samples. The ash behaviour of the untreated and the pretreated samples was characterized by simple sintering tests performed in a laboratory muffle furnace, whereas a number of the thermally treated samples were further characterized with the use of SEM-EDX techniques.

1. INTRODUCTION

Biomass fuels for generating heat and power are of interest because biomass is a renewable form of energy with low ash and sulphur content, which can contribute significantly to the reduction of the greenhouse effect and to increase the energy independence of the user countries. Thermochemical technologies, especially in the form of combustion and gasification, are considered to be promising solutions for energy production from biomass; the most advanced forms are considered to be fluidised bed combustion and gasification. Among the various types of biomass fuels that can be used for energy production are agro-residues resulting as byproducts of agricultural or agro-industrial activities (e.g., straws, pits, hulls, pods, cobs). Such residues are important especially in the underdeveloped areas of the planet, where the use of these biofuels for energy production could fill a substantial gap in the energy production of local communities.

Table 1
Proximate and elemental analysis of wheat straw samples (% dry basis)

	Wheat Straw	Wheat Straw Fraction L>1 mm	Wheat Straw Leached Fraction L>1 mm
Proximate analysis			
Moisture	6.5	7.5	17.95
Ash	7.5	5.04	3.43
Volatiles	76.0	77.23	80.00
Fixed carbon	16.5	17.73	16.6
Ultimate analysis			
Nitrogen	0.79	0.63	0.29
Carbon	43.7	47.8	48.7
Hydrogen	5.08	6.023	4.42
Sulfur	0.43	0.25	0.00
Oxygen	42.4	40.3	43.2
Gross calorific value (kcal/kg)	4520	4699	4850

Table 2
Proximate and elemental analysis of olive kernels samples (% dry basis)

	Olive Kernels	Olive Kernels Fraction L>1 mm	Olive Kernels Leached Fraction L>1 mm
Proximate analysis			
Moisture	5.5	8.76	9.33
Ash	4.6	2.58	1.53
Volatiles	72.0	77.9	80.84
Fixed carbon	23.4	19.5	17.63
Ultimate analysis			
Nitrogen	1.36	1.00	1.1
Carbon	50.7	51.31	54.1
Hydrogen	5.89	5.82	5.98
Sulphur	0.3	0.28	0.22
Oxygen	37.2	39.00	37.1
Gross calorific value (kcal/kg)	5071	4742	5069

However, the use of these "low quality" biomass fuels in sophisticated conversion systems can cause serious problems during operation and even lead to shutdown of the operating units, thus increasing operational costs and discouraging the setup of new biomass units. The majority of these problems is associated with the low melting point ash of these biofuels, which contain large quantities of inorganic elements. The alkali metals (K, Na), alkali earth metals (Ca, Mg), silicon, chlorine, and sulphur are considered to be the main troublesome elements. The problems include agglomeration of the inert bed material, slagging, deposition, fouling, and corrosion or erosion of heat exchange surfaces or of the reactor walls.[1,2,4,5]

2. MATERIALS AND METHODS

Two types of residues were used in this work, wheat straw and olive kernels, each one being pretreated according to two different techniques, fractionation and leaching. The different samples studied include wheat straw, wheat straw fraction L > 1 mm, wheat straw leached fraction L > 1 mm, olive kernels, olive kernels fraction L > 1 mm, and olive kernels leached fraction L > 1 mm. The analysis and characterization of the materials and of the materials ash are shown in Tables 1, 2, and 3. Leaching included the treatment of the materials with tap water at room temperature in order to reduce the ash content or change the ash chemistry to avoid the ash melting behaviour of the materials. The experimental methodology for studying the pretreatment techniques, fractionation and leaching, can be found elsewhere.[6,9]

Representative ash samples of each material were heated using a laboratory muffle furnace from 750°C to 1000°C at intervals of 50°C to 100°C for a period of one hour each time. At the end of each heating period the samples were weighed and inspected to determine changes in their physical state. SEM-EDX analysis technique was used for elemental analysis of the thermally treated ash samples.[3,7,8]

Table 3
Ash elemental analysis of wheat straw & olive kernels samples

Ash basis %	K_2O	Na_2O	CaO	MgO	SiO_2	Al_2O_3	SO_3	Cl
Wheat Straw	14.06	4.3	14.4	4.3	39.2	3.85	5.27	2.73
Wheat Straw Fraction L > 1 mm	13.12	3.3	4.43	0.92	38.5	1.73	5.9	2.98
Olive Kernels	15.10	8.90	21.30	7.90	32.60	6.03	4.97	1.43
Olive Kernels Fraction L > 1 mm	41.82	3.94	9.79	3.32	18.86	2.52	0.68	1.42
Olive Kernels Leached Fraction L > 1 mm	5.89	1.44	22.89	2.54	22.72	3.70	0.2	<0.005

3. RESULTS AND DISCUSSION

3.1. Wheat straw

Ash thermal treatment of samples of wheat straw and wheat straw fraction L > 1 mm changed the heated ash from a loose material to a completely fused, hard, material at the end of the heating process. At 750°C the samples started to sinter, and they were fully fused after 900°C and 800°C respectively. According to Table 3 the materials ash contain mainly silicon and also large amounts of alkali metals, such as potassium, and elements such as chlorine and sulphur, which are characterized as troublesome constituents since they have the tendency to react, in the temperature range of 750-900°C, with silica to form low-melting-point silicates.

Table 4
SEM-EDX elemental analysis of various wheat straw & olive kernels ash samples

	Olive Kernel Fraction L>1 mm		Olive Kernels Leached Fraction L>1 mm		Wheat Straw Fraction L>1 mm			Wheat Straw Leached Fraction L>1 mm	
	White grain	Black grain	White grain	Black grain	White grain	Black grain	Black&White grains	White grain	Black grain
K2O	4.64	22.37	0.00	0.27	45.5	43.77	16.37	1.01	0.38
Na2O	2.21	0.99	0.00	0.00	3.57	3.44	6.85	0.57	0.00
CaO	83.15	14.86	94.51	59.23	2.09	2.28	9.47	33.76	38.57
MgO	0.00	0.55	0.9	6.99	0.54	0.46	3.93	6.35	9.94
FeO	0.00	1.64	1.18	9.31	0.00	0.00	0.33	1.09	1.43
SiO2	4.63	46.35	0.83	17.07	4.01	3.36	60.7	55.86	49.12
Al2O3	0.00	12.46	0.00	3.63	0.00	0.00	0.00	0.00	0.00
Cl	0.37	0.00	0.00	0.36	0.00	0.00	0.00	0.00	0.00
SO3	5.01	0.78	2.58	3.13	44.23	46.68	2.34	1.37	0.57

In addition, the thermally treated ash of the wheat straw fraction L > 1 mm sample was further characterized with the use of SEM-EDX techniques. As it seen on Photo 1 the material is composed of black, white and black, and white grains. Table 4 shows that the areas of white and black grains contain mainly potassium and sulphur at about equal amounts, which is a strong indication of the presence of a potassium sulphate coating on the surface. On the contrary, elemental analysis of the black and white grains revealed that the grains are composed mainly of potassium and silica in a ratio approximately 1:4. This analysis indicates the existence of alkali silicates especially in the form of potassium silicate in the region.

 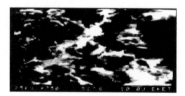

Photo 1. SEM images of wheat straw ash samples.
a) Wheat straw fraction L > 1 mm.   b) Wheat straw leached fraction L > 1 mm.

Nevertheless, the ash of the material wheat straw leached fraction L > 1 mm appears to have an improved thermal behaviour compared with the two former straw samples. Ash changed from a loose orange-white material to a white sintered material at the end of the heating process. The change started at 750°C and ended at 900°C when the sample sintered. The material is expected to be depleted in alkali metals, chlorine and sulphur as a result of the leaching process, which explains to a great extent the improved ash thermal behaviour during thermal treatment. SEM-EDX analysis of the resultant material showed that is composed of black and white grains containing mainly silicon and calcium and traces of other elements such as magnesium, potassium, sulphur, iron and sodium.

3.2. Olive kernels

As in the case of the former straw samples the ash of the materials olive kernels and olive kernels fraction L > 1 mm appeared to start to sinter at 750°C (where the transformation seemed to be concluded for the first sample), while the second sample completely fused above 900°C. This result correlates well with the ash elemental analysis of the specific olive samples depicted in Table 3. Both samples contain mainly alkali metals, chlorine and sulphur, which are thought to be the main troublesome constituents associated with problematic ash thermal behaviour. Table 4 depicts the results obtained from the SEM-EDX analysis of the thermally treated ash of the sample olive kernels fraction L > 1 mm. The results show that the main mineral phase present in the white grains is calcium rich (up to 83%), probably in the form of calcium oxide and to a lesser extent calcium silicate. To the contrary, the major constituent of black grains is silica (up to 46%), in combination with substantial quantities of potassium, calcium and aluminium. This is clear evidence for the presence of a complex potassium-calcium aluminosilicate formed during the heating process on the material's surface.

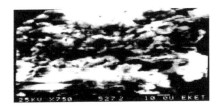 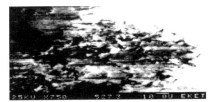

Photos 2. SEM images of olive kernels ash samples.
   c) Olive kernels fraction L > 1 mm.   d) Olive kernels leached fraction L > 1 mm.

Finally, the ash of the material olive kernels leached fraction L > 1 mm changed only slightly during the thermal treatment, taking a darker color, while there were no changes in its physical state. These results agree with the ash elemental analysis of the material depicted in Table 3 and also with the results from the SEM-EDX analysis of the ash sample depicted in Table 4. Elemental analysis of the white grain area shows that grains contain mainly calcium (up to 90%) and traces of other elements, while the analysis of the black grains shows that they contain mainly calcium, followed by substantial amounts of silicon, iron, magnesium and traces of aluminum, chlorine and sulphur.

4. CONCLUSIONS

The two pretreatment techniques, fractionation and leaching, appeared to have totally different effects on the ash thermal behaviour of the studied biomass materials. Fractionation deteriorates the ash thermal behaviour of the treated samples, by increasing the amounts of the "problematic" elements (K, Na, Cl, S) in the ash of the resultant material. On the other hand, leaching reduces the treated material's ash content, while improving the material's ash composition by substantially reducing the amount of K, Na, Cl, S present in the ash of the resultant materials. Finally, the experimental technique applied for the evaluation of the ash thermal behaviour of these biofuels shows that is possible to predict to a great extent the ash behaviour of various biomass materials during combustion and gasification processes.

REFERENCES

1. Baxter, L.L. (1993). Ash deposition during biomass and coal combustion: a mechanistic approach, Biomass and Bioenergy, 4, p. 85.
2. Miles, T.R., P.E., T.R. Miles, Jr., L.L. Baxter, R.W. Bryers, M.M. Jenkins, and L.L. Oden (1995). Alkali Deposits Found in Biomass Power Plants, Volumes I, II. Summary Report for National Renewable Energy Laboratory, NREL Subcontract TZ-2-11226-1.
3. Benson, S. et al. (1996). Ash Formation, Deposition, Corrosion, and Erosion in Conventional Boilers. Applications of Advanced Technology to Ash-Related Problems in Boilers, Plenum Press, New York, p. 1.
4. Moilanen, A., M. Nieminen, K. Sipila, and E. Kurkela (1996). Ash behaviour in thermal fluidised-bed conversion processes of woody and herbaceous biomass, Proceedings of the 9th European Bioenergy Conference, Copenhagen, 2, p. 1227.
5. Baxter, L.L. et al. (1996). Inorganic material deposition in biomass boilers, Proceedings of the 9th European Bioenergy Conference, Copenhagen, 2, p. 1114.
6. Koukios, E.G. and S. Arvelakis (1998). NTUA Group, FIBEGAR Final Report, European Commission, JOULE Programme (JOR3-CT95-0021).
7. Folkedahl, B.C. et al. Inorganic phase characterization of coal combustion products using advanced SEM techniques, 5th Engineering Foundation Conference on 'Inorganic Transformations and Ash Deposition During Combustion', Palm Coast, p. 399.
8. Nordin, A. et al. (1996). Agglomeration and Defluidization in FBC of Biomass Fuels-Mechanisms and Measures for Prevention. Applications of Advanced Technology to Ash-Related Problems in Boilers, Plenum Press, New York, p. 353.
9. Jenkins, M.B., R.R. Bakker, and J.B. Wei (1996). On the properties of washed straw, Biomass and Bioenergy, 10, p. 177.

EXTRUSION CORK POWDER BRIQUETTES

L. Gil and C. Nascimento

INETI, Estrada das Palmeiras, Q. Baixo, 2745-578 Barcarena, Portugal

We studied the production of cork powder briquettes by extrusion using the self-binding capacity of the cork particles or "condensed resins" as the binding and lubrication agent. Cork powder is the major cork industry waste. Several operational parameters and types of cork powder were studied.

1. INTRODUCTION

Cork is the bark of the cork oak tree (*Quercus suber* L.), which is periodically extracted to make items such as cork stoppers, coverings, and insulation products for buildings.

Operations such as grinding, screening, cutting and sanding (granulates and composition cork) and perforation and finishing (cork stoppers) produce a large amount of cork powder. This amount is estimated at 30,000–40,000 ton/year in Portugal alone and globally may reach 50,000 ton/year.

The production of insulation cork board gives rise to a different waste, the "condensed resins" of the backing steam used in the agglomeration process, in amounts of about 1,000 ton/year.

2. MATERIALS USED

The three types of cork powder tested were obtained at three different processing steps:

1 – finishing (sanding) of composition cork tiles
2 – finishing of cork stoppers
3 – cutting and sanding of insulation cork boards

Condensed resins blocks were ground to < 0.25 mm, and this powder was characterised and used in some briquettes' formulations. The characteristics of these materials are summarised in Table 1.

Table 1
Characteristics of the materials

Powder	Density (Kg/m³)	Moisture Content (% w/w)	Wax Content (% w/w)	Tannin Content (% w/w)	Ash Content (% w/w)
Type 1	104	3.1	7.9	3.7	0.9
Type 2	68	3.2	6.6	7.8	2.0
Type 3	61	2.9	8.3	6.1	1.6
Resins	518	---	77.2	4.5	---

3. EXPERIMENTAL PROCEDURE

A laboratory prototype of an extrusion device was developed and adapted to an universal testing machine. A regulated heating belt was used for heating during extrusion. The material was compressed by a piston in a discontinuous process.

Several operational parameters were studied: heating temperatures, extrusion angles, cork types and content of condensed resins.

These parameters ranged as follows:
Extrusion angles $\Rightarrow$ 0.5°; 1.0°; 1.5°
Heating temperatures $\Rightarrow$ 120-150°C
Volume ratio cork:resins $\Rightarrow$ 20:1 to 80:1

Some experiments studied only cork powder and others studied a blend of cork powder and condensed resins in powder form.

4. RESULTS AND DISCUSSION

The briquettes obtained were characterised for the following characteristics whose values ranged as follows:

Moisture content = 1.9 – 7.9 % (w/w)
Higher heating value (HHV) = 6380 – 6970 Kcal/Kg
Ash content = 0.9 – 3.4% (w/w)
Oxygen index = 18 – 21%

Some handling characteristics were also determined.
- In the range of operational temperatures (T), when T increased, the compression force (F) decreased;
- Higher compression velocity increased F;
- The use of condensed resins decreased F and increased the briquettes' HHV;

- Briquettes of cork powder type 3 have higher HHV;
- Most of the briquettes could be handled without breaking apart.

5. CONCLUSION

Cork powder can be easily transformed by extrusion into a briquette that can be used as a combustible material with useful characteristics for industrial or domestic applications.

BIBLIOGRAPHY

Chen, P.Y.S., J.G. Haygreen, M.A. Graham (1987). Forest Products Journal, pp. 53–58.
Nascimento, C. (1994). Novos materiais à base de cortiça – Fabrico de briquetes de pó de cortiça por extrusão, Graduation thesis, ISA.
Gil, L. (1996). Proceedings 9th European Bioenergy Conf., Copenhagen, pp. 960–962.
Gil, L. (1997). Biomass and Bioenergy, 13, pp. 59–61.
Gil, L., J. Santos, M.I. Florêncio (1986). Bol. IPF - Cortiça, 575, pp. 255–261.
Gil, L. (1987). Bol. IPF-Cortiça, 586, pp. 141–142.

Biomass Production and Integration With Conversion Technologies

Feedstocks: Harvesting, Handling, and Storage

DEVELOPMENT OF PRODUCTION METHODS, COSTS, AND USE OF WOOD FUELS IN FINLAND

Ismo Nousiainen and Veli-Juhani Aho

VTT Energy, P.O.Box 1603, FIN-40101 Jyväskylä, Finland
Tel. +358 14 672 611, Fax. +358 14 672 597, e-mail: Ismo.Nousiainen@vtt.fi
e-mail: Veli-Juhani Aho@vtt.fi

At present the share of wood fuels in the Finnish energy supply is large and higher than in any other industrialized western country. The energy content of wood-derived fuels used in 1997 was equivalent to 5.7 million toe, which was 19% of total energy consumption.

The development of new production methods, production costs and use of wood fuels in Finland are described in this paper.

Owing to inadequate development work in the 1980s, production costs of wood fuels from logging residues increased from 58 FIM/MWh to 64 FIM/MWh, which was much higher than the price of alternative fuels. In the beginning of the 1990s the Finnish government set as an objective for the year 2005 to increase the use of wood fuels to about 125% of the 1994 level. The R&D work is needed to achieve this goal.

Since than three new production methods have been developed for commercial use: Evolution-chipper, Moha-chipper and Chipset-terrain chipper. The production costs of wood fuels have decreased to 46 FIM/MWh, which is competitive with the price of alternative fossils fuels.

The change in production costs shows that R&D-work is very important in the promotion of bioenergy.

1. INTRODUCTION

Even today Finland is one of the leading countries in the use of wood fuels. The energy content of wood-derived fuels used in 1997 was 5.7 million tons of oil equivalent (toe). More than half of the wood-derived fuels was waste liquors from pulping industry (3.1 million toe) and the rest was mainly waste wood from forest industry (1.5 million toe) and firewood (1.1 million toe). The share of wood-derived fuels was 19% of the total primary energy consumption, 30.6 million toe.

Finland still has excellent wood resources that can increase the use of wood fuels in energy production. The main barrier to the utilization of these resources is a production cost higher than the price of alternative fuels.

In 1994 the Finnish Government set an objective for the promotion of bioenergy. The aim is to increase the use of bioenergy by about 25% from the present level by 2005. The increment corresponds to 1.5 million toe per year. The research and development work to decrease the production cost has been considered as an important factor in achieving this goal.

2. PRODUCTION POTENTIAL OF WOOD FUELS

The annual growth of stemwood in Finnish forest is over 80 million m^3 (including bark). The growth of all above-ground biomass, including branches, is 110 million m^3 per year. On the other hand, the annual withdrawals of stemwood in 1997 was 66 million m^3, which consisted of 59 million m^3 of commercial timber, 2 million m^3 of natural deadwood and 5 million m^3 of waste stemwood.

The main resources of wood fuels are trees removed during precommercial and early commercial thinnings, logging residues from regeneration fellings and poor-quality, small-sized trees from under-productive stands (Figure 1).

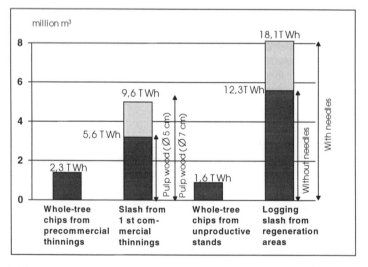

Figure 1. The harvestable forest energy reserve (Hakkila, Nurmi, 1997).

Logging residues, unmerchantable stemwood and branches, left in regeneration felling areas after the harvesting of stemwood, form the main wood fuel potential. They are also the most attractive from the standpoint of harvesting techniques and harvesting costs. According to Hakkila & Nurmi (1997), the technical logging residue harvesting potential is 8.6 million m^3/a including needles and 5.6 million m^3/a excluding needles.

3. NEW PRODUCTION METHODS

During the 1990s five new harvesting methods for the production of logging residue chips were developed. Three of them were commercial by 1998 (Figure 2).

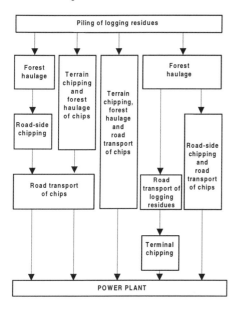

Figure 2. Various chains of the production methods of logging residue chips.

In the road-side chipping method, logging residues are piled in conjunction with the cutting of timber. By adapting the cutting method slightly, the piling does not even complicate the timber harvesting. After piling, logging residues are hauled by a forwarder equipped with a logging-residues grapple and an enlarged load space. Logging residues are chipped at a site by a truck-mounted drum-chipper and transported to the power plant. An alternative is to transport unhandled logging residues and chip them at the terminal or at the power plant.

Terrain chipping methods also demand piling of logging residues. The Moha-chipper method also demands a second piling, after piling in conjunction with cutting. The Moha-chipper is a multi-purpose chipper lorry, which could be used both as a chipping and transportation unit. For terrain chipping method we developed a completely new type of chip harvester, which produces wood chips in the stand and also forwards the chips to a landing site, where they are dumped into a truck container.

4. THE DEVELOPMENT OF PRODUCTION COSTS

In the begining of 1980s the production cost of wood fuel from logging residues was 58 FIM/MWh, when the average transport distance was about 100 km.

Throughout the 1980s and in the begining of the 1990s development work was inadequate. Rising production costs were restricted by the increasing productivity of forwarders that harvested industrial wood and the decline of forest machinery rates, which affected also the production cost of wood fuel. In 1992 the production cost was 64 mk/MWh, which was significantly higher than the price of alternative fuels (for example, coal sold for 44 mk/MWh inland).

After 1992 three new methods were developed for commercial use. The production costs of these methods are presented in Figures 3 and 4.

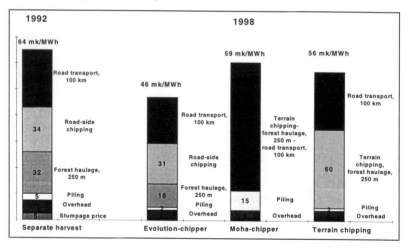

Figure 3. The production costs of logging residue chip by developed methods. Road-transport distance 100 km.

In Figure 5 is presented the development of production cost of logging residue chips in 1985–1998. New production methods decreased production cost about 28%. The total funding for work to develop wood fuel production was at that time about 86 million FIM.

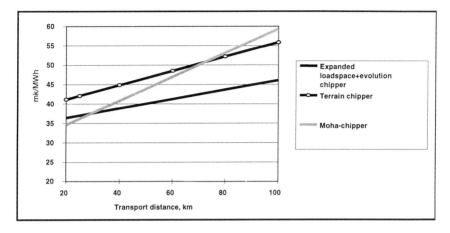

Figure 4. The production costs of logging residues chips by developed methods as a function of transport distance.

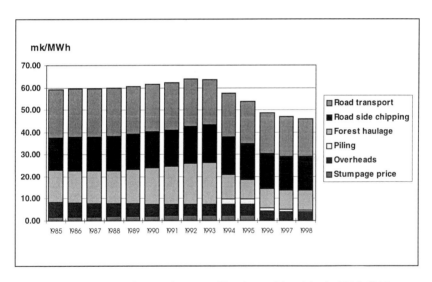

Figure 5. Development of production cost of logging residue chips in 1985–1998.

This means that funding demand was 4.7 million FIM to decrease production cost 1 FIM/MWh.

4. THE USE OF WOOD FUELS IN FINLAND

In 1997, the total consumption of primary energy was 30.6 million toe in Finland. The share of wood-derived fuels was as high as 19% (5.6 million toe). More than half of wood-derived fuels was waste liquors from the pulping industry (3.1 million toe) and the rest was mainly waste wood from the forest industries (1.1 million toe) and firewood (0.85 million toe). The use of forest chips was still modest, about 600,000 m^3 (0.12 million toe).

The most important source of forest chips was whole-tree chips made from small trees (about 400,000 m^3). In spite of the high costs, the use of chips made from delimbed stems is still quite common. The reason for this is the outdated conveyer mechanisms at the small plants which require stick-free chips to function properly. In comparison, the consumption of logging residue was still modest when considering its availability and low cost. In 1997, the figure was 200,000 m^3, which amounts to less than 4% of the harvestable reserve. However, owing to the decreased production costs and energy taxes of alternative fuels the use of logging residues will increase strongly during next few years.

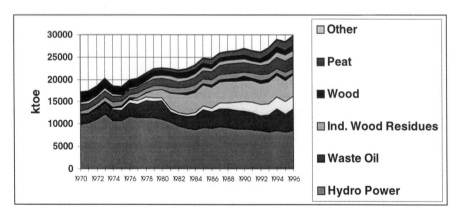

Figure 6. Primary energy consumption in Finland in 1970–1996.

5. CONCLUSIONS

The production potential of wood fuels enables a large increase in use of wood fuels in Finland. The main barrier to the utilization of these resources has been production costs higher than the price of alternative fuels.

With new production methods, it is possible to produce logging residue chips at a competitive price. The development of production costs has shown that R&D-work is very important in the promotion of bioenergy.

SUPPLY SYSTEMS FOR BIOMASS FUELS AND THEIR DELIVERED COSTS

A. Hunter,[a] J. Boyd,[a] H. Palmer,[b] J. Allen[c] and M. Browne[c]

[a]Scottish Agricultural College, Environmental Division
[b]Management Division, Bush Estate, Penicuik, EH26 0PH, UK
[c]Transport Studies Group, University of Westminster,[*] 35 Marylebone Road, London, NW1 5LS, UK

This paper describes a project on the logistics of supplying biomass fuels to power stations in UK. Four types of biomass fuel are covered: forest residues, straw, short rotation coppice, and *Miscanthus*. The paper explains the approach used to model the delivered costs of these fuels, taking straw as the main example. The results indicate that straw supply systems are capable of producing the lowest delivered costs per tonne of dry matter, while the highest costs are likely to occur with short rotation coppice and *Miscanthus*. A significant proportion of total delivered costs is represented by the logistics costs of transport, storage, and handling. The main benefits of the analysis are to quantify sensitivities to variations in parameter value, such as transport distance, and to indicate targets for future cost reduction.

1. INTRODUCTION

Government targets for supplying renewable energy in UK are being progressively raised. In 1994, the national target was 1,500 megawatts of electricity generating capacity by 2000;[1] in 1999, the new target is 10% of electricity supplies by 2010,[2] which is approximately five times as high as before. In both cases, biomass fuels are recognised as important because the existing resource is large, in the form of wastes from forestry and agriculture, and the potential resource from energy crops is even larger. Municipal and industrial wastes, which may be seen as secondary biomass resources, are outside the scope of this paper.

At present, biomass electricity supplies are not competitive with conventional supplies because the technologies involved are immature and costs need to fall; biomass heating and CHP systems are more competitive. In order to obtain objective information on biomass fuel

---

[*] This project was led by Professor Michael Browne and his team from the Transport Studies Group of the University of Westminster, in close collaboration with the Scottish Agricultural College. The work was commissioned and funded by the Department of Trade and Industry of UK Government, and managed by ETSU on behalf of DTI.

supply costs, this major UK study was commissioned.[3] It covered wood residues from forestry, straw from cereals farming, willow or poplar grown as short rotation coppice, and *Miscanthus*. The paper indicates the approach used to carry out the analysis, with examples and results based on straw. The comparative costs for all four biomass fuels are also illustrated. Additional parts of the project, on animal slurries and on strategic siting of power stations, are not discussed in this paper.

2. STRAW SUPPLY SYSTEMS

2.1. Methodology

The starting point for analysing straw supply systems was a set of supply chain option models. These models represented alternative systems for delivering straw from the field, where it is a byproduct of the cereals harvest, to the power station, where it must meet specifications as a fuel. The defining activities, which differentiated systems from each other, were harvest, handling, transport, and storage. Some examples of the factors in these activities for straw are as follows.
- harvest – size of bale, type of baling machine
- handling – collecting system for bales, stacking system for bales, loading system onto transport vehicle, total number of handling operations, wastage of bales that disintegrate
- transport – transport system on the farm, size of road transport vehicle, transport distance
- storage – at farm or at intermediate depot between farm and power station, losses due to spoilage by rain or rotting, delivery schedule for the power station

In order to select five representative systems as supply chain options, the research team drew on their own collective experience, which was already extensive, and on a wide range of information sources including interviews with operators and potential operators in the industry. The main features of the five systems are summarised in Table 1: system A is based on small rectangular bales, system B on large roll bales, and systems C, D and E on alternatives for large rectangular bales.

Table 1
Five straw supply systems selected for study

	A	B	C	D	E
Type of bale and dimensions (m)					
Small rectangular, 0.90 × 0.45 × 0.35	#				
Roll, 1.22 × 1.50 dia		#			
Large rectangular, 2.44 × 1.22 × 1.22			#	#	#
Transport machinery on-farm					
Farm tractor, bale accumulator, trailed bale carrier	#				
Farm tractor, farm trailer		#			
Self-propelled bale carrier "Transtacker"			#		
High speed farm tractor "Fastrac", mounted bale carrier				#	#
Storage location					
On-farm	#	#	#	#	
Intermediate					#

The supply chain option models were constructed as a series of spreadsheets for analysing the costs of the activities, and the costs of the system elements within the activities, using a "bottom-up" approach. In addition to the cumulative activity costs, the cost of purchase from the original straw producer was included to complete the supply chain models.

An identical approach was used for forest residues, coppice, *Miscanthus*, and straw, thus providing a sound basis for direct comparison of system costs between the four types of biomass fuel.

2.2. Delivered costs

All costs were analysed on the basis of one tonne of dry matter delivered to the power station, and are given in this paper as UK pounds per oven dry tonne (£/odt). Because the lower calorific value of biomass fuel depends on dry matter content, while the biomass fuels in the study could have very variable moisture contents, this provided a rational basis for calculation in which cost elements could be correctly added together, and for cost comparison.

One important implication is that many of the cost elements within a single supply chain are interdependent. For example, if there are losses in storage, the quantity of dry matter at purchase and harvest must be proportionately increased to compensate, so that the calculation is for one tonne ultimately delivered as a net quantity to the power station. The quantities at handling and transport must also be adjusted, together with the storage quantity itself. The increased quantities for each activity element will then be reflected in increased elemental costs and in increased total cost.

The base case results for system costs had a number of common assumptions for all the biomass fuels, e.g., transport distance 40 km (80 km round trip), maximum size road transport vehicle, storage period six months, no unproductive downtime, costs exclude profit.

The five system cost totals for straw are given in Table 2. System C with large rectangular bales has a lower cost (£28/odt) than system A with small bales (£37/odt) or system B with roll bales (£36/odt). The equivalent costs per tonne of fuel at 15% moisture content, and per gigajoule, are also shown.

Table 2
Summary of delivered costs for five straw supply systems

	A	B	C	D	E
Delivered cost, £/t (15% moisture content)	37	36	24	24	26
Delivered cost, £/odt	43	43	28	28	31
Delivered cost, £/GJ	2.61	2.57	1.67	1.67	1.88

The elemental costs of these systems are given in Table 3. Each Figure in the Table is the result of a calculation using data and algorithms to take account of fuel quantity and cost inputs for labour, capital, machinery operation, maintenance, workrate, etc. It is assumed for storing straw that the entire top layer of a straw stack will be spoiled by rain and must be counted as a loss. This effect will be proportionately greatest for System B with roll bales, because stack height needs to be limited for safety, and is revealed as an increase in purchase cost for the system (Figure 1). Both systems A and B suffer from high handling costs compared with system C, which has lower handling costs. System E is slightly higher cost than the other two large bale systems because of using intermediate storage rather than on-farm storage.

Table 3
Analysis of delivered costs for five straw supply systems

Operation	Item	A £/odt	B £/odt	C £/odt	D £/odt	E £/odt
Purchase		7.64	8.75	7.88	7.88	7.88
Harvest	Baling + twine	6.73	6.84	6.77	6.77	6.77
	Building field stacks	2.15	-	0.54	-	-
Transport on-farm	Loading	0.36	3.86	0.29	0.86	0.86
	Roping / sheeting	-	0.3	-	-	-
	Trip time	7.26	1.93	3.93	2.62	2.62
	Unroping / unsheeting	-	0.29	-	-	-
	Unloading	0.36	3.86	0.29	0.34	0.34
Storage on-farm	Stacking	2.15	-	0.47	0.47	-
	Rent, interest, insurance	1.64	1.97	1.36	1.29	-
Transport from farm	Loading	4.24	5.09	1.05	1.05	1.18
	Roping / sheeting	0.41	0.49	0.27	0.35	0.40
	Container / trailer transfer	-	-	-	-	-
	Trip time	5.18	6.22	3.45	4.48	3.17
	Unroping / unsheeting	0.20	0.25	0.14	0.18	0.20
	Sampling / weighing	0.61	0.74	0.41	0.53	-
	Unloading	4.24	2.03	0.88	0.88	0.98
Storage intermediate	Rent, interest, insurance	-	-	-	-	1.91
Transport from store	Loading	-	-	-	-	1.05
	Roping / sheeting	-	-	-	-	0.35
	Trip time	-	-	-	-	1.87
	Unroping / unsheeting	-	-	-	-	0.18
	Sampling / weighing	-	-	-	-	0.53
	Unloading	-	-	-	-	0.88
Total		43.18	42.60	27.72	27.69	31.17

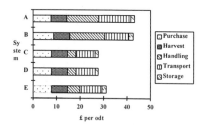

Figure 1. Breakdown of costs for five straw supply systems.

2.3. Sensitivity to distance

Transport distance affects transport costs only and will not affect other activity costs. For example, handling costs are affected by straw quantity, bale size, and loading system, but not by transport distance. For all systems, transport distance has a significant effect (Table 4). In system D, the lowest-cost system, reducing the transport distance by half from 40 km to 20 km will reduce transport cost from 28 to 23% of total cost;

conversely, doubling the transport distance will increase percentage cost to 33%. Transport cost is highest as a percentage of total cost for System E, because of additional transport to the intermediate storage depot.

Table 4
Sensitivity of straw supply costs to transport distance

		A		B		C		D		E	
	km	£/odt	%	£/odt	%	£/odt	%	£/odt	%	£/odt	%
Transport	20	10.5	25	8.0	20	6.5	24	6.0	23	8.5	28
	40	12.0	27	10.0	24	8.0	28	8.0	28	10.0	32
	80	16.0	34	14.0	30	10.0	33	10.0	33	13.5	39
Total	20	42.0	100	40.5	100	27.0	100	26.5	100	30.0	100
	40	44.0	100	42.5	100	28.5	100	28.5	100	31.5	100
	80	47.0	100	46.5	100	30.5	100	30.5	100	35.0	100

3. COMPARISON OF BIOMASS FUELS

The lowest cost systems for each of the four biomass fuels are compared in Table 5, with a breakdown by activity as £/odt and per cent of total cost. In order of ascending cost, the totals are straw (£28/odt), forest residues (£32/odt), short rotation coppice (£47/odt), and *Miscanthus* (£54/odt). The most important cost is the purchase cost for energy crops (coppice, *Miscanthus*) that must be paid to the producer: at 55% of total cost, it is markedly higher than that for forest residues (7%) and straw (28%), which are waste products.

Table 5
Comparison of lowest cost systems for four biomass fuels

	Forest		Coppice		Straw		*Miscanthus*	
	£/odt	%	£/odt	%	£/odt	%	£/odt	%
Purchase	2.0	7	26.3	55	7.9	28	29.0	54
Harvest	14.5	46	5.0	11	6.8	24	11.3	21
Handling	6.2	20	4.0	9	3.6	13	3.6	7
Transport	7.5	24	8.8	19	8.2	30	7.3	14
Storage	1.3	4	3.4	7	1.3	5	2.5	5
Total	32	100	47	100	28	100	54	100

The ranges of costs across the alternative systems for the four fuels are shown in Figure 2. Clearly, at the present stage of development in systems and technology for biomass fuels, the energy crops are more costly than the wastes. The supply chain option models are able to provide objective analysis of costs using existing data, and they will be able to use new data as developments occur. They provide a valuable tool for sensitivity analysis and for indicating targets for cost reduction.

The harvest cost of forest residues is high (£14.5/odt), reflecting the high capital cost and relatively low workrate of forest machinery. The harvest cost of *Miscanthus* (£11.3/odt) is higher than straw (£6.8/odt) because *Miscanthus* must be harvested as a standing crop before baling, unlike straw which is simply baled; it is also higher than coppice (£5.0/odt) where there is the benefit of new machinery for direct harvest and chipping at high workrates. Handling costs are highest for forest residues (£6.2/odt) mainly because of the awkward bulk of the material, while transport costs are comparable for all four fuels (£7 to £9/odt). Storage costs are highest for coppice (£3.4/odt), accounting for expected dry matter losses of 4% per month from chipped material stored in bulk over a six month period: by contrast, the lowest cost forest residue system is based on bulk storage of the raw residue material, with chipping to be done later at the power station.

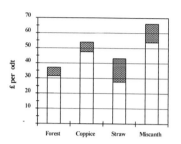

Figure 2. Cost ranges from minimum to maximum for supplying four biomass fuels.

REFERENCES

1. Anon (1994). New and renewable energy: future prospects in the UK, Energy Paper Number 62, DTI, HMSO, London, UK.
2. Anon (1999). New and renewable energy: prospects for the 21st century, DTI, HMSO, London, UK.
3. Allen, J., M. Browne, A. Hunter, J. Boyd, and H. Palmer (1996). Transport and supply logistics of biomass fuels: Volume 1—Supply chain options for biomass fuels, Report No. ETSU/B/W2/00399/REP/1), ETSU, Harwell, UK. p. 143.

FIELD CHOPPING AS AN ALTERNATIVE TO BALING FOR HARVESTING AND HANDLING SWITCHGRASS

David I. Bransby

Agtec Development LLC, 2668 Wire Road, Auburn, Alabama 36832

Current plans to harvest and handle switchgrass for energy production on a commercial scale involve conventional hay-making procedures, including baling into big round bales. However, in order to facilitate feeding into an energy plant and efficient conversion to energy, bales need to be ground up into a particulate format. Grinding is an energy-intensive operation that also requires control of particulate pollution. Although forage choppers are used for harvesting moist (40%+ moisture content) forage, such as silage or haylage, nothing is documented on using field chopping to harvest and handle dry herbaceous material. Furthermore, no research has been done to determine the bulk density of this material, and therefore, the amount that can be transported in a truck, and the associated transport costs. Neither has work been done to determine how best to store dry, chopped material.

The objective of this study was to evaluate field chopping as an alternative to baling for harvesting and handling switchgrass for energy. Switchgrass was successfully mown with a mower-conditioner, picked up and chopped with a pull-behind forage chopper after it had dried to less than 20% moisture in the windrow, and stacked in a pile with a front end loader. The pile of chopped switchgrass shed rain water due to a thatching process, and therefore did not appear to need artificial protection from rain. A front end loader with a large scoop took about 30 minutes to load 13 tons of chopped switchgrass onto a 45-ft walking floor trailer, which offloaded the cargo in about 20 minutes. Therefore, field chopping appears to be a feasible alternative to baling for harvesting and handling switchgrass.

HARVESTING AND HANDLING OF *MISCANTHUS GIGANTEUS*, *PHALARIS ARUNDANICEA* AND *ARUNDO DONAX* IN EUROPE

W. Huisman

Farm Technology Group, Department of Agricultural, Environmental and Systems Technology, Wageningen Agricultural University, Bomenweg 4, NL 6703 HD Wageningen, The Netherlands

A survey is given of harvest, storage and pretreatment operations of rhizamatous perenial grasses for use as energy and fibre crop. Methods are given for mowing, size reduction, densification, storage and pretreatment. Important aspects to be considered are elaborated and relevant experimental results are given. It was concluded that many production chains are possible and that optimization can be done only for given local conditions, including historical weather data.

1. INTRODUCTION

In Europe various crops are under investigation for use as industrial crops. A group with similar characteristics, called rhizomatous perennial grasses (PRG crops) are recognized as possible energy or fibre crops. *Miscanthus x Giganteus* GREEF et DEU or Japanese Elephant Grass is a triploid variety of a group of *Miscanthus* plants introduced in Europe in 1935 as ornamental plants. It is a C4 plant originally from Eastern Asia. It grows well under subtropic conditions and moderate climates. The crop is established by rhizomes or meristem plants since the seeds are not fertile. In Europe it grows from the southern countries up to Germany and Denmark as Northern boundary. *Arundo donax* L or giant reed is native to southern Europe and susceptible to low temperatures. It has a C3 photosynthesis and is also established by rhizomes. Both crops look like reed and bamboo and can reach heights of 3-5 m. *Phalaris Arundanicea* L. or reed canary grass is also a C3 grass, native in wet areas in Europe, and formerly used in cattle grazing mixtures. In Sweden and Finland it has been investigated for use for paper pulp production and energy. Another PRG crop would be *Panicum virgatum* L. or switchgrass. This grass is a C4 grass well known in the USA. It has a high drought resistance. About this crop only very little information is available about production in Europe. In this paper it therefore will not be mentioned although there will be many similar aspects especially as compared with reed canary grass. Both last grasses can reach a height of 1.5 metre and can easily be established by seeding.

All crops mentioned here can be used as fibre crop for various applications but mainly for paper making as well as an energy crop.

2. HARVEST CONDITIONS

The crops will be harvested only once a year. In most cases harvest will be done in spring time (so called "delayed harvest") because in that way the conditions for harvest are more favorable. The crop properties depend on harvest time and location. Important properties are as follows:

Moisture content. The moisture content is decreasing during wintertime an springtime. When waiting for harvest until spring the crop is dried naturally. For *Phalaris*, in Finland and Sweden after the snow has melted it is 10-20%. For *Arundo*, 30-40% in Italy in March (Pari, 1996). For *Miscanthus* in The Netherlands it is from 40% in March till 10% in late April. (Huisman, Kortleve, 1994)

Dry matter losses. Dry matter is lost during the period of drying from September to April next year. For *Phalaris* loss is caused by leaves that fall during winter (Pahkala, 1997). Compared to harvest in autumn, a yield about 30% higher will be reached when harvest is done in spring. This is because more nutrients are translocated in the rhizomes after summer (Pahkala, 1997). For *Miscanthus* in The Netherlands, harvestable yield decreases from an average level of 155% at October 1st to the standard level of 100% in April. The decrease is due to leaves and tops of the plant fallen on the ground and is roughly linear with 0.28% per day from October 1st.

Translocation of minerals. At the end of the growing season the plants start to allocate minerals to the rhizomes. In this way the content of these minerals in the harvested matter decreases.

Workable days. The available period for harvest, defined by the "harvest window," is the period between the first day at which harvest can start (the required moisture content is present) and the day before the regrowth is too large, which will result in harvest of fresh material. For *Phalaris* (Hemming, 1997) after snow has melted *and* soil can carry machines (trafficability) there are 2-6 days available before new shoots are too large. For *Miscanthus* it depends also on the moisture content level at harvest and the local weather. A moisture content of 15% in The Netherlands is available for only one or two weeks. The higher the acceptable moisture content the longer the "harvest window."

3. MOWING

The first action at harvest is generally mowing, i.e., separating the stems from the roots and putting the material in good condition for the following operation.

Cutting height defines the losses. Data about *Phalaris* are given by Pahkala (1998) for the conditions in Finland. Harvested mass is 40% lower for 10 cm cutting height compared to 5 cm cutting height in the first week of may. This is 20% for the last week in May.

Data about *Miscanthus* were collected in The Netherlands. Cutting height losses are about 0.55% per cm cutting height for the first 30 cm above the soil.

Possible machines for mowing are (Venturi & Huisman 1998) swath mower, cutterbar, disc mower, disk mower conditioner, flail mower, maize mower and special mowers. Important aspects when comparing these systems are
- minimum possible cutting height. In order to harvest as much material as possible special mowers are required for *Miscanthus* and *Arundo* since the plant can be cut just above the soil, without damaging the roots.
- loss of small pieces. In case *Phalaris* is mowed when it is very dry and so very brittle, broken pieces fall on the ground and cannot be collected. A mower conditioner is not advised in these conditions. Mowing and baling in a combined machine is favorable then.
- forming a good swath. When additional drying is required after mowing the swath must be loose and open for high penetration of sun radiation and wind. A swath mower or mower conditioner make good swaths.
- handling lodging crop. Because the crop can be lodged due to heavy snow layers or storms, lodged crops should be harvestable too. Flail mowers can take up any crop but give short pieces that can be lost when the material is put in a swath.
- handling fallen leaves. When harvesting *Miscanthus* or *Arundo* and maximum yield is wanted, the fallen leaves can be gathered by mowing low. In that case moisture content will be higher and the content of soil and minerals in the collected matter will be higher, which can give problems at energy conversion. In that case less minerals are recycled by decomposition, so more fertilizer should be applied.
- combination with next operation. If possible, mowing should be combined with the following operation in one machine.

4. SIZE REDUCTION

Reduction of size can be necessary or beneficial. After size reduction density in storage and in bales will be higher, reducing cost for handling and storage. Handling of whole stems of *Miscanthus* and *Arundo* is difficult. If length of stem pieces is below 50 cm the bulk can be handled as a flowing material. Size reduction can be performed by various machines as explained below.

Conditioners. The conditioners usually are used to increase drying speed after cutting because of their "crushing" action. After conditioning the particle length has also decreased but the length distribution is very wide.

Choppers/forage harvesters. In these machines the particle length can be adjusted between roughly 0.4 and 5 cm. The shorter the chopping length the more dust will be produced and the more losses can occur in case of strong wind while loading the trucks.

Hammer mills. These machines can be used stationary to get lengths below 0.5 cm. The resulting length distribution depends on the use and size of screens.

Chunking machines. Chunks are stem pieces in lengths between about 10 and 50 cm. A bulk mass of chunks has a low density, permitting natural ventilation for drying. The material can be picked up easier than whole stems. Chunks can be made by a sugar cane harvester.

A Claas prototype *Miscanthus* harvester also shortens the material in lengths of about 50 cm. For harvest of hemp a machine was designed by Hempflax in The Netherlands that cuts the crop by a row-independent maize mowing header after which a cutting device derived from a forage harvester cuts the stem in lengths of 50 cm. This machine was also used in *Miscanthus*.

5. DENSIFICATION

The main reason for densification is to make units for easier handling and with higher density than loose material. The higher the density, however, the fewer possibilities for natural or mechanical drying are available. The density in storage will differ from the density in the units if these are not square. There are several principles as given below.

Baling. Using standard farm machines for straw or hay the following densities in kg dry matter per m^3 will be reached in *Miscanthus*: round bales 100-130 (*Phalaris* 170), square bales 130-160, compact rolls 200-350 (for ø: 1.0- 0.25 m respectively). These compact rolls are made by prototype machines, designed in Germany. In stationary machines higher densities can be reached. For instance the density of chopped (2 cm) *Miscanthus* densified in a recycled paper compactor was 250 kg d.m./m^3. Tests with a big baler for compacting small bales of *Phalaris* in Finland showed also a density of 250 kg d.m./m^3.

Pelleting. Hartmann (1997) reports on a prototype mobile pelleting machine used to harvest whole grain crops for energy. The density in the pellets was 875 kg d.m./m^3 and in bulk 380 kg d.m./m^3. The same densities are expected for the PRG crops. With a stationary pelleting machine in *Miscanthus* a bulk density of 500 kg d.m./m^3 was reached.

Briquetting. In briquets a density of 600 kg d.m./m^3 was reached (Heuvel, 1996).

Bundling. In The Netherlands *Miscanthus* was harvested with a self-propelled reed mowing and bundling machine. In the bundles with a diameter of 25 cm the density was about 100 kg d.m./m^3. On this machine these small bundles are collected in big bundles with a diameter of 1.5 m held tight by big ropes. Then the overall density reaches 120 kg d.m./m^3.

6. CONSERVATION AND STORAGE

During storage the quality and quantity of the product should be maintained by conservation. Drying is a common method for conservation, but also sealed storage (ensiling) at which lactic acid fermentation occurs, can be applied to wet biomass.

6.1. Drying

For dry storage moisture content should be below 15% to prevent fungal growth (Huisman and Kortleve, 1994). Self heating will occur at moisture contents above 25% unless continuous ventilation is possible. The methods for drying are field drying by sun radiation, natural drying in storage with ambient air, mechanical ventilation with a fan using ambient air, thermal ventilation with fan and heated air and combinations of these methods, successively. When the biomass will be applied for energy conversion heated air is too expensive except when waste heat can be used, for instance at the energy plant.

6.2. Ensiling

This is natural lactic acid fermentation in anaerobic conditions. If sealing is done on the same day of harvest, losses are less than 1%. If sufficient sugars are available, pH level will decrease to 4.3 at high moisture content (60% wet base) or 4.7 at moisture contents of about 30% in two weeks. This was proven with *Miscanthus* (Huisman and Kortleve, 1994). When continuously kept in an anaerobic condition, *Miscanthus* will last for years without additional loss.

6.3. Coverage

During dry storage it is necessary to cover the harvested material against precipitation. Except for bales of *Miscanthus* and *Arundo* this cover could be plastic sheeting. *Miscanthus* and *Arundo* bales need cloth to prevent damage by the sharp stems. Plastic sheeting and cloth are cheap, but they are laborious methods that require much inspection and repair after storms (Venturi and Huisman, 1998). Storage in structures requires more investments, but just covering and simple roof structures are sufficient and even recommended for natural ventilation. Such roofs built for large quantities are not much more expensive than plastics and are much safer for year-round storage.

7. PRETREATMENT

Depending on the application and processing techniques some pretreatment can be necessary. This can be done combined with harvest, after storage on the farm or at the plant.

7.1. Leaf removal

The leaves have less fibre content and strength as well as higher mineral, SiO_2 and ash content. For building material and paper pulp application the leaves should be removed. For *Miscanthus* a leaf loosening and separation machine has been designed. For *Phalaris* Hemming (1997) described a system of chopping followed by classification in an air stream of the light leaf fraction from the heavier stem pieces.

7.2. Reducing moisture

Fresh biomass or ensiled material has a high moisture content. In order to improve energy conversion efficiency, or decrease drying costs, water can be squeezed out. *Miscanthus* silage was pressed in a laboratory piston press with a pressure of 60 N/m^2. This caused a decrease in moisture content from 51% to 39% w.b. In the pressed material the mineral content decreased also slightly.

8. CONCLUSION

Many different production chains are possible and even more chains when the relevant transport means and storage locations are also varied. The choice of the best system depends on the local conditions and chosen optimization criteria. These criteria can be financial yield minus costs, energy input, energy output/input ratio, environmental aspects, social aspects, or organization structure. The optimization of choice of methods and chains is possible only by modeling for local conditions. Such a model is being developed for biomass crops and can be applied for given local conditions, including historical weather data, in order to get a good estimation of the optimum taking into account workability (Huisman, Venturi, and Molenaar, 1996; Venturi and Huisman, 1998)

REFERENCES

Hartmann, H. (1997). Comparing logistical chains for biofuel delivery- Features and costs of pelleting compared to conventional systems, in Book of Abstracts, International conference, Sustainable Agriculture for Food, Energy and Industry, Braunschweig, June 1997, Federal Agricultural Research Center, Braunschweig, Germany.

Hemming, M. (1997). Options for harvesting and preprocessing of Reed Canary Grass, in Conference proceedings of SusAgrCon97.

Heuvel, E.v.d. (1996). Pretreatment technologies for energy crops. Biomass Technology Group, Enschede, The Netherlands.

Huisman, W. and W.J. Kortleve (1994). Mechanisation of crop establishment, harvest and post harvest conservation of *Miscanthus Sinensis Giganteus*. Industrial crops and products, 2, pp. 289–297.

Huisman, W., G.J. Kasper, and P. Venturi (1996). Technical and economical feasibility of the complete production-transport chain of *Miscanthus x Giganteus* as an energy crop, in Proceedings of First European Energy Crops Overvieuw Conference, 1996, Enschede, The Netherlands. Biomass Technology Group, Enschede, pp. 1–8.

Huisman, W., P. Venturi, and J. Molenaar (1997). Costs of supply chains of *Miscanthus giganteus*. Industrial crops and products, 6, pp. 353–366.

Pahkala, K. (1997). Delayed harvest of reed canary grass in Finland; Cutting time in relation to the new growth, in COST 814-II Meeting of the WG "Alternative fibre crops" 1997 at IACR Rothamsted, ed. by T. Mela, Agricultural Research Center of Finland, Institute of Crop and Soil Science, Jokioinen, pp. 70–77.

Pahkala, K.A. (1998). The timing and stubble height of delayed harvest of Reed Canary Grass grown for energy and fibre use in Finland, in Biomass for energy and industry, 10th European conference and technology exhibition, Würzburg, 1998, ed. by H. Kopetz et al., CARMEN, Rimpar, Germany, pp. 204–206.

Pari P. (1996). Harvesting, storage and logistics of herbaceous biomass crops for fuel production, in Biomass for energy and the environment, Proceedings of the 9th European Bioenergy Conference, Copenhagen, Denmark, 1996, ed. by P. Chartier et al., Copenhagen.

Venturi P., W. Huisman, and J. Molenaar (1998). Mechanization and costs of primary production chains for *Miscanthus x giganteus* in the Netherlands, Journal of Agricultural Engineering Research, 69, pp. 209–215.

Venturi, P. and W. Huisman (1998). Modeling the optimization of primary production costs of *Miscanthus*, in: Biomass for energy and industry, 10th European conference and technology exhibition, Würzburg, 1998, ed. by H. Kopetz et al., CARMEN, Rimpar, Germany, pp. 806–809.

EQUIPMENT PERFORMANCE AND ECONOMIC ASSESSMENTS OF HARVESTING AND HANDLING RICE STRAW

R. Bakker-Dhaliwal,[a] B.M. Jenkins,[a] and H. Lee[b]

[a]Dept. of Biological and Agricultural Engineering
[b]Dept. of Agricultural and Resource Economics
All of the University of California, Davis, California, USA 95616

Equipment performance studies were conducted to evaluate costs and capabilities of existing rice straw harvesting operations in the Sacramento Valley of California. These studies reveal limitations in equipment mobility, transfer of straw from field to road transport, number of machines and timeliness of operations, and straw storage, as well as substantial differences in straw yield. Costs of harvesting and roadsiding vary from $16 to 34 Mg^{-1}. Owing to the limited amount of straw harvested, operators have so far been able to select for fields with better soil conditions. Larger scale harvesting operations will likely require improved field equipment.

1. INTRODUCTION

California generates approximately 1.5×10^6 metric tons of rice straw each year. Recent California legislation requires a substantial reduction in open-field burning of rice straw by the year 2000. As a result, industrial uses for straw are under development, including energy, chemicals, paper, animal feed, structural materials and other manufactured products. To support this development, reliable harvesting and handling systems must be available that can operate under difficult field conditions and with short harvesting seasons. Equipment performance studies of existing rice straw harvesting operations and related economic assessments were conducted during the first two years of a multi-year project to characterize system efficiencies and associated costs. Increasing straw utilization to account for a larger share of production will require improved equipment and systems. Straw grading techniques will also require development to meet the needs of the diverse emerging industries seeking to use straw.

2. OBJECTIVES

The objectives of the equipment performance study were to document the types of machinery currently utilized in the straw harvesting process, the types of field operations performed, the straw yields harvested in relation to previous field operations performed, and finally the field capacities of the machines during various operations. The equipment performance studies and field observations were further used to corroborate field data with survey data, to generate system cost assessments, and to some degree test the assumptions developed under a geographic information system (GIS) model created as part of the project modeling objectives.[1] In subsequent years, these results will be used to develop alternative straw harvesting and handling concepts including recommendations for new equipment design, to identify storage requirements and processing facilities, and to determine transportation effects.

3. FIELD STUDY SUMMARY

During 1998, notable as an El Nino year, the straw harvesting season commenced late (first week of October) and ended early (second week of November). Approximately 27,210 dry Mg of straw from 5,667 hectares was harvested in 1998. This amount is similar to that harvested in 1997 even though there were 81 harvesting days in 1997 and half that in 1998.

Equipment performance studies of rice straw harvesting operations from the first two years represent observations of 17 different systems. For all systems observed, operations on a single field were monitored and recorded for the full complement of operations from raking or swathing and continuing through off-site haulage. In addition, straw yields based on measured bale weights as well as yield relationships to stubble height, grain yields, and rice varieties were documented. Additional data were obtained regarding soil strength and mobility, spring harvest, survey of growers, balers, and straw users, and short and long-term storage.

Time and motion studies were conducted after the rice grain harvest primarily on fields that had been harvested using conventional cutter bar combines. Stripper harvested fields were scarce during fall 1998 owing to widespread areas of lodged rice related to the late planting and harvesting season. In addition to raking, swathing, baling, and roadsiding operations, data were gathered on loading, distance and speed of transport on different road types (as classified by the USGS), and unloading at off-site storage facilities.

In-field harvesting and handling operations were observed on a total of approximately 176 hectares, hauling data gathered on 942 dry Mg representing approximately 197 hectares and 50 loads, and yield data including grain yield and variety gathered on over 1,667 hectares. To date varieties represented in the field efforts include Akitakomachi, Koshihikari, Calmochi 101, M201, M202, M204, and L204. Including all varieties, grain yields in 1998 ranged from 7.6 to 10.1 dry Mg ha^{-1}, the harvested straw yields ranged from 2.5 to 7.0 dry Mg ha^{-1} and an average of 5.0 dry Mg ha^{-1}, and straw to grain ratios varied between 0.26 and 0.60, dry basis. Preliminary grain ratios on some of the varieties are reported in Table 1.

Table 1
Average straw:grain yield ratios of some varieties monitored in 1998

Variety	Calmochi 101	M201	M202	M204	L204
Average straw:grain ratio	0.41	0.49	0.52	0.52	0.33

Table 2 summarizes the types of machinery observed during the first two years of the study. Table 3 presents a sample from the 17 systems observed and arranges the machines in the different types of systems that were used in the cost accounting. The cost accounting was done on a system-by-system basis. Each system in our economic study includes either a raking or a swathing operation, followed by baling and roadsiding, respectively.

Table 2
List of machinery observed during harvesting of rice straw, 1997 & 1998*

Tractors	Rakes	Swathers	Balers	Roadsiding Equipment
Case 2390, 2590, 2690	Allen 8827	MacDon 9000	Freeman 1592	Freeman 4400
Case 5250	Darf 950	(turbo)	Freeman 330	Freeman 5000
FarmAll 560	New Holland	New Holland 1116	Freeman 370	New Holland 1075
FarmAll 806	216	New Holland 2550	Hesston 4900	(modified)
Fastrac JCB 185-65		Speedrower	Hesston 4910	New Holland 1085
Ford 4000			New Holland	Roadrunner, Hyster, &
Ford 5610			505	Mack squeezes
Ford 8870			New Holland	CAT 910 forklift
John Deere 4955/4960			595	Bale Skoop

*Disclaimer: Machine brand names are listed by operation in the following Tables. The performance of each machine is given for completeness to show the actual range in performance observed. The data should not be used to compare different machines. Performance is highly dependent on operator skill, field condition, weather, and other uncontrolled factors. This study does not constitute a formal test of any of the machines, and the reporting of machine data constitutes neither a critical evaluation of machine performance nor an endorsement of any machine type.

3.1. Raking and swathing

Raking operations take place when the straw stubble is left at a height between 5 to 25 cm or an average of 15 cm after the rice grain harvest. This normally occurs when the cutter-bar grain combine operates in lodged (fallen) rice, and the stem is cut short yielding high straw loadings. Swathing operations take place when the stubble height is in the range from 25 to 91 cm. During swathing operations the stubble is further cut to a height of 5 to 10 cm above the ground to increase straw yield. In 1998, 90% of the fields observed in Colusa and Butte counties had lodged rice. The grain combines had to keep the header low to the ground to maximize grain yields. Therefore, only raking operations without swathing were needed in most cases. In 1997 the situation was reversed, most fields had standing rice, and swathing was the norm.

Table 3
System specifications

System	Raking	Swathing	Baling		Roadsiding	Hauling
	Model (rake & tractor)	Model (swather)	Model (baler & tractor)	Bale size	Model (stackwagon or hydraulic fork)	Model (stackwagon or truck-trailer))
A	NH 216 & Farmall 560	NH 2550	NH 505 & Farmall 560	16"×24"×46"	NH 1085	NH 1085
B	Allen 8827 & Case5250		Freeman 1592 & JD 4955/ 4960	36"×48"×96"	Freeman 5000	2 trailer flatbed w/ Roadrunner squeeze
C		MacDon 9000 Turbo	Heston 4900 & Case 2690	48"×48"×96"	Case 584 C	2 trailer lowboy w/ JCB loadall 505-19

Raking was observed on a total of 134 hectares in 1997 and 119 hectares in 1998. Raking capacities were similar in both years ranging from 2 to 9 ha h^{-1} and had an average of 5 ha h^{-1}. Differences in raking capacity and speed of operation were related primarily to heavy straw loadings and straw moisture, generally when greater than 30%, than to differences in soil moisture or operator performance. Average material capacity averaged 16 and 24 Mg h^{-1}, in 1997 and 1998, respectively. High moisture straw tended to amass together resulting in clogged wheels and reel bars. Although straw moisture problems associated with raking are currently dominant during the start of the harvesting season, this problem is likely to increase with increased straw harvesting in wetter field conditions. Operators then must frequently stop to unplug their rakes, decreasing field efficiency. If operators did not have to stop to unclog their machines, raking capacities could be improved by approximately 13%. Raking operating speeds ranged from 4.5 to 18 km h^{-1}. Raking was observed in fields with soil moisture contents ranging from 8 to 35% (wet basis).

Swathing was observed on a total of 53 hectares in 1997 and 5.2 hectares in 1998. Average field capacities on an area basis were 2 ha h^{-1} in both years. Average material capacities were 5 and 10 dry Mg h^{-1} in 1997 and 1998, respectively. Overall straw yields in the swathed fields averaged 2.9 dry Mg ha^{-1} in 1997 and 5 dry Mg ha^{-1} in 1998. Although straw volume and moisture were also problematic in swathing operations, soil moisture was of more concern with swathers than with rakes. The distance between the tires on swathers does not allow the machine to follow previously made tracks by the grain combines. Instead swathers often traverse softer, wetter soils in non-compacted, un-tracked field areas. In addition, swathers are heavier than rakes.

Based primarily on 1997 raking data, costs were approximately \$5 ha^{-1} and \$2.20 Mg^{-1}. Swathing costs ranged from \$14 to 26 ha^{-1} and from \$3.31 to 8.82 Mg^{-1}.

3.2. Baling

Baling was monitored on approximately 83 and 167 hectares, in 1997 and 1998, respectively. Overall average yield in both years was 5.0 dry Mg ha^{-1}. Overall field capacities were 2 and 3 ha h^{-1} and material capacity was 7 and 14 dry Mg h^{-1} for small and large balers, respectively. In 1998, average soil moisture was 29% and 20% and straw moisture was 15% and 10% for small and large balers, respectively. The overall average straw yields, field capacities, and working speeds for the large balers were generally higher than for the small balers suggesting higher overall efficiencies for the large balers. The performance of the small balers was highly affected by unfavorable field conditions in 1998. However, in 1997 the field conditions were reversed, large balers performed in high moisture soils (30% average) and small balers in relatively dry fields (14% average). The large balers nevertheless had higher overall capacities in 1997.

Baling costs ranged from $22 to 60 and from $19 to 48 ha^{-1} for small and large bales, respectively. On a material basis, the range was from $5.50 to 17.60 Mg^{-1} for small bales and $4.40 to 7.70 Mg^{-1} for large bales.

3.3. Roadsiding

Roadsiding was monitored on approximately 79 hectares. The roadsiding capacities on an area basis were 1 and 5 ha h^{-1} for small and large bales, respectively. The roadsiding capacities on a material basis were 5 and 26 dry Mg h^{-1} for small and large bales, respectively. The capacities for the small bale stackwagons are only somewhat reflective of the wet field conditions in which they operated, but are mainly due to the long distances the stacks were roadsided. The small wagons transported the bales distances ranging from 0.4 to 6.4 km, directly to the long-term storage facility. Speeds for the large and small bale stackwagons were substantially different during the second year. In-field speeds ranged from 4.8 to 8 km h^{-1} for the small bale wagons and from 13 to 21 km h^{-1} for the large bale wagons. On-road speeds for the small bale stackwagon, fully loaded with 80 bales, ranged from 32 to 64 km h^{-1}. The fork-lifts had lower efficiency in clearing bales off the field when compared to the large bale stackwagons mainly due to the large paddy size in which the fork-lifts operated. The fork-lift carried half the load, two bales, to the edge of the paddy compared to four bales carried by the large bale wagon. In general, speeds increased with increasing paddy size and distance to the roadside stack.

Roadsiding costs ranged from $19 to 29 and $14 to 26 ha^{-1} for small and large bales, respectively. On a per ton basis, costs ranged from $7.22 to 8.82 Mg^{-1} and $4.40 to 5.15 Mg^{-1} for small and large bales, respectively.

3.4. Hauling bales, off-field-loading, transportation, and unloading of trucks

Substantial data on hauling, including loading on truck-trailer combinations, transportation, and unloading of trucks at a storage facility, were collected in 1998. A total of 285 dry Mg were observed during loading operations, 392 dry Mg were monitored in round trip transportation to and from a storage or processing facility, another 265 dry Mg observed in one way (either to or from storage site) transport, and 431 dry Mg during unloading operations. Ten trucks were timed from time of arrival at weigh-in scales to arrival at the unloading site. Scales were approximately 0.32 km from the unloading area. Total weigh-in time ranged from 3 to 11 minutes depending on number of trucks ahead of the straw truck. Actual weigh time averaged one minute. Straw bales were not sampled for moisture content by the processing facility at the scales. If incoming loads were to be monitored and sampled for moisture by a processing facility, the total weigh-in time would be expected to increase depending on the number of samples collected.

A majority of the trucks monitored provided official scale weights of their loads and were sampled for straw moisture content. Flatbed trailers transporting 30 large bales of the Hesston variety ranged from 15 to 16 dry Mg (0.5 to 0.6 Mg bale^{-1}). Flatbed trucks transporting 48 Freeman bales ranged in weight from 17 to 19 dry Mg (0.35 to 0.40 Mg bale^{-1}) and averaged 18 dry Mg (0.38 Mg bale^{-1}). Trailers with lowered mid-sections ("low-boys") hauling 36 Hesston bales ranged from 16 to 22 dry Mg (0.44 to 0.61 Mg bale^{-1}) and for 38 bales ranged from 19 to 22 dry Mg (0.5 to 0.59 Mg bale^{-1}). Average weights transported for 30, 36, and 38 Hesston bales were 15, 19, 21 dry Mg (0.50, 0.53, 0.55 Mg bale^{-1}), respectively.

Loading straw bales on trucks was accomplished by the hydraulic fork loaders or by squeeze type loaders. Loading times, including securing bales on trailers, varied from 13 to 40 minutes and were only partly dependent on the number of bales loaded. Other factors influencing loading rates included operator skill, distance of stack from truck, and number of workers in the field. Unloading times ranged from 17 to 45 minutes and were primarily a function of operator skill and the number of bales unloaded. During a 10-hour work day, four loads per truck within a 32 km radius were typical, while five to six loads per truck were more common within an 8 km radius.

4. SUMMARY AND CURRENT WORK

Total system costs, without distinguishing between small and large bales and not including transportation and storage, ranged from $46 to 110 ha^{-1} and $12 to 34 Mg^{-1} in 1997. Costs have yet to be determined on equipment performance data gathered during the 1998 season. Cost accounting underway includes analyses of direct costs related to transportation and storage. Preliminary survey data indicate storage costs to range from $2.2 to 3.3 Mg^{-1} yr^{-1} for tarping and approximately $4.40 to 6.60 Mg^{-1} yr^{-1} in barns under permanent cover.

For each field activity discussed above, in addition to the actual field capacities on both an area and a material basis, potential machine capacities and field efficiencies are being determined. Data on short and long term storage, soil strength and mobility information, as well as time and motion on industry treatment of straw at the storage or processing facility (i.e., grinding straw bales, are currently being analyzed).

REFERENCE

1. Bernheim, L.G., B.M. Jenkins, R.E. Plant, and L. Yan (1999). Geographic information system modeling of rice straw harvesting and utilization in California, Proceedings Fourth Biomass Conference of the Americas.

343

RAPESEED AS AN ALTERNATIVE CROP FOR NON-FOOD USES IN PORTUGAL

S. Ferreira-Dias,[a] A.C. Correia,[a] J.V. Mazumbe,[a] E.V. Lourenço[b] and J.E. Regato[c]

[a]Instituto Superior de Agronomia, Tapada da Ajuda, 1349-017 Lisboa, Portugal
[b]Universidade de Évora, Dep. Fitotecnia, Apartado 94, 7001 Évora, Portugal
[c]Escola Superior Agrária de Beja, Pr. Rainha D. Leonor, 1, 7800 Beja, Portugal

1. INTRODUCTION

In the South of Portugal, Alentejo is a flat region with a Mediterranean climate of hot, dry summers and mild, wet winters, where small grains have been the traditional crops. Consequently, farmers feel the negative effects of the Common Agriculture Policy (CAP). So, in order to maintain land fertility, there is a need to find alternative crops to include in rotation with cereals and to grow in set-aside lands, such as crops for oil production. In fact, the high petroleum prices related to unstable Middle East politics and environmental concerns have been recently drawn attention to the use of vegetable oils for fuel production (biodiesel). However, the direct use of vegetable oils as a fuel may cause problems such as poor ignition and combustion as well as clogging of fuel lines and filters. These problems are mainly ascribed to the high viscosity, low volatility and incomplete combustion of the vegetable oils, and to the presence of highly reactive unsaturated fatty acids and phosphatides (Johnson & Myers, 1995). The conversion of vegetable oils into their fatty acid methyl esters (FAME) by transesterification with methanol has been carried out to improve combustion characteristics. Blending conventional fuels with vegetable oils or FAME has proved to be a feasible option (Chang et al., 1996). Current interest in biodiesel is increasing. In fact, the use of biodiesel blends for fueling urban buses has been encouraged in Europe in order to reduce atmospheric pollution in big cities (Johnson & Myers, 1995).

The aim of this study was to evaluate the feasibility of non-traditional crops for the Alentejo Region, namely rapeseed (*Brassica napus* var. *oleifera*) for oil production. Three different sites (Beja, Evora and Elvas) (Figure 1) were selected for the culture of 13 varieties of rapeseed. In addition to edible purposes, the use of rapeseed oil (unmodified or converted to FAME) for biodiesel fuels production was investigated.

Figure 1. Selected sites for the testing of rapeseed crop (Evora, Elvas and Beja).

2. MATERIALS AND METHODS

2.1. Materials
Thirteen varieties (Starlight, Briol, Drakar, Star, Bristol, Karat, Pactol, Kabel, Desirée, Ester, Galaxy, Eurol and Loreto) were grown in 1995-96, in three different sites in Alentejo (Beja, Evora and Elvas).

2.2. Oil extraction
Seed preparation. Immediately after harvest, seeds were cleaned by sieving, dried at 60°C for 12 hours and stored in a cold place until oil extraction.

Seed Grinding. In order to select the most suitable mill, seeds from Beja (Pactol variety) were crushed in a household coffee mill (blade-type mill) or in a ball mill (Lampen mill from Lhomme & Argy) and the extraction yield in oil was compared. Experiments were carried out in triplicate.

Oil extraction. The oil was obtained from the seeds by solvent (*n*-hexane *p.a.*) extraction in a Soxhlet apparatus. To establish the extraction time, the solution was recovered in a flask, at different extraction times, the solvent evaporated under reduced pressure and the extracted amount of oil quantified by weighting. Experiments were carried out in triplicate and the results were expressed in a dry basis.

2.3. Oil characterization
Physical properties determinations. Density at 15°C was assayed by the hydrometer method (ASTM D 1298-85); dynamic viscosity at 40°C was assayed in a Brookfield viscometer, model DV-I; kinematic viscosity was calculated as the ratio between dynamic viscosity and density.

Chemical properties. The acidity, i.e., the amount of free fatty acids (%) was assayed by titration with 0.1N NaOH solution; cetane number was estimated as a function of both saponification value and iodine number, according to Nag et al. (1995). The sediment was evaluated by hot filtration (EN 210307.1) and the water content was assayed with a Metrohm 684 Karl Fischer coulometer.

The fatty acid composition of the oils was evaluated as their methyl esters (Ferreira-Dias and Fonseca, 1993) in a gas chromatograph (Carlo Erba, vega 2000 GC) equipped with a SUPELCO capillary column (SP™-2380, 0.2 µm, 60 m x 0.25 mm; fused silica). Both detector and injector (FID) were heated at 250°C. Temperature was programmed as follows: 175°C for 25 minutes, a slope of 5°C/min from 175°C to 220°C and 220°C for 10 minutes. Hydrogen was the carrier gas at a column head pressure of 60 kPa.

2.4. Preparation of fatty acid methyl esters

The FAME were prepared by transesterification between the crude rapeseed oil and anhydrous methanol at a molar ratio of 1:6 (twice as much alcohol as is stoichiometrically required) (Johnson and Myers, 1995). The reaction was carried out in a thermostated cylindrical glass vessel closed with a rubber stopper, at 40°C, under magnetic stirring. Sodium methylate was used as a catalyst (2% of the mass of oil) (Erickson, 1995).

To monitor the reaction, samples were taken out during the time-course of the reaction, and added to equal volumes of water to stop the reaction. The mixture was vigorously stirred and centrifuged at 2000 rpm for 10 minutes. The supernatant was diluted in n-hexane and spotted on silica gel thin-layer chromatography plates. Elution was carried out in petroleum ether/diethyl ether/acetic acid (85/15/1; v/v/v) as the mobile phase. Plates were sprayed with 2% 2':7'-dichlorofluorescein in 95% ethanol, observed under U.V. at 366 nm and the various groups of compounds identified by comparing with standards.

After the equilibrium was attained, the ester separation was achieved by adding water to wash both the catalyst and hydrophilic compounds, followed by an extraction by n-hexane. The residual solvent was removed at 40°C under reduced pressure.

2.5. Statistical analysis

Data concerning oil content of seeds from different varieties and sites, as well as the corresponding oil productivities of the fields were submitted to an analysis of variance (ANOVA) by using the program "Statistica"™, version 5, from Statsoft, USA.

3. RESULTS AND DISCUSSION

3.1. Oil extraction

Comparison of the performance of the two mills used for seed grinding on the efficiency of oil extraction is represented in Figure 2. A slightly higher yield in oil was obtained with the samples ground in the coffee-mill. In addition, after about 10 hours extraction time, no significant increase in extracted oil was observed. Therefore, every sample was ground in the coffee mill prior to a 10 hour solvent extraction.

The yield of oil per hectare (oil productivity) was estimated by multiplying grain yield (kg/ha) by the respective oil content in seeds, for every variety and site. Data were submitted to ANOVA. No significant differences were found between varieties from the same location. On the contrary, significant differences (LSD test, $p < 0.05$) were found for oil content of the seeds and for oil productivity (kg/ha) between samples from different sites (Figure 3). The highest oil contents were obtained on seeds from Beja, followed by the seeds from Evora. However, owing to differences in grain production on the various crops, the highest average oil production was 1280 kg/ha in Evora, followed by Beja (1030 kg/ha) and Elvas crops (373 kg/ha) (Figure 3). The observed behaviour may be ascribed to soil and climate differences at the selected sites.

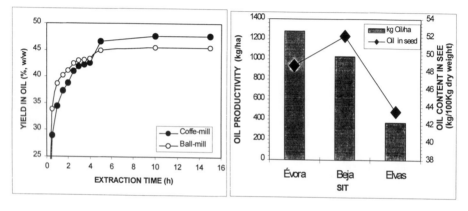

Figure 2. Selection of milling conditions and solvent extraction time.

Figure 3. Average oil content observed on 13 varieties of rapeseed cultivated in three different sites and respective oil productivities of the crops.

3.2. Oil characterization

The average values for the most important physico-chemical properties of crude rapeseed oil and the range values for diesel, according to the European Union legislation are displayed in Table 1. The fatty acid composition for the crude oil is shown in Table 2.

Table 1
Average values for some important properties assayed for crude rapeseed oil (39 samples) and diesel specifications, according to the EU legislation

OIL	Cetane Number	Acidity (%)	Water (mg/kg)	Density (kg/m^3)	Dynamic viscosity (mPa.s)	Kinematic viscosity (mm^2/s)	Sediment (mg/kg)
Rapeseed	25.2	0.4	470	904.3	25.5	31.5	0.01
Diesel	Min. 46	-	Max. 200	820-860		2.0-4.5	Max. 24

Table 2
Major fatty acids of crude rapeseed oil (average of 39 samples)

Fatty Acids	C16:0	C18:0	C18:1	C18:2	C18:3
Average value (%)	4.4	1.8	62.7	18.5	9.3
St. Deviation	0.5	0.2	1.1	0.9	1.5
Range	3.9-5.8	1.5-2.2	60.1-64.0	17.3-18.5	5.9-11

Concerning the potential use of these crude oils as "biodiesel," low acidity values and the absence of sediments are encouraging results (Table 1). However, viscosity values about 10 times the viscosity of diesel, and too-low cetane numbers restrain direct use of rapeseed oil as a fuel.

With respect to fatty acid composition (Table 2), no traces of erucic acid were found in the fatty acids profile for any sample. This was to be expected, since all the varieties tested are edible. The rapeseed oil is mainly composed of unsaturated fatty acids being oleic acid (C18:1) the major fatty acid, followed by linoleic (C18:2) and linolenic acids (C18:3).

3.3. Preparation of fatty acid methyl esters

Under the previously described methylation conditions, nearly all triglycerides were converted into their FAME after 4 minutes reaction time. Dynamic and kinematic viscosities of the recovered ester were assayed: values of 4.0 mPa.s and 4.5 mm^{2}s^{-1} were found, respectively. These values fit well the requirements for diesel.

4. CONCLUSIONS

The direct use of crude rapeseed oil suggests potential drawbacks. However, the obtained results on transesterification of rapeseed oil with methanol are rather encouraging. The rapeseed crop for biodiesel production may be an option for farmers in the South of Portugal.

5. ACKNOWLEDGMENTS

The authors are grateful to Mrs. Natalia de Melo Osorio and Mrs. Sandra Borges, from the Instituto Superior de Agronomia, Lisbon, for their invaluable help with this experiment. This study was supported by a Portuguese grant (PAMAF-IED, projecto n° 1016).

REFERENCES

1. Annual Book of ASTM Standards (1994). American Society for Testing and Materials, Philadelphia.
2. Chang, D.Y.Z., J.H. Van Gerpen, I. Lee, L.A. Johnson, E.G. Hammond, and S.J. Marley (1996). Fuel properties and emissions of soybean oil esters as diesel fuel, J. Am. Oil Chem. Soc., 73, pp. 1549–1555.
3. Erickson, M.D.E. (1995). Interesterification, in Practical Handbook of Soybean Processing and Utilization, ed. by D.R. Erickson, AOCS Press and United Soybean Board, pp. 277–296.
4. Ferreira-Dias, S., and M.M.R. Fonseca (1993). Enzymatic glycerolysis of olive oil: a reactional system with major analytical problems, Biotechnol. Techn. 7, pp. 447–452.
5. Johnson, L.A. and D.J. Myers (1995). Industrial uses for soybeans, in Practical Handbook of Soybean Processing and Utilization, ed. by D. R. Erickson, AOCS Press and United Soybean Board, pp. 410–417.
6. Nag, A., S. Bhattacharya, and K.B. De (1995). New utilization of vegetable oils, J. Am. Oil Chem. Soc., 72, pp. 1591–1593.

SUITABILITY OF CHAIN-FLAIL DEBARKING TECHNOLOGY FOR THE INTEGRATED PRODUCTION OF WOOD FUEL AND PULPWOOD—TECHNOLOGY TRANSFER FROM USA TO FINLAND

Veli-Juhani Aho and Ismo Nousiainen

VTT Energy, P.O. Box 1603, FIN-40101 Jyväskylä, Finland
Tel. + 358 14 672 669, Fax. + 358 14 672 799, e-mail: Veli-Juhani.Aho@vtt.fi
e-mail: Ismo.Nousiainen@vtt.fi

Experimental research carried out at VTT Energy has modified chain-flail delimbing-debarking technology to respond better to Finnish conditions, especially in the utilisation of first-thinning wood for bioenergy and pulp-chips. The current national Bioenergy Research Programme is studying the profitability and costs of chain-flail delimbing-debarking, the raw material balance, and the accumulation and the properties of pulpwood and delimbing chips. The objective of this method is to produce chips of branchwood and bark for fuel and good quality barkless pulpwood chips for the pulping industry at a profitable price. Pulpwood chips should contain < 1.0% bark. The target price of the fuel fraction has been 45 FIM/MWh. Comparisons were made on the basis of different production methods. The chain-flail delimbing-debarking technology was originally imported to Finland from the USA, but it is not entirely suitable for Finnish conditions.

This paper describes modifications of chain-flail delimbing-debarking units that improve their operability in full-scale test equipment. The results will be applied to full-scale production equipment.

The results can be used to size delimber-debarker rollers, chains, revolution speeds and power of processing equipment for both the summer-wood and frozen first-thinning wood. By using the whole-wood processing equipment it is possible to optimise the smallest degree of fragmentation of wood to obtain a defined bark content with a given equipment combination. The targeted 1.0% bark level was easily obtained with the test equipment both in the case of frozen and summer wood. About 3% of the wood in softwood trees and about 1% of birch were lost.

1. BACKGROUND

Flail delimbing-debarking is a new integrated harvesting method for the integrated recovery of clean chips for pulping and residual biomass for energy. During the past ten years, this technique has been adapted widely in the USA and Canada for recovering small-sized timber for sulphate pulping.

In Finland trees are felled by chainsaw operators using the so-called felling-piling method. Felled trees are hauled by conventional forwarders equipped with a grapple saw, which is used for bucking the undelimbed trees into 5-m sections in conjunction with loading. The trunks are delimbed and debarked at the roadside or at the terminal by whipping with steel chains. The delimbed pulpwood, passed through the process, can be chipped directly on the truck, and the wood waste thus formed, such as branches, needles and bark, can be used as fuel. Some chain-flail debarking units are used in Finland.

Because chain-flail delimbers used in the USA are not directly applicable to Finnish conditions, it has been essential to modify the method, so it can better serve the suitable parts of the Finnish forest sector. The research and development environment, developed especially for the research and development work, serves both new equipment as well as that already in use. This kind of product development allows full-scale development and testing of different kinds of methods.

The national Bioenergy Research Programme investigated the productivity and costs of chain-flail delimbing-debarking method, the raw material balance and the accumulation and properties of pulpwood and delimbing wood chips, and the applicability of the method for Finnish conditions. The goal of the research has been to develop an integrated production method based on delimbing-debarking technology. This method, by which it is possible to produce branch and bark chips for fuel purposes and good quality barkless chips for pulping industry at a competitive price, must then be put into production. The research objective was to reduce the residual bark content of the chips below 1.0%. The target price of the fuel fraction has been 45 FIM/MWh. The cost of different production methods is presented in the Table 1.

The project's objective is to develop a chain-flail delimbing-debarking method that keeps the residual bark content of the wood < 1.0 % before chipping, to reduce the portion of sticks and impurities going into the chipper, and to use the removed sticks and branches directly for energy production.

2. IMPLEMENTATION OF THE EQUIPMENT

The delimber-debarker consists of three operational modules: a feeding table, a delimber-debarker and an output pocket. Under the delimber-debarker, trays collect delimbing and debarking residues. In addition, a control centre controls the equipment and collects computerized data. The equipment is 23.4 m long, 5.2 m high, and 3.3 m wide. The width of the control centre is 3.0 m. The width of the feeder conveyor is 1.2 m,

Table 1
Production costs of different production methods. First thinning pine, minimum top diameter of longwood and partial trees 7 cm. Forest haulage distance 250 m, lorry transportation distance 80 km

		Procurement as longwood	Procurement as partial trees	Procurement as whole-trees
		Single-tree processing	Chain-flail delimbing	Chain-flail delimbing
Pulpwood chips production costs	Fl/m³	270	247	252
Price of the fuel fraction	FIM/MWh	45	45	45
Price of the pulpwood chips	FIM/m³	270	270	270
Production costs of fuel fraction	FIM/MWh	45	12	28
Bark content of the pulpwood chips	p-%	0,1	2	2

and the height of the inlet opening 0.7 m. The total weight of the equipment is about 30 tonnes (Figure 1).

The rotation speed of the debarking drums can be adjusted within the range 0-1400 r/min and the debarking angle within ± 45 degrees; and the present maximum timber feed rate is 40 m/min.

The sides of the unit (adjacent to the chains and brushes) have been covered with thick glass-sheet, so the chains and brushes can be viewed as they strike the wood. The striking spots of the brushes are recorded with high-speed video-camera, so it is much easier to analyse the debarking process.

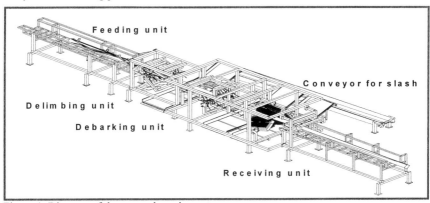

Figure 1. Diagram of the research equipment.

The following measurements were taken by a data collection system: the wood feeding speed (the rotation speed of the feeding chain of the feeding rolls and feeding table), rotation speed and the elevation of the chain rolls, and the electric power demand of the rolls. Graphs of the measurements can be produced on screen using different time-axes for the measuring targets.

3. EXAMPLES OF THE MAIN RESEARCH RESULTS

The measurements can be used both for dimensioning the delimber rollers to be constructed both for summer-wood and frozen first-thinning wood, the chain outfits, and to determine the rotation speeds and powers. By using whole-tree processing device and Unscrambler software it is possible to optimise the smallest degree of fracturing needed for a given type of equipment and the desired bark content.

This test equipment can obtain the targeted 1.0% bark content when debarking single frozen or defrosted single-trees. When debarking first-thinning wood as single trees, the smaller chains break the thinner layer on the surface of the stems, and the debarking result is similar to that of the debarking drum. On the other hand, delimbing of the trees requires two stages and thick chains are used for delimbing efficiency and to prolong the life of the chains. Two-stage debarking improves the debarking result, but the high rotation speed of the delimber equipped with 16 mm chains rapidly increases fragmentation of the wood (Figure 2).

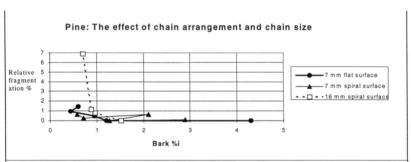

Figure 2. The effect of chain arrangement and chain size.

In a frozen stem there is no clear boundary between the bark and the tree along which the bark can be loosened from the tree. In debarking of frozen timber the bark is hence loosened as small fragments and it is difficult to adjust the debarking depth so that no timber is loosened along with the bark. The larger power demand required to debark the sides of the trees increases the wood losses both from the upper and lower surfaces of the trees. The smaller power required in the processing of the sides of the trees, caused by the

trees. The smaller power required in the processing of the sides of the trees, caused by the geometry of the trees, should be compensated for by higher peripheral velocity of the chains (longer chains) or by using thicker chain-links in the chains hitting on the sides of the trees. These measures, however, require the directioning of the trees into a defined debarking place and the use of chains of different size on the same delimbing roll. Aiming the hit of the chain at a small angle toward the side of the tree increases the debarking efficiency.

In the tests with frozen softwood it was possible to obtain 1.0% bark content with about 3.0% relative wood losses. The corresponding losses in the tests with birch were about 1.0%, which means that the losses with birch are considerably lower than with softwoods (Figure 3).

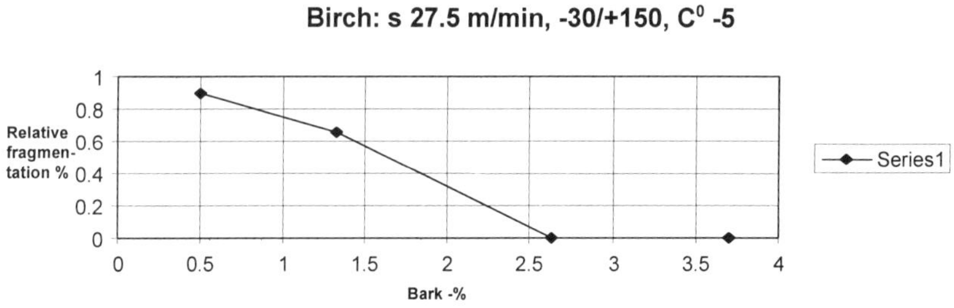

Birch: s 27.5 m/min, -30/+150, C^0 -5

Figure 3. Bark-losses diagram of birch.

Frozen wood does not bend elastically as does non-frozen wood when the chains hit, and hence frozen wood fragments more easily. Using an elastic base for such timber might decrease fragmentation. This might be an interesting research topic.

The life of the chains is decreased considerably by hitting each other. In addition, there is enough space in a chain link to hit towards another link bound to it, so that the surface of the inner link can be worn. If the loose space of a chain link is removed, it might affect the debarking result. The hitting force of the chain can be directed more effectively on the surface of the timber and the wear of the chain might be divided more evenly between the links of the chain, and the links of the chain might not hit towards each other so easily.

4. SUMMARY

The result of the chain-debarking depends on the properties of the raw material, on the process technical parameters and on the weather conditions. In good unfrozen conditions the tested unit can obtain the bark content level of the industry—the bark less than 1% of the dry mass. This bark content was obtained both in summer and winter conditions in the case of debarking single trees of different species. Debarking the sides of the trees and the uncontrolled sliding of the trees between the rollers are problems with debarking tree-bundles. Overcoming these difficulties will require further development of the stem transfer and clamping devices. If the trees are very small or they have dried before processing, the debarking result is still unsatisfactory. The bark content of pine chips has usually been on the level 1-2%. The bark content of chips produced at the -13 to $-30°C$ can rise to 3.0%.

Controlled transfer of the trees between the delimbing-debarking rollers is essential for the debarking of wood bundles so that trees are debarked evenly on all sides of the stems. Additional development is still required for feeding and clamping constructions in order to ensure the controlled transfer of the trees.

The whole-tree processing equipment of VTT Energy appeared to be as good as expected. The cameras and high-speed cameras used in the device assist in the evaluation of the test runs, but they do not yet assist the measurement of the results. High-speed camera and image processing systems might be developed for the bark content and power measurements.

Chain-flail delimbing-debarking technology is applicable in Finnish industry as a special processing line, for instance for first-thinning trees. This use enables the controlled processing of even the smallest trees, hence large wood losses are avoided. The durability of the chains has to be improved and the bundle-feeding phenomenon to be controlled in order to make this kind of technology more commonly used.

REFERENCES

1. V-J. Aho, L. Nikala, and H. Laitinen (1997). Ketjuharja-menetelmä puun karsinnassa ja kuorinnassa. Julkaisussa: Bioenergian tutkimusohjelman julkaisuja 14. Vuosikirja 1996. Osa I: Puupolttoaineiden tuotantotekniikka. (Chain-brush method in delimbing and debarking of trees, In Bioenergy Research Programme, Yearbook 1996. pt. 1 Wood Fuels Production Technology). Jyväskylän Teknologiakeskus Oy (Jyväskylä science park), Jyväskylä,1997, pp. 163–175.
2. K. Rieppo, A. Poikela, P. Hakkila, V-J. Aho, and L. Nikala (1996). Puupolttoaineen ja selluhakkeen integroitu tuotanto ketjukarsintakuorintatekniikalla. Loppuraportti 1995. Moniste (Integrated production of wood fuels and pulpwood chips with chain-flail delimbing-debarking technology, Final report 1995). Helsinki 26.6.1996. 66 s.

THE RECOVERY OF POPLAR ROOTWOOD

Raffaele Spinelli and Riccardo Spinelli

CNR – Ist. per la Ricerca sul Legno, via Barazzuoli 23, I-50136 Firenze, Italy

Poplar roots are an interesting source of biomass, available in large quantities and easy to harvest. As an average, the mature poplar stand can yield 18 fresh tonnes/ha of rootwood. Working togheter, manufacturers and contractors have developed a harvesting system that is simple and effective. Extractors and cleaners can be attached to general-purpose prime movers that are easy to find. Under favourable conditions these machines can reach a very high productivity: 150 roots/h for the extractor and 180 roots/h for the cleaner. Rootwood can be extracted, cleaned and delivered to the factories for a price of about 20 USD/tonne (metric). This is the full cost and includes contractor profit.

1. INTRODUCTION

The roots of harvested trees represent an interesting source of industrial wood biomass. Roots represent a substantial portion of the tree mass, and their wood is more compact and dry than in the stem. In plantations, the removal of the root system is considered as a service rendered to the landowner. Therefore, harvesting tree roots does not require the payment of a concession, and more often involves an additional remuneration paid by the landowner himself. Of course roots are removed only in plantation forests, where favourable site conditions facilitate harvesting.

Poplar plantations offer such conditions. Poplar roots are an interesting source of biomass, available in large quantities and easy to harvest. As an average, the mature poplar stand can yield 18 fresh tonnes/ha of rootwood.

Records exist of poplar root extraction since the early 1900s. In Italy, root extraction was already mechanised by the 1960s, but Italian experts did not participate in the intense research activities carried out in the 1970s by the Finns – and later by the Swedes, the Danes and the French. These were the times of the excavator-mounted *Pallari* uprooting attachment and of his omologues: the *Cranab* and the *ARMEF* uprooters, the *Ösa* stumpharvester and the *TX-1600* whole-tree extractor.

2. ROOT WOOD RECOVERY IN ITALY

Since the 1930s, the Italian furniture industry has based its supply of raw material on the fast-growing poplar plantations established in the plains. Clonal plantations are established in rows, with trees spaced 5 to 6.5 m. Rotation reaches 10-14 years, according to the clone, the site and the market. Poplar grows on agricultural land, which must be cleared after harvesting. Therefore, root systems must be removed. The need to re-cultivate the soil as soon as possible requires the immediate extraction of fresh roots, which have a strong hold to the ground. In 1960 the firm *Elléttari* bought the CCM extractor patent and began producing its root extractors. The machine consisted in a pipe-auger fitted to the tractor rear and powered through the PTO. The pipe would be lowered over the stump and driven into the ground to the depth of 1-1.5 m. Then the pipe was raised with the "carrot" inside it. An ejection ram would push the "carrot" out of the pipe, dropping it to the ground. *Ellèttari* developed two extractor models that can be mounted on any farm tractor with sufficient power. These are still produced today, and they incorporate a number of innovations. On request, *Ellèttari* also build a complete extraction machine, the *Elefante*.

In the early 1980s a second manufacturer began building his own root extractor, very similar to the one made by *Ellèttari*.

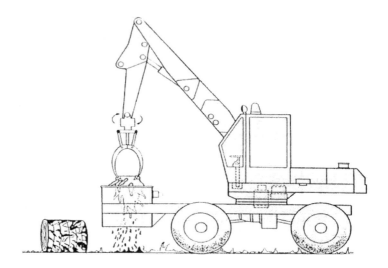

Figure 1. The Maséra chain-flail cleaner.

Whatever the manufacturer, the core-sampler system is ideal for trees with a strong taproot (poplar, pine etc.). In all cases, a pair of propeller blades is welded to the terminal crown of the pipe. These blades crush the remaining side roots so that they do not hinder recultivation.

In the early 1970s, the *Masèra* workshops developed a simple chain-flail cleaner. The machine was mounted on a wheeled chassis and towed by an hydraulic loader. The two flail axles were powered by independent hydraulic motors, connected to the pump of the loader. Each axle carried 24 flails, each made of a chain segment of seven links. Hardened steel chains were used. The loader would travel along the rows of extracted roots, pick up the "carrots" one by one, and dip each one a few seconds between the rotating flails. The clean roots would be thrown 5-6 m away, to form small heaps. *Masèra* sold the patent to *Ellèttari*, who is still producing the same chain-flail trailer without any major modifications.

Clean roots are loaded directly onto 10-ton trucks driven into the field and along the heaps. The truck is loaded by a hydraulic loader driving along—often the same loader that cleans the roots. If the soil is wet, the truck is assisted by a farm tractor or a small bull-dozer that pulls it through soft patches. Recently, someone has turned to high-speed farm tractors with 10-ton trailers. These have the same road speed as the truck, but enjoy better off-road mobility.

3. CASE STUDIES

We studied seven operations, each performed by a different contractor. All seven contractors extracted, cleaned, loaded and transported the root wood. All operated on the banks of the Po river, close to the factories that buy the root wood. Transportation distance never exceeded 30 km, and it was generally much shorter.

Site description is provided in Table 1. All operations were carried out between late January and late March. In fact, numbering follows chronological order.

Table 1
Site description

Case	n°	1	2	3	4	5	6	7
Locality		Coriano	Roncaglia	Torricella	Casalmag.	Dosolo	Senna	Torre
Clone		I-214	I-214	I-214	Canadian	Canadian	Canadian	Canadian
Age	years	11	13	13	10	11	12	11
Density	trees/ha	333	333	278	278	370	333	237
Terrain		Rutted	Even	Rutted	Even	Even	Even	Even
Ground		Mud	Solid	Mud	Solid	Solid	Solid	Solid
Soil		Clay	Sandy	Sandy	Sand	Sandy	Sand	Sandy
Stump size	tons	0.052	0.068	0.065	0.061	0.039	0.060	0.052

In principle, all operations involve three teams: the extraction team, the cleaning team and the transportation team. They generally work separately, assembling only during loading. In fact, loading requires that all three teams stop their work and assemble to carry out the new task. The loader attached to the chain-flail stops cleaning and comes to load the transportation vehicle. This vehicle is driven into the field, and therefore its driver is also kept busy. The extraction unit is often attached to the transportation vehicle, and tows it through difficult terrain.

For the extraction there are basically two options: farm tractor or Elefante, each coming in a number of variants (Table 2).

Cleaning is done with the ubiquitous *Masèra* chain-flail trailers, attached to a wheel loader or to a farm tractor. The hydraulic system of the farm tractor is not powerful enough to drive both the chain-flail and the loader mounted under the tractor cab. Therefore, the flail motors must be driven by an independent pump. A contractor went so far as to fit the chain-flail trailer with its own hydraulic system and with a salvaged 59 kW diesel engine.

Table 2
Resources used in seven operations, separated by work task

Operation	n°	1	2	3	4	5	6	7
				Extraction				
Tractor	type	Elefante	tractor	Tractor	tractor	tractor	Elefante	tractor
Power	kW	118	107	132	92	59	132	59
Extractor	type	Ellèttari 200	Ellèttari 200	Ellèttari 200	Ellèttari 200	Ellèttari 200	Ellèttari 200	Ellèttari 200
Diam.pipe	cm	45	50	50	45	50	45	45
Crew	n°	1	1	1	1	1	1	1
				Cleaning				
Loader	type	loader	loader	Loader	tractor	loader	loader	tractor
Power	kW	80	88	110	99	78	110	92
Cleaner	type	Masèra	Masèra	Masèra	Masèra	Masèra	Masèra	Masèra
Crew	n°	1	1	1	1	1	1	1
				Transport				
Vehicle	type	tractor & trailer	10-ton truck	10-ton truck	tractor & trailer	tractor & trailer	10-ton truck	tractor & trailer
Payload	T	10	10	10	10	10	10	10
Crew	n°	1	1	1	1	1	1	1

Transport vehicles are either 10-ton trucks or farm tractors with a 10-ton trailer. In one case, the farm tractor was a high-speed model (JCB Fastrac), capable of reaching 80 kmh on asphalt road. This is considered an intermediate solution, since the high-speed tractor performs as a truck on the asphalt road, while enjoying the terrain mobility of the farm tractor.

4. RESULTS AND DISCUSSION

Table 3 shows the productivities recorded in each operation.

There are several explanations for the variability recorded in the study. The poor performance of operation n.5 largely depends on the small size of the average stump: even handling many stumps per hour, it is difficult to produce a large tonnage under these conditions. Concerning operation n.1, the responsibility falls partly on the loader and partly on the soil type. The loader was too slow and the soil had a dominant clay texture, which made it stick to the roots and prolonged cleaning times.

Table 3
Productivity of root wood recovery

Case	n°	1	2	3	4	5	6	7
				Extraction				
Productivity	Stumps/d	561	1162	729	809	598	634	535
Productivity	T/d	29.2	78.8	47.4	49.4	23.3	38.0	27.8
				Cleaning				
Productivity	Stumps/d	462	1067	1354	679	587	1302	613
Productivity	T/d	24.0	72.6	88.1	41.4	22.9	78.2	31.9
				Loading				
Productivity	Stumps/d	995	2261	3161	1894	2154	2912	1240
Productivity	T/d	51.7	153.7	205.5	115.5	84.0	174.8	64.5
				Transport				
Payload	T	10.5	10.5	11.0	10.7	6.42	10.8	9.7
Productivity[1]	trips/day	2.3	3.8	4.1	2.6	2.7	4.1	2.3
Productivity	T/day	24.29	39.61	44.65	27.43	17.32	43.78	21.84

Notes: 1 - Equalised over a 20 km distance and a 18 min. unloading time

As to extraction, the modern 100 kW farm tractor seems the best option, especially if fitted with integrated electro-hydraulic controls. Such a machine can reach a gross productivity of 150 stumps per hour, twice that of the old models and 50% more than a tractor without improved controls. Old machines are cheap to operate, but they are too slow and they are good for part-time business only.

Cleaners attached to a farm-tractor are less manoeuvrable than loaders, and their crane is not as quick. Hydraulic loaders can move while performing other tasks, especially if they are handled by a skilful operator Specialised forest loaders are not deterred by muddy, rutted terrain. With a good operator, the loader produces twice as much as the tractor: 180 stumps per hour, when operating on sandy soil.

The hydraulic loader is also the best option for truck-loading. Specialised forestry-versions can load about 400 stumps per hour, filling a 10-ton truck in 15 minutes. Tractor-based units fall far behind, with a maximum output of 250 stumps per hour. Loading speed is particularly important, since long loading times are not paid by the loader only, but also by the transportation unit and by any other unit eventually assisting them.

In principle, transportation is more effective when done by road trucks, even if one has to use an extra tractor to assist them for the terrain transport during loading. If loading is quick enough, the use of an extra machine does not inflate too much the overall cost of loading and transportation. Once on the road, trucks are much faster than farm tractors, both empty and loaded. What's more, farm-tractors are not any cheaper. A 10-ton high-side trailer costs $38,000 US, which adds to a tractor cost already in the range of $60,000 US.

5. CONCLUSIONS

Poplar roots are an interesting supply of energy biomass, which is available in large quantities and is easy to tap. The average poplar cut can yield 10 tons of clean root biomass, endowed with a higher dry matter content than found in tops and branches.

Italian contractors have developed recovery systems that are both efficient and cheap. Extraction and cleaning units are based on general-purpose prime movers, easily available on the market. Under favourable conditions these units can achieve a very high productivity: 150 stumps per hour for the extraction unit and 170 for the cleaning unit.

Loading and transport is the bottleneck. Limited road speed works against farm-tractors, while poor terrain mobility restricts the use of standard trucks. The use of high-speed tractors may not be an option if traffic regulations will impose on them the same speed limit prescribed for farm-tractors. A container-shuttling system may be the solution, but it requires substantial investments that are not justified by the current market situation.

BIBLIOGRAPHY

Czereyski K., I. Galinska, and H. Robel. Rationalization of Stump Extraction. Genève (CH) ECE, FAO, ILO, LOG 158, 30 p.

De Simiane, C., J.L. Bertrand, B. Doubliez, and C. Artigue (1976). Extraction et récupération des souches. Nangis (FR) Annales Armef,-Cermas, pp. 229–267.

FAO (Nov.1962). Roma, Forestry Equipment Notes, B.33.62, 2 p.

Hakkila, P. and M. Makela (1973). Harvesting of stump and root wood by the Pallari stumpharvester. A sub-project of the joint Nordic research programme for the utilization of logging residues. (FI) Helsinki, Communicationes Instituti Forestalis Fenniae, 77, 5, 56 p.

Jonsson, Y. (1978). Comparisons of harvesting systems for stumpwood and rootwood. (SE) Skogarbeten, Report n.1, 31 p.

Nylinder, M. (1977). Harvesting of stump- and rootwood. (SE) Skogsarbeten report n.5, 19 p.

Pottie, M.A. and D.Y. Guimier. Harvesting and transport of logging residuals and residues. (CA) FERIC - IEA, pp. 62–65.

CROP RESIDUES AS FEEDSTOCKS OF CHOICE FOR NEW GENERATION COOPERATIVE PROCESSING PLANTS

Donald L. Van Dyne & Melvin G. Blase

A relatively new form of business organization – a new generation cooperative – is appropriate for investing in a biomass refinery for the production of ethanol and other higher-value chemicals. As a hybrid organization, similar to both a conventional cooperative and a for-profit corporation, part of the assured supply of feedstocks for the refinery is guaranteed by the shareholders. This is because shares in the cooperative carry both the right and the requirement for farmer-members to deliver feedstocks to the plant. From their perspective, farmer-shareholders view the cooperative as a means of adding value to their agricultural production. Consequently, a strong incentive to deliver the crop residues to the processing plant is a built-in feature of this form of organization. This is especially important when the harvesting window may be relatively short, e.g., harvesting cornstalk residue before the onset of winter in the Corn Belt.

This paper considers the advantages and economics of using crop residues in a refinery-type processing plant producing ethanol and furfural, and the role that new generation cooperatives could play in initiating and sustaining this type of plant.

1. INTRODUCTION

Most energy-producing firms are owned as profit-making private firms or as public sector-owned utilities. The exception are systems owned by cooperatives such as rural electric cooperatives.

The cooperative form of organization has been characterized by two primary attributes: 1) one-patron-one-vote, and 2) profits distributed as patrons' dividends in accord with the volume of business transacted by the patron with the cooperative. In spite of their special tax status due to the latter, this form of business organization in the energy industry has been used sparingly with respect to volume of energy produced. That is, relative to the volume of production of the entire industry, this form of organization has been a minor player. This may change in the future.

A radically different form of cooperative has emerged recently. Known as a new generation cooperative (NGC), this organizational form is a hybrid of a private for-profit firm, and a conventional cooperative. Distribution of "profits" depends on shares (delivery rights) owned by a member, and the one-person-one-vote form of governance is used. These and other characteristics of NGCs are worthy of elaboration. Subsequently, an applicable ethanol and chemical plant feasibility study that would lend itself to an NGC is reviewed. Finally, the advantages of ownership and operation of this type of plant by a new generation cooperative will be summarized.

2. CHARACTERISTICS OF NEW GENERATION COOPERATIVES

NGCs' characteristics concern its 1) basic philosophy, 2) membership, 3) delivery rights, 4) value-added objectives, 5) equity entrance requirements, 6) free riders, 7) required equality of a member's patronage and a member's equity, and 8) negotiability of equity shares. Each deserves elaboration (Developed in Van Dyne and Blase).

2.1. Basic philosophy

An NGC is basically an extension of the farm firm. That extension provides a value-added function that the individual firm cannot efficiently perform. As an extension of the farm business, incentives are provided to the firm to invest in, provide quality raw products for, and make active inputs into the management of a partly-owned business. Moreover, the objective of the business is to provide a substantial portion of a farm firm's income from the NGC's residual revenue after all of its expenses have been paid at the end of the year.

2.2. Membership

The above alludes to this distinguishing difference, i.e., closed membership. Closed membership means just that. Either the farm firm takes an equity position in the NGC or it is not a player. There is no middle ground. A firm is either in or out.

At the other extreme is the conventional cooperative. Anyone who does business with it (albeit in some cases a minimal threshold volume is required) is a member in most open cooperatives. Hence, the customer becomes eligible for patronage dividends on the basis of the volume of business transacted during the year.

The difference in membership dimensions clearly segregates NGCs from other cooperatives. In some respects, NGCs are very similar to privately held firms with regard to this characteristic. If this were the only distinguishing difference, as a matter of fact, NGCs might not be called cooperatives at all.

2.3. Delivery rights

This section could really have the heading of delivery rights and obligations. That is to say, the firm not only buys these rights to have some of its output processed, but also is obligated to deliver a predetermined volume of the NGC's feedstock. This feature is one of the keys to the success of an NGC. An assured supply of raw material is essential for any value-added enterprise. NGCs are no exception. This assured supply is obtained as part of the fund-raising drive. The maximum number of delivery rights is determined by the size of the plant needed to be efficient. That size of investment is divided by the volume of feedstock the NGC will need to operate profitably in order to determine the per share price of memberships with their delivery right obligations.

2.4. Value-added objectives

As stated above, NGCs exist to add value, usually to agricultural commodities. One common theme runs through all of them: the basic driving force is economics. NGCs are a deliberate creation to enable firms to capture additional revenues from successive stages in the marketing channel. NGCs are an effort by firms to reap the benefits of vertical integration and gain additional control in the marketing of their commodity with the expressed goal of increasing profits.

2.5. Equity entrance requirements

Up front equity investment requirements of NGCs make them more similar to for-profit firms than conventional cooperatives. Since a given number of units of feedstock to be processed is associated with a unit of equity investment, firms purchase as many units as fit their operations. Usually 30% to 50% of the investment cost of, and feedstocks for, the NGC are raised in this manner.

2.6. Free riders

As indicated above, there are no free rides with an NGC. To participate, everyone must buy shares. Although they are tradable, shares cannot be cashed in at the NGC. This assures the cooperative of a constant equity by its members. At the same time they can be sold, thereby not locking an investor in to the NGC indefinitely. In fact, their value is likely to fluctuate in direct relationship with the profitability of the NGC.

2.7. Required equality of a member's patronage and a member's equity

As mentioned above, a member's equity in an NGC entitles that member to deliver a specific quantity of the commodity to which value will be added, usually by processing. This constitutes a given proportion of the plant's capacity and, hence, its potential residual earnings. Since a member's patronage dividend is a proportional amount of those residual earnings, equality is maintained between patronage and equity.

2.8. Negotiability of equity shares

Most NGCs are likely to outlive their members. Hence, some method of disposing of their equity shares must be available to investors. This has been done by making them negotiable, as indicated above. Over time the share prices can be expected to reflect the earning capacity of the NGC for providing its value added function. In light of the fact that the equity shares carry with them delivery rights and obligations, the potential field of buyers likely is going to be limited to producers of the selected feedstock. Although limited experience has been obtained with regard to equity share transactions, initial sales suggest an appreciation in value as the prospects for future residual earnings of the NGC increase.

2.9. Summary of concepts

In many respects NGCs are unique organizations. Nevertheless, NGCs bear some similarities to both for-profit firms and conventional cooperatives. Regardless, their value-added focus and the equity investments required of a closed membership make them somewhat unusual in concept.

3. ECONOMIC FEASIBILITY OF AN ETHANOL/CHEMICAL PLANT

Previous research investigating the use of dilute acid hydrolysis in converting ligno-cellulosic residues into ethanol and higher value chemicals reported at BioEnergy '98 indicated the following (Van Dyne, Kaylen and Blase):
- the production of ethanol alone is not economically feasible without continued subsidies,
- the plant is profitable if it produces ethanol and furfural as coproducts,
- economies of size require a world-class size plant to attain optimum profitability,
- crop residues are the feedstock of choice,
- the optimum size plant would process 4,360 tons of residue daily (1.44 million tons annually),
- 47.5 million gallons of ethanol and 323 thousand tons of furfural would be produced annually, and
- annual net income would be an estimated $108 million, with about three-fourths of revenues coming from the sale of furfural.

The optimum size plant in terms of net present value over a 15-year lifetime versus volume of feedstocks used daily is represented as the top of the revenue curve in Figure 1. Also important is that a plant processing less than 1,300 tons of LCF resources daily and one processing more than about 8,500 tons daily generate returns of less than 15 percent. With less than 1,300 tons processed daily the plant is too small to be efficient. With a plant larger than the optimum size of 4,360 tons per day, the marginal cost of moving to a larger size plant (largely caused by high transportation, collection and delivery costs from long distances) is greater than the marginal revenues from product sales. Thus, the overall plant profitability continually declines (after peaking at the 4,360 tons per day capacity) and becomes negative after about 8,500 tons of feedstock processed daily.

4. ADVANTAGES OF NEW GENERATION COOPERATIVE ORGANIZATIONAL FORM FOR MULTI-PRODUCT ETHANOL AND CHEMICAL FIRMS

An overwhelming need of the above-described energy/chemical firm is a huge volume of feedstock in order to take advantage of economies of size to gain maximum plant profitability. Moreover, crop residues are widely dispersed, and very bulky. Finally, there are limited harvesting windows, as illustrated by the onset of winter soon after corn harvest. Clearly, both a strong desire to furnish these feedstocks and careful management of their provision are essential to realizing the profitability of the plant. These make the ownership of an NGC worthy of special consideration.

As stated above, owners of an NGC acquire that status by investing in delivery rights. In this case, farmers who produce crop residues and are interested in gaining income from them would acquire these rights as a means of adding value (by off-farm processing) to a byproduct of little value at present. On the other hand, plant managers see these as obligations to deliver a portion of the plant's feedstocks that must be assured for sustained plant operation. In fact, if the farmer

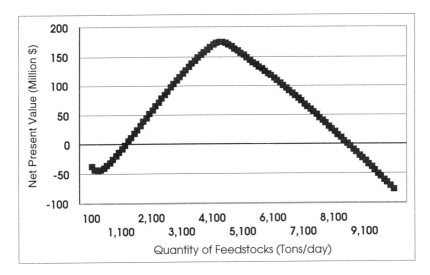

Figure 1. Estimated net present value of a ligno-cellulose-to-ethanol and furfural plant with varying levels of feedstock use, Central Missouri, 1998 (Van Dyne and Blase).

fails to deliver his obligated amount of crop residues the management of the cooperative is authorized to purchase them on the open market and charge them to the shareholder's account. Regardless, at the end of the business year each shareholder participates in the NGC's profits (or losses) in accord with the percent of the business owned, and the shareholder's delivery performance. Obviously, there is a powerful incentive for each farmer-shareholder to deliver his obligated volume of residues to the plant. This mechanism for assuring the delivery of the absolutely essential supply of feedstock for plant profitability makes the NGC an organizational form of ownership for this type of plant that deserves careful consideration.

NGCs can also facilitate the financing of the business. Clearly, the capital requirements for this type of plant to gain the efficiencies that size economies yield are very formidable. Part of these requirements are met in the instance of NGCs by the proceeds of the sale of shares, i.e., delivery rights. Past experiences with this form of organization suggest that when 33% to 50% of the capital requirements for plant construction have been met by this means, debt financing is likely to be available for the remaining. In the formation of existing new generation cooperatives, the Bank for Cooperatives has been a major lender. With regard to operating capital, NGCs also have a financing advantage. Since many farmer-members of the cooperative consider it to be an extension of their farm businesses (a value-added activity that must be performed off the farm), they are willing to accept partial rather than full payment for their share of the plant's feedstocks at the time of delivery. The remainder value, plus or minus the plant profits or losses, is paid to the shareholder at the end of the NGC's business year. Clearly, this significantly reduces the amount of operating capital to run the plant, much of which is frequently borrowed.

One final potential benefit of NGC ownership and operation of the ethanol/chemical plant described is noteworthy. It is the built-in demand for ethanol represented by farmer/shareholders. Two major uses are substantial. First, farmers are major users of energy, especially liquid fuels, in the production of crops from which the residues result, as well as others that do not yield marketable residues. Second, the fact that the cost of transporting the residues exceeds the cost of the residues themselves suggests that those who haul the residues comprise a major market for the ethanol. Splash blending at the plant resulting in direct sales to truckers and farmer/shareholders will not only provide a ready market for part of the ethanol but also minimize the transaction costs of the fuel for both participants in the industry.

NGCs appear to be an attractive alternative as an organizational form for an ethanol/chemical plant based on renewable crop residues. The fact that cooperatives are being used in the grain ethanol industry suggests that NGCs, with their unique features, are likely to play a major role in the lignocellulose to ethanol/chemical industry as well.

REFERENCES

Van Dyne, D.L., M.S. Kaylen, and M.G. Blase (1998). Estimating the optimum biomass refinery size for converting lignocellulosic feedstocks into ethanol and furfural, Proceedings, BioEnergy '98: Expanding BioEnergy Partnerships, Madison, Wisconsin, USA, pp. 194–201.

Van Dyne, D.L., M.S. Kaylen, and M.G. Blase (1998). The Economic Feasibility of Converting Ligno-Cellulosic Feedstocks into Ethanol and Higher Value Chemicals. Report to The Missouri Division of Energy and Department of Natural Resources and to The Consortium for Plant Biotechnology Research, Inc. of the National Ethanol Research Institute, Published by Department of Agricultural Economics, University of Missouri-Columbia.

Van Dyne, D.L. and M.G. Blase (1998). Evaluating New Generation Cooperatives as an Organizational Structure for Methyl Ester/Biodiesel Production, Report to The Southeastern Regional Biomass Energy Program, Tennessee Valley Authority, Muscle Shoals, AL., June 1998.

Biomass Production and Integration With Conversion Technologies

Biomass: The Carbon Connection

CONTRIBUTION OF BIOMASS TOWARD CO_2 REDUCTION IN EUROPE (EU)

Ausilio Bauen and Martin Kaltschmitt

Div. of Life Sciences, King's College, University of London, Campden Hill Road, London, W8 7AH, United Kingdom

The consequences of the anthropogenic greenhouse effect caused by the use of fossil energy sources and actions to prevent it are debated worldwide. This paper analyses the possible contribution of heat or electricity derived from solid biomass to reduce anthropogenic CO_2 emissions in the European Union (EU). The results and discussion are based on the already used biomass and the existing potentials as well as cost considerations for the provision of useful energy from biomass. About 2–8% more CO_2 would be emitted within the EU without the current use of solid biomass. An estimate of the remaining solid biomass potential indicates that CO_2 emissions could be reduced by a further 7–28%. Therefore, biomass could make a significant contribution to meeting the Kyoto Protocol targets.

1. INTRODUCTION

The use of biomass, instead of fossil fuels, as a source of energy can substantially reduce greenhouse gas emissions. The EU agreed to the CO_2 reduction aim in Kyoto, by which EU member states have promised to reduce greenhouse gas emissions between 2008 and 2012 by 8% compared with 1990 levels. CO_2 emissions will differ among various member states. In all cases, the use of renewable energy, biomass in particular, is an important measure to meet the aims.[1,2]

This paper presents the current and potential use of biomass from forestry and agricultural activities for heat and electricity generation and its contribution to avoided CO_2 emissions by replacing of fossil fuels. Also, the costs and benefits of biomass use as part of a CO_2 reduction strategy are briefly discussed. Energy consumption statistics relate to 1995 as it is the latest year for which a consistent set of data is available for the EU member states considered.

2. PRIMARY ENERGY CONSUMPTION

The total primary energy consumption of conventional energy carriers in the EU amounted to 57.8 EJ in 1995, of which 46.4 EJ were provided by fossil fuels (Coal: 21%, Oil: 55%, Natural Gas: 24%) and 9.8 EJ by nuclear power and hydropower. This means that most of the energy demand is met by fossil fuels, especially crude oil, a large part of which is used in the transport sector (Table 1). Table 1 shows also that the energy mix in different EU member states varies considerably. This mix significantly affects the CO_2 reduction potential owing to the different specific CO_2 emissions characteristic of the different fuel cycles.

In 1995, the overall CO_2-emissions based on the primary energy consumption amounted to approximately 3,700 million t (Table 1). This figure includes emissions due to the extraction processing and transport of the energy carriers.

Table 1
Primary energy consumption and CO_2 emissions in the EU in 1995[3]

	Primary Energy Consumption						CO_2 Emissions
	Coal	Crude Oil	Natural Gas	Fossil Fuels (total)	Hydro and Nuclear Power	Total	
	in PJ	in PJ	in PJ	in PJ	in PJ	in PJ	in Mt
Austria	104.7	464.7	251.2	820.6	142.4	963.0	63.25
Belgium & Lux.	389.4	1084.4	448.0	1921.7	427.1	2348.8	153.63
Denmark	276.3	443.8	121.4	841.5	0.0	841.5	70.62
Finland	171.7	418.7	121.4	711.8	251.2	963.0	58.39
France	544.3	3726.3	1239.3	5509.8	4345.9	9855.7	433.57
Germany	3872.8	5656.4	2805.2	12334.3	1741.7	14076.0	1007.08
Greece	355.9	711.8	0.0	1067.6	12.6	1080.2	93.48
Ireland	83.7	242.8	87.9	414.5	4.2	418.7	33.35
Italy	464.7	3973.3	1800.3	6238.3	150.7	6389.1	478.92
Netherlands	406.1	1591.0	1398.4	3395.5	41.9	3437.4	252.45
Portugal	163.3	540.1	0.0	703.4	29.3	732.7	60.59
Spain	791.3	2348.8	309.8	3449.9	686.6	4136.6	289.35
Sweden	87.9	715.9	29.3	833.2	967.2	1800.3	69.56
United Kingdom	2001.3	3420.6	2754.9	8176.8	983.9	9160.7	637.49
Total	9713.4	25338.5	11367.2	46419.1	9784.6	56203.6	3701.73

3. BIOMASS USE AND POTENTIAL

The biomass resources that are commonly used in the EU are fuelwood, wood residues from the wood processing industry, used wood products (e.g., demolition wood), and straw (in some countries). The main unexploited biomass resources are residual wood from forest management measures, woody residues from agriculture, public parks, road margins etc., herbaceous (e.g., straw) and other agricultural residues (e.g., olive stones), organic waste from industry and households (e.g., vegetable residues, sewage sludge, waste from the food processing industry), and energy crops. Marked differences exist between different countries in the exploitation of biomass resources (e.g., straw).

3.1. Current use of biomass

Table 2 shows the current use of biomass resources for electricity and heat generation. According to Table 2 the use of biomass amounts to approximately 1,825 PJ/yr. This represents a share of about 4% of the consumption of fossil energy in the EU in 1995. The use of fuelwood for domestic heating purpose contributes almost 75% of the overall biomass use in the EU. The use of biomass for district heating is significant in a few countries such as

Table 2
Current use and potential of biomass in the EU[4,5]

	Current Use			Potential			
	Electricity in PJ/yr	Heat in PJ/yr	Total in PJ/yr	Woody residues in PJ/yr	Herbaceous residues In PJ/yr	Energy crops in PJ/yr	Total in PJ/yr
Austria	15.6	111.4	127.0	356.7	22.4	62.7	906.6
Belgium & Lux.	6.7	10.5	17.3	54.4	12.9	37.5	104.8
Denmark	31.1	23.7	54.8	29.2	45.7	60.2	135.0
Finland	51.0	154.1	205.1	634.0	18.5	34.3	1651.0
France	38.4	371.2	409.5	71.9	308.5	708.6	204.6
Germany	71.6	111.5	183.2	70.7	197.1	352.9	576.2
Greece	0.0	58.5	58.5	17.5	27.3	105.4	149.0
Ireland	0.0	6.8	6.8	183.5	9.3	122.1	586.8
Italy	13.4	135.1	148.5	15.6	109.5	293.8	82.0
Netherlands	23.3	15.8	39.1	131.4	8.4	58.0	165.7
Portugal	5.8	93.3	99.1	265.4	7.7	26.6	656.1
Spain	21.6	140.7	162.3	494.0	96.0	294.8	546.9
Sweden	65.3	209.5	274.8	164.5	29.9	59.1	249.7
United Kingdom	26.5	12.6	39.1	655.9	108.0	397.5	744.9
Total	370.3	1454.8	1825.2	3144.8	1001.1	2613.4	6759.2

Austria, Finland and Sweden where mainly fuelwood and wood residues from the forestry and wood processing industry are used, and in Denmark where straw is used to some extent. In comparison, the use of biomass in industry and for electricity generation is modest. In some countries, such as Sweden, electricity is generated in combined heat and power plants used in the supply of district heating.

3.2. Biomass potentials

The EU can increase its use of biomass (Table 2). The biomass categories considered in the present analysis are wood and wood residues, straw and other agricultural residues, and energy crops.

Wood and Wood Residues. Residual wood (i.e., the by-product of stemwood production) can be estimated based on stemwood production, which is about 235 million m^3 in the EU. Assuming a share of residual wood from timber production of 15% for softwood and 20% for hardwood, an energy potential of 414 PJ/yr is derived. This represents 0.9% of the fossil energy consumption (i.e., coal, oil and gas) in the EU in 1995.

Forests in the EU are mainly managed for timber production, a part of which (about 360 PJ/yr) is used as fuelwood mainly in private households. This is about 0.8% of the fossil energy consumption. Since the wood price is presently low, only limited thinning is carried out. Assuming additional thinning to assure an optimal management of the forests, a potential in the range of 496 PJ/yr is estimated (i.e., 1.1% of the fossil energy consumption).

Wood processing and manufacturing industries generate wood waste that, if not used as a raw material, can be used as a source of energy. This is also true for wood products at the end of their lifetime (e.g., demolition wood). Assuming that about 80% of the overall produced timber can be used energetically either as industrial wood waste or as demolition wood, this totals approximately 1,712 PJ/yr or 3.7% of fossil energy consumption.

Wood residues are also available from road margins, public parks, fruit or olive plantations, viticulture, private gardens, etc. The estimation of these wood resources is very difficult due to a lack of data. Therefore, these biomass fractions are estimated roughly at approximately 164 PJ/yr or 0.4% of fossil energy consumption.

These different potentials add up to 3 145 PJ/yr and represent 6.8% and 5.6% of the fossil and total primary energy consumption in the EU in 1995 (Table 1 and 2), respectively.

Straw and other agricultural residues. In the EU, about 181 million t/yr of cereal are produced on 35 million ha. The specific grain yields vary between 1.9 t/(ha yr) in Portugal and 8.1 t/(ha yr) in The Netherlands.[6] Based on this, the total amount of straw produced can be estimated. Not all of the available straw can be used for energetic purposes, as part of it needs to be left in the fields to maintain the nutrient content of the soil, and part is used for animal bedding or other purposes. Assuming that only 30% of the straw is available for energy use results in a potential of 834 PJ/yr (1.8% of the fossil consumption).

Many other agricultural residues are produced besides straw (e.g., olive stones, peels, shells, husks). These fractions are assessed roughly at about 167 PJ/yr (0.4%). These sources of agricultural biomass represent 1,001 PJ/yr (cf. Table 2) or 2.2% and 1.8% of the fossil and total primary energy consumption in the EU in 1995 (cf. Table 1), respectively.

Energy crops. A significant potential for energy crop cultivation strongly depends on future agricultural policies. It is assumed that about 15% of the agricultural land currently used in the EU (145 million ha including grasslands) could be planted to energy crops. Assuming an average country-specific biomass yield for a mixture of annual and perennial crops of about 80% of the grain and straw yield currently achieved, due to an assumed less intensive energy crop production, an energy potential of 2,613 PJ/yr can be calculated for the EU (Table 2). This represents a share of about 5.6% and 4.6% of the fossil and total primary energy consumption in the EU in 1995, respectively.

Total potential. The biomass resources considered represent about 6,759 PJ/yr or 14.6% and 12.0% of the fossil and total primary energy consumption in the EU in 1995 (Table 1), respectively. Additionally, Table 2 shows the country distribution of the described energy potential in the EU. The energy potential from wood, for example, plays a major part in those countries where forests cover a considerable part of the land area and are of significant economic importance (e.g., Finland, Austria, Sweden, France). Similarly, for the biomass potential from agriculture, the potentials are high in those regions with large agricultural areas (e.g., France, United Kingdom, Italy). Therefore, the potentials are high in countries with extensive agricultural areas (e.g., France, United Kingdom, Germany) and low in countries with limited agricultural land areas (e.g., Finland, Belgium, Portugal).

4. AVOIDED CO_2 EMISSIONS

Biomass already contributes to avoided CO_2 emissions by meeting part of the energy demand in the EU, which would otherwise be mainly satisfied by fossil fuels. The unused biomass potential could make a substantial additional contribution to reductions in CO_2 emissions and to the fulfilment of the Kyoto Protocol targets. The analysis of the current and potential CO_2 reductions is based on a life cycle approach (LCA). For the biomass fuel cycle, no CO_2 emissions are assumed to result from the conversion stage due the fact the CO_2 released during combustion is removed from the atmosphere during the growing of the biomass. The only CO_2 emissions result from the use of fossil fuels during biomass production and transport.

4.1. Current avoided emissions

To estimate the current avoided CO_2 emissions (Table 3) it has been assumed that electricity and heat from biomass replaces electricity and heat from coal and gas. High and low avoided CO_2 emissions estimates are provided by the replacement of coal and gas, respectively. Coal conversion is characterised by specific CO_2 emissions of about 92 t/TJ and natural gas of 56 t/TJ. Electricity and heat from biomass assumes a conversion efficiency range between 20 and 30% and between 70 and 80%, respectively. Electrical conversion efficiencies for coal and gas are 35 to 40% and 35 to 50%, respectively. Heat conversion efficiencies for coal and gas are 75 to 85% and 80 to 90%, respectively. Also the respective fuel supply chains are taken into consideration. Current use of biomass, assumed to replace fossil fuels, is then estimated to avoid 71 to 288 million t CO_2/yr. The wide range results from differences in the specific emissions of fossil fuels and the ranges of biomass and fossil fuel conversion efficiencies. The current CO_2 emissions in the EU would be approximately 2 to 8% higher without the currently used biomass. Therefore, biomass already contributes significantly to limit CO_2 emissions within the EU.

Table 3
Current and potential avoided CO_2 emissions from biomass use

	Current avoided CO_2 emissions						Potential for additional avoided CO_2 emissions	
	Electricity		Heat		Total			
	min in Mt	max in Mt	min in Mt	max in Mt	min in Mt	max in Mt	in Mt	in%[a]
Austria	0.36	2.84	4.80	16.89	5.17	19.73	9.07 – 36.63	14 – 58
Belgium + Lux	0.13	1.04	0.43	1.50	0.56	2.54	4.00 – 16.16	3 – 11
Denmark	0.72	5.65	1.02	3.59	1.75	9.25	5.26 – 21-24	7 – 30
Finland	1.19	9.27	6.64	23.37	7.83	32.64	20.86 – 84.25	36 – 144
France	0.89	6.98	16.00	56.29	16.90	63.27	59.82 –256.52	14 – 59
Germany	1.67	13.01	4.81	16.91	6.47	29.92	34.68 – 140.05	3 – 14
Greece	0.00	0.00	2.52	8.87	2.52	8.87	7.39 – 31.70	8 – 34
Ireland	0.00	0.00	0.29	1.03	0.29	1.03	5.99 – 24.19	18 – 73
Italy	0.31	2.44	5.82	20.49	6.14	22.92	20.79 – 89.17	4 – 19
Netherlands	0.02	0.18	0.03	0.09	0.05	0.27	3.23 – 13.04	1 – 5
Portugal	0.54	4.24	0.68	2.40	1.22	6.63	6.18 – 26.50	10 – 44
Spain	0.13	1.05	4.02	14.15	4.16	15.20	24.62 – 105.59	9 – 36
Sweden	0.50	3.93	6.07	21.34	6.57	25.26	28.23 – 114.02	41 – 164
United Kingdom	1.52	11.87	9.03	31.77	10.55	43.64	23.13 – 93.41	4 – 15
Total	8.61	67.31	62.72	220.62	71.33	287.93	253.26 – 1052.4	7 – 28

[a] as a share of CO_2 emissions from fossil energy consumption in 1995

4.2. Additional avoided emissions

To estimate the potential additional avoided CO_2 emissions it is assumed that the biomass potential (Table 2) will be used to replace coal and gas to satisfy the given energy demand. The additional avoided CO_2 emissions are calculated in a similar way to the current avoided CO_2 emissions. The efficiency rates of the additional electricity- and heat-generating capacity from biomass is estimated to range between 25 and 45% and between 75 and 85%, respectively. It is also assumed that the share of biomass for heat generation varies between 60 and 70% for different countries according to climatic considerations. The remainder of the biomass potential is allocated for electricity generation. A range of potential avoided CO_2 emissions is obtained that reflects the substitution of fossil fuels characterised by low (i.e., natural gas) and high (i.e., coal) CO_2 emissions (Table 3). The estimated additional avoided emissions indicate that the use of the biomass potential could save between 253 and 1,053 million t CO_2/yr. This represents a share of 7 to 28% of the 1995 CO_2 emissions from fossil energy use in the EU countries. It is then obvious that biomass could make a significant contribution towards the fulfilment the Kyoto Protocol targets.

5. OUTLOOK

This paper considers the current use and potential solid biomass use for heat and electricity in the EU to assess current and additional avoided CO_2 emissions. The results show that solid biomass already contributes significantly to avoid CO_2 emissions, which are estimated at between 2 and 8% of the 1995 CO_2 emissions from fossil energy use. Also, the remaining biomass energy potential could make a considerable contribution to additional avoided CO_2 emissions, estimated at between 7 and 28% of the 1995 CO_2 emissions from fossil energy use, for achieving the targets of the Kyoto Protocol.

An important question is what additional costs, if any, would be incurred to avoid CO_2 emissions due to an increased use of biomass. The BioCosts study[6] estimated current abatement costs for different biomass energy case studies compared to reference conventional energy fuel cycles. They are defined as the ratio of the specific cost difference between the biomass fuel cycle and a conventional reference fuel cycle over the difference of specific CO_2 emissions. The study shows that CO_2 abatement can be achieved at no additional cost in the case of the combustion of forestry residues in circulating fluidised bed boilers for combined heat and power generation in Sweden compared to a reference coal combustion case. This may be also the case where biomass is used for heat generation in small scale plants for replacing stoves fired with coal or light oil. Heat and electricity generation from gasification-based combined cycle systems possess positive abatement costs, up to about 80 US$/t CO_2 for the current state of the technology. However, gasification-based fuel cycles have significant potential for cost reductions and abatement costs could be reduced to about 30 US$/t CO_2.

The use of biomass for heat and power generation could then be a cost-effective option in policies aimed at the reduction of CO_2 emissions, in particular in relation to potential damage costs and alternative CO_2 abatement options.

REFERENCES

1. Hall, D.O. (1999). Biomass energy versus carbon sinks: Trees and the Kyoto Protocol, Environment, 41, pp. 5–39.
2. European Commission (1997). White Paper on Renewable Energy, European Commission, Brussels, Belgium.
3. British Petroleum Company (1996). BP Statistical Review of World Energy, BP, London, UK.
4. Kaltschmitt, M.and A.V. Bridgwater (eds) (1997). Biomass Gasification and Pyrolysis - State of the Art and Future Prospects, CPL Scientific, Newbury, U.K.
5. Kaltschmitt, M., C. Rösch, and L. Dinkelbach (eds) (1998). Biomass Gasification in Europe; European Commission, DG XII, Brussels, Belgium.
6. European Commission DG XII (ed.) (1998). Total costs and benefits of biomass in selected regions of the EU, BioCosts Project, Contract JOR3-CT95-0006, Brussels, Belgium.

379

THE NET CO_2 EMISSIONS AND ENERGY BALANCES OF BIOMASS AND COAL-FIRED POWER SYSTEMS

Margaret K. Mann and Pamela L. Spath

National Renewable Energy Laboratory, 1617 Cole Blvd., Golden, Colorado 80401 USA

To determine the environmental implications of producing electricity from biomass and coal, life cycle assessments (LCA) were conducted on systems based on three power generation options: 1) a biomass-fired integrated gasification combined cycle (IGCC) system, 2) three coal-fired power plant technologies, and 3) a system cofiring waste biomass with coal. Each assessment was conducted in a cradle-to-grave manner to cover all processes necessary for the operation of the power plant, including raw material extraction, feed preparation, transportation, and waste disposal and recycling. Each study was conducted independently and can therefore stand alone. However, the resulting emissions, resource consumption, and energy requirements of each system can ultimately be compared. Although the studies quantified resources consumed, as well as several air, water, and solid waste emissions, this paper will pay particular attention to net CO_2 emissions and energy balances. The biomass IGCC system emits only 4.5% of the CO_2 produced by the average coal power system. This low amount is due to the absorption of CO_2 from the power plant by the growing biomass. Cofiring residue biomass at 5% and 15% by heat input reduces greenhouse gas emissions on a CO_2-equivalent basis from the average coal system by 6.7% and 22.4%, respectively, per unit of electricity produced. The life cycle energy balance of the coal systems is significantly lower than the biomass system because of the consumption of a non-renewable resource. Not counting the coal consumed by these systems, the net energy produced is still lower than that from the biomass system because of energy used in processes related to flue gas clean-up. Cofiring biomass reduces the energy consumed by the total system by 6.4% and 19.9% for the 5% and 15% cofiring cases, respectively.

1. INTRODUCTION

The march of advanced biomass power technologies toward commercialization has provided a more complete set of data for conducting economic analyses and writing operating procedures. These data can also be used to better define the environmental consequences of producing electricity from biomass. Life cycle assessment (LCA) is a systematic analytic method used to quantify the emissions, resource consumption, and energy use of a manufacturing process. LCAs were conducted on a biomass power system, and for comparison purposes, on three coal-fired technologies and a power plant cofiring biomass with coal. Even though the results of each LCA can be compared to highlight the environmental benefits and drawbacks of each process, each study was conducted independently so that the total environmental picture of each process could be evaluated irrespective of any competing process. Material and energy balances were used to quantify the emissions, resource depletion, and energy consumption of all processes between transformation of raw materials into useful products and the final disposal of all products and by-products.

2. DESCRIPTION OF SYSTEMS STUDIED

2.1. LCA of a biomass gasification combined cycle power plant

An LCA on the production of electricity from biomass in a combined cycle system based on the Battelle/FERCO gasifier was completed in 1997. The overall system consists of the production of biomass (hybrid poplar) as a dedicated feedstock crop, its transportation to the power plant, and electricity generation. Upstream processes required for the operation of these sections are also included. The primary purpose of conducting this LCA was to answer many of the questions that are repeatedly raised about biomass power in regard to CO_2 and energy use, and to identify other environmental effects that might become important once such systems are more widely used. For details about the methodology and results of this biomass-to-electricity LCA refer to Mann and Spath (1997).

2.2. LCA of coal-fired power production

In order to examine the environmental aspects of current and future pulverized coal boiler systems, three systems were studied: 1) a plant that represents the average emissions and efficiency of currently operating coal-fired power plants in the U.S. (this tells us about the status quo), 2) a new coal-fired power plant that meets the New Source Performance Standards (NSPS), and 3) a highly advanced coal-fired power plant using a low emission boiler system (LEBS). The overall systems consist of coal mining, transportation, and electricity generation. In keeping with the cradle-to-grave concept of LCA, upstream processes required for the operation of these three subsystems were also included in this study. All three cases use the same type of coal (Illinois No. 6), and both surface and underground mining were examined. The coal is transported via rail, truck, or a combination of rail and barge by one of four cases tested: average user by land, average user by river, farthest user, and mine mouth. The main modes of transportation were barge and train, although some diesel-fueled trucks were required for transporting items such as chemicals, catalysts, and ash. Major results will be presented here; for a more detailed description of the methodology and complete results, refer to Spath and Mann (1999).

2.3. LCA of a plant cofiring biomass with coal

An LCA was conducted on the production of electricity from a coal-fired power plant cofiring waste (primarily urban) biomass. The power plant is similar in design to the average case studied in the coal LCA, since currently operating coal-fired boilers can obtain the most benefit from cofiring biomass. Overall changes in emissions, resource consumption, and energy use were quantified for systems cofiring at rates of 5% and 15% by heat input, compared with a baseline system firing only coal. Cofiring was assumed to take place in the course of normal power plant operation. Thus, no construction or decommissioning of the plant is included in the assessment, although plant modifications required for cofiring were assessed.

3. MAJOR RESULTS

Although each LCA examined many different air, water, and solid waste emissions, plus numerous natural resources, only CO_2 and energy balance results will be presented here. Because of increasing concerns about the role of man-made gases on global climate change, special attention is directed toward CO_2. Quantifying CO_2 emissions from the power plant are not as much of a concern as looking at the net CO_2 produced by the entire life cycle system. This is especially obvious when biomass systems are being studied, because CO_2 is absorbed during photosynthesis, greatly reducing CO_2 emissions per unit of energy produced. Similarly, the net amount of energy produced by the system is more important than the amount of energy that is produced by the power plant. Upstream processes such as feedstock production, transportation, and chemical manufacture consume substantial quantities of energy, resulting in less energy produced by the system overall.

3.1. CO_2 emissions

Figures 1 and 2 illustrate the major sources and amounts of CO_2 emissions for the biomass IGCC and average coal systems. In terms of total air emissions, CO_2 is emitted in the greatest quantity from all systems examined. Net CO_2 emissions from the biomass IGCC system account for approximately 67% by weight of all air emissions. From the coal systems, CO_2 accounts for 98-99% of the total air emissions. However, note that in the case of the biomass IGCC system, because carbon dioxide emitted from the power plant is recycled back to the biomass as it grows, net emissions from this system are only 4.5% of those from the average coal system. Net CO_2 emissions for the NSPS and LEBS coal cases are 941 g/kWh, and 741 g/kWh of net electricity produced, respectively.

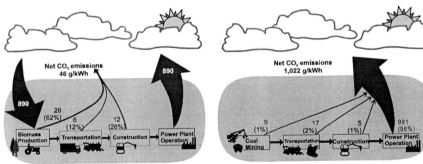

Figure 1: Biomass Power System
95% carbon closure

Figure 2: Average Coal Power System
0% carbon closure

The carbon closure of a system can be defined to describe the net amount of CO_2 (as carbon) released from the system in relation to the total amount of carbon circulating through the system. Referring to Figure 1, the carbon closure of the biomass IGCC system is:

$$1 - \frac{net}{Ctotal} = 1 - \frac{46}{46 \cdot 890} = 95\%$$

In addition to CO_2, two other greenhouse gases, methane and N_2O, are produced by these systems. The capacity of methane and N_2O to contribute to the warming of the atmosphere, a measure known as the global warming potential (GWP) of a gas, is 21 and 310 times as high as CO_2 (Houghton, et al, 1995). Thus, the GWP of a system can be normalized to CO_2-equivalence to describe its overall effect on global warming. Because biomass is diverted from landfills for the cofiring cases, methane and CO_2 that normally would be produced at the landfill are avoided. These avoided emissions are taken as a credit against the emissions from the cofiring systems. Therefore, the reduction in the GWP of the cofiring systems is higher than the rate of cofiring on an energy input basis. The 15% cofiring case reduces the GWP of the no-cofiring case by 22% on a per kWh basis. A 7% reduction is obtained by cofiring at 5%. Table 1 shows the carbon closures, net GWP on a CO_2-equivalence basis, and net CO_2 emissions, for all of the systems studied. Because no CO_2 is removed from the atmosphere by the coal systems, their carbon closures will always be zero. The carbon closure of the biomass IGCC system could be higher than 95% if the soil on which the biomass is grown is able to permanently sequester carbon.

Table 1
Carbon closure, global warming potential, and net CO_2 emissions

	Biomass IGCC	Average coal	NSPS coal	LEBS coal	15% cofiring	5% cofiring	0% cofiring
Carbon closure	95.1%	0%	0%	0%	15.1%	5.1%	0%
Net GWP (g CO_2 equivalent / kWh)	49	1042	960	757	816	981	1,052
Net CO_2 (g/kWh)	46	1,022	941	741	927	1,004	1031

3.2. Energy production and consumption

Given that the systems being studied exist for the purpose of producing electricity, the net energy balance was examined carefully. Energy is consumed in two ways: 1) in upstream processes that create an intermediate feedstock (e.g., fertilizer or limestone) or affect an operation (e.g., transportation), and 2) by destroying a material that has the potential to be converted to energy (e.g., natural gas). The net energy of the system is the energy produced as electricity by the power plant, minus the energy consumed throughout the system. In the case of the coal-fired systems, because coal is a non-renewable resource, it is said to be consumed by the process; thus, its energy content is subtracted from the energy produced by the plant. Biomass, on the other hand, is both created and consumed within the boundaries of the system, and so its energy is not subtracted from the net. In addition to the standard power plant efficiency, which is the energy delivered to the grid divided by the energy in the feedstock to the power plant, four other measures of efficiency were defined in Table 2.

Table 2
Measures of net energy production

Life cycle efficiency (%) (a)	External energy efficiency (%) (b)	Net energy ratio (c)	External energy ratio (d)
$\dfrac{Eg-Eu-Ec-En}{Ec \cdot En}$	$\dfrac{Eg-Eu}{Ec \cdot En}$	$\dfrac{Eg}{Eff}$	$\dfrac{Eg}{Eff \cdot Ec \cdot En}$

where:
Eg = electric energy delivered to the utility grid
Eu = energy consumed by all upstream processes required to operate power plant
Ec = energy contained in the coal fed to the power plant
En = energy contained in the natural gas fed to the power plant (LEBS system only)
Eff = fossil fuel energy consumed within the system (e)

The net energy ratio describes the amount of energy produced per unit of energy consumed. Although this is a more accurate and rigorous measure of the net energy balance of the system, the external measures are useful because they expose the rate of energy consumption by upstream operations. Table 3 gives the energy efficiency and ratio results for all systems studied.

Table 3
Energy results

	Biomass IGCC	Average coal	NSPS coal	LEBS coal	15% cofiring	5% cofiring	0% cofiring
Power plant efficiency	37%	32.0%	35.0%	42.0%	31.1%	31.5%	32.0%
Life cycle efficiency	35%	-76%	-73%	-66%	-60%	-70%	-74%
External energy efficiency	35%	24.3%	26.8%	34.1%	25.5%	25.4%	25.6%
Net energy ratio	15.61	0.29	0.31	0.38	0.34	0.31	0.30
External energy ratio	15.61	4.97	5.09	6.72	5.51	5.13	5.03

Because the energy contained in the coal is greater than the energy delivered as electricity, the life cycle efficiencies of the coal systems are negative. Another way to view this is that because a non-renewable resource is expended, the coal systems consume more energy than they produce. The net energy ratio likewise indicates that only about one-third of every unit of energy put into the coal-fired systems is obtained as electricity. Cofiring at 5% and 15% reduces the net energy consumption of the average coal system by 6.4% and 19.8%, respectively. This is due almost exclusively to the reduction in coal consumption. The biomass IGCC system results demonstrate that far more energy is produced than is consumed, because the process is based on a renewable resource.

The external energy efficiency and external energy ratio indicate that upstream processes are large consumers of energy in the coal systems. In fact, the operations related to flue gas clean-up and those associated with coal transportation account for between 3.8% and 4.2% of the total system energy consumption, and between 67.4% and 70.5% of the non-coal energy. Processes involved in the gas clean-up operations include the production, transport, and use of limestone and lime in the average and NSPS systems; and the production, distribution, and combustion of natural gas in the LEBS system. These operations consume between 35.3% and 38.5% of the non-coal energy, and between 2.0% and 2.4% of the total energy of the systems. Transportation of the coal uses similar amounts: between 30.1% and 32.2% of non-coal, and 1.8% of total system energy.

4. CONCLUSIONS

The net CO_2 emissions of the biomass system are substantially lower than those of any of the coal systems because of the uptake of CO_2 during biomass growth. Biomass IGCC can obtain carbon closures of 95% or greater, depending on the amount of carbon that is sequestered in the soil. Coal power systems, because they do not remove from the atmosphere any of the CO_2 they produce, have carbon closures of zero. Cofiring biomass offers the opportunity to reduce the net GWP of coal-fired systems. The reduction in GWP is higher than the rate of cofiring (on a heat-input basis) because of the avoided landfill methane. Net GWP reductions are 7% and 22% when biomass is cofired at 5% and 15% by heat input. Therefore, cofiring waste biomass helps coal-fired power plants reduce greenhouse gas emissions in two ways: 1) the well-known cycling of carbon between the power plant and growing biomass, and 2) avoiding emissions that would have been produced at the landfill if the biomass were not used at the power plant.

The net energy balance of the biomass IGCC system shows that 16 units of energy are produced for every unit of energy consumed. Because of the use of a non-renewable resource, the coal systems consume more energy than they produce. Cofiring biomass with coal reduces net energy consumption by 20% and 6.4% for the 15% and 5% cofiring cases; however, the net energy balance is still negative.

REFERENCES

Houghton, J.T., L.G. Meira Filho, B.A. Callander, N. Harris, A. Kattenberg, and K. Maskell, eds. (1996). Climate Change 1995 - The Science of Climate Change. Published for the Intergovernmental Panel on Climate Change. Cambridge University Press, New York.

Mann, M.K. and P.L. Spath (1997). Life Cycle Assessment of a Biomass Gasification Combined-Cycle System. NREL/TP-430-23076. National Renewable Energy Laboratory, Golden, Colorado.

Spath, P.L. and M.K. Mann (1999). Life Cycle Assessment of Coal-fired Power Production. NREL/TP-570-25119. National Renewable Energy Laboratory, Golden, Colorado.

387

PRODUCING A LIFE-CYCLE ENERGY BALANCE FOR AN INTEGRATED SWEET SORGHUM / SUGARCANE SYSTEM IN THE SEMI-ARID SOUTHEAST REGION OF ZIMBABWE

J. Woods and D.O. Hall

Division of Life Sciences, King's College London, London, W8 7AH, United Kingdom

During the last eight years we have carried out research to establish the potential for energy production from sweet stemmed varieties of the C_4 crop sorghum [*Sorghum bicolor (L.)* Moench] that are integrated with sugarcane. Sweet sorghum is a versatile crop capable of producing large amounts of liquid fuels (i.e., ethanol by fermentation), and electricity or thermal heat (by combustion of fibrous stem residues). It is an annual energy crop that is tolerant to a wide range of environments and it efficiently utilises water, nitrogen and light.

A life-cycle energy balance methodology has been designed as a part of the development of a systems analysis model (AIP). The methodology will be used in the AIP to generate site-specific and time-specific estimates of energy balances using both the integrated mechanistic sorghum-specific plant growth model (a modified CERES model) and the harvesting, transport and energy conversion modules of the AIP. This paper summarizes a worked example of an energy balance calculation based on a 1997/98 sweet sorghum productivity trial and the utilisation of sweet sorghum and sugarcane for energy production in the Triangle Ltd. sugar mill in south-eastern Zimbabwe.

1. METHODOLOGY

Energy use is estimated for each step of the crop growth and conversion process—
- tillage, including land preparation, planting, weeding, and harvesting (manual),
- fertiliser and pesticide application,
- irrigation,
- harvesting,
- transport, and
- conversion.

Both direct and indirect energy use is estimated. 'Direct energy' use is defined as the actual energy used in carrying out the various operations during crop growth and conversion, whilst 'indirect energy' use estimates the energy requirements for the manufacture, delivery, and repair of the equipment used.

The energy output side of the balance is calculated by using the sweet sorghum productivity levels achieved during the 1997/98 trial (located near Triangle).[1] Sugarcane productivity is based on the long-term average commercial cane productivity of 115 t_{cane} stems ha^{-1} as delivered to the Triangle mill. Energy inputs during juice separation, ethanol and sugar production are calculated from 1997 Triangle Ltd data.[2,3] Energy balances are calculated for the existing sugar- and ethanol-producing mill configuration and for a postulated no-sugar (ethanol-only) configuration. They are based on Triangle's existing conversion technologies. (See description below.)

2. BIOENERGY PRODUCTION (OUTPUTS)

Useful products derived from sweet sorghum and sugarcane are the sugars and the fibre ('bagasse'[a]). The sugars' value is derived through the production of crystalline sugar (sucrose only) and ethanol (all fermentable sugars). The bagasse value is derived from the generation of energy through combustion; all the bagasse is currently required at Triangle Ltd. to provide the energy for crystalline sugar and ethanol production. The composition of sweet sorghum and commercial sugarcane are given in Table 1.

2.1. In-field sorghum and sugarcane biomass productivities

Sugarcane: Data for sugarcane are based on Zimbabwe average yields of delivered cane (115 t_{cane} ha^{-1} yr^{-1}) using conversion factors from Hall et al., and a sugarcane growth period of 12 months; it does not include detached leaves.[4] The delivered stem mass of 1 t_{cane} is composed of 145 kg fibre + 168 kg BRIX + 687 kg water; the BRIX is composed of 140 kg sucrose, 7 kg reducing sugars (predominantly glucose and fructose), and 21 kg other dissolved solids. Harvesting and transport losses are assumed to equal 5% of in-field stem biomass.

Delivered biomass (see Table 1) = 115 t_{cane} ha^{-1} delivered.
= 158*0.75*0.95*0.25 = 28.14 odt delivered.

Sweet sorghum: The productivity of sweet sorghum is estimated at 80 fw t ha^{-1} crop^{-1} of total above ground biomass. Of this biomass (75% moisture content, wet weight basis), the tops and leaves represent 20% (wt/wt) at physiological maturity. Harvesting, transport and handling losses represent approximately 5% of the biomass (minus tops and leaves). The delivered stem mass of 1 t_{cane} is composed of 126 kg bagasse + 174 kg BRIX + 700 kg water; with the BRIX being comprised of 124 kg sucrose, 14 kg reducing sugars (predominantly glucose and fructose), and 36 kg other dissolved solids. Therefore the mass of delivered biomass to the mill can be calculated as:

Delivered biomass (see Table 1) = 60 t_{cane} ha^{-1} delivered.
= 80*0.8*0.95*0.25 = 15.2 odt delivered

---

[a]Bagasse is the fibre of delivered stems at 50% moisture content (i.e., fibre x 0.5): 7.5 GJ t^{-1} LHV.

Table 1
Composition of sweet sorghum (cv. Keller) and commercial sugarcane

Genotype	Keller [a]		Sugarcane [b]	
Total above ground standing biomass: t ha^{-1} (%)				
Fresh	80	(100)	158	(100)
Dry	20	(25)	47	(30)
Moisture	60	(75)	111	(70)
Main stem biomass: t ha^{-1}				
Fresh Weight	64		121	
Dry Weight	18		36	
Biomass components: % dry weight				
Stems	68		77	
Attached Leaves	13		23	
Panicle + Seeds	9		0	
Fermentable sugars: % stem dry weight				
Sucrose	29		44	
Glucose	4		3	
Fructose	3		2	
Total	35		48	
Total t ha^{-1}	8		18	
Bagasse	43		47	
t ha^{-1} (oven dry)	8		17	

[a] Sweet sorghum var. Keller data from Chiredzi trials 97/98- 3.5 month growth period (1203 Growing Degree Days)
[b] Data for sugarcane are based on Zimbabwe average yields of delivered cane (115 t_{cane} ha^{-1} yr^{-1}) using conversion factors from Hall et al. Harvesting and Transport losses are assumed to equal 10% of in-field stem biomass. The sugarcane growth period is 12 months. It does not include detached leaves. (Associated with 1 t_{cane} are 140 kg bagasse, 160 kg BRIX, 92 kg attached tops + leaves; not included are 188 kg detached material, i.e., dead leaves)

2.2. Production of ethanol, steam, electricity, and crystalline sugar

Triangle's cane processing capacity is 490 t_{cane} h^{-1}, resulting in 145 t bagasse, 60 t sugar and 4,370 l of ethanol being produced per hour. In order to meet the power requirements of the existing mill, sugar factory, ethanol plant, and estates, a total of 130 t bagasse h^{-1} is burnt in the boilers producing 279 t high pressure (HP) steam h^{-1} (3.2 MPa at 400°C) with an energy content of 820 GJ (i.e., 230 MW$_{th}$). Seventy percent of this steam is fed through the mill turbines and back-pressure turbo-alternators to provide virtually all of the mill, factory, ethanol plant and estate requirements for electricity and direct power. The exhaust steam from the turbines is combined with 'let-down' steam (30% of the HP steam) and passed through the de-superheater where make-up water is added. The resulting production of

301.6 t of 'exhaust steam' has an energy content of 760 GJ (190.1 kPa at 118.6°C). This exhaust steam is then used for direct heat applications, e.g., juice heating, evaporation, distillation.

3. ENERGY CONSUMPTION (INPUTS)

3.1. Agronomic

Tillage & harvesting: The primary energy input for this section results from the diesel fuel requirements of the machinery required to prepare and maintain the land for crop growth, i.e., discing, ploughing and the movement of sprayers. The energy sequestered in the machinery during manufacture, transport to their location from the factory and repairs is also accounted for. A tillage and harvesting energy requirement of 2262 MJ ha^{-1} for sweet sorghum and 3082 MJ ha^{-1} for sugarcane is assumed. This energy analysis is based on Bowers.[5] A manual harvesting rate of 5 t h^{-1} is assumed (a rate similar to that of sugarcane).

Fertilisers & pesticide energy use: Direct and indirect energy requirements for fertiliser and pesticide use are based on Bhat and actual use during sweet sorghum trials and commercial sugarcane growth.[6] Energy requirements of 5865 MJ ha^{-1} for sweet sorghum and 10,317 MJ ha^{-1} for sugarcane are assumed.

Irrigation: Detailed energy accounting for irrigation use is very complex due to problems associated with estimating indirect energy use. The methodology used to calculate irrigation energy input is derived from Sloggett, including direct and indirect energy application.[7] Energy requirements of 1534 MJ ha^{-1} for sweet sorghum and 11672 MJ ha^{-1} for sugarcane are assumed.

Transport: Transport costs for sorghum and sugarcane are calculated as 7.1 MJ t.km^{-1} derived from Lewis. An average transport distance of 20 km is assumed.[8]

3.2. Conversion

The process of separating the sugars (juice) from the fibre (bagasse) requires large amounts of electricity (pumps and motors), direct power, and heat, with virtually all of this energy being provided from the combustion of the bagasse.

The bagasse is burnt in 10 boilers (of various capacity, ages, and efficiency; average 82%), some of which are capable of supplying high pressure steam (3.1 MPa and 380°C) for use in four direct power turbines and six turbo-altenators (total capacity 23 MW$_e$). The exhaust steam from the turbines is then used for heat applications in both the crystalline sugar and ethanol production processes. Old back-pressure turbines that allow a pressure of 0.1 to 0.15 MPa to be maintained after the turbine provide low pressure exhaust-steam usable for all heat requirements. These turbines provide nearly 20% of the final energy requirements and lose 3.3% of the bagasse energy.

More than 80% of the bagasse energy is used to provide heat and in the process lose over 10% of the bagasse energy input. The energy requirements for crushing (juice separation) and crystalline sugar extraction are roughly equal at just over 40% each (approx. 1 GJ t$_{cane}^{-1}$ each).

4. ENERGY BALANCES & CONCLUSIONS

As can be seen from Table 2, the current configuration at Triangle Ltd. does not allow the net production of bioenergy when both ethanol and crystalline sugar are produced, because use of fossil fuel outweighs the potential energy benefits from ethanol. Even if all the sugars are diverted to ethanol production the positive energy balance shown does not allow the export of electricity, all of which is necessary for the mill and estate use.

Only with the introduction of newer technologies that reduce the overall steam requirements for sugar and ethanol production will allow Triangle to become a substantial net exporter of bioenergy in the future. Retrofitted sugar mills are capable of reducing their steam consumption to around 300 kg steam per t_{cane} as compared to Triangle's current rate of steam consumption of 570 kg steam per t_{cane} processed. The investment required to achieve these efficiencies in steam use will only occur in an economic environment which encourages the production of environmentally sustainable bioenergy.

Table 2
Ethanol-only and ethanol + sugar energy balances (GJ per delivered t_{cane})

	Ethanol Only		Ethanol + Sugar	
	Sweet Sorghum	Sugarcane	Sweet Sorghum	Sugarcane
Agronomic (inc. harvesting)	0.181	0.223	0.181	0.223
Transport	0.149	0.143	0.149	0.143
Crushing	0.970	0.970	0.970	0.970
Ethanol production	0.888	0.934	0.128	0.128
Sugar production	-	-	1.03	1.03
'Energy In' Sub-Total	2.19	2.27	2.46	2.49
Bagasse energy content	1.793	2.175	1.7925	2.175
Ethanol energy content	1.307	1.375	0.198	0.173
'Energy Out' Sub-Total	3.10	3.55	1.99	2.35
'Energy Out':'Energy In'	1.42	1.56	0.81	0.94

Notes: Values for sweet sorghum represent mean values for cv.s Keller and Cowley for both estimated ethanol and fibre production. Source: Based on J.Woods.[9]

REFERENCES

1. Mvududu, E., L. Nyabanga, J. Gopo, et al. (1998). 1st year report: CFC sweet sorghum project. CFC Sweet Sorghum Project Progress Reports. SIRDC. Harare, Zimbabwe.
2. Wenman, C. (1999). Triangle Limited: Weekly factory performance summary.
3. Hoekstra RG. (1997). Steam load overall balance (SLOB): Triangle runs [computer program], Mount Edgecombe, SA: Tongaat-Hulett Sugar Ltd.

4. Hall, D.O., F. Rosillo-Calle, R.H. Williams, et al. (1993). Biomass for energy: supply prospects, Chap. 14 in Renewable Energy: Sources for Fuels and Electricity, 2nd ed., ed. by T.B. Johansson, H. Kelly, A.K.N. Reddy, and R.H. Williams. Washington, D.C., pp. 593–652. Island Press.
5. Bowers, W. (1992). Agricultural field equipment, Chap. 10 in Energy in Farm Production., ed. by R.C. Fluck, Elsevier, Amsterdam, pp. 117–130.
6. Bhat, M.G., B.C. English, A.F. Turhollow, et al. (1994). Energy in Synthetic Fertilizers and Pesticides: Revisited. ORNL/Sub/90-99732/2. pp.1–45. ORNL. Oak Ridge, Tennessee, USA.
7. Sloggett, G. (1992). Estimating energy use in world irrigation, Chap. 14 in Energy in Farm Production, ed. by R.C. Fluck, Elsevier, Amsterdam, pp. 203–218.
8. Lewis, G. (1984). Some aspects of the production of ethanol from sugar cane residues in Zimbabwe, Solar Energy, 33(3/4), pp. 379–82.
9. Woods, J. (1999). A modular model for designing and optimising environmentally sustainable bioenergy systems: Integrating sweet sorghum growth and conversion for both electricity and ethanol production, PhD dissertation,. King's College, London.

FIRST TOTAL EVALUATION OF THE ECOLOGICAL COMPARISON: BIOFUELS VERSUS CONVENTIONAL FUELS

G. Zemanek and G.A. Reinhardt

IFEU-Institut für Energie- und Umweltforschung Heidelberg GmbH, Wilckensstr. 3, D-69120 Heidelberg, Germany

Early comprehensive life cycle assessments (LCAs) that compared biofuels with fossil fuels appeared in the beginning of the nineties. Since then the number of assessed parameters has continuously increased and the methodology improved. One of the most extensive studies on this topic was completed in 1997. A large variety of ecological parameters was examined—some of them for the first time—over the whole life cycle by using a unified methodical approach on a high level of differentiation.[1] The results of this study constitute the basis for the present project. They allow for the first time an overall replacement of the ecological advantages and disadvantages of the replacement of fossil fuels by biofuels in accordance with the methodological demands on LCAs.[2] Based on the determination of the "specific contributions" and the ecological importance of the various evaluated parameters the final evaluation shows that under certain assumptions some biofuels—especially wood chips from short rotation forestry and residues—have definite environmental advantages compared to fossil fuels and other biofuels.

1. OBJECTIVE AND PROCEDURE

It is the objective of this study to determine if (and if so, then which) biofuels perform ecologically better than their fossil counterparts and how biofuels perform in a comparison among themselves. By a comparative LCA we examine biofuels that are potentially cultivable in Germany or that result as residues from agriculture, forestry or preservation measures. They are compared with fossil fuels that they can replace (see Table 1). One basic life cycle is determined for every biofuel. We select this basic life cycle according to the present or at present foreseeable conditions, under which the biofuel is or will most probably be used.

We balance all energy carriers over their whole life cycles and consider agricultural reference systems. We also include all additives and co-products (see Figure 1). The latter are balanced as credits. The residues are not balanced as co-products, i.e., we do not assume any expenditures for their production. The biofuels are compared with their fossil counterparts on the basis of the same amount of useful energy. Details of the scope of the study and the underlying assumptions are published elsewhere.[1,2,4] Inventory analysis and impact assessment prepare the evaluation. The latter includes two steps: firstly, the parameters that are finally assessed are determined; secondly, the final assessment is carried out.

Table 1
Selected basic life cycles

Basic life cycles	
Grain vs. fuel oil in heat plant	Grass residues vs. fuel oil in CHP plant [1]
Fast growing grass [2] vs. fuel oil in CHP plant [1]	Fuelwood from residues vs. fuel oil in heat plant
Orchard grass vs. fuel oil in heat plant	Ethanol [3] vs. gasoline in passenger car
Fuelwood from short rotation forestry vs. fuel oil in heat plant	Rapeseed oil methyl ester (RME, biodiesel) vs. diesel oil in passenger car
Straw vs. fuel oil in heat plant	Rapeseed oil vs. diesel oil in tractor

[1] Combined heat and power plant
[2] *Miscanthus, Arundo donax*
[3] Ethanol from sugar beet

2. INVENTORY ANALYSIS AND IMPACT ASSESSMENT

Table 2 shows the parameters that were examined quantitatively in the LCI (life cycle inventory). After a first analysis of the results, unreliable parameters such as benzene were excluded from further considerations. Sensitivity analyses are carried out in order to examine the impact of uncertain basis data or variations of the life cycles on the results of LCI and LCIA (life cycle impact assessment).

For the final assessment we select the impact categories "Resource demand," "Greenhouse effect," "Acidification" and "Eutrophication;" "Noise," "Land use" and "Photo smog" are not considered because of missing balancing procedures. In the category "Ozone depletion," nitrous oxide (N_2O) is the only relevant substance that appears on a significant scale. Therefore this

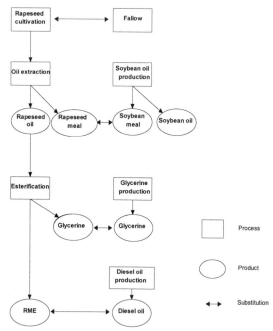

Figure 1. Schematic life cycle comparison rapeseed oil methyl ester (RME) vs. diesel oil.[3]

category can be described by N_2O as a LCI parameter (see discussion below). The category "Human and eco toxicity" is included on the basis of an individual assessment of LCI parameters because no aggregation models are available. Column 1 in Table 3 shows the finally evaluated parameters. They are described by category indicators or LCI parameters.

Table 2
LCI results for the life cycle comparison, fuelwood from short rotation forestry vs. fuel oil

LCI parameters	Results [1]	LCI parameters	Results [1]
Finite primary energy [2]	– 144 GJ/(ha*a)	CO	3,3 kg/(ha*a)
Limestone	145 kg/(ha*a)	NMHC	– 1,7 kg/(ha*a)
Phosphate ore	93,3 (kg/ha*a)	Diesel particulates	0 g/(ha*a)
Sulphur total	6,2 kg/(ha*a)	Dust	319 g/(ha*a)
Potash ore	504 kg/(ha*a)	HCl	495 g/(ha*a)
Sodium chloride	0 kg/(ha*a)	NH_3	0,2 - 2,0 kg/(ha*a)[3]
Clay	0 kg/(ha*a)	Formaldehyde	– 6,3 g/(ha*a)
N_2O	0,89 - 2,30 kg/(ha*a)[3]	Benzene	74,4 g/(ha*a)
NO_X	18,2 kg/(ha*a)	Benzo(a)pyrene	– 3,8 mg/(ha*a)
SO_2	– 4,5 kg/(ha*a)	Dioxines	1.129 ng/(ha*a)

[1] The unit "ha*a" indicates how much energy and emissions is/are saved or additionally used/emitted when the amount of wood that is harvested per hectare and year substitutes fuel oil in a heat plant; positive figures mean a result in favor of the fossil fuels, negative figures in favor of the biofuels.
[2] Crude oil, natural gas, uranium ore, mineral and brown coal
[3] Results of a minimum-maximum-evaluation

3. FIRST ASSESSMENT STEP

The objective of this step is to exclude those parameters from further consideration that are of minor ecological significance and to distinguish significant parameters by ranking according to their ecological importance. Thus we determine the ecological importance of the evaluated parameters on a scale from "very low" to "very high." Then we exclude all those parameters from further consideration that show the following:
- neutral results, i.e. neither significant advantages nor disadvantages for one fuel,
- no significant ecological importance, i.e., that are rated "very low" or "low." For instance neither CO nor SO_2 (as parameters for human and eco toxicity) particularly affect human health anymore.

Table 3 indicates which parameters have qualified for the final assessment. The ranking is based on value choices that are subjective by nature. Therefore we do not focus on the derivation of the individual ecological importance of each parameter but consider if and how different rankings influence the results of our final assessment.

As already mentioned, no significant amounts of other ozone-depleting substances other than N_2O occur during the life cycles. An aggregation to OD-potentials according to the LCIA method is not possible due to lack of characterisation factors. Therefore the LCI parameter N_2O is used in order to indicate the stratospheric ozone-depletion. A clear statement concerning the ecological importance of N_2O is not possible (neither absolute nor in relation to other ozone depleting substances). Therefore the whole range is considered when the ecological importance of N_2O is determined. Regarding the impact category "Human and eco toxicity" only one of the parameters that have individually been considered remains for the final assessment, i.e., the LCI parameter "diesel particulates."

Table 3
Finally evaluated parameters and their ecological importance

Evaluated parameters[1]	Category indicators	LCI parameters	Ecological importance
Resource demand	Total of finite primary energy	Crude oil, natural gas, uranium ore, brown and mineral coal	High (very high)[2]
Greenhouse effect	CO_2-equivalents[3]	CO_2, N_2O, CH_4	Very high
Acidification	SO_2-equivalents[4]	SO_2, NO_X, NH_3, HCl	Medium (very high)[2]
Eutrophication	Nitrogen total	NO_X, NH_3	Medium (very high)[2]
N_2O[5]	_[5]	N_2O	Very low – very high[5]
Diesel particulates[5]	_[5]	Diesel particulates	High

[1] We combine the resulting parameters from LCI and LCIA under the term "evaluated parameters"
[2] Range of variation of the ecological importance
[3] According to [5]
[4] According to [6]
[5] For details see text.

4. FINAL ASSESSMENT STEP: BIOFUELS VERSUS FOSSIL FUELS

This part of the final assessment is based on the evaluation of the "specific contributions" and the ecological importance of the selected parameters. The determination of the specific contributions is a means to measure the importance of the individual ecological advantages and disadvantages relative to the overall situation in Germany. The so-called equivalent value per capita is a suitable reference unit. Average per capita energy consumption and emissions of a German citizen are related to the corresponding consumption and emissions that result from the production and use of the examined fossil fuels and biofuels. Reference year is 1996.

The qualitative results are very uniform: for all life cycle comparisons the results for the parameters "Resource demand" and "Greenhouse effect" favor the biofuels; all other parameters favor of fossil fuels. All values have approximately the same scale as the results for the life cycle comparison "Fuelwood from short rotation forestry vs. fuel oil" (see Figure 2). Due to this similarity of the results we give higher priority to the parameters with high to very high ecological importance than to the others. This leads to the following interpretation of the results: both evaluated parameters with high to very high ecological importance ("Resource demand" and "Greenhouse effect"—N_2O is rated "medium", "low" or "very low" particulates add as a third parameter for the biodiesel life cycle comparison (for all other comparisons the diesel particulate balance is zero[7]). Even if the unfavorable N_2O-emissions are rated "high" or "very high" the advantages outweigh this disadvantage. Except from N_2O two parameters with only medium ecological importance are in favor of the fossil fuels ("Acidification," "Eutrophication"). Therefore an overall final assessment in favor of the biofuels can be justified. However, this assessment is not inescapable. In particular when a precautionary environmental approach is preferred the above argument can be reversed as long as there is interpretational ambiguity concerning N_2O.

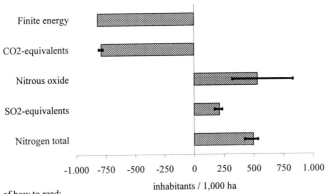

Examples of how to read:

- the amount of finite energy that is saved when the harvest of short rotation wood from 1,000 ha replaces fuel oil equals the average energy consumption of 816 inhabitants of Germany.
- the amount of N_2O-emissions that are additionally emitted when the harvest of short rotation wood from 1,000 ha replaces fuel oil equals the average N_2O-emissions of 532 inhabitants of Germany.

Figure 2. Ecological advantages and disadvantages for the life cycle comparison "Fuelwood from short rotation forestry vs. fuel oil" by means of the "specific contributions" (positive values indicate contributions in favor of the fossil fuel, negative values in favor of the biofuel; the small bars indicate the results of a minimum-maximum-evaluation).

5. FINAL ASSESSMENT STEP: BIOFUELS VERSUS BIOFUELS

This part of the final assessment is also based on the ranking according to the ecological importance (see Table 3). N_2O is again rated "low" or "very low." As a clear statement on the ecological importance of N_2O is not possible, we consider also the results if N_2O is rated "very high" (see above). Thus we cover the whole range of possible results. We also look at further ranking possibilities for the parameters in order to analyze how important the determination of the parameter with highest importance is. The overall assessment leads to the following interpretation of the results:

- If all biofuels are compared the cultivated solid biofuels come off best when the main objective is to reduce the greenhouse effect. Greenhouse effect and resource consumption are closely related. If the reduction of energy consumption is taken as being more important than the reduction of the greenhouse gases, the results do neither change here nor for other considered ranking variations. The liquid biofuels obtain better results than the solid biofuels from residues. But if the minimization of the N_2O-emissions is main objective, the energetic use of residues has to be favored. As a further result it turns out that these evaluations depend to a very high degree on the determination of the parameter of highest ecological importance.
- If only the cultivated biofuels are compared, fuelwood from short rotation forestry is clearly the best biofuel. In general liquid biofuels compete badly compared to most solid biofuels. Only when for example the reduction of acidification is main objective they are better than all other solid biofuels (except from fuelwood). Here the results depend only partly on the determination of the parameter of highest ecological importance.
- If only the biofuels from residues are compared residues from forestry are best.

6. CONCLUSION AND OUTLOOK

The present assessment is based on the results of life cycle comparisons. These comparisons are made on numerous assumptions. Although scientifically reliable results can be derived, these results and the resulting interpretations cannot be generalized because other assumptions and system boundaries lead to other results. Therefore the results must be explicitly discussed within the frame of the underlying assumptions. On the other hand, the evaluated results from LCI and LCIA are—in this frame—very stable because data that are not reliable at all are excluded from the final assessment and other uncertain data have been examined by sensitivity analyses.

Finally, it is necessary to mention that the present statements are obtained on some assessment procedures that are *a priori* not completely scientifically objective but include subjective elements. This fact requires that the assessment results must always be discussed only on the basis of the underlying value choices. Furthermore, it requires the documentation of the complete evaluation process in order to allow a comprehensible review. Other value choices can lead to very different interpretations. Therefore, it is necessary to define clear positions concerning the evaluation of the ecological importance of the considered parameters. Additionally, we point out that some parameters like heavy metal emissions have been excluded due to missing data for the life cycles or aggregation models for environmental factors. For example misuse of fertilizers and biocides or impacts on biodiversity or ecological functions of the soil cannot yet be quantified. And finally, the results of this study allow no conclusions on economy and technical practicability of the use of biofuels.

However, this study has proved that the applied evaluation method—based on and in correspondence with the ISO norm requirements—gives good results and is practicable. Therefore this study presents a data source and a methodical example for future LCA projects. But it is up to every user to determine his own value choices and define the suitable scales, scopes and parameter conditions for his project.

REFERENCES

1. Kaltschmitt, M. and G.A. Reinhardt (eds.) (1997). Nachwachsende Energieträger: Grundlagen, Verfahren, ökologische Bilanzierung, Vieweg Verlag, Braunschweig/ Wiesbaden.
2. Reinhardt, G.A. and G. Zemanek (1999). Ökobilanz Bioenergieträger. Bewertung von Lebenswegvergleichen Bioenergieträger versus fossilen Energieträgern, Erich Schmidt Verlag, Berlin, in preparation.
3. Reinhardt, G.A. (1998). First total ecological assessment of RME (biodiesel) versus diesel oil, Proceedings, 1998 Biomass for Energy and Industry International Conference, Rimpar.
4. Borken, J., A. Patyk, and G.A. Reinhardt (1999). Basisdaten für ökologische Bilanzierungen: Einsatz von Nutzfahrzeugen für Transporte, Landwirtschaft und Bergbau, Vieweg Verlag, Braunschweig/Wiesbaden, in preparation.
5. IPCC (1996). Climate Change 1995. The Science of Climate Change, Cambridge University Press, Cambridge.
6. CML & TNO & B&G (1992). Environmental Life Cycle Assessment of Products. Guide (Part 1) and Backgrounds (Part 2), prepared by CML, TNO and B&G, Leiden.
7. Reinhardt, G.A., J. Borken, A. Patyk, R. Vogt, and G. Zemanek (1997). aktualisierte Fassung von 1999). Ressourcen- und Emissionsbilanzen: Rapsöl und RME im Vergleich zu Dieselkraftstoff, Heidelberg.

401

BIOENERGY AND CARBON CYCLING—THE NEW ZEALAND WAY

Ralph E.H. Sims

Director, Centre for Energy Research, Massey University, Palmerston North, New Zealand
e-mail: R.E.Sims@massey.ac.nz

The New Zealand Government has been a strong advocate of plantation forests as sequesters of atmospheric carbon dioxide. Coupled with credits for the tradeable absorption of greenhouse gas emissions, the concept has been presented internationally with some success. However, there is a limit to how much land can or should be covered in forest owing to competition for land use to produce food and fibre products. Already in New Zealand some public resistance to establishing further extensive areas of monoculture plantation forests is evident, the current area being 1.4 million hectares (over 5% of the total land area). This has increased by approximately 60,000 ha/y for the past five or six years coming out of pasture. This land change process cannot continue indefinitely to offset fossil fuel use since sooner or later the whole country would be covered in trees. Forests as carbon sinks can only be a short-term measure at the best.

The preferred option is to grow forest sinks but harvest the entire aboveground biomass and utilise it as an energy resource, thereby displacing fossil fuels. This paper analyses the concept accounting for variations in soil carbon, choice of forest regime, and nutrient cycling to provide a sustainable system.

1. INTRODUCTION

Biomass is a renewable energy source arising from a range of organic products such as forest residues, wastes or purpose-grown crops that can be converted into heat, power or transport fuels. Technically biomass can be a total substitute for fossil fuel products since it can also be used as feedstock to produce virtually any petrochemical or material currently obtained from oil, coal or gas.

In New Zealand biomass already provides over 30 PJ per year or 6% of the primary energy supply (EECA, 1996). This will double within the next decade and provide around 10 times the energy contributed from all the existing or planned wind, small hydro (< 10MW), and solar projects. This dominance of new and emerging renewables by biomass is also the case in Australia where significant biomass resources exist (ERDC, 1994). In Austria and Scandinavia, biomass contributes around 30% towards their total primary energy, and in the USA more electricity is generated from biomass annually than the total power demand of New Zealand. Furthermore, biofuels for transport are commercially available at service stations in several countries including USA, Brazil, France, Germany and Austria. At present these biofuels such as ethanol tend to be heavily subsidised in order to compete with cheap oil products but research investment is increasing worldwide in order to reduce their production and processing costs. Since a more detailed assessment of biofuels has recently been presented elsewhere (Sims, 1998), this paper will outline examples of bioenergy used to supply heat and power projects where these are commercially competitive with fossil fuels without the need for subsidies. Future opportunities to develop bioenergy projects to obtain tradeable carbon credits will also be discussed.

2. COMMERCIAL PROJECTS

The forest product and sugar processing industries can become largely energy self sufficient, using process residues to meet their on site heat and power demands. Export of power off-site is feasible in many instances where surplus fuel exists. For example most Australian sugar cane processing plants "waste energy efficiently" simply to dispose of the bagasse by-product. In 1996 over 11 million wet tonnes of bagasse were produced with an energy content of around 120 PJ. Annual volumes vary with the growing season and are produced only during the cane crushing period from mid-June to mid-December. Currently all of the 33 sugar factories are virtually self sufficient in energy with around 250MW$_e$ of total capacity installed, generating over 500 GWh/y and an abundance of medium grade heat suitable for sugar processing. When more modern and efficient cogeneration plants are installed, as has been achieved at the CSR Invicta sugar factory in 1998, the available bagasse could then supply over 1000 MW$_e$ capacity and generate nearly 4000 GWh/y, much of which could be exported off-site.

Hence there is opportunity to expand the current power export capacity considerably using the existing biofuel resources and particularly if the additional bagasse currently either burnt off or left in the field at harvest could also be collected and used for fuel. It is encouraging to note that the Australian Greenhouse Office recently awarded the Rocky Point Green Energy Corporation a Showcase grant of $3M to install a 30 MW$_e$ cogeneration plant at their Brisbane sugar mill. The innovative concept of this project is to use locally obtained wood waste outside the 20-week crushing season to enable the power plant to provide green power to the Queensland grid all year round. No doubt other mills will follow suit where wood resources exist nearby. Eventually 10,000 GWh/y of electricity generation from biomass may become feasible.

Of the $10M made available for Showcase grants, a further $2M was awarded to another biomass project to enable EDL to design and install a solid waste-to-energy conversion facility. This facility will reuse and recycle resources prior to gasifying the residual organic components and generating heat and power from the gas. Such waste-to-energy processes are sensible conversion routes because the waste disposal costs are avoided and environmental impacts can be minimised. Many other commercial waste-to-energy projects exist including landfill gas, biogas from organic wastes, and use of wood process residues to provide heat, power and transport fuels. For example, in New Zealand a 39 MW$_e$ cogeneration plant using bark and other wood wastes was recently commissioned at Carter Holt Harvey's pulp and paper plant at Kinleith. Previously around 200,000 t of residues were dumped in landfills at high economic and environmental costs to the company. Combusting the material to recover the energy through the steam cycle for use mainly on site made commercial sense and some excess power is exported to the grid.

3. FUTURE OPPORTUNITIES

The proportion of primary energy supplied from biomass is likely to continue to increase in the foreseeable future in spite of new and improved wind and PV technologies being continually developed. The most modern biomass conversion plants also incorporate improved designs and performance efficiencies, particularly when designed for cogeneration of heat and power. New concepts such as integrated combined cycle gasification plants giving clean gas, improved conversion efficiencies, and resulting low air emissions are rapidly becoming commercially viable. Plant capacities currently range from 50 MW_e down to 50 kWe. The latter are perhaps better suited to rural communities, particularly in developing countries where abundant and sustainable supplies of biomass exist (such as old coconut trees, rice husks, and copra) and labour is relatively cheap and readily available.

In many developed countries biomass fuel sources that arise from the harvesting of plantation forests, wood process residues, and specifically grown energy crops are all available in relatively large volumes. The use of such fuels can be fully sustainable and environmentally acceptable assuming the overall production system is designed and managed correctly. The main problems associated with utilising biomass are the following:
- competition with food and fibre production for land use;
- the removal of additional nutrients from the soil thereby depleting reserves;
- the transport of large volumes of low energy density biomass fuels to the conversion plant (Sims and Culshaw, 1998).

The key to overcoming these problems and to ensuring the use of such biofuels remains an economic and environmentally acceptable option in competition with cheap coal, oil and gas, as well as with other renewables, is to obtain additional benefits from their use. Such benefits include:
- improving the fertility of unproductive saline soils (of which there are 4 million ha in Western Australia alone) by putting a proportion of the area into short rotation energy plantations to lower the water table and hence reduce the overall salinity problem for cereal production;
- providing easier re-planting of a commercial forest after stemwood harvest by clearing the residual trash traditionally left to decompose;
- avoiding the disposal costs for waste organic products by utilising them on site;
- treating effluents from municipal sewage works, food and fibre processing industries and farms by irrigating them on to energy crops rather than on to food crops or discharging into waterways; and
- sequestering carbon thereby gaining tradeable credits when used to displace fossil fuels.

Considerable recent debate has been undertaken internationally by the IPCC and other environmental organisations concerning utilising forests as carbon sinks (Figure 1).

Assuming sinks are ultimately accepted by the international signatories to the Kyoto Protocol, growing more plantation forests can provide only a short term solution to reducing the ever-increasing atmospheric carbon levels until eventually all the available land is covered by trees. So the concept makes sense as a long-term solution only if the accumulated carbon stored in the biomass can be utilised for energy purposes, either for cogeneration of heat and power or for the production of transport fuels or materials. Then the carbon is recycled within the overall biomass production and bioenergy utilisation system and the "renewable carbon sink can last forever" (Hall, 1998).

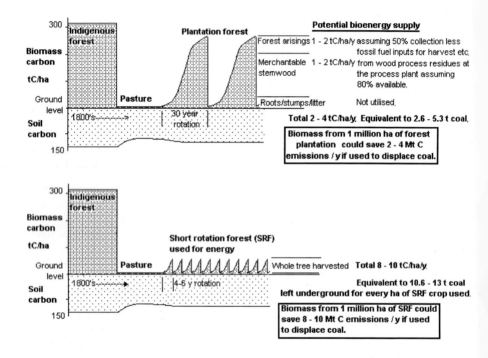

Figure 1. Carbon storage in biomass and soil fluctuates with land use over time as illustrated when indigenous forest is cleared for pasture then later replanted in plantation or short rotation energy forests.

Calculations of land availability have shown that sufficient suitable land exists to supply all food and fibre needs of the growing world population together with enough energy to meet a significant proportion of the global demand. There is much yet to learn about growing the biomass, transporting it and operating highly efficient conversion plants. It is exciting to note that the Shell Oil company has recently established a fifth division of the Royal Shell Group "International Renewables," as part of their long term strategy to become an energy company rather than an oil company. Growing short rotation forests to supply woody biomass for gasification features highly in their core business plan for the future (van der Veer and Dawson, 1997). With the high level of investment made available for this initiative, rapid advancements in the current knowledge relating to the biomass production, harvesting, processing and utilisation chain can be expected.

4. CONCLUSIONS

Growing forests to sequester carbon can be only a partial and temporary solution to mitigating for greenhouse gas emissions from using fossil fuels. The main benefits result from using all or part of those trees as fuelwood to displace fossil fuels. Then the carbon is cycled thorough the combustion/photosynthesis cycle so further atmospheric increases do not occur as any carbon released is reabsorbed. The concept of tradeable carbon permits will benefit forest owners in that they will receive credits which can then be sold on the international market to firms and industries as offsets against their excess greenhouse gas emissions. It is anticipated that more bioenergy projects, currently not commercially viable at present, will become so once their carbon credit value is included.

REFERENCES

Billins, P. (1998). Power from the land, Proc. 36th Annual conference of the Australian and New Zealand Solar Energy Society, Christchurch, pp. 3–6.
EECA (1996). New and emerging renewable energy opportunities in New Zealand, Centre for Advanced Engineering and Energy Efficiency and Conservation Authority, Wellington, p. 266.
ERDC (1994). Biomass in the energy cycle study. Energy Research and Development Corporation, Canberra, Report ERDC, pp. 234–272.
Hall (1998). The role of bioenergy in developing countries, in Proc. 10th European Conference on Biomass for Energy and Industry, Wurzburg, Germany, pp. 52–55. CARMEN.
Sims, R.E.H. (1997). Energy sources from agriculture. Proc. International conference Sustainable Agriculture for Food, Energy and Industry, Braunschweig, Germany, pp. 748–752. James and James.
Sims, R.E.H. and D. Culshaw (1998). Fuel mix supply reliability for biomass-fired heat and power plants, in Proc. 10th European Conference on Biomass for Energy and Industry Wurzburg, Germany, pp. 188–191. CARMEN.
Van der Veer, J. and J. Dawson (1997). Shell International Renewables—bringing together the group's activities in solar power biomass and forestry. Media release, London, 6 October.

PEAT AND FINLAND, AN EXAMPLE OF THE SUSTAINABLE USE OF AN INDIGENOUS ENERGY SOURCE

T. Nyrönen,[a] P. Selin[a] and J. Laine[b]

[a]Vapo Oy, P.O. Box 22, FIN-40101 Jyväskylä, Finland
[b]University of Helsinki, Department of Forest Ecology, P.O. Box 24, FIN-00014 Helsingin yliopisto, Finland

Finland is a cold country in northern Europe. The indigenous sources for the generation of power and heat are hydro, wood and peat. Almost one third of the country's area are peatlands totalling some 10 million hectares.
Currently the area reserved for peat production is 132,000 hectares and half of it is in active use. During the short summer season about 7 million tonnes of peat is produced. This amount covers 6,5% of the total primary energy demand in Finland. Peat has the specific CO_2 emission of 104-106 g/MJ, which is of the same order of magnitude as wood. The accumulation of peat and hence the sequestering of carbon in the Finnish peatlands exceeds the current use of peat for energy and horticultural purposes. The total CO_2 emissions of peat utilization are 8 million tonnes while the sequestering of the CO_2 in pristine and for forestry ditched peatlands is 11-20 million tonnes.
Peat production is conducted at about 300 individual sites all over the country. These sites are located in rural areas where the unemployment rate is usually high. Peat creates directly and indirectly about 6000 job opportunities in Finland.
Peat is used for power and heat generation in CHP plants of the inland cities. As long as the annual accumulation of peat exceeds the use of peat, the resource is used in a sustainable manner.

1. IMPORTANCE AND USE OF PEATLANDS IN FINLAND

Over a third of Finland's surface area, in other words approximately 10 million hectares, was originally peatland. Some of this peatland has been put to other forms of use by means of ditching. About 9% of Finland's peatlands have already been turned over to peatland conservation programmes and the Natura 2000 programme will further increase the total under conservation. Peat producers have 1.4% of Finland's mires, roughly 132,000 hectares reserved for production purposes. Of this about half has been fully or partly developed for production. About 50% of Finland's mires, or approximately 5.3 million hectares, has been ditched for forestry purposes.

Peat production benefits peatlands whose natural state has already been changed by man's intervention, since almost half the peatlands used for peat production have already been used previously for forestry purposes. Approximately 93% of Finland's peatlands are not suitable for peat production due to their location or the characteristics of the peat deposits. On the basis of current peat utilization estimated technically and economically viable peat resources are sufficient to last 350-500 years depending on the production technique employed.

Peat production in an individual site lasts about 15-25 years. During this time the area is ditched, vegetation is removed from the surface and thin layers are milled from the peat stratum, which are then dried by means of solar energy. The theoretical number of production days during the Finnish summer is 90 but the number of effective days ranges from 40 to 50. The production sites are located at reasonable distances from the power plants, usually less than 100 kilometers. The whole logistic chain from bog development to the stack of the power plant gives 6000 direct and indirect job opportunities.

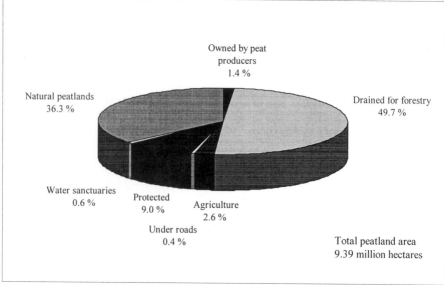

Figure 1. Use of peatlands in Finland.

In 1998 utilization of fuel peat amounted to 19.8 TWh. In combined production of heat and power the efficiency factor may be as high as 85-90% of the fuel's energy content. Calculated on the basis of the combustion efficiency factor, the specific emission of carbon dioxide for fuel peat is 104-106 g CO_2/MJ. For the purposes of comparison it should be mentioned that the corresponding figure for wood fuel is the same, i.e., 104-106 g CO_2/MJ.

The combustion techniques developed for the use of fuel peat also permit other solid fuels to be burned in addition to peat. In practice using fuel peat as the main or secondary fuel in existing boilers also supports the increased use of wood fuels. In addition peat can be used in conjunction with sorted waste.

2. PEATLANDS AND CLIMATE CHANGE

According to the Silmu (Research Program of Finnish Climate Change) research project financed by the Academy of Finland peat is bioenergy, since a peat bog in its natural state functions in the same way as a forest. Carbon dioxide emissions associated with wood are assumed to become embodied in the growth of new wood biomass over a period of roughly 100 years, whereas quantities of carbon dioxide released when peat is burned are bound up significantly more slowly in the formation of new peat, over a period of time 20-50 times longer. It is clear that the carbon released from peat combustion does not remain in the atmosphere but becomes embodied in new vegetation. The principle has not met with undisputed acceptance, however.

Finnish emissions of carbon dioxide amounted to approximately 0.3% of total carbon dioxide emissions worldwide for survey year 1990 in accordance with the agreement on global warming. Finland is a northerly country, thus energy consumption per capita is higher than in the more southerly countries of the EU. Fuel peat accounts for 6.5% of total energy consumption and 9% of emissions of greenhouse gases from energy production. Emissions of greenhouse gases from energy production are well documented in comparison to many other functions and sources of emissions. The current estimate of the total quantity of carbon contained in Finnish peatlands is almost 5 billion tonnes (5 Pg), while the annual carbon sink of peatlands in their natural state is estimated at roughly 1.1 million tonnes. According to Silmu studies the binding up of carbon in the growth of plants that form peat and the extenuating effect on methane emissions caused by ditching serve to reduce Finland's global warming potential.

Ditching alters the water regime of a bog. The bog's growing-season water level falls by an average of 30 - 60 cm measured from the surface and the bog's oxygen-containing surface layer extends deeper and facilitates efficient, aerobic decomposition in peat layers even deeper down.

The increase in biomass and forest litter production caused by ditching and, on the other hand, the increasing ratio between oxidation and decomposition of the peat decide whether a bog changes from a binder to a source of carbon subsequent to ditching. In the Silmu project the carbon sink of peatlands ditched for forestry purposes in Finland was estimated at 1.9-4.5 million tonnes of carbon per year, which corresponds to a CO_2 sink of 7-16 million tonnes. The annual carbon sink of the biomasses of bogland timber was estimated at 3 million tonnes of carbon, which corresponds to a CO_2 sink of 11 million tonnes. When reduced emissions of methane are also taken into account in calculations of peatlands' greenhouse effect, the result after ditching is a lower global warming potential which would seem to last for at least several hundred years.

In Finland the average effect of ditching on the carbon reserve in peat has been investigated relatively well even by international standards. Where carbon binding and especially the decomposition of organic matter are concerned there are clear differences between the EU countries, because the growing season in Finland is only 3-4 months in length, while in the United Kingdom, for instance, it is 12 months. On the other hand, we have insufficient knowledge of how practices used in peat forests, such as felling and improvement ditching, affect CO_2 sinks.

3. SITES RELEASED FROM PEAT PRODUCTION AND THEIR USE

In 1990 cutaway areas totalled approximately 3,500 hectares and the current figure is now in excess of 10,600 hectares. In 2010 the total area of sites released from peat production will amount to 45,000-50,000 hectares. Cutaway areas are suited to forestry or agricultural use as well as special cultivation (berries, herbs) or then by restoring the water level subsequent to the production phase they can be turned into bird sanctuaries, fish farming ponds or wetlands. In all probability the commonest post-production use will be reforestation or restoration.

If we compare the reforestation and restoration of cutaway areas over a 100 year period, the quantity of carbon dioxide sequestered in the afforestation option begins to decrease after the 100 year mark as the timber generation ages. In the restoration option a 100 year period is short, since carbon dioxide continues to be sequestered for thousands of years. The strongest uptake of carbon dioxide occurs in young forest and young wetlands, so in both options the binding of carbon dioxide per year is at its strongest 15-25 years after reforestation or restoration.

4. CONCLUSIONS

The use of peatlands for energy production has increased in Finland since the 1970s. Finland does not have resources of fossil energy and the transportation distances in the country are long. On the other hand, Finland possesses extensive reserves of peat. The utilization of fuel peat at a good efficiency factor is both justified and sensible in a country like Finland. The re-use of human-induced cutaway areas released from peat production will increase during the remainder of the 1990s and first decade of the next century. Direct human-induced measures used in sink calculations in Finland from the 1990s onwards have been taken in forests and peatlands.

However, when reporting Finland's carbon balance the starting-point should be the following facts:
- All definitions associated with control of the greenhouse effect and decisions based on them must be made in accordance with scientific principles.
- The carbon uptake of peat and wood is the same, on natural sciences principles, but the time scale is different.
- About 20 TWh of peat is used in Finland each year, with a resulting carbon dioxide emission of 8 million tonnes. The total carbon dioxide uptake of Finland's peatlands is 11-20 million tonnes annually, a quantity larger than the average amount of carbon dioxide released as a result of peat utilization each year.

- All measures associated with growth utilization of cutaway areas can thus be made use of as Finnish sinks and these are in harmony with the Kyoto principle of reforestation. Where peatlands are concerned Finland can take the following measures to increase the quantity of sinks:
- Cutaway areas released from production to be mainly afforested or restored, in which case their ability to bind carbon as regards the afforestation option is strong for about 100 years while for the restoration option the carbon binding effect may last for thousands of years even.
- Peatland arable fields to be used for peat production purposes, in which case the total global warming potential is reduced.
- Other measures to include boosting the efficiency of energy production by increasing combined heat and power production.

BIBLIOGRAPHY

Kuusisto, E., P. Kauppi, and P. Heikinheimo (ed.) (1996). Climate change and Finland—Silmu report, Helsinki University Press, Helsinki, 265 p.
Laiho, R., J. Laine, and H. Vasander (ed.) (1996). Proceedings of the International Workshop on Northern Peatlands in Global Climatic Change, Hyytiälä, Finland, Publications of the Academy of Finland 1/96. EDITA.
Laine, J. and K. Minkkinen (1996). Effect of forest drainage on the carbon balance of a mire: A case study. Scandinavian Journal of Forest Research, 11, pp. 307–312.
Laine, J., J. Silvola, J. Alm, H. Nykänen, H. Vasander, H., T. Sallantaus, I. Savolainen, J. Sinisalo, and P.J. Martikainen (1996). Effect of water level drawdown in northern peatlands on global climatic warming, Ambio, 25, pp. 179–184.
Laine, J., P.J. Martikainen, M. Myllys, T. Sallantaus, J. Silvola, K. Tolonen, and H. Vasander (1996). Peatlands, in Climate Change and Finland, ed. by E. Kuusisto, L. Kauppi,. and P. Heikinheimo, Helsinki University Press, Helsinki, pp. 107–126.
Lappalainen, E. and P. Hänninen (1993). The peat reserves of Finland. Geological Survey of Finland. Report of Investigation, 117, pp. 1–118.
Vasander, H. (ed.) (1996). Peatlands in Finland, Finnish Peatland Society, Helsinki, Finland.

# HYDROGEN PRODUCTION AND BIOFIXATION OF CO_2 WITH MICROALGAE

John R. Benemann,[a] JoAnn C. Radway[b] and Anastasios Melis[a]

[a]Department of Plant and Microbial Biology, University of California Berkeley, Berkeley, California 94720-3102
[b]School of Ocean and Earth Science & Technology, University of Hawaii, Honolulu, Hawaii 96822

Research into H_2 production and fixation of CO_2 into fuels by microalgae has been supported for over two decades in the U.S. (by the Department of Energy) and abroad (most recently as part of a very large Japanese R&D effort). Although several systems and economic analyses over the years have suggested that such processes are feasible, practical achievements require a more than two-fold increase in solar conversion efficiencies and over ten-fold cost reduction, compared with current commercial microalgae production processes. One major problem in both biohydrogen production and biofixation is the inefficiency of the photosynthetic apparatus at high light intensities. This is being addressed by a research project at the University of California Berkeley, whose aim is to reduce the number of chlorophyll molecules in the light-gathering "antenna" complexes of the photosynthetic apparatus of microalgae. This would more than double solar conversion efficiencies by microalgae cultures. The ultimate goal is to approach in algal mass cultures the theoretical maximum 10% solar conversion efficiencies of photosynthesis. Research at the University of Hawaii is evaluating the use of low-cost closed photobioreactors for microalgal H_2 production and CO_2 fixation. Although practical applications of such processes still require long-term R&D, some near-term applications are possible, through fossil CO_2 utilization in microalgae wastewater treatment ponds and in the commercial production of higher value microalgae products.

1. INTRODUCTION

Microalgae are microscopic plants that carry out the same process of photosynthetic O_2 evolution and CO_2 fixation as higher plants. Microalgae have been studied for some fifty years as a source of foods, feeds, fertilizers, fuels, high value products, and as a means of wastewater treatment (Burlew, 1953; Benemann and Weissman, 1993). Commercial production of microalgae was first achieved in Japan with *Chlorella* (a small unicellular green alga), then in Mexico and the U.S. with *Spirulina* (a filamentous cyanobacterium), followed by *Dunaliella* (in

Australia and the U.S.) and, most recently, *Haematococcus* (in Hawaii). *Chlorella* and *Spirulina* are produced for sale as food supplements ("nutriceuticals"), *Dunaliella* (a unicellular green algae that grows at high salinity) is a source of natural beta-carotene, and *Haematococcus* (a freshwater green alga) is a source of the pigment astaxanthin, used in aquaculture feeds. In addition, controlled microalgae systems are being used in wastewater treatment in the U.S. and other countries (Benemann et al., 1980; Benemann and Oswald, 1996).

Despite this steady record of technological development, the microalgae industry is still quite small, with less than a score of significant (> 50 t/y) production facilities worldwide. On a biomass basis, total world production is below 10,000 tons, plant gate revenues are less than $100 million, and unit costs roughly $10/kg, to provide order-of-magnitude estimates. In the area of wastewater treatment, only a handful of engineered, so-called "high-rate," microalgal treatment systems are currently operating. Compared with higher plant agriculture, microalgae are a minor industry, with small production facilities, high unit costs of production and limited markets.

Microalgae have also been investigated since the early 1950s for the production of methane fuel (by anaerobic digestion), later for H_2 production, and since the early 1980s, as a source of vegetable oils for use as liquid fuels such as biodiesel. The latter was the main objective of the U.S. Dept. of Energy (DOE) sponsored "Aquatic Species Program" (reviewed by Sheehan et al., 1998). Most recently, microalgae have been considered for the "biofixation" of CO_2, i.e., the capture of fossil CO_2 emissions from power plants and other stationary sources (Benemann, 1993). The largest R&D effort has been in Japan, where over $250 million has been spent during the 1990s on microalgal biofixation and H_2 production R&D.

Considering the limited applications and very high costs of current commercial microalgae production processes, the question arises whether microalgae will ever become sources of low-cost fuels. In brief, microalgae technology is still very much in its infancy, with great potential for process and economic improvements. Microalgae can directly utilize and recycle CO_2 emissions from power plants. Microalgae can use waste, saline and brackish waters, as well as "hard pan" and other land resources otherwise unusable by conventional crop plants. Their short life cycles would allow relatively rapid R&D. Most important perhaps, microalgae could achieve productivities almost an order of magnitude higher than higher plant annual crops or short rotation forests.

2. MICROALGAE FUELS PRODUCTION

Microalgae production systems are essentially continuous hydraulic production processes, as these microscopic plants grow suspended in water and their fast growth rates require frequent harvesting, essentially on a daily or at most weekly basis. There are several types of microalgae production processes, with one major distinction being between open ponds and closed systems, or "photobioreactors" (discussed in the next section). Open ponds can be of various designs, including circular, shallow inclined, large unmixed, and paddle-wheel mixed-raceway ponds. The latter design has become the industry standard, as it is easily scaled, very flexible and of relatively low cost. Commercial raceway production ponds are typically 0.2 - 0.5 ha in size, though a few commercial production ponds of up to about 5 ha have been constructed, and scales of up to 10 ha appear possible. In such paddle wheel mixed raceways, the algal culture is typically 15-30 cm deep and mixed at a velocity of some 20 - 30 cm/sec, fast enough to achieve turbulence but requiring relatively small energy inputs (which increase as the cube of mixing velocity).

Several engineering and economic analyses of increasing sophistication and detail have estimated the capital and operating costs of large-scale (> 1,000 ha) production processes for microalgae fuel production using paddle wheel mixed raceway ponds (Benemann et al., 1982, Weissman and Goebel, 1987; Benemann and Oswald, 1996). The basic design is of a single long raceway (channel) earthwork pond of about 10 ha in size, with a single paddle wheel and one or more CO_2 injection sumps spanning the width of the channel. CO_2 supply is a major design issue and requires balancing pH, alkalinity, mixing velocities, sump depth and other factors to maximize CO_2 utilization efficiencies. An overall efficiency of utilization of over 90% has been demonstrated in practice (Weissman and Tillett, 1992). For utilization of CO_2 from power plants and other stationary emission sources of fossil CO_2, the ponds must be located in the immediate vicinity of the power plant to avoid excessive transportation costs. Alternatively, flue gas CO_2 could be concentrated and piped some considerable distance, though at somewhat higher costs. Algae harvesting would be accomplished by allowing the cells to flocculate and settle in large sedimentation ponds ("bioflocculation," Benemann et al., 1980).

The capital cost projections for large open-raceway ponds are based on earthwork construction methods similar to those used for the large ponds used in rice fields and on agricultural engineering practices generally (with minimal engineering and other overhead costs). Total projected capital costs of about $100,000/ha, including mixing, dilution, CO_2 supply and harvesting, were estimated by several prior and recent studies (reviewed and updated by Benemann and Oswald, 1996). Operating costs, assuming a productivity of somewhat above 100 t/ha/y of biomass and a high oil content of about 60% by weight, were estimated at roughly $100/t. Both these projections are based on numerous, very favorable, assumptions about location, site, and process parameters, but suggest the potential for achieving major cost reductions compared to current technology. However, for fuel production, even higher productivities, and thus lower costs, must be achieved, as discussed in Section 5 below.

3. BIOHYDROGEN PRODUCTION

An alternative to the production of algal biomass with a high oil content, which would then be processed to liquid fuels (e.g. biodiesel), is the production of H_2 by microalgae cultures (Benemann et al., 1980). The attraction here is that such a process requires no net inputs of CO_2 and directly produces a high-value fuel from only sunlight and water. The microalgae cultures would essentially act as catalysts in such a conversion process. Despite the attractiveness of the concept, and a quarter century of applied R&D in the U.S., Europe and Japan (Zaborsky, 1998), practical advances have been limited and economic analyses derivative of microalgae fuels production processes (e.g., Benemann, 1998; Benemann, 1999).

One issue is the best process: direct biophotolysis, H_2 production by nitrogen-fixing cyanobacteria, or two-stage processes, using microalgae followed by photosynthetic bacteria or using only microalgae, to list only the main examples. Most such concepts suffer from major drawbacks: direct biophotolysis, in which O_2 and H_2 are produced simultaneously without intermediate CO_2 fixation, is only observed in laboratory experiments at very low partial pressures of O_2, with scale-up requiring a thus-far-unknown O_2-resistant H_2 evolution reaction. Nitrogen-fixing cyanobacteria readily produce H_2 gas (in the absence of N_2, the substrate of nitrogenase), but do so very inefficiently. Processes using photosynthetic bacteria, widely studied in Japan and Europe (see Zaborsky, 1998, for examples), have similar problems. Two-stage processes using microalgae capable of both CO_2 fixation and H_2 production appear to be the most promising, but require demonstration of a sustainable process and scale-up to outdoor systems. This process has become the focus of ongoing research, sponsored by the U.S. DOE, at the University of California Berkeley, the University of Hawaii and the National Renewable Energy Laboratory, using green algae (*Chlamydomonas, Dunaliella*) and cyanobacteria (*Spirulina*). The concept is to produce microalgal biomass high in fermentable carbohydrates in open ponds followed by anaerobic H_2 production, first in the dark and then in the light.

The final light-driven H_2 production stage requires closed photobioreactors to capture the H_2 gas. There are many different designs, of which the two major options are being internal and external gas exchange (Benemann, 1999). Microalgal cultures require large amounts of CO_2 and produce equivalent amounts of O_2, which must be supplied and removed to avoid, respectively, starving or poisoning the algal cells. Gas transfer is also a major issue in open ponds, and it becomes the critical factor in closed systems (Benemann, 1999). A "near horizontal tubular reactor" (NHTR) has been developed by Tredici and co-workers in Italy, which appears to be relatively inexpensive (as low as $50/m^2$), and potentially applicable to H_2 production (Tredici et al., 1998). Two such reactors, each consisting of eight 20-m-long tubes have been set up at the University of Hawaii and are being used to develop and demonstrate a two-stage process for biohydrogen production (Radway et al., 1998). Closed photobioreactors would also find applications in the scaleup of algal cultures from the laboratory to large outdoor ponds, required for the highly efficient, genetically engineered microalgal strains that must be developed for such applications.

4. MICROALGAE PRODUCTIVITIES

The above-reviewed processes and cost analyses suggest, optimistically, that relatively low-cost microalgae production for liquid fuels and H_2 is plausible. However, the major underlying assumption is the ability to achieve very high productivities, that is solar energy conversion efficiencies approaching the 10% theoretical limit of photosynthesis. In practice, however, such high efficiencies are not observed except under very low light intensities in the laboratory.

Outdoor pond operations, or even closed photobioreactors, typically exhibit maximum sustained solar conversion efficiencies of only 2 to 3%. The reason for such low efficiencies, even under optimal conditions, is that under the high intensities of sunlight, more photons are captured (absorbed) by the pigments of microalgae (typically chlorophyll) in the so-called antenna complexes than can be processed by the reaction centers of the photosynthetic apparatus. The surplus photons are wasted as heat or fluorescence. Although this problem is also present to some extent in higher plants, it is most severe and pronounced in microalgae cultures.

Several solutions have been proposed over the years to this fundamental limitation, including rapid mixing of the cultures and even use of optical fibers to transmit light uniformly into the cultures. Indeed the latter has been the major approach by the recent and very large Japanese effort in this field (see Zaborsky, 1998 for examples). However, neither of these solutions can be considered practical. A potentially practical approach is to genetically engineer microalgae with a greatly reduced content of chlorophyll (and other pigments) in the antenna complexes. This would reduce the number of photons captured by the reaction centers and, consequently, reduce photon wastage at high light intensities. Although long recognized, such an approach has become practical only with recent advances in our knowledge of the biochemistry and molecular biology of photosynthesis (Melis, 1991, 1998). Recent work at the University of California Berkeley has demonstrated that cultures of the green alga *Dunaliella* shifted from high to low light intensities exhibit small chlorophyll antenna sizes, and also both high rates and efficiencies of photosynthesis at high light intensities (Melis et al., 1998). Although sustained for only a short period, this work demonstrates the inherent feasibility of the goal of permanently truncating the chlorophyll antenna sizes in microalgae using genetic engineering techniques. Such strains could achieve the very high productivities required for solar energy conversion and fuels production.

The efficient and economical production of fuels and biofixation of CO_2 with microalgae is still a long-term goal. However, some near-term applications are possible. For example, in Hawaii a commercial microalgae production company has built a small power plant to produce both electricity for sale and CO_2 to feed their algal production process—a small-scale but present-day example of the longer-term goal. Wastewater treatment systems provide additional opportunities for near-term applications, combining CO_2 utilization with fuel production. The development of closed photobioreactors for H_2 production could find uses in microalgae specialty chemicals production. Such near-term applications and the long-term potential justify a sustained and expanded U.S. research effort in microalgae technology.

REFERENCES

Benemann, J.R. (1993). Utilization of carbon dioxide from fossil fuel-burning power plants with biological systems, Energy Conserv. Mgmt., 34, pp. 999–1004.

Benemann, J.R. (1998). The technology of biohydrogen, in BioHydrogen, ed. by O. Zaborsky et al., Plenum Press, New York pp. 19–30.

Benemann, J. (1999). Photobioreactors for Hydrogen Production, Report to the National Renewable Energy Laboratory, February 1999.

Benemann, J.R. and J.C. Weissman (1993). Food, fuel and feed production with microalgae Proc. First Biomass Conference of the Americas, Burlington, Vermont, September, 1993 NREL/CP-200-5768, pp. 1427–1440.

Benemann, J.R. and W.J. Oswald (1996). Systems and Economic Analysis of Microalgae Ponds for Conversion of CO_2 to Biomass, Final Report to the Federal Energy Technology Center under Grant No. DE-FG22-93PC93204.

Benemann, J.R., K. Miyamoto, and P.C. Hallenbeck (1980). Bioengineering Aspects of Biophotolysis. Enzyme and Microbial Technology 2, pp. 103–111.

Benemann, J.R., B.L. Koopman, J. C. Weissman, D. E. Eisenberg, and R. P. Goebel (1980). Development of Microalgae Harvesting and High Rate Pond Technology, in Algal Biomass, ed. by G. Shelef and C.J. Soeder, Elsevier Biomedical Press, Holland, pp. 457–499.

Benemann, J.R., R.P. Goebel, J.C. Weissman, and D.C. Augenstein (1982). Microalgae as a Source of Liquid Fuels, Final Report U.S. Department of Energy, 202 pp.

Burlew, J., ed. (1953). Algae Culture: from Laboratory to Pilot Plant, Carnegie Inst.,Washington, D.C.

Melis, A. (1991). Dynamics of photosynthetic membrane composition and function. Biochim. Biophys. Acta (Reviews on Bioenergetics), 1058, pp. 87–106.

Melis, A. (1998). Photostasis in plants: mechanisms and regulation, in Photostasis, ed. by T.P. Williams and A. Thistle, Plenum Press, New York, pp. 207–221.

Melis A., J. Neidhardt, I. Baroli, and J.R. Benemann (1998). Maximizing photosynthetic productivity and light utilization in microalgae by minimizing the light-harvesting chlorophyll antenna size of the photosystems, in BioHydrogen, ed. by O. Zaborsky, Plenum Press, New York pp. 41–52.

Radway, J.C., B.A. Yoza, J.R. Benemann, M.R. Tredici, and O.R. Zaborsky (1998). Evaluation of an outdoor tubular photobioreactor system in Hawaii, Annnual Meet. Amer. Soc. Microbiol., May 17 - 22, 1998, Atlanta, Georgia.

Sheehan, J., T. Dunahay, J.R. Benemann, J.C. Weissman, and P. Roessler (1998). Close-Out Report for the U.S. D.O.E. Aquatic Species Program, 1976 - 1996, Report to the National Renewable Energy Laboratory, Golden, Colorado.

Tredici, M.R., G.C. Zittelli, and J.R. Benemann (1998). A tubular internal gas exchange photobioreactor for biological H_2 production: Preliminary cost analysis, in BioHydrogen, ed. by O. Zaborsky, Plenum Press, pp. 391–402.

Weissman, J. C. and R. P. Goebel (1987). Design and Analysis of Pond Systems for the Purpose of Producing Fuels, Solar Energy Res. Institute, Golden, Colorado, SERI/STR-231-2840.

Weissman, J.C. and D.T. Tillett (1992). Design and Operation of an Outdoor Microalgae Test Facility: Large-Scale System Results, ed. by L.M. Brown and S. Sprague, Aquatic Species Project Report, FY 1989-1990, NREL, Golden Colorado, NREL/MP-232-4174. pp. 32–56.

Zaborsky, O., ed. (1998). BioHydrogen, Plenum Press, New York.

A SUSTAINABILITY ANALYSIS OF BIOMASS USE BY INFORMATION ENTROPY THEORY

M.X. Ponte

Visiting Professor at the University of Missouri – Columbia, from Federal Univ. of Para-Br, 508 Stalcup, Columbia, Missouri, 65203, e-mail:mximenes@gte.net

Some important developments in the industrial use of biomass are changing rural activities. Sophisticated methods are emerging to evaluate how resources can be captured and converted to useful commodities, especially the goods and services that yield significant on-site benefits.

Information theory is a mathematical approach that describes the structure of some agroecosystems as a network of flows of energy. This paper reports an analysis using this theory to investigate changes in the structure of traditional agriculture in the Brazilian Amazon. The analysis uses empirical data from "cut and burn" agricultural system to evaluate the potential of biomass as an agricultural commodity that also produces great energetic and environmental gains. Two possibilities for use of biomass are taken into account: generation of the electrical energy and the production of ethanol and chemicals by acid hydrolysis. The results suggest that the potential for economic and environmental sustainability are improved.

421

ENVIRONMENTAL COSTS OF ENERGY FROM TWO BIOGAS PLANTS IN DENMARK

P. S. Nielsen

Department of Buildings and Energy, Technical University of Denmark, Building 118, 2800 Lyngby, Denmark

Biogas presently produces 2.4 PJ—less than one percent—of the Danish primary energy supply. The physical potential is, however, 15 times higher, due to the intensive animal production in Danish agriculture. The paper assesses the environmental impacts and external costs of two biogas installations in Denmark producing electricity and heat. Environmental impacts of the biogas plants are compared with the environmental impacts of a natural gas system. The methodology used is a bottom-up impact pathway method that addresses the environmental impacts throughout the entire fuel chain.

The results show that the largest estimated non-global damages (effects on human health) are due to the emission of NO_x during combustion of both biogas and natural gas. The largest environmental impact of natural gas systems is the emission of greenhouse gases. It depends, however, on the valuation of CO_2 damages. The largest benefit of a biogas plant is the avoided emission of methane for the traditional storage tanks. Furthermore a reduction of the leakage of nitrate to the groundwater is expected in the biogas energy systems.

1. INTRODUCTION

In order to reach a CO_2 reduction goal of 20% in 2005 and 50% in 2030, biomass is an important fuel in the future energy system in Denmark.[1] The biogas potential is around 30 PJ, depending on the future developments in technology and in the agricultural sector. Biomass produces around 64 PJ—7.7%—of the total primary energy supply, including energy recovery from waste incineration. The future potential of biomass is expected to be 145 PJ.[2]

In 1985 the Danish government launched a national biogas programme to stimulate the development of centralised joint biogas plants. Today there are 19 large joint biogas plants established in Denmark. The biogas programme supports up to 30% of investment costs. Despite advances in the last 15 years, the technology has not up to now been competitive when only the traditional costs are included in the energy prices.

2. THE ENERGY SYSTEMS

The biogas plants assessed are Ribe Biogas Plant (RBP) and Hashøj Biogas Plant (HBP). RBP, one of the largest biogas plants in Demark, has been in operation for eight years. Seventy-nine farms supply the biomass, which is mixed with industrial organic waste to increase biogas production. The total biomass supply is 410 tonnes of biomass/day, where industrial organic waste covers around 20%. The total biogas production is 12,000 m^3/day (gross) and about 10,000 m^3 is transmitted to a CHP plant in Ribe per day.

The slurry to HBP is supplied by 19 farmers with a total of around 140 tonnes per day of which 30% is industrial organic waste. The total biogas production is approximately 5,500 m^3 biogas per day. In both cases the energy conversion systems are gas engines.

At HBP there are two gas engines, one using biogas and the other natural gas. Furthermore it is possible to switch between the fuels at the individual unit. At RBP a natural gas engine was recently installed; no estimated environmental costs exist for that plant. The environmental damage of the natural gas units are estimated for the HNP.

2.1. The biogas fuel chain

The biogas fuel chain starts at the farm where it is collected by a truck from the biogas plant and transported to the biogas plants. At the biogas plant the slurry is mixed with industrial organic waste before it enters the biogas reactor. The produced biogas is transmitted 2 km to underground pipelines in a CHP plant which uses it to produce electricity and heat.

After leaving the reactors the fermented biomass is stored for few days at the plants and then transported back to the farms. In some cases the slurry from a number of farmers is stored at intermediate storage tanks, which are established at strategic sites. The farms can collect the fermented slurry when it is going to be spread on the fields as fertilisers during March and May.

2.2. The natural gas fuel chain

The natural gas used in Denmark is domestic and extracted in the Danish part of the North Sea. It is transmitted 300 km through sea pipelines to Esbjerg. The natural gas is stored in two large stores, one in the northern part of Jutland and the other in the middle of Zealand, before it is delivered to various power plants or directly to private consumers for heating and cooking.

3. METHODOLOGY

Impact assessments and valuations used the 'damage function' or 'impact pathway' approach. Methods ranged from the simple statistical relationships, as in the case of occupational health effects, to complex models and databases, as in the cases of acid rain and global warming effects. The approach required detailed definition of both the fuel cycle and the system within which the fuel cycle operates, with respect to both time and space.

The study used a unified approach, to ensure compatibility between results. This is achieved through the use of the EcoSense software package, which is an integrated computer system developed at the University of Stuttgart.[3,4] It assesses the environmental impacts and resulting external costs from electricity generation systems. The system has an environment database at both a local and regional level including population, crops, building materials and forests. The system also incorporates two air transport models, allowing local and regional scale modelling. It also includes a set of impact assessment modules, based on the dose-response relationships used in the ExternE Study, and a database of monetary values are included for different impacts.

In a life cycle assessment the definition of the system boundaries is important, and it is especially important in the biogas fuel cycle. To establish the biogas fuel cycle it was important to study the traditional handling of slurry because the use of slurry for biogas production reduces the emission of CH_4 from the large storage tanks at the farms.[5] This emission increased during the last decade as stricter regulations on spreading of slurry directly on fields has caused nitrate to leak into the groundwater. Therefore it has been important for this study to establish the total balance of emissions of greenhouse gases (GHG) for the entire fuel cycle, taking this benefit of reducing CH_4 into account.[6] A GHG balance was also prepared for the natural gas fuel cycle.

Both cases involve a combined heat and power system. District heating system was not included as it is similar in the two cases. Both systems also produce electricity, which is sold to the public grid, which means that any effect on grid is not taken into account.

4. IDENTIFICATION AND SELECTION OF PRIORITY IMPACTS

A wide variety of impacts were considered for the two fuel cycles: the effects of air pollution on the natural and human environment, consequences of accidents in the workspace, and the effects on climate change arising from the release of greenhouse gases. More than 200 impacts were identified for the life cycle of the fuel cycles of natural gas.[3] The prioritised impacts related to the natural gas chain and the biogas fuel chain are shown in Table 1.

Table 1
Identification of priority impacts a natural gas CHP plant and a biogas CHP plant

Impact	Natural gas	Biogas
Visual amenity	x	-
Atmospheric emissions	x	x
Greenhouse gases	x	x
Public accidents	x	x
Occupational accidents	x	x
Natural gas storage	x	-
Land use changes	x	x
Emissions to soil	-	x
Road damage	-	x

5. QUANTIFICATION

For the biogas fuel cycle the largest effects are produced by emissions from the combustion process, and mainly the release of NO_x (Table 2). NO_x emissions of 290 mg/Nm^3 were measured at RBP, and 400 mg/Nm^3 at HBP were measured. Both values are below the national limit of 650 mg/Nm^3 for natural gas systems. In Table 2 the emission is given per kWh electricity. The difference in NO_x emissions between HNP and HBP is related to emissions from transportation of slurry for the biogas fuel cycle; they are higher at RBP because more slurry arrives there. The emission of SO_2 at RBP and HBP reflect both the transportation process and the generation of energy. The emission of SO_2 from the engine running on natural gas is negligible. Except for the figures on leakage of nitrate the figures are transferred from [6] and [7]. The allocation between electricity and heat is done on an energy basis.

Table 2
Quantification of damages for a natural gas CHP plant and a biogas CHP plant

Impact	Natural gas, HNP	Biogas, HBP	Biogas, RBP
Atmospheric emissions	~0 g SO_2/kWh$_{el}$ 1.8 g NO_x/kWh$_{el}$, ~0 g TSP/kWh$_{el}$	0.04 g SO_2/kWh$_{el}$ 2.0 g NO_x/kWh$_{el}$	0.14 g SO_2/kWh$_{el}$ 3.3 g NO_x/kWh$_{el}$
Greenhouse gases	480 g CO_2/kWh$_{el}$	-550 g CO_2/kWh$_{el}$	-450 g CO_2/kWh$_{el}$
Public accidents	Negligible	0.04 minor 0.06 major 0.01 death	0.2 minor 0.3 major 0.03 death
Occupational accidents	1 minor 0,1 major 0.01 death	2 minor 0.3 major 0.004 death	5 minor 0.8 major 0.01 death
Avoided leakage Of nitrate		0.6 kgN/ton slurry	0.6 kgN/ton slurry
Natural gas storage	200 houses	-	-
Road damage	-	42,000 km	210,000 km

Impacts of the natural gas fuel cycle are mainly related to emissions (NO_x and CO_2) directly from the combustion process. Gas storage poses a relatively large risk. Only minor effects were recorded on occupational health, and other effects assessed are regarded as negligible.

Public accidents for the biogas fuel cycle are due to the transportation process alone. Average data for Danish industry were used, except for exploration natural gas in the North Sea where actual data are used.[8]

The concentration of NH_4^+-N in fermented slurry is increased by around 20% compared with raw slurry. Studies show an increased uptake of NH_4^+ from fermented slurry compared with raw slurry (10-15%), but it is possible that the efficiency of utilising the NH_4^+ could be even higher.[9] This indicates that if farmers are able to fully utilise the NH_4^+ the use of fermented slurry will reduce the leakage of nitrate by up to 0.6 kgN/ton slurry. The actual figure depends on weather and is very site specific. But the figure shows to an order of magnitude the avoided environmental impact.

The main difference between the two fuel cycles is the emission of greenhouse gases. As stated above, a total balance of greenhouse gases has been established for both case studies.[6] For the biogas fuel cycle there is a large benefit of reducing the emission of CH_4. In total this benefit is larger than the emission CO_2 from the transportation of slurry, emission of unburned CH_4 passing through the gas engine and emission from storage of fermented biomass. The emission of CH_4 passing the engine unburned is measured to be 1.8% of the total amount of in-going CH_4 for engines below 1 MW_{el}.[10] The unburned CH_4 passing through the engine is in this comparison similar for the two case studies, but it is included in the figures for CO_2 emission in Table 2.

6. VALUATION

Figure 1 shows the most important results of the calculations related to the biogas and natural gas fuel cycles. The results show that the local and regional externalities are lower for the biogas fuel chain (HBP and RBP) than the local and regional externalities for the natural gas fuel chain (HNP). The local and regional damages are due to emissions of especially NO_x to air. The reason why the local and regional damages are lower for the biogas fuel chain is the avoided damage due to the avoided nitrate leaching; using a value of 3 ECU/kgN was used.

Total damage of fuel chains

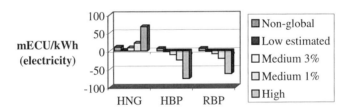

Figure 1. Externalities related to the fuel chains.

By including global damages, the difference in damages between natural gas and biogas increases, as natural gas emits large amounts of greenhouse gases, whereas biogas does not disturb the greenhouse gas balance. The low damage cost estimates for the greenhouse effect is 3.8 ECU/tCO$_2$.[4] The medium estimates include both a 1% and 3% discount rate and are 18 and 46 ECU/tCO$_2$, respectively. The high estimate is 139 ECU/t CO$_2$.

7. CONCLUSIONS

Atmospheric emissions are the most dominant damages produced by all fuel cycles. Such emissions stem predominantly from the generation process in the natural gas fuel cycle, but from a number of different parts of the biogas fuel cycle.

Damage due to the emission of greenhouse gases predominate in the natural gas fuel cycle even when a discount rate of only 3% is taken (the medium estimate). Avoided greenhouse gas emission is the main advantage of the biogas fuel chain. However, the figures are still uncertain. To get a more precise picture of total damages the damages due to the emission of NO$_x$, nitrate leaching and the GHG balance should be studied in more detail.

REFERENCES

1. Danish Energy Agency 1996. Energy 21, Copenhagen, Denmark.
2. Ministry of Environment and Energy (1995). Denmark's Energy Futures (In Danish), Copenhagen, Denmark.
3. ETSU and Metroeconomica (1995). ExternE - Externalities of Energy. Vol. 2 Methodology, ECE, DG-XII, Luxemburg.
4. Schleisner, L. and P.S. Nielsen (eds) (1998). ExternE National Implementation. Appendix, (DG-XII) and Risø National Laboratory, Roskilde, Denmark.
5. Nielsen, P.S. and J.B. Holm-Nielsen (1996). CO$_2$ balance in production of energy based on biogas. Proceedings from the European Conference: Environmental Impact of Biomass for Energy, November 4-5, 1996. CLM, Utrecht, The Nederlands.
6. Schleisner, L. and P.S. Nielsen (1998). ExternE National Implementation - Danish Fuel Cycles. ECE (DG-XII) and Risø National Laboratory, Roskilde, Denmark.
7. Almeida, A., A. Bauen, A., F.B. Costa, S.O. Ericsson, J. Giegrich, G. Gosse, N. von Grabczewski, H.M. Groscurth, D.O. Hall, O. Hohmeyer, K. Jörgensen, C. Kern, I. Kuehn, B. Leviel, R. Löfsted, J. da S. Mariano, P.M.G. Mariano, N.I. Meyer. P.S. Nielsen, C. Nunes, A. Patyk, E. Poitrat, G.A. Reinhardt, F. Rosillo-Calle, I. Scrase, C. Vergé, and B. Widmann (1998). Total Costs and Benefits of Biomass in Selected regions of the European Union. The European Commission, Brussels.
8. Danish Energy Agency (1997). Oil and Gas Production in Denmark 1996, Danish Energy Agency, Copenhagen, Denmark.
9. Holm-Nielsen, J.B., N. Halberg, and S. Huntingford (1993). Biogasfællesanlæg landbrugsmæssige nytteværdier (In Danish), Danish Energy Agency, Copenhagen.
10. Nielsen, M. (1996). UHC/metan-emission fra gasmotorbaserede kraftvarme-installationer, Dansk Gasteknisk Center, Hørsholm.

CARBON CREDITS—HOW TO MEASURE THEM FOR THE PETROLEUM REFINING INDUSTRY

J.J. Marano and S. Rogers

Federal Energy Technology Center, P.O. Box 10940, Pittsburgh, Pennsylvania 15236-0940, USA

Implementation of the Kyoto protocol in the United States, an outcome far from certain, will require the establishment of a mechanism for estimating greenhouse-gas reduction—that is, the calculation of carbon credits. In addition, if these credits are to be bought, sold or traded, a means of valuing them is required. Clearly, establishment of such metrics involves a total lifecycle, "well-to-wheel" approach. For example, the use of biomass components as fuel additives reduces emissions from "upstream" petroleum subsystems, as well as from combustion of the fossil-based fuels. Further, a baseline or temporal reference for the credits was specified for the U.S. at Kyoto. Carbon emission reductions are to be measured relative to a base year of 1990, with reductions below this baseline (7%) required by the years 2008 to 2012.

This scenario is strikingly similar to the requirements laid down in the Clean Air Act Amendments (CAAA) of 1990 for the production of reformulated fuels, aimed at reducing air emissions (toxics, VOCs and NO_x) from mobile sources. The CAAA90 involved roughly the same implementation period (10 years), employed a 1990 baseline, and specified mandatory reductions from baseline. The baseline can be different for each petroleum refiner since each refiner's emissions are different owing to differences between crude oils, product slates, and processing equipment. The problem then facing the refiner is one of measurement or estimation of carbon emissions, and analysis to identify ways of reducing these emissions. The carbon content and emissions associated with producing various petroleum-based fuel components and biomass-derived substitutes have been compiled and used to develop models for predicting refinery emissions. These sub-models have been incorporated into a refinery linear programming model, a common planning tool used in most refineries, to investigate carbon emission reduction strategies for the transportation sector, allowing both the size and value of carbon credits for biomass to be estimated. Preliminary results indicate that the production of hydrogen and liquid fuel additives are cost-effective approaches for introducing biomass into the transportation fuel market.

429

TOTAL COSTS OF ELECTRICITY FROM BIOMASS IN SPAIN

Sáez, R, Lechón, Y., Cabal, H. and Varela, M.

CIEMAT. Avda Complutense, 22.E28040 Madrid, Spain

In the White Paper on Renewables, the European Commission has set the goal of achieving a 12% penetration of renewables in the Union by 2010, of which biomass could be the main contributor. Spain is developing a Biomass National Plan to achieve that objective. Therefore, the objectives of this study are to select the sites for and size of possible biomass power plants, which are needed to contribute at least 2% to the current national energetic mix, and to assess the electricity costs.
The site selection is carried out by use of a geographic information system, and the direct cost of the electricity as well as the external costs and benefits are assessed. The study describes the species selected for Mediterranean edaphoclimatic conditions in each site and the biomass production cost is included in the electricity cost analysis. Environmental externalities are assessed applying the Extern-E methodology and socio-economic externalities using the Spanish input/output tables. Results of total costs of electricity from biomass are obtained.

1. INTRODUCTION

An increasing concern about the environmental and socio-economic consequences of our current energy use drives a move towards the use of renewable energies (RE). To date wind energy is the most developed, and it is competitive in some cases with conventional sources. However, wind energy could not possibly provide as large amount of energy as biomass does. The biomass contribution, for final thermal uses, in the Spanish national energy mix was 3.57 Mtoe, that is, 3.5% of the total energy consumption during 1996. Spain is currently developing a Biomass National Plan to achieve the objective set by the White Paper on Renewables of the European Commission, which is to increase the penetration of RE to 12% in the European Union (EU) by 2010. Although all renewable energies will be developed, biomass could be the main contributor.

Biomass and other renewable energies are not yet implemented on a full commercial basis mainly owing to their high cost. This obstacle may be overcome if externalities are assessed and the total/full cost of the electricity is considered in the electricity price. The external costs are imposed on society and not accounted for by the producer. The quantification of the externalities would permit us to know the true cost of electricity and what price society is paying for it. Producing electricity leads to both negative externalities such as environmental damage and positive externalities such as employment, increased economic activity, reduced dependence on fossil fuels. These benefits are associated with some sources of energy, mainly renewables such as biomass.

Because biomass has been generally used for thermal purposes in Spain, this work studies how to add 2% to the current national power capacity (which in 1997 was 43.55 GW) fueled by biomass from energy crops. The biomass scheme studied is based on woody biomass as short rotation coppicing (SRWC) in arable land, taking into account the edaphoclimatic conditions, land use and the Spanish agriculture infrastructure. The electricity production costs are calculated including the energy crops production, storage, transport and grid connection costs. The environmental externalities (health and agriculture damages caused by air emission, erosion and non-point-source pollution caused by biomass production and global warming effects) have been calculated applying the ExternE project methodology (EC, 1995), also named the "damage function." Employment and macroeconomic effects have been assessed using the Spanish input/output tables.

2. SITE, SPECIES AND TECHNOLOGY SELECTION

2.1. Site selection

Land use, topographic data, road and hydrology network, municipalities, land prices, electricity grid and rainfall data were introduced into the geographical information system (ARC-INFO) in order to develop the site selection process. The criteria for both plantations and biomass power plants location were defined in terms of availability of fresh water, of enough nonirrigated land areas with more than 400 mm rainfall, roads, minimum distance from population centres, slope and altitude (Varela et al., 1999).

Owing to the huge amount of information to be treated, a two-step process was used. Two intermediate outputs were obtained: the potential sites for the power plants and the potential available land. These two coverage were overlaid in order to obtain the final sites for power plants (Varela et al., 1999). This two-step process allowed the identification of 28 potential sites for biomass power plants with a total power capacity that reaches around $1GW_e$. Land suitable to produce the needed biomass to fuel each power plant surrounds all of them. The size of each plant has been determined according to two main criteria. The biomass plantations have to be located inside a circle of 40 km of radius around the power plant, and the total land needed for each plant should be a maximum of 10% of the available non-irrigated arable land in that circle (Varela et al., 1999).

2.2. Species considered

Only woody species have been considered in this study. Selection criteria have been based on the country's edaphoclimatic conditions and empirical data obtained in several places of Spain. The selected species in this study are *Populus sp., Eucalyptus sp., Robinia pseudoacacia, Sophora japonica and Gleditsia triacanthos*. Plantation irrigation was not considered for economic and resource-optimization reasons. Therefore SRWC would be established only in areas with a minimum rainfall of 400 mm/y. The distribution of species in each area depends on minimum temperatures in winter and frequency and volume of rainfall (average of the 30 last years) (Varela et al., 1999). Of the total area, 5% would be planted with poplar, 33% with eucalyptus and 62% with acacias.

In this analysis the potential production of biomass per hectare has been estimated taking into account the concept of water use efficiency (Kramer and Boyer, 1995) and empirical results obtained in Spain (San Miguel and Montoya, 1983). Therefore, the biomass harvested per hectare has been calculated for all 28 sites to be an average of 13.3 tonnes of dry matter/ha and a total of 6.3 million of tdm/yr. of gross biomass. CO_2, matter and energy balances were calculated. In addition to the aerial parts of the plants, i.e., biomass harvested, there are stump roots that should be taken into account since they act as a CO_2 sink in the biomass scheme. A study of Lodhijal and Singh estimates that the amount of roots in a high-density plantation of poplar is 17.6% of the entire (aerial and underground) biomass per hectare and year (Lodhijal and Singh, 1994). Assuming this figure for all plantations and considering the average harvested biomass 562.4 thousand t/yr of carbon would be permanently sequestered. That means that 272 g of CO_2 would be sequestered per kWh produced. Further research should be done in order to confirm this assumption. Leaves, twigs and thin roots on the soil that decay in a short period of time are considered as neutral regarding CO_2 balance.

Losses in chipping (2%), handling (1%), and storage (2%) have been estimated from the biomass yield and taken into consideration in order to calculate final costs. Other losses such as pest and diseases, long periods of drought or other eventual losses have already been included since the concept "biomass harvested" used in the study embodies those losses.

2.3. Technology

Fluidised bed combustion (FBC) and biomass gasification integrated in a combined cycle (BIGCC) have been selected as the most appropriate technologies according to the possible state of the art of the year 2000. In order to avoid technical risks, the range of plant size considered has been 60–25 MW_e for FBC and 35–30 MW_e for BIGCC. Because this technology has been used less than FBC, only 6 BIGCC plants would be set up. The efficiency would range from 25% (FBC) to 40% (BIGCC), including parasitic consumption by the plants. The electricity produced by each plant, minus the losses during the distribution stage, results in the total energy output. Emissions of CO_2, SO_2, NOx and particulates have been calculated for the power plants. The emissions of CO_2 would be of 1.4 t CO_2/MWh. It is estimated that one job per MW is produced, so 1,000 jobs per year would be created during the generation stage.

3. EXTERNALITIES ASSESSMENT

The methodology used for the assessment of the environmental external effects is the one developed within the ExternE project (EC, 1995) in the European Commission R&D Programme Joule II. The ExternE methodology is a bottom-up methodology with a site-specific approach; that is, it considers the effects of an additional fuel cycle located in a specific place. Effects are quantified through the "damage function," or effect pathway approach. This approach uses logical steps to trace the effect of an activity that creates it to the damage it produces, independently for each effect and activity considered. The stages of the methodology are site and technology characterisation, identification of fuel chain burdens, identification of effects, prioritisation of effects, quantification of effects and economic valuation. The term 'burden' relates to anything that is, or could be, capable of causing any effect.

Environmental externalities assessed include the following:
- Health effects of atmospheric emissions produced throughout the whole fuel cycle. Detrimental effects on crops have not been considered, because the main pollutant emission harming crops is SO_2, which is negligible in biomass power plants.
- Global warming effects produced by the overall CO_2 and CH_4 emissions inventory, considering the fixation of CO_2 by the photosynthesis process in the cultivation stage.
- Erosion and non-point pollution effects produced by the crops considered.

Health effects of atmospheric emissions have been estimated based on a previous study of externalities of the whole Spanish electricity system in which the average damages per tonne of pollutant emitted were calculated at different locations of Spain (CIEMAT, 1998). These damages are independent of the technology and are site specific. Ten representative power plants were analyzed, eight of them FBC plants and two of them BIGCC plants. The damages were estimated using the emission factors calculated for these plants and the damage factors of the closest power plants of the ones analyzed in the study above mentioned. Results obtained are summarised in Table 1 as weighted averages considering the proposed participation of each type plant in the scheme. For global warming, two cases were considered, the first case ignoring the possible sequestration of CO_2 by the roots, and the second one considering that there is effectively a permanent sequestration of CO_2 in the roots of the crop. The damages were estimated following the recommendations of the ExternE updated methodology (EC, 1998) in which a range of 18–46 ECU/t CO_2 was estimated as an illustrative restricted range (Table 1). CO_2 sequestration leads to positive external effects in the sense that they would be avoided damages. They are expressed as negative external costs.

Erosion from the cultivation stage was estimated based on a previous study of a hypothetical biomass power plant located in Valdecaballeros (CIEMAT, 1996); those results were adjusted for the different energy crop considered in the present study. Results obtained ranged from 0.004 to 0.010 c$/kWh.

Effects from non-point pollution produced in the crop stage were also calculated based on the previously cited study and the results ranged from 0.002 to 0.03 c$/kWh.

Subtotal computed externalities ranged from 1.04 to 1.31 c$/kWh when sequestration of CO_2 was ignored. If an actual sequestration of CO_2 by the roots is admitted the results ranged from negative values (-0.46 c$/kWh) to slightly positive values (0.67 c$/kWh). These results are only preliminary estimations because they were extrapolated from other studies and therefore are only proximate figures. Furthermore, they are only subtotals because other potential effects have not been considered.

Socioeconomic externalities considered were those due to the employment generated— either direct or indirect. Direct employment refers to jobs needed to build the plants, operate them and produce the fuel. Indirect employment is that created in other sectors of the economy because of the demand for goods and services generated by the project. Direct employment has been estimated based on the ratios obtained in previous analysis of a biomass scheme in Spain (CIEMAT, 1996; Varela et al., 1999). Results for the whole life cycle of the scheme proposed are 1071 man/year in the agricultural sector and 2052 in other sectors of the economy. Indirect employment was calculated based on the results previously obtained for a biomass power plant in Spain using input/output tables of the 1990 Spanish economy. This indirect employment has been estimated in 2592 man/year.

Positive externalities from job creation arise only when the workers were previously unemployed and would remain so otherwise. When unemployment rates are higher than the natural rate (around 5%) social benefits are produced by job creation. In order to evaluate in monetary terms the benefits of job creation, the reduction in the amount of unemployment subsidies paid by the government has been used. In the Spanish agricultural sector, the unemployment rate is very high, around 45% of the active population. Therefore, it is reasonable to assume that all workers employed by the scheme proposed in the agricultural sector will come from the unemployment pool. However, in this sector unemployment subsidies exist only in two regions of Spain. The number of direct jobs created in these regions has been estimated at 321 and each job created saves $1638/yr to the government.

Employment generated in the other sectors of the economy, both direct and indirect, are 4644 man/year. In these sectors employment subsidies exist under certain conditions. Since most of the sectors have unemployment rates above 5% (CIEMAT, 1996), it is assumed that all of the workers were receiving these subsidies that range from 424 to 1243 $/month.

4. ECONOMIC ANALYSIS

Certain financial assumptions are part of the economic analysis of the overall biomass cycle studied. The main assumptions are as follows (all costs are expressed in 1997 values).

- The power plant size was limited to a minimum of 15MW.
- The design and construction period of a biomass power plant was established as four years and the capital payments during construction were scheduled as 10%, 50%, 30% and 10% during the four years.
- The useful lifetime of the biomass power plants for electricity generation was set at thirty years, with two fifteen-year crop cycles.
- The load factor for power plants was set at 85% of their rated capacity (7,500 h/y).
- All capital requirements during the biomass cycle were treated as equity and the annual cost of equity was set at the same rate used to discount the annual cash-flows of the project. A general discount rate of 5% was applied.
- During the one-year commissioning period no incomes from electricity sales were considered. The fuel storage must supply the plant for 15 days. The costs associated with the shutdown of power plants have been calculated considering that their salvage value could be 1% of the initial capital investment and that it is equivalent to the decommissioning costs.

- A 1% and 2% share of the initial investment have been estimated for insurance costs and fee payments of the biomass power plants, respectively, and 8%, 5% and 2% of the initial investment have been estimated for contingence requirement, routine and maintenance and operating labour costs, each in the order given.

Investment costs were also estimated, and a final analysis of the conversion stage (i.e., power generation and distribution, decommissioning) was included in a specific global worksheet, taking into consideration the intrinsic variations between different locations. The model used in the economic analysis is based on "discounted cash-flow analysis" (IEA, 1991). The net present value (NPV), revenue requirements (RR), levelized energy costs (LEC) and levelized required revenues (LRR) were used as indicators. For FBC, the specific cost ranges from 1947 to 1589 $/kW considering the size interval of 25-60 MW. For BIGCC, these costs correspond to 1877-1792 $/kW in a size range of 30-35 MW.

The average electricity costs, considering all plants, would be 8.91 c$/kWh. If the percentage of participation of FBC and BIGCC technologies in the whole biomass scheme is considered, a weighted average cost of electricity of 8.93 c$/kWh is obtained. The main participant to the energy cost is the fuel cost (50.7%), followed by capital cost and O&M cost.

5. CONCLUSIONS

The study evaluated the integration of energy crops on a large-scale within an energy supply system, the benefits obtained in economic terms, and the effects from the environmental point of view. The scheme proposed considers an additional power of 1GW$_e$. Weighted average results of total costs obtained considering the proposed participation of both types of plants (FBC and BIGCC) (Table 1). Direct electricity costs calculated for the whole scheme proposed were 8.93 c$/kWh, environmental external costs ranged from 1.04 to 1.31 c$/kWh if sequestration of CO_2 is not considered and from –0.46 to 0.67 c$/kWh if CO_2 sequestration by roots is admitted, and socioeconomic externalities from –0.93 to –0.32 c$/kWh. The scheme is therefore feasible from an economic point of view, and it is very favourable from both the environmental point of view (especially when sequestration of CO_2 is included) and from the socioeconomic point of view.

Table 1
Total costs of electricity in the biomass scheme proposed

	c$/kWh
Direct costs	
Fuel costs	4.53
O&M costs	1.88
Capital costs	2.52
Total direct costs	8.93
Environmental external costs	
Atmospheric emissions	1.01 to 1.22
Global warming effects without CO_2 sequestration	0.03 to 0.06
Global warming effects with CO_2 sequestration	-0.58 to -1.48
Erosion	0.004 to 0.010
Non-point source pollution	0.002 to 0.033
Subtotal environmental externalities without CO_2 sequestration	1.04 to 1.31
Subtotal environmental externalities with CO_2 sequestration	-0.46 to 0.67
Socioeconomic external costs	
Employment effects	-0.93 to -0.32
Total costs without CO_2 sequestration	9.04 to 9,92
Total costs with CO_2 sequestration	7.54 to 9.28

REFERENCES

1. EC (1995). ExternE: Externalities of energy, Vol.1. Summary, EUR 16520 EN.
2. EC (1998). ExternE: Externalities of energy, Vol 7. Methodology, 1998 Update (in press)
3. CIEMAT (1996). Assessment of the externalities of biomass energy for electricity production.
4. CIEMAT (1998). ExternE National Implementation Spain. Colección Informes Técnicos. EUR 18269.
5. IEA (1991). Guidelines for the economic analysis of renewable energy technology applications. OECD/IEA, Paris, France.
6. Varela, M., R. Sáez, and H. Audus (1998). Large scale economic integration of electricity from short rotation woody crops, Solar Energy Paper 98-04-24-0050R (in press).
7. Kramer, P.J. and J.S. Boyer (1995). Water relations of plants and soils. Academic Press, 385 p.
8. Lodhijal, L.S. and R.P. Singh (1994). Productivity and nutrient cycling in poplar stands in central Himalaya, India, Can. J. Far. Res., 24, pp. 1199–1209.
9. San Miguel, A. and J.M. Montoya (1983). Resultados de los primeros cinco años de producción de tallares de chopo en rotación corta (2-5 años). Anales del INIA (Serie Forestal), 8, pp. 73–91.

437

GREENHOUSE GAS EMISSIONS OF BIOENERGY SYSTEMS COMPARED TO FOSSIL ENERGY SYSTEMS

Methodology and First Results

G. Jungmeier, L. Canella, and J. Spitzer

JOANNEUM RESEARCH, Institute of Energy Research, Elisabethstrasse 5, A-8010 Graz, Austria.

Increasing the use of bioenergy is a promising way to reduce greenhouse gas emissions. Hence it is important to know the greenhouse gas emissions of bioenergy systems in comparison with fossil fuel systems. A life cycle analysis of biomass and fossil fuel energy systems compared the overall greenhouse gas emission of both systems for heat and electricity supply. Different bioenergy systems to supply electricity and heat from various biomass sources were analysed for Austria in year 2000 and 2020. Total emissions of greenhouse gases (CO_2, N_2O, CH_4) along the fuel chain, including land use change and by-products, are calculated. Various conversion technologies and various fuels from forestry, agriculture and industry were considered. The methodology was developed within the IEA Bioenergy Task 25 on "Greenhouse Gas Balances of Bioenergy Systems" and is orientated on ISO 14 040. In this paper the results of selected bioenergy heat supply systems are shown as emission of CO_2-equivalents per kWh heat for bioenergy systems as compared with fossil fuel systems, and as a percentage of CO_2-equivalent reduction. The results demonstrate that some bioenergy systems already reduce greenhouse gas emissions because of avoided emissions of the reference biomass use or because of certain substitution effects of by-products. In general the greenhouse gas emissions of bioenergy systems are lower compared to the fossil systems. Therefore a significant reduction of greenhouse gases is possible by replacing fossil energy systems with bioenergy systems. This comparison may help policy makers, utilities and industry to identify effective biomass options in order to reach emission reduction targets.

1. GOAL AND SCOPE

1.1. Goal of the study

The goal of the study was to analyse different bioenergy systems that produce electricity and heat supply from various biomass sources. Emissions of greenhouse gases over the entire life cycle, including land use change and by-products, were calculated. The results are compared with those of fossil energy systems. Because of the increased use of bioenergy as one promising way to reduce greenhouse gas emissions, this comparison may help policy makers, utilities and industry identify effective uses for biomass in order to reach emission reduction targets.

1.2. Scope of the study

A life cycle analysis of biomass and fossil fuel energy systems compared the overall greenhouse gas emissions (CO_2, N_2O, CH_4) for heat and electricity supply systems. The systems considered for the Austria in the year 2000 and 2020 were combinations of biomass production, processing, fuels and conversion (Figure 1). In this paper selected bioenergy systems for heat supply are compared with oil, natural gas and coal for the year 2000.

The methodology used, developed mainly within the IEA Bioenergy Task "Greenhouse Gas Balances of Bioenergy Systems"[1] compared greenhouse gas balances of bioenergy with those of fossil energy systems (Figure 2) by considering the following: time period and changes of carbon stocks; reference energy systems; energy inputs required to produce, process and transport fuels; mass and energy losses; energy embodied in the infrastructure; distribution systems; cogeneration systems; by-products; waste wood and other biomass waste for energy; reference land use and other environmental issues. In this paper the above methodology is applied to selected systems for heat supply.

1.3. Function and functional unit

The function of the energy systems is to supply heat to the consumer at a temperature of 35-60°C for space heating. The functional unit is the greenhouse gas emissions of CO_2, CH_4 and N_2O in CO_2-Eq. for 100 yr/kWh of heat (g CO_2-eq.$_{100}$/kWh$_{heat}$).

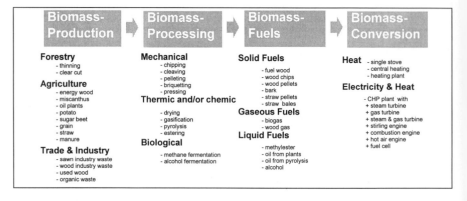

Figure 1. Elements of bioenergy systems.

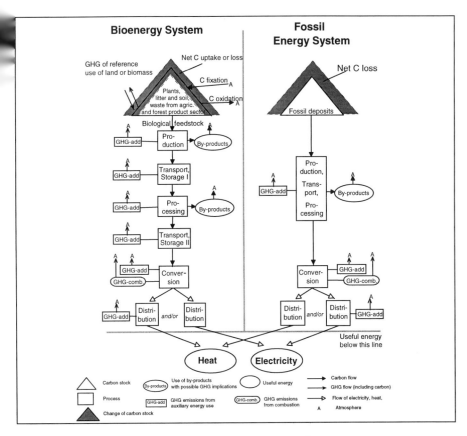

Figure 2. Comparison of bioenergy and fossil energy systems.[1]

1.4. System boundaries

The system boundaries include the entire life cycle including by-products, land use change and the reference use of biomass.

1.5. Data quality requirements

Data are considered to be "default values" for best average calculation for Austria in the year 2000. The data quality is proven with plausibility criteria.

1.6. Comparison between systems

Bioenergy systems are compared with fossil energy systems with the same functional unit for heat supply in Austria in the year 2000, and the percentage of greenhouse gas reduction is calculated.

1.7. Critical review considerations

For a critical review the project is attended by a national advisory board of Austrian experts and by members of IEA Bioenergy Task 25.

2. INVENTORY ANALYSIS

2.1. General description of the life cycle inventory

For space heating, supply data for each stage and process of the entire life cycle were collected, especially for the bioenergy systems. For the fossil fuel systems many data were taken from the Austrian database contained in TEMIS-A (total-emission model of integrated systems — the Austrian situation).[2] The life cycle inventory covered the greenhouse gas emissions of all stages throughout the energy system's life from raw material acquisition to end of life (recycling, landfill or incineration of materials).

2.2. Data collection and calculation procedures

The data were collected from literature and as far as possible from practical experience. The data calculation procedure is done with TEMIS 3.1.[3] and an additional computer tool for the formatting of the results. Table 1 shows the inventory analysis of greenhouse gas emissions for selected heating systems.

3. IMPACT ASSESSMENT

The inventory analysis is used for impact assessment by using the Global Warming Potential for 100 years (1 CO_2 = 1 CO_2-Eq.$_{100}$, 1 CH_4 = 21 CO_2-Eq.$_{100}$, 1 N_2O = 310 CO_2-Eq.$_{100}$[4]) to calculate the cumulative effect on global warming (Table 1, Figure 3), details for biogas from cow manure in a central heating system are shown in Figure 4.

4. INTERPRETATION

The reduction of greenhouse gas emissions from bioenergy systems as compared with fossil fuel systems based on the emission of CO_2-equivalents during the entire life cycle per kWh heat are shown in Table 2 and Table 3. Depending on the systems compared, greenhouse gases can be greatly reduced by replacing fossil with bioenergy systems in the

Table 1
Inventory analysis of greenhouse gas emissions for selected heating systems

	CO_2	CH_4	N_2O	CO_2	CH_4	N_2O	CO_2-eq.
	[g/kWh$_{heat}$]			[g CO_2-eq.$_{·100}$/kWh$_{heat}$]			
wood chips CH	29	0,12	0,017	29	2,5	5,2	36
wood chips DH	45	0,17	0,027	45	3,6	8,2	56
fuelwood spruce ST	16	0,13	0,044	16	2,7	14	32
fuelwood beech CH	25	0,15	0,017	25	3,1	5,4	33
wood chips poplar CH	51	0,16	0,047	51	3,4	15	69
wood chips willow CH	45	0,15	0,047	45	3,2	15	63
rape seed oil CH [1]	113	0,29	0,36	113	6,1	111	231
rape methylester CH [1]	26	-10	0,36	26	-216	111	-79
biogas cow manure CH [1,2]	26	-29	-0,10	26	-617	-32	-624
biogas co-digestion CH [1,2]	24	-6,7	-0,015	24	-141	-4,5	-122
coal briquetts CH	673	3,8	0,019	673	80	5,8	759
lignite briquetts CH	809	0,30	0,076	809	6,4	24	839
extra light oil CH	397	0,13	0,0055	397	2,6	1,7	401
natural gas CH	251	2,1	0,0049	251	44	1,5	297
natural gas DH	330	1,7	0,0035	330	36	1,1	367
natural gas electr. HP	129	0,67	0,0006	129	14	0,20	143

ST....stove, CH.....central heating, DH....district heating, HP.....heat pump
1) includes the GHG benefits from the use of by-products
2) includes the avoidance of methane from manure storage

year 2000, the tables show which of the selected biomass heating systems will maximally reduce greenhouse gases.

The results of the life-cycle greenhouse gas emissions, in this case for heat supply systems, demonstrate that some bioenergy systems have no net emissions of greenhouse gas or are even associated with "negative" emissions, as shown in Figure 3 for biogas and methylester. In the case of biogas this is because emissions from the reference biomass use are avoided (the reference use of manure is storing the manure—associated with uncontrolled emissions of methane). In the case of methylester it is due to substitution effects of by-products (for example, by-products of methylester are glycerine that replaces conventionally produced glycerine for chemical use, and rape cake that replaces soybean feed).

A comparison of bioenergy systems with fossil energy systems shows that a significant reduction of greenhouse gas emissions is predicted in all possible combinations of bioenergy and fossil energy systems.

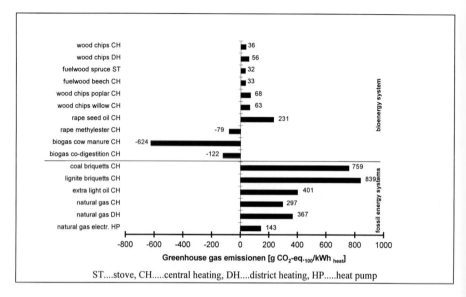

Figure 3. Impact assessment of selected heating systems on global warming.

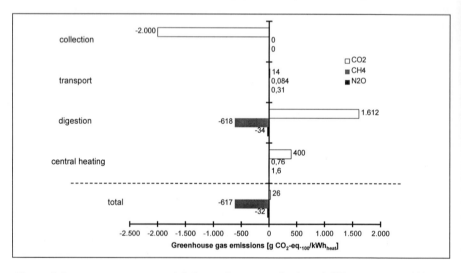

Figure 4. Impact assessment on global warming – contribution of different stages of biogas from cow manure for central heating.

Table 2
Differences between greenhouse gas emissions of bioenergy and fossil energy systems (g CO_2-eq.$_{100}$/kWh$_{heat}$)

	coal briquetts CH	lignite briquetts CH	extra light oil CH	natural gas CH	natural gas DH	natural gas electr. HP
wood chips CH	-722	-803	-365	-260	-330	-107
wood chips DH	-702	-783	-345	-241	-310	-87
fuelwood spruce ST	-726	-806	-369	-264	-334	-111
fuelwood beech CH	-726	-806	-368	-264	-334	-110
wood chips poplar CH	-690	-770	-333	-228	-298	-75
wood chips willow CH	-696	-776	-338	-234	-304	-80
rape seed oil CH [1]	-528	-608	-171	-66	-136	87
rape methyelster CH [1]	-838	-918	-481	-376	-446	-222
biogas cow manure CH [1, 2]	-1.380	-1.460	-1.030	-921	-990	-767
biogas co-digestition CH [1, 2]	-880	-961	-523	-419	-488	-265

ST....stove, CH.....central heating, DH....district heating, HP.....heat pump
1) includes the GHG benefits from the use of by-products
2) includes the avoidance of methane from manure storage

Table 3
Reduction of greenhouse gas emissions from heat supply using bioenergy instead of fossil energy systems (percentage-reduction based on fossil emissions)

	coal briquetts CH	lignite briquetts CH	extra light oil CH	natural gas CH	natural gas DH	natural gas electr. HP
wood chips CH	-95%	-96%	-91%	-88%	-90%	-75%
wood chips DH	-93%	-93%	-86%	-81%	-85%	-61%
fuelwood spruce ST	-96%	-96%	-92%	-89%	-91%	-77%
fuelwood beech CH	-96%	-96%	-92%	-89%	-91%	-77%
wood chips poplar CH	-91%	-92%	-83%	-77%	-81%	-52%
wood chips willow CH	-92%	-92%	-84%	-79%	-83%	-56%
rape seed oil CH [1]	-70%	-73%	-43%	-22%	-37%	61%
rape methyelster CH [1]	-110%	-109%	-120%	-127%	-122%	-155%
biogas cow manure CH [1, 2]	-182%	-174%	-255%	-310%	-270%	-536%
biogas co-digestition CH [1, 2]	-116%	-115%	-130%	-141%	-133%	-185%

ST....stove, CH.....central heating, DH....district heating, HP.....heat pump
1) includes the GHG benefits from the use of by-products
2) includes the avoidance of methane from manure storage

5. ACKNOWLEDGEMENTS

The work for this paper was mainly carried out within the project "Greenhouse Gas Balances of Bioenergy Systems in Comparison to Fossil Fuel Systems" funded by Austrian Federal (Ministries of Economics, Science and Environment) and State research funds, the Austrian Power Association (VEOE) and the Austrian oil and gas company (OMV).

REFERENCES

1. Schlamadinger, B., M. Apps, F. Bohlin, L. Gustavsson, G. Jungmeier, G. Marland, K. Pingoud, and I. Savolainen (1997). Towards a standard methodology for greenhouse gas balances of bioenergy systems in comparison with fossil energy systems, Biomass and Bioenergy, 13.
2. TEMIS-A: Total Emission Model of Integrated Systems—Austria (1998). Austrian Environmental Agency, Vienna.
3. TEMIS 3.1: Total Emission Model of Integrated Systems, developed by the Ökoinstitut Darmstadt, http://www.oeko.de/service/gemis.
4. Climate Change 1995 - The Science of Climate Change, ed. by J.T. Houghton, L.G. Meira Filho, B.A. Callander, N. Harris, A. Kattenberg, and N. Maskell (1995). Contribution to WGI to the second Assessment Report of the Intergovernmental Panel on Climate Change, Cambridge University Press.

445

THE NEED FOR AND THE BENEFITS OF A DOMESTIC PROTOCOL TO APPROVE AND ACCREDIT GHG EMISSION OFFSETS FROM U.S. PROJECTS

Charles E. Parker III, Principal, JI Services

The International Series of Conference of Parties (COPs) is moving towards establishment and ratification (by governments) of a mandatory cap (by country) on GHG emissions by the year 2008. A vehicle (Protocol, Procedures, Requirements & Guidelines, and an Approving Body) exists for approving and accrediting GHG emission offsets from international projects. This will shortly be split into two protocols, one for developing nations and one for developed nations. No such vehicle or protocol exists for domestic projects. GHG Emission Offsets from domestic projects cannot be approved or accredited and these offsets cannot be counted towards meeting caps nor certified for trading or sale. American companies are at a huge disadvantage compared with foreign counterparts. A domestic protocol, enabling legislation, a governing body and incentive for action ("Early Credits for Early Action") are all needed together to redress this situation.

A recent letter from the California Biomass Energy Alliance to the Western Regional Biomass Energy Program of U.S. DOE identifies a grave risk facing the biomass energy industry. Deregulation of the electric utility industry has brought about the loss of a crucial federal subsidy of the price of biomass-generated electricity. Without this support, all existing biomass will become uneconomic and be forced to shut down. Appendix 1 shows how the projected value of a ton of offsets (by Council of Economic Advisors) will offset the loss of the subsidy within five years' time.

1. SUMMARY OF THE PROJECT

The summary of the project includes the steps we will take for each segment of the process in establishing the domestic protocol.
1. Define and construct a draft JI Proposal for a first targeted biomass project, to be submitted under the new domestic protocol, once it and the governing body are established and functioning.
2. Structure and detail the governing body charter, procedures, process, guidelines for proposers, philosophy, steps in approval and accreditation, staffing and budget for operation.
3. Construct the protocol and procedures for:
   a. depositing and registering approved and accredited offsets in an offsets facility,
   b. monitoring, tracking and verifying deposits and offsets in their own accounts,

c. certifying and then transacting Certified Tradable Offsets (CTO's) in the marketplace.
4. Assist in the creation of enabling legislation for the protocol, the governing body and 'Early Credit for Early Action' by working to educate and convince a critical mass of senators (12) to back S.2617.
5. Pass into Phase II, where appropriations are distributed, the governing body is staffed and goes on line and a stream of domestic proposals is prepared and submitted for approval and accreditation.

2. PROCESS

We will use available models to construct the process the procedures, the guidelines, the requirements, the standards or criteria of judgement, the agreements necessary to implement the process, and the structure, makeup and rules which govern the approving body. The models we will use for this make up the resource documents for project and proposal development, application submittal and procedures for responding to the evaluation process of the United States Initiative on Joint Implementation (USIJI) for approving and accrediting international project GHG emission offsets. We will also use the Environmental Defense Fund (EDF) papers, documents, treatises, and outline for creating a global GHG emissions trading market (including domestic) plus the Hopp & Associates creation of the protocols, procedures and agreements for registering, depositing, monitoring, tracking, verifying, certifying and transacting international project GHG emission offsets for the International Offsets Facility, LLC.

The use of the above models will be in creating the domestic protocol for approving and accrediting domestic GHG emission offsets from U.S. projects and in creating the domestic Offsets Facility, LLC for depositing, registering, monitoring, tracking, verifying, certifying and transacting domestic project GHG emission offsets. These creations will proceed in concert with and support the Senate in their creation of enabling legislation to allow the administration to give 'Early Credit for Early Action'.

3. TECHNICAL DESCRIPTION AND FEASIBILITY

3.1. Project design

This project will follow the existing market commercialization pathway of setting up the International Offsets Facility LLC (IOF) to handle GHG emission offsets from international projects approved and accredited by the USIJI, except that in this case, the projects will all be domestic and be located in the U.S.A. As in the international example, this domestic protocol will provide for the registering and depositing of the offsets in a Facility, endorsed by the Congress and meeting the criteria of the Department of Energy. The Facility will entertain projects seeking 'Early Credit for Early Action' in its initial years. The Facility will use an offsets tracking system, modified from the IOF model and as structured and proposed by the Environmental Defense Fund. It will adopt a certification process, suggested by the IOF model, but sufficiently modified so as not to be infringing. Transactions of offsets, whether they be simple sales, pooled sales, trades or holding for appreciation or later internal use, will similarly follow the guidelines as set up by the IOF (but not too closely). Guidelines and Procedures, as set up in the USIJI publications "USIJI Sept. 1998—Reducing Greenhouse Gas Emissions Through International Partnerships" and "Resource Document on Proposal and Project Development under USIJI" will similarly be formulated for the domestic protocol. On the separate recommendations of the Department of Energy's Offices of Fossil Energy and of Energy Efficiency/Renewable Energy, the Federal Energy Technology Center (FETC), the White House Climate Challenge Office, and the EPA, we will propose that the protocol be administered by and be under the governance of the U.S. DOE, Climate Challenge Program, Larry Mansueti, Director. This entity, when picked, will be endowed with all of the structure, procedures for supplicants, criteria for evaluation, and other necessary ingredients of the USIJI, to function as it does, on domestic projects. In addition, the concept of 'Early Credit for Early Action' will be supported and encouraged as Senate Bill S.2617, to passage and fruition, as the enabling legislation and necessary support for the Administration's adoption and employment of this incentivizing reward to parties that act now.

3.2. Project plan

The project will construct the following deliverables for the Domestic Protocol:
1. The procedures and necessary agreements for proposal preparation, submittal and process formation.
2. The criteria and their application for approval and accreditation.
3. The procedures and agreements for registering and depositing offsets.
4. The protocol and procedures for monitoring, tracking, verifying, certifying and conducting transactions of the offsets.
5. Establishing a charter, operating procedures, the guiding philosophy of, and the necessary staffing for, the governing body.
6. Drafting the first domestic project (biomass) for submittal under the domestic protocol established by this project. This will be the California Biomass Alliance Project. It, and all other Western biomass projects brought forward during the conduct of the subject proposed project, will be given first priority of proposal preparation for submittal to the governing body for its accreditation, under the protocol, once the protocol is approved and enabling legislation is passed to empower the Executive Branch to establish the process and the governing body.

3.3. Milestone dates for completion of tasks

1. Formulate proposal preparation, submittal and process procedures by the end of the third month of the project.
2. Define the criteria for proposal approval and accreditation by the governing body by the end of the sixth month of the project.
3. Establish the procedures and agreements for registering and depositing offsets by the end of the third month of the project.

4. Construct, review and refine (with the cooperation and the participation of EDF) the protocol and procedures for monitoring, tracking and verifying the offsets by the end of the fifth month of the project.
5. Construct the protocol and procedures for the various transactions possible with the offsets (solo sales, pooled sales, holding for appreciation or internal application, and trading) by the end of the fifth month of the project.
6. Establish the charter, operating procedures and guidelines, etc., for the governing body's setup by the end of the sixth month of the project.
7. Draft the first submittal of the first domestic project under domestic protocol by the end of the sixth month of the project. Subsequent answers to questions, modifications, resubmittals and ensuing discussion and negotiations with the governing body (leading to approval and accreditation) will be conducted under Phase II of the project and will last at least six months time.

Phase II of the project will last six to twelve months and will be composed of the following activities:

1. Lobby and persuade the Senate to support the protocol, as well as the Senate Bill S.2617 "Early Credit for Early Action" together with an FY00 Appropriation in the Energy & Water Bill (106th Congress 1st Session) of $1.5 to 2.0 million (tbd) to the Climate Challenge Program.
2. Outline and oversee the creation of the department (as a governing body) with the appropriate staff, protocol, procedures, process, review schedule and qualifying criteria.
3. Set up and run the first projects through the Climate Challenge Program, including at least the following fields of application.
   a. Biomass project life extensions and new additions.
   b. Separation of carbon from flyash and substitution of the ash for cement in making concrete.
   c. Substitution of hydraulic for Portland cement in concrete.
   d. Waste heat recovery and utilization in replacement of combustion of solid fuels.
   e. Coal cleaning, slurrying and pipeline delivery for clean solid fuel generation.
   f. Coalbed methane recovery and generation of power on site from the methane gas.

4. CAPABILITIES AND EXPERIENCE OF OUR TEAM

Skills of key individuals needed to accomplish the domestic protocol.

Robert Tamaro—Project Developer, Project Manager and Consultant, Small Business Manager, Engineer, Corporate Executive with degrees in finance and engineering.

Charles Parker—Product and Project Developer, Small Business Manager, Consultant, Corporate Executive with degrees in finance and engineering, extensive background in marketing.

Rachel Hopp—Environmentalist, Lawyer with expertise in governmental contracts - Environmental and General Law, expert in offsets trading.

Experience—Messrs. Tamaro and Parker have worked together on the preparation, submittal and negotiation of proposals for international projects achieving GHG emission reductions under the USIJI International Protocol. Countries have included the Philippines, Brazil, Poland, China, India and England. Mr. Parker and Ms. Hopp have worked together on preparation of all the procedures, process, documents, agreements and the international protocol for registering, depositing, monitoring, tracking, verifying, certifying and transacting the international offsets. They have also conducted discussions on setting up a comparable domestic protocol in the future.

The principals have a strong background in power plants (boilers, turbines and pollution control systems), the utility industry, international projects, various forms of financing, contracting for and owning capital equipment and systems. The team has worked with the Department of Energy, the EPA, and the Departments of State, Commerce and Treasury. Members have been involved in development, completion and operation of approximately twenty energy efficiency and renewable energy projects. In addition, members have prepared over fifty feasibility studies of projects involving these technologies. Members have been involved in successful USIJI Projects in forestation in Bolivia, biomass power in Central America and in a nearly completed rice hulls project in the Philippines.

5. DOMESTIC PROTOCOL SUPPORT

The following organizations and individuals have indicated their active support.

5.1. Pew Center on Climate Change

Asea Brown Boveri	Air Products	American Electric Power
Baxter International	Boeing	BP Amoco
CH2M Hill	DuPont	Euron
Entergy	Holnam, Inc.	Intercontinental Energy
Intn'l Paper	Lockheed Martin	Maytag
Shell Intnl	The Sun Company	Toyota
United Technols	U.S. Generating Co.	Weyerhauser
Whirlpool		

5.2. Trade associations

Cement	APCA
Coal	National Coal Council & World Coal Institute
Biomass	National Bio Energy Industrial Association
Elec Utility	Edison Electric Institute

5.3. National environmental organizations
Environmental Defense Fund
Van Ness Feldman
ENSR

5.4. Governmental (executive branch) bodies
DOE
EPA

5.5. U.S. Senate
Sens.: Chafee, Mack, Lieberman, Kerrey, Lugar, Kerry (more tbd)

6. POSITIVE IMPACTS OF THE PROJECT

The creation of the domestic protocol will slow, stop and then reverse the shutdown of biomass projects in California and the West by replacing the lost government subsidy of electricity price with the ability to sell (annually) the tons of certified offsets accredited under the domestic protocol. It will advance the objectives of the biomass industry.

1. Activities that reduce the cost of biofuels or biopower. By supplanting the government's subsidy of the price of biomass power with the sale or valuation of the GHG emission reductions, the protocol keeps alive biomass projects which would otherwise shut down and, at the same time, countenances the reduction in cost of power to consumers from these projects.

2. Encourage economic development through public and private sector investments in biomass energy technologies. By creating the vehicle for realization of domestic GHG emission reduction offsets, this project and effort will directly spur the final commercialization of the EUA BIOTEN technology and the TAZCOGEN proprietary system improvement, which makes the technology competitive with any other form of power generation. Then, because it's 'green,' it will then become the system of choice for technology application.

3. Activities that reduce or eliminate market barriers. By accomplishing objectives (1) and (2) above, this project will allow the California Biomass Alliance and its associates to keep existing projects open and to, in fact, open up new ones.

4. It will result in expanded use or reduction of negative impacts. Letters from the California Biomass Alliance state that, without the federal government subsidy of their price, they will be forced too shut down. This represents $.015/kWh and pushes them below the break-even line. When the realization of the GHG emission offsets valuation is applied to the price they charge, the following are the results:

Today—informal price of GHG offset of $3 to 4/ton results in an equivalent price support of $.003/kWh.

5 Years—Council of Economic Advisors projection of $14 to 23/ton results in an equivalent price support of $.009 to .0146/kWh.

5. Price will expand jobs, with a positive impact on economic development. As a result of this project and the Domestic Protocol, biomass projects will not shut down and jobs will not be lost with all the attendant impacts on economic development.

REFERENCES

Cooperative Mechanisms and the Kyoto Protocol--The Path Forward, by the Environmental Defense Fund.
Council of Economic Advisors July 1998 Paper The Kyoto Protocol—Administration Economic Analysis.
Early Action and Global Climate Change, Bob Nordhaus of Van Ness Feldman.
Market Mechansims and Global Climate Change, Dan Dudek and Joe Goffman of EDF.
Overview of International Offsets Facility, Presentation by C.E. Parker and R.M. Hopp.
Paper—Potential Benefits of Biomass Energy Conversion Systems by California Energy Commission.
Paper—Why a Domestic Protocol for GHG Emission Credits is Necessary to Save the Biomass Industry, Biomass Power Industry Information Sheets.
Pew Center for Climate Change - Papers.
Procedures for Registration, Deposit and Transfer of Domestic Offsets and CTO's by Industrial Offsets Facility.
Senate Bill S.2617 'Early Credit for Early Action', sponsored by Senators Chafee, Mack and Liberman.
The International Offsets Facility Flyer.
The JI Services Flyer and accompanying memo Reasons for Offset Accreditation.
USIJI Brochure on their International Program.

Appendix A

Table 1
Value of Offsets

Emissions Rate Tons/Y	/MW 5000	/kW 5
Tons/Hr @ 90% CF	0.63419584	0.0006347
Value Based on $/Ton	$/MWh	$/kWh
14	$8.879	$0.0089
15	9.513	0.0095
16	10.147	0.0101
17	10.781	0.0108
18	11.415	0.0114
19	12.050	0.0120
20	12.884	0.0127
21	13.318	0.0133
22	13.952	0.0140
23	14.587	0.0146

Biomass Production and Integration With Conversion Technologies

Systems Integration

455

EKOKRAFT™ PROGRAM—MUNICIPALITY OF HEDEMORA, SWEDEN

Mats O. Wilstrand

Salix Maskiner AB, Hamre 1, S – 776 90 Hedemora, Sweden

Short rotation coppices (SRC), when applied to the solution of water and soil-related environmental problems, generate unique added value that no other energy source can achieve. The fundamental idea behind EKOKRAFT™, a US$ 30 million economically and ecologically sustainable development program started in 1998 and run by the Municipality of Hedemora, is the conversion of "glocal" environmental problems into local bioenergy assets. The program is based on the full-scale implementation of newly developed SRC cultivation methods, willow clones, harvesting technology and logistics. The results expected are primarily minimising of eutrophication on watercourses the improvement of the biomass energy balance at lower biofuel production costs, the creation of new high-quality biofuels and the further reduction of CO_2 emissions. The program is divided into various main projects: KOMBIBRUK™ is a new method for sustainable food and energy farming; use of urban and industrial effluents as a nutritional resource for willow production; the establishment of a biomass cover on today's sterile mining soil as a new biofuel source; and installation of two new district heating systems and the conversion of 15 small oil-fuelled boilers, all to be fuelled with a new SRC-based biofuel, EKOPELLETS™. The program is also intended as a model for replication. An EKOKRAFT™ international network is established.
Español o Portugués: Este resumen o artículo entero, podrá usted recibirlos en su idioma.

1. BACKGROUND

1.1. Global concern: from Kyoto to Hedemora 1998

After Kyoto 1998, the Commission of the European Union (EU) set up as its prime objective the doubling of renewable energy sources by 2010,[1] a process in which biomass is fully expected to play a major role. Today, member countries are working out and planning their own development programs. The goal is the production from biomass of 500 TWh/a in the EU within approximately ten years.

In Sweden there is an old tradition of using biomass for energy purposes. The bio-energy sector has dramatically expanded since 1986 when eco-taxation of fossil fuels came into force. In 1996, the Swedish Government stated as a prime objective that: The most important environmental problems must be solved within a generation. In August 1997, gradual replacement of nuclear power began. In 1998, the Government allocated US$ 1.12 billion to an energy policy program[2] to change to a sustainable energy system. US$ 675 million (of a planned total of US$ 1 billion) have also been allocated to support local investment programs and to help municipalities extend their use of district heating. The grants also aid the change to renewable energy sources, clean up contaminated soil, and reduce the discharge of nutritive salts into the sea. Swedish government grants normally pay up to one third of a project's total investment costs. The size of these environmental investments must be related to the size of the Swedish population: 9.2 million.

1.2. Turning "Glocal" environmental problems into local assets

Within the framework of EU and Swedish policies and goals and particular local conditions, Hedemora mounted an economically and ecologically sustainable development and demonstration program called EKOKRAFT™ (Ecopower).

The Municipality of Hedemora, which has a population of 16,000 and a long tradition of mining, industry, forestry and agriculture, decided to invest in the future. To help launch this US$ 30-million program, the Swedish government approved a grant of US$ 8 million in April 1998.[3] Investments will be scheduled over a four-year period.

The fundamental idea behind EKOKRAFT™ is the conversion of "glocal" environmental problems into local assets. Of course, it is of vital importance that politicians, civil servants, administrators, companies, developers, researchers, financial institutions and associations interact—think "globally" and work "locally," hence "glocally."[4] As a natural consequence of this school of thought and action, EKOKRAFT™ is a very concrete local investment program for Hedemora while also serving as a model for economically and ecologically sustainable development in other regions. An international network called the EKOKRAFT™ FORCE has been established. Players dealing with environmental and bio-energy projects are welcome to join the network![5]

1.3. Latest biologic and technical R&D—the basis for biomass take-off

Replacing fossil fuels with biomass leads to global benefits such as reduced emissions of greenhouse gases; this is a well-known fact. Biomass has another, unique, characteristic: when a plantation is applied to the solution of water or soil-related environmental problems it generates added value that no other energy source can achieve.

Agriculture is reported to be responsible for about half the total volume of nitrogen and other nutrient salts discharged to seas causing eutrophication and a number of problems for the maritime eco-systems.[6] Human urban and industrial activity produces the other half. The solution to this and other serious environmental problems opens up a major opportunity for the biomass sector.

A national survey on the potential environmental and economic benefits of short rotation coppices in ten applications was conducted on behalf of the biggest power utility in Sweden.[7] The most important environmental improvements can be achieved when energy crops are used as vegetation filters for the treatment of municipal waste water, as buffer strips for the interception of fertiliser run-off from food crops to watercourses and as plantations for the recirculation of sewage sludge.

The reported environmental values for energy crops can in some cases rise to the same level as the production costs for willow biofuel, which is, with conventional direct chipping harvesting technology, around 13 US$/MWh.[8] Moreover, if new cultivation methods, *Salix* clones, direct bundling harvesting technology and logistics as well as a re-organisation of the *Salix* business are applied the expected production cost of biofuel is estimated at 8–10 US$/MWh.[9] These constitute an ideal situation for the biomass sector.
- With environmental values higher than production costs, the competitiveness of bio-fuels as against fossil fuels increases dramatically.
- A short-rotation coppices-based energy sector can be started even in countries exempted from eco-taxation of fossil fuels (the vast majority of countries in the world).

In this context, EKOKRAFT™ is an important full-scale investment and demonstration program.

2. INTRODUCTION TO EKOKRAFT™ PROGRAM

The main material flows resulting from both nature and interrelated human activities such as urban life, agriculture and industry must be re-directed. Today, this is technically and economically feasible.[10] As an example, the brief analysis of the Hedemora program shows that the main material flows causing environmental problems are these:
- urban and industrial sewage: Effluent is discharged into river, unbalancing the aquatic ecosystem.
- agricultural fertiliser: Surplus fertiliser from food crops drains into watercourses, unbalancing the ecosystem.
- mining soils: Big areas leak heavy metals into groundwater and surface water.
- pulp and paper mill wastes: The need to dispose of very large volumes of mainly fibre sludge and furnace ashes.

The re-direction of these flows to strategically placed biomass plantations will have these results:
+ a durable solution to the above-mentioned problems in the medium and long terms;
+ local production of high-quality biofuels for the substitution of fossil fuels and nuclear power at lower costs than today;
+ higher production and consumption of locally produced agricultural food and energy crops as a result of increased awareness among consumers of improved quality, leading to less transportation;
+ local job creation and higher income or less expenditure for the farmer, the municipality, the inhabitants, plus minimum costs for the disposal of pulp industry wastes.

3. PROJECTS WITHIN THE EKOKRAFT™ PROGRAM

There is about 200 ha of conventionally established energy forestry (*Salix spp*) in the Municipality used already for district heating. In the projects involved, new cultivation methods and technologies will be implemented, but also the newest developed varieties of high frost resistant willows will be introduced in full-scale.[11] The latitude of Hedemora is considered as the northernmost agricultural border for the establishment of short rotation coppices. Latitude influences the program in a double sense. It is a challenge for all the players involved in the optimisation of invested resources, as investments must pay back. For those players interested in replicating EKOKRAFT™ south of Hedemora (that is most of them), the main conclusion of the demonstration part of the program will probably be, "If it works in northern Hedemora it should even work in our latitude...."

3.1. KOMBIBRUK™, a new method for the sustainable food and energy farming

No one can bathe in Brunnsjön, the lake closest to the town of Hedemora. Its flora and fauna have been seriously affected by the discharge, during the last decades, of surplus fertilisers from intensive agriculture around the main stream leading into the lake. Positive experiences on the use of surplus fertilisers from drained agricultural land on 3 ha of willow plantations are reported in southern Sweden.[12] In Hedemora, soil erosion of 2500 ha of farmland and the further severe eutrophication of Lake Brunnsjön would be greatly reduced by the establishment of willow buffer strips to be situated on strategically selected topographic areas; about 10-20% of the arable area will be planted. Depending on a number of factors, the density of the plantation will vary as well as the pattern of plantation. The farmers will benefit from increased turnover and in order to suit specific biofuel demands, and the willow buffer strips will be harvested in different rotation periods. The further transformation of the bundles into consumer specific biofuels will depend on the location of the farm, the distance to the nearest boiler and, of course, the type of boiler. The project will demonstrate how the farmer can derive increased benefit in an environmentally friendly agriculture with the very compatible combination of food and bioenergy production.

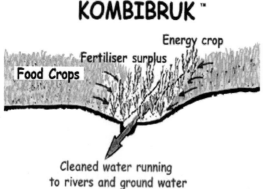

3.2. Urban and food industry effluents, a nutritional resource for willow production

The establishment of vegetation filters for the caption of nutrient salts and other components in urban and other sewage waters has been long known. During the last decades much research and implementation has been carried out.[13] The principle of waste water filtration is today used in South Sweden in ten different willow establishments, for the treatment of urban sewage and food industry waste water and landfill lecheate waters.[14] In Hedemora, sewage sludge is normally recycled to the existing energy forestry plantations. The sewage water, after treatment in the sewage water plant, which also belongs to the Municipality, flows into the River Dalälven. If the nutrients are used as fertiliser instead, the sewage water will irrigate about 70 ha of biomass and at least halve the discharge of nutrients into the river. Probably the first big vegetation filter in Europe will be established in a latitude where the biomass vegetative cycle is no longer than 3 to 4 months.

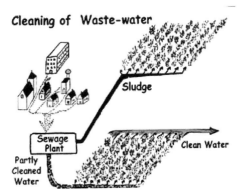

At present, the effluent contains the discharge from the 7,000 inhabitants of the town of Hedemora and the waste water from a big milk products factory, equivalent to the discharge from a town of 17,000 inhabitants. Various types of vegetation filters and irrigation techniques will be implemented. Considerable savings in investment costs and chemicals for the sewage treatment plant are expected and at the same time the biofuel cost for various heating purposes will diminish.

3.3. Green cover on sterile mining soil—new biofuel sources

The main aim of the project is to further develop an optimum method to create biotopes on spoiled mining areas. A mining company and a pulp and board mill will cooperate in the remediation of very old environmental transgressions. In order to prevent the leaching of heavy metals into the watercourse and groundwater systems a biotope will be created on sterile mining soil. The project will start with a first trial on 22 ha. A layer of 0.5-1.0 m of fibre sludge and furnace ashes from the pulp and board mill will be spread on the area. Willow and other species will be planted in these sterile areas. This will serve to minimise leaching of heavy metals, generate cost savings in sludge and ash disposal and produce biofuels. Provided that results are satisfactory, the area of demonstration will be increased and the method will be replicated in many places in Sweden.[15]

3.4. EKOPELLETS™—possibly the twenty-first century's biofuel

New cultivation methods, new willow clones and a new direct bundle harvesting technology also developed in Hedemora[16] for implementation within EKOKRAFT™, together make for a dramatic increase in energy balances for biomass. These can be as much as 1:60—as compared with current values of 1:20 in direct chipping harvesting systems—at about 20% lower harvesting costs. Homogeneous bundles of willow stems are produced as "virtual timber." These can be handled as such, and stapled for storage. After the summer at this latitude, the moisture content of the bundles will halve and their heating value increase by 30%. Bundle harvesting technology allows the year-round supply of biofuel to the bigger power plants and opens up a new market for short rotation coppices: the rapidly expanding stove and small boiler markets. Biomass now in the form of stored drier bundles facilitates product diversification and combustion technical development, in, for example, the EKOPELLETS™ project. The latter involves the technical development and installation of processes for the production of biofuel pellets directly out of willow stems and their combustion.[17] With 3 to 10 times higher energy balances and 25% to 40% lower production costs in comparison with standard pellets,[18] EKOPELLETS™ will enable the Municipality to change 15 smaller heating systems (in schools and hospitals) from oil and electricity. Approximately 5 GWh/a of locally produced EKOPELLETS™ will be supplied at a substantial savings for the Municipality of Hedemora.

3.5. District heating systems: expansion and new district nets

The Hedemora Energy company, also owned by the Municipality, already operates two 10-13 MW biofuel-based district heating plants. The project involves extension of the actual net in the town of Hedemora and the installation of new, smaller bio-fuelled district heating systems in two villages. With their production based on all the new *Salix* clones, technologies as well as a *Salix* business re-organisation to be implemented in EKOKRAFT™, biomass fuels make even small district heating systems economically feasible. This also is new for the bio-energy sector.

REFERENCES

1. Commission of the European Union (1996). DG XVII. White Book.
2. Swedish Government (1998). Commission on Ecologically Sustainable Development Fact Sheet.
3. Swedish Government (1998). Commission on Ecologically Sustainable Development Approval of Grant.
4. Grundström, G. (1997). PM. Secretary of Trade and Industry, Municipality of Hedemora.
5. Lindqvist, R. and B. Alriksson (1998). Invitation Letter. EKOKRAFT™ FORCE Network. Rodolofo@salix.se, Hedemora and Borje.Alriksson@lto.slu.se, Uppsala.
6. Askling, U. (1998). Pressmeddelande, Jordbruksdepartamentet.
7. Miljöeffekter vid odling av energigrödor (1997). Vattenfall AB, Projekt Bioenergi.

8. Börjesson, P. (1998). Biomass in a Sustainable Energy System, Department of Environmental and Energy Systems Studies Lund University, Sweden.
9. Wilstrand, M. (1997). Calculation run, BMC-program, Salix Maskiner AB, Hedemora.
10. Modée, Å. (1997). Project Leader for "Local production of consumption of biofuels" projects, EKOKRAFT™ program, SKG AB, Stockholm.
11. Larsson, S. (1998). PM, Frost resistent willow clones, Svalöf Weibull AB.
12. Elowson, S. and U. Johansson (1995).Cleaning of drainage water from agricultural land. Report 54. SLU-Swedish University of Agricultural Sciences, Dep. of Ecology and Environmental Research, Section of Short Rotation forestry, Uppsala.
13. Aronsson, P. and K. Perthu (1994). Willow vegetation filters for municipal wastewaters and sludge, Report 50. SLU-Swedish University of Agricultural Sciences, Dept. of Ecology and Environmental Research Section of Short Rotation Forestry, Uppsala.
14. Rosenqvist, D. (1999). PM. Rosenqvist Mekaniska Verkstad, Kristianstad.
15. Lindström, A. (1999). PM. Dalarna University College in Garpenberg, Boliden Mineral AB and STORA Fors kartongfabrik AB, Garpenberg, Sweden.
16. Wilstrand, M. (1998). Salixodling, anläggning och skörd. Bioenergi från lantbruket. Temadag. Sveriges Utsädesförening and Kungl. Skogs- och antbruksakademien. Stockholm.
17. Wilstrand, M. (1997). Projet Ekopellets. PM. Wilstrand Innovation AB, Hedemora.
18. Hadders, G.and B. Sundell (1996). Pelletspärmen, Energibalans, JTI–Swedish Institute of Agricultural Engineering, Uppsala.

463

IEA BIOENERGY FEASIBILITY STUDIES

Y. Solantausta,[a] D. Beckman,[b] E. Podesser,[c] R.P. Overend,[d] and A. Östman[e]

[a]VTT Energy, PO Box 1601, FIN - 02044 VTT, Finland
[b]Zeton Inc, 5325 Harvester Road, Burlington, Ont L7L 5K4, Canada
[c]Joanneum Research, Institute of Energy Research, Elisabethstraße 5, 8010 GRAZ, Austria
[d]NREL, 1617 Cole Blvd, Golden, Colorado80401-3393, USA
[e]Kemiinformation, Birkagatan 35, S - 113 39 Stockholm, Sweden

The International Energy Agency (IEA) Bioenergy aims at advancing bioenergy utilisation. In one of the current tasks, Techno-Economic Assessments for Bioenergy Applications 1998-99, the objectives are 1) to analyze bioenergy systems to support organisations working with products and services related to bioenergy, 2) to build and maintain a network for R&D organisations and industry, and 3) to disseminate data on advanced biomass conversion technologies. These IEA resources have been maintained and improved since the original studies in 1982. State-of-the-art, performance, and feasibility analyses of biomass to (heat and) electricity (CHP), and liquid, gaseous, and solid fuel systems have been carried out.

This paper presents technical and economic feasibility studies carried out for several advanced biomass power and fuel conversion technologies in Austria, Canada, Finland, Sweden and the United States during 1998-99. The core technologies analysed include
- a flue gas condensing system for increased heat production integrated to a biomass boiler,
- fast pyrolysis for chemicals,
- small-scale steam boiler power plant compared with proposed new bio-power concepts,
- fast pyrolysis liquid production for district heat production within a city.

1. INTRODUCTION

IEA Bioenergy is an international collaboration in bioenergy within the International Energy Agency (IEA). IEA is an autonomous body within the framework of the Organisation for Economic Co-Operation and Development (OECD) working with the implementation of an international energy programme.

Techno-Economic Assessments (TEA) is a task within the IEA Bioenergy, in which several studies are carried out for national authorities together with private industry. The IEA TEA resources have been maintained and improved since the original studies in 1982. State-of-the-art, performance, and feasibility analyses of biomass to (heat and) electricity (CHP), and liquid, gaseous, and solid fuel systems have been carried out (Kjellström, 1985; Beckman, 1990; Solantausta, 1994, 1996, 1998).

The following general objectives for the current task are defined:
- to promote companies commercialising new bioenergy technologies and products by carrying out site specific pre-feasibility studies,
- to support the development of new technologies for appropriate bioenergy applications. Based on the expertise within the task, concepts will be selected that appear feasible for commercial applications.
- to summarise experiences from the studies in such a manner that they may be employed as guidelines for R&D,
- to report opportunities, solutions and problems in bioenergy business in participating countries.

TEA on bioenergy applications will be produced for selected sites. The results of the studies will be utilised by industry, funding agencies, and research organisations. The following specific objectives have been defined by various project participants:
- to estimate the competitiveness of flash pyrolysis liquid as an alternative boiler fuel,
- to evaluate a small scale power production concept,
- to improve the current flue gas condensing system by using heat pump,
- to assess on a techno-economic basis the potential for further development of proposed technologies employing fast pyrolysis to produce chemicals, fuels, and fertilisers.

The role of an IEA task in the collaboration is
- to determine rigorous performance analyses for the systems being proposed, and together with the industry, to carry out economic assessments, and
- to promote the pre-feasibility studies to a wider audience (users and decision makers) than possible for the companies alone.

It is believed that this approach of helping the industry to make their products and services known would benefit the whole bioenergy community.

2. METHODS

Performance (mass and energy balances) of systems will be determined rigorously, and the economic assessment will be carried out with companies supplying or planning to use these systems. The companies, whose technologies or sites are considered, have been selected by the funding agencies in the countries participating in this project. The performance analysis, a most critical topic in a techno-economic assessment, will be carried out using AspenPlus™ as a modelling tool as in two previous IEA Bioenergy activities (Solantausta, 1996, 1998).

Cost assessments for the systems will be estimated in collaboration with industry.

3. CASES ASSESSED

3.1. Pyrolysis liquids as boiler fuel

Stockholm Energi Ab, Sweden, is currently using wood pellets and tall oil pitch as a renewable fuel for district heating within the Stockholm city area. Pyrolysis liquid is a potential substitute for petroleum fuel oil. A technical, economic, and environmental assessment for the whole utilisation chain is being carried out.

In Sweden a major test with pyrolysis liquid is planned to be carried out by Stockholm Energi AB. It started in 1997 with test combustion of pyrolysis liquids in amounts of tonnes; larger amounts are scheduled to be used in 1999 followed by a possible commercial application in 1999 and thereafter.

Stockholm Energi AB is using several biofuels such as wood pellets and tall oil pitch. Thus, an opportunity is established for a techno-economic comparison between different ways of utilising biofuels from wood.

The Swedish forests are used as a source for several wood products. By-products are obtained in the forests (light thinning, clearing, etc.) as well as in the subsequent manufacturing processes (sawdust, tall oil, black liquor, etc.). To some extent they are already in use for energy production. However, further utilisation is regarded feasible.

With the support of Stockholm Energi AB and some other companies involved in fuel manufacture, basic data on energy consumption, investment, transportation cost are collected either as specific figures in actual cases or as estimates with an acceptable accuracy.

The techno-economic evaluation will be carried out in a conventional way. Two cases are considered for boiler combustion, one with an existing mineral oil fired boiler, and one with a modified boiler.

A major uncertainty for biofuels concerns prices and taxes. They will be handled as sensitivity calculations.

3.2. Small scale power production from biomass

The first BioPower co-generation power plant (0.9 MW power, 6 MW heat) suitable for sawmill and district heat operation will be commissioned by Sermet Oy, Kiuruvesi, Finland, during the fall of 1999. A comparison between the conventional steam boiler power plant and two new concepts proposed (gasification–gas engine, pyrolysis–diesel engine) is carried out to study the competitiveness of the BioPower concept.

Small-scale electricity production is often proposed a potential market for biomass. Especially, advanced cycles are promoted for this market. The comparison in this work is carried out in a power-only production mode. Operating small power plants in combined heat and power mode would be more economic. However, sufficient heat loads are not always available.

It should be noted that only the conventional system is ready for commercial operation. There are several technical uncertainties related to the new cycles proposed. The comparison is carried out to define at which conditions the conventional cycle is preferable.

Electricity production costs of the three concepts are compared as a function of annual operation time in Figure 1. Owing to the small scale, electricity costs are rather high for all the cases. It may be seen that with more than 3 000 annual operating hours the Rankine cycle offers the lowest power production cost, around 0.14 US$/kWh at 5 000 h/a. This may be compared to 0.04 or 0.06 - 0.08 US$/kWh, which are typical costs of power produced in a large natural gas combined cycle, and a large wood-fired Rankine cycle power plant, respectively.

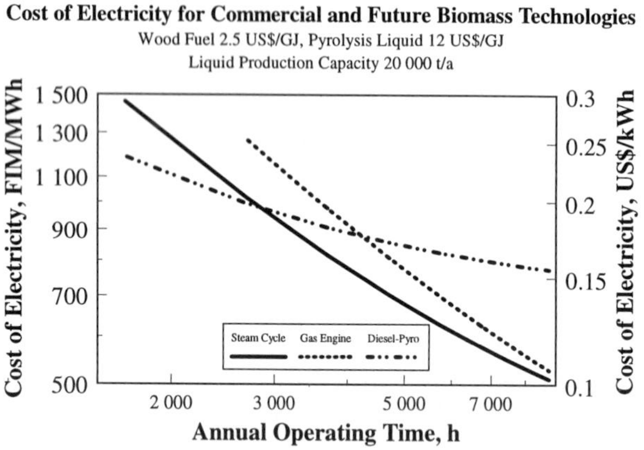

Figure 1. Power production costs at 2 MWe for three technologies. Investment costs: Steam boiler power plant (PP), 6 MUS$; Gasification–gas engine PP, 9 MUS$; Pyrolysis–diesel PP, 3 MUS$; Pyrolysis liquid production plant, 13 MUS$.

3.3. Improved heat recovery in biomass district heat plants

An improved version of the current flue gas condensation concepts is envisioned, in which a larger fraction of the available condensing heat of flue gas is used for district heating. Technical aspects with real site data are employed, together with economic and environmental aspects. The concept has been proposed by Joanneum Research, Graz, Austria.

Heat recovery from flue gas in biomass furnaces of district heat plants increases efficiency, because of the high water content of wood chip and bark fuel. Due to the water content of the biofuel, the low heating value is commonly reduced to 50% of dry wood. However, if the flue gas is cooled to about 30°C, large quantities of heat (30 to 50% of the furnace capacity) may be recovered by condensation (Figure 2).

If a heat pump is used, the low temperature condensation heat recovered from the flue gas may be increased to the temperature of the district heat return level. For this purpose a resorption heat pump with a mechanic compression unit should be employed due to the high coefficient of performance (Podesser, 1997). The mechanical compression unit is powered either by a grid-connected electrical motor, or by a flue-gas powered Bio-Stirling engine. Detailed calculation and design of a resorption heat pump plant following the Lorenz Process for heat recovery from the flue gas is carried out. The preliminary economic assessment without real plant operation data shows that due to the high coefficient of performance the amortization times of such plants range from four to six years. Research and development work should be started to investigate and improve this special type of heat pump technology.

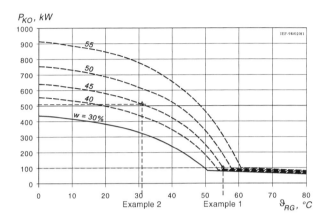

Figure 2. Two examples of heat recovery by flue gas condensation; example 1: Conventional condensation plant, example 2: Active condensation with resorption heat pump; (w ... biofuel water content, P_{KO} ... Condensation capacity, boiler capacity 1000 kW$_{th}$; Combustion air ratio 1,75; Flue gas temperature at boiler outlet: $\vartheta_{RG} = 140°C$).

3.4. By-products from fast pyrolysis liquid

Production of fertilisers, chemicals and fuel from fast pyrolysis liquid is evaluated. Each of these processes was developed by Resource Transforms International, RTI, of Waterloo, Canada. The objective is to estimate which of the proposed concepts would be most competitive for further research and development.

1. Production of slow release fertilizers from bio-oil produced by fast pyrolysis. An analysis of the process producing slow-release nitrogenous fertilizers starting with a bio-oil feed was carried out. Comparison was made to the alternative ammoxidation process for nitrogenous fertilizer production. The process has been patented (Radlein et al., 1997).

2. Upgrading of bio-oil by reaction/blending to improve bio-oil properties for fuel applications. The value added to the bio-oil fuel versus the costs of the upgrade will be analysed. An European patent application for this process has been made (Radlein et al., 1995).

3. Production of sugars (levoglucosan) by fast pyrolysis of wood. The process to produce sugars from wood via fast pyrolysis will be analysed. The wood is pretreated by acid hydrolysis prior to conversion to bio-oil by fast pyrolysis. Water is added to the bio-oil, producing a solid phase, consisting of lignin and an aqueous phase, and containing soluble

sugars. The sugar solution can be used to produce chemicals, or to produce ethanol by fermentation.

4. PRELIMINARY CONCLUSIONS

In Sweden a major test with pyrolysis liquid is planned to be carried out followed by a possible commercial application in 1999 and thereafter.

Rankine cycle appears competitive against new power plant concepts, if the annual peak operation time is above 3,000 h. Competitiveness of new cycles will be improved, if fuel costs will be higher than today. The pyrolysis–diesel engine concept may become competitive, if pyrolysis liquid is produced in large scale.

A resorption heat pump with mechanic compression appears interesting coupled to co-generation plants. The amortization times of such plants may range from four to six years. Research and development work should be started to investigate and improve this special kind of heat pump technology.

The production of by-product chemicals is believed to be important to the development of economically feasible pyrolysis liquid fuel applications. Some approaches to chemical development are included in the Canadian case studies.

5. ACKNOWLEDGMENTS

This work is carried out within the IEA Bioenergy, and has been financed by the funding agencies in participating countries. Countries participating are Austria, Brazil, Canada, Finland, Sweden, and the United States of America, whose support is gratefully acknowledged.

REFERENCES

Kjellström, B., (ed) (1985). A Study of a Biomass Liquefaction Test Facility, Final Report of IEA Cooperative Project Biomass Liquefaction Test Facility, Statens Energiverk R:1, Stockholm.

Beckman, E., D.C. Elliott, B. Gevert, C. Hörnell, B. Kjellström, A. Östman, Y. Solantausta, and V. Tulenheimo (1990). Techno-Economic Assessment of Selected Biomass Liquefaction Processes. Final Report of IEA Cooperative Project Direct Biomass Liquefaction. Espoo: Technical Research Centre of Finland. 169 p. + app. 95 p. (VTT Research Reports 697).

Podesser, E. Resorptionswärmepumpen zur Nutzung der Kondensationswärme an Biomassefeuerungsanlagen zur Fernwärmeversorgung, JOANNEUM RESEARCH, IEF-B-06/97.

Radlein, D., J. Piskorz, and P. Majerski (1997). Method of producing slow-release nitrogenous organic fertilizer from biomass. United States Patent 5,676,727, Oct. 14, 1997.

Radlein, D., J. Piskorz, and P. Majerski (1995). Method of upgrading biomass pyrolysis liquids for use as fuels and as a source of chemicals by reaction wih alcohols. European patent application EP 0 718 292 A1, Dec. 22, 1995.

Solantausta, Y., J.P. Diebold, D.C. Elliott, T. Bridgwater, and D. Beckman (1994). Assessment of liquefaction and pyrolysis systems. Espoo: Technical Research Centre of Finland. 123 p.+ app. 79 p. (VTT Research Notes 1573).

Solantausta, Y., A. Bridgwater, and D. Beckman (1996). Electricity production by advanced biomass power systems. Espoo: Technical Research Centre of Finland. 120 p.+ app. 61 p. (VTT Research Notes 1729).

Solantausta, Y., T. Koljonen, E. Podesser, D. Beckman, and R. Overend (1998). Feasibility Studies on Selected Bioenergy Concepts Producing Electricity, Heat, and Liquid Fuel. Espoo: Technical Research Centre of Finland. Vols I - VI.

Stiglbrunner, R. Beurteilung der Kondensationsanlage der Biomasseheizanlage in Pfarrwerfen. JOANNEUM RESEARCH, Bericht Nr.: IEF-B-11/95.

471

ECONOMIC EVALUATION OF TECHNICAL, ENVIRONMENTAL AND INSTITUTIONAL BARRIERS ON BIOMASS RESIDUE COLLECTION COST IN CALIFORNIA

Prab Sethi,[a] Valentino Tiangco,[a] Ying Lee,[b] Virak Dee,[a] David Yomogida,[c] and George Simons[a]

[a]California Energy Commission, Research and Development Office, Sacramento, California, 95814
[b]Lawrence Livermore National Laboratory, Livermore, California 94551
[c]Henwood Energy Services, Inc., Sacramento, California 95833

This paper deals with the economic impacts of the technical, environmental, and institutional barriers in the production, harvesting, and processing of various biomass fuels in California. For each biomass fuel, a base case scenario was developed to characterize the procedures of harvesting, processing, and transporting. An economic model was utilized to project the biomass production costs supplied to direct-combustion power plants for each technical, environmental, and institutional barrier. These results will enable the California Energy Commission (Energy Commission) to identify the most significant barriers to economical biomass energy production (production, harvesting, and processing).

1. INTRODUCTION

Currently, California's biomass power facilities are facing the 11-year price cliff, the time when their Standard Offer Number 4 contract will expire. Most biomass projects signed 30-year power purchase contracts. Generally, the first 10 years of the contract provided fixed, escalated, or levelized energy prices based on the early 1980s prevailing utility forecasts of avoided costs. Projects committing to firm capacities received levelized avoided capacity prices ranging from $140 to $200 kW-year, depending on what year the project became operational. The impending end of the fixed energy price period will cause a substantial number of these projects to become uneconomical. Many of the contracts will possibly be terminated, forcing some projects into bankruptcy. In addition, the future is uncertain due to the deregulation of electric utility industry in California. Thus, the Energy Commission has been seeking methods to mitigate the financial losses incurred by the expirations of these contracts.

Model, technical, environmental, and institutional barriers associated with fuel harvesting, production, collecting, and transporting can be economically quantified. By identifying the major barriers to economical fuel collection, the Energy commission can accurately target and prioritize specific biomass projects (production, harvesting, and processing) for future research funding to aid the biomass industry.

2. DESCRIPTION OF THE BIOFUEL MODEL

The BioFuel spreadsheet Model, written in EXCEL, is a collection of several economic sub-models that are descriptive of the biomass residue production process. BioFuel (version 0.11) can estimate the biomass production costs for energy crops, forest slash or stand improvement, urban wood wastes, rice straw, and food processing wastes (pits and shell). Furthermore, the BioFuel Model was specially designed so that other biomass residue collection processes (e.g., collection of sorghum for ethanol production) can be easily incorporated to the model to meet future modeling needs.

2.1. Conventional processes

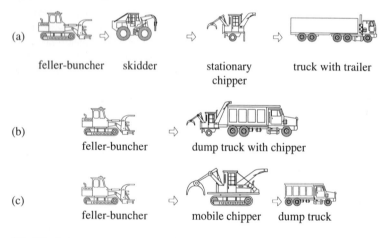

2.2. Whole tree process

Figure 1. Possible methods of harvesting, processing, and transporting energy crops

One important feature of BioFuel is that the biomass collection process can be customized at the user's request. Up to eight machines may be selected during the harvesting and processing of a biomass residue. For example, Figure 1 shows several possible ways to harvest and process *Populous* on an energy plantation. By selecting different combinations of machinery for the collection process in the BioFuel Model, almost any harvesting system can be selected. Although only eight machines can be stored in the BioFuel Model, built-in databases for establishment machines, harvesting machines, and transportation machines were added to the BioFuel Model so that an unlimited number of equipment can be stored and retrieved through the database manager of the BioFuel Model.

3. A CASE STUDY OF ENERGY CROPS

For the base case energy crop scenario, fifty thousand acres of marginal farm land in Northern California is used to plant *Populous* hybrid NE-388 (*Populous maximowiczii* crossed with *trichocarpa*) or commonly called poplar. The yield for poplar is based on an experimental study done by Strauss and Wright (1990). The length of the whole project is assumed to be 18 years, and the rotation period of the crop is 6 years. Furthermore, coppicing is allowed to re-establish the whole crop, and low-intensity management (i.e., minimum fertilizer, weed control, and irrigation during the whole life span of the tree) is employed to minimize the operation cost. In addition, the amount of rainfall is assumed to be sufficient for the crop so that irrigation is not necessary. The harvesting process utilizes feller-bunchers, skidders, and chippers. Traditional trucks with semi-trailers are used to transport the wood chips to a local biomass facility, which is 25 miles from the plantation.

The resulting base case cost for energy crops, found by BioFuel, is about $45/BDT in 1994 current dollars (Table 1). The BioFuel Model found that the irrigation cost had the greatest economic impact of all the technical and environmental barriers since irrigation water in California is expensive in most of the state where marginal land is available. A 74% increase in the base case biomass cost is expected if the water cost is $100/acre-ft (compared to no water fee in the base case). To reduce the irrigation cost, energy farms should be established where rainfall is more abundant. Additionally, waste water from animal farms or other agricultural facilities (agroforestry) may be used to meet water demands.

4. A CASE STUDY OF FOREST SLASH—STAND IMPROVEMENT

The timber stand improvement process involves feller-bunchers, skidders, and chippers. The base case data was adapted from a study done by Huyler (1981). The transportation distance is assumed to be a 50-mile round trip.

According to the BioFuel Model, the base case cost for timber stand improvement residue is about $31/BDT in 1994 current dollar as shown in Table 1. The technical, environmental,

Table 1
Estimated costs of different biomass residues by the BioFuel Model (in 1994 current dollar)

	Energy Crops	Forest Slash	Urban Wood Wastes	Rice Straw	Orchard Prunings
Land	7.1	N/A	0.0	N/A	N/A
Establishment	7.2	N/A	N/A	N/A	N/A
Cultural Management	34.3	N/A	N/A	N/A	N/A
Annual Operating	N/A	N/A	4.4	N/A	N/A
Harvest/Collect/Process	21.1	23.9	9.7	24.4	19.6
Transportation	5.1	7.0	7.7	19.9	6.5
TOTAL ($/BDT)	44.8	30.9	21.8	44.3	26.1

*N/A: Not Applicable

and institutional barriers to timber stand improvement residue production were studied. BioFuel found that the processing (chipping) cost had the greatest economic impact of all the quantifiable technical and environmental barriers. The chipping cost can contribute almost 40% of the total production cost. To eliminate this cost, whole-tree harvesting could be employed to circumvent the chipping process. Whole-tree burning, however, is not commonly practiced in industry but is ready for full-scale demonstration (Wiltsee, 1993).

5. A CASE STUDY OF URBAN WOOD WASTES

Urban wood waste facilities serve as local dump sites for wood wastes. On the facility, field workers hand-separate wood into different piles, according to the wood waste quality. Loaders are used to move wood wastes from the landing to conveyors by which wood is sent to hammermills for processing. The end product is stored in a silo or piled outside waiting for eventual transportation by trucks.

According to BioFuel, the base case cost for processed urban wood waste is about $22/BDT in 1994 current dollar as shown in Table 1. BioFuel found that the high transportation cost had the greatest economic impact of all the quantifiable technical and environmental barriers. The most successful way to minimize this cost is to minimize the transportation distance or to consider the back-hauling practice. Another barrier which may have a significant impact on the processed urban wood waste cost is the initial cost for unprocessed urban wood waste. Assembly Bill AB688 mandated a 50% resource reduction for each county in California. Accepting urban wood waste (with a tipping fee) may be a sensible alternative for urban wood waste facilities in California.

6. A CASE STUDY OF RICE STRAW

The rice straw base case scenario is based on an experiment performed by Jenkins et al. (1985). Jenkins' collection process included forming the rice straw into windrows with harvesters, baling the rice straw with square balers, roadsiding the bales with tractors, and transporting the bales with flat bed truck. In the base case, however, the rice straw is cut and formed into windrows during the harvesting of the rice. As a result, only the baling and transportation costs were included in the overall collection cost.

According to the BioFuel Model, the base case cost for rice straw delivered to a direct-combustion facility is about $44/BDT in 1994 current dollar as shown in Table 1. BioFuel found that the high processing cost due to high moisture content had the greatest economic impact of all of the quantifiable technical and environmental barriers. For example, if the rice straw is baled immediately after the harvesting, the moisture content may be as high as 60%. With this high moisture content, the processing cost of rice straw is expected to increase by 73% compared to the base case, where the rice straw is dried for a few days until the moisture content is 30%. As a result, the most successful way to minimize this cost is to allow the straw to dry in the fields after windrowing.

7. A CASE STUDY OF ORCHARD PRUNINGS

The collection of the pruning residues is assumed to start at the roadside of an orchard, where farmers typically pile the residues after pruning. Since pruning is required for all orchards, the cost of pruning is not a biomass collection cost in this scenario. Before transporting the prunings to a biomass facility, the prunings are chipped at the roadside by a chipper with two field workers. The chipped wood is blown directly to a dump box or a trailer. Once the trailer is full, it is hauled away by a truck to a biomass facility, which is located 25 miles from the orchard.

According to the BioFuel Model, the base case cost for processed orchard pruning residue is $26/BDT in 1994 current dollar as shown in Table 1. The BioFuel Model found that the on-site labor costs for chipping the prunings had the greatest effect on the orchard pruning fuel cost. If the number of field workers can be reduced by one person, the cost of fuel would reduce by 29%. This may be achieved by hiring experienced field workers or by improving the chipper efficiency.

8. CONCLUSION

Through the use of the Energy Commission's BioFuel Model, the costs (in 1994 current dollar) for various processed biomass fuels were projected for California: energy crops are about $45/BDT, timber stand improvement residues are about $31/BDT, urban wood waste residues are about $22/BDT, rice straw residues are about $44/BDT, and orchard prunings are about $26/BDT. These figures depend heavily on the machine operating conditions, management systems, financial arrangements, and resources of a particular biomass processor.

To determine the effects of various technical and environmental barriers on the base case fuel cost, different scenarios were developed by varying the operating parameters of the base case. For energy crops, the irrigation cost had the greatest economic effect on the cost of fuel. The irrigation cost is capable of increasing the base case cost of fuel by as much as 74%. For timber stand improvement residue, eliminating the chipping process by adopting whole-tree harvesting reduced the base case fuel cost by 31%; to reduce the production cost for urban wood waste, minimizing the transportation distance had the greatest effect (5%); for rice straw, drying the straw in the field before collection can reduce the processing cost by as much as 73%; and for orchard prunings, reducing the number of field workers in the chipping process can reduce the cost of fuel by as much as 29%. For a complete fuel production analysis refer to "Investigation and Economic Evaluation of Technical, Environmental, and Institutional Barriers to Biomass Residue Production/Collection in California" (CEC, 1995).

REFERENCES

California Energy Commission (1995). Investigation and Economic Evaluation of Technical, Environmental, and Institutional Barriers to Biomass Residue Production/Collection in California, Research and Development Office, ETAP Biomass Program, Sacramento, California.

Huyler, N.K. (1992). The cost of thinning with a whole-tree chip harvesting system, Northern Logger and Timber Processor, July.

Jenkins, B.M., D.A. Toenjes, J.B. Dobie, and J.F. Arthur (1985). Performance of large balers for collecting rice straw, Transactions of ASAE, 28, pp. 360–363.

Strauss, C.H. and L.L. Wright (1990). Woody Biomass Production Costs in the United States: an economic summary of commercial populus plantation systems, Solar Energy, 45, pp. 105–110.

Wiltsee, George (1993). Strategic Analysis of Biomass and Waste Fuels for Electric Power Generation, report for the Electric Power Research Institute, EPRI TR-102773, Final Report, December.

LONG TERM PERSPECTIVES FOR PRODUCTION OF FUELS FROM BIOMASS: INTEGRATED ASSESSMENT AND RD&D PRIORITIES[*]

A.P.C. Faaij and A.E. Agterberg

Department of Science, Technology and Society, Utrecht University, Padualaan 14, 3584 CH, Utrecht, The Netherlands, E-mail: a.e.agterberg@chem.uu.nl

The purpose of this project is to give better insight in the potential performance of different bio-energy systems for the production of fuels from biomass in a timeframe of about 30 years. The focus is on biomass for liquid transport fuels and as a feedstock for chemicals. The results should give insight into the energetic performance as well as the costs of the bio-energy systems and are used for directing long term research, development and demonstration policies.

In this project different bio-energy systems including biomass-conversion and end-use are analyzed and compared.

The conversion routes considered are:

Conversion Route	Feedstock	Resulting Fuel
Hydro-thermal-upgrading	Any biomass feedstock	Biocrude (possibly upgraded)
Pyrolysis	Any biomass feedstock	Pyrolysis oil (possibly upgraded)
Fermentation	Sugar- and starch-crops	Bio-ethanol
Hydrolysis+Fermentation	Lignocellulosic biomass	Bio-ethanol
Extraction+esterfication	Oil-containing crops	Bio-diesel
Gasification	Any biomass feedstock	Hydrogen or methanol

Various end-use options are also included in the chain analyses and cover among other things distribution and future transport systems like fuel-cell vehicles and advanced internal combustion engine vehicles.

The bio-energy systems are compared integrally on the basis of feedstock requirements, energetic performance and costs. Furthermore the technological bottlenecks and

---

[*]The project is funded by a Dutch foundation for technological sciences (STW) and is supported by an expert-panel of the biomass conversion platform of the foundation for sustainable chemical development (DCO).

uncertainties that may occur in the further development of the bio-energy systems are pointed out. Data are collected through literature reviews and personal communication with various experts, institutions and industries.

The results so far consider hydro-thermal-upgrading and pyrolysis. Currently, inventories and overviews are made of the production of hydrogen and methanol through gasification and ethanol from lignocellulosic biomass. The final results are expected end of April 1999.

OPPORTUNITIES FOR EFFICIENT USE OF BIOMASS IN THE PULP AND PAPER INDUSTRY

Katarina Maunsbach, Viktoria Martin, and Gunnar Svedberg

Department of Chemical Engineering and Technology/Energy Processes
Royal Institute of Technology
SE-100 44 Stockholm, Sweden

1. INTRODUCTION

The pulp and paper industry is of great importance for Sweden as it earns the country's largest net export income. Furthermore, black liquor and other biomass-based byproducts represent 13% of Sweden's primary energy supply. However, to maintain its worldwide competitiveness, the Swedish pulp and paper industry must make better use of its resources. More efficient use of forest residue for power production would minimize the use of external energy sources. New technology may allow some mills to produce a surplus of electricity, refined fuels or heat to be sold within a deregulated market. Efficient use of biomass lessens environmental impacts and thus makes the industry more ecocyclic.

In 1996, the total amount of electricity consumed in the Swedish pulp and paper industry was 19 TWh; of this, 4 TWh was generated internally mainly from biomass.[1] A large portion of the total fuel consumption was internally supplied fuels such as black liquor and bark. Common technologies used for energy conversion are the Tomlinson boiler for black liquor recovery and boilers fired by oil and bark. The steam generated in these boilers is used in back-pressure steam turbines to generate electricity. Typically, the efficiency of electricity generation is low in these systems—less than 10%. Within the next 20 years some of these systems are to be retired,[2] creating an opportunity to introduce more energy efficient systems for chemical recovery, steam and power generation. Also, modern, efficient pulp mills tend to have a lower steam demand as compared to today's average mill. Hence, a larger portion of the energy content of the internally supplied fuels may be used to generate power. Therefore, a combined heat and power system should operate at a higher ratio of power to heat production as compared with the systems used today.

This paper explores two possibilities: whether the Swedish chemical pulp industry and the integrated pulp and paper industry can become self-sufficient with respect to energy use, and whether excess power can be produced for sale. This analysis was based on the total amount of fuel, heat and power generated and utilized by the Swedish pulp and paper industry. Two scenarios for heat and power generation have been considered: (1) conventional with a

Tomlinson boiler, biomass boiler, and back-pressure steam turbine; (2) advanced, which emphasizes high efficiency power generation.

2. BIOMASS AVAILABILITY

The main biomass-based fuels used by the pulp and paper industry are felling waste, bark, and black liquor generated in the chemical pulping process. In Sweden, the internal biomass fuels used by the industry amounts to 30 TWh black liquor and 8.5 TWh felling waste and bark.[3] The total amount of biomass-based fuel used in Sweden is around 39 TWh excluding black liquor.[3]

It is possible for the pulp and paper industry to increase the amount of wood-based energy available for steam and power production partly by utilizing a larger part of the tree than is used today. Table 1 shows typical fractions of a pine tree's energy content in various parts of the tree.

Table 1
Typical energy content of various tree parts.[4]

Tree part	Energy content [%]
Tree trunk	57
Crown of the tree and branches	30
Bark	6
Tree stump	7

As shown in Table 1, the tree trunk and bark represent 63% of the energy content of a pine tree. Some of the felling waste (crown of the tree and branches) is left behind in the forest. As compared with using only bark to generate energy, adding branches and the crown of the tree would increase the amount of available biomass by 500%. Ecological restrictions requires that if a large part of a tree is removed for heat and power production, the ashes must be returned to the forest ecosystem to prevent the depletion of important nutrients. Naturally, this regulation influences the economics of using felling waste and bark as a fuel. Estimates of how much additional biomass fuel could be utilized in Sweden by the year 2020 range between 93 TWh (The Swedish Department of Forest-Industry-Market Studies, or SIMS) and 17 TWh (Swedish Forest Industries Association).[5] Hence, by the year 2020, the amount of available wood-based fuel (excluding black liquor) could increase by 40 to 240% as compared to today, depending on the economic feasibility, among other things.

3. ENERGY CONVERSION TECHNOLOGIES IN A PULP MILL

There are a number of ways to utilize wood-based biomass in conjunction with a pulp and paper mill. For instance, the felling waste and bark can be used to generate steam and power, transportation fuels (e.g., methanol), upgraded biomass such as wood pellets, or any combination of these three types of products. In this study, the potential for increased power generation is examined.

Several advanced thermal processes and system configuration have been studied with the aim of achieving high efficiency for power generation and high overall efficiencies. One interesting alternative for increased power generation employs gasification of black liquor and biomass and then uses the resulting gaseous fuel in
- a gas turbine cogeneration system (a so-called combined cycle or CC) whose efficiency for power generation is 40-60%[6,7]
- an evaporative gas turbine process (EvGT) whose power efficiency is 50-60%[8,9]
- a steam injected gas turbine (STIG) whose power efficiency is 40-50%.[9,10]

As an alternative to gasification, externally fired gas turbines can be used where heat is transferred to the turbine working fluid through a high temperature heat exchanger. Here the efficiency for electricity generation varies between 30-50%.[11] Note that the variations in power efficiencies as listed above are due to varying system operating conditions.

Recently, performance modeling by Larsson et al.[12] showed that a pulp mill using integrated black liquor and biomass gasification in combination with a gas turbine cogeneration system (IGCC) increased the net power efficiency from 10% to 25% as compared with a mill using conventional boilers. Aside from improvements in efficiency, black liquor gasification can be used as a supplement to an existing boiler to increase the capacity of a pulp mill. Ahlroth and Svedberg[13] found that by using an IGCC system to handle the additional black liquor resulting from an increased pulp production, twice as much electricity could be generated as compared with sending the additional black liquor to a Tomlinson boiler system. Although there is an increased potential for power production when using gasification integrated with a combined cycle, supplemental biomass may be needed to meet the process steam demand for the pulp and paper making processes.[2,12]

4. BIOMASS-BASED COMBINED HEAT AND POWER PRODUCTION

Biomass-based combined heat and power production within the Swedish chemical pulp industry, as well as the pulp and paper industry as a whole (including mechanical pulp mills), has been studied to examine the industry's potential for becoming self-sufficient with respect to energy use. Statistics on the total annual amount of fuel, heat, and power generated and used by the industry is shown in Table 2, along with the amount of pulp and paper produced. This data has been used as the basis for this study. Table 2 also shows the current net power efficiency in back-pressure turbines (BPT). This efficiency is obtained by dividing the amount of power produced in the BPT by the total amount of fuel supplied to the BPT.

Data from Table 2 was used to compute the average net power efficiency of a combined heat and power process required to make both the chemical pulp industry, and the pulp and paper industry as a whole, self-sufficient with respect to electricity. This net power efficiency then was compared to the efficiencies of the advanced thermal processes previously mentioned. Assuming a constant total thermal efficiency (here it is approximately 91% based on data in Table 2), increasing the power efficiency means that either the steam demand for pulp and paper making must be reduced through process improvements, or additional wood-waste must be utilized as compared with today. Therefore, we examined the effect of an increased net power efficiency on the required steam demand or the biomass fuel requirement. The need for additional wood-waste was also compared to the estimates on the

Table 2
Summary of energy use by Swedish pulp and paper mills in 1994 (based on Wiberg[14]).

	Market pulp		Integrated pulp and paper production			By-products	Total
			Pulp				
	Mechanical	Chemical	Mechanical	Chemical	Paper		
Pulp/paper production [Tonne]	401000	3381000	2464000	3878000	9322000		
Fuel consumption:							
Internal [GWh]	86	17312	32	10692	8500	335	36958
Total [GWh]	464	19192	32	12280	17334	583	49885
Fuel for BPT:							
Internal [GWh]	86	16880	32	10346	8500	335	36180
Total [GWh]	335	17615	32	10962	17183	583	46710
Power generation in BPT [GWh]:	20	1650	0	780	1240	60	3750
Net power efficiency	6.0%	9.4%	0.0%	7.1%	7.2%	10.3%	8.0%
Total power consumption [GWh]	740	2880	5760	2430	6700	170	18680

potential for increased biomass availability by Swedish Forest Industries Association (17 TWh) and SIMS (93 TWh) previously mentioned. Finally, using a process with high net power efficiency can in certain cases enable production of surplus electricity that can be sold within a deregulated market. The amount of electricity surplus (or deficiency) is also examined as a function of the net power efficiency. Results from the study are given below.

4.1. The Swedish chemical market pulp mill

Figure 1 shows the amount of electricity produced by an average Swedish chemical pulp mill as a function of the net power efficiency, using internal fuels only. Also shown is the

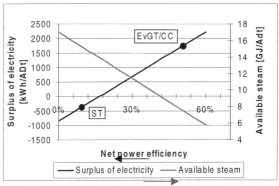

Figure 1. Net power efficiency vs. surplus of electricity and steam demand.

steam demand that would be satisfied as a function of the net power efficiency, assuming that the total system efficiency is constant and the amount of internal fuels is the amount listed in Table 2 (17312 GWh). Estimates of the net power efficiencies for a combined cycle (CC) and an evaporative gas turbine process (EvGT) as reported in the literature are indicated, along with the efficiency of back-pressure steam turbines (ST) used today.

As shown in Figure 1, a 17% net power efficiency would make the average chemical pulp mill self-sufficient with respect to electricity. However, to avoid intake of supplemental biomass, the process steam demand must then be lowered from today's 15.4 GJ/ADt[14] to 13.9 GJ/ADt. When using CC/EvGT-processes with estimated net power efficiencies around 50%, it would be necessary to drastically reduce the steam demand (Figure 1). If this is not possible, supplemental biomass must be used. Figure 2 shows the supplemental biomass required as a function of the net power efficiency when the steam demand is kept at 15.4 J/ADt. As shown, the estimates by the Swedish Forest Industries Association and SIMS on the amount of additional biomass available by the year 2020 will be more than enough to increase the net power efficiency to 60%.

Figures 1 and 2 show that there is a large potential for the chemical pulp industry to generate surplus power that may be sold on the market. This also suggests that an integrated chemical pulp and paper mill may have the potential to be self-sufficient with respect to electricity.

4.2. The total Swedish pulp and paper industry.

Figure 3 shows the calculated power surplus and the required supplemental biomass intake as a function of the net power efficiency for the Swedish pulp and paper industry as a whole. Here, the industry's steam requirement is assumed to be at today's levels. Estimated net power efficiencies for EvGT/CC, together with the potential increase in biomass availability deemed feasible by Swedish Forest Industries Association, are indicated in the figure. Results indicate that to satisfy the power requirements, the net power efficiency has to reach 50%, which is feasible with advanced gas turbine technologies. So doing would provide not only

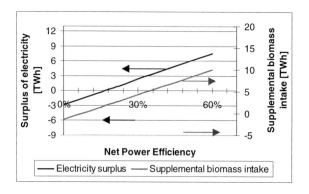

Figure 2. Net power efficiency vs. surplus of electricity and supplemental biomass intake.

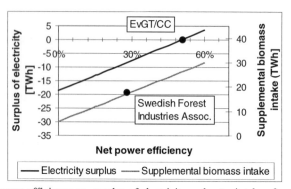

Figure 3. Net power efficiency vs. surplus of electricity and extra intake of wood waste fuel.

the chemical pulp industry with power, but also integrated pulp and paper mills and mechanical pulp mills which have very small amounts of internal fuels. However, the amount of additional biomass that could be available according to Swedish Forest Industries Association would not be sufficient to cover the process steam demand at this power efficiency. The significantly larger estimate by SIMS covers the need for additional biomass.

5. CONCLUDING REMARKS

The Swedish chemical pulp industry, and the pulp and paper industry as a whole, can become self-sufficient with respect to electricity by increasing their net power efficiency to 17% and 50%, respectively. These power efficiencies are possible by using advanced gas turbine processes. By increasing the net power efficiency in the chemical pulp mill to 50%, it is possible to generate a power surplus of 1708 kWh/ADt.

To increase the electricity generation significantly and still satisfy the industry's steam demand with wood-based fuels, it may be necessary to use a larger quantity of biomass fuels, or to reduce the steam demand. Although certain estimates suggest that it is possible to sufficiently increase the amount of available biomass, it needs to be pointed out that wood-based fuels may be interesting to heat and power generating companies other than the pulp and paper industry.

The proposed techniques for increasing the net power efficiency have to be fully developed before practical opportunities and economical incentives for investments will exist.

ACKNOWLEDGEMENT

This work is a part of the Ecocyclic Pulp Mill research program financed by MISTRA, the Swedish Foundation for Strategic Environmental Research.

REFERENCES

1. Swedish Forest Industries Association (1997). The Swedish forest industry—facts and figures 1997. Published by Swedish Forest Industries Association, Stockholm, Sweden.
2. Consonni, S., E.D. Larson, and N. Berglin (1997). Black liquor-gasifier/gas turbine cogeneration. Proceedings of the 1997 International Gas Turbine & Aeroengine Congress & Exposition, ASME, New York, NY. Paper: 97-GT-273.
3. NUTEK, Swedish National Board for Industrial and Technical Development (1997). Energy in Sweden 1996. Published by NUTEK, Stockholm, Sweden.
4. Marklund, L.-G. (1988). Biomass functions for pine, spruce and birch in Sweden. Published by The Swedish University of Agricultural Sciences, Umeå, Sweden.
5. NUTEK, Swedish National Board for Industrial and Technical Development (1996). Effects of increased biomass utilization—state of the art. (Report in Swedish.) Published by NUTEK, Stockholm, Sweden. Report No. R 1996:37.
6. Sipilä, K., A. Johansson, and K. Saviharju (1993). Can fuel-based energy production meet the challenge of fighting global warming - a change for biomass and cogeneration? Bioresource Technology: Biomass, Bioenergy, Biowastes, Conversion Technologies, Biotransformations, Production Technologies, 43, pp. 7–12.

7. Smith, D. J. (1994). Advanced gas turbines provide high efficiency and low emissions. Power Engineering, 98, pp. 23–27.
8. Ågren, N., A. Cavani, and M. Westermark (1997). New humidification concept for evaporative gas turbine cycles applied to a modern aeroderivative gas turbine. Proceedings of the 1997 ASME International Mechanical Engineering Congress and Exposition. ASME, Fairfield, NJ, 37, pp. 223–230.
9. Gallo, W.L.R. (1996). A comparison between the HAT cycle (humid air turbine) and other gas-turbine based cycles: efficiency, specific power and water consumption. Proceedings of the 1996 International Symposium on Efficiency, Costs, Optimization, Simulation and Environmental Aspects of Energy Systems, ECOS'96. Elsevier Science Ltd, Oxford, England, 38, pp. 1595–1604.
10. Ågren, N., H.U. Frutschi, and G. Svedberg (1994). A parametric study of steam injected gas turbine with steam injector. Proceedings of the 8th Congress & Exposition on Gas Turbines in Cogeneration and Utility, Industrial and Independent Power Generation. ASME, International Gas Turbine Institute, 9, pp. 177–184.
11. Yan, J. (1998). Externally fired gas turbines. Published by the Royal Institute of Technology, Stockholm, Sweden. Technical Report No. TRITA-KET R87, ISSN 1104-3466
12. Larsson, E.D., T.G. Kreutz, and S. Consonni (1998). Combined biomass and black liquor gasifier/gas turbine cogeneration at pulp and paper mills. Proceedings of the 1998 International Gas Turbine & Aeroengine Congress & Exhibition. ASME, Fairfield, NJ. Paper: 98-GT-339.
13. Ahlroth, M. and G. Svedberg (1998). Case study on simultaneous gasification of balck liquor and biomass in a pulp mill. Proceedings of the 1998 International Gas Turbine & Aeroengine Congress & Exhibition. ASME, Fairfield, NJ. Paper: 98-GT-350.
14. Wiberg, R. (1995). Energy consumption in the pulp and paper industry 1994. (Report in Swedish.) Published by ÅF-IPK, Stockholm, Sweden.

ENERGY CROPS VERSUS WASTE PAPER: A SYSTEM COMPARISON OF PAPER RECYCLING AND PAPER INCINERATION ON THE BASIS OF EQUAL LAND-USE

Marko Hekkert, Richard van den Broek, and Andre Faaij

Utrecht University, Department of Science, Technology and Society, Padualaan 14, 3584 CH Utrecht, The Netherlands, email: M.P.Hekkert@chem.uu.nl

Due to the renewable nature and CO_2 neutrality of biomass, one may expect a large future demand for both biomass energy and biomass-based materials. Because land availability is limited, at some point choices need to be made about the type of biomass that is grown and for which purposes it is used. In this paper we compare the related CO_2 emissions of two biomass-based product cycles. In the first system, biomass is used for the production of paper which is directly used for energy recovery after consumption. In the second system, paper is largely produced from recycled fibers. The land area that becomes available by paper recycling is then used for energy crop production. Preliminary results show that the second system leads to lower energy use and reduced CO_2 emissions.

1. INTRODUCTION

There is a growing interest for both biomass-based energy and materials, due to the renewable nature and CO_2 neutrality of biomass. Current research in, for example, biofuels, biomass gasification and bioplastics supports this interest. Also, materials are shifting from oil-based products to biomass-based products, such as replacement of plastic shopping bags by paper bags in many countries.

Because land is a limited resource, at some point in the future choices may have to be made about the use of land for biomass production for various purposes. These choices are complex because insight into the whole production and consumption chain is needed in order to be able to judge the influences of choices made. An example of possible choices is the choice between short rotation biomass for energy production or long rotation biomass for material production.

Besides the type of biomass used, in many stages in the product life cycle choices need to be made about the use these biomass-based products. One of the major questions related to these choices is whether to recycle biomass-based materials or to use them directly for energy recovery? The relevance of this question is revealed by the large number of studies that are dedicated to this subject. Most studies focus on the matter of paper recycling versus energy recovery from waste paper.[1,2,3,4] Comparison of the results of these studies does not give an unambiguous answer about whether paper recycling is more environmentally advantageous than paper incineration. Causes for these differences involve the assumptions made and differences in system boundaries.[5]

Up till now it is not common to explicitly integrate land use considerations with the analysis of environmental impacts of material use. A reason for this may be that Life Cycle Assessments, which are suitable to compare environmental impacts of different life cycles, use separate impact categories for land use and other environmental effects. It is not common to integrate these impact categories.

In this paper we try to integrate land use and environmental impacts of biomass-based materials by using a broader system comparison. When all environmental impacts are taken into account such a system comparison is basically an LCA with expanded system boundaries. For this paper we just focus on land use and climate change in order to keep the comparison comprehensible. Furthermore, we use newsprint as a biomass-based material case. Such a system comparison should lead to preliminary insights in whether biomass production should be focused on material or energy production and whether newsprint should be recycled or used for energy recovery.

In this paper we compare two systems. In the first system biomass is used for the production of paper which in turn is directly used for energy recovery after consumption. In the second system newsprint is largely produced from recycled fibers. The land area that becomes available through this paper recycling is now used for energy crop production. We compare the two systems on energy use and CO_2 emissions.

2. SYSTEM DESCRIPTION

The system comparison is based on current practices in The Netherlands. The newsprint is assumed to be produced and consumed in this country. Because the forest industry in The Netherlands is small, wood pulp is imported from other countries like Sweden. In Figure 1 the two systems that are compared in this paper are depicted. In Table 1 basic input data for the analysis are presented. In this analysis the production of 1000 tons of newsprint is used as functional unit.

System A is a representation of the situation where no waste paper is recycled. Newsprint is produced and collected with other municipal solid waste after being used. It is disposed of in a waste incineration facility with energy recovery. The pulp used for newsprint production is mechanical pulp and is imported from Sweden. Mechanical pulping is an electricity-intensive process.

In order to avoid an even more extended system analysis in this preliminary study we assumed hybrid poplar plantations for fiber production because current harvest practices in Central Europe result, for example for pine, in 45% sawlogs, 34% pulpwood, 8% fuelwood and 8% waste.[8] This multi-product output makes is difficult to allocate the reduced demand for pulpwood to land use. For hybrid poplar the relation between land use and demand for pulpwood is straightforward because poplar is solely used for pulpwood production. For poplar production in Sweden we assumed a rotation period of 15 years. In system A, transport of wood and pulp by truck is needed in Sweden, transport of pulp by boat is needed to The Netherlands and within The Netherlands both transport of pulp and waste paper by truck is necessary.

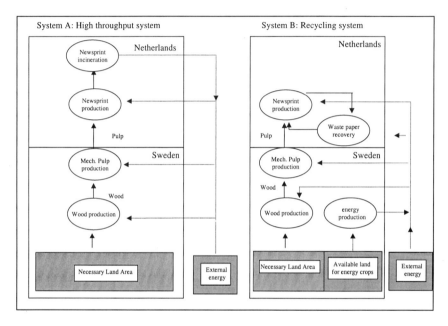

Figure 1. Schematic overview of system comparison (the dashed lines represent energy flows and the solid lines represent product flows).

Table 1
Description of input data

	Unit	Value System A+B	Value System A	Value System B	Source
Energy use truck	MJ/ton/km	2.9			(10)
Energy use boat	MJ/ton/km	0.23			(10)
International transport	km	1000			(14)
Transport in The Netherlands	km	100			
Transport in Sweden	km	80			(14)
CO_2 emissions electricity prod. Sweden	kg/GJ_{el}	15			(12)
CO_2 emission elec. prod. The Netherlands	kg/GJ_{el}	158			(12)
CO_2 em. fuel mix district heating Sweden	kg/GJprim	53			(13)
Yield hybrid poplar production	t.d.m./ha/yr		4.7		(9)
Yield willow production	t.d.m./ha/yr			9.3	(10)
Energy use willow production	GJ_{prim}/ha/yr			6.9	(10)
Energy use poplar production	GJ_{prim}/ha/yr			2.6	(14)
Energy use mechanical pulping	GJ_{el}/ADMT	8			(7)
Electricity use waste paper recovery	GJ_{el}/ADMT			2.2	(11)
Heat use waste paper recovery	GJ_{th}/ADMT			1	(11)
Electricity use newsprint production	Gjel/ADMT	1.5			(6)
Heat use newsprint production	GJ_{th}/ADMT	2.3			(6)
Electric efficiency waste incinerator	%		24		

System B represents the situation where a maximum amount of newsprint is collected and recycled after use. The recycling process is assumed to be a flotation deinking facility and is located in The Netherlands. Less wood pulp is imported from Sweden which results in reduced land use. The available land is used for short rotation willow production. The willow chips are used as fuel for district heating in Sweden. By doing so, it replaces a mixture of energy carriers, i.e., wood fuels, peat, oil, blast furnace gases and electricity, for heat pumps and boilers. For (international) transport we used the same figures as in system A.

For both systems external energy is needed. The type of energy used and the related CO_2 emissions is largely influenced by the country where the processes are located. In The Netherlands a large part of the energy supply is fulfilled by natural gas while this energy source is hardly used in Sweden. Furthermore, the CO_2 emissions related to electricity production are much lower in Sweden than in The Netherlands due to the large input of nuclear and hydro-energy in Sweden and the large input of coal and natural gas in The Netherlands.

3. RESULTS

In Table 2 the material use, energy use and CO_2 emissions of both systems are presented. Moreover, Table 2 shows the division of land use. It shows that in system B more wood on the same acreage is produced than in system A. This is the result of the higher productivity of willow for energy purposes compared to poplar for fiber purposes. Table 2 also shows that system B uses much less energy than system A; in system B even more energy is produced than consumed. The large difference in energy use is also reflected in the difference in CO_2 emission.

4. DISCUSSION

The results of the system comparison are influenced by a number of assumptions and system choices. Three aspects are important in this matter: selection of technologies, selection of countries and determination of reference energy sources.

The selection of technologies is important, the efficiency of the processes as well as fuel input. In this paper we have chosen paper incineration in a municipal solid waste incinerator and combustion of willow for district heating. Other possibilities are conversion of waste paper and biomass to electricity using incineration in biomass incinerators, co-firing in pulverized coal boilers and biomass with combined cycle.

In addition to choices of conversion technologies, choices of pulp processes will greatly affect the results. In this paper we have analyzed the situation for newsprint using mechanical pulp. For chemical pulp that uses lignin for energy production the outcome may differ.

The effect of the country-specific data on the system comparison is shown in Table 2 by the large consumption of CO_2-extensive electricity in Sweden. The differences in carbon intensity of electricity between Sweden and The Netherlands not only influences the CO_2 emissions of processes located in Sweden or The Netherlands but also the avoided CO_2 emissions when electricity is produced.

CO_2 emissions are avoided when electricity or primary energy sources are produced that replace fossil fuels or fossil fuel-based electricity. For the analysis in this paper we have assumed that biomass production in Sweden would be used for district heating since biomass is already used intensively for this purpose.[13] The question remains: what type of fuel is replaced when extra biomass is used for district heating? In this analysis we assumed that the biomass produced in system B replaces the mixture of energy carriers that is now used for district heating in Sweden. The choice of alternative energy sources greatly influences the outcome of this system analysis. If we had chosen that the produced biomass be converted into electricity to replace Swedish electricity, then the total CO_2 emissions of system B would amount to +587 tons. In this case system A should be preferred over system B from a CO_2 emission point of view. The choice of alternative energy sources is also greatly dependent on country specific conditions. However, if the entire production system is located within one country, then differences in CO_2 intensity of electricity would not exist and the replaced primary energy sources would be equal in both systems. In that case the difference in CO_2 emissions between system A and B would be from the same order of magnitude as the difference in energy use. Recycling is then preferred from a CO_2 point of view when the created land space is used for energy purposes.

Table 2
Land use, material use, energy consumption and CO_2 emissions of the two systems compared in this paper for a fixed surface of land and 1000 tons of paper

	Process	Quantity (ADMT)	Energy use (TJ_{prim})[1]	CO_2 emissions (tonnes)[1]	Land use (ha.yr)
System A	Paper making	1000	6.3	338	
	Pulp making	1100	21.8	128	
	Poplar production	1198	0.6	45	231
	Transport		0.6	49	
	Paper incineration		-10.9	-691	
	Total of processes		18.4	-131	
	Total land use				231
System B	Paper making	1000	6.3	338	
	Pulp making	267	5.3	31	
	Deinked pulp making	833	5.5	326	
	Poplar production	290	0.7	51	56
	Willow production	1788	1.2	90	175
	Transport		0.3	21	
	Biomass incineration		-29.6	-1534	
	Deinking sludge incineration		-2.2	-140	
	Total of processes		-5.1	-816	
	Total land use				231

[1]negative figures represent energy production or avoided CO_2 emissions.

5. CONCLUSIONS

Under the standard assumptions made in this paper, maximum recycling of newsprint has smaller CO_2 emissions and less energy use than direct incineration of waste paper. However, other definition of the reference system may turn the results regarding CO_2 emissions around. From an energy point of view it is always best to close the paper cycle as much as possible and use the created land space for energy crops.

Including land use in system comparisons of different product life cycles is useful. It can lead to insights in how to combine both land and material flow management.

This study was a preliminary approach in which land use was explicitly integrated in an environmental system comparison. Follow-up work will elaborate on the sensitivity of the regional and technological choices made and on the methodological questions regarding the choice of reference energy systems.

REFERENCES

1. Virtanen, Y and S. Nilsson (1993). Environmental Impacts of waste Paper Recycling, International Institute for Applied Systems Analysis, Earthscan Publ. Ltd., London, U.K.
2. British Newsprint Manufacturers Association (1995). Recycle or Incinerate? The Future for Used Newspapers: an Independent Evaluation, U.K.
3. Byström and Lönnstedt (1995). Waste Paper Recycling: Economic and Environmental Inpacts, Proceedings, LCA - A challenge for Forestry and Forest Industry, Hamburg, Germany.
4. Kärnä et al. (1994). Life cycle analysis of newsprint: European scenarios, Paperi ja Puu, 76, pp. 232–237.
5. Teeuwisse, S. (1998). A Comparison of Life Cycle Analyses about Paper and Board Products, Depart. of Science, Techn. and Society, Utrecht University, The Netherlands.
6. Nilsson, L.J. et al. (1995). Energy Efficiency in the Pulp and Paper Industry, American Council for an Energy Efficient Economy, Washington DC, USA.
7. Komppa, A. (1993). Paper and Energy: A Finnish view, Proceedings, Int. Conference on Energy Efficiency in Process Technology, Commission of the European Communities, Brussels, Belgium.
8. Muiste, P. (1997). Actual Forestry Market in Estonia, Import of Biomass, Workshop document of a study trip to Estonia and Sweden, Biomass Technology Group, Enschede, The Netherlands.
9. Christie, J.M. (1994). Provisional Yield Tables for Poplar in Britain, Forstry Commission technical Paper 6, Forestry Commission, Edinburgh, U.K.
10. Börjesson, P.I.I. (1996). Energy analysis of biomass production and transportation, Biomass and Bioenergy, 11, pp. 305–318.
11. McKinney, R.W.J. (1995). Technology of Paper Recycling, Glasgow, U.K.
12. OECD (1997). Energy Balances of OECD countries, 1994–1995.
13. Nutek (1997). Energy in Sweden 1997, Stockholm, Sweden.
14. Agterberg, A.E and A.P.C. Faaij (1998). Bio-Energy Trade, Possibilities and constraints on short and longer term, Department of Science, Technology and Society, Utrecht University, The Netherlands.

OPTIMIZING USED RECYCLED PULP IN PAPER PRODUCTION

M.D. Berni and S.V. Bajay

State University of Campinas—UNICAMP*, Interdisciplinary Centre for Energy Planning – NIPE, Department of Energy/School of Mechanical Engineering, CEP 13-083-970, P.O. Box 6122, Campinas, SP, Brazil, Fax. 55.019.239.3722, email: bajay@fem.unicamp.br

In recent years interest in paper recycling has grown, as part of the increase in popular awareness of environmental quality problems. Paper makes up about half of the weight of municipal refuse and municipal refuse disposal is an increasingly difficult problem. Landfill sites near large cities, which offer the cheapest disposal method, are filling up at an increasing rate as the per capita solid waste generation continues to increase. Costs of acquiring more distant sites also continue to rise, as do transportation costs, hence reducing the total amount of refuse for disposal is of increasing economic importance, even aside from the environmental quality issue.

This paper presents a case study with application of a linear programming model for production use and recycling of various paper and related products. This model is used to estimate the maximum feasible recycling rate given the current state of paper technology. The objective of this case study is to establish the quantities of each product to be manufactured, according to profit maximization criteria, regarding the market restrictions, capacity of the equipments and of process, quality final product, energy, environmental and social aspects.

497

A LINEAR PROGRAMMING TOOL FOR OPTIMIZING BIO-ENERGY AND WASTE TREATMENT SYSTEMS (*)

B. Meuleman, A. Faaij

Department of Science Technology and Society, Faculty of Chemistry, Utrecht University, Padualaan 14, 3584 CH Utrecht, The Netherlands. Tel. 31-30-2537643/00 fax. 31-30-2537601, e-mail: A.Faaij@chem.uu.nl.

In many regions a virtually unlimited variety of bio-energy chains can be considered given a (complex) mixture of biomass resources, organization of logistics and possible conversion systems. One would like an optimal (either in terms of costs or in terms of energy production) utilization of (often-limited) resources. However, the realization of bio-energy projects do not seem to lead to optimal utilization of available resources in many regions (or countries) either from an economic or especially from an energy point of view. This problem is particularly pressing in the Dutch context where CO_2 mitigation policies have led to a large demand for the limited biomass resources. Maximizing the net energy production from those resources at minimal costs is a highly relevant question (and not only in the Netherlands).

This work has focused on the development of a tool that can facilitate optimal planning of new bio-energy projects in a region, given a local biomass availability (and characteristics), a feasible technology mix (with specific characteristics), existing infrastructure, economic conditions and the like.

1. MODEL DESIGN

After an inventory of modelling approaches, the second step of this study was the creation of a model design. Figure 1 and Figure 2 give a simple impression of this design. The model includes optimization procedures for identifying optimal systems given resources, selected technologies and existing capacity. The model has been developed in a combination of Excel and Visual Basic, which allow for easy access to various parameters and the database.

(*) This work is funded by NOVEM (Dutch Organization for Energy and Environment and Hanze Environment NV., a daughter of the electricity distribution company EDON. Work has been carried out in collaboration with Hanze Environment and the 'Region Twente' organisation.

A large part of the model exists of two databases. The first database includes the case-independent data and the second includes the case dependent information.

The case-independent information contains data on biomass conversion technologies and waste treatment options. This information includes estimates of performance and costs versus scale (described by trend lines). This first phase of the work focused laid on options that result in electricity production. Other examples of case-independent information are the biomass characteristics and the logistic systems, reimbursements levels, interest rate, and lifetime of the plants. Also, a matrix describing whether a given biomass stream can or cannot be converted to energy per technology included is taken up.

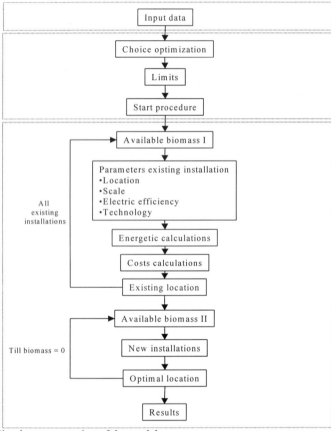

Figure 1. Simple representation of the model.

The case-dependent information exists of a 'distance table' for the region, the road network and the existing capacity for biomass conversion or waste treatment. Crucial case dependent information included into the database is the available biomass in the region. For this study we focused on the "Twente" region in the east of the Netherlands. For this case the spatial distribution of the biomass resources is given by quantities per municipality in the region. Those data were partly supplied through local parties and partly by translating available statistics to the local level.

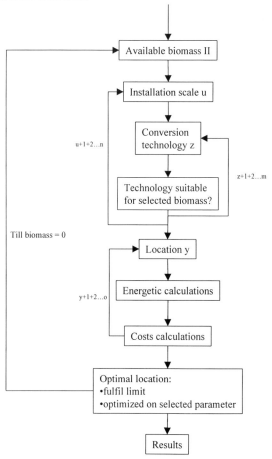

Figure 2. Procedures after calculations with the existing installations.

1.1. Running the model

For a procedure for a certain region, the model can start from an inventory of the existing conversion units in the region itself and the surrounding regions. The model calculates biomass and waste utilised in those units and subtracts these quantities from the total available biomass. For a region existing of x municipalities, the model makes an inventory of all biomass streams (input data) within a specific municipality (y) and all the biomass in the surrounding municipalities. Then the model includes in this municipality (y) a biomass conversion technology (z) with a specific scale (u) and calculates the needed figures. In the following loop the scale is increase by one step (u+1) and this procedures is repeated until all scales get a change (u, u+1, u+2, ...n). Afterwards the conversion technology is changed to the second technology (z+1) and all the scales are recalculated. If all the conversion technologies get a change (z, z+1, z+2, ...m) the following municipality (y+1) is used in the calculations and all the scales and conversion technologies are recalculated. In this way the model fills a 3D matrix with needed figures, as shown in Figure 3. If the matrix is filled the model selects the cell with a location, a scale and a conversion technology that fulfill the criteria selected (such as lowest waste treatment costs or maximum energy production). After subtracting the needed biomass for this new plant the procedures are repeated until all available biomass used. In this way a list is produced with conversion technologies of specific scales and specific locations.

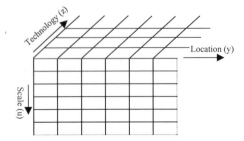

Figure 3. Filling of the 3D matrix by the model.

1.2. Case study of the region Twente

After building the model and including the information into the databases the model was demonstrated for a concrete region in the east of the Netherlands. This region has a diameter of approximately 50 km and it contains 23 municipalities. This area produced about 263 $kton_{1995}$ of biomass—such as organic household waste, pruning, verge grass, sewage sludge. Biomass conversion units in this region exit are two dumping grounds (and three in the surrounding region), a composting unit, municipal incinerator and 13 wood combustion units with a capacity of 38 kton of demolition wood. Table 1 gives an overview of some of the results that can be obtained for the case study region Twente. By the optimization of the biomass streams it was possible to increase the primary energy saved by the given input of 0.2 $GJ/ton_{biomass\ wet}$ to 9.9 $GJ/ton_{biomass\ wet}$. (primary energy saved). A wide range of sensitivity analyses can be executed with the model. The examples given in the table show a high sensitivity to changes in heating values of biomass streams and to the limiting of the waste treatment costs. Other possibilities are varying the supply, existing capacity, technology performance or technology mix considered.

Table 1
Some modelling results of the case study Region Twente. The first column gives the results of the calculations of the model. The following columns give a sensitivity analysis executed with the model by multiplying the transport distances by a factor 2 and 4 (1 Dfl. = 1.90 U$)

		Multiplying transport distances by a factor 2 and 4	
	1	2	4
Location	Borne	Borne	Borne
Collecting biomass (GJ)	-32732	-32732	-32732
Transfer biomass (GJ)	-2546	-2546	-2546
Transport (GJ)	-8158	-16317	-32633
Conversion (GJ)	2572622	2572622	2572622
Total (GJ)	2529185	2521027	2504710
Share collecting biomass (%)	1.27	1.27	1.27
Share transfer (%)	0.10	0.10	0.10
Share transport (%)	0.32	0.63	1.27
Available biomass (ton)	263319	263319	263319
Needed biomass (ton)	254636	254636	254636
Remaining biomass (ton)	8684	8684	8684
Scale (MWe)	44.00	44.00	44.00
Efficiency (%)	43.54	43.54	43.54
Conversion technology	BV/STEG	BV/STEG	BV/STEG
Produced electricity (MWh)	308000	308000	308000
Annual depreciation (Dfl/year)	16316399	16316399	16316399
O&M (Dfl/year)	163281	163281	163281
Fuel costs (Dfl/year)	15278155	15278155	15278155
Logistic (Dfl/year)	13207846	13526023	14162375
Costs* (Dfl)	44965681	45283858	45920210
Costs (Dfl/GJ)	17.78	17.96	18.33
Costs (Dfl/kWh)	0.15	0.15	0.15
Reimbursement produced electricity (Dfl/year)	16016000	16016000	16016000
Reimbursements installed capacity (Dfl/year)	9240000	9240000	9240000
Waste treatment costs ** (Dfl/year)	19709681	20027858	20664210
Waste treatment costs (Dfl/ton)	77.40	78.65	81.15
Energy production (GJ/ton)	9.93	9.90	9.84

* annual depreciation+O&M+fuel costs+logistics
** annual depreciation+O&M+fuel costs+logistics- reimbursements

Some main assumptions for the case study of region Twente:
- technologies included in the database: digestion, composting, combustion, gasification;
- waste streams included: verge grass, organic domestic waste, sewage sludge, public garden waste, organic components household waste;
- reimbursement for produced electricity: 0.052 Dfl/kWh;
- reimbursements for installed capacity: 210 Dfl/kW.

2. CONCLUSIONS AND FURTHER WORK

The developed tool is able to calculate the performance of thousands of potential bio-energy chains for a given area in short time. It seems to be an ideal instrument for analyzing various scenarios for organizing waste treatment and production of energy from available biomass in a given region. Optimization procedures allow for identification of least-cost options, highest net energy production and optimal location selection. Furthermore, the tool allows for complex sensitivity analyses for evaluating the impact of different options. Existing waste treatment and conversion capacity can be included, as well as different technology characteristics, biomass characteristics and biomass (and waste) supply.

A concrete case study has been carried out for the Twente region in the east of the Netherlands to illustrate the functionality of the model. Given sufficient data, the tool is usable for optimal planning in any region.

However, the model also lacks a number of aspects that are to be included in an improved version: utilization of heat on a local level, and increasing the data set describing the performance of various technologies, and including a larger number of logistic options.

The work has raised various new questions for further improvement and sophistication of the model. A next phase is foreseen for the coming year. Heat production, supply and demand, more conversion options and logistic systems as well as the use of GIS are being considered. Furthermore, attention is paid to the influence on costs in relation to applicable emission standards. A second phase is currently planned to create an extended version of the model on a national scale.

REFERENCE

1. Meuleman, B. and A. Faaij (1998). An optimization model for organic waste streams, Department of Science Technology and Society, Faculty of Chemistry, Utrecht University, The Netherlands.

THE CONCEPT OF AN FAO-INTEGRATED ENERGY FARM: A STRATEGY TOWARDS SUSTAINABLE PRODUCTION OF FOOD AND ENERGY

N. El Bassam and W. Bacher

Institute of Crop Science and Grassland Research; FAL, Bundesallee 50, D – 38116 Braunschweig, Germany, Tele: 0049-531-596-605; FAX: 0049-531-596 365, e-mail: elbassam@pf.fal.de

More than two billion people worldwide, especially in rural areas, lack access to modern energy resources, and their current approaches to energy are unsustainable and not renewable. In order to face this challenge a concept of an Integrated Energy Farm (IEF) is being developed within the frame work of the SREN, FAO, and UN activities in this field. The IEF is a farming system model with an optimal energetic autonomy including food production and, if possible, energy export. Energy production and consumption are based mainly on non-polluting renewable sources, such as modern biomass, wind and solar electricity and heat generation systems.

1. INTRODUCTION

Energy is directly related to the most critical social issues which affect sustainable development: poverty, jobs, income levels, access to social services, gender disparity, population growth, agricultural production, climate change and environment quality, and economic and security issues. Without adequate attention to the critical importance of energy to all these aspects, the global social, economical and environmental goals of sustainability cannot be achieved. Indeed, the magnitude of change needed is immense, fundamental and directly related to the energy produced and consumed nationally and internationally.

The importance of energy in agricultural production, food preparation and consumption is evident and essential. The world still continues to seek energy to satisfy its needs without giving due consideration to the social, environmental, economic and security impacts of its uses. Current approaches to energy are unsustainable. It is the responsibility of political institutions to ensure that the research and the development of technologies supporting sustainable systems be transferred to the end users. Scientists must bear the responsibility in understanding the earth as an integrated whole and the impact of our actions on the global environment in order to ensure sustainability to avoid disorder in the natural life cycle.

The key challenge to realizing these targets is to overcome the lack of commitment and to develop the political will to protect people and the natural resource base. Failure to take action will lead to continuing degradation of natural resources, increasing conflicts over scarce resources and widening gaps between rich and poor. We must act while we still have choices. Implementing sustainable energy strategies is one of the most important levers humankind has for creating a sustainable world. Most present trends in energy indicate a deteriorating situation. Furthermore, current energy patterns are aggravating this process by an over-preoccupation with centralized energy supply and fossil fuels to the detriment of energy efficiency, decentralized supply and renewable energy. Business-as-usual energy patterns and conventional approaches to energy are contributing to the sustainability. Thus any attempt to tackle the social, environmental, economic and security issues, as was done by the United Nations conference in Rio, must pay full attention to the energy aspects.

In order to meet challenges, the future energy policies should put more emphasis on developing the potential of energy sources, which should form the foundation of future global energy structure. In this context the FAO, in support of the Sustainable Rural Environment and Energy Network (SREN), is developing a concept model for the optimization and evaluation of an Integrated Energy Farm (IEF) and the preparation and verification of joint projects.

2. THE INTEGRATED ENERGY FARM

2.1. Concept

An integrated energy farm (IEF) is a farming system model with an optimal energetic autonomy including food production and, if possible, energy exports. Energy production and consumption at the IEF has to be environment-friendly, sustainable and eventually based mainly on renewable energy sources. It includes a combination of different possibilities for non-polluting energy production, such as modern wind and solar electricity production, as well as the production of energy from biomass.

The IEF concept includes a decentralized living area from which the daily necessities (food and energy) can be produced directly on-site with minimal external energy inputs. The land of an IEF may be divided up into compartments to be used for growing food crops, fruit trees, annual and perennial energy crops and short rotation forests, along with wind and solar energy units within the farm.

An integrated energy farming system based largely on renewable energy sources would seek to optimize energetic autonomy and ecologically semi-closed system while also providing socio-economic viability and giving due consideration to the newest concept of landscape and bio-diversity management. Ideally, it will promote the integration of different renewable energies, promote rural development and contribute to the reduction of greenhouse gas emission.

2.2. The objectives

The project is an activity of the SREN of the European System of Cooperative Research Networks in Agriculture (ESCORENA). Under the leadership of FAL and the SREN

working group on Biomass Production and Conversion of Energy, a group of SREN member scientists and others will collaborate to
- develop the conceptual framework of the IEF model,
- gather the required information,
- elaborate the results into a functional model,
- publish the results, and
- prepare a project proposal for experimental verification of the model.

The overall objective is that the IEF concept be successfully introduced into agricultural production systems which have to be completely sustainable.

As a contribution to the above, this project seeks to
- develop a workable scientific model for estimating the requirements and feasibility of an integrated energy farm ready for experimental verification or demonstration in Europe, and
- prepare a project proposal for experimental verification (demonstration farms) of the model under various conditions.

The demonstration farms will also offer possibilities for demonstration and education. Data will be collected and models developed in close collaboration with the Sustainable Development Department Group of the regional office for Europe (REUS), the Environment and Natural Resources Services of the Sustainable Development Department (SDRN) and other technical departments of FAO. This collaboration will assure adaptability of the model to non-European regions and the mutual benefit of this project from and to ongoing FAO programs (e.g., World Food Submit Action Plan, Strategy 2000) and to future FAO initiatives such as "Energizing the Food Chain" and in organic farming.

3. RESULTS

3.1. Global approach

A minimum of basic data should be available for the verification of an IEF. Various climatic constraints, water availability, soil conditions, infrastructure, availability of skills and technology, population structure, flora and fauna, common agricultural practices and economic and administrative facilities in the region should be taken into consideration.

From a global point of view, the major assumptions have been established in order to evaluate the possible contribution of the major renewable energy sources to support food production for a given area.

Climatic conditions prevailing in a particular region are the major determinants of agricultural production. In addition, other factors, such as local and regional needs, availability of resources and other infrastructure facilities also determine the size and the product spectrum of the farmland. The same requisites also apply to an IEF. The climate fundamentally determines the plant species and the intensity of cultivation needed for energy production on the farm. Moreover, climate also influences the energy-mix (consisting of biomass, wind and solar energies) at a given location; and what single technology can be installed also depends decisively on climatic conditions of the locality in question. For example, cultivation of biomass for power generation is not wise in arid areas. Instead a larger share can be allocated to solar energy techniques in such areas. Likewise, coastal regions are ideal for wind-power installations.

Taking these circumstances into account, a scenario was developed for an energy farm of 100 ha in definite climatic regions of northern and central Europe, southern Europe, northern Africa and Sahara and Equatorial regions. It was presumed that one unit of this size needs about 200 megawatt-hours (MWh) heat and 100 MWh power per annum for its successful operation. A need for approximately 8,000 liters of fuel per annum is calculated. The possible shares of different renewable energies are presented in Table 1.

Table 1
The possible share of different renewable energies in diverse climatic zones

Climate Region	Energy source	Power production (% of total need)	Heat production (% of total need)	Biomass need (t/a)	Biomass area (% of the total area)
North and Central Europe	Solar 200 m^2	7	15		
	Wind 100 kW	100	-	60	12
	Biomass	100	105		
South Europe	Solar 250 m^2	12.7	40		
	Wind 100 kW	100	-	36	4.8
	Biomass	70	65		
North Africa Sahara	Solar 300 m^2	21	90		
	Wind 100 kW	75	-	14	1.2
	Biomass	25	25		
Equatorial region	Solar 200 m^2	18.2	37.5		
	Wind 100 kW	45	-	45	10
	Biomass	70	80		

It is evident that in Europe, wind and biomass energies contribute the major share of the energy-mix, while in North Africa and the Sahara the main emphasis lies with solar and wind energies. Equatorial regions offer great possibilities for solar as well as biomass energies and little share is expected of wind energy in these regions. Under these assumptions, in southern Europe, Equatorial regions and north and central Europe, a farm area of 4.8, 10 and 12 percent, respectively, would be needed for cultivation of biomass for energy purposes. This area corresponds to annual production of 36, 45 and 60 tons for the respective regions.

In North Africa and Sahara regions, in addition to wind and solar energy, 14 tons of biomass from 1.2 percent of the total area would be necessary for energy provisions.

Moving ahead, in order to broaden the scope and seek the practical feasibility of such farms, the dependence of local inhabitants (end users) is to be integrated in this system. Roughly 500 persons (125 households) can be integrated with one farm unit. They have to be provided with food as well as energy. As a consequence, the estimated extra requirement of 1,900 MWh of heat and 600 MWh of power have to be fulfilled from extra sources. Under the assumption that the share of wind and solar energy in the complete energy provision remains at the same standard as without the households, 450 tons of dry biomass must be produced to fulfill this farm demand at a location in north and central Europe. For the production of this quantity of biomass, 20 percent of farm area is to be dedicated to cultivation. In southern Europe and the Equatorial regions, 15 percent of the land area should be made available for the provision of additional biomass.

4. LAYOUT OF AN IEF

An ideal IEF consists of adequate buildings and other infrastructure facilities. At the smallest level the IEF is divided into several compartments that are used for different purposes. The nucleus of the farm could be dedicated to a main building and other storage and energy installations. The compartment may be used for cultivation of crops for daily necessities such as fruits and vegetables, spices and medicinal herbs, floricultural plants and for animal husbandry. Some portion of this segment can be allotted to workshops and parking purposes. Further rings may be used for cultivation of annual energy crops, cereals, oilseeds, sugar and starch crops. The perennial energy crops and energy forest are proposed in the outermost compartment so that they may provide shelter to the inner ring crops. The energy farm is also intended to have provisions for fish and bee farming (apiculture).

In order to generate its maximum output, it is necessary to determine the most appropriate equipment and facilities to be located on the farm. These would probably include
- wind mills
- solar collectors (thermal)
- solar cells (PV)
- briquetting and pelleting machines for biomass
- bio-gas production units
- power generators (bio-fuels, wind and PV operated)
- bio-oil extraction and purifying equipment
- fermentation and distillation facilities for ethanol production
- pyrolysis unit
- Stirling engines
- water pump (solar wind and bio-fuel operated)
- vehicles (solar, bio-fuels operated or draught vehicles)
- monitoring system
- cooking stoves

The overall objective is to promote a complete sustainable farming system. Depending upon this objective, the verification of demonstration farms in various regions of the world has to be prepared taking into account the following influential factors:
- impact, influence and needs of climate, soil and crops
- ratio of required food and bio-fuel production
- input requirement for cultivation, energy balance and output : input ratio
- equipment choices (wind, solar and biomass generation and conversion technology)

The information so collected from different regions would support decisions about the optimal farm size, its regional adaptation and would solve other related constraints.

5. REGIONAL IMPLEMENTATION

In an attempt to verify the implementation of the IEF in practical sense at a regional level taking into consideration the climatic and soil conditions, planning work has been started at Dedelstorf (northern Germany). An area of 260 ha has been earmarked for this farm, which would satisfy the food and energy demands of 700 respondents. For the settlement purposes, old military buildings are being renovated. A period of three years will be needed to complete the project. The main elements of heat and power generation will be solar generators and collectors, wind generator, biomass combined heat and power generator, Stirling engines and a bio-gas plant. The total energy to be provided amounts to 8,000 MWh heat and 2,000 MWh power energy. The cultivation of food and energy crops will be according to ecological guidelines. The energy plant species foreseen are short rotation coppice, willow and poplar, *Miscanthus*, polygonum, sweet and fibre sorghum, switchgrass and reed canary grass and bamboo. A research, training and demonstration centre will accompany this project.

ECONOMIC EVALUATION OF TECHNICAL, ENVIRONMENTAL AND INSTITUTIONAL BARRIERS FOR DIRECT-COMBUSTION BIOMASS POWER PLANTS IN CALIFORNIA

Prab Sethi,[a] Valentino Tiangco,[a] David Yomogida,[b] Virak Dee,[a] Ying Lee,[c] and George Simons[a]

[a]California Energy Commission, Research and Development Office, Sacramento, California 95814
[b]Henwood Energy Services, Inc., Sacramento, California 95833
[c]Lawrence Livermore National Laboratory, Livermore, California 94551

This paper describes with the economic effects of various technical, environmental, and institutional barriers in the direct-combustion of biomass fuels for power generation. An economic model was used to project the cost of electricity for a "base case" direct-combustion biomass facility for each barrier. Based on these calculations, the California Energy Commission will be able to identify the greatest barriers to economical biomass energy production and target those for future research funding.

Biomass Transformation into Value-Added Chemicals, Liquid Fuels, and Heat and Power

Value-Added Products: Lignins, Pyrolytic Oils, and Carbons

513

APPLICATION OF THE SLOW PYROLYSIS EUCALYPTUS OIL TO MAKE PF RESINS

J.D. Rocha,* S.S. Kelley, and H.L. Chum

National Renewable Energy Laboratory, 1617 Cole Boulevard, Golden, Colorado, 80401-3393, USA

A crude bio-oil produced during the slow pyrolysis of eucalyptus wood was used to partly replace the petroleum-derived phenol used for preparation of phenol formaldehyde wood resins. Large quantities of this bio-oil are currently produced by the Brazilian charcoal industry and are available as a low-cost feedstock. In the local market this bio-oil is priced at US $200 per ton while phenol is priced in the range of US $750 to 900 per ton. A series of bio-oil modified phenol formaldehyde resins were prepared using different resin formulations. The levels of phenol replacement (25, 50 and 75%), the formaldehyde to phenol ratio, (F/P = 1.0 and 1.2) and the base to phenol ratio (N/P = 0.3 to 0.5) were all studied. The resins were prepared using a standard temperature profile and cooked to a target viscosity. Small wood composites were used to test the curing and performance properties of these resins before and after accelerated aging. Both the bond strength and the wood failure values were measured. These results indicate that phenol formaldehyde resins containing up to 50% eucalyptus bio-oil can be used to prepare resins with commercially attractive properties.

1. INTRODUCTION

Slow carbonization is widely practiced in Brazil. The main product is charcoal (average yield of 25% wt of initial wood) to supply the pig iron, steel and iron-alloy industries. Owing to its sulfur-free characteristic and low ash content, charcoal is a better iron reduction agent than coke, and it produces high purity final metals. The charcoal technology based on masonry kilns is still very traditional and inefficient. Recent process improvements allow for recovery of the pyrolitic liquids.[1]

---

*Visiting Researcher at NREL from Sept. 97 to Sept. 99 sponsored by FAPESP (026260/97), the São Paulo State Research Foundation in Brazil.

Currently, pyrolysis vapors are emitted to the atmosphere resulting in air pollution problems and a lost opportunity cost. Identification of commercial uses for these bio-oils can improve the profitability of the charcoal manufacturing process, decrease the emissions, and help develop the economy in rural areas. The current production rate for the bio-oil is about 200 metric tons a month. This production is planned to increase up to 2,000 tons a month by the next year. Thus there is an ever increasing need to identify uses for these bio-oil.

Both in Brazil and internationally there has been a considerable amount of research into finding new uses for these bio-oils, including uses as fuels or chemicals. Because of the inherent phenolic character of these bio-oils, phenolic resins appear to be a reasonable alternative use. With the high local price of petroleum-derived phenol, it appears that the replacement of pure phenol in phenol formaldehyde (PF) resins with phenolics derived from bio-oil can play an important role in the Brazilian market.[2,3]

Research conducted at the National Renewable Energy Laboratory (NREL) on uses for fast pyrolysis oil strongly supports the concept of replacing petrochemical phenol with phenolics derived from biomass pyrolysis oil.[4,5,6] While this work demonstrates the feasibility of this technology where fast pyrolysis oils are the source of the natural phenolics, there is much less known about the use of the slow pyrolysis oils produced by the charcoal production process.

Phenolic resins are used worldwide for wood composites, foundry and friction materials, insulation and abrasives. The worldwide consumption was estimated to be 2,250 thousands of metric tons in 1993. The USA is the biggest consumer of PF resins at about 30% of the total, followed by European countries with 20%; Brazil consumes about 3% or 60,000 metric tons per year. An overview of the Brazilian market indicates that phenol production is 115,000 metric tons per year and about 50,000 metric tons is used to produce phenolic resins.[7,8,3] Thus there is a good match between the potential production volumes for bio-oil phenolics and the volumes of phenol used for PF resins.

2. EXPERIMENTAL

Eucalyptus oil was produced by slow carbonization, at a maximum temperature of 600°C, with an oil recovery system connected to a masonry kiln. The sample was supplied by Biocarbo Ind. Com. Ltda in Brazil. This sample is the water-insoluble organic fraction that was decanted from the aqueous components. The as-received sample had the following properties: pH 2.75 at 25°C, viscosity of 132 cP at 25°C and 43.5 cP at 40°C (Brookfield), water content of 13.7% and $M_n = 260$, $M_w = 470$ by GPC.

The PF resin formulation variables included the phenol-formaldehyde mole ratio (F/P), sodium hydroxide-phenol mole ratio (N/P), percentage of organic solids and percentage of phenol substitution by the pyrolysis oil. In calculating the mole ratios the bio-oil was treated as phenol. The resin solids were 45%, which is typical for a PF plywood resin. A 50% sodium hydroxide was used to neutralized the acidic functionalities in the bio-oil and raise the pH of the mixture to around 11. High purity paraformaldehyde and phenol were used as received. Table 1 shows the values for all variables examined in this study.

Table 1
Variables in resin formulation

(F/P) ratio	(N/P) ratio	Phenol replacement by wood oil (%)
1.0	0.3	25
1.2	0.4	50
1.2	0.5	75

All components were charged to a 500-ml three-neck round bottom flask equipped with a reflux condenser, a thermometer, and a mechanical stirrer. The heat was supplied through an external heating mantle connected to the thermocouple. The temperature profile was the same for all samples. The mixture was heated to 70°C for 90 minutes, and then to 90°C until the solution viscosity was equivalent to stages to U-V as measured with Gardner-Holt viscosity tubes, about 1,000-1,200 cP. The resin was then quenched and stored in a freezer until needed for preparation of the wood composites.

The wood composites used to measure the resin quality were made according to the European Standard EN 301 for adhesives, phenolic and aminoplastic, using hard maple wood as the substrate. Between 100-200 mg of resin were applied to a 25.4 × 25.4 mm area at one end of each stick. After an open assembly time of 10 minutes at ambient conditions, the sticks were lapped over the length of their coated ends. The over-lapped stick assembly was then hot pressed at 200 psi and 160°C for 2, 2.5, 3.5 and 6 minutes, using a Carver Laboratory Press, Model-C, 12-ton capacity. Five wood composite samples were prepared for the shorter press times and 10 composite samples were prepared for the 6 minute press time. The samples pressed for 6 minutes were separated into two groups; the first was tested dry and the second was subjected to an accelerated aging sequence. After bonding, all specimens that were tested "dry" were reconditioned at 51% relative humidity for at least 24 hours prior to mechanical testing. Samples that were tested "wet" were boiled in water for 4 hours, followed by drying for 20 hours at 60°C and 15 mm Hg and boiled again for 4 hours and kept in plastic bag until the mechanical tests. Mechanical properties were measured with an Instron Testing machine. Maximum tensile load was recorded for each specimen and averaged for the five specimens prepared for each resin press time combination. A commercial plywood PF resin obtained from Georgia Pacific Resins (GPR) was used as a standard.

The wood failure was visually estimated based on the ASTM D 5266-92 standard procedure. The final viscosity, and for same samples, the viscosity in all measurement during the cooking time, was made in a Brookfield viscometer using appropriate spindle. The pH was for each sample measured in an Orion Research 601A pH meter.

3. RESULTS AND DISCUSSION

Three levels of phenol replacement were examined, 25, 50 and 75 wt.%. The individual resin formulations are shown in Table 2. Several of the formulations "failed" due to phase separation of the sample during the cook. The main cause of the phase separation appeared to be related to the low solubility of the phenolic polymers as the molecular weight increased.

A series of resins, BC-01 through BC-05, were prepared with different phenol/oil, F/P and N/P ratios. A standard cooking profile was used and the cooks were terminated when the resin reached the target viscosity of 2,000 cP. These resins were then used to bond lap-shear samples. The mechanical properties of the lap-shear samples were measured as described above. These initial tests showed that high quality resins could be prepared using 50% bio-oil and F/P ratios of 1.0 to 1.2. The N/P ratio was also an important factor that influenced the quality of the bonded composite and needed to be between 0.4 and 0.5.

Table 2
Samples conditions and properties

Resin Sample	Phenol/Oil ratio	F/P	N/P	pH	Viscosity (cP)
GPR (Control)	100/0	n.a.	n.a.	11.94	1,100
BC-01	75/25	1.2	0.3	10.95	2,700
BC-02	50/50	1.2	0.5	11.98	1,700
BC-03	50/50	1.0	0.5	12.48	1,000
BC-04	50/50	1.0	0.4	11.36	1,800
BC-05	25/75	1.2	0.5	11.45	2,300
BC-06 (30 min.)	50/50	1.2	0.5	12.11	1,900
BC-07 (60 min.)	50/50	1.2	0.5	11.90	2,500
BC-08 (120 min.)	50/50	1.2	0.5	11.90	2,000
BC-09 (120 min.)	50/50	1.0	0.5	12.08	1,370
BC-10 (150 min.)	50/50	1.2	0.5	11.93	1,420
BC-11 (150 min.)	50/50	1.0	0.5	12.06	1,800

Note: Time in parenthesis refer to bio-oil delayed addition. n.a. means not available.

Based on these initial tests a second series of resins where prepared, i.e., BC-06, BC-07, BC-08, BC-09. The resins used the same phenol/oil, F/P and N/P ratios, but they varied the time between the start of the resin cook and the addition of the bio-oil, e.g., the "delay time" (see Table 2). Again the cooks were terminated when the resin reached the target viscosity of 2,000 cP. These resins were then used to bond lap-shear samples and the mechanical properties were measured. For the fully cured samples, e.g., those pressed for 6.0 minutes, the mechanical properties and wood failure values of all of the bio-oil based resins were better than or equal to those of the commercial control, except for sample BC-06 where the bio-oil was added at the start of the resin cook. For the samples that were cured for only 3.5 minutes the mechanical strength of BC-06 was again markedly lower than any of the other bio-oil based resins or the commercial control.

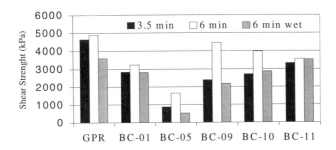

Figure 1. Shear strength in kPa for five different eucalyptus oil resins and GPR.

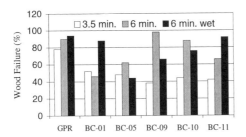

Figure 2. Wood failure for GPR and five eucalyptus oil resins.

Finally a third set of resins was produced to compare small variations in the F/P ratio on the properties of the bio-oil based resins. For these resin, BC-08 through BC-11 the F/P ratio was varied between 1.0 and 1.2. Again the cooks were terminated when the resin reached the target viscosity of 2,000 cP. These resins were then used to bond lap-shear samples and the mechanical properties were measured. In general the mechanical strength and wood failure values are comparable to the commercial control. The actual values are shown in Figures 1 and 2. The exception to this trend is the wood failure values for the samples that were pressed for 3.5 minutes. For this one measure all of the bio-oil samples are lower than the commercial control. However, for the key set of samples that was subjected to the critical boil-dry-boil test, the mechanical strength and wood failure properties lap-shear samples bonded with the bio-oil resins are as good or even better than the commercial control.

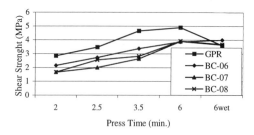

Figure 3. Shear strength in MPa versus press time for GPR and three eucalyptus oil resins.

Figure 3 shows a plot of shear strength as a function of press time for four different resin formulations (including GPR). Not too surprisingly the shear strength for all of the samples increases as the press time increases from 2 minutes to 6 minutes. The shear strength for all of the samples also decreases following exposure to the accelerated aging conditions. There are no statistical differences between any of the three bio-oil based PF resin formulations. While the commercial control has higher shear strength values at 2 and 2.5 minutes, there are no statistical differences between the control and BC-06. The most promising results for the bio-oil PF resin formulations are seen following the accelerated aging. While the shear strength of the commercial control decreases significantly, the shear strength of all three bio-oil formulations remains virtually unchanged. The bottom line result is that there are no statistically significant differences between the shear strengths of the four resin formulations following the critical boil-dry-boil test.

These results highlight the importance of understanding the subtle chemical differences between phenol and bio-oil that control the formation of the polymeric network. Since the bio-oil has a different functionality and pH the amount of formaldehyde and NaOH must be adjusted to allow the polymerization reactions to proceed in a controlled manner. These results also highlight the importance of how and when the bio-oil molecules are incorporated into the polymer network. Since they have different functionalities and reactivities they can not simply be added to a standard PF resin recipe, but the recipe has to be modified to accommodate these differences.

4. CONCLUSIONS

Phenolic oils produced from the slow pyrolysis of biomass can be an alternative feedstock for phenolic resin production. The resin formulation and cooking profile must be modified to account for the chemical differences between phenol and the bio-oil but substitution in the range of 50% it is feasible. Biomass oil resins worked better in severe conditions like in wet tests. The price for pyrolysis oil is extremely attractive.

5. ACKNOWLEDGEMENT

We thank Ms. Emília Antunes from Biocarbo Company for sending the eucalyptus oil. J.D. Rocha thanks FAPESP for the scholarship to research at NREL.

REFERENCES

1. Resende, M.E.A., V.M.D. Pasa, and A. Lessa (1994). In Advances in Thermochemical Biomass Conversion, 2, pp.1289–1298.
2. ABIQUIM (1994). Anuário da Indústria Química Brasileira, SP, Ano 21, 272 p.
3. ABIQUIM (1997). Guia da Indústria Química Brasileira, SP, Ano 13, 256 p.
4. Chum, H.L. and R.E. Kreibich (1992). U.S. Patent 5,091,499.
5. Chum, H.L., S.K. Black, J.P. Diebold, and R.E. Kreibich (1993). U.S. Patent 5,235,021.
6. Kelley, S.S., X.M. Wang, M.D. Myers, D.K. Johnson, and J.W. Scahill (1997). In Developments in Thermochemical Biomass Conversion, 1, pp. 557–572.
7. Gorbaty, L. (1996). Phenol. Chemical Economics Handbook Marketing Research Report. 48 p.
8. Gorbaty, L. (1994). Phenolic Resins. Chemical Economics Handbook Marketing Research Report. 34 p.

WOOD COMPOSITE ADHESIVES FROM SOFTWOOD BARK-DERIVED VACUUM PYROLYSIS OILS

C. Roy,[a] L. Calvé,[b] X. Lu,[a] H. Pakdel[a] and C. Amen-Chen[a]

[a]Pyrovac Institute Inc., 333, rue Franquet, Sainte-Foy (Québec) Canada G1P 4C7
[b]Forintek Canada Corp., division de l'Est, 319, rue Franquet, Sainte-Foy (Québec) Canada, G1P 4R4

The Pyrocycling^{TM} process developed by Pyrovac involves the thermal decomposition of softwood bark residues to phenolic-rich pyrolysis oils which can be used in the manufacture of phenol-formaldehyde resol resins. Different adhesive recipes were formulated and tested in oriented strand board (OSB) panels at 40% phenol and 24-30% formaldehyde replaced by pyrolysis oil. The mechanical test results showed that the OSB panels prepared with the pyrolysis oil-modified resins performed similarly to the commercial surface resin used as a reference. DSC analysis confirmed the OSB panel test results.

1. INTRODUCTION

Large volumes of bark residues are produced worldwide annually as a by-product of the forest industry. For example, the province of Quebec alone generates more than 3 Mt of dry bark residues annually.[1] In Quebec, approximately 50% of the volume of bark generated is used to produce energy while the remainder is incinerated or discharged in landfills or dumps creating potential air, water, and soil pollution problems.[2] An alternate solution is fast pyrolysis that enables the conversion of various biomass materials to biooil and wood charcoal. Chum et al.[3] extracted a phenol/neutral (PN) fraction from a wood pyrolysis oil by a solvent extraction technique. The pyrolysis process was performed in a vortex reactor and yielded approximately 12 to 21 wt.% (anhydrous biomass basis) of a PN fraction, depending on the type of biomass treated. The PN fraction was mixed (50:50) with pure phenol and polymerized with formaldehyde. The final wood composite resin compared very well with a commercial resin, Cascophen 313^{TM}. However, the complexity of the lengthy solvent extraction method associated with a relatively low PN yield negatively influence the process economics. Himmelblau[4] has formulated an adhesive made of raw pyrolysis oil derived from mixed hardwoods and reported successful replacement of 50% or more of the phenol in a typical phenol-formaldehyde adhesive formulation. The pyrolysis oil yield was 10.3%. This paper describes the Pyrocycling^{TM} of softwood bark to produce a phenolic-rich oil used as a petroleum phenol substitute in resol resins.

2. EXPERIMENTAL

2.1. Bark residues

The bark sample was provided by Daishowa, Inc., a large pulp and paper plant in Quebec City. The feedstock originated from processed softwood and was composed of fir (70% v/v) and spruce (28%). The remainder was composed of pine, hemlock spruce and larch (2%). The as-received wet sample was air-dried at room temperature to reduce its moisture content below 12%. It was then ground and sieved with a 0.5" U.S. standard sieve.

2.2. Pyrolysis runs

The pyrolysis experiments (runs # H40, H41, H42) were performed in a process development unit at an average throughput capacity of 50 kg/h. The bark particles were fed semi-continuously into the reactor and were moved and stirred over two horizontal superimposed heating plates using a molten salt eutectic mixture at 530°C as heat carrier. The enclosure of the reactor was maintained at 500°C under a total pressure of 18 kPa. The bark organic material was decomposed into vapours which were rapidly withdrawn from the hot chamber by means of a vacuum pump. The vapours were directed towards two successive packed towers in which the pyrolytic vapours were cooled by direct contact with the recirculating pyrolyzates. The first tower operated at such a temperature as to avoid both water and low-molecular-weight compounds condensation. The second tower was operated at a lower temperature to recover the light oil and to condense water. The noncondensable gases were removed by a vacuum pump and subsequently burnt. The charcoal produced was removed from the reactor using a screw conveyor.

The pyrolytic oil was produced using a proprietary process leading to an overall phenolic-rich oil yield of 34 wt. % (anhydrous biomass feedstock basis) for the vacuum pyrolysis process. This yield is higher than any of the previously mentioned processes. The phenolic-rich oil obtained can be used directly in the formulation of PF resin.

2.3. Analysis

The whole pyrolysis oil was fractionated into fourteen subfractions by sequential elution liquid chromatography on silica gel. The oil sub-fractions obtained were analyzed by GC/MS following a method which has been published previously.[5] The analytical results are summarized in Table 1.

A series of differential scanning calorimetry (DSC) tests were applied on the experimental resins. The kinetic data derived from these tests were obtained with a Mettler DSC-20 apparatus equipped with a TC11TA processor (Wood Science Dept., Université Laval). The sample was sealed in a high-pressure capsule pan and heated to 250°C at 10°C/min. The treatment of data was performed with a software provided with the apparatus, which contained the Bochard and Daniels (B/D) kinetic model.

Table 1
Chemical classes of compounds found in the pyrolysis oil (wt. %)

Hydrocarbons	3	Ketones	2
Sugars	9	Hydroxy ketones	1
Low-molecular weight acids	1.5	Cyclic alcohols	1.5
High-molecular weight acids	10	Steroids	2
Alcohols	1	Triterpenoids	2
Esters	1	Lignins / tannins-based compounds	46
Phenols	10	Total	100

2.4. Adhesive formulation methods

The adhesives were prepared by Forintek Canada Corp.'s laboratories (Québec City) with the vacuum pyrolysis oil sample at 40% phenol and 24-30% formaldehyde replacement levels using commercial PF resin formulations currently employed in the manufacture of oriented strand board (OSB) panels. Two commercial core-powder-adhesive recipes and three commercial surface-liquid-adhesive recipes were tested for the manufacture of two experimental powder adhesives (Oil PF_A, Oil PF_B) and three experimental liquid adhesives (Oil PF_C, Oil PF_D, Oil PF_E). The content of alkaline catalyst was adjusted for each resin to obtain the same final pH as the commercial recipe selected as a basis for its formulation. The commercial powder surface control resin was obtained directly from a resin manufacturer. The other PF control resins were also prepared in Forintek laboratory. The general parameters for panel manufacture were as follows:

Replicates: 2 per press cycle Resin content: 2.0%
Strands: 3 inches poplar Moisture content 4.9%
Press cycle: 30 s closing and 30 s opening Press temp.: 215 °C
Panel type: Homogeneous, 11.1mm Wax: 1.5%

2.5. OSB testing

Internal bond tests (IB) were performed according to the test methods and requirements described in CSA 0437 resins-93.[6] In addition, wet torsion shear strengths after 2-hour boil treatments were evaluated as an indication of the degree of resin adhesive cure.[7]

3. RESULTS

The OSB panel test results for the liquid and powder experimental resins and the commercial control resins are shown in Table 2. The two experimental powder-core resins Oil PF_A and Oil PF_B prepared by replacing 40% phenol and 24-30% formaldehyde with pyrolysis oil, exhibited internal bond (IB) and wet torsion shear values higher than that of the commercial control surface PF powder resin. With regards to the core resin, the IB of the experimental resins was close to that of the commercial PF control; however, poor wet torsion shear results were obtained. Among the various liquid-surface experimental resins containing 40% phenol and 24-30% formaldehyde replaced by pyrolysis oil, the Oil PF_E resin produced superior IB and torsion shear values at 210 s press cycle compared with the control. At 150 s press cycle, the quality of the experimental resins was close to that of the control. All the resins tested (powder or liquid) had IB values above the CAN 3 0437 Series-93 minimum requirement.

Table 2
Strength of OSB panel made with Oil PF adhesive at 40% phenol and 24-30% formaldehyde replacement

Resin Type	Press Cycle (sec)	Density (kg/m^3)	Internal Bond (MPa)	Torsion Shear (N.m)
Powder:				
PF Core	150	637	0.495	2.9
Commercial	210	643	0.595	4.3
Oil PF_A	150	638	0.396	1.1
	210	642	0.515	2.6
Oil PF_B	150	637	0.289	1.1
	210	637	0.487	2.4
PF Surface	150	639	0.048	0.2
Commercial	210	644	0.446	1.6
Liquid:				
Oil PF_C	150	666	0.429	2.4
	210	548	0.512	3.4
Oil PF_D	150	637	0.468	2.3
	210	647	0.481	2.7
Oil PF_E	150	654	0.416	2.2
	210	647	0.566	4.7
PF Surface	150	648	0.472	2.5
Commercial	210	639	0.529	3.5
CAN 3 0437 series-93	-	-	0.345	-

Table 3 summarizes the kinetics data from the DSC tests for the commercial control resin and the experimental Oil PF_C liquid resins. The DSC results shown in Table 3 confirm the OSB panel performance results. The Ea (obtained from a simple Bochard and Daniels, or BD, single peak evaluation) and ΔH values are generally associated with the rate of reaction and the extent of cross-linking in the resins, respectively. Very similar Ea and ΔH values were obtained for the commercial surface resin and the experimental Oil PF_C liquid resins. Interestingly, the peak temperature for the exotherm of the experimental resin was found to be lower than that of the commercial resin (see data in Table 3 and Figure 1). Conversion α for the isothermal plot vs. time was higher for the experimental resin in comparison to the control (Figure 2). These results are encouraging as they suggest that the experimental resin may polymerize faster than the commercial control resin. The results also suggest that the formulation may need further optimization to take advantage of this fast polymerization reaction.

Table 3
Kinetic analysis of the DSC data for experimental liquid Oil PF_C resin and commercial liquid surface PF control resin

Kinetic parameters	PF Control	Oil PF_C	Kinetic parameters	PF Control	Oil PF_C
Solids content (%)	47.8	47.2	Peak Temp (°C)	151	131
Ea (kJ/mol)	180	186	α (4 min 140 °C)*	0.47	0.81
ΔH (J/g)	197	195	α (2 min 150 °C)*	0.63	0.87

* From isothermal plot α vs time (Figure 2)

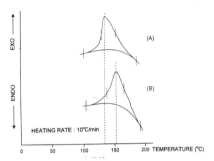
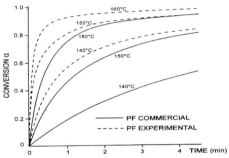

Figure 1. DSC thermogram of (A) Oil PF_C experimental liquid PF resin and (B) liquid commercial surface PF resin.

Figure 2. Isothermal plot for α vs time for Oil PF_C liquid experimental surface PF resin and liquid commercial surface PF resin.

4. DISCUSSION

IB and torsion shear values similar to the commercial resin were obtained for both powder and liquid surface resins made of 40% phenolic oil derived from bark residues. The DSC tests confirm the OSB panel performance tests. The energy of activation and the exothermic heat of reaction of some experimental and commercial surface resins were very similar. There is also some indication that the curing rate of the experimental resin could be further improved by improving the resin formulation because higher α (α vs. time for the isothermal curve) was obtained for the experimental resin. For the more reactive core resin applications the IB value of the experimental resin approaches that of the control. However, poor torsion shear results were obtained with powder core resin at 40% phenol substitution. Further work is needed to modify and optimize resins formulated by resins manufacturers using pure phenol. A fundamental study of the properties of resins prepared with pyrolysis oil is in progress in our laboratory.

5. CONCLUSION

The vacuum pyrolysis process converts softwood bark to a reactive phenolic-rich pyrolysis oil at high yields. The vacuum pyrolysis oil can be directly used to replace 40% of pure phenol and 24-30% of formaldehyde in PF surface resol resin formulations with no reduction in the reactivity of the resin. The preliminary OSB panel performance tests are very promising. The resin formulations tested so far can be further modified to take full advantage of the vacuum pyrolysis oil.

6. ACKNOWLEDGMENTS

This study has been supported by Pyrovac Technologies Inc., Province of Québec. Thanks are due to Dr. B. Rield, Wood Science Dept., Université Laval, for useful technical discussion and permission to use the DSC equipment. This paper is dedicated to the memory of Dr. Louis Calvé who passed away in February 1999.

REFERENCES

1. Gouvernement du Québec (1993). Ministère des forêts Direction du développement industriel. Potentiel de développement de la cogénération à partir de biomasse industrielle au Québec.
2. Anonymous (1990). Les déchets dangerereux au Québec. Une gestion environnementale. Les publications du Québec, pp. 265–71.
3. Chum, H.L. et al. (1993). Resole resin products derived from fractionated organic and aqueous condensates made by fast-pyrolysis of biomass materials. US patent 5,235,021. Aug. 10.
4. Himmelblau, D.A. (1998). Production of wood composite adhesives with air-blown, fluidized-bed pyrolysis oil, 52nd Forest Products Society Annual Meeting.
5. Pakdel, H., H.G. Zhang, and C. Roy (1994). Detailed chemical characterization of biomass pyrolysis oils, polar fractions. Advances in Thermochemical Biomass Conversion, Vol 2, ed. by A.V. Bridgwater, Blackie Academic, Glasgow, Scotland, pp. 1068–85.
6. Canadian Standards Association 0437 series-93:1993. Standard on OSB and Waferboard. Forest Products.
7. Shen, K.C. and M.N. Carroll (1975). Measurement of layer-strength distribution in particleboard. Forest Prod. J., 25, pp. 32–37.

FOAM CONCRETE USING BAGASSE PYROLYSIS TAR

L.A.B.Cortez,[a] L.E. Brossard Perez,[b] E. Izquierdo,[b] E. Olivares,[a] G. Bezzon[c]

[a]Faculty of Agricultural Engineering, University of Campinas, Brazil
[b]Faculty of Chemical Engineering, University of Oriente, Santiago de Cuba, Cuba
[c]Interdisciplinary Energy Planning Center (NIPE), University of Campinas, Brazil

This work presents some technical characteristics of light foam concrete obtained by adding alkaline tar solutions (ATS) to mixtures of water and Portland cement.
The low thermal conductivity of this type of concrete allows its use as a thermal insulator improving the comfort of buildings with significant energy saving. Comparative data are also given for acoustic absorption of the studied material and a commercial one known as "Siporex."
Foam concrete elements were tested for compression resistance as a function of its apparent density. Both studied parameters, thermal conductivity (C) and compression resistance (R), are expressed density (D) functions of foam concrete in the form of the following mathematical models:

$$C = -0151 + 2.3742 \cdot 10^{-4} D \quad (W/m.°C)$$

and $R = -0.0015D + 8.376 \cdot 10^{-6} D^2$ (*MPa*)

It is concluded that slow pyrolysis tar made soluble by alkaline treatment constitutes an excellent foaming agent for light concrete construction elements.

1. INTRODUCTION

Ancient pyrolysis processes gave charcoal as the only product destined to be used as a fuel. Long afterwards, dry distillation (Nikitin, 1966) expanded pyrolysis possibilities by obtaining additional products such as methanol, acetic acid, and eventually acetone. However, this process also became obsolete and slow pyrolysis was reduced to charcoal making at industrial and primitive levels. Usually, carbon-making installations either burn or throw away vaporized tar and other effluents. However, those effluents may constitute an important source of products with practical applications (Brossard and Cortez, 1997).

Thus, when slow pyrolysis tar, obtained from sugarcane bagasse carbonization, is transformed in a water-soluble substance by reaction with an alkaline hydroxide, the resulting alkaline tar solution (ATS) can be used as an excellent foaming agent for the production of light foam cellular concrete.

This paper deals with the production and technical characteristics of foam concrete using ATS.

2. MATERIALS AND METHODS

2.1. ATS production method

Air dried sugarcane bagasse was pyrolyzed in a bench scale continuous installation at a heating rate of 8°C/min up to 450°C.

The condensed effluents were separated into two fractions. The tar fraction was neutralized with NaOH solution and finally filtered to eliminate solid impurities. The resulting ATS was prepared with approximately 20% mass fraction and 1.02 g/ml density at 28°C.

2.2. Preparation of foam concrete at laboratory level

P-350 Portland cement from cement factory Hermanos Calderon located in Santiago de Cuba (265 g), was mixed with water in a mass ratio water:cement between 0.67 and 0.70.

After the cement paste was thoroughly homogenized, ATS was added in an ATS volume (ml) to cement mass (kg) ratio of 0.015 and submitted to the action of a paddle cement paste stirrer (80 rpm) for 2 minutes. At this time, the apparent density of the foamed paste is approximately 450-500 kg/m^3.

The resulting cement foam is poured in 10 × 10 × 10 centimeter molds and cured in a wet chamber for 28 days.

The air-dried foam cement cubes were tested for compression strength, apparent density, thermal conductivity and acoustic absorption coefficients.

Laboratory preparations of ATS foam concrete as well as the employed characterization methods were carried out according to Cuban regulations RC-3190 and RC-3182.

3. RESULTS

Table 1 shows foamed ATS concrete apparent density in dried state (D) and its correspondent compression strength (R). It can be observed that in this type of concrete a little lowering of D causes a large diminution in R.

Table 2 presents values of thermal conductivity (k) for ATS-foamed concrete having different apparent densities in the dried state (D).

The acoustic absorption coefficients (λ) at different frequencies of sample of ATS-foamed concrete and another sample of Siporex having the same apparent density (i.e., 500 kg/m^3), were compared. Their respective acoustic absorption coefficients (λ) at different frequencies are presented in Table 3.

Table 1
Apparent density (D) and compression strength (R) of dried foamed ATS concrete elements

D (kg/m^3)	R (MPa)	D (kg/m^3)	R(MPa)
380	0.431	450	0.866
240	0.156	316	0.231
317	0.463	401	0.899
232	0.085	321	0.335
315	0.204	397	0.823
282	0.121	353	0.378
341	0.449	408	0.863
243	0.202	282	0.282

With the data presented above, the following statistical regression model was obtained.

$$R = -0.001585D + 8.3758.10^{-6} D^2$$

Table 2
Thermal conductivity, k, for ATS-foamed concrete elements at different apparent densities in the dried state

D (kg/m^3)	k (W/m. °C)
282	0.056
383	0.074
418	0.085
474	0.093
480	0.094
489	0.099
503	0.105
573	0.127
577	0.123

Former exposed parameters were correlated by following mathematical model:

$$C = -0.0151084 + 2.37428.10^{-4} D$$

The acoustic absorption coefficients (λ) at different frequencies of sample of ATS-foamed concrete and another sample of Siporex having the same apparent density (i.e., 500 kg/m^3), were compared. Their respective acoustic absorption coefficients (λ) at different frequencies are presented in Table 3.

Table 3
Acoustic absorption coefficients (λ) for ATS foamed concrete and Siporex at different frequencies (Hz) and at the same apparent density in dry state (500 kg/m^3)

Freq.	400	500	630	800	1000	1250	1600	2000	2500	3150	4000
ATS-foamed concrete	0.12	0.02	0.22	0.28	0.28	0.28	0.32	0.22	0.29	0.24	0.12
Siporex	0.08	0.02	0.10	0.22	0.24	0.44	0.15	0.03	0.13	0.13	0.19

For the specific case of using ATS foam concrete as roof conditions previous to impermeabilization and as flooring, steel savings were calculated in comparison with traditional earth filler. Thus, for a five-floor building, steel savings owing to decreased loads associated with use of ATS foam concrete is:

Floors and roof 29%
beams 13%

columns 17.5%
concrete 28.5% (for loads < 0.2 MPa)

The costs and mass for a 10-cm-high cover for earth filler and for foam ATS concrete are presented in Table 4.

A difference of only one cent per m^2 yields a significant saving on steel consumption because the load, introduced when using foam concrete, is lowered by 80%.

On the other hand, the low thermal conductivity of foam concrete allows better insulation behavior in comparison with earth fillers.

Table 4
Costs and mass per m^2 for a 10 cm height cover of earth filler and foam concrete

	Costs (US$)	Mass (kg)
Earth filler	5.45	173.62
Foam concrete	5.46	35.00

Assuming, for example, a building covering 216 m^2, and considering a thermal conductivity of 0.58 W/m.°C for the earth filler and of 0.08 W/m.°C for foam concrete with 400 kg/m^3 (in dry state), the resulting heat loads are shown in Table 5.

Table 5
Heat loads for building roof preconditioned for impermeabilization with earth filler and with foam concrete according to several concepts

Concept	Heat Load, Q (W)	
	Earth Filler	Foam Concrete
Opaque surfaces	10655.36	2062.36
Transparent surfaces	1806.00	1806.00
Occupants	3875.00	3875.00
Lighting	2112.00	2112.00
Total	18448.36	9855.36

One aspect that should not be neglected is related to the productivity in applying both materials. Around 31 m^3 of foam concrete can be applied in an 8-hour shift, while only 2.5 m^3 of earth filler in the same period.

For impermeabilization preconditioning purposes the ATS foam concrete should have densities in the dry condition between 300 and 400 kg/m^3. Compression strength as well thermal conductivity data for this range is given in Table 6.

Table 6
Compression strength (R) and thermal conductivity (k) for ATS foam concrete in the range of densities from 300 to 400 kg/m^3 (dry condition)

D (kg/m^3)	R (MPa)	k (W/m.°C)
300	0.28	0.0541
350	0.47	0.0669
400	0.71	0.0797

4. CONCLUSIONS

Slow pyrolysis tar can be easily transformed in a water soluble substance (ATS) which behaves as an excellent surface-active agent. It can be used to produce foam cement for roof impermeabilization purposes as well as to repair old damaged buildings.

ATS foam concrete also has low heat and acoustic transmission properties that could be utilized for specialized construction. In this way a highly polluting vegetable tar can be turned into a value-added industrial product.

5. NOMENCLATURE

ATS	Alkaline tar solution
R	Resistance to compression (MPa)
C	Thermal conductivity (W/m.°C)
λ	Acoustic absorption coefficient (dimensionless)
rpm	revolutions per minute

REFERENCES

1. Nikitin, N.I. (1966). The Chemistry of Cellulose and Wood, Ed. MIR, Moscow.
2. Brossard, L.E. and L.A.B. Cortez (1997). Potential for use of pyrolytic tar from bagasse in industry, Biomass & Bioenergy, 12, pp. 363–366.
3. RC-3190 (1990). Ejecucion de obra. Production de Hormiter. Ministerio de Construcciones, Cuba.
4. RC-3182 (1982). Hormiter en cubiertas. Requisitos para su ejecucion. Ministerio de Construcciones, Cuba.

533

PLASTICIZERS THAT TRANSFORM ALKYLATED KRAFT LIGNINS INTO THERMOPLASTICS[*]

S. Sarkanen and Y. Li

Department of Wood and Paper Science, University of Minnesota, 2004 Folwell Avenue, St. Paul, Minnesota 55108, USA

Under alkaline conditions with bisulfide at elevated temperatures, the industrial conversion of wood chips into pulp for making paper is accompanied by the formation of kraft lignins in huge quantities. As far as possible, this byproduct is burned as a fuel, but a surplus of kraft lignin can sometimes be generated as a result of maximizing pulp production. Among the possible ways that such a surplus could be used, the creation of lignin-based polymeric materials has been a particularly appealing prospect for more than two decades. Nevertheless, it had been almost an article of faith that incorporating over 25-40% (w/w) of a lignin derivative into polymeric materials is difficult to achieve without seriously compromising their mechanical properties. Yet the noncovalent interactions between kraft lignin components are very powerful and, as such, should have a substantially favorable impact on cohesion in lignin-based polymeric material domains. Indeed, about four years ago 85% kraft-lignin-containing thermoplastics were reported for the first time; their tensile strengths and Young's moduli bore a direct relationship to the degree of intermolecular association. The past two-year period has witnessed a much more dramatic development, however: polymeric materials composed exclusively of alkylated kraft lignin components have been produced that are very similar to polystyrene in terms of their tensile behavior. Now extensive plasticization of alkylated 100% kraft-lignin-based polymeric materials has been achieved with single commercially available blend components costing as little as US$ 3.75 per kg.

---

[*] Paper No. 994436801 of the Scientific Journal Series of the Minnesota Agricultural Experiment Station, funded through Minnesota Agricultural Experiment Station Project No. 43-68, supported by Hatch Funds. Acknowledgment for support of this work is made to the United States Environmental Protection Agency through the National Center for Clean Industrial and Treatment Technologies (although it does not necessarily reflect the views of the Agency or the Center, so no official endorsement should be inferred).

1. GROUNDS FOR LIGNIN-BASED PLASTICS

One of the most demanding challenges that has confronted lignin chemists for the past two-and-a-half decades has been the creation of lignin-based thermoplastics. It should be borne in mind that such a goal cannot realistically be achieved simply by introducing progressively greater quantities of some lignin derivative or other into perfectly good thermoplastic polymeric materials and documenting how the mechanical properties of the resulting blends begin to deteriorate. Thus, while an interesting range of formulations has been investigated, few if any of the materials produced have been genuinely lignin-based.[1] Nevertheless, enthusiasm for the idea has wavered little because of the enormous tonnage in lignin derivatives that could be converted to useful polymeric materials if only the means for doing so reliably and profitably had been realized.

Industrially, the alkaline pulping of wood chips in the presence of bisulfide each year generates enormous quantities (over 25 million tons in the United States alone) of byproduct lignins. It is a sad commentary on how imperfectly the physicochemical properties of these kraft lignins are understood that less than 0.1% of them are employed outside the recovery furnaces of pulp mills where they serve as a relatively low value fuel.[2] In certain mills the maximization of production causes the capacity of the recovery furnace to be surpassed so that the kraft lignin cannot all be burned any longer.[2] Apart from such traditional sources of surplus byproduct lignin, the economic viability of environmentally friendlier pulping methods, such as the steam explosion and Organosolv processes, for example, has been tied to the development of high value-added lignin-based products.[3] Furthermore, if lignocellulose from plants and woody tissues is employed as the raw material for producing bioethanol[4-6] and chemicals[7,8] from sugars, lignins will again materialize as substantial byproducts that cannot be ignored.

2. TOWARDS AUTHENTIC LIGNIN-BASED PLASTICS

The creation of authentic lignin-based thermoplastics, with mechanical properties that can be predictably tailored to different applications, would constitute one of the most worthwhile vehicles for effective byproduct lignin utilization. There appear to be few alternative ways of converting lignins to high-volume commodity products in the near future. An attribute of lignin-containing polymeric materials, which may merit serious consideration, is that many such formulations are biodegraded by white-rot fungi.[9] Be that as it may, an imperfect understanding of their physicochemical properties has presumably been the chief obstacle to the widespread use of lignin derivatives as polymeric materials. For about four decades, lignins have been generally viewed as three-dimensional network polymers composed of (p-hydroxyphenyl) propane units randomly linked together in about ten different ways.[10]

Aliphatic and aromatic hydroxyl groups are usually the most frequent reactive functional groups in lignins, and in this respect it is perhaps not surprising that they have provided the most common points of departure for attempting to make lignin-containing polymeric materials. Thus byproduct lignins have been incorporated into phenol-formaldehyde resins, polyurethanes, epoxies and acrylics; moreover other synthetic polymer chains have been successfully grafted onto byproduct lignin components through free-radical polymerization reactions.[1]

Outside the University of Minnesota, there seems to have been little significant progress in attempts to utilize lignin derivatives as polymeric materials in their own right. These efforts have inevitably suffered from the shortcomings pervading current understanding of lignins as far as their structural characteristics and physicochemical properties are concerned. It is a sobering thought that adequate analytical degradative methods have yet to be developed for deducing the sequences of interunit linkages (namely, the primary structures) in polymeric lignin components. Certainly it must be acknowledged that, if the distributions (although not frequencies) of primary structural elements along macromolecular lignin chains were random, there would be little point in (or prospect of) determining their sequences explicitly. However, the physicochemical properties of lignin derivatives, to the extent that they have been discriminatingly investigated, have not supported a description of the native biopolymer as a random three-dimensional network.[11]

Of particular relevance to the development of kraft lignin-based thermoplastics, the noncovalent interactions prevailing between kraft lignin components seem to be remarkably selective in their effects; this specificity has been interpreted as arising from vestiges of native macromolecular configurational motifs which are arranged along the polymeric chains in a highly ordered way.[12-18] Moreover, as far as lignin biosynthesis is concerned, recent indications have tended to emphasize that the dehydrogenative polymerization of monolignols leading to the assembly of lignins in plant cell walls is a highly regulated process. Actually a working hypothesis has been advanced to describe how the formation and replication of specific macromolecular lignin primary structures might be brought about *in vivo*.[19-21] Thus, if native lignins are characterized by well-defined primary structures (as are almost all other biopolymers), vestiges of these features will be preserved to varying extents in lignin derivatives. The recognition of such prescriptions could eventually exert a profound influence on the effective development of authentic lignin-based polymeric materials.

3. CLUES FROM 85% KRAFT LIGNIN-BASED THERMOPLASTICS

The strong noncovalent interactions between the individual molecular components in lignin preparations[12-18] will be instrumental in organizing macromolecular domains in the solid state that could play a dominant role in producing cohesion in lignin-based polymeric materials. Thus there is no obvious reason *a priori* why thermoplastics cannot be fabricated with lignin contents far beyond a supposed incorporation limit of 25-40% (w/w).[22] The search for such materials was first vindicated by successfully blending underivatized softwood kraft lignin at 85% (w/w) levels with poly(vinyl acetate) and two plasticizers.[23,24]

The key finding from these new polymeric materials lay in the observation that their tensile behavior was strongly affected by the degree of association between the individual molecular components of which the kraft lignin was composed. The polymeric material incorporating the most associated kraft lignin preparation was difficult to deform, while that containing the most dissociated kraft lignin preparation was able to withstand strains greater than 65%. Indeed these results suggested that plastic deformation may be encountered in true lignin-based materials when the individual components are separated from one another enough for the polymer chains to be drafted past each other in response to external stresses. Thus brittleness need not be an inherent characteristic of true lignin-based materials, but their mechanical properties will inevitably be strongly influenced by the extent to which the tendency for association between the individual molecular components has been satisfied.

4. ALKYLATED 100% KRAFT LIGNIN-BASED THERMOPLASTICS

The question arises as to what kinds of polymeric materials would be most worthwhile as standards for comparative purposes in judging the performance of thermoplastics with the highest attainable lignin contents. In the absence of any indications to the contrary, it intuitively appeared that a single (simple) kind of lignin derivative, if properly handled, could ultimately be the most telling. It was considered that the thermal properties of such materials might be improved if intermolecular hydrogen bonding were completely eliminated, and thus the kraft lignin preparations to be employed in the quest for incipient standards were alkylated completely. Interestingly, the resulting kraft lignin derivatives did indeed exhibit quasi-melting behavior.

The inherent aptness of kraft lignin derivatives for producing polymeric materials is a question that cannot altogether be dismissed since substantial proportions of the molecular components present in such preparations are of low and intermediate molecular weights.[12-18] In order to address the matter, the parent kraft lignin was exhaustively ultrafiltered through a 10,000 nominal molecular weight cutoff membrane in aqueous 0.10 M NaOH. Both the higher molecular weight kraft lignin fraction and the parent preparation were treated with diethyl sulfate at pH 11-12 in aqueous 60% dioxane and then completely methylated with diazomethane in chloroform. Through solvent-casting from DMSO, the resulting ethylated methylated kraft lignin derivatives were transformed into cohesive sheets with quite promising mechanical properties (Figures 1 and 2). This may represent the first occasion on which measurable tensile strengths have been recorded for polymeric materials composed exclusively of a simple lignin derivative.

The Young's moduli and tensile strengths, respectively extending to ~ 1.9 GPa and ~ 37 MPa, are much higher than anything previously reported for the mechanical properties of materials with very high lignin contents. Although their ultimate strain and tensile energy absorption to fracture remain low, these alkylated kraft lignin sheets show undeniable promise as thermoplastic materials in their own right: it would not be surprising if relatively unsophisticated changes in composition could affect their mechanical properties to a considerable extent. Certainly the tensile behavior of the material embodying the ethylated methylated higher molecular weight kraft lignin fraction constituted a significant improvement over that of the material based on the parent preparation (Figures 1 and 2).

Most remarkable of all, however, is that the tensile properties of the polymeric material composed of the alkylated higher molecular weight kraft lignin fraction are almost exactly the same as those of polystyrene.[25] Both are brittle and thus both require plasticization if tough components are to be produced from them through injection-molding. The problem was that 100% lignin-based materials had never before been plasticized even though plasticization of

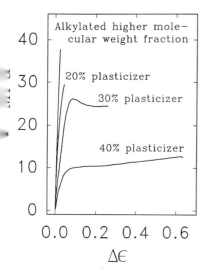 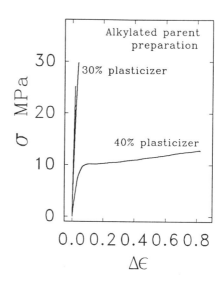

Figure 1. Plasticization of ethylated methylated higher molecular weight kraft lignin-based polymeric material by poly(1,4-butylene adipate). (Stress-strain curves delineated by Instron Model 4026 Test System employing 0.05 mm min^{-1} crosshead speed with 9 mm gauge lengths.)

Figure 2. Plasticization of ethylated methylated parent kraft lignin-based polymeric material by poly(1,4-butylene adipate). (Stress-strain curves delineated by Instron Model 4026 Test System employing 0.05 mm min^{-1} crosshead speed with 9 mm gauge lengths.)

polystyrene is notoriously easy. The powerful noncovalent interactions between the kraft lignin components are probably the main obstacle to plasticizing the alkylated kraft lignin-based polymeric materials. These interactions must be overcome to the point of permitting sufficient chain segmental mobility without destroying the cohesiveness of the material. That such a goal can be attained was previously substantiated by the tensile characteristics of the 85% kraft-lignin-based thermoplastic incorporating the most dissociated kraft lignin preparation examined.[24]

Indeed, effective plasticization of alkylated 100% kraft-lignin-based polymeric materials is readily accomplished with commercially available aliphatic polyesters costing about US$ 3.75 per kg (Figures 1 and 2). For example, the material based on the alkylated higher molecular weight kraft lignin fraction exhibits progressively more plasticity as a result of blending with poly(1,4-butylene adipate) at levels extending to 40% (w/w) (Figure 1). Interestingly, materials based on the alkylated parent kraft lignin preparation require higher proportions of poly(1,4-butylene adipate) to reach the threshold of plasticization, but thereafter plastic deformation becomes somewhat more extensive (Figure 2). The tensile strength of the poly(1,4-butylene adipate) alone fell below measurable limits and thus there is little doubt that such aliphatic polyesters function as true plasticizers. Achieving sufficient degrees of plasticization with 10-15% blend component levels has now become the next step in the quest to produce tough lignin-based thermoplastic components by injection-molding techniques.

REFERENCES

1. Glasser, W.G. and S. Sarkanen (eds.) (1989). Lignin—Properties and Materials, American Chemical Society, Washington D.C., ACS Symp. Ser., p. 397.
2. Kirkman, A.G., J.S. Gratzl, and L.L Edwards (1986). Tappi J., 69, p. 110.
3. Vanderlaan, M.N. and R.W. Thring (1998). Biomass and Bioenergy, 14, p. 525.
4. Gregg, D. and J.N. Saddler (1996). Appl. Biochem. Biotechnol., 57/58 p. 711.
5. Cao, N.J., M.S. Krishnan, J.X. Du, C.S. Gong, N.W.Y. Ho, Z.D. Chen, and G.T. Tsao (1996). Biotechnol. Letters, 18, p. 1013.
6. Vinzant, T.B., C.I. Ehrman, W.S. Adney, S.R. Thomas, and M.E. Himmel (1997). Appl. Biochem. Biotechnol., 62, p. 99.
7. Dorsch, R. and N. Nghiem (1998). Appl. Biochem. Biotechnol., 70-72, p. 843.
8. Donnelly, M.I., C.S. Millard, D.P. Clark, M.J. Chen, and J.W. Rathke (1998). Appl. Biochem. Biotechnol., 70-72, p. 187.
9. Milstein, O., R. Gersonde, A. Hüttermann, M.-J. Chen, and J.J. Meister (1992). Appl. Environ. Microbiol., 58, p. 3225.
10. Sakakibara, A. (1980). Wood Sci. Technol., 14, p. 89.
11. Goring, D.A.I. (1989). ACS Symp. Ser., 397, p. 2.
12. Dutta, S. and S. Sarkanen (1990). Mat. Res. Soc. Symp. Proc., 197, p. 31.
13. Garver, Jr., T.M. and S. Sarkanen, in Renewable-Resource Materials—New Polymer Sources, (ed. by C.E. Carraher, Jr. and L.H. Sperling) (1986). Plenum Publishing Corp., New York, p. 287.
14. Himmel, M.E., J. Mlynár, and S. Sarkanen (1995). In Handbook of Size Exclusion Chromatography, ed. by C.-s. Wu, Marcel Dekker, New York; Chromatographic Science Series, 69, p. 353.
15. Dutta, S., T.M. Garver, Jr., and S. Sarkanen (1989). ACS Symp. Ser., 397, p. 155.
16. Sarkanen, S., D.C. Teller, C.R. Stevens, and J.L. McCarthy (1984). Macromolecules, 17, p. 2588.
17. Sarkanen, S., D.C. Teller, E. Abramowski, and J.L. McCarthy (1982). Macromolecules, 15, p. 1098.
18. Connors, W.J., S. Sarkanen, and J.L. McCarthy (1980). Holzforschung, 34, p. 80.
19. Guan, S.-y., J. Mlynár, and S. Sarkanen (1997). Phytochemistry, 45, p. 911.
20. Lewis, N.G., L.B. Davin, and S. Sarkanen (1998). ACS Symp. Ser., 697, p. 1.
21. Sarkanen, S. (1998). ACS Symp. Ser., 697, p. 194.
22. Kelley, S.S., W.G. Glasser, and T.C. Ward (1989). ACS Symp. Ser., 397, p. 402.
23. Li, Y., J. Mlynár, and S. Sarkanen (1989). Proc. 8th Internat. Symp. Wood Pulp. Chem., I, p. 705.

24. Li, Y., J. Mlynár, and S. Sarkanen (1997). J. Polym. Sci. B: Polym. Phys., 35, p. 1899.
25. Rubin, I.I. (ed.) (1990). Handbook of Plastic Materials and Technology, Wiley, New York.

PRODUCTION AND PERFORMANCE OF WOOD COMPOSITE ADHESIVES WITH AIR-BLOWN, FLUIDIZED-BED PYROLYSIS OIL

D. Andrew Himmelblau[a] and George A. Grozdits[b]

[a]Biocarbons Corporation, 71 Cummings Park, Woburn, Massachusetts 01801, USA
[b]Louisiana Forest Products Laboratory, Louisiana Tech University, P.O.Box 10138, Ruston, Louisiana 71272, USA

The ability to substitute lower-cost phenolics from biomass pyrolysis oil for pure phenol in adhesives for exterior-grade plywood and oriented strandboard is attractive for the potential cost savings in a commodity product. Beyond meeting adhesive certification requirements, commercial success depends on convincing adhesive and panel manufacturers that replacing homogeneous phenol with heterogeneous pyrolysis oil will not complicate processing. Analysis of pyrolysis oil made in a bubbling, air-blown fluidized bed shows about 40 compounds present at greater than 1 mole % representing about 82 mole % of the material. Identified compounds, all polymerizable, have, on average, two positions available for methylene linkages, versus three for phenol. Consequently, some phenol (50% by weight or less) must still be used to provide sufficient linkages for a water-resistant adhesive. Biocarbons Corporation has made adhesives with 50% substitution for phenol with pyrolysis oil that perform nearly as well in making three-ply plywood from southern yellow pine as a commercial adhesive. The pH of a successful pyrolysis oil adhesive should match that of commercial adhesives. Adjustments of pH should be done mainly during adhesive making via multistage NaOH addition, rather than after adding fillers and extenders.

1. THE PATH FOR COMMERCIALIZATION OF PYROLYSIS OIL ADHESIVES

Pyrolysis oil from biomass can be produced at a lower cost than the phenol that is used to make adhesives for water-resistant structural wood panels such as plywood and oriented strandboard (Himmelblau, 1995). Consequently, there has been considerable interest in using pyrolysis oil as a phenol substitute in phenol-formaldehyde adhesives (Kelley et al., 1997). A successful adhesive made with pyrolysis oil must meet conventional press times to avoid reducing panel mill output, not complicate the adhesive or panel manufacturing processes, and pass the certification procedures required by any new or modified adhesive used to make structural panels. For plywood, this procedure is NBS-PS 1/83.

Technical success alone will not guarantee commercial application. Major hurdles facing commercial use of any pyrolysis oil include the following:
- the ability of the three major North American manufacturers to pass through their costs. Adaptation will require panel mill (customer) push or a manufacturer attempting to increase market share by offering a lower cost, pyrolysis-oil adhesive;
- previous problems with pyrolysis oil adhesives such as odor, especially in panel pressing, and unimpressive results, or results with "easy" furnishes such as hardwoods;
- skepticism that a heterogeneous raw material will not ultimately cause problems (such as extra VOCs) not currently experienced with homogeneous or pure raw materials;
- skepticism that the composition and use of pyrolysis oil products will not be affected by variations in biomass feedstock,
- in the United States, Toxic Substances Control Act registration of pyrolysis oil and notification of change requirements for manufacturing processes to make adhesives.

This paper presents recent work that Biocarbons Corporation has done to commercialize adhesives with pyrolysis oil (for prior work see Himmelblau and Grozdits, 1997 and 1999).

2. BIOCARBONS CORPORATION'S PYROLYSIS OIL PRODUCT AND COMPOSITION

While all pyrolysis oil processes produce phenolics suitable for adhesive use, the air-blown, bubbling fluidized bed that Biocarbons Corporation uses produces an oil that can be used directly to make adhesives. No product separation is required to remove any oil fraction that is non-reactive. Biocarbons Corporation has focused on making pyrolysis oil around 590°C, well above the oil-maximizing temperatures typically used by fast pyrolysis processes to maximize fuel production, accepting lower oil yields, but with 100 % selectivity.

Like other pyrolysis oil processes, Biocarbons Corporation's air-blown, bubbling fluidized bed produces a large number of oil compounds. Mixed hardwood (maple, birch and beech) pyrolyzed at typical operating conditions and used in the Trial 7 adhesive (below) produced 69 peaks by GC/MS analysis. Of these, 14 peaks present at > 2 mole %, represented 45 mole % of the product. The 27 peaks between 1 and 2 mole %, represented an additional 37 mole % of the product. Compounds that were identified are listed in Table 1, in order of appearance (time). Several of the #4-position groups could also be occurring at the #3 position. All are reactable to make a phenol-formaldehyde type adhesive. Pyrolysis oil from pine that was made at the same operating condition (but has not yet been tested for adhesive use) had essentially the same compounds present at > 1 mole %, but at different relative concentrations.

Although some have three or more, most of the compounds in Table 1 have two available positions for methylene linkages, compared to three for phenol. Some have only one site and are polymer chain breakers. Most of the aldehydes are capable of linking at positions in addition to the HC=O group. The overall "linkability" of the compounds present helps explain why complete substitution for phenol does not produce a suitable thermoset, and why 50% or less phenol is still needed to provide an adequate network of methylene linkages.

Table 1
Compounds identified in air-blown, bubbling bed pyrolysis oil at > 1 mole % in order of appearance from the column (time)

Phenol	Phenol, 2-(1-methylethyl)	Vanillin
1,2-Cyclopentanedione, 3-methyl	or Phenol, 2-ethyl-4-methyl	Phenol, 2-methyl-6-(2-propenyl)
	1,2 Benzenediol, 3-methyl	Phenol, 4-ethyl-2-methoxy
Phenol, 2-methyl	1,4 Benzenedicarboxalde-hyde-2-methy	Phenol,2-methoxy-4-(2-propenyl)
Phenol, 4-methyl		1,3-Benzenediol, 4-propyl
Phenol, 2-methoxy	1H-Inden-1-one,2,3-dihydro	Ethanone, 1-(4-hydroxy-3-methoxyphenol)
Phenol, 2,4-dimethyl	1,2-Benzenediol, 4-methyl	
Phenol, 4-ethyl	2-Methoxy-4-vinylphenol	Benzaldehyde, 4-hydroxy-3,5-dimethyl
Phenol, 2-ethyl	Phenol, 2-methoxy-3-(2-propenyl)	
Phenol, 2-methoxy-4-methyl		4-Hydroxy-2-methoxycinna-maldehyde
1,2-Benzenediol	4-Ethylcatechol	

3. PYROLYSIS-OIL ADHESIVE EXPERIMENTAL WORK

3.1. Pyrolysis oil production

The experimental equipment used to make pyrolysis oil from mixed hardwoods has been described elsewhere (Himmelblau and Grozdits, 1997, 1999). Operating conditions used to make the oil samples for adhesives have been the same, to test reproducibility of results. Reactor conditions were bed feedpoint average temperature of 591°C with a standard deviation of about 7°C. Bed bottom temperature just above the distributor plate averaged 15°C higher than the feedpoint. Freeboard temperature was below 450°C. Feed rate was about 3.3 kg/hr of +0.50 mm mixed-hardwood coarse sawdust with a wet-basis moisture content of 7 weight %. The product oil was water insoluble with water emulsified into the oil. Owing to scrubber modifications, the water content of the oil has increased to about 20 to 23 weight %. Char and ash in the oils measured by filtration was less than 1 weight %.

3.2. Adhesive formulation

The adhesive made for Trial 3 was an attempt to replicate the adhesive formulation for Trial 1 and 2 (Himmelblau and Grozdits, 1997), substituting pyrolysis oil for 50 weight % of the phenol. The adhesive formulation for Trial 4 was a reasonably successful attempt to increase the pyrolysis oil substitution to about 60 weight %, but the results from both the experimental and control adhesives were affected by the overly rough and dry veneer used. Trials 3 and 4 are described elsewhere (Himmelblau and Grozdits, 1999). The adhesive made for Trial 7 (reported here) had 50 weight % substitution for phenol but was a multistage (pH) cook ending at a higher pH than the Trial 3 adhesive. For Trial 7 adhesive cooking, phenol was first added to pyrolysis oil, water was added, the pH was increased to 10 to 10.5 by NaOH addition, and the mixture was cooked at 75-77°C for 80 minutes. Next, paraformaldehyde and additional water were added and the mixture was cooked at 75-77°C (with NaOH added in stages to boost the pH) until the target viscosity (measured at room temperature) was reached. The final composition is shown in Table 2.

Table 2
Composition of the Trial 7 adhesive as charged

	Grams	Weight %	
Net Pyrolysis Oil	81.05	14.60	
Water in Oil	24.85	04.48	
Char in Oil	1.00	00.18	
Phenol	81.04	14.60	
Water	216.18	38.95	
CH_2O Net	76.03	13.70	
Water in CH_2O	7.52	01.35	
NaOH	33.68	06.07	
Water in NaOH	33.68	06.07	
Total	555.03	100.00	
Weight % Solids		49.15	Measured: 49.15 Wt%

3.3. Board pressing conditions

The adhesive samples were kept refrigerated or frozen until used. Three-ply, 22.8 cm × 22.8 cm (9 in. × 9 in.) sample boards made with fresh southern yellow pine veneer taken directly from the dryer at a plywood mill. The veneer for Trial 7 was 3.2 mm (1/8 in.) thick with moisture content of 4.0%, and fairly smooth.

The adhesive was applied with a putty knife, first to one side of the core, then after attaching the core to the bottom, to the top of the core. Assembly time was 2-3 min. The gluelines reached 100°C approximately 45 s after press closure. The pressed boards were not hot stacked. Control panels were made with commercial adhesive from the mill that contained filler and extender. Twelve weight % Glue-X□ extender and 20 weight % unknown filler (supplied from the mill) were added to the pyrolysis oil adhesive, maintaining the same weight % solids. The pH of the adhesive used was 11.34 before and 11.03 after filler and extender were added. This pH was increased to 12.34 for some boards by adding 30 weight % NaOH to the adhesive. Platten temperature was 177°C, and platten pressure was 11.9 bar. Referring to Table 3, boards 7A to 7C were made without filler and extender, with an adhesive pH of 11.34; all other boards had filler and extender. The pH of the adhesive for boards 93 to 98 was 11.03. The pH of the adhesive for boards 99 to 105 was 12.34. The pH of the control adhesive for boards RA to RF was 13.3. Glue loading for boards 7A, 7B, 93 to 95, 103 to 105 and RD to RF was 0.224kg/m^2 (double glueline). Loading for boards 7C, 96 to 102 and RA to RC was 0.264kg/m^2.

3.4. Board testing

Each board was cut into twelve 8.25 × 2.54 cm (3.25 × 1.0 in.) specimens according to NBS/APA PS 1-83, after the outer 3.18 cm of the board was removed. Four specimens from each board were sheared dry, four were sheared after the vacuum-pressure soak (VPS) test, and four were sheared after the boil test. For a commercial board to "pass," the average wood failure (versus adhesive failure) of all specimens tested must be at least 85%. For experimental small-scale board making, relative wood failure, especially against a commercial control adhesive, is relevant especially without hot stacking after pressing (for additional cure), and because of greater edge effects in small versus production boards. Results from the trials are shown in Table 3.

Table 3
Trial 7 board testing results—averages of four specimens per test

Board	Shear Pressure (bar)			% Wood Failure			Press Time
	Dry	VPS	Boil	Dry	VPS	Boil	minutes
7A	19.6	17.2	10.3	54	65	68	3
7B	15.1	12.1	11.8	75	85	64	4
7C	20.8	12.5	11.8	75	79	82	5
93	16.0	11.0	6.4	80	70	78	3
94	19.2	10.2	9.3	88	68	72	4
95	17.9	13.6	8.8	86	69	62	5
96	16.4	9.5	6.6	80	75	67	3
97	12.1	9.1	3.7	70	65	62	4
98	15.2	5.4	5.9	79	54	58	5
99	17.3	6.0	5.3	63	35	60	3
101	16.4	10.7	8.0	92	58	56	4
102	14.5	11.6	10.9	80	76	69	5
103	20.8	13.9	10.5	76	61	59	3
1-4	15.1	10.6	9.4	75	60	70	4
1-5	18.5	12.2	9.3	70	65	74	5
RA	17.5	10.9	9.3	81	69	78	3
RB	20.0	12.1	11.6	84	62	71	4
RC	19.5	13.3	11.3	91	88	81	5
RD	14.7	8.6	7.6	88	74	76	3
RE	15.1	9.1	10.2	72	75	66	4
RF	18.9	8.0	6.7	85	68	72	5

4. DISCUSSION OF RESULTS

There was not much difference in board performance at the different glue loadings. There was not much improvement in board performance with increasing press times; 3 min seems adequate. There was some improvement by adding filler and extender; more pronounced improvement would be expected with rougher veneer that requires more gap filling. The adhesive made with pyrolysis oil could readily use conventional filler and extender and obtain better performance with them.

While it is not clear from the Trial 7 data, increasing the pH of the adhesive with pyrolysis oil while it is being made, through staged cooking/NaOH addition, improved adhesive performance. The absolute results and results relative to the control are better for Trial 7 than for Trials 1 to 3, whose neat adhesives had lower pH. Boosting pH by adding NaOH after mixing with extenders and fillers did not have much effect in Trial 7. Attempting this in Trial 5, with the Trial 3 adhesive, actually reduced adhesive performance. Also, the final pH (for southern pine) should be higher than for Trial 7, nearer to the pH of commercial adhesives.

The averaged results of all boards with filler and extender are shown in Figures 1 and 2. As can be seen, the wood failures and shear pressures for the pyrolysis-oil adhesive are close to those for the control commercial adhesive. Reduction in property values with wet testing is also similar to that of the control. It is likely that by boosting the cook pH and optimizing the viscosity of the applied adhesive, the performance of adhesives with pyrolysis oil can be further improved, both absolutely and relative to commercial adhesives.

Future work will continue to investigate the effect of glue loadings, press times, press temperatures, and filler and extender levels on adhesive performance. Near-term work will focus on the effect of veneer moisture content on performance, raising pyrolysis-oil adhesive pH, pyrolysis oils from pine rather than hardwood, and lower temperature, higher yield oils.

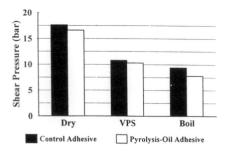

Figure 1. Average shear pressures for pyrolysis-oil and control adhesives

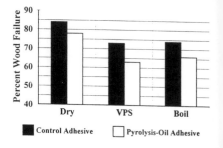

Figure 2. Average wood failure for pyrolysis oil and control adhesive

REFERENCES

Himmelblau, D.A. (1995). Phenol-Formaldehyde resin substitutes from biomass tars. Proceedings, Wood Adhesives 1995, Forest Products Society, Madison, Wisconsin, pp.155–162.

Himmelblau, D.A. and G.A. Grozdits (1997). Production of wood composite adhesives with air-blown, fluidized-bed pyrolysis oil, Proceedings, 3rd Biomass Conference of the Americas. Addendum.

Himmelblau, D.A. and G.A. Grozdits (1999). Production of wood composite adhesives with air-blown, fluidized-bed pyrolysis oil, Proceedings, 1998 Forest Products Society Annual Meeting. Forest Products Society, Madison, Wisconsin (to be published).

Kelley, S.S., et al. (1997). Use of biomass pyrolysis oils for preparation of modified phenol formaldehyde resins, in Developments in Thermochemical Biomass Conversion, ed. by A.V. Bridgwater and D.G.B. Boocock, Blackie Academic, London, pp. 557–572.

PREPARATION OF ACTIVATED CARBONS USING POPLAR WOOD AND BARK AS PRECURSORS

A. Bahrton, G. Horowitz, G. Cerrella, P. Bonelli, M. Cassanello and A.L. Cukierman

PINMATE, Dep. Industrias, FCEyN, Universidad de Buenos Aires, 1428, Ciudad Universitaria, Buenos Aires, Argentina. FAX: +54-1-576-3366, e-mail: analea@di.fcen.uba.ar

Residues from fast growing tree species used in the wood industry have been recognized as suitable precursors for preparing activated carbons (AC). In this work the feasibility of preparing AC from poplar wood and its bark is experimentally examined. The method employed involves pyrolysis of the raw materials and subsequent activation of the resulting solids with CO_2 under mild operating conditions, which involve relatively low temperatures, from 873 to 1003 K, and reaction times up to 2 hours.

The work determined the surface areas developed by the reaction and the characteristics of the solid products obtained at different conversion levels. The effect of activation conditions on AC features is examined by several techniques, including proximate and ultimate analyses, adsorption measurements employing N_2 at 77 K and CO_2 at 298 K as adsorbates and mercury porosimetry. Activation temperature has a marked influence on the adsorption capacity of the prepared AC, as inferred from the developed surface area.

Results indicate that, for certain conditions, poplar wood can be a convenient precursor for preparation of AC with surface areas higher than 700 m^2/g. Although AC prepared from poplar bark shows moderate surface areas, about 450 m^2/g, the product obtained from this precursor is comparable to some others commercially available.

1. INTRODUCTION

Activated carbon has been widely used as an adsorbent since the early part of the 20th century for water purification, and for volatile organic compound (VOC) and pesticide residual removal due to its ability to adsorb large quantities of contaminants from air and water. In recent years, activated carbons (AC) manufacture has gained renewed interest because of their wide range of applications, particularly for solving environmental problems. In this context, the search for renewable, inexpensive materials for producing high quality AC has increased

Wood and coconut shell are the primary materials used to manufacture activated carbon. Furthermore, some agricultural by-products, such as fruit stones and some fast growing tree species have been recognized as suitable precursors for the attainment of high quality products.[1-3]

The method most frequently used to prepare activated carbon from several raw materials involves two stages:
- a first step in which pyrolysis, or thermal degradation of the precursor material, is carried out in an inert atmosphere, in order to obtain an intermediate product with a high carbon content commonly named "char," and
- a subsequent step in which the char is activated by partial gasification using a mild oxidant agent such as steam or CO_2.

Operating variables of both steps strongly affect the adsorptive properties of the obtained activated carbons.

This work analyzed the characteristics of activated carbons prepared from a fast growing tree species, poplar wood and bark, by CO_2 gasification of devolatilized samples, to assess their suitability as precursors. It focuses on the effect of activation conditions on activated carbons features, as examined by several techniques. The evolution of porosity upon pyrolysis and further activation is also investigated.

2. EXPERIMENTAL

Poplar wood and bark from *Populus euramericana*, a fast growing hardwood mostly used in pulp and paper manufacture, are chosen as precursors for preparing activated carbons.

The raw material is first pyrolyzed, without a previous treatment, in a fixed bed reactor at 1073 K under N_2 flow during one hour. Chars prepared in this way are then milled and sieved. Standard proximate and ultimate analyses are performed on the virgin and devolatilized precursors.

The activation process of the pyrolyzed samples is then performed in a laboratory scale fixed-bed differential reactor, heated by an electrical furnace, using fractions of particle diameter between 1 and 2 mm and CO_2 (99.99%) as the activating agent. Operating conditions for the activation comprise temperatures between 873 and 1003 K and different activation times, up to 2 hours maximum. A high CO_2 flow rate is employed to ensure negligible external mass transfer resistance.

Chars conversions are evaluated from weight records at different reaction times. The reactor is set up with a special device to measure the solid conversion and to take samples at different conversion levels for determining the activated carbons features, as detailed in a previous paper.[4]

Adsorption experiments are performed in Micromeritics Accusorb 2100E and Gemini 2360 instruments. Surface areas of virgin and pyrolyzed samples and of the obtained activated carbons are calculated from adsorption isotherms of N_2 at 77 K using the conventional BET procedure, S_{N2}, and from CO_2 isotherms at 298 K applying the Dubinin-Radushkevich equation, S_{CO2}. Solids porosity and mesopore–macropore size distribution are determined using a Quantachrome Autoscan-60 porosimeter. Scanning electron micrographs (SEMs) are obtained with a 515 Philips Microscope. They were made to visualize the structure of the samples and changes by the preparation procedure.

3. RESULTS AND DISCUSSION

Proximate and ultimate analyses of the virgin and devolatilized precursors, expressed on a dry basis, are detailed in Table 1. They show typical values generally found for woods and their chars.[5,6] Both precursors have a high volatile content which sensitively decreases as a result of the thermal degradation. Ash content of virgin and pyrolyzed precursors is low, making these materials adequate for activated carbons preparation.

Table 1
Proximate and ultimate analyses* of virgin and pyrolyzed precursors

Precursor	Volatile %	Fixed Carbon %	Ash %	%C	%H	%N	%O **
Wood	71.4	28.1	0.5	48.3	5.8	0.1	45.8
Bark	67.0	31.9	1.1	47.8	5.8	0.8	45.6
Wood char	7.4	88.4	4.6	89.6	0.8	0.3	9.3
Bark char	10.1	83.7	6.2	83.3	0.7	0.4	15.6

* expressed on a dry basis; ** estimated by difference

Surface areas evaluated from N_2 adsorption isotherms are presented in Figures 1 and 2. They characterize the activated carbons prepared from wood and bark chars, respectively, at different temperatures and activation times.

The results point to activated carbons with higher adsorption capacity as temperature and time increase, although temperature exerts a predominant effect.

Comparison of values for both precursors indicates that larger surface areas are obtained from poplar wood. According to SEMs, this can be attributed to a more ordered and open structure of the original material in which horizontal and longitudinal channels and conducting vessels of the virgin wood are preserved upon thermal treatment; as a consequence, the development of the area by pyrolysis is much more significant for this precursor than for the bark, which originally has a more uniform structure showing less fluid conducting elements.[7]

Although activation time favors areas development, values longer than 1 hour are not advisable as energy consumption increases and product yield decreases.[4] Furthermore, for high temperatures, the value of surface areas trends to a maximum at conversion levels around 50-60%.[7]

As it can be inferred from the Figures, surfaces areas larger than 700 m^2/g are obtained for AC from poplar wood, whereas values around 450 m^2/g characterize those prepared from the bark. These values are characteristic of activated carbons which are suitable for liquid effluents treatment.[8]

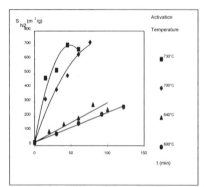
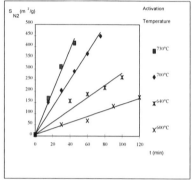

Figures 1 and 2. Effect of the activation conditions on the surface area development of activated carbons prepared from wood and bark chars, respectively.

N_2 and CO_2 surface areas, evaluated for the virgin and devolatilized precursors and for two activated carbons obtained from both precursors at 973 K and 49% conversion level, are listed in Table 2.

Although similar values of S_{N2} are found for both virgin precursors, the values of S_{CO2} point to a more microporous structure of the bark. After pyrolysis, the wood char sample shows larger surface areas indicating a more pronounced porosity development as a result of the complex network of reactions which take place during the thermal treatment and the higher volatile matter release of poplar wood compared to its bark.

Significant increases in S_{CO2} for both chars point also to microporous development induced by pyrolysis. This is slightly more pronounced for the poplar wood char, in agreement with the larger volatile matter content of the virgin wood, mentioned above.

In addition, S_{N2} and S_{CO2} values for both activated carbons indicate that poplar wood char, upon activation, develops larger surface areas than bark char, under similar operating conditions. Nevertheless, activation of both chars leads also to further enhancement of micropores, as reflected in general in the values of S_{CO2}.

Results obtained from mercury porosimetry measurements are associated with the presence of macropores and mesopores in the samples. Table 2 also lists these results.

Both precursors show similar amounts of mesopores; macropore volume is larger for the poplar wood than for the bark sample. Total pore volumes increase considerably upon pyrolysis of precursors and also due to subsequent activation of derived chars. Decreases in mesopores volume of activated carbons might be attributed to their transformation into macropores as a consequence of activation.

Table 2
Surface areas and total, meso and macropores volumes for poplar wood and bark, their chars and prepared activated carbons

Sample	S_{N2} (m^2/g)	S_{CO2} (m^2/g)	V_{Hg} (cm^3/g)	$V_{macropores}$ (cm^3/g)	$V_{mesopores}$ (cm^3/g)
Poplar Wood (PW)	1.3	82	0.135	0.121	0.014
PW char	21	490	0.211	0.182	0.029
PW activated carbon	720	730	0.289	0.271	0.018
Poplar Bark (PB)	1.1	128	0.091	0.077	0.014
PB char	6.9	447	0.264	0.199	0.065
PB activated carbon	447	590	0.316	0.291	0.026

Pore development is also reflected in the SEMs obtained for AC prepared from both precursors, depending strongly upon the features of the original botanical structure. An oriented structure, formed by large pores and small pits in the walls characterize the AC obtained from poplar wood. In addition, it shows a slightly more ordered and open structure compared with AC prepared from bark.[7] These features are in close agreement with textural parameters determined from nitrogen adsorption measurements and mercury porosimetry.

4. CONCLUSIONS

Porosity development in AC prepared from poplar wood and bark depends markedly upon activation conditions. Higher temperatures and longer activation times induce larger surface areas, although the influence of temperature is dominant. Activation of the chars is found significant at temperatures equal to or higher than 973 K. Activated carbons prepared from both chars, at this temperature for one hour, show the largest adsorption capacities, macropores and micropores mainly compose porous networks.

Activated carbons from poplar wood show S_{N2} larger than 700 m^2/g, whereas those obtained from poplar bark have S_{N2} of 450 m^2/g. Differences attest to the strong dependence on the nature of the starting material. Particularly, the constituent structures of the precursors exert a remarkable influence on the resultant activated carbon.

Results indicate that wood from fast growing trees such as poplar wood, may be considered as a convenient supply to prepare activated carbon with adequate adsorptive properties under mild activation conditions. Even though activated carbon prepared from bark under the same conditions show moderate S_{N2}, they are comparable to some others available commercially.

5. ACKNOWLEDGMENTS

The authors gratefully acknowledge CONICET, UBA and ANPCYT for financial support.

REFERENCES

1. Heschel, W. and E. Klose (1995). Active carbon from agricultural by-products, Fuel, 74 pp. 1786–1791.
2. Tancredi, N., T. Cordero, J. Rodríguez-Mirasol, and J.J. Rodríguez (1996). Activated carbons from Uruguayan eucalyptus wood, Fuel 75 N°15, pp. 1701–1706.
3. Rodriguez-Reinoso, F., M. Molina- Sabio, and M.T. Gonzalez (1995). The use of steam and CO_2 as activating agents in the preparation of activated carbons, Carbon, 33 N°1, pp. 15–23.
4. Bahrton, A., G. Horowitz, P. Bonelli, E. Cerrella, M. Cassanello, and A. Cukierman (1998). Residuos forestales como precursores para la preparación de carbones activados, ASADES, 2 N°1, pp. 93–96.
5. Cukierman, A., P. Della Rocca, P. Bonelli, and M. Cassanello (1996). On the study of thermochemical biomass conversion, Trends in Chemical Engineering, 3, pp. 129–144.
6. Ghetti, P., L. Ricca, and L. Angelini (1996). Thermal analysis of biomass and corresponding pyrolysis products, Fuel, 75, N° 5, pp. 565–573.
7. Bahrton, A. (1998). Preparation of Activated Carbons from a Fast Growing Tree Species, Master's Thesis, KTH, Sweden.
8. Bansal, R., J. Donnet, and F. Stoeckly (1988). Active Carbon, Marcel Decker, New York.

LIGNIN—THE RAW MATERIAL FOR INDUSTRY IN THE FUTURE

P. Zuman and E. Rupp

Department of Chemistry, Clarkson University, Potsdam, New York 13699-5810, USA

Lignin is a promising raw material for syntheses of aromatic organic compounds in the future, when petroleum and coal reserves will be exhausted. Economic cleavage of lignin under mild conditions (pH 8-12, 25°C) has been demonstrated. Cleavage follows first order kinetics and rate constants (k') increases with increasing pH. The plot of k' = f (pH) indicates the presence of a rapidly established acid-base equilibrium with pK_a about 11. Phenolate formation is considered to precede the unzipping of the polymer. Rate determining addition of OH^- ions can be excluded. Similarity in cleavage pattern of natural (rot wood) lignins and kraft lignins indicates similar cross-linking.

1. INTRODUCTION

The major source of raw materials for industrial manufacture of organic compounds has been first coal and more recently petroleum. Eventually these resources, which are not renewable, will be exhausted and it will be necessary to find for industrial organic syntheses other, renewable raw materials. In addition to carbohydrates, lignin, which represents 10-25% w/w of wood, straw and related natural resources, offers the possibility to become such material. The first condition for the use of lignin as a source of low-molecular-weight organic compounds is to find an economic way for the cleavage of lignin.

Currently lignin (an unwanted compound) is separated from cellulosics in the manufacture of paper and synthetic polymers under extreme conditions: strongly alkaline solutions are used, containing sodium hydroxide at concentrations usually higher than 1 molar, and it is treated at high temperatures and high pressures. In addition to a considerable use of energy in such processes, consecutive chemical reactions of the low-molecular-weight molecules can occur under such conditions. Primary reaction products can for example undergo Cannizzaro reactions, aldol formation and Michael additions, resulting in a complex mixture of oligomers. Such mixtures are precursors of humic acids, formed after acidification, and are unsuitable for obtaining synthetic intermediates. Moreover, such processes using high pressures and high temperatures are costly.

No information about cleavage of lignin under mild conditions, such as pH 8-12 at 25°C, has been found in the literature. Attempts were made[1,2] to investigate the cleavage of model compounds, but the relationship to the cleavage of three-dimensional protolignins is tenuous.

In our preliminary studies of the alkaline cleavage of lignins, the possibility to use measurements of changes in polarographic current with time was demonstrated. This technique allows measurements of concentration changes of electroactive species directly in the reaction mixture. The measurement of the current, which is a linear formation of concentration of the electrolyzed species, can be obtained by placing the mercury-dropping electrode directly into the suspension of lignin in a buffer. Polarographic currents remain practically unaffected by dispersed solid or colloidal particles. Such real-time analysis is not possible using other analytical techniques, involving quenching the reaction and carrying out the analysis after removal of solid and colloidal particles. Presence of colloids and gels makes use of HPLC difficult. UV-visible spectra recorded following ultrafiltration do not show characteristic absorption peaks as they are merged with the background due to light scattering by colloidal species. Absorbance measured at a constant wavelength, e.g., 260 nm or 280 nm, shows a time dependence as obtained for polarographic currents. As polarographic measurements allow following the time changes of concentrations of cleavage products directly in the reaction mixture and enable determination of individual functional groups, the polarographic technique was used for obtaining reported data.

2. EXPERIMENTAL

2.1. Materials

Three natural lignins and one commercial one were used in this study. Natural lignins were obtained from a rotten maple log (P), from a rotten maple stump (Q) and from a rotten gray aspen log (R). The rotted wood was ground using a mortar and a pestle, passed through a screen with 1.4 nm openings and washed with distilled water. The commercial lignin was Indulin ATR-CK1 from Westwaco (64.5% C, 5.8% H, 27.7% O, 1.6% S, 1.65% ash), contents of COOH 0.42 mol kg^{-1}; OCH$_3$ 4.50 mol kg^{-1} and phenolic OH 4.2 mol kg^{-1}. After washing twice with water and drying, this material was sieved using a 120 opening 1 inch screen.

2.2. Procedures

To follow kinetics of the cleavage reaction as a function of pH, 15 ml of a buffer was deoxygenated by a stream of nitrogen for 2 min, then 0.1 g of the natural or 0.02 g of the kraft lignin were added and after stirring by gas stream the current-voltage curves were recorded using a dropping mercury electrode and a differential pulse polarograph. Such current-voltage curves were recorded in phosphate and borate buffers after selected time-intervals. The solution containing suspended lignin was periodically stirred using a stream of nitrogen. Individual peaks on recorded I-E curves were identified by spiking with vanillin, syringaldehyde, isovanillin, 4-hydroxybenzaldehyde and trans-cinnamaldehyde. Concentrations of benzaldehydes and α,β-unsaturated carbonyl compounds in reaction mixture were determined by standard addition.

2.3. Kinetics

The current of infinite time was estimated by extrapolation. The cleavage with a half-time of the order of tens of minutes was followed. Plots of the natural logarithms of the difference between the current at infinity and individual reaction times were a linear function of time. Rate constants were obtained from slopes of $\ln(i_p^t - i_p^\infty) = f(t)$ plots at different pH values. Alternatively, after quenching the reaction by adjusting the pH to about 6 followed by ultrafiltration, absorbance at 280 nm was measured and treated similarly as polarographic currents. Similar kinetics and similar values of rate constants were observed.

3. RESULTS AND DISCUSSION

Current-voltage curves obtained in suspensions of lignin in buffers pH 8-12 as a function of time, show an increase of a predominant peak at about -1.6 V (Figure 1). By spiking, this peak was identified as belonging to hydroxybenzaldehydes. In pure aqueous buffers, vanillin, isovanillin and syringaldehyde can be analyzed polarographically. In a mixture in the presence of surface active species in the suspension of lignin, their peaks overlap and only the total concentration of aromatic aldehydes can be determined. The analytical results were confirmed after extraction of buffered solutions with ether. The extraction was most effective between pH 8 and 10, as at higher pH-values a dissociation of phenolic groups occurs.
Slowly increasing peak at -1.2 V was identified by spiking as corresponding to species bearing a CH=CH-CO grouping. The nature of the small peak at -1.4 V which is rapidly formed and remains practically time-independent was not identified.

Changes of the peak current at -1.6 V with time were used for the kinetic study. At pH 9.8 the changes in this current followed a similar pattern as that of the absorbance at 280 nm, measured after quenching and ultrafiltration.

The kinetics of the alkaline cleavage of all studied lignins—natural and kraft—follows a similar pattern (Figure 2). The cleavage involves a fast reaction, which is complete within first few minutes with a relatively small yield. This reaction is followed by a reaction with half-times of the order of tens of minutes, which was studied in some detail.

Both for natural and commercial lignins this slower reaction follows first order kinetics at least up to 70% conversion (Figure 3). As the time changes for other lignins, for example the expansion lignin or lignosulfonates, follow a different pattern, it is possible to conclude that not only the nature of functional groups but the three-dimensional structure of kraft lignins resembles that of protolignins. The rate of cleavage and the degree of conversion increases with increasing pH (Figure 4) in a similar way both for natural lignins and kraft lignins. Due to a better homogeneity of the kraft lignin, its samples were chosen for a more detailed study of the dependence of the rate of cleavage on pH.

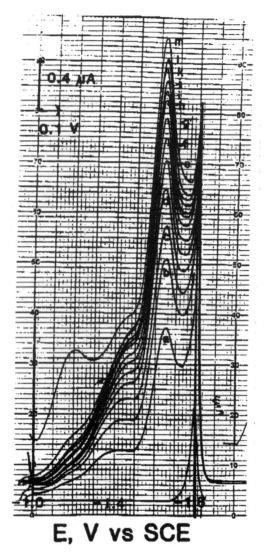

Figure 1. Current-voltage curves obtained by DPP for the hydrolysis product of 0.02 g indulin ATR-CK1 lignin in 25 mL phosphate buffer pH 10.5, $\mu = 0.5$. Time in minutes until start of recording DPP (a) 0.37, (b) 2.17, (c) 3.87, (d) 6.30, (e) 8.08, (f) 10.87, (g) 14.12, (h) 19.10, (i) 22.85, (j) 32.20, (k) 48.78, (l) 63.73, (m) 1338 (22 hours 20 min). $\Delta e = 50$ mV, curves starting at -1.0 V vs SCE.

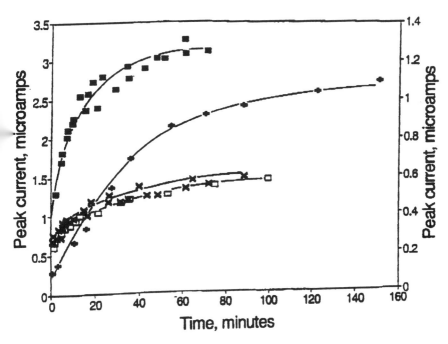

Figure 2. Comparison of the time dependence of DPP currents in the course of the hydrolysis of various lignins (0.1 g/15 mL) in borate buffer, pH 9.0, $\mu = 0.5$. Manistee lignin (Q), at -1.5 V (■) and Clarkson maple (P_1), at -1.6 V (□) with current scale on left axis, grey birch (R_1), at -1.5 V (X) and indulin ATR-CK1 (L_1), at -1.6 V (+) with current scale on right axis.

Over the entire pH-range studied (pH 8 to 12) the cleavage follows first order kinetics, enabling us to obtain pseudo first order rate constants k'. The values of rate constants k' increases with increasing pH (Figure 5). The slope of the dependence corresponds to an expression $k' = kK_a/(K_a + [H^+])$ which corresponds to a cleavage preceded by a rapidly established acid base equilibrium:

$$\text{ArOH} \xrightarrow{K_a} \text{ArO}^- + \text{H}^+$$

$$\text{ArO}^- \xrightarrow{k} \text{Products}$$

where k_a is an acid dissociation constant and k the rate constant of the rate determining step. As the slope of the k' = f (pH) plot corresponds to a reaction with $pK_a \approx 11$, it is possible to conclude that the unzipping of the polymer is initiated by a dissociation of a phenolic hydroxyl.

An alternative possibility considering a nucleophilic catalysis involving addition of OH^- ions either to a C=O or C=C seems less probable, as in such reactions the rate constant k' would be a linear function of $[OH^-]$.

In glycine buffers or ammonia–ammonium ion buffers the kinetics indicates participation of Schiff base formation.

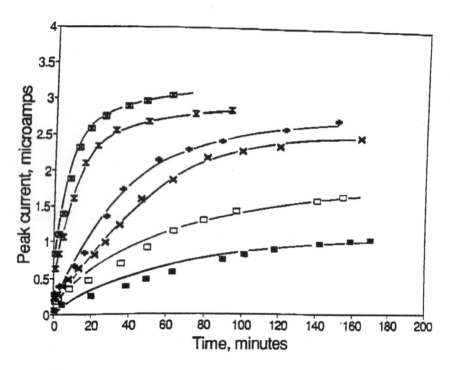

Figure 3. First order kinetic plot for DPP currents obtained for the hydrolysis of 0.1 g indulin ATR-CK1 in 15 ml buffer. Phosphate buffer, 0.2 M, pH 7.4, E_p -1.46 V (□) (right scale); borate buffer, $\mu = 0.25$, pH 9.0, E_p - 1.54 V; and phosphate buffer, $\mu = 0.5$, pH 10.3, E_p -1.6 V (left scale).

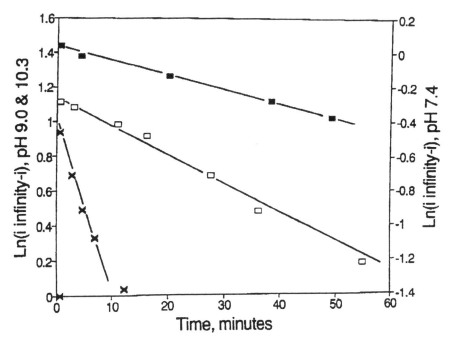

Figure 4. The time dependence of DPP currents in the course of the hydrolysis of indulin ATR-CK1, 0.1 g/15 mL in various buffers. Phosphate 0.2 M, pH 7.4 (∎); borate, μ = 0.5, pH 8.3 (□), pH 8.7 (X), pH 9.0 (+), pH 10.0 (X); and phosphate 0.2 M, pH 10.3 (⊠).

4. CONCLUSIONS

Lignin can undergo cleavage at pH 8-12 at 25°C, resulting in formation of hydroxy substituted aromatic aldehydes and α,β-unsaturated carbonyl compounds. Such compounds are promising as synthetic intermediates in manufacture of aromatic organic compounds. Investigation of the nature of individual products and their yields is in progress. Remarkable similarity of the cleavage pattern of natural and kraft lignins and its dependence of pH, indicates that crosslinking in kraft lignin results in structures rather similar to those in natural lignin samples. Rotwood lignins used in this study can be considered to have structures close to those of protolignin, as they are obtained with minimal handling and operations which could alter the structure of the natural product.

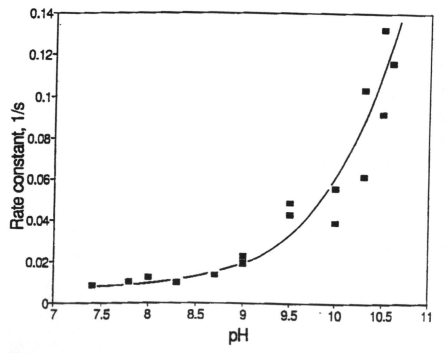

Figure 5. Dependence of the first order rate constant for the hydrolysis of indulin ATR-CK1 on pH.

REFERENCES

1. Gierer, J., B. Lenz, and N.H. Wallin (1964). Acta Chem. Scand., 18, p. 1469.
2. Kratzl, K., E. Risnyovsky-Schäfer, P. Claus, and E. Wittman (1966). Holzforschung, 20, p. 21.

MODERN TECHNOLOGIES OF THE BIOMASS CONVERSION FOR CHEMICALS, CARBON SORBENTS, ENERGY, HEAT AND HYDROCARBON FUELS PRODUCTION

V. Anikeev

Boreskov Institute of Catalysis, 630090 Lavrentieva, 5 Novosibirsk, Russia

This paper discusses biomass conversion processes in supercritical solvents to produce various chemicals. The common method and algorithm for calculating the phase state of complex biomass-containing mixtures in subcritical and supercritical conditions, using thermodynamic models, are developed. We studied chemical technological processes (CTP) of converting solid biomass to methanol, carbon sorbents, energy and heat, light and heavy hydrocarbons for industry application. Then, models of such processes were created, and mathematical modeling, thermodynamic analysis, optimization and calculations of the exergy efficiency of CTP were performed. The relationship between the main operation parameters of biomass gasification, their properties and the target products yield were studied. The Fischer-Tropsch reactor and process scheme were calculated. Productivity and product selectivity are given as the function of biomass gasification and hydrocarbon synthesis parameters.

1. INTRODUCTION

Complex biomass processing to produce various chemical products is increasingly supported, because it is not wise to use biomass (organic raw materials, wastes of wood-working industry and agriculture) as a fuel for both energetic and environmental pollution reasons. Cellulose of the conventional formula $C_6H_{10}O_5$ served as biomass in our studies, since it makes up about half of the main biomass materials. If cellulose content influences the gasification process, then studies of pure cellulose gasification will promote understanding of the gasification of cellulosic wastes.

Supercritical fluids (SCF) near the critical point possess such unique thermodynamic and transport properties as low viscosity, high solubility and diffusivity. SCF technologies used are homogeneous and heterogeneous catalysis, waste treatment, and conversion of biomass and coal to valuable chemicals or fluid fuels. Supercritical water (SCW) acts as a solvent and is also an important reactant for the biomass conversion. SCW facilitates hydrolysis and oxidation of organic compounds very effectively.

2. PRODUCING MULTINUCLEAR ALCOHOLS FROM BIOMASS HYDROLYSIS PRODUCTS IN SUPERCRITICAL WATER

Biomass hydrolysis products (cellobiose, glucosyl-erythrose, glucose, fructose, erythrose and other) are the initial raw material for production of multinuclear alcohols (sorbite, xylite, erythrit, its isomers, glycerin, ethylene glycol) by catalytic hydrogenolysis in a hydrogen flow. The reaction occurs in the water solution under pressure (150-200 atm) at temperatures of 175-230°C, in the presence of heterogeneous hydrogenation catalysts. The homogeneous co-catalyst $Ca(OH)_2$ plays the role of "cracking agent." Hydrogenation of aldehyde groups in the sugar molecules under these conditions produces six-nuclear alcohols (sorbitol and its isomers). Besides, C-C bonds in the six-nuclear alcohols break forming the lowest polyols, such as xylite, erythrit, glycerin, ethylene glycol, 1,2-propylene glycol. The selectivity of reactions towards lowest polyols depends on the selected catalyst and operation conditions.

The above products are valuable for medicine, pharmacology, and the manufacture of cosmetic preparations. The cheap vegetative raw material and ecological purity of products are attractive.

However, the developed "classical" variant of the proposed process lacks a low hydrogenolysis reaction rate. The residence times both at hydrolysis and hydrogenolysis stages change within 15-60 min. The reason for such low intensity of the hydrogenolysis process is extremely low solubility of hydrogen in a reaction mixture in the reaction conditions. Recent publications on the first process stage (hydrolysis) in supercritical water report that hydrolysis was finished in 1-1.5 s even without catalyst.[1,2]

We suggest it is possible to realize the second stage (hydrogenolysis) in supercritical conditions also. For this purpose is enough to increase the process pressure from 150-200 atm to 240 atm. The water solution of sugars obtained at the first stage is processed in a hydrogen flow. Hydrogen solubility increases by orders of magnitude in the near-critical water. Such effects are known for hydrogenation processes in supercritical solvents. As a result, we have an opportunity to reduce reactor volumes and increase the productivity more than tenfold. In a variant of the concurrent process the hydrogen is added at the hydrolysis stage with simultaneous hydrogenolysis reactions.

Design and operation of modern biomass conversion technologies require appropriate thermodynamic and vapor-liquid-solid equilibria data. Our phase equilibrium calculations are based on the Redlich-Kwong-Soave (RKS) state equation, which is widely used for the description of P-V-T properties of any fluid phase. However, we modified the coefficients of binary interaction k_{ij} and c_{ij} based on experimentally obtained correlation equations for the equilibrium in a two-phase system as a function of the temperature and pressure.[3]

In the determination of phase stability it is very important to localize the critical point of the reactant mixture, because the unique properties of SCFs change dramatically with small changes in pressure and temperature near the critical point.

An algorithm[4,5] developed in this work is based on a well-known homotopic method. Unlike conventional techniques, parameters for solution continuation in our calculations were temperature and pressure. The method exploits a two-step algorithm "predictor-corrector" using the Euler linear step predicator corrected by the Newton method.

3. CRITICAL POINT CALCULATION; COMPARISON WITH EXPERIMENTAL DATA

When modelling a technological process that occurs in the reactor under subcritical and supercritical conditions, one should necessarily calculate the mixture's critical point. Therefore we focused special attention on comparing the calculated results with the published data[6,7] as well as with experimental results.

An example calculation was carried out for model mixture CH_4 - 0.9430, C_2H_6 - 0.0270, C_3H_8 - 0.0074, n-C_4H_{10} - 0.0049, n-C_5H_{12} - 0.0010, n-C_6H_{14} - 0.0027, N_2 - 0.0140; this mixture is used for comparative analysis in some works.[6-8] Experimentally obtained values of critical parameters for this mixture are given in the latter work.[8]

Table 1
Coordinates of critical point for the model mixture

Source	This work	Michelsen M.L[6]	Peng and Robinson[10]	Heidemann R.A[9]	Experiment[8]
T_C, K	202.77	203.13	202.44	202.20	201.09
P_C, atm.	59.20	58.11	59.04	58.89	55.78

The calculated results agree well with experiments and different models.

4. BIOMASS GASIFICATION TO PRODUCE METHANOL, HEAT AND ENERGY

Let us consider a principal scheme of biomass gasification to syn-gas to synthesize methanol and to produce heat accumulated in some chemical products (Fig. 1). The gas produced (B1), which contains H_2, CH_4, CO, CO_2, H_2O at temperature T_1, comes to a heat-exchanger, is cooled to temperature T_h and enters a unit for methanol synthesis (B3). The solid biomass residue, carbon, is removed before the heat exchanger. In the methanol synthesis unit (B3) the gas is compressed adiabatically to pressure P_2, cooled in a heat exchanger to the methanol synthesis temperature T_2 and comes into reactor.

The synthesis products are condensed to methanol and water. Liquid methanol is one of target products. All gases left as well as methanol and water vapors under the saturation pressure come to a combustion chamber (B2), in which air is also supplied by a compressor. Combustion products under pressure P_2 go to a turbine, where they produce work.

In order to estimate the useful work produced by the turbine, we take the work produced by the turbine at gas expansion to atmospheric pressure, subtract energy consumed by compressors and relate the difference to the inlet biomass exergy. The values of this criterion (reduced turbine work) allow an estimate of whether this scheme can support itself with energy.

We performed calculations using parameters of real gasification and methanol synthesis processes. The total exergy efficiency of the scheme, its productivity towards methanol and reduced turbine work as the functions of gasification temperature and pressure, biomass wetness (moles of water per biomass mole, w) represent the calculation results.

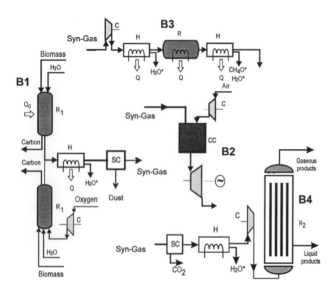

Figure 1. Fragments of schemes of complex energy-chemical processing of solid biomass. B1—gasification unit; B2—electric energy production in gas turbine; B3—methanol production unit; B4—unit for synthesis hydrocarbons fuels. R1—gasification reactor, H—heat-exchangers, C—compressors, SC—scrubbers, R2—"slurry-reactor" for hydrocarbons synthesis.

5. THERMOCHEMICAL CONVERSION OF BIOMASS TO PRODUCE LIGHT AND HEAVY HYDROCARBON FUELS IN THE FISCHER-TROPSCH SYNTHESIS

It seems very promising to synthesize motor fuels, alcohols and other chemical products from biomass during its thermochemical processing. Using mathematical modeling and thermodynamic analysis, we have considered the principal aspects of a possible thermochemical processing of biomass to hydrocarbon fuels by gasification followed by the Fischer-Tropsch (FT) synthesis. We propose to study the technological scheme, consisting of a biomass gasificator and reactor for FT synthesis.

In order to solve the problem, we take a simplified technology of hydrocarbon fuels synthesis from the products of biomass gasification (Figure 1 [B1, B4]). A $CO + H_2$ mixture, which contains small amounts of methane and water vapour (about 3%) is first compressed in the compressor and fed to a "slurry-reactor" for the FT synthesis.

We take for a mathematical model a perfectly mixed reactor, operating under isothermal conditions. The model is formulated by the following system of nonlinear algebraic equations:

$$\mathbf{M}^0 - \mathbf{M} + (g_{cat})\mathbf{Z}^T \mathbf{r}(\mathbf{y}, \mathbf{K}) = 0 \qquad (1)$$

Where:

$\mathbf{M}^0 = [m_1^0, m_2^0, ..., m_{Ns}^0]^T$, $\mathbf{M} = [m_1, m_2, ..., m_{Ns}]^T$ —Ns-vectors of molar flows of components at the inlet and outlet of the reactor, Z—the stoichiometric matrix of ($Nr \times Ns$) dimensions, $\mathbf{r} = [r_1, r_2, ..., r_{Nr}]^T$ —Nr-vector of reactions rates, $\mathbf{y} = [y_1, y_2, ..., y_{Ns}]^T$ —Ns-vector of components given molar portions; K- Np-vector of kinetic constants. Ns is the number of components, Nr is the number of reactions, and Np is the number of constants.

In equation (1), the sample weight is given in brackets, assuming that the rate and rate constants are expressed in mol h/kg$_{cat}$. The function r(y, K) is a kinetic model of the FT process. This model was obtained using published experimental data.[11] The methodological aspects of kinetic models identification, algorithms and some other details are thoroughly considered by Anikeev et al.[12] This report also shows the rate constants found by the model identification.[12]

6. RESULTS AND DISCUSSION

The technological scheme (Figure 1) was calculated at fixed temperature, 533K, and pressure, 2.0 MPa in the FT reactor. The residence time of the reaction mixture in the reactor (the ratio of the catalyst weight to the inlet total molar gas rate) changed solely. Temperature and water feeding to the reactor of biomass gasification were varied, providing CO/H_2 variations within 0.53-1.38 at the FT reactor inlet. The goal of our calculations was to find out the process conditions providing the maximum yield of wanted fractions, such as C_5-C_{11} and C_{12}-C_{18}, C_{19}-C_{24}, C_{25}-C_{40} ones.

In order to estimate a range of conventional residence time variations of the reaction mixture in the FT reactor, which provides the preset total conversion of $CO + H_2$, we calculated the technological scheme at fixed parameters of the biomass gasification process, which secured the basic ratio $CO/H_2 = 1$ at the inlet of the gasification reactor, when residence time varied solely. Following from the dependence of $CO + H_2$ total conversion and curves of the hydrocarbon fraction yields on the residence time, the required conversion degree of > 0.75 is obtained at the residence time equal 0.2-0.35 kg h/mol.

The biomass gasification process was estimated at temperature variations from 1000 to 1300K. The amount of water added varied from 0 to 0.5 kg per 1 kg of biomass. The conditions of biomass gasification that can govern the coke formation were not considered, since it sharply decreases the overall gas yield at the gasification reactor outlet, thus decreasing process efficiency towards the desired hydrocarbon fractions, C_5-C_{11}, C_{12}-C_{18}, C_{19}-C_{24}, and C_{25}-C_{40}.

Conditions ensuring maximum yields of wanted fractions, C_5-C_{11}, C_{12}-C_{18}, C_{19}-C_{24}, were then optimized. Diagrams showing the yield of fractions as the function of gasification temperature and wetness were created. Maximum yield of the light fraction is observed when the temperature equals 1100K and the wetness is 0.5. The heavy fraction (C_{19}-C_{24}) yields are maximum at the biomass gasification temperature equals 1300K. At fixed wetness, the yield of fraction C_5-C_{11} passes over its maximum at a low wetness and a temperature of 1150-1200K, while at a higher wetness the yield of fraction C_5-C_{11} passes over its maximum at a temperature of 1050K. When the wetness is fixed, the yields of C_{12}-C_{18} and C_{19}-C_{24} fractions increase at all values of wetness.

As the temperature of gasification increases and the amount of water added decreases, owing to an increase in the ratio of CO/H_2, the portion of light hydrocarbons decreases and that of heavy ones increases. These regularities are observed both inside the fraction and between fractions.

7. CONCLUSIONS

Using analysis results and selected types of biomass, one can choose the biomass processing technology suitable for regional demands, products obtained, economics and ecology.

The advantages of biomass processing to syn-gas with the following synthesis of hydrocarbons are evident. Thus, a lower ratio of CO/H_2 in syn-gas obtained by gasification allows higher productivity of all hydrocarbons at the same gas flow rates; the process is practically wasteless, since biomass contains only small amounts of minerals and sulfur, and its is highly efficient.

It is expedient to perform this process on a small-scale energy-chemical setup sited near a biomass deposit.

This paper developed a simplified flash procedure for complex mixtures and an algorithm for calculation of the phase equilibrium and critical point of the mixture.

REFERENCES

1. Kabyemela, B.M. and T. Adschiri et al. (1997). Ind. Eng. Chem. Res. 36, p. 5063.
2. Kabyemela, B.M. and M. Takigawa et al. (1998). Ind. Eng. Chem. Res. 37, p. 357.
3. Anikeev, V.I. and A. Yermakova (1998). Theoretical Foundations of Chemical Engineering, 5, p. 508.
4. Yermakova, A. and V.I Anikeev (1998). Journal of Physical Chemistry, 12, p. 2158.
5. Yermakova, A. and V.I. Anikeev (1999). Journal of Physical Chemistry, 1, p. 140.
6. Michelsen, M.L. (1982). Fluid Phase Equilibria, 9, p. 21.
7. Savage, E.P., S. Gopalan, T.I. Mizan, Ch. J. Martino, and E.E. Brock (1995). A.I.Ch.E. J., 7, p. 1723.
8. Karapetiyans, M.Ch. (1975). Chemical Thermodynamic, Chemistry, Moscow. (in Russian).
9. Heidemann, R.A. and A.M. Khalil (1980). A.I.Ch. E. Journal, 5, p. 769.
10. Peng, D.Y.and D.B. Robinson (1977). A.I.Ch. E. Journal, 2, p. 137.
11. Novel Fischer-Tropsch Slurry Catalysts and Process Concepts for Selective Transportation Fuel Production (1986) Quarterly Technical Report DOE/PC/70030--T7.
12. Anikeev, V.I., A. Yermakova, and A.V. Gudkov (1996). Chemistry for Sustainable Development, 4, p. 121.

571

STUDY ON GLUABILITY OF COPOLYMER RESINS OF BIOMASS EXTRACTS FOR LAMINATING CCA-TREATED LUMBER UNDER ROOM TEMPERATURE

Chia-Ming Chen and David L. Nicholls

School of Forest Resources, The University of Georgia, Athens, Georgia 30602

Alkaline extracts of various forest and agricultural residues were used to synthesize copolymer resins containing phenol, resorcinol, and formaldehyde. CCA-treated Southern pine, untreated Southern pine, and oak lumbers were laminated with these copolymer resins under room temperature and at 190 psi for an overnight period of approximately 18 hours. The gluelines were evaluated with the vacuum/pressure water soaked method for testing exterior gluelines of structural gluelam.
In laminating CCA-treated Southern pine lumbers, several copolymer resins containing approximately 12.5% of resorcinol provided more than 85% wood failures and 1,000 psi shear strength. Among different biomass materials, pecan shell flour provided the highest wood failures for gluelines of CCA treated lumbers.

1. INTRODUCTION

Given the environmental strain caused by fossil fuels and chemicals, efforts to develop renewable sources of chemicals and energy must be accelerated. Georgia, along with other southeastern states, has tremendous biomass resources in forest and agricultural residues. At the present time, the wood products industry is virtually dependent on chemicals derived from petroleum and natural gas for adhesives. An effective utilization of these biomass residues as a source of adhesive chemicals would not only be good for the environment, but also enable the forest products industry to secure a dependable and renewable source of raw materials for adhesives and binders.
The total utilization and low processing costs of agricultural and forest biomass residues are the keys to a feasible and practical use of these residues for industrial chemicals and adhesives. As the author has reported in previous papers, a family of fast-curing phenolic copolymer resins containing extracts of forest and agricultural residues was developed for exterior grade plywood, particle boards, flakeboards, and room temperature curing of wood laminating adhesives (Chen, 1982a; 1982b; 1994; 1995; 1997).
Resorcinol, phenol-resorcinol, phenol, and melamine are defined in the ASTM D 2559 Standard Specification as suitable for bonding wood (including treated wood) into a

structural laminated wood product where a high strength, waterproof adhesive bond is required (ASTM 1995). The American National Standard for structural glued laminated timber ANSI/AITC A190.1 prohibits the use of adhesives containing urea and melamine modified urea resins for gluing hardwoods or chemically treated woods before or after gluing (ANSI/AITC 1992). The resorcinol-formaldehyde (RF) and phenol-resorcinol-formaldehyde (PRF) resins are the primary adhesives for bonding these difficulty to bond woods. However, Vick reported in 1994 that virtually no CCA-treated southern pine lumber was used in adhesive bonded structural lumber products (Vick 1994). One problem that has slowed progress in the CCA-treated structural glued wood products is the difficulty encountered in gluing of CCA-treated lumbers.

Previous investigations by many researchers have clearly pointed out that chemicals of CCA preservatives cause mechanical strength deterioration and difficulty in bonding with RF and PRF adhesives (Sellers and Miller 1997, Shaler, et al., 1988; Winandy, 1986; Vick, 1994; Zhang et al., 1997). After studying the physical and chemical nature of the surfaces of CCA-treated southern pine and the effect of these characteristics on adhesion, Vick stated that there are two reasons that cause lack of CCA-treated structural lumber products: (1) the CCA chemicals in wood interfere with the adhesion of treated wood and (2) glued wood products treated with CCA preservatives severely degrade from warps, splits, and checks with subsequent drying after CCA treatment (Vick, 1994). Vick reported in 1995 that a new hydroxymethylated resorcinol coupling agent, when used as a dilute aqueous primer on lumber surfaces before bonding, may physicochemically link the phenol-resorcinol-formaldehyde resin with unsolvable chromium, copper, and arsenic oxides on the surfaces of CCA-treated southern pine. Thus the coupling treatment enhanced the durability of adhesion (Vick, 1995).

Resorcinol is an expensive chemical and its price is the determining factor in the high cost of RF and PRF adhesives. Pizzi reported that the percentage (by mass) of resorcinol in liquid resins of current commercial PRF is on the order of 16 to 18% (Pizzi, 1994). As the senior author reported previously, alkaline extracts of forest and agricultural residues have been used to successfully replace phenol in the synthesis of copolymer resins with phenol and formaldehyde. Grafting approximately 10 to12% resorcinol to copolymer resins with 40% of their phenol replaced with extracts of biomass provided more than 80% of wood failures in laminating Southern pine and oak wood blocks at room temperature (Chen 1995, 1997). Thus, it is of interest to examine whether the biomass copolymer resins could provide structural grade gluelines for gluing structural lumber of CCA-treated woods.

Thus, the objective of this study was to examine the gluability of biomass copolymer resins in laminating CCA-treated Southern pine wood blocks at room temperature. The gluelines were evaluated with block shear test specimens for dry and vacuum/pressure soaked samples.

2. EXPERIMENTAL PROCEDURES

2.1. Copolymer resin preparation
Alkaline extracts of peanut hulls and pecan shell flour were used in this study. The ground

biomass materials were extracted with a 22% solution of sodium hydroxide at 95°C for an overnight period of approximately 17 hours. The alkaline extracts of these biomass materials were then used to synthesize experimental copolymer resins.

All experimental copolymer resins, with 40% by weight of their phenol replaced by the extracts, were prepared by loading the biomass extracts, phenol, and formaldehyde into a resin reaction kettle. Water was then added to the mixture to adjust the target nonvolatile content of the resin to 45%, by weight. Stepwise addition of a 50% sodium hydroxide solution was then carried out with moderate heating to enhance the chemical addition of formaldehyde to the phenolic ring while incurring minimal formaldehyde loss due to the Cannizzaro reaction (Chen, 1989).

The mixture was then heated and polymerized to a target viscosity of 500 cps at 25°C. The mixture was then cooled to 50°C and the prescribed amount of resorcinol, equal to 12.5% of liquid resin, was added. The reaction temperature was raised, and the reaction was allowed to continue at 70°C for 45 minutes to graft the resorcinol onto the copolymer resin. The molar ratio of formaldehyde to all phenolic materials (the combination of phenol, biomass extracts, and resorcinol) was 0.86.

2.2. Wood blocks

Kiln-dried tangential above ground grade CCA-treated Southern pine, untreated Southern pine, and red oak were planed to a 3/4 inch thickness and then cut into 2-1/2 inch by 12 inch blocks. The average equilibrium moisture content of wood blocks was 11.24% for CCA-treated pine, 9.86% for untreated pine, and 8.5 1 % for red oak.

2.3. Wood block gluing

The resorcinol grafted copolymer resins were mixed with a hardener, composed of 50% paraformaldehyde and 50% pecan shell flour, immediately before gluing the wood blocks. The molar ratio of formaldehyde to all phenolic materials, after the addition of the hardener, was 1.5. Glue was applied to one surface of the wood blocks with a hand roller with a spread rate of 7-8g/30 square inch single glue line. Two blocks were then glued together with the grain oriented in a parallel fashion at a pressure of 195 psi. The open assembly time ranged from 2 to 5 minutes, and total assembly time ranged from 7 to 30 minutes. The glued blocks were then kept under pressure at a room temperature of 18°C to 20°C for approximately 17 hours to ensure complete resin cure. Two duplicate wood blocks were glued for each wood and resin combination.

2.4. Block shear testing

Each duplicate glued wood block was cut into five standard block-shear test specimens according to ASTM D-905 standard test methods. The ten test specimens were then divided into two groups: one group for testing dry specimens and another group for testing after vacuum/pressure water soaking. The block-shear specimens were tested to failure at a speed of 0.2 in./min. Both shear strength and percentage of wood failures were recorded.

3. RESULTS AND DISCUSSION

The composition and characteristics of the copolymer resins, along with their gel times, are listed in Table 1. Average shear strengths and percent wood failures for laminated wood blocks from the bonding tests are presented in Table 2.

Table 1
Characteristics of copolymer resins

Resin composition			Resin analysis			
Type of biomass extracts	Extracts/ Phenol	pH	NaOH Contents (%)	Nonvolatile Contents (%)	Viscosity @ 25°C (cps)	Gel time @60°C (min:sec)
Peanut hull	40/60	10.1	3.86	44.74	880	2:33
Pecan shell Flour	40/60	9.92	3.68	45.00	790	2:25

Table 2
Shear-Block test results of copolymer resins

Type of wood	Specimens tested	Peanut hull copolymer		Pecan shell flour copolymer	
		psi*	WF(%)*	psi*	WF(%)*
CCA treated Southern pine	Dry	1702	92.5	1898	85.3
	Soaked	992	92.5	1010	92.5
Untreated Southern pine	Dry	1526	90.0	1641	85.8
	Soaked	883	85.0	956	50.4
White oak	Dry	2720	55.0	2749	34.2
	Soaked	1957	67.1	1651	85.00

*The shear strength and wood failures are averaged from five specimens, with two or three specimens randomly selected from each of the two glued blocks.

As can be seen from the Table, copolymer resins containing either peanut hull extracts or pecan shell flour extracts exhibited a good bonding quality for gluing CCA-treated Southern pine wood blocks. The copolymer resins provided a wood failure of more than 80% after vacuum/pressure water soaking. Somewhat surprisingly, the glueline quality of CCA-treated southern pine was better than that of untreated Southern pine wood.

In conclusion, the biomass extracts copolymer resins containing approximately 12.5% resorcinol are capable of gluing the difficult-to-bond CCA-treated Southern pine wood blocks laminated under room temperature. The study on the optimization of the copolymer resins is underway and will be published in forthcoming articles.

REFERENCES

ANSI/AITC (American National Standards Institute/American Institute of Timber Construction) (1992). American National Standard for wood products—structural glued laminated timber, ANSI/AITC A190.1-1992.
ASTM (American Society for Testing and Materials) (1995). Standard specification for adhesives structural laminated wood products for use under exterior (wet use) exposure conditions. D 2559-92.
Chen, C.M. (1982a). Copolymer resins of bark and agricultural residue extracts with phenol and formaldehyde: 40 percent weight of phenol replacement. Forest Products Journal. 32, 11/12, pp. 14–18.
Chen, C.M. (1982b). Bonding particle boards with the fast curing phenolic-agricultural residue extract copolymer resins. Holzforschung, 36, pp. 109–116.
Chen, C.M. (1994). Wood adhesive copolymer resins made of biomass residue components. Proceedings of Second Pacific Rim Bio-Based Composites Symposium, pp. 304–315.
Chen, C.M. (1995). Copolymer resins made of agricultural and forest residues extracts for wood laminating adhesives. Proceedings of Second Biomass Conference of the Americas: Energy, Environment, Agriculture, and Industry, Portland, Oregon, pp. 1210–1218.
Chen, C.M. and D.L. Nicholls (1997). Room-temperature curing of copolymer resins derived from forest and agricultural residue extracts and used for wood laminating adhesives. In Making A Business from Biomass in Energy, Environment, Chemicals, Fibers, and Materials, ed. by R.P. Overend and E. Chornet, Elsevier Science, Kidlington, Oxford, United Kingdom, 2, pp. 905–913.
Pizzi, A. (1994). Resorcinol adhesives. In Handbook of Adhesive Technology, ed. by A. Pizzi and K.L. Miftal, Marcel Dekker, Inc., New York, pp. 369–380.
Sellers, Jr., T. and G.D. Miller, Jr. (1994). Evaluations of three adhesive systems for CCA-treated lumber. Forest Products Journal, 47, 10, pp. 73–76.
Shaler, S.M., et al. (1988). Strength and durability of phenol-resorcinol joints of CCA-treated and untreated southern pine. Forest Products Journal, 38, 10, pp. 59–63.
Winandy, J.E. and B.H. River (1986). Evaluation of a method for testing adhesive-preservative compatibility. Forest Products Journal, 36, 1, pp. 27–32.
Vick, C.B. (1994). Preliminary findings on adhesive bonding of CCA-treated southern pine. Proc. of Symp., Adhesives and Bonded Wood Products, Proc. No. 7315, Forest Products Society, Madison, Wisconsin, pp. 158–176.
Vick, C.B. (1995). Coupling agent improves durability of PRF bonds to CCA-treated southern pine. Forest Products Journal, 45, 3, pp. 78–84.
Zhang, H.J. et al. (1997). Surface tension, adhesive wettability, and bondability of artificially weathered CCA-treated southern pine. Forest Products Journal, 47, 10, pp. 69–72.

AGRONOMIC EVALUATION OF ASH FOLLOWING GASIFICATION OF FIVE BIOMASS FEEDSTOCKS

D.I. Bransby,[a] G.R. Mullins,[a] and B. Bock[b]

[a]Department of Agronomy and Soils, 202 Funchess Hall, Auburn University, Alabama 36849, USA
[b]TVA, P. O. Box 1010, CEB 3A, Muscle Shoals, Alabama 35662-1010, USA

The objective of this project was to determine ash composition and the agronomic value of ash following gasification of switchgrass, broiler litter, sugarcane bagasse, rice straw and paper min sludge. Gasification tests were conducted by Primenergy Inc. of Tulsa, Oklahoma, using the gasification technology licensed from PRM Energy. The by-products (ash) following gasification of the feedstocks were evaluated in greenhouse pot experiments using rye and forage sorghum as test crops. Results indicated that land application of the ash generally increased soil pH, thus showing some liming value. In addition, ash from broiler litter increased plant growth, but ash from paper mill sludge depressed growth. The ashes tested did not appear to immobilize nitrogen in the soil, and the nutrients contained in them appeared to be available for plant uptake. Therefore, except for paper mill sludge, ash following gasification of these feedstocks can be safely applied to crop land at reasonable rates, usually with no effect, or with some benefit to crop growth.

1. INTRODUCTION

Agricultural enterprises and forestry generate a wide variety of biomass feedstocks that could be used to produce energy in the southeastern United States. Furthermore, the relatively low cost of land, low income from major existing enterprises (forestry and production of beef cattle) and high yields from test plots of switchgrass make the Southeast one of the most promising regions for the commercialization of switchgrass as an energy crop. Although some biomass (mainly forestry-related residue) is already used to generate energy in the region, relatively little is known about the properties of ash which results from combustion or gasification of these and other feedstocks. With the increased focus on compliance of industry with federal, state and municipal environmental standards, and continued competition from low-cost fossil fuels, cost of ash disposal may influence the economic feasibility of producing energy from biomass. Therefore, the objective of this study was to determine ash composition and agronomic value of ash following gasification of switchgrass, broiler litter, sugarcane bagasse, rice straw and paper mill sludge.

2. PROCEDURE

The residual material or by-products from gasification of the feedstocks in this study still contained some carbon, indicating that gasification was not entirely complete. Therefore, strictly speaking, this material does not represent true ash. However, for convenience, it will be referred to as "ash" in this paper. The five feedstocks were gasified by Primenergy, Inc., of Tulsa, Oklahoma, and results of the gasification tests were reported by McQuigg and Scott (1998). On receipt of the ash materials from Primenergy, samples were ground and analyzed for total C, N, S, Ca, K, Mg, P, Cu, Fe, W Zn, Ba, and Pb. A greenhouse trial was then conducted from March to June, 1998, using a Marvyn loamy sand (Typic Paleudult) that was collected near Auburn University. Prior to the test, initial properties of the soil were determined and it was treated with 100 mg K/kg of soil, 40 mg P/kg of soil and 12 mg Mg (as Mg-sulfate)/kg of soil.

Treatments involved application of the ash from gasification of the five feedstocks at four rates, in combination with three rates of N. Based on initial analyses, each ash was applied at 0, 10, 20, and 40 g/kg of sod, which is approximately equivalent to 0, 10, 20, and 40 tons/acre. Nitrogen rates were a total of 0, 45 and 90 mg N/kg of soil, applied in three equal dressings split over the growth of the two test crops. Each pot contained 5 kg of treated soil and the treatments were arranged in a randomized complete block design with three replications per treatment. Soil in the pots was maintained near field capacity by daily addition of deionized water.

In the initial trial rye *(Secale cereale;* variety Bonel) was seeded at a rate of 50 seeds/pot and thinned to 20 plants/pot one week after planting. Forage yields were determined 35 days after seeding by clipping the plants at approximately one inch above the soil surface. Harvested plant material was dried and weighed. After harvest, soil samples were collected from each pot for chemical analysis. In the second trial (May-June, 1998), sorghum-sudangrass *(Sorghum* hybrid; variety Sugargraze II) was seeded at a rate of 50 seeds/pot and thinned to 20 plants/pot one week after seeding. Forage yields were determined 28 days after planting and plant material was processed in the same way as rye in the previous test.

3. RESULTS AND DISCUSSION

Analysis of the Marvyn loamy sand soil used in this study indicated a soil pH of 6.1, and Mehlich extractable nutrient contents of 37 lb P/acre, 94 lb K/acre, 480 lb Ca/acre, and 48 lb Mg/acre. Results from analyses of the residual ash material supplied by Primenergy after gasification indicate that their composition varied widely (Table 1). Analysis of variance showed that the three and two way interactions and main effects were all significant ($P<0.05$) for forage yield in the rye experiment. This indicated a complex set of results which was difficult to interpret. Perhaps the most distinct feature of these data was the depressed yields caused by paper mill sludge ash. Application of ash from the other feedstocks resulted in relatively little change in yield, although ash from the broiler litter at all levels increased yield at all N levels, and application of ash from rice straw increased yield at the low and medium N levels.

Table 1
Chemical analyses of the residual ash material following gasification of the five feedstocks

Component	Paper Mill Sludge	Switchgrass	Broiler Litter	Sugarcane Bagasse	Rice Straw
C	13	62	25	8	25
N	0.54	0.57	1.15	0.27	0.78
S(%)	0.57	0.11	0.20	0.00	0.21
Ca	10.2	1.17	5.86	1.46	0.82
K (%)	0.09	2.16	2.85	0.73	2.93
Mg	11.1	0.64	0.90	0.30	0.47
P (%)	0.63	0.53	3.40	0.30	0.23
Cu (mg/kg)	180	11	408	37	11
Fe (mg/kg)	3666	1456	5541	4419	2683
M-n (mg/kg)	820	394	640	266	606
Zn (mg/kg)	121	84	292	51	31
Ba (mg/kg)	132	126	129	105	121
Pb (mg/kg)	110	19	35	25	24

Yields in the sorghum test were considerably higher than in the rye test, and this probably allowed greater resolution among treatments. However, patterns were similar, with all interactions and main effects being significant. Once again, ash from paper mill sludge depressed yield (Fig. 1). The low and medium levels of ash from broiler litter increased yield when N was applied, but resulted in no difference when a high rate of ash was applied. Ash from rice straw at the low and medium levels increased yields when N was applied, but at the high level of application it depressed yield. All levels of switchgrass and sugarcane bagasse ash application increased yield at the low level of N, but there was no response to the ash when no N was applied, or at the high level of N application.

After the harvest of rye, soil sample analyses showed that Mehlich I extractable nutrients in the soil were largely related to the concentration of those nutrients in the different ashes. Calcium was increased most by ash from paper mill sludge, followed by that from broiler litter. Ash from the other feedstocks had little effect (Fig. 2). Magnesium was affected only by ash from paper mill sludge, and sodium by ash from rice straw and broiler litter. Potassium was increased by ash from rice straw, broiler litter and switchgrass, in that order, but was influenced very little by ash from sugarcane bagasse and paper mill sludge. This corresponds strongly to the concentration of potassium in these materials as determined in the initial analysis (Table 1). Phosphorus was increased only by ash from broiler litter. These results suggest that the nutrients contained in the ashes are available for plant uptake.

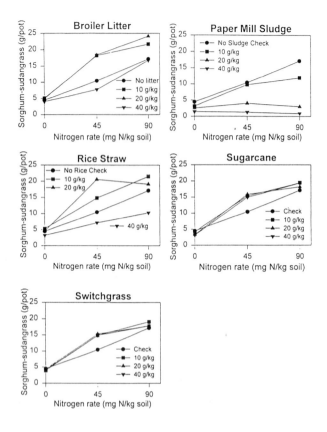

Figure 1. Sorghum-sudangrass yields 28 days after planting, as affected by feedstock ash and nitrogen application.

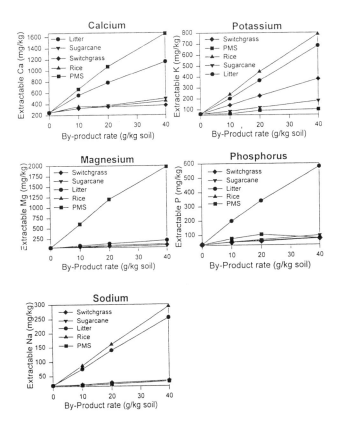

Figure 2. Melich I extractable nutrients as affected by feedstock gasification by-product (ash) rate (PMS = paper mill sludge)

Soil carbon was increased by all feedstock ashes, approximately in the order in which carbon concentration was determined in the original analyses (Table 1). All the ashes had some neutralizing or liming effect due to the addition of basic cations. However, this effect from the ash of paper mill sludge was excessive, probably causing salinity problems and depressed plant growth. Following the harvest of sorghum, Mehlich I extractable nutrient contents were still almost identical to those recorded after the first experiment with rye. Mehlich I extractable Cu, Mn, AL Fe, and Zn were influenced by ash from the different feedstocks, again, mainly in relation to the concentration of these elements determined in the original analyses.

4. CONCLUSIONS

Agronomic tests demonstrated that land application of the ashes from the gasification of the five feedstocks tested in this study generally increased soil pH. In addition, ash from broiler litter increased plant growth, but ash from paper mill sludge depressed growth. Results suggest that the ashes tested did not immobilize N in the soil, and the nutrients contained in them were available for plant uptake.

5. REFERENCES

McQuigg, K. and W. N. Scott. (1998). Starved air gasification tests on five biomass feedstocks. Proceedings of Bioenergy '98, Madison, Wisconsin, pp. 443–452.

ALKALINE BAGASSE TAR SOLUTIONS AS FOAMERS IN COPPER MINING*

E. Olivares,[a] L.E. Brossard,[b] L.A.B. Cortez,[a] N. Varela,[b] G. Bezzon[c]

[a]Faculty of Agricultural Engineering, University of Campinas, Brazil
[b]Faculty of Chemical Engineering, University of Oriente, Santiago de Cuba, Cuba
[c]Interdisciplinary Energy Planing Center (NIPE), University of Campinas, Brazil

Alkaline tar solutions (ATS) obtained from slow pyrolysis of sugarcane bagasse were used as foamers in the flotation process of sulphurized copper numerals carried out in Mina Grande El Cobre in Santiago de Cuba. The ATSs used differed only in age (0.5 and 3 years) and had 0.22 mPa.s viscosity and a 25% mass fraction of neutralized tar in aqueous solution. Copper recovery (%) was studied as a function of following variables: χ_1 = mineral granulometry; χ_2 = mineral pulp's pH; χ_3 = collector dose; χ_4 = flotation time; χ_5 = foamers dose; χ_6 = type of mineral; χ_7 = age of foaming agent. Experiments were conducted according to a 2_{IV}^{7-3} fractional factorial design. The mathematical model in coded form thus was

$$\% \text{ Copper Recovery} = 72.56 + 1.27\chi_1 + 2.69\chi_2 + 3.37\chi_4 - 6.34\chi_6 - 0.98\chi_7$$
$$- 2.77\chi_5\chi_6 + 2.18\chi_2\chi_4 - 1..11\chi_1\chi_7$$

Behavior of 0.5 year old ATS is similar to commercially used pine oil and represents an opportunity for the development of added value products from pyrolysis effluents.

1. INTRODUCTION

Slow pyrolysis of lignocellulosic materials is usually regarded only as a source of vegetable carbon. Recently, Brossard and Cortez (1997) presented practical applications in different industries of slow pyrolysis tar obtained from sugarcane bagasse: alkaline tar solutions (ATS). The present paper describes a more detailed study of the use of ATS as a flotation foamer in the beneficiation sulphurized copper mineral, replacing commercial pine oil.

---

*The authors thank the State of São Paulo Research Foundation-FAPESP for the financial support, which made this publication possible.

The experimental study deals with a rather large number of independent variables, looking for a complete description of the flotation process and specially of the role of ATS's age and dosage in connection with copper recovery.

2. MATERIALS AND METHODS

Previously conditioned copper mineral coming from two different mine zones, labelled M-1 and M-2, were mixed with water in a liquid solid ratio (v/v) = 3:1. In order to regulate the mixture's pH, the calcium carbide hydrolysis residue from acetylen's production (C_2C_a "ash") was added.

Next, potassium amyl xanthale, acting as collector agent, was added to the alkalinized slurry.

Finally, the flotation mixture was completed by the addition of the specified foamer (i.e., 0.5- and 3-year-old ATS) (See Table 1).

All experiments were conducted in a bench scale flotation cell, located at Mina Grande El Cobre in Santiago de Cuba.

The detailed sequence of operations as well as the analythical techniques employed and ATS preparation is described elsewhere (Varela, 1998).

Two blocks of experiments were planned according to a 2_{IV}^{7-3} fractional factorial design (Box, 1993). The studied variables and their respective levels are shown in Table 2.

Table 1
Characteristics of ATS

Age (years)	Density (g/ml) 26° C	Mass Fraction (%)	Viscosity 26° C (mPa.s)
0.5	1.04	25	0.22
3.0	1.06	25	0.22

The experiment's matrix belonging to block N°.1 (Table 3) shows the selected associations for χ_4; χ_5; χ_6; and χ_7. The experiments 9-16 are replicates of the first eight experiments and were planned in order to obtain an estimate of $S_{exp.}^2$.

The complementary block N°.2 appears in Table 4 with negative interaction associations for χ_4; χ_5; and χ_6. The measured response for these experiments was copper recovery (%).

3. RESULTS AND DISCUSSION

The obtained copper recovery (%) for each experiment in block No. 1 is given by the same Table 3. For this set of experiments the pooled experimental error variance, S_{exp}^2, was 4.94 with 8 degrees of freedom.

Table 2
Studied variables and their levels

Natural Variables	Coded Variables	Levels	
		-1	+1
Particle size (%)	χ_1	40	60
Pulp's pH	χ_2	9	12
Collector dosage (g/t of mineral)	χ_3	130	230
Flotation time (min)	χ_4	6	16
Foamer dosage (g/t of mineral)	χ_5	150	250
Type of mineral	χ_6	M-1	M-2
Type of foamer	χ_7	E-1 (0.5 years old)	E-2 (3 years old)

Table 3
Block N°.1—Matrix of experiments

Run No.	Factors							% Cu Recovery
	χ_1	χ_2	χ_3	$\chi_4 = \chi_1\chi_2$	$\chi_5 = \chi_1\chi_3$	$\chi_6 = \chi_2\chi_3$	$\chi_7 = \chi_1\chi_2\chi_3$	
1	-1	-1	-1	1	1	1	-1	60.94
2	1	-1	-1	-1	-1	1	1	64.46
3	-1	1	-1	-1	1	-1	1	81.30
4	1	1	-1	1	-1	-1	-1	85.35
5	-1	-1	1	1	-1	-1	1	71.44
6	1	-1	1	-1	1	-1	-1	75.62
7	-1	1	1	-1	-1	1	-1	70.59
8	1	1	1	1	1	1	1	73.37
9	-1	-1	-1	1	1	1	-1	66.07
10	1	-1	-1	-1	-1	1	1	59.06
11	-1	1	-1	-1	1	-1	1	78.12
12	1	1	-1	1	-1	-1	-1	86.55
13	-1	-1	1	1	-1	-1	1	69.54
14	1	-1	1	-1	1	-1	-1	77.43
15	-1	1	1	-1	-1	1	-1	72.01
16	1	1	1	1	1	1	1	75.06

Table 4
Block N°.2—Matrix of experiments

Run No.	Factors							% Cu recovery
	χ_1	χ_2	χ_3	$\chi_4 = -\chi_1\chi_2$	$\chi_5 = -\chi_1\chi_3$	$\chi_6 = -\chi_2\chi_3$	$\chi_7 = \chi_1\chi_2\chi_3$	
1	-1	-1	-1	-1	-1	-1	-1	72.04
2	1	-1	-1	1	1	-1	1	84.88
3	-1	1	-1	1	-1	1	1	69.28
4	1	1	-1	-1	1	1	-1	59.89
5	-1	-1	1	-1	1	1	1	57.35
6	1	-1	1	1	-1	1	-1	72.46
7	-1	1	1	1	1	-1	-1	86.70
8	1	1	1	-1	-1	-1	1	74.95

A Cochran test (Box, 1993) applied to pooled experimental variances confirmed the homoscedasticity hypothesis:

$G_{exp} = 0.37$

$G_{(1,8;\ \alpha = 0.05)} = 0.68$

As $G_{exp} < G_{critical}$, homoscedasticity is accepted.

For Block No. 2, Table 4 also shows copper recovery (%) for planned experiments.

Results from Blocks N°. 1 and N°. 2 were processed by means of the methodology described by Box (1993), in which the confounding effects appearing in each experimental block are processed in a way as to obtain coefficients of considered factors free of bias. In relation to significance groups of interactions, a careful examination of each one gave what was considered the significant interaction responsible for the noted effect.

In this way, a mathematical model was obtained for copper recovery (%) as a function of the following coded factors and factor interactions:

% Copper Recovery = $72.56 + 1.27\chi_1 + 2.69\chi_2 + 3.37\chi_4 - 6.34\chi_6 - 0.98\chi_7 - 2.77\chi_5\chi_6 + 2.18\chi_2\chi_4 - 1.11\chi_1\chi_7$

at a significance level $\alpha = 0.05$

So, copper recovery is larger when mineral M-1 (-1 level of χ_6) is used at a bigger particle size (+1 level of χ_1), the pulp's pH = 12 (+1 level of χ_2), at the highest flotation time (+1 level of χ_4) and also when ATS with 0.5 years is used (-1 level of χ_7).

It is important to note that although ATS dosage (variable χ_5) is not significant by itself, it does influence copper recovery as it can be seen from $\chi_5\chi_6$ interaction term. In other words, copper recovery is favored by a high ATS dosage when M-1 mineral is employed.

In addition, the interaction term $\chi_1\chi_7$ (mineral's granulometry - age of ATS) must be considered when predicting copper recovery. In this case, both factors appear to work at the levels observed previously when describing each significative individual factor.

Interactions $\chi_2\chi_4$ (pulp's pH - flotation time) indicate that higher pH (i.e., pH = 12) and higher flotation time (i.e., 16 minutes) contribute to a greater copper recovery.

4. CONCLUSIONS

Even though phenolates are considered weak foamers (Dudenkov, 1980), ATS, which is doubtless a complex mixture of various phenolates, has a quite satisfactory behavior as a foamer flotation agent. This behavior could probably be caused by the presence in ATS of "heavy phenolates," i.e., phenolates having rather large alkyl functional groups attached to the benzene ring.

During slow pyrolysis, initially formed light phenols coming from lignin decomposition are submitted to a polycondensation reaction due to prevailing high temperature and also long residence times. This could be the way in which biomass is partially transformed into the substances exhibiting surface-active properties present in ATS.

In relation to the performed industrial tests, it is important to note that no changes were needed in the technology routine of the factory and also that no special dosage device was necessary.

On the other hand, although C_u recovery (%) was the studied response, another important parameter, % C_u in enriched mineral, was also favorable when using ATS.

Thus, for flotations with ATS, mean value of C_u % in the concentrated mineral was around 21%. Pine oil correspondent values were not statistically different.

The possibility of using slow pyrolysis tar is confirmed, in the form of an alkaline soluble solution (ATS), as a good substitute for pine oil in sulphurized copper minerals' enrichment.

Flotation carried out with ATS as a foamer gives copper recovery values of the same order as other commercial foamers.

ATS dosage depends greatly on the type of copper mineral being processed.

Rather recently prepared ATS is preferred when using a mineral's larger particle size in order to obtain a larger copper recovery.

The study of other simple transformation procedures for integral slow pyrolysis tar as well as additional applications of ATS seems to be a promising way of finding added value products from actual pollutants in carbonization process.

5. NOMENCLATURE

ATS = alkaline tar solution
χ_i = coded symbol for independent variables or factors
-1; +1 = coded inferior and superior levels respectively
M-1; M-2 = coded qualitative levels for two copper minerals
E-1; E-2 = coded qualitative levels for two ATS of different age
G = Cochran statistic
α = significance level

REFERENCES

1. Box, G.E.P., W.G. Hunter, and J.S. Hunter (1993). Estadistica para investigadores. Introduccion al diseño de experimentos, análisis de datos y construcción de modelos, Editorial Reverte S.A., España.
2. Brossard, L.E. and L.A.B. Cortez (1997). Potential for the use of pyrolytic tar from bagasse in industry, Biomass & Bioenergy, 12, pp. 363–366.
3. Dudenkov, S.V. (1980). Fundamentos de la teoria y la práctica del empleo de reactivos de flotación, Edición MIR (en español), Moscú.
4. Varela, N. (1998). Estudio de la flotacion de mineral sulfuroso de cobre com solucion alcalina de alquitrán de bagazo, M.Sc. Thesis, Universidad de Oriente, Santiago de Cuba, Cuba.

589

EFFICIENCY TEST FOR BENCH UNIT TORREFACTION AND CHARACTERIZATION OF TORREFIED BIOMASS

F. Fonseca Felfli, C. A. Luengo, P. Beaton and J. A. Suarez

Grupo Combustíveis Alternativos (GCA), IFGW/DF, Universidade Estadual de Campinas
C.P 6165, CEP 13083-970, Campinas SP, Brasil. Fax: (19) 289 2421

This work evaluates results of a bench unit for biomass torrefaction. Torrefaction improves the energy properties of biomass by use of a short residence time and temperature less than 300°C. The unit was evaluated with wooden briquettes as raw material, reporting efficiency around 80%, and weight yield between 45 and 95%. Analysis of torrefied briquettes showed fixed carbon contents ranged from 19 to 50% and higher heating value between 21 and 23 MJ/kg. It was determined from analysis of results that the temperature of torrefaction has greater influence on the torrefied products than residence time. For this reason, a temperature range of 250 to 300°C with residence times less than 60 minutes is recommended for torrefaction. The torrefied briquette has low moisture, hydrophobic character and good properties for combustion and gasification, and it is able to replace firewood or charcoal in some cases.

1. INTRODUCTION

Direct use of agricultural and wood residues as fuel is usually difficult, because they have poor energy characteristics (low heating value, high moisture content, hygroscope nature, low density and polymorphism) that cause high costs during transportation, handling and storage. In this context, the main objective of this research was to investigate the torrefaction process as an alternative treatment to biomass residues that could increase the competitiveness of biomass residues as an energy resource. Torrefaction is a pyrolysis process with a low heating rate at a temperature lower than 300°C. Torrefied biomass achieves a hydrophobic character, low moisture, higher calorific value and less smoke when burnt, and a nonfermentable nature.[1,2,3]

A bench unit for biomass torrefaction was developed; this unit was evaluated using wooden briquettes as raw material. The bench unit consists of a combustion chamber and a torrefaction chamber; the combustion chamber supplies the energy necessary for the torrefaction process by burning biomass. The hot gases from combustion circulate in a pipe to the torrefaction chamber, where they exchange heat with the biomass. In the torrefaction

gases with torrefaction vapors, ensures that these vapors are recycled and burned in the combustion chamber, and avoids environmental pollution.[4]

2. EFFICIENCY TEST OF UNIT

Efficiency can be determined by direct or by indirect methods. In this work we used the indirect method,[5,6] because it provide information about the principal heat losses, facilitating the energy analysis of the system. The efficiency can be express by follow equation:

$$\eta = 100 - \sum_{i=2}^{5} q_i \qquad (1)$$

Where:
η : Gross efficiency of unit %
q_2 : losses by sensible enthalpy of outlet gases %
q_3 : losses by unburned fuel gas %
q_4 : losses by unburned particles fuel in grate ash and fly ash %
q_5 : losses by radiation and convection %.

Several efficiency tests were carried out in the bench unit, using wooden briquettes as raw material. The heat losses by unburned particles fuel (q4) were determined by ash balance, reporting a value around 0.9% for all conditions. On the other hand, the losses by unburned fuel gas (q3) ranged from 0.4 to 4% of the total energy fed to a process, depending on the excess air ratio. With a larger amount of excess air ratio, about 1.6, the loss q3 achieved the maximum value (4%) as a result of the low combustion temperature and short time in the hot zone of fuel gases; with an excess air about 1.3 the loss q3 achieved the lowest value (0.4%).

The major loss was determined in the sensible enthalpy of outlet gases (q2), which ranged between 12 and 14% in function of outlet gases temperature (from 250 to 252 C). It was determined that the loss by radiation and convection is constant for all conditions with a value about 4.3% of total energy losses. The gross efficiency ranged from 77 to 82%, but can be increased using the enthalpy of outlet gases in an previous drying process of biomass.

3. CHARACTERIZATION OF TORREFIED BRIQUETTES

The torrefied briquettes were analysed in term of proximate analysis, ultimate analysis, and high caloric value;[7,8,9] the results are summarized in Table 1 and 2. The fixed carbon and volatile matter of torrefied briquettes depend on torrefaction temperature and residence time (Table 1). Temperature had greater influence on the torrefied briquettes than residence time.

Table 1
Proximate analysis and heating value. 1 hr processing time

Temperature °C	Volatile Mater %	Fixed carbon %	Ash content %	High Heating Value MJ/kg
220	74.6	19.0	6.4	20.99
250	65.0	27.2	7.8	22.06
270	52.1	38.2	9.7	22.98

Table 2
Ultimate analysis. 1 hr processing time

Temperature °C	C %	H %	O%	H/C	O/C
220	52.02	6.50	35.08	1.49	0.50
250	55.81	6.60	29.79	1.40	0.40
270	59.82	5.26	25.22	1.05	0.31

The temperatures higher than 250°C present a larger effect in the high calorific value, because the volatile emission is more intense at temperature range 250-300°C (Table 1). The yield weight of torrefied briquettes ranged from 45 to 95%; it decreased with increasing temperature and residence time (see Figure 1).

During torrefaction, the briquettes undergo changes in chemical composition. The carbon content increases at the expense of oxygen and hydrogen content, provoking decreases in H/C and O/C ratios (Table 2). Torrefied briquettes remained unaffected on immersion in water for 17 days, whereas a normal briquette quickly disintegrates in 10 minutes.

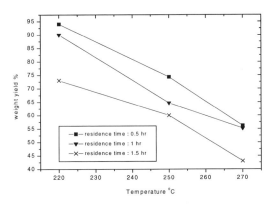

Figure 1. Weight yield of torrefied briquettes at different conditions.

4. CONCLUSIONS

Torrefaction of briquettes is a feasible alternative to improve their energy properties, increase calorific value and avoid moisture absorption, facilitating handling and storage.
The gross efficiency of a bench unit can be increased by using the energy of outlet gases in a previous drying process of biomass or preheating the air to combustion process. The torrefaction temperature has greater influence on the torrefied briquettes than residence time. A temperature range of 250 to 300°C with residence time less than 60 minutes is recommended for torrefaction.

5. ACKNOWLEDGMENTS

Authors thank FAPESP for the financial support (process N^0 96/2661-7).

REFERENCES

1. Bhattacharya, S C. (1990). Carbonized and uncarbonized briquettes from residues. In : Workshop On Biomass Thermal Processing, London: Proceedings. Shell, 1990. V. 1, pp. 1–9.
2. Bourgeois, J.P. and J. Doat (1985). Torrefied wood from temperate and tropical species, advantages and prospects, in Bioenergy 84, London: Proceedings, ed. by Egneus and Ellegard, pp. 153–159.
3. Antal, M.J. and Willian S.L. Mok (1990). Review of methods for improving the yield of charcoal from biomass, Energy & Fuel, 4, pp. 221–226.
4. Fonseca Felfli, F.E, Carlos A. Luengo, and P. Beaton (1998). Bench unit for biomass residues torrefaction, in Biomass For Energy And Industry, 8-11 June, Wurzburg: Proceeding: 10th European Conference and Technology Exhibition, 1998, pp. 1593–1595.
5. Cortez, L.A.B. and E. Silva (1997). Tecnologias de conversão energética da biomassa. Manaus: EDUA/EFEI, p. 540.
6. Beaton P. and E. Silva (1991). Pruebas de balance térmico en calderas para bagazo. Santiago de Cuba: ed ISPJAM, 1991, p. 94.
7. A.S.T.M. (1985). Philadelphia D 3286-85: Standard Test Methods for Gross Calorific Value of Coal and Coke by the Isoperibol Bomb Calorimeter. Philadelphia, p. 5.
8. A.S.T.M. (1984). Philadelphia. D 3178-84: Standard Test Methods for Carbon and Hydrogen in the Analysis of Coal and Coke. Philadelphia, p. 3.
9. A.S.T.M. (1984). Philadelphia.D 1762-82:Standard Test Methods for Proximate Analysis of Wood Charcoal. Philadelphia, p. 2.

Biomass Transformation into Value-Added Chemicals, Liquid Fuels, and Heat and Power

Value-Added Products: Bioproducts and Fibers

PRODUCTION OF LEVULINIC ACID AND USE AS A PLATFORM CHEMICAL FOR DERIVED PRODUCTS

D.C. Elliott,[a] S.W. Fitzpatrick,[b] J.J. Bozell,[c] J.L. Jarnefeld,[d] R.J. Bilski[e]
L. Moens,[c] J.G. Frye, Jr.,[a] Y. Wang,[a] and G.G. Neuenschwander[a]

[a]Pacific Northwest National Laboratory, Battelle Boulevard, P. O. Box 999, MSIN K2-12, Richland, Washington 99352, USA
[b]Biofine, Inc., 300 Bear Hill Road, Waltham, Massachusetts 02154, USA
[c]National Renewable Energy Laboratory, 1617 Cole Boulevard, Golden, Colorado 80401, USA
[d]New York State Energy Research and Development Authority, Corporate Plaza West, 286 Washington Ave. Ext., Albany, New York 12203, USA
[e]Chemical Industry Services, Inc., 1420 Fawn Ridge Drive, West Lafayette, Indiana 47906, USA

Levulinic acid (LA) can be produced cost effectively and in high yields from renewable cellulosics. The technology to convert cellulosic biomass to LA is being demonstrated on a 1 ton/day scale using paper mill sludge and municipal solid waste as the feedstock. Low cost LA has great possibilities as a platform chemical for the production of a wide range of value-added product chemicals. For example, a process developed at Pacific Northwest National Laboratory produces methyltetrahydrofuran (MTHF) from LA in > 80% molar yield by a single-stage catalytic hydrogenation process. MTHF may be used as a solvent and as a fuel. In other work, the National Renewable Energy Laboratory has developed a new preparation of δ-aminolevulinic acid (DALA), a broad spectrum herbicide, from LA. Each reaction step proceeds in high (> 80%) yield and affords DALA in greater than 90% purity, giving a process that could be commercially viable. LA is also being investigated at Rensselaer Polytechnic Institute as a starting material for the production of diphenolic acid, a direct replacement for bisphenol A in several commercial polymers.

1. BACKGROUND

Biofine, Inc., of Waltham, Massachusetts, has developed a thermochemical process to convert cellulosic biomass into levulinic acid (LA) using high-temperature, dilute-acid hydrolysis.[1,2] At the reaction conditions used, cellulose in the biomass is converted through several steps into LA ($CH_3COCH_2CH_2CO_2H$). Furfural, formic acid, and a solid residue (which could be used as fuel) are co-products. The LA is made from low-cost and abundant waste feedstocks. Wet feedstocks can be used without drying, thereby saving energy. Paper mill wastes appear to be ideal feedstocks for the process because they are finely divided and easy to handle.

In August 1997, Biofine, the U.S. Department of Energy, the New York State Energy Research and Development Authority (NYSERDA), and Biometics, Inc. began manufacturing LA from paper mill sludge at a 1 ton/day demonstration plant in South Glens Falls, New York. Biofine's process had already been demonstrated at a smaller scale with a variety of cellulosic feedstocks, including municipal solid waste, unrecyclable municipal waste paper, waste wood, and agricultural residues.

LA's niche markets provide excellent small-scale opportunities. LA's worldwide market is about 1 million lbs/yr at a price of $4-6/lb. Full-scale commercial plants are feasible at 50 dry ton/day of feedstock. At this scale, LA could be produced at $0.32/lb, and converted into commodity chemicals such as succinic acid and diphenolic acid, which sell for $2/lb or less, or acrylic acid, which sells for $0.50/lb. Large-scale opportunities also exist which are even more economical. Larger plants to convert 1,000 dry ton/day of feedstock into LA at $0.04-0.05/lb allow economical production of fuel additives, such as methyltetrahydrofuran (MTHF). The worldwide commercial market for LA and its derivatives could reach one 1 trillion lbs/yr. Full-scale plant opportunities are being assessed for several locations in the U.S. and worldwide. One full-scale commercial plant using 1,000 dry ton/day of feedstock could manufacture more than 160 million lbs/yr of product. The broad range of economical plant sizes means even the 1 ton/day demonstration plant is self-sufficient at LA's existing price.

Because LA is a platform chemical, it need not be sold as a commodity chemical. Derivatives are the key to marketability, and markets for such LA derivatives as tetrahydrofuran, 1,4-butanediol, γ-butyrolactone, succinic acid, and diphenolic acid exist. The National Renewable Energy Laboratory (NREL), Pacific Northwest National Laboratory (PNNL), and Rensselaer Polytechnic Institute (RPI) are developing market applications and production methods for other derivatives, including a biodegradable herbicide, a gasoline fuel additive, and new biodegradable polymers.

2. PROCESS DESCRIPTION

Cellulosic biomass is converted to LA through high-temperature, dilute-acid hydrolysis. The continuous two-stage system combines a short-residence-time tubular reactor with a longer-residence-time backmix reactor. Cellulose is first converted to intermediate sugars and furfurals, which are then converted to LA. High product yield is achieved by the patented reactor configuration, which was developed based on reaction kinetics and thermodynamic considerations. The controlled hydrolysis proceeds at temperatures in the range of 200°C and pressures of several hundred psig. Typical yields of LA range from 50-70% of theoretical, or approximately 0.5 lb of product per pound of cellulose. A key advantage of this technology is that it can utilize dirty, wet cellulosic biomass streams and produce a single chemical product in high yield and high purity.

To facilitate storage and handling at the demonstration plant, feedstock is dried to about 60-70% solids content. (Note: feedstock does not need energy-intensive drying in a full-scale commercial facility.) It is fed into the first stage as an aqueous acidified slurry (using sulfuric acid), with about 15% solids. The feed is heated in the first stage by injecting high-pressure steam to maintain a pressure of 3.06 MPa at 215°C. The contents of the first-stage reactor (constructed of Teflon-lined zirconium) flow to a lower-pressure (1.52 MPa at 200°C) second stage through a pressure let-down valve. Hydrolysis to LA is completed in the second stage, and volatiles produced in the reaction are allowed to vaporize. LA remains in the aqueous solution at a concentration of 6-12% by weight. LA is then extracted and concentrated. LA is separated and purified by distillation. The raffinate containing sulfuric acid is recycled to the reactor feed.

2.1. By-products

Formic acid and condensed tars are by-products of the process. The reactor system also produces furfural in high yields if hemicellulose is present in the feed. In the demonstration plant, furfural and formic acid are not recovered. The formic acid is neutralized, and the aqueous stream is discharged to a local wastewater treatment plant. The condensed tars are filtered out of the process and sent to a local incinerator. These tars have fuel properties similar to those of brown coal. Full-scale options include selling the tars as a solid fuel product, or using the tars onsite to generate steam for the process. All by-products have value in this process, and at commercial scale would be recovered and sold as commodities, or used as fuel or to produce LA derivatives.

2.2. Product Potential

Although LA's primary market is small, it would expand greatly if LA could be used to economically produce other chemicals. With Biofine's new technology, LA can be made cheaply enough to allow inexpensive production of higher-value derivatives such as diphenolic acid, succinic acid, pyrrolidinones, and agricultural chemicals such as δ-aminolevulinic acid. Over the longer term, Biofine will produce larger-volume chemicals such as 1,4-butanediol, γ-butyrolactone, and tetrahydrofuran. Ultimately, the largest-volume, lowest-cost LA could be used to produce the gasoline oxygenate additive, methyltetrahydrofuran (MTHF). The LA or LA derivatives produced at future commercial facilities could become a feedstock for a new chemical industry based on biomass wastes.

An industry–government consortium was established to convert LA into a larger suite of chemical products. The consortium brings together two industrial partners, Biofine and Chemical Industry Services (a custom and contract chemical manufacturing agency representing more than 40 U.S. chemical companies), with NREL, PNNL, and NYSERDA. In addition, Merichem Company has provided financial support and technical advice regarding operations at the demonstration plant.

Research at NREL and PNNL sponsored by U.S. DOE's Office of Industrial Technologies has resulted in two potentially high-profile products: the MTHF, a fuel additive with a huge potential market, and δ-aminolevulinic acid (DALA), a broad-spectrum herbicide with a projected market of 200-400 million lb/yr. These two products alone have the potential to expand LA's market into hundreds of millions of pounds annually. Commercial scale-up work is planned at NREL and PNNL for both these products. In NYSERDA-sponsored work, RPI is exploring large-volume, short-term applications for diphenolic acid (DPA) as a monomer for producing polycarbonates, epoxy resins, etc.

Future plans foresee commercial facilities manufacturing LA derivatives onsite, but work is needed to scale up commercial production methods for some derivatives. Meanwhile, the market for new products like MTHF and DALA may develop slowly. Therefore, ongoing research is balanced between derivatives with short- and long-term potential.

3. MTHF

Laboratory-scale continuous reactor work at PNNL converts Biofine's LA into MTHF with a single-bed catalytic hydrogenation process that uses a proprietary catalyst at elevated temperature and pressure.[3] The LA undergoes multiple hydrogenations (three moles of hydrogen per mole of LA) and two dehydrations in a single reaction step in a tubular reactor with a fixed bed of palladium-rhenium catalyst. Operating conditions are 240°C and 1500 psig using liquid hourly space velocities around 1 liter of LA/liter of catalyst bed/hour. Laboratory tests have indicated an 83% yield on a theoretical (molar) basis. On a weight basis, the yield is 63 lb of MTHF for every 100 lb of LA. Previous literature values using competing processes suggested low yields (3%) of MTHF as a minor by-product.[4] Plans are under way to scale up production of MTHF from LA in a mobile processing unit at Biofine's demonstration plant. MTHF production there is projected at about 20 gallons per day.

The major projected use of MTHF is as a transportation fuel. Because it is miscible with gasoline at all proportions and hydrophobic, MTHF could be blended at the refinery and transported by pipeline. In contrast, ethanol must be added later in the distribution process because contamination with water can cause a phase separation. MTHF can be blended in gasoline up to 70% by volume without adverse engine performance. MTHF has a higher specific gravity than gasoline; therefore, mileage from MTHF-blended fuel would not decrease. As a component of "P-series" fuels (recently approved by the U.S. Department of Energy), MTHF can be used to meet the requirements for alternative-fuel fleet vehicles stipulated by the Energy Policy Act of 1992. The "P-series" fuels are blended such that they have a minimum anti-knock index of 87 and a maximum vapor pressure of 15 psi.[5] P-series fuel emissions are generally below those for reformulated gasoline using methyltertiarybutyl ether (MTBE) and are well below federal emissions standards. And, because MTHF is essentially a high-density (liquid) storage system for hydrogen as fuel, it could help establish hydrogen as a practical transportation fuel.

4. DALA

DALA is a broad-spectrum biodegradable herbicide that shows high activity toward dicotyledonous weeds while showing little activity toward monocotyledonous crops such as corn, wheat, or barley. DALA exerts its effects by stimulating overproduction of tetrapyrroles in the plant at night. In daylight, the accumulated tetrapyrroles photosensitize the formation of singlet oxygen in the plant, leading to its death.[6,7] More recently, DALA has been found to be useful as an insecticide[8] and as a component in photodynamic therapy as a cancer treatment.[9]

DALA is a component of all living cells and a metabolic precursor to the biosynthesis of porphyrin. A variety of synthetic routes towards DALA have appeared during the last five decades. The obvious starting material for the preparation of DALA is, of course, LA, which requires the formation of a C-N bond at the C5 carbon. By far, the most common approach towards activating the C5 position toward amination is by selective bromination of LA in an alcohol medium.[10] This leads to mixtures of 5-bromo- and 3-bromo-esters that can be separated by distillation. The 5-bromolevulinate can then be aminated through the use of nucleophilic nitrogen species such as sodium azide.[11]

The key areas of study at NREL for DALA manufacture have been improvement of each of these synthesis steps, with particular focus on overall cost effectiveness. The work is the basis of a patent application whose claims have been allowed. The synthesis begins with LA, which is brominated in MeOH to give 5-bromomethyllevulinate. This ester is treated with diformylamide anion (easily prepared from NaOMe and formamide[12]). Acid hydrolysis of this intermediate leads to DALA. The only side product observed from the hydrolysis is formic acid, which can be easily removed and used in several applications. Each of the steps proceeds in high (> 80%) yield and affords DALA (as the hydrochloride salt) in greater than 90% purity, fostering a process that could be commercially viable. The material prepared by this process has been submitted for herbicide testing at the University of Illinois. The first samples were 85% as effective as a highly purified control sample of DALA.

More recently, the process has been improved further by demonstrating the use of a different amination reagent for the second step of the process. This new reagent generates much less waste, and can be used in several different solvents.

5. DPA

Work at Rensselaer Polytechnic Institute has just begun on near-term applications for DPA, particularly ones that displace currently marketed bisphenol-A (BPA) products. Researchers are exploring DPA/BPA copolycarbonates that could partially or wholly displace BPA formulations. Within these formulations, researchers are also studying the use of dibrominated DPA in fire retardants. Brominated DPA also has some promise as an environmentally acceptable marine coating that could replace the toxic tributyltin. Further development of DPA uses will concentrate on highly crosslinked polymers and charged polyesters, or "ionomers."

DPA is used as a component in protective and decorative finishes. It also can be used as a substitute for BPA, the primary raw material for epoxy resin, which is a potential threat to human health because it is an endocrine disrupter. Indeed, before BPA was developed, DPA was used more widely in coating applications.

6. FUTURE PLANS

A small (1 ton/day feedstock) plant to convert low-grade cellulosic biomass to LA has been successfully designed, procured, constructed, and operated. Operating and economic data thus collected are helping to optimize the demonstration, as well as plan for a commercial-scale plant. Biofine is partnering with Pencor Environmental Ventures, Inc., to assess sites in the U.S. and worldwide for commercial development of large-scale plants. Work on commercializing new production methods and applications for an integrated family of LA and LA derivatives and by-products continues at NREL, PNNL, RPI, and elsewhere.

Today's chemical and fuel industries are based primarily on fossil fuels, but Biofine's demonstration promises to lead to a full-scale biomass waste industry. During the next two decades, this technology has the potential to transform major sectors of the chemical and fuel manufacturing industries in the U.S. and abroad from petroleum-based to renewable feedstock.

REFERENCES

1. Fitzpatrick, S.W. (1990). Lignocellulosic degradation to furfural and levulinic acid, US Patent No. 4 897 497.
2. Fitzpatrick, S.W. (1997). Production of levulinic acid from carbohydrate-containing materials, US Patent No. 5 608 105.
3. Elliott, D.C. and J.G. Frye, Jr. (1999). Method of hydrogenating a 5-carbon compound, US Patent No. US5883266.
4. Christian, Jr., R.V., H.D. Brown, and R.M. Nixon (1947). Jour. Amer. Chem. Soc., 69, p. 1961.
5. Paul, S.F. (1997). Alternative Fuel, US Patent No. 5 697 987.
6. Rebeiz, C.A., A. Montazer-Zouhoor, H.J. Hopen, and S.M. Wu (1984). Enzyme Microb. Technol., 6, pp. 390–401.
7. Rebeiz, C.A., J.A. Juvik, and C.C. Rebeiz (1988). Pestic. Biochem. Physiol., 30, pp. 11-27.
8. Rebeiz, C.A., S. Amindari, K.N. Reddy, U.B. Nandihalli, M.B. Moubarak, and J.A. Velu, (1994). δ-Aminolevulinic acid based herbicides and tetrapyrrole biosynthesis modulators, in Porphyric Pesticides: Chemistry, Toxicology, and Pharmaceutical Applications, ACS Symposium Series 559, Washington, DC.
9. Rebeiz, N., S. Arkins, C.A. Rebeiz, J. Simon, J.F. Zachary, and K.W. Kelley (1996). Photodynamic Therapy, Cancer Res., 56, pp. 339–344.
10. MacDonald, S.F. (1974). Methyl 5-bromolevulinate, Can. J. Chem., 52, pp. 3257–3258.
11. Ha, H.-J., S.-K. Lee, Y.-J. Ha, and J.-W. Park (1994). Synth. Commun., 24, pp. 2557–2562.
12. Yinglin, H. and H.A. Hongwen (1990). Synthesis, pp. 615–618.

DECENTRALISED PRODUCTION OF BIODEGRADABLE LUBRICANTS, RENEWABLE FUELS, FODDER, AND ELECTRICITY FROM OIL CROPS ON THE PRODUCTION-SITE, IN MOBILE UNITS

Dr. U.C. Knopf

Agrogen Foundation, P.O. Box 21 CH-1701 Freiburg, Switzerland. Fax: + 26 670 46 51
Agricultural Institute, CH-1725 Grangeneuve, Switzerland. Fax: + 26 305 55 68

Biomass represents an enormous yet largely unused energy source. Its development and use is a great challenge for the coming millennium if we aim for sustainable economic development on our globe. For this, not only new concepts and technical procedures must be developed, tested, and introduced, but in addition a new consciousness for the potential of renewable energy resources must created in our society, so that the politico-economical frame necessary for its development will be established.

1. THE AGROGEN CONCEPT: PLANT OIL FOR THE SUBSTITUTION OF MINERAL OIL

The principal goal of our Bio-Energy project is to develop a technology that allows the gradual replacement of mineral oil and products made therefrom by plant oil and its derived products. This goal further develops agricultural resources and their use in the energy sector and improves the environmental situation. Our technology is based on the principle that biomass must be transformed and its products consumed as close as possible to the site where the biomass is grown. This requirement minimises transportation of energy over long distances, as is done currently for mineral oil.

2. REALISATION OF THE CONCEPT

2.1. Establishment of a politico-economical frame

Canola is the main oil crop in Switzerland. Canola oil is suited for consumption, but it also has excellent properties for technical purposes. In 1993 the Swiss government established a politico-economical framework that allowed us to make contracts with farmers all over Switzerland for the production of canola to be transformed for technical purposes. The basis of the frame consisted essentially (but not exclusively) in the financial support for the farmers, allowing them to cultivate canola and us to buy it at a relatively low price.

2.2. Construction of mobile production units

Currently in Switzerland canola is transformed exclusively in large industrial complexes. The location of these industrial plant oil production complexes again imply relatively long transportation distances for the seeds on one hand and the products therefrom (oil and meal) on the other. Furthermore, the transformation processes used have been very energy intensive. Finally the quality requirements of oil for consumption purposes and the ones of oil for technical purposes are not identical. For these reasons we decided to develop relatively small mobile production units allowing the decentralised production of a specific quality of plant oil suited for different technical applications. In 1993 we constructed the first mobile unit which we started to test extensively in 1994. We optimised the unit and the procedures during the following years. These production and transformation units have the following characteristics:
- small and mobile, therefore usable on different production sites; can be transported without special permits on the roads, on the railway, ships, and even with air-planes;
- energetically optimised: optimal isolation, use of the gravity force and solar energy wherever possible, use of the internal process energy generated during the transformation process of the bio-mass; use of part of the plant oil produced as process energy (electricity);
- directed by a microprocessor and ergonomically optimised; operated by only one part-time worker;
- annual production and transformation capacity of one unit: 500 (1,000) tons of plant oil, lubricant or fuel and 1,000 (2,000) tons of meal;
- relatively low cost for one production unit: US $ 400,000.

2.3. Development and testing of various substitutes for mineral oil products, technical accessories in particular combustion engines

The pilot plant allowed us to develop and to test a number of products made from plant oil able to substitute for products made from mineral oil; biodegradable lubricants and renewable fuels for different kinds of combustion engines (cars, trucks and other machinery), and machines able to co-generate heat and electricity using canola oil as a primary energy source and different kinds of fodder. The production procedures were chosen in order to minimize energy input. During the past four years we introduced and tested these products in the market, with positive results. In addition to the technical results, we were able to make a number of interesting sociological and psychological observations.

3. CONCLUSIONS

All this leads us to the following conclusions. We have today the technology available allowing us to produce and gradually introduce plant oil and products therefrom as substitutes for mineral oil and products derived from it. The technology has a high potential for further industrial and agricultural development. Thus it is of great interest to industrialised countries, but also to countries of the third world, with a strong need and will to develop their own energetic and agricultural resources. There is a wide-spread lack of information on substitutes of mineral oil. Important communication work has to be done before establishing a sound politico-economical frame if one aims to introduce this technology.

605

AGRI-PULP™ NEWSPRINT*

Al Wong

Arbokem Inc., P.O. Box 95014, Vancouver V6P 6V4, Canada

Using agricultural fibre for newsprint in North America is a technical and economic challenge. The principal obstacle is the current availability of relatively low-price wastepaper and virgin wood fibre for the manufacture of newsprint.

In 1995, Arbokem began the development of a novel newsprint which contains substantial agricultural pulp fibres. The goal was to make a high-performance newsprint using cereal straw and seed grass straw on a modern paper machine.

About 163 tonnes (180 short tons) of standard (49 g/m^2) newsprint was made successfully on the No. 1 Bel-Baie paper machine of Smurfit Newsprint Corp. (Oregon City, Oregon). The final furnish of the Agri-Pulp™ newsprint contained 20% Agri-Pulp™, 12% thermomechanical wood pulp (TMP) and 68% old newspaper pulp (ONP). The Agri-Pulp™ was made from Oregon rye grass straw, California rice straw and British Columbia red fescue straw.

In the weeks following the papermaking test run, the Agri-Pulp™ newsprint was test-printed commercially by the Los Angeles Time, Santa Rosa Press-Democrat, San Jose Mercury News, Paradise Post, Sacramento Bee, The Oregonian and The Orange County Register. The pressroom operators generally found the printing quality and physical strengths of the test Agri-Pulp™ newsprint to be somewhat better than those of standard newsprint.

1. BACKGROUND

Agri-pulp is a papermaking pulp made from agricultural plant fibre that is grown on-purpose or is recovered as cropping residues. Developing countries such as Indonesia, India, Mexico and Cuba now make some agri-pulp newsprint.[1] Bagasse (sugar cane residue) is the principal fibre used.

A kenaf-based newsprint in North America was tested in the early 1980s.[2-7] The test kenaf newsprint contains nearly 20% expensive kraft wood pulp. Newsprint currently made in North America contains virtually no kraft wood pulp. Moreover, kenaf is an expensive on-purpose fibre crop; newsprint is a relatively low-priced paper product.

---

*Agri-Pulp™ is a registered trademark of Arbokem Inc.

What is new? There is a renewed interest in North America in using agriculture-based fibres in place of wood, for the production of pulp and paper.[8,9] The growing shortage of economical wood supply can be alleviated partially by the increased use of waste paper. But ultimately, virgin fibre is needed to compensate for the quantitative and qualitative losses of fibre in the paper recycling system.

From an ecological viewpoint, growing paper demand increases the pressure to harvest virgin forests.[10] From an environmental viewpoint, profitable utilization of straw for pulp and paper production would eliminate the hazardous and polluting practice of stubble burning in heavily populated regions such as the Spokane Valley (Washington) and the Sacramento Valley (California).[11] From a social viewpoint, use of agricultural cropping residues (i.e., straw) would provide a significant new income for American farmers. For example, if an agri-pulp mill purchases straw from eastern Washington farmers, the net farm income could be increased by at least 20%, on a "$ per acre" basis, even at the unusually high price level of wheat in past years.[12] The problem of declining income for grain farmers is real as government subsidies to farmers are being reduced through legislative mandates.

2. FIBRE AVAILABILITY

Pulp production from agricultural cropping residues appears to be the most practical economic means to supplement the fibre needs of the paper industry. Unlike on-purpose fibre cropping of kenaf and hemp, agricultural cropping residues are readily available.

Table 1 shows the enormous availability of surplus straw. It is the largest single uncommitted fibre supply in North America and is truly renewable in real time.

Figure 1 illustrates the geographic occurrence of agricultural cropping residues in North America. With careful soil management practices, a substantial amount of straw can be used for pulp and paper production without any adverse long-term effect on soil structure and fertility.

3. AGRI-PULP™ FIBRE USAGE STRATEGY

In the case of agricultural cropping residues, e.g., cereal straw, there is an optimum combination of such pulp with other pulps for use in papermaking. As given in Figure 2, example wheat straw Agri-Pulp™ has the basic physical strengths required for use in newsprint production and subsequent paper runnability in the press room. Note the superior tensile strength of wheat straw pulp in comparison with that of softwood chemithermo-mechanical pulp (CTMP).

Table 1
Estimated availability of surplus straw in the United States[11]

Commodity and state	Average annual production, 1990-96 1,000 tons	Grain-to-residue ratio	Estimated crop residue 1,000 tons	Harvestable fraction Percent	Estimated availability 1,000 tons
Winter wheat					
Kansas	10,903	1:1.7	18,535	43	7,970
Oklahoma	4,301	1:1.7	7,312	51	3,729
Washington	3,700	1:1.7	6,290	50	3,145
Texas	2,912	1:1.7	4,950	33	1,634
Colorado	2,446	1:1.7	4,158	13	541
Montana	2,285	1:1.7	3,885	20	777
Nebraska	2,168	1:1.7	3,686	34	1,253
Ohio	1,896	1:1.7	3,223	50	1,612
Illinois	1,832	1:1.7	3,114	50	1,557
Idaho	1,816	1:1.7	3,087	50	1,544
Missouri	1,676	1:1.7	2,849	50	1,425
South Dakota	1,481	1:1.7	2,518	26	655
Arkansas	1,295	1:1.7	2,202	50	1,101
Indiana	1,060	1:1.7	1,802	50	901
U.S. Total	49,086	1:1.7	83,446	-	-
Spring wheat					
North Dakota	10,745	1:1.3	13,969	73	10,197
Montana	3,094	1:1.3	4,022	21	845
Minnesota	2,699	1:1.3	3,509	50	1,754
South Dakota	1,789	1:1.3	2,326	36	837
U.S. Total	20,930	1:1.3	27,209	-	-
Rice					
Arkansas	3,505	1:1.5	5,258	100	5,258
California	1,746	1:1.5	2,619	100	2,619
Louisiana	1,327	1:1.5	1,991	100	1,991
U.S. Total	8,530	1:1.5	12,795	-	-
Total	78,546	-	123,450	-	51,342

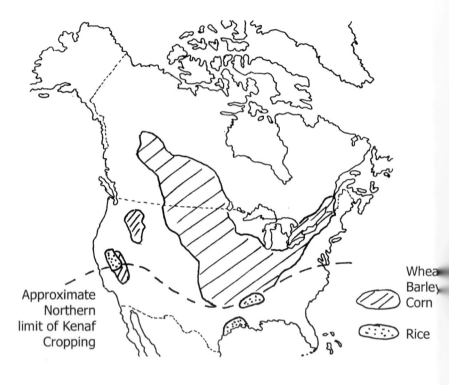

Figure 1. Occurrence of agricultural cropping residues in North America.

Typically, straw would be cooked with a mixture of K_2SO_3 and KOH, and followed with H_2O_2 bleaching. The spent chemicals consisting of mainly K_2SO_4 and dissolved straw organics would be collected and processed for blending into conventional fertilizers. The inflow of straw silica into the Agri-Pulp™ mill can thus be managed profitably.

4. THE AGRI-PULP™ NEWSPRINT PROJECT

Arbokem Inc. and Smurfit Newsprint Corp. collaborated to demonstrate that Agri-Pulp™ can be mixed with standard newsprint pulp to produce standard 49 g/m^2 newsprint.[13,14] In compliance with prevailing California post-consumer wastepaper (PCW) content rule for newsprint, the ultimate furnish was 80% deinked old newspaper (ONP) and 20% Agri-Pulp™. The ONP was made at the Smurfit's own processing facility in Oregon City.

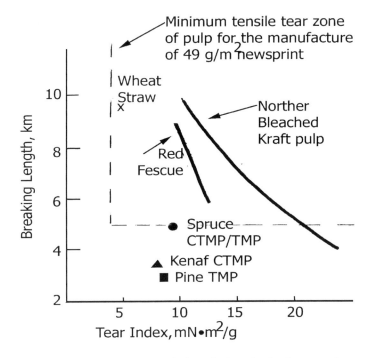

Figure 2. Tear-tensile strength data of selected pulps for newsprint manufacture.[9]

Several months before the papermaking trial, the Rice Producers of California (Colusa, Calif.) had shipped 30 tonnes of rice straw bales to the Arbokem Agri-Pulp™ mill in Vulcan, Alberta. Additionally, several truckloads of creeping red fescue (*Festuca rubra* L.) straw bales and rye grass (*Lolium multiflorium* Lam.) straw bales were procured from grass seed farmers in British Columbia's Peace River Valley and in Oregon's Willamette Valley, respectively. Each type of agricultural fibre was pulped separately. About 10 tonnes of Agri-Pulp™ were made for shipment to Smurfit Newsprint's Oregon City newsprint mill.

The test newsprint was made on Smurfit's Paper Machine #1 (105-in trim; 2,600 feet per minute) which has a modern Bel-Baie former wet end and full cross directional profiling controls. The experiment was started using normal newsprint blend of softwood thermomechanical pulp (TMP) and deinked pulp. As the Agri-Pulp™ was added in steps of 10% and 20%, both TMP and deinked pulp were reduced proportionally. The test newsprint with the highest amount of agri-pulp used has the following composition:

Deinked ONP		68%
TMP		12%
Agri-Pulp™:	red fescue straw	3%
	rice straw	6%
	rye grass straw	11%
		100%

About 163 tonnes (180 short tons) of Agri-Pulp™ newsprint, viz., paper with > 10% agri-pulp content, was made during the 24-hour trial. These major operational aspects were observed during the papermaking trial under non-optimized conditions:
- The number of paper breaks was normal.
- Loss of paper bulk with the usage of Agri-Pulp™ could be compensated by using fewer stack rolls.
- As the Agri-Pulp™ substitution was increased, the wet end drainage was decreased, which could be compensated for, to a certain extent, by reducing paper machine speed or increasing steam for drying. For example, at the 20% Agri-Pulp™ level, the paper machine speed was reduced by about 5% to maintain satisfactory draw.

The quality of the Agri-Pulp™ newsprint was generally satisfactory. The physical strengths, particularly tensile and stretch, of the Agri-Pulp™ newsprint were found to be generally higher than those of standard newsprint. Paper smoothness, opacity and porosity were within the values of standard newsprint. As expected, the Agri-Pulp™ provided higher ash content in the finished newsprint. The higher ash content, probably from the field dirt debris, of the Agri-Pulp™ would need to be reduced to minimize undesired wear on papermaking equipment.

The following dailies participated in the subsequent printing trial of the Agri-Pulp™ newsprint: Los Angeles Times, Sacramento Bee, San Jose Mercury-News, Santa Rosa Press-Democrat, Orange County Register, The Oregonian, and Paradise Post.

Pressroom feedback indicated that both 10% Agri-Pulp™ and 20% Agri-Pulp™ newsprint ran well without any appreciable paper-related difficulties. In each printing test run, no special press adjustments were made. As reported by the participating dailies, the print results (including colour) were considered normal.

The future commercial direction of Agri-Pulp™ newsprint depends, among other things, on the relative pricing of ONP and virgin wood fibre for TMP production. ONP prices reflect the severe cyclic demand of Asian newsprint mills. Availability of economical virgin wood fibre (as wood chips) is tied closely to the domestic production of sawn timber. Straw-based Agri-Pulp™ could provide a practical buffer against the extremes of economic sourcing of ONP and virgin wood fibres.[15]

5. CONCLUDING REMARKS

Commercial trial production of an experimental newsprint containing up to 20% Agri-Pulp™ was completed satisfactorily at the Oregon City mill of the Smurfit Newsprint Corp. There were no major paper machine operating problems. The physical strengths of the Agri-Pulp™ newsprint were noted to be better than those of reference standard newsprint which contains 85% ONP and 15% TMP.

About 163 tonnes of Agri-Pulp™ newsprint were made in the 24-hour trial. The paper was shipped to seven dailies in California and Oregon for commercial printing trials. Feedback from all participating pressrooms indicated that the Agri-Pulp™ newsprint ran well without difficulties. The print results were considered comparable to standard newsprint.

The technical feasibility of Agri-Pulp™ newsprint has been demonstrated successfully. Future commercial manufacture of Agri-Pulp™ newsprint is dependent, on among things, the relative supply of economical ONP and virgin wood fibres.

6. ACKNOWLEDGMENTS

The cooperation of Smurfit Newsprint Corporation and its sales, technical and production staff at the Oregon City newsprint mill is greatly appreciated. (AK17708).

REFERENCES

1. Hernandez, J.R. and J.R.J. Calatayud, eds. (1990). Proc. International Seminar on Newsprint from Bagasse, CUBA 9- UNIDO, Havana, Cuba, October.
2. Hodgson, P.W. et al. (1980). ANPA, International Paper Sales Kenaf Newsprint Project, Tappi Non-wood Plant Fiber Pulping Progress Report No. 11.
3. Hodgson, P.W. et al. (1981). Commercial paper machine trial of CTMP kenaf in newsprint, Tappi, 64, p. 161.
4. Lawford, W.H. and G. Tombler (1982). Kenaf pulp for newsprint, Pulp Paper Can., 83, p. 99.
5. Klugler, D.E. et al. (1988). Kenaf - A Fiber Source for Newsprint, Proc. First International Non-Wood Fiber Pulping and Papermaking Conference, CTAPI, Beijing, China, July.
6. Anon. (1990). Proc. Workshop on Kenaf Pulping and Alkaline Papermaking, UNIDO, Bangkok, Thailand, December.
7. Rosenberg, J. (1996). Alternative fiber sources for newsprint, Editor & Publisher, January 20, pp. 22-26.
8. Wong, A. (1995). Agri-Pulp Newsprint - Why, How and When, paper presented at the Newspaper Association of America Conference, Tysons Corner, Virginia, October 30.
9. Wong, A. (1996). The Agri-Pulp Newsprint Alternative, paper presented at the Newspaper Association of America SuperConference, Miami, Florida, March 6.

10. Anon. (1998). Report says world pulpwood supply must increase; industry urged to think globally, Pulp & Paper Week, August 12, p. 5.
11. Anon. (1997). Straw and Kenaf Make Inroads in Building Materials and Paper, in Industrial Uses of Agricultural Materials: Situation and Outlook Report IUS-7, Economic Research Service, U.S. Dept. of Agriculture, July, pp. 17-25.
12. Wong, A. (1996). Agri-Pulp Industry Faces Challenge, Paper Asia, May, pp. 23–25.
13. Anon. (1996). Smurfit to test rice straw in newsprint, Pulp & Paper Week, August 19, p. 3.
14. Tisdale, J.R. (1997). Agri-Pulp Content Newsprint, paper presented at the Newspaper Association of America SuperConference, Miami, Florida, January 15.
15. Wong, A. (1997). Agri-Pulp - The Third Papermaking Fibre, Proc. Financial Times World Pulp and Paper Conference, London, UK, December 6.

POLYMERS FROM THE EXPLOITATION OF BIOMASS

Alessandro Gandini

Ecole Française de Papeterie et des Industries Graphiques (INPG), BP 65, 38402 St. Martin d'Hères, France

A brief survey reviews recent advances in the use of biomass components to synthesize novel polymeric materials. Polysaccharides such as cellulosics and chitosans were grafted and crosslinked in order to obtain polymer electrolytes and fibres for composites. Lignins were employed as macromonomers to prepare polyesters and polyurethanes. Furan derivatives were the source of macromolecular structures with special properties. Vegetable oils substituted for fossil counterparts in lithographic inks, and suberin was extracted from cork in order to assess use of its components as macromonomers or additives. This overview emphasises that renewable resources constitute a viable alternative to petrol and coal as building blocks for polymeric materials.

1. INTRODUCTION

Biological activities produce an incessant amount of raw materials: the biomass. All these substances are renewable resources, which can participate in ecological biodegradation and regeneration cycles, as opposed to the single-use character of fossil counterparts. Several years ago, our laboratory took up the challenge of studying the use of these renewable resources as starting points for the elaboration of original polymeric materials.[1-3] This communication is aimed at bringing the subject up to date by summarising a number of recent studies that illustrate the scope of our strategy. Two points should be stressed from the start: (1) agricultural and sea *wastes* can be used quite profitably in this general context, i.e., there is no need to exploit more valuable raw materials such as wood; and (2) the properties of the materials obtained are quite comparable to those of polymers derived from petrol or coal—there is no loss of quality or diversity associated with switching from the extinguishing to the renewable resources.

2. CHEMICAL MODIFICATION OF POLYSACCHARIDES

Two different types of materials are sought. Polymer electrolytes, in which the bulk of the polysaccharide is modified while preserving its skeleton, takes the important role of forming films. Cellulosic fibres to be used in composites are modified only at their surface in order to become more compatible with polymeric matrices. In both instances, specific chemical modifications are applied to the natural substrate, all based on the reaction of hydroxy and amino functions. However, whereas in the former case the material as a whole is involved in the modification, in the second, the bulk crystallinity of the fibres is preserved and only their surface is altered appropriately.

2.1. Polymer electrolytes

Our long-standing research in this field[4] led to materials with very satisfactory mechanical and electrochemical properties, but which required a specific improvement in terms of the possibility of preparing films as thin as possible, preferably in the few micrometer range. In fact, the manufacture of solid-state batteries, sensors or electrochromic devices requires optimally the use of *thin* films of polymer electrolytes. Cellulose ethers and chitosan, appropriately modified by oligoethylene oxide (OEO) and propylene oxide (OPO) branches and crosslinks, gave an adequate solution to that quest.[5,6]

The synthesis of these networks was achieved by different procedures including the reaction of mono- and di-functional OEO and OPO isocyanates with the OH and NH_2 functions of the polysaccharide and the oxypropylation of the latter. The resulting structures, as in the chitosan-based scheme shown below, insured both the essential role of ionic solvation (provided by the polyether moieties) and the film-forming aptitude (provided by the natural polymer). The introduction of lithium salts of strong Brønsted acids into these fully-amorphous elastomers, possessing Tg around -50°C, produced polymer electrolytes with good ionic conductivities and mechanical properties.

2.2. Modified cellulosic fibres

The elaboration of composites in which a polymeric matrix is associated with cellulosic fibres requires good compatibility between these components leading to a highly adhesive interface. When glass fibres are used in this type of materials, specific reactions are carried out at the surface of the glass in order to achieve those two aims. A similar approach was followed in our laboratory, by adapting it to the specific context of natural fibres.[7] We investigated different types of cellulosic materials, so that the characterization of the modified surface could be as complete as possible. Thus for example, pure tracing paper was used as a substrate to measure contact angles before and after the specific chemical treatment of its surface; a semicrystalline cellulosic powder was the best starting substrate for inverse gas chromatography; and high-modulus regenerated fibres was best for the preparation and mechanical testing of composites.

A number of reagents was selected for studying the chemical alteration of the surface of the substrates. Some of them were commercially available, such as alkenyl succinic anhydrides (ASA) and styrene-maleic anhydride copolymers, whereas others were synthesized in the laboratory, namely the already-mentioned OPO and OEO monoisocyanates and a set of copolymers of styrene with 3-isopropenyl-α,α'-dimethylbenzyl isocyanate (TMI). Thus, two distinct kinds of reactions occurred on the superficial OH groups, their esterification or their transformation into urethane moieties. Moreover, oligomeric or polymeric branches were attached onto the fibres in order to study both the changes in surface energy and the formation of adhesive entanglements with the macromolecules of the matrix.

The encouraging outcome of this study prompted its pursuit according to a different strategy which consists in appending polymerizable functions onto the cellulosic fibres, e.g., by the reaction of styrenic isocyanates or acrylic anhydrides.[8] The resulting materials were then introduced into a monomer and its ensuing polymerization also involved the reactive unsaturations appended on the fibres. In other words, the matrices were prepared *in situ* and were linked covalently to the reinforcing elements, thus providing good interfacial adhesion.

3. LIGNINS AS MACROMONOMERS

After a thorough study of the synthesis of polyester networks based on the condensation of both aliphatic and phenolic OH groups from different types of lignin with difunctional acidic reagents,[9] attention was switched more recently towards the preparation of new materials based on urethane and oxirane chemistry.

The possibility of modifying the surface and solubility characteristics of lignins by treating them with monofunctional isocyanates revealed that attaching relatively short "hair" possessing hydrophilic properties (e.g., oligomeric ethylene oxide chains (OEO)), did indeed modify those properties quite dramatically, even when only about one half of the available OH groups were involved.[10] These materials showed a marked solubility in water and displayed glass transition temperatures that varied inversely with the higher the extent of grafting; 60°C for the pristine lignin to -60°C for 80% substitution. In the same vein, the use of the OEO and OPO diisocyanates produced networks which gave a swelling behaviour and Tg values reflecting the high hydrophilicity and segmental mobility of the bridging chains. Various gels were prepared by changing the relative proportion of oligodiisocyanate with respect to the total OH groups borne by the lignin core molecules.[10]

A schematic visualization of these two types of macrostructures is shown below:

L=lignin

4. THE ISOLATION, CHARACTERIZATION AND USE OF SUBERIN

Cork is a unique natural material in which the classical ligneous composition is drastically altered by the almost total absence of lignin and the presence of a different polymer called suberin, which can reach contents of up to 50%. Native suberin is a crosslinked polyester bearing long aliphatic chains which impart the well-known hydrophobic character to cork. The traditional industrial exploitation of cork to make stopcorks, insulating panels, etc., generates a large amount of rejects in the form of irregular pieces, particles and sawdust that are usually burnt. An alternative way of making good use of these cork wastes consists in extracting the suberin and utilising it as a macromonomer or a specific polymer additive.

The chemical exploitation of suberin, preceeded by a thorough study of its composition and physical properties,[11] was initiated by using it as a macromonomer in the synthesis of polyurethanes, given the presence of OH groups in its oligomeric components, using conventional aliphatic and aromatic diisocyanates.[12] These reactions proceeded in a characteristic fashion without any anomalous behaviour and gave mixtures of thermoplastic and crosslinked materials (the individual OH functionality of the oligomeric species in suberin varies from 0 to 4). The glass transition temperature of these polyurethanes depended on the isocyanate used and, as expected, was lower with hexamethylene diisocyanate than with aromatic counterparts. The intervention of all available OH groups in the polycondensations was proved by the facts that both the extent of crosslinked material (as measured by extraction) and its Tg went through a maximum for an [OH]/[NCO] ratio of unity.

As for the use of suberin as a non-reactive additive, its role in improving certain properties of inks and varnishes was tested using various commercial and experimental all-vegetable compositions. Suberin does not dissolve in these formulations and can therefore migrate to the surface of the drying ink or varnish and improve its optical and tribological features through partial crystallization and to increase its hydrophobic character.

5. FURAN POLYMERS

Our recent survey on the general topic of macromolecular materials bearing furan moieties[3] covers the literature up to 1996. The work discussed here deals with only two recent additions to the state of the art concerning studies conducted in our laboratory.

5.1. Conjugated oligomers and polymers

Following our discovery of a simple system permitting the synthesis of poly(2-fury-lene vinylene) (PFV),[13] the scope of this work has been extended to the realm of high-tech materials. The reaction involves the self-condensation of 5-methyl-2-furancarboxaldehyde in a strongly nucleophilic medium followed by a growth that can take place only between an aldehyde end-group and a monomer molecule:

$$(n+1) \text{ [furan-CHO]} \xrightarrow[\text{benzene - } n\,H_2O]{t\text{-BuOK}} \text{ [furan-CH=CH]}_n\text{-furan-CHO}$$

The fact that this novel synthesis also applied to the thiophene analogue, initiated another study which consisted in preparing model oligomers in which the O and S heterocycles were placed in preestablished sequences. Thus, random and block copolymers, as well as dimers, trimers (as shown below) and tetramers involving homo- and heterostructures were prepared and thoroughly characterized. Whereas the polymers displayed high conductivity after doping, the shorter structures produced luminescence at different wavelengths in the visible spectrum, when excited in the near uv. Their electroluminescence is now being investigated.

X=O Y=O Z=O
X=S Y=O Z=O
X=S Y=O Z=S
X=O Y=O Z=S

These compounds were also used as reagents with various aromatic amines to synthesize Schiff bases in search of liquid crystal features. It is well known that furanic derivatives with substituents at C2-C5 do not give rise to mesogenic structures, even when the aspect ratio seems correct, because the bond angle resulting from that substitution pattern is too small and the molecular structure is therefore distorted with respect to a rod-like shape. Thus, among the numerous combinations explored, only a few gave rise to mesophases.[13]

The other interesting property of these oligomers and polymers, whether furanic, thiophenic or mixed, is their photosensitivity. When the dimers were submitted to a detailed photophysical and photochemical study in the solid state, cyclodimerization was the only relevant reaction taking place, with high regiospecificity.[14] Thus, for example, the simple furanic dimeric aldehyde gave a single cyclobutane derivative:

These photochemical reactions were found to be thermally reversible.

5.2. Polymers based on the reversible Diels-Alder cycloaddition

The furan heterocycle is among the classical structures capable of responding to the Diels-Alder (DA) reaction because of its pronounced dienic character. Its condensation with typical dienophiles such as maleic anhydride or mono- and bis-maleimides are textbook examples. We recently applied this property to the (thermoreversible) cross-linking of polymers bearing furan rings[15] in different amounts, e.g., copolymers with a few percent or polyurethanes with one heterocycle per unit. Their reaction with bis-maleimides gave networks by intermolecular coupling. These gels were then heated in the presence of an excess of 2-methylfuran, bis-maleimide that was liberated in the ensuing retro-DA reaction was therefore trapped by this additive upon cooling, and the starting furanic polymer was recovered as a soluble decrosslinked product.

The implications of this study are that it is possible to elaborate crosslinked structures arising from the DA reaction which can be readily returned to their linear (thermoplastic) origin by heating. The application of this procedure to recyclable tyres seems attractive.

Another application of the DA reaction calls upon the use of the polycondensation of difuranic monomers with bis-maleimides or of a single monomer bearing each function.[15]

6. CONCLUSION

The few examples discussed above illustrate, albeit in a limited breadth, the interest and potential development associated with using renewable resources from the vegetable and animal realms to elaborate a rich variety of polymers including high-added-value materials.

REFERENCES

1. Gandini, A. (1992). In Comprehensive Polymer Science, Suppl. Vol. 1, ed. by S.L. Aggrawal and S. Russo, Pergamon Press, Oxford, p. 527.
2. Gandini, A. and M.N. Belgacem (1996, 1998). In Polymeric Materials Encyclopedia, ed. by J.C. Salamone, CRC Press, New York, 11, pp. 8518, 8530 and 8541; Polym. Int. 47, p. 267.
3. Gandini, A. and M.N. Belgacem (1997). Prog. Polym. Sci., 22, p. 1203.
4. Gandini, A. and J.F. Le Nest (1996). In Polymeric Materials Encyclopedia, ed. by J.C. Salamone, CRC Press, New York, 11, p. 7809.
5. Gandini, A., J.F. Le Nest, and C. Schoenenberger (1994). Trends Polym. Sci., 2, p. 432.
6. Velazquez-Morales, P., J.F. Le Nest, and A. Gandini (1998). Electrochim. Acta, 43, p.1275.
7. Trejo, J.A., J.Y Cavaillé, and A. Gandini (1997, 1998). Cellulose, 4, p. 305; J. Adhes., 67, p. 359.
8. Botaro, R.V. and A. Gandini (1998). Cellulose, 5, p. 65.
9. Evtughin, D. and A. Gandini (1996). Europ. Polym. J., 47, p. 344 and refs therein.
10. Montanari, S. (1996). Doctoral thesis, University of Sao Paolo at Sao Carlos, Brazil.
11. Cordeiro, N., N.M. Belgacem, C. Pascoal Neto, and A. Gandini (1998, 1998). Int. J. Biol. Macromol., 22, p. 153; Biores. Biotechnol., 63, p. 153.
12. Cordeiro, N., N.M. Belgacem, C. Pascoal Neto, and A. Gandini (1997, and in press). Ind. Crops Prod., 6, p. 163; Ind. Crops Prod.
13. Coutterez, C. and A. Gandini (1998). Polymer, 39, p. 7009; C. Coutterez, A. Gandini, G. Costa, and B. Valenti (1998). J. Chem. Res. M 2966, p. 679.
14. Baret, V., A. Gandini, and E. Rousset (1997). J. Photochem. Photobiol., A103, p. 171.
15. Goussé, C., A. Gandini, and P. Hodge (1998). Macromolecules, 31, p. 314; C. Goussé and A. Gandini (1998). Polym. Int., 40, p. 389; (in press) Polym. Int.

UTILIZATION OF *CHLORELLA* SP. FOR PLASTIC COMPOSITE AFTER CO_2 FIXATION USING HIGH DENSITY POLYETHYLENE

Toshi Otsuki,[a] Masatada Yamashita,[a] Zhang Farao,[b] Yuji Ikegami,[b] Masami Yoshitake,[b] Hiroshi Tsutao,[b] Hiroshi Kabeya,[c] Ryoichi Kitagawa[c] and Takahiro Hirotsu[c]

[a]RITE IHI Laboratory in Research Institute, Ishikawajima-Harima Heavy Industries Co., Ltd., 1, Shin-Nakahara-cho, Isogo-ku, Yokohama 235-8501, Japan
[b]Research Institute of Innovative Technology for the Earth (RITE), 2-8-11, Nishi-shinbashi, Minato-ku, Tokyo 105-0003, Japan
[c]Shikoku National Industrial Research Institute, Agency of Industrial Science and Technology, MITI, 2217-14, Hayashi-cho, Takamatsu 761-0395, Japan

The purpose of this study is to make ecofriendly building materials using micro-algae and high density polyethylene (HDPE). *Chlorella* sp. (hereafter chlorella), a kind of green micro-algae, is being studied for fixing CO_2 because of its high growth rate compared with common plants. After chlorella has been grown in the photobioreactor to fix CO_2 emitted from the power plants, etc., carbon should stay in chlorella bodies for a long period. While chlorella is incorporated into materials such as floor tiles and laminated plastic boards, CO_2 could not appear in the atmosphere again.
The mixture of chlorella powder and powdered HDPE was set in the mold frame and then pressed and heated simultaneously using a thermocompressor to make a rigid form. After being cooled the sample was removed from the frame, and its physical properties were examined.
Chlorella-HDPE composites showed the same strength as commercial plasticized PVC compounds[1] and we suggest that it is applicable to building materials.[2]

1. INTRODUCTION

Increasing CO_2 in the atmosphere is so serious a problem for mankind that many studies are implemented to reduce CO_2 emission. One is a biological CO_2 fixation using the photosynthetic function of micro-algae (RITE project[3]). One of the systems to utilize the micro-algae (chlorella in this case) for ecofriendly materials is shown in Figure 1.

This work was performed under the management of Research Institute of Innovative Technology for the Earth (RITE) as a part of the biological Fixation and Utilization Project supported by New Energy and Industrial Technology Development Organization (NEDO).

By using the collector with a transmission system of solar light, photobioreactor, and highly efficient photosynthesis, a great amount of chlorella can be produced. However, it is difficult to use the produced chlorella without releasing CO_2 into the atmosphere again.

In order to utilize chlorella effectively, various studies are being conducted. One of these studies uses chlorella as a filler in thermoplastics such as polyethylene, in order to develop ecofriendly materials (Figure 1).

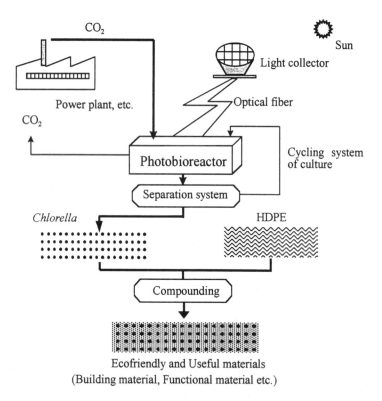

Figure 1. Concepts of biological CO_2 fixation and utilization for ecofriendly materials.

2. EXPERIMENTAL

2.1. Materials

Dried chlorella powder (Yaeyama Ltd.) as an alternative to real photobioreactor products and an HDPE pellet (7000F, MW = 2×10^5, Mitsui Chemicals, Inc.) were used. The grain of chlorella, which was constituted of cells (about 3 ~ 5 μm in diameter), was a nearly spherical aggregate (about 50 μm in average size) as shown in Figure 2. HDPE had a density of 0.92 g/cm^3. It was used as pellet in the early stage and then used after it was shivered mechanically into a spherical powder with a size of about 1 mm.

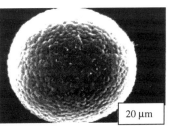

Figure 2. SEM photograph of *chlorella* aggregate, overall appearance.

2.2. Molding methods

Use of the HDPE pellet required a complicated method as shown in Figure 3(a), but after we succeeded in shivering the pellet to powder at low temperature, the method was simplified as shown in Figure 3(b).

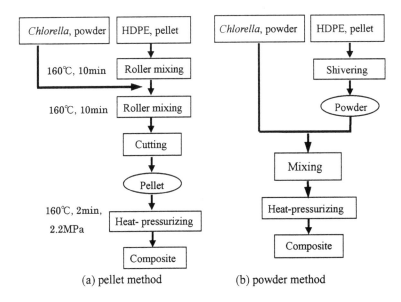

Figure 3. Chlorella-HDPE composite molding method, (a) pellet of HDPE, (b) powder of HDPE

Basically the aluminum mold frame with an inner size of 120 × 20 × 7mm was used to measure tensile strength. A thermocompressor (NSF-37, Shinto Metal Industries Ltd.) for molding was used with an oil pressure apparatus and heating device. A sample of mixture filled in the frame and was molded under selected condition. When molding was finished, the frame was cooled to room temperature and the sample was peeled off the frame.

2.3. Measurement

The test piece (sheet sample) was conditioned at 20 and 65% relative humidity in the room for 24 h before tensile strength and elongation were measured. A test piece for the measurement was made according to the method.[4] A tensile tester (AG-100A, Shimadzu Ltd.) was used to measure tensile strength and elongation at break in accordance with the testing method for tensile properties of plastics (cross-head speed of 5 mm/min). The microstructure of the materials was observed by a scanning electron microscope (S-2460N, Hitachi Ltd.).

3. RESULTS AND DISCUSSION

3.1. Optimum molding condition of chlorella-HDPE

The effect on the tensile strength of the molding temperature, pressure, time, water content of chlorella, and chlorella/HDPE ratio in the mixture were examined.

Figure 4 showed the relation between molding temperature and tensile strength at the ratio of chlorella/HDPE = 20/80 in weight, pressure of 2.2 MPa, time of 2 min in which both pellet and powder were tested. It was decided the suitable molding temperature was 160°C and the powdered HDPE was preferable to the following works.

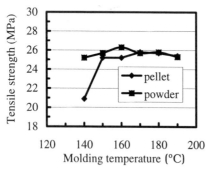

Figure 4. Effect of molding temperature on tensile strength. chlorella/HDPE = 20/80, Pressure: 2.2 MPa, Time: 2 min

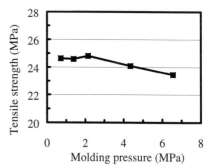

Figure 5. Effect of molding pressure on tensile strength. chlorella/HDPE = 20/80, Temp: 160°C, Time: 2 min

Figure 5 showed the effect of pressure on the strength at chlorella/HDPE = 20/80, 160°C, and 2 min. The tensile strength decreased gradually as pressure increased, and the best pressure was 2.2 MPa.

Figure 6 showed the effect of molding time on the strength at chlorella/HDPE = 20/80, 160°C, and 2.2 MPa. The tensile strength did not change from 2 min to 30 min, so it was decided the best molding time was 2 min.

Figure 7 showed the effect of water content of chlorella on the strength at chlorella/HDPE = 20/80, 160°C, 2.2 MPa and 2 min. It was important to determine how much water could be retained because the drying of wet chlorella takes much energy. The water content of chlorella was found to be 10 ~ 20%. As the tensile strength did not change up to 13% of water content, it seemed 13% water was permissible.

Figure 8 showed the effect of chlorella/HDPE ratio in the mixture on the strength at 160°C, 2.2 MPa and 2 min. The tensile strength decreased as the ratio of chlorella increased, and it crossed the line of the standard[1] (15 MPa of strength) at 20% of chlorella.

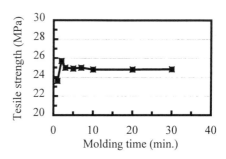

Figure 6. Effect of molding time on tensile strength. chlorella/HDPE = 20/80, Pressure: 2.2 MPa, Temp: 160°C

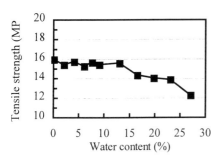

Figure 7. Effect of water content of chlorella on tensile strength. chlorella/HDPE = 20/80, Pressure: 2.2 MPa, Temp: 160°C, Time: 2 min

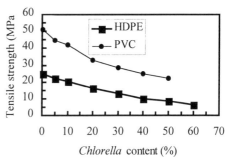

Figure 8. Effect of chlorella content in the mixture on tensile strength.
HDPE Pressure: 2.2 MPa, Temp: 160°C, Time: 2 min
PVC Pressure: 4.4 MPa, Temp: 180°C, Time: 5 min

3.2. Evaluation for floor tile

The chlorella-HDPE composite as a floor tile was evaluated as shown in Table 1. The chlorella-HDPE composite can be a good floor tile because the characteristics of the sample almost came up those of the standard.

Table 1
Evaluation of the chlorella-HDPE composite as a floor tile

Items	Unit	Standard[5]	Sample
Rate of length change after heating	%	0.2	0.09
Rate of length change after dipping water	%	0.2	0.04
Rate of weight decrease after heating	%	0.5	0.23
Dent at 20	mm	0.25	0.19
Dent at 45	mm	0.8	0.29
Rate of dent residual	%	8.0	0.31

Molding condition: chlorella/HDPE = 20/80, 160°C, 2.2 MPa, 2 min
Sample: 150 × 150 × mm

3.3. Improvement of tensile strength of the composite—chemical modification of HDPE

As shown in Figure 8, the tensile strength of chlorella-HDPE was lower than that of chlorella-PVC. It seems that the adhesion between chlorella and HDPE is so weak because of an air gap around the chlorella grain which prevents the reaction of chlorella and HDPE. So we modified HDPE with maleic anhydride and benzoyl peroxide to change its surface from hydrophobic to hydrophilic. Chlorella can be strongly combined with HDPE after this chemical modification, and as much as 40% can be used as filler without a sharp drop of tensile strength.

4. CONCLUSION

We succeeded in making a useable sample of floor tile using chlorella-HDPE composite. The tensile strength after HDPE chemical modification was much higher. An effective process to treat a great amount of chlorella could be developed. It will extend the possibilities of micro-algal CO_2 mitigation.

REFERENCES

1. JIS K6723-1995. Plasticized polyvinyl chloride compounds, Japan Standard Society.
2. Otsuki, T. and M. Yamashita, et al. (1998). Utilization of micro-algae for building materials after CO_2 fixation, in Advances in Chemical Conversions for Mitigating Carbon Dioxide, Elsevier Science B.V., 114, pp. 479–482.
3. Usui, N. and M. Ikenouchi (1997). The biological CO_2 fixation and utilization project by RITE (1), Energy Convers. Mgmt., 38, pp. S487–S492.

4. JIS K7113-1995, Testing method for tensile properties of plastics, Japan Standard Society.
5. JIS A5705-1992, Floorcovering -PVC, Japan Standard Society.

VALORISATION OF FLAX-BYPRODUCTS BY MEANS OF MYCELIAL BIOMASS PRODUCTION IN DIFFERENT INDUSTRIAL APPLICATIONS

P.V. Vilppunen,[a] O.K. Mäentausta,[a] and P.I. Kess[b]

[a]Department of Process Engineering, University of Oulu, P.O. Box 444, FI-90571 Oulu, Finland
[b]Department of Economics, University of Oulu, P.O. Box 444, FI-90571 Oulu, Finland

This project addresses a novel use of flax-byproducts as a substrate material for growing lignivorous mushrooms. In the process flax waste is degraded and can be used as a cattle feed or as a compost in agriculture.

The flax textile industry mainly aims at high-value linen textiles, applied in fashion. This market segment holds two disadvantages: dependence on fashion and sensitivity to economic factors. An important economic consideration and environmental necessity is the use of the by-products as innovative products such as substrates for "exotic mushrooms." Valorisation of the by-products and development of other markets for them will reduce the costs of flax processing for textiles.

The production of "exotic mushrooms" has rapidly expanded in the last decades. Because the availability of conventional substrate material is restricted, mushroom growers are looking for valuable and environmental friendly alternatives.

The work is based on studying thoroughly the bioconversion rate of different types of flax waste as related to different exotic mushroom species, while following their biological characterisation: incubation, maturation or induction and fructification behaviour on different substrate compositions alone and in combination with other raw materials. In addition the left-over substrate will be evaluated on its composting qualities.

1. OBJECTIVES OF THE STUDY

This study was conducted by University of Oulu, FinFlax Ltd and Mycelia bvba (Belgium) in order to evaluate the suitability of flax by-products in mushroom substrate production at laboratory and pilot scale using *Pleurotus ostreatus* and *Lentinula edodes* as examples.

The means to achieve the objectives of the study were the following:
1) test in the laboratory growth properties of different exotic mushrooms in flax-waste based substrate materials,
2) test the production of uniform quality substrate material from flax-waste remaining from bio-technical retting of flax,
3) test the production of substrates in pilot scale using commercially available mushroom spawn,
4) test the production of mushrooms in pilot scale, as an example, *Pleurotus ostreatus*, and
5) develop microbiological methods to follow up and control common microbial contamination in mushroom production.

2. PLEUROTUS OSTREATUS

2.1. Production of substrate material from flax

An enzymatic flax retting process was developed in FinFlax Ltd together with Department of Process Engineering in University of Oulu. The aim was to achieve controlled biotechnical retting process to produce high quality flax fibres and a consequence clean flax shives to be used as a mushroom substrate material. Shive is a bulky woody by-product of flax.

Substrate was prepared from flax waste, mixed shives and flower bottoms, left over from the enzymatic retting process. The substrate material (250 kg/batch) was neutralised with chalk (5%) before pasteurisation. The substrates for *Pleurotus ostreatus* were pasteurised in pilot scale using a 1-cubic-meter vessel connected to a steam generator (12 kW) for 7 hours at temperature 85°C. The pasteurised substrate mass was let to cool to 25°C and mixed with *Pleurotus ostreatus* spawn (3%). The mixture was packed manually into Microsac bags in 10 kg quantities with dimensions of height 50 cm and diameter 35 cm. After this the bags were sealed tightly and moved to further processing.

The inoculated substrates were kept for two weeks in an incubation room (temperature 25°C, humidity 95%, lightning 200 lux). Conditions were kept constant using an air conditioner and ultrasound humidifier.

Twelve later substrate batches were prepared in pilot scale in similar conditions in order to investigate reproducibility of substrate production and to evaluate the production costs and technical risks.

2.2. Production of the *Pleurotus ostreatus* mushrooms

After incubation and maturation phases the substrates were removed into the growing room where temperature (15°C), humidity (85%rh) and lighting (200 lux with circadian rhythm of 12 hours) were kept constant.

2.3. Results

An average yield of mushrooms was 18% of wet substrate weight and 60% of dry weight material. A cumulative yield of 65% was reached after a two-week growth period. An additional 25% of harvest was obtained in two weeks after the first harvest. This means that a one-month cycle is likely the most economical way to grow *Pleurotus ostreatus* on these substrates compared to 2-month cycle when a conventional substrate, such as wheat straw, was used.

2.4. Analysis of the results

Production of mushroom substrate material from flax shives and flower bottoms was shown to be reproducible in terms of yield. In addition, this part of the production of fibres and raw material needs no additional processing which means cost efficiency in the production of the substrate.

The mushroom substrates were produced in 12 later batches under similar conditions. The quality of substrate was tested by growing the mushrooms. The results were quite similar in each batch; the average yield was 18% mushrooms of substrate mass.

The production of *Pleurotus ostreatus* mushroom in pilot scale was reproducible in terms of quantity and sensory quality. The produced mushrooms were distributed to end users for field trials and verbal feedback was positive. No negative comment was received concerning the sensory characteristics of these mushrooms.

Using conventional microbial culture assay techniques the microbial quality was tested before and after pasteurisation of substrate mass. Further studies concentrate on the developing a controlled pasteurisation process. This needs further development of microbial quality control protocols.

3. *LENTINULA EDODES*

3.1. Substrate experiments

Used materials and methods for the incubation tests of *Lentinula edodes* are in Table 1. Results of the induction tests on the substrates are in Table 2 and results of the fructification tests on the substrates are in Table 3.

The difference substrates were filled into special bags, named "Microsacs." These autoclavable bags are provided with special filters sealed between strips of plastic folio. The latter allows the necessary gas exhange for mycelium growth without any risk of infection.

Table 1
The materials and methods used for the incubation tests of *Lentinula edodes*

Recipe	Quantity (kg)	Description	Origin
A	9.6	wood particles	Blaak - Netherlands
	0.2	corn flour	Voeders Laroy - Belgium
	0.2	bran	Voeders Vermeulen - Belgium
	12	water	
B	5.6	wood particles	Blaak - Netherlands
	4	flax-shives	Bentex - Belgium
	1.2	corn flour	Voeders Laroy - Belgium
	1.2	bran	Voeders Vermeulen - Belgium
	12	water	
C	4	wood particles	Blaak - Netherlands
	4	flax-flowers	FinFlax - Finland
	2	flax-fruits	FinFlax - Finland
	2	bran	Voeders Vermeulen - Belgium
	12	water	
D	8	flax-shives	FinFlax - Finland
	2	flax-flowers	FinFlax - Finland
	2	bran	Voeders Vermeulen - Belgium
	12	water	
E	8	flax-shives	FinFlax - Finland
	4	flax-flowers	FinFlax - Finland
	12	water	

Table 2
Results of the induction tests on the substrates of *Lentinula edodes*

Strain	Recipe	End of incubation (days)	Formation of knobs (days)	Start of browning (days)	End of browning (days)	Formation of primordia (days)
3770	A2	25	41	54	64	89
3770	B2	25	39	51	60	85
3770	C2	25	40	52	62	86
3770	D2	25	41	52	61	85
3770	E2	25	40	52	62	86

Table 3
Results of the fructification tests on the substrate of *Lentinula edodes*

Strain	Recipe	Substrate weight (g)	Product. (g)	EF (%)	Production (%/day)	Picking time (days)
3770	A2	3486	556	15.95	0.20	80
3770	B2	3271	497	15.19	0.19	79
3770	C2	3807	603	15.84	0.19	82
3770	D2	3047	431	14.15	0.18	78
3770	E2	2911	485	16.66	0.21	81

Weight (g) = the initial humid weight of the substrate in grams
Produce (g) = the total weight of the fresh mushrooms, picked on the different flushes
EF (%) = the efficiency in percent = (Produce × 100)/weight
Production/day (%/day) = EF/harvest time

Remarks:
a) The mycelium growth rate on the different substrates was examined and compared for different mushroom-species.
b) The fructification experiments were exclusively carried out on the recipes and strains, on which the best incubation results were obtained.
c) Each test was performed in threefold, in order to avoid mistakes and wrong conclusions.

4. CONCLUSIONS

The preliminary studies led to the following conclusions:
1) Clean, uniform quality flax shives, which are left over from the enzymatic retting process, may be considered as an interesting raw material for the production of lignivorous mushrooms.
2) The production of substrates for *Pleurotus ostreatus* and a series of exotic mushrooms, using flax shives and flower bottoms, showed promising results and was reproducible in laboratory and pilot scale.
3) The production of *Pleurotus ostreatus* and *Lentinula edodes* mushrooms was reproducible, and the observed results in terms of quantity and quality were encouraging. The tests show that the induction time decreased 3 or 4 days when flax-based substrate was used. This is very important for mushroom growers. The most important advantage of the use of flax-waste for the cultivation of *Pleurotus ostreatus* is the standardisation of the ingredients for the substrates.

The economic and industrial opportunities in this project are for the flax producers the possibility to upgrade the flax-waste to a valuable food-product through mushroom cultivation and to animal feed or compost in a second phase. For the mushroom growers the economic and industrial opportunities are as follows:
- the creation of new production opportunities for lignivorous mushroom,
- the elimination of product risks while offering a standard raw material,
- the reduction of production costs,
- the utilisation of the spent substrates in animal feed and/or composting.

REFERENCES

1. Van Dam, J.E.G., G.E.T. van Vilsteren, F.H.A. Zomers, W.B. Shannon, and I.T. Hamilton (1994). Increased applications of domestically produced plant fibres in textiles, pulp and paper production and composites materials, Directorate-General XII, Science, Research and Development, EUR 16101 EN.
2. Houdeau, G., J.M. Olivier, S. Libmond, and H. Bavadikji (1991). Improvement of *Pleurotus* cultivation, Science and Cultivation of Edible Fungi, ed. by Maher, Balkema, Rotterdam.
3. Sharma, H.S.S. (1997). Comparative study of the degradation of flax shives by strains of *Pleurotus* Appl. Microbiol. Biotechnol, 25, pp. 524–546.

UTILIZATION OF *CHLORELLA* SP. FOR PLASTIC COMPOSITE AFTER CO_2 FIXATION USING PVC

Toshi Otsuki[a], Masatada Yamashita[a], Zhang Farao[b], Yuji Ikegami[b], Masami Yoshitake[b], Hiroshi Tsutao[b], Hiroshi Kabeya[c], Ryoichi Kitagawa[c] and Takahiro Hirotsu[c]

[a] RITE IHI Laboratory in Research Institute, Ishikawajima-Harima Heavy Industries Co., Ltd., 1, Shin-Nakahara-cho, Isogo-ku, Yokohama 235-8501, Japan
[b] Research Institute of Innovative Technology for the Earth (RITE), 2-8-11, Nishi-shinbashi, Minato-ku, Tokyo 105-0003, Japan
[c] Shikoku National Industrial Research Institute, Agency of Industrial Science and Technology, MITI, 2217-14, Hayashi-cho, Takamatsu 761-0395, Japan

The purpose of this study is to make ecofriendly building materials using micro-algae and thermoplastics polyvinyl chloride (PVC). *Chlorella* sp. (hereafter *chlorella*), a kind of green micro-algae, is being studied for fixing CO_2 because of its high growth rate compared with common plants. After *chlorella* was grown in the photobioreactor to fix CO_2 emitted from the power plant etc., carbon should stay in *chlorella* bodies for a long period. While *chlorella* is in materials such as floor tiles and laminated plastic boards, CO_2 could not appear in the atmosphere again.

The mixture of *chlorella* powder and PVC powder, with some chemicals for stabilizing the composite, was set in the mold frame. A thermocompressor then pressed and heated it simultaneously for a certain period to make a rigid form. After being cooled, the sample was removed from the frame, and its physical properties were examined.

Chlorella-PVC composites showed the same strength as commercial plasticized PVC compounds[1,6] and we suggest that they are applicable to building materials.[2]

1. INTRODUCTION

Increasing CO_2 in the atmosphere is so serious a problem that many studies seek ways to reduce CO_2 emission. One way is biological CO_2 fixation using the photosynthetic function of micro-algae (RITE project[3,4]); the micro-algae (*chlorella* in this case) can then be used in ecofriendly materials.

This work was performed under the management of Research Institute of Innovative Technology for the Earth (RITE) as a part of the Biological Fixation and Utilization Project supported by New Energy and Industrial Technology Development Organization (NEDO).

By using a collector that transmits solar light, a photobioreactor, and highly efficient photosynthesis, a great amount of *chlorella* can be produced. However, it is difficult to use the *chlorella* without releasing CO_2 into the atmosphere again.

In order to utilize *chlorella* effectively, various studies are being conducted. One of these studies is to use *chlorella* as a filler in thermoplastics such as PVC.

2. EXPERIMENTAL

2.1. Materials

Dried *chlorella* powder (Yaeyama Ltd.), an alternative to real photobioreactor products, and PVC powder (degree of polymerization of 1100, Wako Chemicals Inc.) were used. The grain of *chlorella* was constituted of cells about 3~5 μm in diameter and was a nearly spherical aggregate (about 50 μm in average size) as shown in Figure 1.

Some chemical additives were used as stabilizers. These chemicals were barium–zinc fatty acids complex (PSE, Mizusawa Chemicals, Ltd.), complicated compounds of organic tin (KM, Katsuta Ltd.), and paraffin.

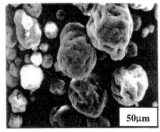

Figure 1. SEM photograph of *chlorella* aggregate, overall

2.2. Molding methods

The *chlorella*-PVC composite was prepared by a heat pressurizing method. The mixture (about 8 g) was set in the aluminum mold frame (press area is 20 × 120 mm); it was then pressed and heated simultaneously for a certain period to make a rigid form using the thermocompressor (NSF-37, Shinto Metal Industries Ltd.). After molding was finished, the frame was cooled to the room temperature and the sample was peeled off of the frame.

2.3. Measurement

The test piece (sheet sample) was conditioned at 20 and 65% relative humidity in the room for 24 h before measuring tensile strength and elongation. The test piece for the measurement was made according to the method.[5] The tensile tester (AG-100A, Shimadzu Ltd.) was used to measure tensile strength and elongation at break in accordance with testing method for tensile properties of plastics (cross-head speed of 5 mm/min.). The TG-DTA analyses of *chlorella* were carried out using the thermal analyzer (MC-2000, MCA Science Co.). Microstructure of the materials was observed by a scanning electron microscope (S-2460N, Hitachi Ltd.).

3. RESULTS AND DISCUSSION

3.1. Optimum molding condition of *chlorella*-PVC

The effect of molding temperature, pressure, time has been discussed.[2] In this section, we examined the thermal behavior of *chlorella*, and the effect of water content of *chlorella* and the effect of *chlorella* content in the mixture on the tensile strength, respectively.

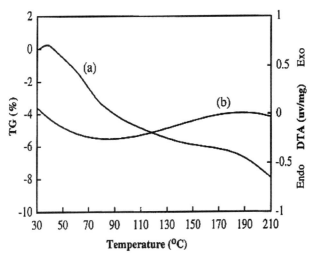

Figure 2. TG-DTA analysis of *chlorella*. (a) Thermogravimetric analysis (TG). (b) Differential thermal analysis (DTA). Heating rate: 5°C/min in air.

Figure 2 shows TG-DTA results for *chlorella*. Weight loss was greater above 180°C than between 110~180°C. The loss might reflect volatilization and decomposition of some substances such as chlorophyll, considering the color change (from deep green to brown) when the molding temperature was greater than 190°C. Thus the molding temperature should be kept below 190°C so as not to incur the deterioration of *chlorella* during heat-pressurizing. It was already decided the suitable molding temperature was 180°C.[2]

Figure 3 shows the effect of water content of *chlorella* on the strength at chlorella/PVC = 20/80 in weight, pressure of 6.6 MPa, and time of 5 min. Because tensile strength decreased sharply when the water content was above 13%, much water might weaken the bond between *chlorella* and PVC in the composite. It seemed that the permissible water content was about 10%.

Figure 4 shows the relation between *chlorella* content and tensile strength at 180°C, 4.4 MPa, 5 min.

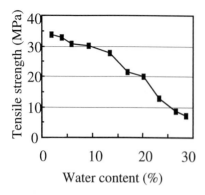
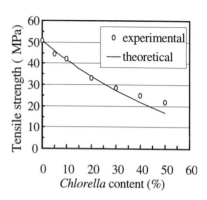

Figure 3. Effect of water content of *chlorella* on tensile strength. Chlorella/PVC=20/80, Pressure: 6.6Mpa, Temp:180°C, Time:5min Additives:3% of PSE-227, 0.5% of Paraffin in PVC

Figure 4. Relationship between *chlorella* content and tensile strength. Pressure: 4.4Mpa, Temp:180°C, Time:5min Additives:KM-55, 0.5% in PVC

We considered the interaction of the *chlorella* grain and PVC molecule by both theoretical and experimental aspects. The total section area (A) could be equal to the sum of the section areas of PVC matrix (A_p) and *chlorella* part (A_c). The strength of the composite was given by equation (1).

$$\sigma = F/A \tag{1}$$

where F was the tension load on the composite. If the strength of the composite was contributed only by the PVC matrix, the tension load could be expressed by equation (2).

$$F = \sigma_p A_p \tag{2}$$

where $_p$ was the tensile strength of the PVC matrix. Introducing equation (2) into equation (1) yielded equation (3).

$$\sigma = \sigma_p A_p / A \tag{3}$$

If there were no air gaps in the composite, the volume of the composite (V) was the sum of the volume of PVC matrix (V_p) and *chlorella* (V_c). V_p and V_c are defined as

$$V_p = W_p/D_p \tag{4}$$

$$V_c = W_c/D_c \tag{5}$$

where W_p and D_p were the weight and density of PVC, and W_c and D_c were the weight and density of *chlorella*, respectively. Assuming that all the cross sections of the composite had the same values of A_p and A_c,

$$A_p/A = V_p/V \tag{6}$$

By using $V = V_p + V_c$, equation (4) and (5), equation (6) could be written as

$$A_p/A = 1/(1 + D_p W_c / W_p D_c) \tag{7}$$

As the experimental D_c and D_p values were 0.7 and 1.4, respectively, the weight content of *chlorella* in the composite (expressed as C%), then Wc/Wp = (C/100)/(1-C/100). Thus by introducing equation (7) into equation (3), equation (8) was obtained.

$$\sigma = \sigma_p [1/\{1 + 2(C/100)/(1 - C/100)\}] \tag{8}$$

Equation (8) shows the relation between *chlorella* content and tensile strength of the composite, and the theoretical tensile strength could be calculated by equation (8).

Figure 4 showed the theoretical curve according to equlation (8) and the experimental results. The experimental values matched well the theoretical curve when the *chlorella* content was lower than 20%, but the experimental values were larger than the theoretical curve when the *chlorella* content exceeded 20%. It could be inferred that there was an effective bonding of *chlorella* and PVC, and if *chlorella* content was higher (> 20%), the contribution of *chlorella* to the strength gradually increased (that is, some partial interaction were produced as shown in Figure 5).

The composite of 20% *chlorella* content had tensile strength higher than 30 MPa which was the value of the lower limit of the standard designation of rigid polyvinyl chloride compounds.[6]

Thus, the optimum molding conditions could be as follows:
a. *Chlorella*/PVC = 20/80 in weight was suitable.
b. The pressure was 2.2~6.6 MPa.

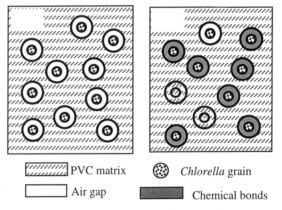

PVC matrix   *Chlorella* grain
Air gap   Chemical bonds

Figure 5. A model of *chlorella*-PVC composite.
(a) *chlorella* content is low (<20%)
(b) *chlorella* content is high (partial interaction produced)

c. The temperature and time were 180°C and 5 min, respectively.
d. Small amounts of chemical stabilizer for PVC were needed.
e. The tensile strength of 30 MPa was obtained.

3.2. Evaluation for floor tile

The *chlorella*-PVC composite as a floor tile was evaluated as shown in Table 1.
A *chlorella*-PVC composite can make a good floor tile because the characteristics of the sample almost came up to the standard.

Table 1
Evaluation of the *chlorella*-PVC composite as a floor tile

Items	Unit	Standard [7]	Sample
Rate of length change after heating	%	0.2	0.06
Rate of length change after dipping water	%	0.2	0
Rate of weight decrease after heating	%	0.5	0.2
Dent at 20	mm	0.25	0.1
Dent at 45	mm	0.8	0.12
Rate of dent residual	%	8.0	0

Molding condition: *chlorella*/PVC=20/80, 180, 3.8MPa, 10min, Sample:150×150×3mm

4. CONCLUSION

We succeeded in making a useable sample of floor tile with a *chlorella*-PVC composite. The use of a process that treats a great amount of *chlorella* could promote micro-algal CO_2 mitigation.

REFERENCES

1. JIS K6723-1995, Plasticized polyvinyl chloride compounds, Japan Standard Society.
2. Otsuki, T. and M. Yamashita et al. (1998). Utilization of micro-algae for building materials after CO_2 fixation, in Advances in Chemical Conversions for Mitigating Carbon Dioxide, Elsevier Science B.V., 114, pp. 479–482.
3. Usui, N. and M. Ikenouchi (1997). The biological CO_2 fixation and utilization project by RITE (1), a highly effective photobioreactor system, Energy Convers. Mgmt., 38, pp. S487–S492.
4. Murakami, M. and M. Ikenouchi (1997). The biological CO_2 fixation and utilization project by RITE (2), Screening and breeding of microalgae with high capability in fixing CO_2, Energy Convers. Mgmt., 38, pp. S493–S497.
5. JIS K7113-1995, Testing method for tensile properties of plastics, Japan Standard Society.
6. JIS K6740-1976, Designation of rigid polyvinyl chloride compounds, Japan Standard Society.
7. JIS A5705-1992, Floorcovering -PVC, Japan Standard Society.

NEW GREEN PRODUCTS FROM CELLULOSICS

E.G. Koukios,[a] A. Pastou,[a] D.P. Koullas,[a] V. Sereti,[b] H. Stamatis,[b] and F. Kolisis[b]

[a]Bioresource Technology Unit—Laboratory of Organic and Environmental Technology
[b]Laboratory of Biosystems Technology
All of the Department of Chemical Engineering, National Technical University of Athens, Zografou Campus, GR-157 00 Athens, Greece

The experimental techniques applied in this work involve partial solubilisation or swelling of polysaccharides in aqueous alkaline media and various non-aqueous solvents to activate the cellulosic matrix. In addition, chemical pretreatments aiming at a selective chemical "opening" of the biopolymer chain were employed, whereas acid hydrolysis was used to produce mixtures of low and high molecular weight polysaccharides. The pretreated cellulosics and their various products so obtained were subjected to lipase-catalyzed esterification with varying success, the main factor in which was the accessibility of the cellulose hydroxyl units to enzymatic action. The efficiency of the latter was increased after an etherification treatment of cellulose.

1. INTRODUCTION

Cellulose, in its native form, is known to resist chemical and, particularly, enzymic attack owing to its crystalline areas; moreover the reactivity of cellulose macromolecules is limited by the presence of other plant cell-wall macromolecules, such as lignin and hemicellulose. In this context, in general, the use of appropriate pretreatments enhancing cellulose accessibility and reactivity seems to be imperative. Research has been carried out on (a) the production of a mixture of oligosaccharides from cellulose using following acid hydrolysis and (b) the etherification of an hydroxyalkylcellulose that is generated from cellulose under epoxidation, based on the reaction of an epoxide and the hydroxyls of the anydroglucose unit.

The enzymic esterification of hydroxypropyl cellulose with long chain fatty acids catalyzed by lipase in organic solvents was carried out.

2. MATERIALS AND METHODS

2.1. Pretreatments of lignocellulosics

The raw materials used were i) for hydrolysis: α-cellulose and sulfuric (96%) and hydrochloric acid (37%) (Sigma); ii) for etherification: α-cellulose (Sigma), sodium hydroxide, 1,2- propenoxid and n-hexane (Merck) and *t*-butanol (Fluka).

Acid hydrolysis: α-Cellulose was treated in sulfuric acid, solid-to-liquid ratio 1:5 for 35, 45 min at 30°C. When hydrochloric acid was used the ratio of cellulose to acid varied between 1:5 and 1:10 at 50°C. The reaction was terminated by neutralisation.

Cellulose etherification: α-Cellulose was treated with sodium hydroxide 5N at room temperature for 12 h. After filtration the alkalicellulose reacted with propenoxid in an autoclave in the presence of a mixture of n-hexane and *t*-butanol. The reaction took place at propenoxid to alpha-cellulose molecular ratios, expressed as propenoxid moles to (g of alkalicellulose/162), where 162 is the molar weight of anhydroglucose. The reaction was terminated by rapid freezing followed by washing with hexane and acetone to remove out the by-product in the liquid phase.

Nuclear Magnetic Resonance Analysis (NMR): Nuclear magnetic resonance analysis of cellulose and substituted cellulose samples diluted in $CDCl_3$ were taken in a 300 MHz Varian Gemini 2000 spectrometer. The goal of the analysis is to determine the presence of the alkyl groups in the ethers and to determine the molecular substitution value (Ho et al., 1972).

Esterification of hydroxypropyl cellulose (HPC): 500 mg HPC and 150 mM of lauric acid were dissolved in 20 ml *tert*-butanol. Then 150 mg Novozyme was added and the reaction mixture was shaken at 45°C for six days. After the enzyme preparation was removed by centrifuging the solvent was evaporated under vacuum and the solid residue was extensively washed with n-hexane until no free fatty acid was detected.

Fourier Transform-Infrared Spectroscopy (FTIR): FTIR spectrograms were recorded on a Nicolet Magna 750 spectrophotometer to establish the presence of ester bonds (about 1730-1740 cm^{-1}) and the reduction of hydroxyls groups on the cellulose molecule.

Colorimetric Determination of Free Fatty Acid: Aliquots were taken from the mixture and the amount of the free fatty acid was measured (Lowry and Tinsley, 1976).

2.2. Enzymatic esterification

Immobilised lipase from *Candida antarctica* (Novozyme) was a gift from Novo Nordisk. Cellulose acetate, cellulose acetate butyrate and lauric acid were purchased from Sigma. Hydroxypropyl cellulose and solvents were obtained from Aldrich and Merck, respectively.

Oligosaccharides produced at the hydrolysis stage were used as substrate for the esterification with lauric acid, using Liposyme and Palatase and an immobilised lipase from *Mucor miehei*.

Cellulose acetate and cellulose acetate butyrate: 180 mg cellulose acetate or cellulose acetate butyrate and 150 mM of lauric acid were dissolved in 5 ml acetonitrile. Then 150 mg Novozyme was added and the reaction mixture was shaken at 45°C for five days. The reaction was terminated by removing of the enzyme preparation by centrifugation. The products were isolated by precipitation into cold water (20 ml). The suspensions were stirred vigorously for another 10 min and filtered through a Buchner funnel. The solid residue was extensively washed with methanol until no free fatty acid was detected. In this case, the molecular characterisation of the product was achieved by ^{13}C-NMR spectroscopy.

Hydroxypropyl cellulose (HPC): 500 mg HPC and 150 mM of lauric acid were dissolved in 20 ml *tert*-butanol. Then 150 mg Novozyme was added and the reaction mixture was shaken at 45°C for six days. After the enzyme preparation was removed by centrifuging, the solvent was evaporated under vacuum and the solid residue was extensively washed with n-hexane until no free fatty acid was detected. The molecular characterisation of the product was achieved by FT-IR spectroscopy.

Fourier Transform-Infrared Spectroscopy (FTIR): FTIR spectrograms were recorded on a Nicolet Magna 750 spectrophotometer to establish the presence of ester bonds (about 1730-1740 cm^{-1}) and the reduction of hydroxyls groups on the cellulose molecule.

^{13}C- Nuclear Magnetic Spectroscopy (^{13}C -NMR): The presence of new carbon chains on the cellulose acetate was determined by ^{13}C-NMR. The NMR spectra were recorded at ambient temperature in DMSO-d$_6$ on a Varian -400 MHz spectrometer.

Colorimetric Determination of Free Fatty Acid: Aliquots were taken from the mixture at various time intervals and the amount of the free fatty acid was measured as follows (Lowry and Tinsley, 1976): 50 ml of the reaction mixture was added to screw cap test tubes containing 1 ml of 5% w/v cupric acetate-pyridine (pH = 5.8-6.0), 5 ml isooctane. After mixing and centrifuging for 3 min at 2500 rpm, the free fatty acids (C8 or more only) were determined in the upper organic phase. Absorbance was followed at 715 nm.

Determination of the Amount of Ester (%) (Tanghe et al., 1963): Acyl moieties on the cellulose derivatives were determined by saponification under mild conditions. Approximately 1 g of esterified cellulose acetate laurate and 40 ml of ethanol (75%) were heated at 55°C for 30 min and then 25 ml of 0.5 N NaOH was added. The mixture was heated and left to settle at room temperature for three days. The excess NaOH was measured with a 0.5N HCl solution and phenolphthalein as indicator. An excess of 1 ml of acid was added and the mixture was left to settle for 12 h to avoid any diffusion problems. The excess of added acid was then back titrated with a 0.5N NaOH solution. The ester content (%) was determined as follows:

$$\text{Ester content (\%)} = [(A - B)N_B - (C - D)N_A]M/10w \qquad (1)$$

where A and B = respective volumes of NaOH solution added to sample and blank (ml); N_B = normality of NaOH solution; C and D = respective volumes of HCl solution added to sample and blank (ml); N_A = normality of HCl solution; w = weight of sample (g); M = molar mass of grafted acyl radical.

3. RESULTS AND DISCUSSION

3.1. Pretreatments of lignocellulosics

Acid hydrolysis: The treatment with sulfuric acid caused a complete hydrolysis of cellulose at rather low temperature. The product was a mixture of oligosaccharides, and glucose production was less than 4%. The length of the reaction lead to oligosaccharides mixtures with lower molecular chains. The treatment with sulfuric acid caused a complete hydrolysis of cellulose. The treatment with hydrochloric acid caused only a partial reduction of cellulose and resulted in a mixture of oligosaccharides and glucose in the liquid phase, between 30-40%.

Cellulose etherification: The chemical structure of the product (degree of substitution and the size of the polyoxypropyl side chains) depends on the reaction conditions. Two factors determine the properties of the product: the molecular substitution value (MS) and the distribution of hydroxypropyl substituents. The hydroxypropyl group is relatively hydrophobic. Different HPC samples show diversity in solubility, which affects to different degrees the MS of the samples. The behaviour of the produced HPC molecules in t-butanol and n-hexane was of interest because these substances are used in the enzymic process. The solubility tests show that HPC derivatives are soluble in t-butanol, more soluble in a mixture of it and 10% methanol, and insoluble in n-hexane.

Proton NMR spectra show the changes of the cellulose structure with a peak at δ 1.12 indicating the methyl protons of the hydroxypropyl chains and a broad band from δ 3.0 to 4.0 due to methylene and methyne protons.

Enzymic esterification: Immobilised lipase from *Candida antarctica* (Novozyme) can catalyse the acylation of hydroxypropyl cellulose with lauric acid in t-BuOH but not the acylation of cellulose and the oligosaccharides produced from cellulose. A possible reason for this weakness is cellulose's high crystallinity index and the high content of salt in the oligosaccharides from the neutralisation stage.

For the acylation of hydroxypropyl cellulose, the effect of the molecular weight of hydroxypropyl cellulose was under investigation (FT-IR spectra). We show that molecular weight plays a crucial role on the degree of the esterification. At higher molecular weight, a lower degree of esterification has been observed. A possible explanation for this observation could be the lower solubility of the substrate with higher molecular weight.

3.2. Enzymatic esterification

Immobilised lipase from *Candida antarctica* (Novozyme) can catalyse the acylation of cellulose acetate with lauric and oleic acids in acetonitrile and the acylatin of hydroxypropyl cellulose with lauric acid in t-BuOH, but not the acylation of cellulose acetate butyrate. A possible reason for the inability of Novozyme to catalyse the acylation of cellulose acetate butyrate is its high total degree of substitution (DS = 2.8-2.9). Figures 1 and 2 show the depletion of fatty acids during the course of the acylation reactions of cellulose acetate and hydroxypropyl cellulose. As can be seen, the final conversion of both fatty acids was about

35% after 96 h of incubation at 45°C. However no depletion of fatty acid was observed in the absence of a biocatalyst, indicating that the reaction is enzymatically controlled. The determination of the increase of the ester bonds on the final product by a method based on the chemical hydrolysis of esterified cellulose, as previously described, indicates an increase of about 8% of the ester bonds on the final product compared with the unmodified cellulose acetate. Based on the above observations one can assume that the acylation of cellulose acetate by lauric acid takes place either in the free hydroxyl groups of the substrate molecule by direct esterification or in the acylated hydroxyl groups by transesterification.

Owing to the absence of ester bonds on the molecule of hydroxypropyl cellulose, the acylation of this molecule was determined by using FT-IR spectroscopy. The structure of the enzymatically modified cellulose acetate was investigated by ^{13}C-NMR spectroscopy. The presence of new methylene groups with ^{13}C chemical shifts within a range of (15-30 ppm) indicate the presence of new carbon chains on the cellulose molecule as a result of its enzymatic acylation with lauric acid. ^{13}C-NMR (DMSO-d_6, TMS as the internal reference), for the dodecanoyl moiety: δ 171.5 (C=O), 34.7 $CH_2(CH_2)_9CH_3$, 32.1 $CH_2CH_2CH_3$, 29.9-30.3 6x CH_2, 25.1 $CH_2(CH_2)_8CH_3$, 22.1 CH_2CH_3, 14.5 CH_3; for the acyl moiety δ 171.5 (C=O), 20.5 CH_3.

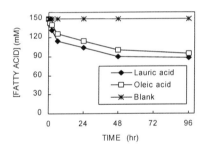
Figure 1. Esterification progress of cellulose acetate with lauric and oleic acid in acetonitrile with Novozyme.

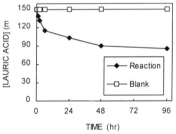
Figure 2. Esterification progress of hydroxypropyl cellulose with lauric acid in t-BuOH with Novozyme.

3.3. Factors affecting the lipase catalytic behavior

As can be seen in Figure 3, the rate of the depletion of fatty acids increased with the amount of immobilised lipase as was expected. At high enzyme concentration the conversion of both fatty acids is about 30% after 96 h of reaction.

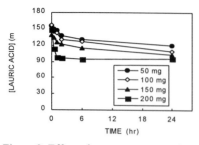 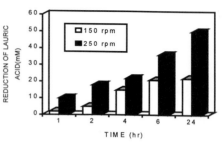

Figure 3. Effect of enzyme concentration on the depletion of lauric acid concentration at the esterification of cellulose acetate in acetonitrile.

Figure 4. Effect of the agitation speed on the depletion of lauric acid at the esterification of cellulose acetate.

For the possible influence of external diffusion limitation on the enzymatic process we performed experiments in a specific reactor where external mass transfer could be controlled. The reaction mixture consisted of equal amounts of cellulose acetate and lauric acid (150 mg in 5 ml of acetonitrile) while the agitation speed was varied from 0 to 250 rpm. Figure 4 shows the effect of the agitation speed on the acylation of cellulose acetate with lauric acid catalysed by Novozyme in acetonitrile. The depletion of fatty acid was increased with time as the speed of agitation was also increased. When the process is carried out without agitation the depletion rate of lauric acid is significantly slower, and the conversion was less than 10% after 96 h of incubation indicating an external mass transfer limitation. This observation, which indicates the low diffusion rate of substrates on the enzyme microenvironment, could be attributed to the high viscosity of the reaction mixture owing to the presence of high molecular weight cellulose acetate.

The effect of the weight ratio of cellulose acetate to fatty acid on the catalytic behavior of Novozyme has also been tested. The experiment was carried out with a constant amount of lauric acid (150 mg in 5 ml of acetonitrile) while the concentration of cellulose acetate was varied from 45 to 225 mg. As can be seen from Table 4, an increase of the weight ratio from 0.3 to 1.7 slightly affected the final conversion yield of the catalytic process. A further increase of the amount of cellulose acetate increased the viscosity of the reaction mixture.

Table 4
Effect of the weight ratio of cellulose acetate on the fatty acid conversion

Weight ratio of cellulose acetate to fatty acid	(%) Fatty acid conversion after 96 hours
0.5	25
1	34
1.7	36

For the acylation of hydroxypropyl cellulose the best experimental conditions were selected based on the acylation of cellulose acetate: a high concentration of fatty acid and hydroxypropyl cellulose, high agitation speed, and a large amount of enzyme. Moreover, the effect of the initial water content of the reaction mixture on the final conversion was examined and the effect of the molecular weight of hydroxypropyl cellulose was investigated. The results are based on the FT-IR spectra of the final product.

We show that molecular weight plays a crucial role on the degree of esterification. A higher molecular weight produced a lower degree of esterification. A possible explanation for this observation could be the lower solubility of the substrate with higher molecular weight.

Moreover, the initial water content of the reaction mixture seems to play a crucial role for the enzymatic esterification of hydroxypropyl cellulose. The initial water content of the reaction mixture was controlled with pre-incubation of the solvent, substrates and enzyme in the presence of a saturated solution of salt giving a known constant of water activity. Three different salts were used in order to achieve three different water activities: LiCl $a_W = 0.1$, $MgCl_2$ $a_W = 0.5$, KNO_3 $a_W = 0.9$. FT-IR spectroscopy showed that the higher production of ester was achieved at initial $a_W = 0.5$.

ACKNOWLEDGEMENTS

This work was supported in part by the European Commission (AIR Programme)

REFERENCES

Asandei N., N. Perju, R. Nicolescu, and S. Ciovica (1995). Cell. Chem. Technol., 29, pp. 261–271.
Ho, F.-L., R.R. Kohler, and G.A.Ward (1972). Analytical Chemistry, 44, p. 178.
R.R. Lowry and J.J. Tinsley (1976). J. Am. Chem. Soc., 53, pp. 470–474.
Tanghe, L.J. (1963). In Methods in Carbohydrate Chemistry, ed. By R.L Whistler, Academic Press, New York, 3, pp. 201–203.

PAPERMAKING PULP FROM *HESPERALOE* SPECIES, AN ARID-ZONE NATIVE PLANT FROM NORTHERN MEXICO

Al Wong[a] and Steve McLaughlin[b]

[a] Arbokem Inc., P.O. Box 95014, Vancouver, V6P 6V4, Canada
[b] University of Arizona, 250 East Valencia Road, Tucson, Arizona 85706

The University of Arizona has discovered in Northern Mexico an unusual native plant, *Hesperaloe* spp., that offers exceptionally balanced properties for the manufacture of high-quality paper products. The ideal combination of 3.3-3.5 mm length and 15-17 µm width of the ultimate fibres of *Hesperaloe funifera* and *Hesperaloe nocturna* is believed to be the key contributing factor.

A field research program on the cropping of Hesperaloe species has been in progress at the University of Arizona's Bio-resources Research Facility in Tucson since 1988.

In cooperation with Arbokem Inc., a laboratory pulping study was undertaken to validate the expected superior papermaking characteristics. The potassium alkaline sulphite pulping method was chosen for the study as this technique is known to be practical for the pulping of a wide variety of agriculture fibres.

Preliminary test results showed that an exceptionally high-strength pulp can be made from *Hesperaloe funifera* fibres. The experimental pulp was also found to have an unbleached brightness in the range of 78%.

Additional tonnage-scale pulping trials are planned at the Arbokem demonstration agri-pulp mill in Vulcan, Alberta.

1. INTRODUCTION

Hesperaloe funifera and *Hesperaloe nocturna* are plants in the Agavaceae family native to the Sonoran Desert of Northern Mexico. These Hesperaloe species are long-lived, perennial acaulescent rosette plants, i.e., they produce a cluster of leaves at the soil surface. Figure 1 shows example stands of mature *Hesperaloe funifera* and *Hesperaloe nocturna*.

The fibre of the leaves of these Hesperaloe species has been identified to have considerable papermaking potential. Typical ultimate fibre dimensions are 3.3-3.5 mm and 15-17 µm width.[1-3] This unusual fibre morphology allows Hesperaloe pulp to provide specific qualities to paper products (e.g., light weight with higher strength) which cannot be achieved technically with any combination of softwood and hardwood pulps. Reeves et al. have reported that Hesperaloe pulp can be used to produce very soft high-strength tissue paper.[4] There are no known North American competitive fibres of natural agricultural origin.

Since 1988, the University of Arizona has been conducting field research studies at its Bioresources Research Facility (BRF) in Tucson, Arizona.[5-7]

Figure 1. Mature stands of *Hesperaloe funifera* (back) and *Hesperaloe nocturna* (front) at the BRF

Physiological studies have determined that Hesperaloe species photosynthesize using Crassulacean acid metabolism, which means that they take up carbon dioxide and lose water principally at night. Transpiration ceases during midday when evaporative demands are highest. This characteristic greatly increases their water-use efficiency and decreases their water requirements for cultivation in arid zones. For example, at nearly the same biomass production, Hesperaloe species require 50% less water than alfalfa.

Agronomic and economic studies using experimental plots at the BRF have demonstrated the potential for high fiber yields: 5 to 7 oven-dry tonnes of fibres per hectare per year, after a stand is established.

A laboratory pulping study assessed if the unique fibre morphology of Hesperaloe fibres would confer exceptional papermaking pulp properties. It should be noted that the pulping conditions used were not optimized to establish, among other things, cooking time, chemical charge, pulp yield and physical strengths.

2. EXPERIMENTAL

The Hesperaloe leaves harvested manually by BRF staff were decorticated partially by pressing lengthwise through twin steel rollers. The pressed fibres were then dried naturally in the sun for several weeks prior to baling for shipment to the Arbokem laboratory in Vancouver, British Columbia. The as-is fibres were cut manually to 10-15 cm length prior to use in the cooking experiments.

An electrically heated 10-litre stainless steel digester equipped with liquor re-circulation was used for the laboratory cooking experiments. The potassium pulping technology[8-10] was selected for testing because this procedure has been found to make a satisfactory papermaking pulp from many agricultural fibres. Technical grades of KOH, K_2SO_3, and anthraquinone catalyst were used.

Typical pulping conditions used were as follows:

K_2O charge	330 kg/oven-dry tonne fibre
K_2O as KOH	10%
K_2O as $K2SO3$	90%
Anthraquinone charge	1 dry kg/oven-dry tonne fibre
Liquid-to-fibre	7 (by weight)
Maximum temperature	175 deg.C
Time to max. temp.	60 minutes
Time at max. temp.	240 minutes

The cooked pulp was routinely screened on a laboratory 10-cut flat screen. The screened pulp were tested according to the Standard Methods of the Technical Section of the Canadian Pulp and Paper Association.

3. RESULTS AND DISCUSSION

Depending on the quality of decortication of the harvested Hesperaloe fibres, the pulp yield based on dry fiber ranged from 42 to 48%. material. The target Kappa number for the alkaline sulphite pulp made under the above operating conditions was 6 to 8. The resulting pulp viscosity was in the range of 10 to 12 mPa.s. The brightness of the unbleached pulp was was 72 to 78%.

The representative physical strengths of the alkaline sulphite-cooked Hesperaloe pulp are given in Table I. It is interesting to note that neither tear nor tensile strengths were "developed" significantly during PFI refining. This finding suggests that the Hesperaloe pulp could be used on an as-is basis in many instances of papermaking. Considerable refining energy saving could be realized.

Figure 2 shows the tear-tensile strength profile of the Hesperaloe pulp to be substantially superior to prime-quality Canadian bleached northern softwood kraft pulp.

Unlike the conventional kraft cooking process, the alkaline sulphite pulping method provides a bright, easy-to-bleach pulp. Indeed, the bleaching of the present Hesperaloe pulp to still higher brightness can be expected to be achieved readily with only one or two stages of H_2O_2 treatment.

Further tonnage-scale pulping tests using freshly harvested Hesperaloe fibres are planned in late 1999. The trial will be made at Arbokem's demonstration agri-pulp mill in Vulcan, Alberta.

Table 1
Physical strength of unbleached hesperaloe pulp. Unbleached Sample V276S (pulp made from well-decortricated *Hesperaloe funifera*)

	PFI revolutions				
	0	1000	2000	3000	4000
Freeness, CSF	392	305	235	170	147
Density, kg/m^3	543	588	592	606	625
Tear Index, mN.m^2/g	20.0	17.0	16.5	17.7	14.3
Burst Index, kPa.m^2/g	11.0	13.3	12.0	13.0	12.9
Breaking Length, km	11.8	12.4	12.0	12.1	12.3
MIT Fold	2001	1961	2100	2177	2097
Porosity, sec.	7	17	20	45	59
Brightness 457μm	77.2				
Opacity (white), %	87.1	85.2	83.9	82.8	81.5

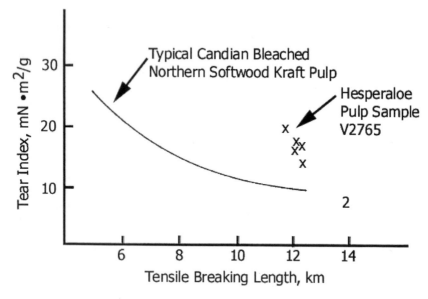

Figure 2. Tear-tensile strength relationship of hesperaloe pulp sample V276S

Follow-up test usage of the Azaloe™ Hesperaloe pulp in the commercial trial production of ultra lightweight coated (LWC) publication paper will be made. The Azaloe™ pulp is expected to have an unique role as a reinforcement fibre for LWC paper. In the present industry practice, the basis weight of LWC paper has apparently reached the technical limit of about 47 g/m^2 at about 30% kraft softwood pulp and 30% filler/coating content. Additional use of kraft softwood pulp would lower base paper opacity unacceptably; further use of mineral filler would lower the physical strength of the finished paper undesirably.

4. CONCLUDING REMARKS

Laboratory-produced unbleached Hesperaloe pulp has been shown to have unusually high physical strengths. These pulp strengths could not be achieved by the use of conventional softwood kraft pulp. This finding provides a commercial opportunity to use Hesperaloe pulp judiciously for the production of ultra lightweight coated publication paper.

Azaloe™ is a registered trademark of Arbokem Inc.

5. ACKNOWLEDGMENT

The 1988-98 agricultural research studies conducted at the University of Arizona was supported in part by funding from the United States Department of Agriculture (AARCC Agreement 93AARC20030, and CSREES Agreement 94-COOP-1-0036).

REFERENCES

1. McLaughlin, S.P. and S.M. Schuck (1988). Evaluation of fibers from native agavaceae—potential for use in paper making, Proc. First International Conference on New Industrial Crops and Products, Riverside, California, USA, October.
2. McLaughlin, S.P. and S.M. Schuck (1991). Fiber properties of several species of Agavaceae from the Southwestern United States and northern Mexico, Econ. Bot., 45, p. 480.
3. McLaughlin, S.P. and S.M. Schuck (1992). Intraspecific variation in fiber properties in *Yucca elata* and *Hesperaloe funifera* (Agavaceae), Econ. Bot., 46, p. 181.
4. Reeves, H. et al. (1994). Soft High Strength Tissue Using Long- Low-Coarseness Hesperaloe Fibers, U.S. Patent 5, 320, 710, June 14.
5. McLaughlin, S.P. (1995). Morphological development and yield potential in *Hesperaloe funifera* (Agavaceae): an experimental fiber crop for dry regions, J. Expt. Agric., 31, p. 345.
6. McLaughlin, S.P. Domestication of Hesperaloe: progress, problems and prospects, Progress in New Crops, ed. by J. Janick, ASHS Press, Alexandria, Virginia, USA. pp. 395–402.

7. Anon. (1996). Hesperaloe has properties that interest papermakers, in Industrial Uses of Agricultural Materials: Situation and Outlook Report IUS-6, Economic Research Service, U.S. Dept. of Agriculture, August, pp. 24–26.
8. Wong, A. (1988). Potassium pulping of industrial fibre crops and agricultural residues, Proc. First International Conference on New Industrial Crops and Products, Riverside, California, USA, October.
9. Wong, A. (1997). The Agri-Pulp alternative, paper presented at the Fiber '97 Conference, California Resource Recovery Association Meeting, Monterey, California, USA, June.
10. Wong, A. (1997). Potassium pulping of straw, paper presented at the 1997 TAPPI Pulping Conference, San Francisco, California, USA, October. (AK17706A).

ANAEROBIC METABOLISM IN THE MARINE GREEN ALGA, *CHLOROCOCCUM LITTORALE*

Y. Ueno,[a*] N. Kurano,[a] and S. Miyachi[b]

[a]Marine Biotechnology Institute, Kamaishi Laboratories, 3-75-1 Heita, Kamaishi, Iwate, 026-0001 Japan
[b]Marine Biotechnology Institute, 1-28-10 Hongo, Bunkyo-ku, Tokyo, 113-0033 Japan

Microalgae can produce various compounds by catabolizing endogenous carbohydrates during fermentation under dark anaerobic condition. *Chlorococcum littorale* is a new species of a unicellular marine green alga that can grow rapidly in a carbon dioxide concentration as high as 60%. In this study, we investigated the carbon metabolism and electron flow under dark and anaerobic conditions.

Cellular starch was degraded under dark anaerobic condition, and ethanol, acetate, hydrogen, and carbon dioxide were detected as fermentation products. The consumption mode of reducing equivalents depended on the incubation temperature and the partial pressure of hydrogen. Ethanol productivity was improved by adding methyl viologen. Enzyme activities involved in anaerobic metabolism are also discussed. Making use of anaerobic metabolism in microalgae could be a useful strategy for converting CO_2 to chemicals.

1. INTRODUCTION

Microalgae, including green algae and cyanobacteria, have been isolated and investigated with respect to their photosynthetic ability to fix CO_2, in the hope of reducing atmospheric CO_2 levels. However, algal photosynthetic products would eventually self-decompose, and CO_2 would be released into the air again unless it could be converted to other compounds that

---

This work was supported by New Energy and Industrial Technology Development Organization (NEDO).
*Present address: Kajima Technical Research Institute, 2-19-1 Tobitakyu, Chofu-shi, Tokyo, 182-0036 Japan

can be stored. The photosynthetic conversion of carbon dioxide to such valuable compounds as polysaccharides,[1] lipids,[2] and pigments[3] is being extensively investigated.

Although microalgae are oxygenic phototrophic organisms, they are often exposed to dark anaerobic conditions in nature. Endogenous substrate that is accumulated during photosynthesis serves as their energy source under such conditions. Fermentative metabolism produces various end-products, while endogenous storage compounds are completely decomposed to CO_2 by aerobic respiration. The following fermentative products have been detected for several microalgae kept in the dark: hydrogen, carbon dioxide, formate, acetate, ethanol, lactate, and glycerol.

Chlorococcum littorale is a newly discovered species of unicellular marine green alga that can grow rapidly at a CO_2 concentrations as high as 60%. Growth rates higher than those of other microalgae have been observed at 5–40% CO_2 concentration.[4] The growth characteristics indicate that this alga is suitable for dense cultivation at an extremely high CO_2 concentration.

In the work reported here, we investigated enzyme activities that regulate anaerobic metabolism in this alga. We also discuss the issue of anaerobic metabolism and its prospect for algal biomass conversion.

2. MATERIALS AND METHODS

Details of the cultivation of the strain, analytical methods, and preparation of cell-free extract for the determination of enzyme activities are described in earlier reports.[5,6] Enzyme activities were measured from the change in absorbance at 340 nm due to reduction of NAD(P) or oxidation of NAD(P)H using a spectrophotometer. The activity of NAD(P)H:ferredoxin oxidoreductase was determined as diaphorase activity by using methyl viologen.[7]

3. RESULTS AND DISCUSSION

3.1. Product formation by dark fermentation[5]

The cellular starch was decomposed immediately after the start of dark anaerobic incubation: approximately 27% of cellular starch was consumed during 24 hours of incubation at 25°C. Simultaneously with this starch decomposition, ethanol, acetate, hydrogen, and carbon dioxide formed. A decrease in starch decomposition rate with prolonged incubation time could be due to product inhibition or to a change in intracellular pH caused by the formation of acetate, although no change in pH caused by reaction buffer was observed.

It is suggested that *C. littorale* can carry out alcohol fermentation, since ethanol is the predominant organic product. The carbon and hydrogen recoveries in the experiment were 98% and 99%, respectively. These values prove that no major fermentation product was overlooked, and that no carbon compound other than cellular starch was used as the major source for fermentation.

Product formation during dark anaerobic incubation is shown in Table 1. Increase of the incubation temperature from 25 to 30°C affected the mode of cellular starch decomposition

Table 1
Product formation from the anaerobic metabolism in *C. littorale*

Incubation temp.	Addition of MV	Initial starch content	Starch decomposition	Productivity			Yield		
(Åé)		(μmol/g-dry wt.)	(%)	(μmol/g-dry wt.)			(mol/mol of glucose)		
				Ethanol	Acetate	Hydrogen	Ethanol	Acetate	Hydrogen
25	none	754	27	296	99	223	1.5	0.5	1.1
30	none	754	47	448	109	86	1.3	0.3	0.2
30	1mM	884	56	938	74	47	1.9	0.1	0.1

Dark anaerobic incubation was conducted for 24 hours.

and increased ethanol productivity. Adding methyl viologen (MV) to the reaction vial markedly decreased hydrogen formation and increased the ethanol productivity. It seems that more reducing equivalents were involved in ethanol formation in the presence of methyl viologen, although a proportion of the reducing equivalents might have been trapped by the added compound itself. At 30°C and with methyl viologen, ethanol productivity was about triple that at 25°C without methyl viologen. A similar effect on the regulation of metabolism by methyl viologen was observed in *Clostridium acetobutylicum*.[8] Methyl viologen may also be effective for product control in other microalgal fermentation processes.

The carbon flow proposed from the yield of fermentative products and the enzyme activities is shown in Figure 1. *C. littorale* seems to have exhibited unique metabolism with dark fermentation when compared with other green algae such as *Chlamydomonas* strains, *Chlorella*, and the fresh water *Chlorococcum* strain that shows the formation of glycerol or formate, unlike the metabolites detected in the present experiment.[9] It has been reported that *Chlamydomonas* strains use the reducing equivalents to produce hydrogen gas as their main metabolite.

3.2. Hydrogenase[7]

Hydrogenase of this alga, which plays an important role in anaerobic metabolism, was purified to homogeneity in five chromatographic steps under strictly anaerobic conditions. Hydrogenases are induced when the green algae are exposed to dark anaerobic conditions.[10,11] The time needed for this process is variable, ranging from a few minutes to several days, depending on the organism and adaptation conditions.[12] The hydrogenase of *C. littorale* was

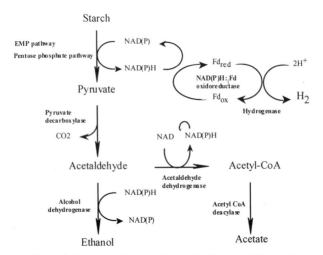

Figure 1. Postulated anaerobic metabolism in *C. littorale*.

also induced during the anaerobic adaptation process, and its activity increased with incubation time. No activity was found in cells without anaerobic adaptation, while the maximum activity was obtained after approximately 12 hours of anaerobic adaptation (Figure 2).

The catalytic properties of the enzyme are similar to those of the green algae, *Chlamydomonas reinhardtii*[10] and *Scenedesmus obliquus*.[11] Two peptide fragments were obtained from the enzymatically digested hydrogenase protein, and their amino acid sequences were determined. No significant homology to any other known sequences of hydrogenases was found.

Hydrogen evolution by the crude extract was observed when NADH, NADPH, or ferredoxin from *Chlorella* (purchased from Wako Co., Japan) was used as the electron donor. On the contrary, neither NADH nor NADPH were used for hydrogen evolution by the partially purified hydrogenase fraction. This indicates that ferredoxin was the physiological electron donor for hydrogen evolution.

3.3. Other enzyme activities related to anaerobic metabolism

The cellular starch in *C. littorale* was decomposed to pyruvate by the EMP pathway and the pentose phosphate pathway.[5] The activities of neither phosphoketolase, the key enzyme of hetero-lactic fermentation, nor of phosphogluconate dehydrase/2-keto-3-deoxy-6-phosphogluconate aldolase, the key enzymes of the Entner-Doudoroff pathway, were detected (data not shown).

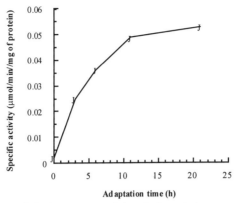

Figure 2. Induction of hydrogenase activity during dark anaerobic adaptation. The hydrogenase showed high affinity to the ferredoxin isolated from the same cell (K_m value = 0.68 µM).

The detection of pyruvate decarboxylase and alcohol dehydrogenase supports the notion that the strain can perform alcohol fermentation in which pyruvate that is formed by starch decomposition is converted to an acetaldehyde before being reduced to ethanol. The activities of several enzymes that were involved in anaerobic metabolism were determined with respect to their cofactor specificity (Table 2). The reducing equivalents formed through starch decomposition are used to form ethanol and hydrogen gas in this alga. NAD(P)H:ferredoxin oxidoreductase of this alga has higher specificity to NADPH than to NADH. It seems that the formation of NADPH promotes formation of hydrogen gas rather than NADH. However, alcohol dehydrogenase was active with both NADP and NAD with high specific activities. This characteristic of the enzyme could have allowed the predominant formation of ethanol. The product distribution of the anaerobic metabolism in this alga might not have been affected by the balance of NADH and NADPH. While hydrogenase was synthesized *de novo* (Figure 2), all other enzymes involved in anaerobic metabolism were constitutively presented in the cell without anaerobic adaptation.

This may also support the finding that ethanol was the predominant product.

In the anaerobic metabolism in *C. littorale*, NAD(P)H:ferredoxin oxidoreductase was crucial to the electron donation from reduced pyridine nucleotides to ferredoxin, although the enzyme usually catalyses in the reverse reaction, as is the case of ferredoxin:NAD(P) reductase in photosystem I.[13]

Table 2
Activities and cofactor specificity of enzymes involved in the anaerobic metabolism in *C. littorale*

Enzyme	Cofactor	Specific activity (nmol/min/mg of protein)
Glucose-6-phosphate dehydrogenase	NAD	1.1
	NADP	3.4
Gluconate-6-phosphate dehydrogenase	NAD	0.8
	NADP	2.8
Glyceraldehyde-3-phosphate dehydrogenase	NAD	0.1
	NADP	0.1
Phosphofructokinase		0.7
NAD(P)H:ferredoxin oxidoreductase	NADH	7.1
	NADPH	208.0
Pyruvate decarboxylase		7.1
Alcohol dehydrogenase	NAD	6.9
	NADP	8.4

4. FUTURE PROSPECTS

The quantity of product generated by dark fermentation in *C. littorale* cells is not comparable with that of the conventional fermentation process for bacteria, in which an exogenous substrate can be supplied. Improvements in the cellular starch content and in its degradation efficiency would be interesting research topics with respect to the enhancement of metabolic productivity. The production of ethanol or other chemicals by anaerobic metabolism of green algae may be the simplest process, since the substrate for end-products is formed from carbon dioxide by photosynthesis and all reactions proceed in the same cell. The enzymes, except for hydrogenase, are constitutively presented in the cell. Improvement of energy metabolism in photosynthesis by genetic engineering would be one option for practical use.

Although anaerobic metabolism has been extensively investigated in green algae and cyanobacteria, there have been few reports concerning anaerobic metabolism in other eukaryotic algae such as Euglenophyceae, Phaeophyceae, or Rhodophyceae. Further investigation is needed to gain deeper insight into anaerobic metabolism. It might also bring about the use of algae for biomass conversion.

REFERENCES

1. Miyashita H., H. Ikemoto, N. Kurano, and S. Miyachi (1995). J. Mar. Biotechnol., 3, p. 136.
2. Wolf, E.R., A. Nonomura, and J. A. Bassham (1985). J. Phycol., 21, p. 388.
3. Ben-Amotz, A., (1991). Bioresource Technology, 38, p. 233.
4. Kodama, M., H. Ikemoto, and S. Miyachi (1993). J. Mar. Biotechnol., 1, p. 21.
5. Ueno, Y., N. Kurano, and S. Miyachi (1998). J. Ferment. Bioeng., 86, p. 38.
6. Ueno, Y., N. Kurano, and S. Miyachi (1999). FEBS Lett., (in press).
7. Bes, M.T., A.L. de Lacey, V.M. Fernandez, and C. Gomez-Moreno (1995). Biochemistry and Bioenergetics, 38, p. 179.
8. Grupe, H. and G. Gottschalk (1992). Appl. Environ. Microbiol., 58, p. 3896.
9. Ohta, S., K. Miyamoto, and Y. Miura (1987). Plant Physiol., 83, p. 1022.
10. Happe, T. and J.D. Naber (1993). Eur. J. Biochem., 214, p. 475.
11. Schnackenberg, J., R. Schulz, and H. Senger (1993). FEBS Lett., 327, p. 21.
12. Schulz, R. (1996). J. Mar. Biotechnol., 4, p. 16.
13. Knaff, D.B. and M. Hirasawa (1991). Biochem. Biophys. Acta, 1056, p. 93.

EVALUATION OF A SOY-BASED HEAVY FUEL OIL EMULSIFIER FOR EMISSION REDUCTION AND ENVIRONMENTAL IMPROVEMENT IN INDUSTRIES AND UTILITIES IN MEXICO[*]

Phillip K. Lee and Bernard F. Szuhaij

Central Soya Company, Inc., P.O. Box 1400, Fort Wayne, Indiana 46801-1400

The purpose of this study was to determine the effects of water-in-oil emulsions, created using an emulsifier derived from soybeans, on the environmental emissions and boiler efficiencies when they were burned in industrial-scale boilers. Because heavy fuel oils do not burn very efficiently, they release large amounts of emissions into the atmosphere. The emulsification of water into heavy fuel oil improves atomization of the fuel in a combustion chamber, which allows greater carbon burnout, lowered emissions, and other benefits. In the past, the use of this technology has been limited and typically used petroleum- or synthetic-based emulsifiers, many of which are toxic. Soybean lecithins have long been used as efficient emulsifiers in food systems and are natural, nontoxic, biodegradable, and renewable.

In this study, a soybean lecithin-based emulsifier was evaluated in several utility and industrial facilities in Mexico, where the use of heavy fuel oil is common. The emulsifier was added to the heavy fuel oil at 0.5% and 1% levels and emulsions of 10% water in heavy fuel oil were prepared and burned in full-scale boilers used to produce steam and electricity. NO_x and other emissions decreased substantially, particulates were substantially reduced, and the excess oxygen requirement decreased when the emulsions were burned as compared with fuel oil alone. It was concluded that the use of a soybean lecithin-based emulsifier may be used to create water-in-oil emulsions that increase the burning efficiency of heavy fuel oils, reduce emissions and particulates, and reduce down time for cleaning boilers in combustion systems.

---

[*]The work was jointly funded by Central Soya Company, Inc., the United Soybean Board, the Indiana Soybean Board, the Ohio Soybean Council, the Indiana Value Added Grant Program, and the Kentucky Soybean Promotion Board.

Biomass Transformation into Value-Added Chemicals, Liquid Fuels, and Heat and Power

Anaerobic Processes

ANAEROBIC BIOCONVERSION OF LIGNOCELLULOSIC MATERIALS

A. Padilla,[a] E. Marcano,[a] and D. Padilla[b]

[a]Universidad de Los Andes, Facultad de Ciencias Forestales, Laboratorio de Bioenergía, Vía Chorros de Milla, Mérida 5101, Venezuela, e.mail: adrianap@forest.ula.ve
[b]Universidad de Los Andes, Facultad de Ingeniería, Ciclo Básico, La Hechicera. Mérida 5101, Venezuela

The desertification of some areas in Venezula is partly due to the very common use of forest biomass for energy, mainly firewood for cooking. The Bioenergy Laboratory works to provide an alternative energy source generated *in situ* in villages that minimizes environmental risks. This research analyzes some variables that influence biogas production.

1. INTRODUCTION

Even though Venezuela has large fossil fuel reserves, it does not escape the consequences of the world energy crisis. One remedy, strongly recommended, is to convert gasoline-powered motors to natural gas.

Recent estimates show that Venezuela has natural gas reserves for 160 years.[1] They reach 5.4 million m^3; 70% are found within its geographical continental limits in eastern Venezuela and 30% in the west.

The above figures indicate a possible natural gas consumption of almost 34 million m^3 per year. However, the construction of enormous gas pipelines to carry fuel the length and breadth of the country is problemmatic.

The anaerobic digestion process provides three great advantages: the *in situ* production of biogas, environmental improvement, and production of organic fertilizers.

Production *in situ* avoids transportation and gas pipeline construction costs. In addition, environmental improvement is shown in the diminishing chemical oxygen demand (COD).

The value or potential fertilizer value depends, in part, on the total nitrogen content in the purines and pyrimidines found in nucleic acids, RNA and DNA, of the vegetable cells.[2,3]

Biogas forms during anaerobic fermentation; the basic biochemical reactions of the microorganisms involved are oxidation and reduction in which a number of organic compounds are oxidized by the removal of hydrogen.

The proportion of chemical components within vegetable material varies and it affects the biomethanation biochemical processes. Cellulose and hemicellulose are rapidly degraded to simple carbohydrates through hydrolysis. However, when these components are associated with lignin, only about half[4] are available for the methanogenic crops because they become only slightly susceptible to hydrolysis. Degradation must than be obtained by submitting lignocellulosic material to alkaline digestive processes.

The increase in fossil fuel prices, among them liquid gas, and the diminished purchasing power of the rural population has caused some rural Venezuelan areas to become deforested. Without stirring up the situation, it can be affirmed that the desertification processes are winning ground. One cause of such circumstances is the widespread use of firewood as a source of energy.

This reason alone justifies studies that seek a short-term solution to an ecological and energy problem by looking at pioneer species of rapid growth and, if possible, fodder. Pastures that grow biomass and rural systems of anaerobic digestion in "batch" operation can be complementary.

The system of batch (or lot) operation, occurs when the fermentation structure is loaded with solid waste, leaving it to ferment until ultimately gas is produced. Users then extract all the material and completely reload the structure. This takes one month.[5]

The *Pinus caribaea* species from the Uverito plantations must be added to those studied for utilization of biomass. This is so because there are large planted areas, and because the wood is used for structures, which leaves great volumes of bark and sawdust that not only is not utilized but obstructs the forest industry.

2. EXPERIMENTAL METHOD

2.1. Raw material

Leaves and pseudostems of the *Musa sapientum* (banana), leaves and branches of the *Titonia diversifiolia* (arnica), leaves and stems of the *Bambusa vulgaris* (bamboo), and sawdust of the *Pinus caribaea* (Caribbean pine) were used.

2.2. Portion of chemical components

Portions of the main components of the test materials are shown in Table 1 and were determined following the method of analysis published in the Universidad de Santander,[6] that uses the TAPPI T 204 om-88 processes for the preparation of extract-free wood; Jaime and Wise to determine the holocellulose K; Seifert to quantify cellulose; and ASTM D-1107 for the solubility in alcohol-benzene. For determining the lignin, the "acid-insoluble lignin in annual plants, developing woods, and forage method" was used.[7]

2.3. Production of biogas

Glass biodigestors with an approximately 400 ml capacity were built especially for laboratory use. The mixture for biogas production consisted in 5% weight suspension. The solid phase was made of 2.5% sawdust or various primary materials and 2.5% mud from a pig-raising oxidation pond. The liquid part (95%) was water. The biodigestors were kept at 36°C for one month, with periodic measurements of gas production. When production was seen to diminish considerably, the pH of the mixtures was measured and the COD was then analyzed.

2.4. Chemical oxygen demand (COD)

The APHA-AWWA-WPCF Standard Method for the examination of wastes and wastewater[8] was used. The liquid part of the degraded suspension was removed after completing the anaerobic process. Then a preliminary determination of COD was made. The liquid part was diluted four times and eight times for successive determinations.

3. RESULTS AND ANALYSIS

Table 1
Chemical composition in an oven dry basis of the raw material used in the anaerobic process.

Raw material	Extracts	Cellulose	Holocellulose	Lignin	Ash
Musa sapientum (pseudostems)	9.54	41.56	56.85	8.74	10.34
Musa sapientum (leaves)	8.67	46.59	60.89	21.91	8.84
Titonia diversifiolia (leaves)	17.76	23.58	46.46	9.96	16.49
Titonia diversifiolia (branches)	16.49	44.59	75.24	17.22	5.80
Bambusa vulgaris (leaves)	8.56	47.71	72.15	19.52	16.60
Bambusa vulgaris (stems)	7.13	39.49	65.84	14.67	2.00
Pinus caribaea (sawdust)	2.08	47.44	73.26	26.39	0.28

Table 2
Total gas volume (m^3N) by kilogram of oven dry raw material, COD (mg/l), consistency (%) of the liquid phase and its dilution, produced after 96 days of anaerobic process.

Raw material	Total gas	C_0	COD_0	C_1	COD_1	C_2	COD_2
Oxidation pond mud	*	0.274	**	0.058	844.17	0.016	224.01
Musa sapientum (pseudostems)	0.281	0.352	**	0.124	149.05	0.023	67.13
Musa sapientum (leaves)	0.160	0.213	**	0.154	288.48	0.010	106.33
Titonia diversifiolia (leaves)	0.159	0.396	**	0.104	592.57	0.022	129.20
Titonia diversifiolia (branches)	0.287	0.223	843.55	0.063	243.04	0.007	103.39
Bambusa vulgaris (leaves)	0.150	0.150	1012.52	0.034	316.54	0.003	125.44
Bambusa vulgaris (stems)	0.148	0.278	888.97	0.081	238.14	0.017	118.09
Pinus caribaea (sawdust)	0.042	*	*	*	*	*	*

C = consistency
0, 1 and 2 = liquid phase and its first and second dilution respectively
* = not determined
** = too high

The consistency determined after the dilutions did not correspond to the exact values expected theoretically. The main cause of this difference is that the mixes are not solutions but suspensions; thus the sample reflects the heterogeneity in the concentration of solids (Table 2).

Although the theoretical values do not correspond with those obtained in the experimental process, a significant correlation between COD values determined in the two liquid parts was established (Tables 2 and 3).

Table 3
Linear regression model of the factors which show significant relation.

Linear equations	Probability level	Correlation coef.	Determination coef.
$COD_2 = 75.86 + 0.11 * COD_1$	0.10902	+0.71669	51.37%
$COD_2 = 159.93 - 261.62 * m^3 gas$	0.06813	-0.77855	60.61%
$COD_1 = 843.56 - 13.28 * cel$	0.07028	-0.77494	60.05%
$COD_1 = 144.255 + 16.0195 * ash$	0.19683	+0.61176	37.43%

COD = chemical oxygen demand in mg/l
$m^3 gas$ = total gas in normal cubic meter
cel = % cellulose
ash = % ash

From Table 3, the following can be deduced: (1) a significant relation exists between the COD of the first dilution and the COD of the second; (2) changes in COD values depend on 60.61% of the total gas values produced for each mix—the probability that the regression model adjusts to real facts is 93.18%; (3) the chemical demand of oxygen in the final anaerobic process depends on the amount of cellulose contained in the primary material before beginning the process (COD diminishes as the percentage of cellulose increases); (4) the increase in the amount of ash of the primary material causes an increase in the COD of the anaerobic process residual waters.

For the regression analysis–correlation (Table 3) *Pinus caribaea* was not used because it is difficult to use in gas production. However, an increase in the pH of the solution was sufficient to overcome any inconvenience.

4. CONCLUSIONS

Even though Venezuela is an OPEC country, it cannot escape the world energy crisis. Evidence of this can be found in the tangible deforestation that has resulted from firewood cutting in rural areas. Mitigating deforestation and subsequent deterioration of the soils can be accomplished by encouraging the rural population to embrace once again their ancestral technology of producing biogas. The optimization of such technology consists, in part, of controlling factors such as pH and temperature and primary materials. Primary material with a high cellulose content, neutral pH, and temperatures close to 36°C all improve the production of biogas.

A comparison of the COD values for mud without anaerobic treatment to those obtained at the end of the biogas production process (Table 2) once more confirms that the anaerobic process contributes to environmental improvement.

The anaerobic process produces a mixture that can be used as a fertilizer. However, the dosage used must be carefully formulated according to culture needs, because the concentration of elements is excessive as it comes out of the biodigestors.

ACKNOWLEDGMENTS

The authors wish to acknowledge the importance of professor Taylhardat's suggestions in the development of this project; the collaboration of professor Lombardo in the determination of COD; and to the CDCHT for its financial support.

REFERENCES

1. González, S. (1993). Conversión directa de metano con catalizadores a base de óxido de lantano. ULA, Facultad de Ciencias, Mérida, Venezuela.
2. Devlin, R. (1976). Fisiología vegetal. Omega S.A., Barcelona.

3. Hernández, R. (1989). Nutrición mineral. Facultad de Ciencias Forestales, ULA, Mérida, Venezuela.
4. Wise, D. (1981). Fuel gas production from biomass Vol II. CRC Press, Inc., Florida.
5. Taylhardat, L. (1992). Comparative analysis on the performance of two anaerobic digester (horizontal and vertical) with filter, in Proceedings of Biotechnologies for Pollution Control and Energy. FAO, CNREE, Braunschweigh, Germany, pp. 91–102.
6. Rodríguez, L. (1978). Métodos de análisis empleados en la industria papelera. Universidad Industrial de Santander, Bucaramanga, Colombia.
7. Stephen, L. and D. Carlton (eds.) (1992). Methods in Lignin Chemistry, Springer-Verlag, Berlin.
8. APHA,AWWA-WPCF (1991). Standard Methods For the Examination of Waste and Waste-water. USA.

CONSTRUCTION AND OPERATION OF A COVERED LAGOON METHANE RECOVERY SYSTEM FOR THE CAL POLY DAIRY

D.W. Williams,[a] M.A. Moser[b] and G. Norris[c]

[a]BioResource and Agricultural Engineering Department, California Polytechnic State University, San Luis Obispo, California 93407
[b]Resource Conservation Management, Oakland, California
[c]USDA Resource Conservation Service, Templeton, California

This paper describes the design, construction and anticipated operation of a lagoon-type methane recovery system for the Cal Poly Dairy. The initial design was based upon the present and anticipated herd size, 300 to 600 cows, heifers and calves. The lagoon design meets USDA-NRCS standards, and accounts for limitations of the site, primarily shallow sandstone bedrock. The new lagoon, which has a liquid volume of 14,000 m^3, was constructed next to an existing lagoon. The new lagoon was covered with a flexible membrane incorporating buoyant material so that the cover floats on the surface, and a gas collection system. The predicted output of the lagoon for the present population of approximately 350 cows, heifers and calves is estimated to average up to 320 m^3 of biogas per day. The biogas will fuel a micro-turbine electric generator, and produce up to 23 kW in parallel with the utility system. Odor control is the most important non-economic benefit. This project will provide environmental benefits—odor control by capturing the odorous gases that result from dairy manure storage; methane, a significant greenhouse gas is kept out of the atmosphere; and water pollution is reduced through the reduction in organic matter in the lagoon. Economic benefits include electricity and process heat, together worth up to $16,000 per year.

1. FARM DESCRIPTION AND BACKGROUND

The Cal Poly Dairy is located adjacent to the Cal Poly campus in San Luis Obispo, California. The dairy presently milks 180 cows; the total population is more than 350 animals. Most of the herd is housed in freestall barns. About 90% of the manure is deposited on concrete and flushed with fresh or recycled water to the lagoon. The remaining 10% of the manure is deposited in the corrals and is collected only seasonally. Solids are separated from the flushed wastewater prior to storage in a single-cell lagoon. This lagoon has a volume of 19,000 m^3, which translates to 50 to 90 days of storage, depending upon the amount of water used by the dairy.

Electricity and natural gas usage was determined from Cal Poly and utility records. These energy forms could be displaced by farm-produced methane. The main electrical service is a 480-volt, 600 amp, 3 phase, Y-connected 4-wire system. Most electric motors are three-phase, and minor uses such as outlets and lights are single phase.

The dairy purchases electricity through one meter. The milking parlor uses the most electricity; the separator system and lagoon recycle pumps use the next largest amount. The milking parlor's use of electric power varies during the day. During the two daily milkings of 4 hours each, electricity use is relatively high. However, it is relatively low during the rest of the day. The monthly consumption is also variable, increasing during months when the irrigation pump is used to empty the lagoon. The average annual power consumption is approximately 234,000 KWh at an average electrical rate $0.09 per kWh, for a total annual cost of $21,000 annually. The natural gas consumption is only that used for water heating, and amounts to approximately 77,000 kJ per year, which at $0.58/100,000 kJ ($.60/therm) would annually cost $450.

2. METHANE PRODUCTION TECHNOLOGIES

A number of methane producing technologies that have been developed and could be considered for dairy manure. The choice of the most appropriate technology depends on specific waste characteristics. Complete mix systems and plug flow digesters are used for concentrated dairy waste and were not considered because of the dilute nature of the Cal Poly dairy waste. Packed bed and upflow anaerobic sludge blanket digesters have been used for soluble organic wastes and are just now being tested for use with dairy flushwater (Wilke, 1998). Covered lagoons have been successful at dairies (Safley and Lusk, 1990), and they are the most appropriate and reliable technology for consideration at this site. Lusk (1998) reported, that in the U.S. covered lagoons had a much higher success rate, 78%, as compared with either plug flow (37%), or complete mix digesters (30%).

Moser (1996) and Williams, et al. (1996, 1998) described the initial design assumptions for this lagoon digester project. This paper compares this original design with the actual lagoon design and its construction details. It was proposed to construct a new, primary lagoon adjacent to an existing lagoon. The existing lagoon is needed for storage, whereas the primary lagoon must be held at a constant volume in order to function as a methane recovery digester.

3. CAL POLY COVERED-LAGOON DESIGN

3.1. Influent manure and flush water

The barn and parlor flush water containing very dilute manure is a feasible digester feed for an unheated covered lagoon. Design of a constant volume methane-producing lagoon must consider all of the volume flowing into the lagoon to avoid hydraulic washout of bacteria. The expected daily volume flowing to the separator includes the manure solids from 180 milk cows, 169 heifers and 8 calves. The inclined screen will capture 15% of the manure volatile solids from the liquids flowing to the lagoon. The estimated daily wastewater is 350,000 L per day at 0.3% total solids (TS) and 0.25% volatile solids (VS).

3.2. Sizing of covered lagoon

Two design variables were analyzed to size the methanogenic covered lagoon: the organic loading rate (OLR) with units of kilograms (kg) of VS per cubic meter per day, and the hydraulic retention time (HRT), with units of days. The limitation on OLR for the climate of San Luis Obispo County, California, is no more than 0.16-kg VS/m^3/day. Using this parameter for the approximate 900 kg of VS per day from the dairy, the calculated lagoon size would be 5,700 m^3. The recommended HRT for this region is 39 days. Based upon this parameter and the 350,000 L of flushwater per day, and allowing for two feet of freeboard, the lagoon should be almost 16,000 m^3 and contain 14,000 m^3 of effective liquid storage. Since this is the larger of the two lagoon design sizes, HRT controls the size of this lagoon. OLR of this lagoon would be only 0.06 kg VS/m^3/day. At this rate, the lagoon will be able to accommodate manure from additional cows or cheese whey from the nearby dairy processing facility, as along as the flush volumes do not greatly increase.

3.3. Components and construction

As shown in Figure 1, the new primary lagoon was located next to the existing lagoon, and it has approximate surface dimensions of 80 m × 80 m. The depth varies from 5.2 m to 3.7 m with 2:1 side slopes. A sandstone bed at the design depths required a relatively shallow lagoon depth. During August 1997, the completed plans for the lagoon were put out to bid. In September a contractor was chosen for the anaerobic lagoon construction at the Cal Poly dairy. Bids were received from 6 companies, and the company chosen was based on a low bid of approximately $60,000.

In November we met with the contractor at the new lagoon site, and got supplies for staking. Construction was delayed because of rain, but the contractor was poised to begin as soon as it stopped. Digester construction was delayed because of rains from December 1997 through most of April 1998. Lagoon construction began on April 27, 1998. The site was cleared of vegetation and the outline of the lagoon was marked with stakes. An elevating scraper and dozer first excavated the "key " of one lagoon wall embankment, that next to the road. Although only 1.5 to 2 m of soil capped a rock layer, the overall lagoon depth was more than 3.7 m because the lagoon was sited in a natural depression requiring less excavation. Construction was again delayed by rain in early May and resumed late May. As of July 1 only one sidewall needed more dirt, and the overflow pipe needed to be installed. The contractors finished the earthen lagoon in August.

Requests for lagoon cover bids were mailed out in late June The manure transfer pump and sump were ordered and were installed in early September. A new 15-cm diameter PVC influent line was installed from the existing lagoon inlet, through a sump and 3.73 kW transfer pump, to the corner of the new lagoon. An overflow outlet pipe 20-cm diameter was located at the corner of the lagoon opposite the inlet pipe, and set to maintain 0.5 m of freeboard.

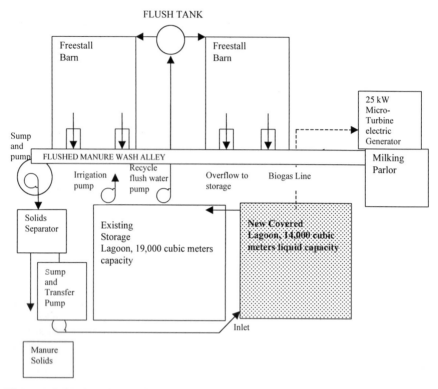

Figure 1. Cal Poly Dairy and lagoon system schematic

Bids were received for the lagoon cover, and two of the three bids were within the budget allocated for the cover. A lagoon cover bid of $41,000 from Sharp Energy, which planned to cover about half of the lagoon surface (2300 m^2) was selected. The bid costs of the cover, $16 to $43/m^2, bids were much higher than the original budgeted amount (less than $10.00/m^2). The cover was to be positioned over the deeper half of the lagoon where the majority of the gas would be generated. The Sharp Energy cover was ordered, delivered in early November, and scheduled to be installed when the lagoon had filled. The electrical hook-up for the manure transfer pump was then completed in early November. The lagoon pump was activated and slurry was loaded into the lagoon starting in mid-November. It reached its capacity in late December 1998, at which point the lagoon cover installation began.

With a crew of 8 to 12 people, the lagoon cover (68 × 34 m) was unfolded, the floats placed in the sleeves on the underside of the cover, and the cover floated out onto the lagoon.

This task took two days; at this point it was discovered that some parts were missing from the lagoon cover: additional sleeves that were to hold the gas collection manifold, and the sucker rod used to secure one edge of the cover to the lagoon bank. Mr. Sharp, who was assisting and directing the lagoon cover placement, contacted the cover fabricator, C.W. Neal, and they sent a crew to weld in the missing sleeves. There was still one task to complete the job—inserting the gas manifold and folding under the 5.2 m skirt at the gas collection edge of the cover; this skirt extends down into the lagoon water and directs the gas between the floating cover and the skirt to the gas manifold. A crew of at least 12 was required to accomplish this task during the first week of January 1999, using BRAE student labor. The rest of the gas handling equipment was ordered—the pipe, the meter, the flare, and associated valves and fittings. The gas handling system, which will be put into place in February 1999, consists of 7.5-cm pipe, a condensation trap, a gas blower, a gas meter and a flare.

According to Farmware, an EPA–developed computer program for farm digester design (EPA, 1997), the output of the lagoon will average 320 m^3 of biogas per day. As gas bubbles form in the lagoon, they are channeled along 23-cm-diameter flotation blocks under the cover to the gas manifold at the bank of the digester. This manifold exits the cover by a 7.5-cm flange, which is connected to a 7.5-cm-diameter PVC buried pipe. The pipe then connects to the rest of the gas handing system as previously described. Eventually the gas system will be piped to a gas handling and utilization building containing the electric generation system. The methane production from the covered lagoon produces 20–25 kW on a continuous basis from the present cow population. After considering an internal combustion-type engine generator and electric generation from biogas, a biogas-fired micro-turbine with 25 kW maximum electrical output was selected. The micro-turbine electric generator will operate in parallel with the utility system at a constant level of output controlled by the biogas supply. The micro-turbine will be supplied at no cost by Reflective Technologies. The actual and estimated cost to date of this methane recovery system including the lagoon construction, flexible cover, influent piping, gas handling, and associated labor and engineering was approximately $150,000 (Table 1).

4. BENEFITS OF METHANE PRODUCTION

4.1. Non-economic benefits

Installation of a covered lagoon and demonstration of its operation to the students and visitors at Cal Poly will serve the California dairy community. Odor control is the most important non-economic benefit. Conversion of volatile solids to biogas, and recovery and use of the biogas limits odor to the immediate area. Two lagoons provides additional storage volume, which will allow the existing lagoon to be properly emptied and over-winter storage of a larger volume of flushwater.

Table 1
Estimated cost of construction of covered lagoon digester

	Cost
Equipment	
Flushed Manure Transfer Pipe and Pump	$6,000
Excavation and Lagoon Construction	$60,000
Digester Cover and Installation	$44,000
Subtotal	$110,000
Gas Handling	$5,000
Micro-Turbine Electric-Generator	$0
Electric Generator Building and Utility Tie-in	$10,000
Subtotal	$12,000
Equipment Subtotal	$125,000
Construction	
Engineering	$25,000
Engineering Subtotal	$25,000
TOTAL	$150,000

4.2. Electricity benefits

The annual dairy electrical use by the dairy operation is approximately 234,000 kWh with a value of $21,000, averaging almost $0.09/kWh. Benefits from using biogas to produce electricity to replace purchased electricity are based upon an average of up to 23 kW, which will be the estimated electrical generation when the lagoon is completely covered. The completed methane recovery system will produce 170,000 kWh of electricity and 77,000 kJ of hot water annually, worth approximately $16,000. These returns will be based upon installation of the remainder of the digester cover and gas handling system, estimated to cost approximately $50,000. This would bring the project total cost to $200,000 and a simple payback of 13 years not including the environmental benefits of odor control and water pollution prevention. Although there is no capital cost of electric generation for this project, the cost of the state-bid lagoon construction was very high, $60,000 compared with typical farm construction of under $25,000 as calculated by Farmware. Since the estimated cost of a 25 kW engine-generator is over $30,000, the higher lagoon cost offsets the absence of an electric generation capital cost for the Cal Poly system.

5. PROJECT STATUS

As of February 1999, the new lagoon is complete, the partial cover installed, and the gas handling system including meter and the flare is installed. The biogas quantity and quality will be monitored, and the micro-turbine generator will be installed in spring 1999, and testing of the system will occur during the following six months. Electricity should be flowing into the utility lines by summer 1999. The remainder of the cover will also be installed during 1999, at which point electricity output will be at its maximum for this dairy.

REFERENCES

Lusk, P. (1998). Methane Recovery from Animal Manures – The Current Opportunities Casebook. National Renewable Energy Laboratory, NREL/SR-580-25145, Golden, Colorado.

Moser, M. (1996). Feasibility of production and use of methane from dairy waste at Cal Poly Dairy, San Luis Obispo, California, prepared for California Polytechnic State University, San Luis Obispo, California, by Resource Conservation Management, Inc., P.O. Box 4715, Berkeley, California.

Safley, L. and P. Lusk (1990). Low Temperature Anaerobic Digester, North Carolina Department of Commerce, Energy Division, Raleigh, North Carolina.

U.S. Environmental Protection Agency (1997). Amstar Handbook, ed. by M.A. Moser and K.F. Roughs, Atmospheric Pollution Prevention Division, EPA-430-B-97-015.

Wilke, A. (1998). Personal communication, University of Florida, Gainesville, Florida.

Williams, D.W., M. Moser, and G. Norris (1998). Covered lagoon methane recovery system for a flush dairy. ASAE Paper Number 984104, Presented at the 1998 ASAE Annual International Meeting, Disney's Coronado Springs Resort, Orlando, Florida.

Williams, D.W., M. Moser, and J. Smith (1996). Design of a covered lagoon methane recovery system for a flush dairy. Proceedings of the Seventh National Bioenergy Conference, The Opryland Hotel, Nashville, Tennessee.

LANGERWERF DAIRY DIGESTER FACELIFT: WHAT WE FOUND WHEN WE TOOK APART A 16-YEAR-OLD DAIRY PLUG FLOW DIGESTER

Mark A. Moser[a] and Leo Langerwerf[b]

[a]Resource Conservation Management, Inc., P.O. Box 4715, Berkeley, California 94704
[b]Langerwerf Dairy, Inc., 1251 Durham-Dayton Highway, Durham, California 95938

Annual digester maintenance has been estimated to cost up to 8% of capital cost. This project found that the long term annual digester maintenance cost was less than 1% of capital cost. The findings will reduce uncertainties that have limited adoption of digesters. Langerwerf Dairy is a 400-cow dairy in Durham, California, with a plug flow anaerobic digester that has been in operation continuously since 1982. The dairy received matching grant assistance from the Western Regional Biomass Energy Program (WRBEP) to refurbish the digester and document the process. The AgSTAR program provided technical assistance. The protective greenhouse was disassembled, gas bag removed, floating crust removed, manure pumped out, heating system examined, and settled solids removed. The digester was basically in good condition. Major costs were removing floating and settled solids, and new greenhouse parts and gas bag. The digester was refurbished and returned to service. Findings include the following: (1) hypalon gas bag material degraded; (2) floating crust accumulated to an average 4 feet thickness; (3) sand accumulated to an average depth of 5.5 feet; (4) a dairy plug flow digester appears to accumulate only 1% of the volume of solids that would be expected in an anaerobic lagoon; and (5) there was minor corrosion of concrete and steel. Further detail including, procedures, costs, and findings, are included.

1. PROJECT HISTORY

Langerwerf Dairy is a 400-cow family run dairy in Durham, California, with a plug flow anaerobic digester that has been in operation since 1981. The farm received a matching funds grant from the Western Regional Biomass Energy Program (WRBEP) to assist in purchasing materials necessary to refurbish the digester, to document the process and to publish a report of findings on the lifecycle condition of digester components. At the time of refurbishing, the digester was 16 years old and needed some parts repaired and replaced. A Caterpillar G3306 engine generator has operated about 90% of the project lifetime, averaging about 40 kW output. In 16 years of operation the digester has produced about 120,000,000 ft^3 of biogas (72,000,000 ft^3 of methane) and the engine converted the biogas into 5,000,000 kWh.

Approximately 23,000 yd^3 of digested fiber have been sold. The original digester, generation and solids separation systems cost $200,000. The value of the electricity produced is about $350,000; the value of the digested fiber sold has been about $138,000; about $75,000 in farm hot water was recovered from the engine; and $135,000 was saved in lagoon cleanout costs. The farm has spent about $160,000 on operation and maintenance including this project. The estimated return for the 16 years of operation is approximately $540,000.

2. DIGESTER DECONSTRUCTION

A 8-foot-high perimeter chain link fence was taken down. The protective greenhouse (a two-layer clear plastic over galvanized steel hoops) enclosing the digester was disassembled and hauled away. The gas collection cover was removed and disposed of. These tasks were performed over a one-week period.

A 3 - 5 foot thick, hard floating scum was found. The scum would easily support the weight of a man. The scum was removed by an articulating trackhoe and a skilled operator. The operator scooped the scum up and deposited it directly into farm dump bed trucks. About 350 cubic yards of matted hard scum were hauled to the fields and applied. Scum removal required about two days and about four people plus the trackhoe operator.

The remaining digesting slurry was pumped out, with about 40,000 gallons reserved in a nearby tank for digester startup. The black iron hot water heating system was examined and found to be in good condition. Five feet of settled solids, mostly fine sand with some organic material, was found covering the bottom and part of the heating system.

Approximately 400 yards of settled sands were in the digester and approximately 330 yards were removed using a hydraulic mining technique. High-pressure hoses using recycled digester water were used to wash settled solids and to pump the mixture to a settling basin near the digester. Clarified liquid flowed back to the digester.

Cleaning out the solids required about 10 days because several different approaches were tried before arriving at the workable solution. The work would have required four men for five days even if the hydraulic mining with recycle had been initially used. Once the digester was emptied all components were inspected for serviceability.

3. COMPONENTS AND MATERIALS AFTER 16 YEARS OF OPERATION

<u>Gas collection cover.</u> The hypalon gas collection cover material degraded and failed due to UV weathering. UV weathering caused pinholes, and subsequently biogas and water infiltrated into the cover thus creating larger holes and leaks.

<u>Gas collection cover attachment.</u> Black flat bar steel, angle iron and galvanized bolts were used to attach the cover to the concrete tank. Sheet metal capping was installed on the top of the digester wall. The exposed wall top galvanized sheet metal cap rusted away. These materials were on the atmosphere side of the gas collection cover and exposed to the same moisture and hydrogen sulfide conditions as the greenhouse hoops. Additionally, some

manure had run along the inside edge between the cover and the digester wall at some time and remained in contact with the bolts. At least a dozen bolt heads had corroded enough to require drilling out. Some of the flat bar steel had corroded because it was left in contact with manure and air. Most of the angle iron was rusty but not corroded significantly. New hardware was used for cover reattachment.

Concrete. Some concrete corrosion was found in the same areas as the cover attachment corrosion and is attributed to manure that had run along the inside edge of the concrete and been exposed to air as it dried. The manure decomposed forming acids that etched away one-eighth to one-quarter of an inch of concrete over a 40-foot length of concrete sidewall. The corrosion presented no problem with the digester operation.

Liner material. Hypalon material was also used for the digester liner. The material was judged to be suitable for continued service. It was aged and grainy in spots but did not exhibit the holes or liquid infiltration. No buildup of struvite was found.

Heat exchanger and pipe. The digester heat exchanger was constructed of black steel. It was found partially buried in sand and judged to be not fully effective. Upon removal of the settled sand no external corrosion was found.

Gas collection pipe. No degradation of any PVC plastic including the gas collection pipe was found. The gas intake T where the cover rested was slightly deformed, probably due to 16 years of the cover resting on it and being exposed to high temperatures. There was some accumulation of manure solids evident in the gas line, probably from startup foaming.

Greenhouse components. Galvanized greenhouse support hoops were corroded at unprotected welds. The corrosion can be attributed to normal condensation mixed with some hydrogen sulfide. The greenhouse plastic was in need of replacement after four or five years.

4. REFURBISHING AND RESTARTING THE DIGESTER

A pipe coupling in the heating system was broken during the cleanout process and later was repaired. The digester was refilled on 10/22 with new and old manure. Heating began on 10/23 with the engine-generator running on propane. The new digester cover was installed using new steel and bolts on 10/28. On 10/29 the digester temperature was at 85 degrees and 5% new manure was added. The digester gas meter was installed and the biogas gas tested in a flame test the same day. The flame was consistent and had good color characteristics. Late on 10/29 the biogas was tested in the engine. Full power was demonstrated and therefore the engine was set at 20 kW and run continuously. Table 1 shows the log of the startup of the digester.

By 11/1 the engine was running full time at 30 kW based on a production of almost 20,000 ft^3/d. The greenhouse was installed mid-November to protect the digester during winter weather. On 11/30 engine output reached 40 – 45 kW at continuous operation.

Table 1
Langerwerf dairy digester startup log

Date	Gas Meter Reading Ft^3	Average ft^3/day since last	Effluent Temp - °F	Effluent pH
November 3	96000	19200	94	7.3
5	161400	32280		
7	209100	23850	101	7.4
9	257700	24300	102	7.4
17	504700	30875		
18	538400	33700	103	7.4
27	864000	36178		

5. CURRENT OPERATIONS

Table 2 shows the farm electricity use before and after the project directly from PG&E invoices. In September 1998 the digester was shut down, showing the real farm requirements for electricity. In October 1998, the digester was heated with propane fueling the engine at low output and reduced the electricity use. November 1998 is the startup month where gas production increased and was burned for electricity production, resulting in a decline of electricity use to 109 kW/d. The digester's electricity production saved the farm $850 in

Table 2
Electricity purchases comparison 1997-1998

Month	1997 kWh/d Use	1998 kWh/d Use
September	194	609
October	118	502
November	221	109

October alone. The 1998 purchases can be compared with the 1997 purchases and it can be seen that even with just startup operation, the farm is buying less electricity than in 1997. At the same time, the digester system produced about 1,000 surplus kWh in November 1998 that were sold to PG&E for $356.27. The farm was not selling nearly as much electricity in 1997.

6. PROJECT EXPENDITURES

The project was completed on time and on budget. Table 3 summarizes the budgeted versus actual expenditures. The situation found upon opening the digester necessitated altering some strategies and planned work. Hired labor was substituted for a contractor when it was obvious that mantime was more necessary than skilled construction assistance. Savings were used to spend more money on rental equipment for removing scum and settled solids. A trackhoe to remove scum saved money which was then used to hydraulically mine solids. AgSTAR assistance substituted for some of the planned farm personnel time. AgSTAR personnel suggested the recycle settling ponds, set up the cover for installation and worked with the cover installation crew.

Table 3
Costs of refurbishing the digester

			Est. Labor Hours	Actual Labor Hours	Costs $ per unit	BUDGET Subtotal	ACTUAL Chargeable
Disassemble system							
Farm Labor	hours		360	213	$ 22.00	$ 7,920.00	$ 4,686.00
Hired Labor	hours		80	210	$ 12.00	$ 2,800.00	$ 2,520.00
Rent	days		10		$ 75.00	$ 750.00	$ 1,084.30
Mixer/Pumps							
Rental crane	days		3		$ 375.00	$ 1,125.00	$ 920.00
Farm truck/tractor	hrs		60	40	$ 25.00	$ 1,500.00	$ 1,000.00
Put system back together							
Farm Labor	hours		250	156	$ 22.00	$ 5,500.00	$ 3,432.00
Hired Labor			80	151	$ 12.00	$ 2,800.00	$ 1,812.00
Consulting							
AgSTAR							$ 9,800.00
Project Manager	hours		100	100	$ 65.00	$ 6,500.00	$ 6,500.00
Materials							
meter	Roots ssm					$ 1,250.00	$ 1,250.00
flare	Varec or equiv					$ 3,205.00	$ 2,833.71
cover	30 mill polypropylene					$ 3,800.00	$ 4,603.71
frame	angle, clips, bolts					$ 740.00	$ 630.46
greenhouse parts						$ 4,550.00	$ 2,750.04
							$ 6,381.96
Contingency purchases							
Subtotal						$ 42,440.00	$ 50,204.18
Contingencies @ 10%						$ 4,244.00	$ 0
TOTAL						$ 46,684.00	$ 50,204.18

7. CONCLUSIONS

The materials originally selected for use were very satisfactory. We are very pleased with the condition of materials in the digester. The cover material did not last 20 years as projected by the manufacturer. Gas collection cover materials may be serviceable only for 10 years; however, the cost of a new cover every 10 years does not materially affect the profitability of the digester. The only unusual finding is that the digester had been operating successfully. The amount of floating scum and settled solids reduced the usable volume of the digester by about 66%. Only a third of the digester was actually hosting active digestion. However, the lack of retention time had caused a drop in performance, leading to the recognition of the need for digester cleanout and refurbishing.

8. OBSERVATIONS FOR INDUSTRY CONSIDERATION

The accumulation of solids and scum in a digester are expected. However, a dairy plug flow digester appears to accumulate only 1% of the volume of solids that would be expected in an anaerobic lagoon. The following section compares the volume of solids found in the plug flow digester with the estimated sludge accumulation volume in an anaerobic lagoon. It may not be fair to compare a digester and a lagoon because a lagoon may also provide storage. However, a separate storage facility for digester effluent requires no sludge volume, because all but 6 inches of sludge is assumed to be pumped out each time the storage facility is emptied.

Approximately 10,000 ft^3 of scum and 8,000 ft^3 of settled solids were found to have accumulated in the Langerwerf plug flow digester over 16 years. Occasional mixing to reduce scum had been attempted with minimal success from years 14–16. For purposes of estimation, we can say that about 20,000 ft^3 of material accumulated over the life of the plug flow digester.

The NRCS *Field Waste Management Handbook* (National Engineering Handbook, *Agricultural Waste Management Field Handbook*, USDA, Soil Conservation Service, April 1992) page 10A-3, contains the method used to calculate solids accumulation in anaerobic lagoon treatment systems. An anaerobic lagoon receiving the waste from 360 cows would accumulate 2,300,000 ft^3 of sludge over the same 16-year period as the cleanout interval of the plug flow digester. A recent revision to Practice Standard 359-1 (*Natural Resources Conservation Service Conservation Practice Standard, Waste Treatment Lagoon Code 359*, January, 1998, Revision 4) distributed through the North Carolina State NRCS office, reduces the dairy sludge accumulation volume and for this comparison would only require 1,500,000 ft^3 of sludge storage volume.

Therefore, in a 16-year cycle, 20,000 ft^3 (740 yd^3) of material accumulates in the plug flow digester and 2,300,000 ft^3 (85,000 yd^3) of material accumulates in a lagoon and must be managed. One can easily calculate that cleaning the digester and hauling off material in 10-yard dump trucks would require 74 truck loads. The same maintenance for a lagoon would require 8,500 truckloads or some other rather large investment to remove sludge. Under North Carolina guidance only 5,555 truckloads would need to be moved.

If a farm has the option of a digester and storage or an anaerobic lagoon, the farm should be aware that sludge storage is not required for a digester while a lagoon will be much larger than the digester with 85,000 yards of excavation required just for sludge storage. Eventually the farm with the lagoon will have to manage 5,000 to 8,000 dump truck loads of sludge that a digester does not accumulate.

MAKING ENERGY RECOVERY FROM ORGANIC WASTES WORK IN A DEREGULATED ELECTRICITY MARKETPLACE

G. Simons,[a] V. Tiangco,[a] R. Yazdani,[b] M. Kayhanian[c]

[a]California Energy Commission, 1516 Ninth Street, Sacramento, California 95814-5512
[b]Yolo County Department of Public Works, Davis, California
[c]University of California at Davis, Davis, California

Californians generate more than 40 million tons of municipal solid waste each year which is disposed of in more than 600 landfills. However, fewer than 60 sites recover landfill gas and generate electricity with it. Those 60 sites account for only 110 megawatts of electrical generating capacity. Low electricity prices, and uncertainties over landfill gas recovery system capitalization and ownership, have slowed development of landfill gas-to-energy projects. Two innovative approaches to digesting and recovering energy from the organic fraction of the municipal solid waste stream have been developed for use in California's deregulated electricity marketplace. By improving conditions for biological processes in landfills, methane generation can be accelerated to up to twice the rate in conventional landfills. As a result, accelerated decomposition processes reduce landfill volume, decrease waste management costs, and offer the possibility of generating electricity at costs approaching 3.5 cents per kilowatt-hour. Two demonstration cells containing 18,000 tons of waste have been tested at Yolo County's Central Landfill since October 1996. Similarly, a unique two-stage, anaerobic digestion and aerobic drying system has been developed by the University of California at Davis that will generate both a humus product and electricity. The UCD system offers a way to help California municipalities meet a mandated 50% recycling goal by the year 2000, while simultaneously generating electricity for on-site use at transfer stations and landfills. The economics, opportunities and benefits associated with each of these innovative energy recovery systems is examined against a backdrop of California's deregulated electricity marketplace.

YOLO COUNTY CONTROLLED LANDFILL PROJECT

D. Augenstein,[a] R. Yazdani,[b] K. Dahl,[b] A. Mansoub,[b] R. Moore,[b] and J. Pacey[c]

[a]I E M, Palo Alto, California 94306
[b]Yolo County Department of Public Works, Davis, California 95616
[c]Emcon, San Mateo, California 94402

A new landfill management approach, "controlled landfilling," is being demonstrated by the Yolo County, California Department of Public Works at the Yolo County Central Landfill (YCCL) near Davis. Overall objectives are to obtain earlier and greater methane energy recovery from landfilled waste and to reduce landfill "greenhouse" gas emissions to near-negligible levels. Methane generation and waste stabilization were accelerated by improving biological conditions within a test cell through carefully controlled additions of water and leachate. A control cell was operated in parallel. Landfill gas capture was maximized, with emissions reduced to minimal levels, by a combination of surface membrane containment, a permeable layer conducting gas to collection points, and operation at slight vacuum. Cells are highly instrumented to determine performance. To date, normalized methane recovery is the highest seen from such a large waste mass, anywhere-- about tenfold that from "conventional" landfill practice. The rationale and details of this project, and first three years' results, are summarized below.

1. INTRODUCTION

The Yolo County, California, Department of Public Works is conducting large-scale demonstration of an advanced landfill management strategy, "controlled landfilling" at its Central Landfill outside of Davis, California. Support has come from sources including the California Energy Commission, Yolo County, and the U.S. DOE (through EPRI). This paper provides a project overview and presents encouraging results that have been obtained to date. Readers interested in more detail should consult Augenstein et al., 1997.

With "conventional" sanitary landfilling (current U.S. regulations), landfilled waste generally remains relatively dry for many years after filling. Such dry waste conditions retard and limit waste decomposition to landfill gas. Long terms of slow landfill gas recovery is well-documented (SWANA 19-landfill study Vogt and Augenstein, 1997), as is findings of un-decomposed legible reading material many decades old, often recoverable from landfill samples (as noted in popular articles by William Rathje). Slow decomposition entails several long-term problems: gas system and containment maintenance, continuing subsidence, and leachate pollutants. Even with all present and expected U.S. regulations, conventional landfilling results in gas recovery less than the potential, inefficient energy use, and often, substantial fugitive methane emissions. In summary, because of these inefficiencies, and fugitive emissions both before and after collection, as little as half of generated gas may be collected by conventional systems. Augenstein and Pacey (1991) estimated that fugitive fractions may be 10% to 60% (i.e., collection system efficiencies between 40% and 90%). California's Air Resources Board, in its Suggested Control Measures (1990), estimated escape at 40% to 60% of gas collected, i.e. 30-40% fugitive gas. Walsh (1994) estimated fugitive gas at 25% to 75%. These estimates are for times when controls are operating. Furthermore, much U.S. waste may, under current regulations, occasionally or always escape gas control.

Surface membranes ("geomembranes") that cover landfills can capture almost all generated gas. However, such membranes as the sole means to maximize gas recovery, absent other measures, also have a serious disadvantage. Membranes retard moisture (precipitation) infiltration thus maintaining waste at initial low moistures, typically 20-25%. This further slows and lessens decomposition beneath impermeable membranes (Kraemer, 1993; Leszkiewicz, 1995) so decomposition may continue much longer, up to a century or more. Dry containment approaches have been termed "dry tomb" technologies (Lee, 1990). Two "dry tomb" problems are common. (1) Poor economics for low-rate gas recovery (overall, as well as low rates per unit area) for extremely long periods because of high fixed annual collection costs per unit of gas collected. (2) Poorer economics of scale for energy use or flaring of gas recovered at lower rates. To avoid such long term problems with entombing of membrane-covered waste it is highly desirable to completely decompose the waste more quickly—i.e., make the landfill a bioreactor. Our approach has been termed "controlled landfilling."

2. CONTROLLED LANDFILLING

Waste decomposition and methane enhancement can be promoted by moisture, temperature, pH, nutrients, and other factors. Elevated moisture (by conventional landfill standards) appears essential for accelerating methanogenesis (Halvadakis et al., 1983). Temperature elevation can also provide major benefits. (Ea. ca. 15 kcal/mol., hence rate doubling for each $\approx$10C increase throughout a span from 10°C to 50+°C [Ashare et al., 1978], also see Hartz and Ham, 1982). Any process for U.S. use must be compatible with current U.S. landfill regulations, and integrate readily with current practice. With this in mind, water and temperature have been the only enhancement techniques applied, even though other techniques are possible. That landfill decomposition times can be reduced and gas generation completed within ten years or less appears well-supported by studies above and is consistent with results of field trials (Pacey et al., 1987)

In combination with methane enhancement, a surface permeable layer beneath a surface membrane, at uniform slight vacuum, provides a good alternative to wells to accomplish near-total gas recovery. In summary, surface membrane containment with management of landfill moisture and temperature can speed completion of methane generation, minimizing fugitive methane emissions (particularly long term) as well as maximizing energy potential.

Anticipated benefits in energy, environmental and landfill operational improvements include the following: (1) reduction to minimal levels of methane emission from landfilled waste masses to which it is applied. Methane is an extremely potent climate-active gas; (2) elimination of almost all organic air pollutants; (3) completing decomposition much sooner, reducing long-term risks to the environment and reducing long-term gas and other management costs; (4) reduced costs for post-closure landfill care and gas system operation and maintenance owing to earlier completion of landfill gas generation and waste stabilization; (5) maximizing rate and yield of methane recovery; (6) more predictable methane recovery, so that landfill-gas–fueled energy equipment may be appropriately sized to fully use gas; and (7) better scale economics for energy use of the greater amounts of captured gas.

The project operates two test cells containing about 9,000 US tons waste each. Cells are large enough (100 x 100 x 40 feet deep) to replicate both compaction and heat transfer of landfilling at normal waste depth. Liquid (well water or leachate) was added to the enhanced cell by 14 scrap-tire-filled "pits." Additions were metered to stay within U.S. regulations limiting base hydrostatic head (to 30 cm). A leachate collection and removal system (LCRS) delivers leachate into a reservoir (prefabricated manhole) from which the leachate can be either recirculated to the cell or (ultimately) disposed of. The control cell receives no liquid. Waste in cells is intensively instrumented to establish performance. Sensors for moisture and temperature were embedded in the waste during placement. In total, waste in the two cells contains 56 moisture sensors and 24 temperature sensors. These are distributed over three layers in the control cell and four in the enhanced cell. Cells also have monitoring of volume reduction, leachate flow and composition, static head on the base liner, gas flow and composition, containment integrity, and other parameters of interest. Sidewalls of compacted clay, also used successfully in an earlier demonstration (Pacey et al., 1987) isolate the demonstration cells' waste from the main body of surrounding waste. The cells were covered with a highly gas-permeable layer of shredded scrap tires. This permeable gas conduction layer is in turn covered with gas-impermeable geosynthetic membrane. For experimental purposes, in large part to assess base lining integrity, the enhanced cell base was double-membrane lined with means to detect leaks. Cells were filled with waste using landfilling approaches that were largely standard. However, greenwaste intermediate cover left waste permeable to later moisture additions and infiltration. Porous cover allowed some limited initial composting which elevated startup temperature as well. Waste was typical residential and commercial material from packer trucks serving households, small businesses, and markets. Tonnages were carefully logged. Loads that were inert were, however, diverted. Much more information can be found in Augenstein et al., 1997.

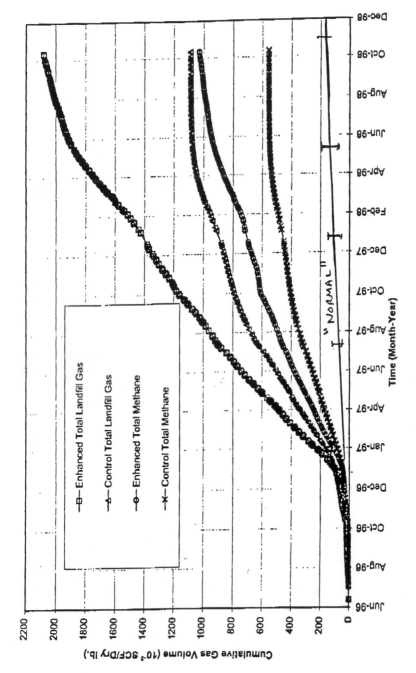

Figure 1. Cumulative Landfill Gas Volume for the Control and Enhanced Cell

Overall performance objectives at the outset included completion of methane generation and biological waste stabilization in under 10 years, and demonstration of fractional gas recoveries of 95% or more (fugitive emissions of less than 5%).

For gas collection, gas is withdrawn to maintain slight vacuum, a < 1 cm water head is uniform throughout the surface permeable layer, and gas moves through perforated pipe to a main collection line. Control and enhanced gas flow are both measured by highly accurate Dresser Industries corrosion-resistant positive displacement meters. Gas composition and particularly the methane of interest is analyzed by gas chromatography.

Waste was brought to the field capacity by liquid (well water) addition to pits. Then outflow recirculated to attain acceptably high readings from emplaced moisture sensors. Makeup well water was used to overcome any moisture deficit, indicated by either dry sensors or minimal or absent outflow. Water was initially added at rate sufficient to bring waste to field capacity in four months.

3. CONCLUSIONS

The most important results may be summarized as follows:

Refuse temperature: Both cells experienced substantially elevated temperatures, 40-55°C in the bulk of the waste upon filling, attributed to limited aerobic composting occurring after waste placement. The combination of heat inputs from methanogenesis and losses has resulted in current temperatures slightly over 40°C in the enhanced cell. The control cell with less biological activity has now cooled to a mean near 33°C.

More than 90% of the enhanced cell waste was wetted within six months after start of liquid addition. Thus, the infiltration approach is successful. For the control cell, waste moisture readings remained dry as expected with the exception of the very bottom layer (where moisture has been detected by other means as well; however, control leachate generation has now ceased).

Cumulated gas recoveries to date are shown in Figure 1. To very briefly summarize, gas recovery has been excellent. Normalized recovery rate from the enhanced cell was 0.7 ft^3 CH_4/lb. waste.yr (50 liters/tonne.yr) during all of 1997. Cumulated enhanced cell methane recovered to October 1998 is already 1.05 ft^3/lb total waste. Cumulated enhanced cell recovery and rate are to this point the highest known to the authors from any waste mass this large anywhere in the world. "Normal" gas recovery that would be expected from this waste mass is also shown in Figure 1 based on a major study of recoveries from 19 landfills (Vogt and Augenstein, 1997), as well as widely applied commercial models such as Emcon and SCS. The demonstration cell recovery exceeds "normal" expectations by a factor of five to ten. Completion of landfill gas should be possible in a few years, possibly substantially complete in fewer than five years. Enhanced cell methane recovery is thought to reflect both moisture and beneficial temperature effects. The control cell exhibited very high early methane recovery as well, more than half that of the enhanced cell. Control cell productivity may be due to temperature effects, which by themselves would result in such enhancement with temperatures > 20°C over normal ambient. After this initial burst, the "dry" control cell productivity fell to near zero.

Based on results to date, incremental energy and greenhouse gas abatement potential from wide application of controlled landfilling to U.S. landfills has been estimated (in preparation for the Federal Energy Technology Center: IEM, 1999). Controlled landfilling can result in capture of 95% of gas generated to maximum yield. It was assumed that controlled landfilling could be applied to about 80% of U.S. waste. If resultant gas fueled electricity, added time-averaged electrical energy could amount to ca. 4000 MWe over and above that available with "conventional" landfill gas collection. This is a significant increment of electrical energy, enough to meet total needs associated with all activities of close to 3 million U.S. citizens. Climate or "greenhouse" benefit from controlled landfilling comes in large part from elimination of vertical well collection inefficiency and long-term methane emissions. Further benefit comes from offset of the fossil CO_2 that would otherwise be emitted. Benefit has been calculated as equivalent to the abatement of about 80 million tons of CO_2 per year. Given uncertainties and variables in such calculations, this suggests possible CO_2-equivalent reductions between 50-100 million tons CO_2 per year for the U.S.

In any case major benefit is possible by extant standards. Wide application of controlled landfilling should be a significant help to the U.S. in meeting any future greenhouse gas reduction commitments.

REFERENCES

Ashare, E., D.L. Wise, and R.L. Wentworth (1977). Fuel gas production from animal residue. Dynatech R/D Company. U. S. Department of Energy/NTIS.

Augenstein, D., D.L. Wise, R.L. Wentworth, and C.L. Cooney (1976). Fuel gas recovery from controlled landfilling of municipal wastes. Resource Recovery and Conservation 2 103–117. Also: Augenstein, D., J. Pacey, R. Moore, and S.A. Thorneloe (1993). Landfill methane enhancement. 16th SWANA Annual Landfill Gas Symposium.

Augenstein, D., R. Yazdani, R. Moore, and K. Dahl (1997). Yolo County controlled landfill demonstration project. Proceedings from Second Annual Landfill Symposium. Solid Waste Association of America (SWANA), Silver Spring, Maryland.

California Air Resources Board (CARB) (1990). Suggested control measure for landfill gas emissions—California Air Pollution Control Officers Association Technical Review Group.

Halvadakis, C.P., A.O. Robertson, and J. Leckie (1983). Landfill methanogenesis: literature review and critique. Stanford University Civil Engineering Report no. 271. Available from NTIS.

Hartz, K.E., R.E. Klink, and R.K. Ham (1982). Temperature effects: methane generation by landfill samples. Journal of Environmental Engineering, ASCE.

Kraemer, T.H., H. Herbig, and S. Cordery-Potter (1993). Gas collection beneath a geomembrane final cover system. Proceedings, 16th Annual Landfill Gas Symposium, SWANA, Silver Spring, Maryland.

Leszkiewicz, J. and P. Macaulay (1995). Municipal solid waste landfill bioreactor technology closure and post closure. Proceedings, U.S. EPA Seminar on Bioreactor Landfill Design and Operation, Wilmington, Delaware, March 23–24. EPA/600/R-95/146.

Pacey, J.G., J.C. Glaub, and R.E. Van Heuit (1987). Results of the Mountain View controlled landfill experiment. Proceedings of the SWANA 1987 International Landfill Gas Conference, SWANA, Silver Spring, Maryland.

Vogt, W.G. and D. Augenstein (1997). Comparison of models for estimating landfill methane recovery. Final report to the Solid Waste Association of North America (SWANA) and the National Renewable Energy Laboratory (NREL). March.

WASTEWATER TREATMENT FOR A BIOMASS-TO-ETHANOL PROCESS: SYSTEM DESIGN AND COST ESTIMATES

Kiran L. Kadam,[a] Robert J. Wooley,[a] Francis M. Ferraro,[b] Richard E. Voiles,[b] Joseph J. Ruocco,[c] Frederick T. Varani[c] and Victoria L. Putsche[d]

[a]National Renewable Energy Laboratory, 1617 Cole Boulevard, Golden, Colorado, USA 80401
[b]Merrick & Co., 2450 S. Peoria St., Aurora, Colorado, USA 80014
[c]Phoenix Bio-Systems, Inc., 1880 S Pierce St, Lakewood, Colorado, USA 80228
[d]7884 Elder Circle, Denver, Colorado, USA 80221

A wastewater treatment (WWT) system was designed for an enzyme-based process for converting lignocellulosic biomass to fuel ethanol. The bioethanol process included the following basic unit operations: biomass pretreatment, hydrolyzate conditioning, fermentation, cellulase production, product recovery, and energy recovery. Anaerobic digestion followed by aerobic treatment was selected as the best WWT strategy, and centrifugation and evaporation of the stillage was employed to minimize wastewater and optimize water recycling. For a 2000 dry tons/day biomass-to-ethanol plant, the capital cost estimate for a WWT system is about $10.4 million. The impact of capital costs and the associated operating expenses of the WWT system corresponds to 2.35¢/L of ethanol or 6.6% of the total ethanol cost. The WWT strategy and cost estimates developed can be generally applicable to similar bioethanol plants.

1. INTRODUCTION

Waste water treatment (WWT) is an important operation of the overall lignocellulosic biomass-to-ethanol processes. The current trend in the similar industries of corn-to-ethanol production and pulp and paper manufacture is to recycle various water streams internally and to reclaim wastewater with appropriate treatment to allow additional recycle. Especially over the past 20 years, once-through water systems have been replaced with minimum discharge systems. This is due not only to the cost of WWT, but also to minimization of environmental impact, cost and availability of make-up water, etc.

A WWT system was designed for an enzyme-based process for converting hardwoods to fuel ethanol (Merrick, 1998). The bioethanol process comprised the following main unit operations: biomass pretreatment, hydrolyzate conditioning, fermentation (simultaneous saccharification and cofermentation), cellulase production, product recovery, and energy recovery. A 2000 dry tons per day (tpd) plant was modeled using the Aspen® Plus™ process simulator (Wooley et al., 1999).

2. WWT ALTERNATIVES

A sound WWT strategy is to concentrate contaminants into a relatively small stream, leaving the major stream sufficiently clean for reuse or discharge. By selecting waste streams that can be recycled separately upstream of the treatment, the WWT system becomes much smaller and overall plant efficiency is greatly increased. These issues were considered in the selection of the best alternative for WWT.

Some of the major WWT alternatives available are anaerobic biological treatment, aerobic biological treatment, evaporation (and incineration), stream discharge, land application, and discharge to a publicly owned treatment works (POTW). The suggested treatment system is a combination of anaerobic biological treatment followed by aerobic biological treatment. This recommendation is based on the calculated flow rates as well as the estimated chemical oxygen demand and biological oxygen demand (COD and BOD). The flows from the bioethanol process average at around 200,000 L/h or 1.3 million gallons per day (MGD) of total flow to the waste treatment block, with an approximate COD loading of 30,000 mg/L. Anaerobic and aerobic facilities in the 1 to 5 MGD range can be obtained in a variety of processes and facility types ranging from custom engineered and constructed "municipal" facilities to vendor-distributed and -installed package-type plants.

3. EFFLUENT CRITERIA

In general, for influents with a COD higher than 1000 mg/L, anaerobic digestion is the preferred first treatment step; anaerobic treatment of soluble organics will average over 90% reduction on a COD basis. For influent to the aerobic treatment step, which is the effluent from the anaerobic treatment, of <400 mg/L, 400–800 mg/L, and 800–1000 mg/L of BOD, the effluent will average <10 mg/L, <20 mg/L, and <30 mg/L each of BOD and total suspended solids (TSS), respectively.

As the site of the proposed facility and therefore the ultimate discharge of the effluents from the WWT facility are unknown, 30 mg/L BOD and TSS each were used as targets for maximum discharge parameters, which are usual stream discharge requirements for the average Western U.S. stream. The fact that a particular project effluent could be higher quality than the regulation of 30 mg/L BOD and TSS each does not typically change the requirement for both an anaerobic and an aerobic treatment step. However, if the typical treatment step appears overdesigned, the design should be evaluated for potential cost savings by reducing the size (residence time) of the equipment to match system performance to the effluent requirement. Other parameters for wastewater discharge requirements such as toxins, metals, nitrogen, and phosphorous will have a bearing on treatment steps in the waste treatment scheme finally selected. However, these generally are not expected to be a problem for a biomass-to-ethanol facility.

4. PROPOSED SCHEME

Figure 1 shows the flow schematic for processing the stillage and treatment of the resultant waste stream. The input to the overall WWT system includes waste streams from ion exchange (hydrolyzate conditioning, i.e., removal of acetic acid), pretreatment flash vents, condensate from the evaporator, and wastewaters from boiler and cooling tower blowdown.

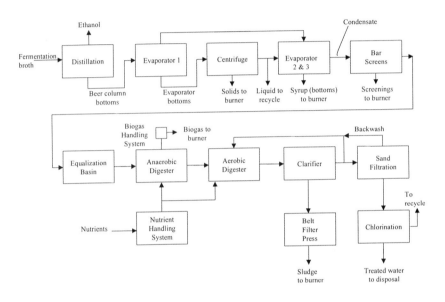

Figure 1. Schematic of stillage processing and overall WWT system.

4.1. Process description

The fermentation broth is fed to the beer column; ethanol is recovered from the top, and the distillation bottoms are sent to the first effect evaporator. The evaporator vapors enter the second effect evaporator and the bottoms (90% of the water and all of the solids, insoluble and soluble) are processed in a centrifuge. Having the first effect evaporator upstream reduces the centrifuge size. The solids-bearing stream from the centrifuge is conveyed to the burner and the centrate is returned to the second effect evaporator. The evaporator bottoms (syrup) from the third effect evaporator is also sent to the burner and burned. A small portion of the condensate is sent to the WWT system, and the rest is recycled to the process.

The feed to the WWT system passes over bar screens, and any large solids are removed. This feed is stored in an equalization basin. In the anaerobic digestor, 90% of the COD is converted via biomethanation to CH_4 and CO_2. Biomethanation is the production of biogas by certain microorganisms using organic (carbonaceous) substances under anaerobic conditions. Part of the energy present in dissolved organic waste is conserved as methane, which is burned in the boiler as fuel. The effluent from this step enters the aerobic digestor, an aerated lagoon. Here 90% of the BOD is converted to cell biomass and CO_2. The effluent from this step enters the clarifier. The settled solids are pressed in a belt filter press, and the resultant sludge is sent to the burner and the liquid is recycled to the aerated lagoon. The clarified liquid from the clarifier is treated by sand filtration and chlorination; part of the treated water is recycled to the process and the rest is sent to disposal.

4.2. Advantages of proposed scheme

The cost of treating the total volume of wastewater from the beer column bottoms is prohibitive. Therefore, evaporation and centrifugation of the stillage was employed to concentrate the waste into smaller streams and maximize recycle. The evaporation step was integrated with the waste treatment process, and an integrated water recycle design was developed. Evaporation of stillage/centrate (the second and third evaporation stages are downstream of centrifugation) is done using heat integration with the distillation section of the process. The resultant WWT system then has significantly reduced flow, making on-site treatment easily achievable with conventional treatment systems using anaerobic treatment followed by aerobic treatment. The resultant wastewater streams are classified as "high strength biologically treatable" and can be economically treated in either package plants of standard designs or in small custom plants.

4.3. Fate of solid-rich streams

The concentrate or syrup from the evaporator can be sent to the boiler directly or to the anaerobic digester. It was assumed that the syrup could be sprayed or mixed with the lignin cake and sent to the boiler as fuel in the first option. If the evaporators use all of the waste heat in the distillation section the syrup is predicted to contain 79% water. The heating value of this stream is positive and contributes 8% of the boiler heat load.

The second option would be to send the syrup to the anaerobic digester. The digester and all downstream equipment, including the aerobic unit, becomes larger, but this is somewhat offset by the production of more methane gas in the anaerobic digester. Various configurations of the anaerobic/aerobic units were considered and judged based on simplicity (ease of operation and maintenance) and cost. The best option was to burn the syrup at approximately 7.7% solids with the ligneous residue in the boiler. The recommended process configuration includes a burner (fluidized bed combustor) to combust suspended solids produced by the centrifuge and can accept the evaporator syrup, screenings from bar screens, and sludge from the aerobic digester.

4.4. Treatment of anaerobic digester off-gas

The wastewater contains dissolved sulfates which will convert to hydrogen sulfide in the digester. The resultant hydrogen sulfide concentration in the anaerobic digester off-gas is approximately 1800 ppm (wt.); furthermore, the boiler stack will emit approximately 1.14 tons/day of sulfur to the atmosphere (2.28 tons/day of SO_2). This emission rate may not be permitted in the United States. EPA regulations are site-specific but a useable rule of thumb is less than 100 tons/year of SO_2 emissions is allowable.

The fluidized boiler will include limestone addition for other sulfurous components in the ligneous fuel. This treatment in the combustion chamber is more economical than any options for removal of sulfur in the WWT area.

5. SYSTEM DESIGN AND COST ESTIMATES

5.1. COD and BOD calculations

COD is a measure of the amount of oxygen required to convert all the carbon in a specific compound to CO_2. For example, the COD of glucose is 1.07 kg oxygen/kg compound and is calculated based on complete oxidation of glucose. COD values were calculated based on the concentrations of cellobiose, glucose, galactose, mannose, xylose, arabinose, ethanol, glycerol, xylitol, acetic acid, lactic acid, and succinic acid. The calculated COD values compared well with the actual laboratory data from simulated waste samples.

Based on the projected wastewater compositions and the proposed treatment system, the estimated BOD/COD ratio is 0.7 for the influent to anaerobic digestion, 0.2 for the influent to aerobic treatment, and 0.1 for the system effluent. Since BOD is based on a laboratory test and cannot be specifically predicted *a priori*, the above ratios are estimates based on authors' experience with other WWT systems and published COD and BOD ranges for similar systems (Tchobanoglous and Burton, 1991; Perry and Green, 1998). Since ammonia is not converted in anaerobic digestion, the contribution of the reduced nitrogen compounds is not included in the overall COD calculation. These compounds (ammonium acetate and ammonium sulfate generated during ion exchange for hydrolyzate detoxification), however, are accounted for in the aerobic step.

5.2. Design basis

Using the Aspen® Plus™ model, the following estimates were developed: total flow to WWT system = 173,980 kg/h and COD = 30,000 mg/L. In the model, a methane yield of 350 liters/kg COD converted (0.229 kg/kg at 25°C) was used, the methane concentration being 75%. In the anaerobic treatment, 90% reduction on a COD basis was assumed with equimolar production of CH_4 and CO_2. (In conventional biomethanation systems, biogas will range from 55% to 70% CH_4, the remainder being CO_2. Hence, some of the CO_2 remains dissolved.) In the aerobic treatment, 90% reduction on a BOD basis was assumed, 60% being converted to CO_2 and 30% to cell mass.

5.3. WWT systems costs

Costs for "off-the-shelf" anaerobic and aerobic units were developed based on the above design data. Cost estimates for package-type anaerobic units were provided by Phoenix Biosystems of Colwich, Kansas and those for an aerobic units—of the sequential cell, aerated, fabric lined earthen pond type—were supplied by Goble Sampson Associates, Englewood, Colorado. Other costs were developed using the Icarus® software. These cost estimates are summarized in Table 1. Added costs at 42% of total project investment (TPI) were used to calculate the TPI of $14.75 million for the WWT system. Using a capital recovery factor of 0.177, the annualized capital cost is $2.6 million/year, which is equivalent to $0.06/gal or $0.016/L of ethanol.

The operating costs for the WWT system were estimated as follows (¢/gal ethanol): chemicals–0.9, electricity–1.4, labor–0.3, overhead and maintenance–0.12, and insurance and taxes–0.09. This yields a total WWT cost of 2.35¢/L ethanol. Hence, the WWT costs represent a non-trivial portion—about 6.6%—of the total ethanol cost.

Table 1
Capital cost estimate for WWT system (all costs in 1998 US$)

Equipment description	Installed cost	Equipment Description	Installed cost
Equalization basin	478,900	Aerobic sludge belt filter press	1,170,000
Anaerobic digestor	3,665,600	Aerobic digestor aerator	700,200
Aerobic digestor	588,700	Screw conveyors	51,700
Nutrient feed system	81,000	Feed cooler	259,800
Biogas handling system	35,000	Agitators	186,000
Clarifier	328,400	Pumps	348,700
Bar screen	138,000	Recycle water tank	20,400
Beer columns bottoms centrifuge	2,372,000	Miscellaneous	4,200
Total capital cost for WWT system = 10,430,000			

6. CONCLUSIONS

The WWT system designed for bioethanol process included anaerobic digestion followed by aerobic treatment. This was considered as best strategy and employed centrifugation of the stillage with evaporation to minimize wastewater and optimize water recycling. For a 2000 dry tpd bioethanol plant, the WWT capital cost was estimated to be $10.4 million. The specific WWT cost is $0.0235/L of ethanol, a material contribution to the total ethanol cost. The WWT strategy and cost estimates generated in this paper are deemed to be generally applicable to similar bioethanol plants and can be useful as a stepping stone.

7. ACKNOWLEDGMENTS

This work was supported by the Biochemical Conversion Element of the Biofuels Energy Systems Program of the US Department of Energy.

REFERENCES

1. Merrick & Company (1998). Wastewater Treatment Options for the Biomass-to-Ethanol Process. NREL Subcontract AXE-8-18020-01, Final Report, National Renewable Energy Laboratory, Golden, Colorado.
2. Wooley, R., M. Ruth, J. Sheehan, H. Majdeski, and A. Galvez (1999). Lignocellulosic Biomass to Ethanol Process Design and Economics, Utilizing Co-current Dilute Acid Prehydrolysis and Enzymatic Hydrolysis: Current and Futuristic Scenarios. NREL Report TP-580-26157, National Renewable Energy Laboratory, Golden, Colorado.
3. Tchobanoglous, G. and F.L. Burton (1991). Wastewater Engineering Treatment, Disposal, and Reuse, 3rd edition, McGraw Hill, New York.
4. Perry, R.H. and D.W. Green (1998). Chemical Engineers' Handbook, 7th edition, McGraw-Hill, New York, pp. 25–62.

PREFEASIBILITY STUDY FOR ESTABLISHING A CENTRALIZED ANAEROBIC DIGESTER IN ADAMS COUNTY, PENNSYLVANIA[*]

P. Lusk[a] and R. Mattocks[b]

[a]Resource Development Associates, 240 Ninth Street, NE, Washington, DC 20002 USA,
[b]Environomics, Suite 17A, 5700 Arlington Avenue, Riverdale, New York 10471 USA

Growth and concentration of population and the agribusiness industry in Adams County, Pennsylvania, creates both problems and opportunities for the disposal of the area's organic residues. Landfill space and transportation costs are increasingly burdensome, and pollutants from decomposing organic residues have the potential to contaminate the environment. Many farm-based digesters have already been deployed in Pennsylvania to process animal manures. However, economies of scale associated with the technology preclude its economic use on smaller farms. Centralized digesters might be used instead of on-farm systems. Centralized digestion systems also allow for the possibility of codigesting manures and the residues from food processing industries, which have their own disposal issues. A centralized anaerobic digestion system could potentially provide a number of economic, environmental, and social benefits. Such a facility in Adams County could convert much of these residues into useful co-products while reducing demand for valuable landfill space and lowering transportation costs. The economic evaluations presented here represent a series of plausible scenarios that demonstrate that the anaerobic digestion of organic residues can be a cost-effective technology. Moreover, the external benefits of such a project could be more important in the long term.

1. INTRODUCTION

In addition to being blessed with its people and outstanding agricultural productivity, Adams County, Pennsylvania, also generates a substantial amount of organic residuum. These materials encompass municipal solid waste, biosolids and septage, through a range of types of animal manures, and many grades of organic industrial residues.

---

[*]Originally prepared for the Adams County Office of County Commissioners and sponsored by the Northeast Regional Biomass Program, Adams County Office of Solid Waste Recycling, and the Adams County Conservation District.

Principal pollutants from decomposing manure and other organic residues are methane, ammonia, excess nutrients and pathogens, and an increased biochemical oxygen demand (BOD). The major pollution problems associated with manure are surface and ground water contamination, and surface air pollution due to odors, dust, volatile organic acids, and ammonia. There is also concern about the contribution of methane emissions to global climate change. Consequently, organic residue management systems such as anaerobic digesters that prevent pollution and produce energy are becoming increasingly attractive.

Many farm-based digesters have already been deployed in Pennsylvania to process animal manures. However, there are economies of scale associated with the technology that preclude its economic use on smaller farms. Centralized digesters could potentially be used instead of on-farm systems. These types of digestion systems would also allow for the possibility of co-digesting manures and the residues from food processing industries which have their own unique disposal issues. Establishing a centralized anaerobic digestion system could potentially provide a number of economic, environmental, and social benefits.

2. RESOURCE AVAILABILITY

A first component to understanding how a centralized anaerobic digestion system could provide some benefit to Adams County is to develop an understanding of the types of organic residues potentially available. One study objective was to carry out a preliminary assessment of the types and amounts of organic residues generated in the county. A total of nearly 70 telephone interviews and physical site visits were conducted to assess sources of organic materials. These sources included animal manures, fertilizer fabrication, food processing, food packers, food service, mills and elevators, residential food, sewage, and slaughterhouses.

The overall volume of organic residues generated in Adams County is quite large, more than 2340 tons per day. However, much of the total volume of organic residues is deposited in the field by livestock is very dilute or is already sold or used. Only about ten percent of the available organic residues, some 230 tons per day, are collected and disposed of at a cost. Still, there are indications that some of the generators of dairy manure and organic industrial residues are open to considering alternative-use strategies, particularly if it could reduce their operating expenses and provide long-term disposal options. Diverting only 15 percent of the amount of organic residues that are now sold or used in the county would add an additional 200 tons per day to the pool of available feedstocks. Other likely sources of organic residues include septage and biosolids.

3. ENVIRONMENTAL BENEFITS

Wise use of the solid and liquid digestate residues from the anaerobic digestion process can have very beneficial environmental impacts. The amount, quality and nature of these products will depend on the quality of the feedstock, the method of digestion and the extent of the post-treatment refining processes. The adoption rate of improved manure management practices will improve if the new practice results in reduced operating costs or, ideally, new profit centers.

Odor from animal manures has become regarded as a nuisance in almost every state in the union. Manure odors are caused principally by intermediate metabolites of anaerobic decomposition. Anaerobic digestion by its very nature converts odor-causing materials in organic matter to methane and carbon dioxide, which are odorless. Generally, that which smells offensive is the prime food source for methane-producing bacteria. Odor reduction using anaerobic digesters can be a very cost-effective alternative when compared to aeration, chemicals, or enzyme treatments. Odor control is probably the main reason livestock farmers install an anaerobic digester, especially those built since 1984.

Animal manure represents a potential source of organisms that may be pathogenic to humans. Therefore, the use of manure must be closely managed to limit the potential for pathogen contamination. The potential to operate digesters at temperatures above 120°F makes the anaerobic digestion process particularly interesting for hygenization. Digesters can be designed to operate entirely at that temperature, or as a post-treatment step following operation at a lower temperature. Along with temperature, the anaerobic chemical environment multiplies the sanitation effect.

4. GENERAL PROCESS DESCRIPTION

A process description of an anaerobic digestion facility in Adams County was developed using a hypothetical 150 ton-per-day system. The initial feedstock focus was on the accessibility of the substantial amount of dairy cow manures and organic industrial residues generated in the county. Other forms of organic residues suitable to anaerobic digestion, such as caged layer or hog manures, were not evaluated. Hog manure was not evaluated because its total solids concentration was too dilute to allow economical transport. Caged layer manure was not evaluated because the amount of dilution required to utilize this material as a feedstock violates fundamental manure management practices.

Digestion products are a medium-Btu biogas and a slurry called digestate. The biogas contains approximately 60-70 percent methane and is water saturated. The balance of the biogas mixture is carbon dioxide, and some parts per million of hydrogen sulfide. The digestate consists of undigested solids, cell-mass, soluble nutrients, other inert materials, and water. All digestate contains a recoverable solid fiber with physical attributes similar to those of a moist peat moss. This product may be used as a soil improver or used as a constituent in potting soils. After the fiber is removed, the main digestion product is a liquid organic substance commonly called "filtrate." Filtrate from manures can be spread directly onto farmland for its nutrient value. Filtrate can also be further processed to provide a liquid material commonly called "centrate" and solid product called "cake."

One aspect of the digestion system worth noting was the specification for a storage tank sized to hold six-months worth of liquid. A storage facility such as this could be a tremendous boon to livestock farmers who participate in the project. Many of them now spread fresh manure on a daily basis, even when climatic conditions may not be favorable. By providing storage, land application can be conducted at favorable agronomic times. Furthermore, storage will allow for a more efficient use of farm labor by doing away with daily spreading. This may save each farm between one-half to one full-time equivalent position. Local food processors are also doing daily hauling of their organic residues, and they may also benefit from the use of the centrate storage tank.

While there are a number of financing possibilities, the economic evaluations were based upon the assumption of private investor ownership because this represents the most conservative outcome. Other ownership possibilities include, but are not limited to, straight county ownership and operation or county ownership with a private company under contract with facility operation. If the county were to fund a centralized anaerobic digestion system its cost of capital would be much lower than any private investor.

5. ECONOMIC EVALUATIONS

A number of *pro forma* economic evaluations for the 150 ton-per-day anaerobic digestion system were conducted using three different feedstock combinations: dairy cow manure only, codigestion of dairy cow manure and organic industrial residues, and organic industrial residue only. Organic industrial residues were defined as a generic mixed vegetable residue from a packing plant. It was assumed that industrial residue generators pay the facility a tipping fee for disposal of their materials.

Four economic evaluations were conducted to establish the parameters most likely to make each project scenario cost-effective. These included a low-efficiency engine with the solids processing plant recovering cake, a high-efficiency engine with the solids processing plant recovering cake, a low-efficiency engine without cake recovery, and a high-efficiency engine without cake recovery. These options were also tested to see whether or not the additional step of cleaning and compressing the biogas to generate electricity on-peak would add economic value to the project.

The digestion system produces biogas that can be converted into electric power, electric energy, and recovered thermal energy. The other main co-product having an economic value at this time is the fiber. After separation in the solids processing plant, it will be combined with appropriate admixtures and stored in windrows until maturation. After maturation, the composted fiber product is hauled off-site for bulk sales. Some consideration was also given to recover cake if it provided an economic advantage to a project scenario. If not, then the centrifuge, cake drying and bagging equipment were not costed into the particular analysis. There are a number of co-products (carbon dioxide gas, grit, and liquid centrate) that had no assumed value in the analyses.

For the dairy-cow-only scenario, it was determined that the greatest profitability would be derived by using a high-efficiency engine, operating that engine off-peak, and using the solids recovery plant to recover cake. The project was costed out to a total of $2,776,976. The value of the digester was established by the amount of salable co-products generated which included electric energy, fiber product, and cake. Total revenue for the first year of the project summed to $964,808. The first year expenses associated with the digester system included thermal energy, manure hauling, fiber admixture, and engine, digester and solids recovery operation and maintenance (O&M). Total expenses during the first year of the project summed to $784,492. First year plant income on the *pro forma* was defined as the sum of depreciation, annual revenue, and annual expenses. The project had a first year income of a positive $1,442. Unfortunately for this scenario, the project had many years in its assumed life-cycle with a negative after-tax income. Given the assumptions used, the conclusion was that the project did not have investment merit. Additional analyses were conducted to determine what it would take to make this scenario cost-effective. Whether these conditions could be met in the real world is a matter for additional discussion.

For the dairy cow and organic industrial residue (OIR) scenario, it was determined that the greatest profitability would be derived by using a low-efficiency engine, operating that engine off-peak, and not using the solids recovery plant to recover cake. The project was costed out to a total of $2,353,124. The value of the digester was established by the amount of salable co-products generated which included electric energy, electric power, thermal energy, and fiber product. Total revenue for the first year of the project summed to $1,225,026. The first year expenses associated with the digester system included manure hauling, fiber admixture, hydrogen sulfide removal, and engine, digester and solids recovery O&M. Total expenses during the first year of the project summed to $823,020. First year plant income on the *pro forma* was defined as the sum of depreciation, annual revenue, and annual expenses. The project had a first year income of a $402,007. This scenario yielded a positive after tax income in every year of the project's life. Given the assumptions used, the conclusion was that the project does have investment merit. Additional analyses were conducted to determine what it would take to make this scenario not cost-effective.

In the OIR only scenario, it was determined that the greatest profitability would be derived by using a low-efficiency engine, operating that engine off-peak, and not using the solids recovery plant to recover cake. The project was costed out to a total of $2,398,387. The value of the digester was established by the amount of salable co-products generated that included electric energy, electric power, thermal energy, and fiber product. Total revenue for the first year of the project summed to $1,506,520. The first year expenses associated with the digester system included effluent disposal, fiber admixture, hydrogen sulfide removal, and engine, digester and solids recovery O&M. Total expenses during the first year of the project summed to $1,220,239. First year plant income on the *pro forma* was defined as the sum of depreciation, annual revenue, and annual expenses. The project had a first year income of a $268,280. This scenario also yielded a positive after tax income in every year of the project's life. Given the assumptions, the conclusion was that the project does have investment merit. Analyses were also conducted to determine what it would take to make the scenario not cost-effective.

One of the objectives of the economic evaluation section was to illustrate the importance of maximizing co-product utilization and other offsets made available by employing anaerobic digestion technology. Two of the scenarios described above achieved true economic merit. There are likely many circumstances not considered that could provide an opposite conclusion. However, while there are a number of conditions and qualifiers that can be incorporated in any economic model, it is also apparent that the true social cost of manure and OIR disposal is not really known.

Table 1
Project scenario economic evaluation summary

	Dairy Cow	Dairy Cow & OIR	OIR
Capital cost	$2,776,976	$2,353,124	$2,398,387
1st year revenues	$964,808	$1,225,026	$1,506,520
1st year expenses, depreciation & taxes	$963,366	$823,020	$1,220,239
1st year net income	$1,442	$402,007	$286,280
Net present value (15% discount rate)	($1,265,327)	$1,897,764	$804,536
Internal rate of return	0.1%	27.6%	20.3%
Cost-effective	No	Yes	Yes

6. CONCLUSIONS

Economics is a science too often criticized for "knowing the price of everything and the value of nothing." The analyses presented here do not provide a quantifiable price impact for some of the more subjective value advantages that can result from adopting anaerobic digesters. Anaerobic digestion can play a role in treating effluents from livestock and industrial operations. After digestion, nutrients are stabilized and pathogens can be reduced by up to 99.99 percent. An additional unquantified externality is the ability of anaerobic digesters to help control odor and flies. Many farms and businesses use a digester specifically installed to control odor. Lastly, conversion of biogas into less odious carbon dioxide can be accomplished through combustion in an engine-generator. Unrecovered methane in biogas produced by the inevitable decomposition of manures and other organic residues is a suspected agent of global climate change.

Since Adams County derives a substantial part of its economy from exportation of agricultural and horticultural commodities, installing an anaerobic digester to process value-added organic solids and liquids could be a tremendous benefit for economic development and balance of trade purposes. The potential value that can be derived by reducing the need to purchase nutrients and soil conditioners could be a significant factor in the decision to encourage the deployment of anaerobic digestion technology. Given that county has a stated public policy to strengthen the agricultural economy, local government could conceivably undertake an evolutionary change in how the manures and OIR generated in the county are treated.

Considering the encouraging results of this prefeasibility study, the further evaluation of the technical and economic merit of centralized anaerobic digestion technologies is warranted. Additional work has been done to bring together a potential system developer and an AD system provider to assess the merit of privately constructing the dairy cow manure and OIR scenario system. These negotiations have revealed a number of new insights into the proposed systems costs and benefits.

PRETREATMENT OF DOMESTIC SEWAGE BY THE MODIFICATION OF AN EXISTING IMHOFF TANK

Santino Di Berardino,[a] Sara Antunes,[a] Michaela Bergs,[a] and Ana Alegria[b]

[a]Departamento de Energias Renováveis, INETI-ITE, Est. do Paço do Lumiar, 1699 Lisboa Codex
[b]Serviços Municipalizados de Sintra, Av. dos Movimentos das Forças Armadas, n° 16, 2710 Sintra

This paper reports the results obtained from an experimental sewage treatment plant to be used as a prototype for domestic wastewater treatment. An existing Imhoff tank was modified; the tank will work as an anaerobic hybrid filter reactor to improve the efficiency and capacity of treatment. The results will be used to recover and upgrade other similar tanks, as well as for the pre-treatment of other wastewaters.

The work program characterized daily and hourly averaged wastewater samples and continuously measured pH, flow-rate, air, and wastewater temperatures. Laboratory continuous experiments were carried out using an anaerobic filter and an UASB reactor. Biomass characterisation and measurements of activity of the Imhoff tank and of laboratory systems were compared.

Laboratory results generally are favourable to the anaerobic filter process, indicating good COD reduction (70-90%) at low hydraulic retention time (1–2 days) and very high sludge activity. The full-scale modified unit is under start-up. The two existing sludge-drying beds have been modified into a submerged macrofite pond and stabilisation pond followed by a polishing plastic filter, to study the possibility of obtaining a high-quality effluent.

1. INTRODUCTION

The use of anaerobic processes for domestic wastewater treatment is very old (since 1896). It is typified by the popular septic tank, which even today is commonly used for sewage treatment in very small communities. In this tank, sewage flows into a sludge bed, losing suspended solids, and undergoes moderate anaerobic degradation. Travis in 1903 and Imhoff in 1904 (McCarty, 1981) separated sedimentation and digestion into distinct compartments, optimising each function. The sewage does not receive any biological degradation and the settled sludge is stabilised by anaerobic methanogenic process. The efficiency of these tanks is greatest in removing suspended solids (60-70%) and relatively low in terms of BOD (30-40%). As a consequence, the sludge digester gained popularity as a separate compartment or tank and it is widely used today. Modern sludge digestion systems are heated, stirred, and processed in a series of two tanks.

From the perspective of sewage treatment, interest focused on aerobic biologic technology to increase BOD removal efficiency to more than 95%.

In 1890 Scott-Moncrieff (McCarty, 1981) placed a bed of stone over an empty space to increase efficiency and constructed the first up-flow anaerobic filter. Despite this promising experiment, direct anaerobic treatment of sewage was not considered realistic. Some 67 years later, Coulter et al. (1957) developed the so-called contact process, which obtained encouraging removal efficiencies (BOD 50–65%; SST 84%). Hemens et al. (1962) inverted the flow of a combined settling-digesting tank, called a "clarigester," transforming it into an anaerobic activated-sludge plant. The discovery of anaerobic filter (Young and McCarty, 1969) and of up-flow sludge blanket reactors (Lettinga, 1971), at the same time as the first world energy crisis in the 1970s, increased interest in this low-energy-demand process and created a new generation of efficient and stable reactors. They are an interesting alternative to aerobic processes for both concentrated and diluted effluents.

The main advantages of anaerobic treatment as compared with conventional system lies in the small, or zero, energy consumption, the production of energetic biogas that can eventually be used, small sludge production and its ability to handle high organic loads. At present, the main limitation of anaerobic systems is their inability to achieve the limits imposed by environmental regulations on wastewater discharges. For this reason, the anaerobic process must be complemented by a final stage of treatment, which can be aerobic.

2. PROJECT DESCRIPTION

2.1. Project objectives

In Portugal some hundreds of Imhoff tanks are used to treat the sewage from small communities. Most of these plants are working at the limits of their capacity. The objective of this project, submitted to an European Life program and approved, was to study one specific Imhoff tank that needed to be upgraded or replaced, and to increase its acceptable load and treatment efficiency. We want to develop a simple and effective anaerobic technology at ambient temperature, adapted to local environmental conditions, which can successfully replace conventional systems and bring substantial advantages in terms of costs, energy consumption and O & M.

This technology can also be used for pre-treatment of facultative or aerated ponds, activated sludge and extended aeration plants, saving on land area and energy. At a national scale, wide spread implementation of anaerobic processes to treat wastes from small and medium sized communities is strategic, allowing substantial electrical energy savings and producing renewable energy.

2.2. Description of the modifications

The experimental plant is situated in Nafarros, a small place near Lisbon, where the "Serviços Municipalizados de Sintra" have in operation one Imhoff tank in reasonably good condition. Its treatment capacity and efficiency are already insufficient. The plant is an excellent candidate for upgrading operations: the plant is small, it needs to be increased, and it does not involve significant economic risks. Also, it is not far away and is easily accessible, which facilitates monitoring.

The modifications to transform the Imhoff tank into an anaerobic hybrid filter were simple and economic. Sewage enters the digestion tank through a vertical PVC pipe with a bottom-end diffuser, which provides mixing and minimizes channelling (Figure 1). Then, it up-flows into the settling compartments where a PVC packing medium improves sedimentation. The treated effluent leaves through the same weir of the original system. The transformations did not require emptying the tank, although the liquid level was lowered 30 cm.

The anaerobic process does not remove many nutrients. So the two existing drying beds were modified into macrofite and facultative pond pilot plants, to study the possibility of obtaining a high-quality effluent. A coarse plastic filter provides additional treatment to the effluent from the facultative pond.

The produced biogas will be treated, stored in a flexible holding tank and burned in the small motor-generator, which furnish electricity to the plant.

The expected investment for the plant modification, for the studies and for dissemination of results was only $16,000 US. Existing plant infrastructures minimized the costs and a grant of 50% was supported by LIFE program. Sintra municipality paid only a low cost to upgrade the plant and gain knowledge about this technology.

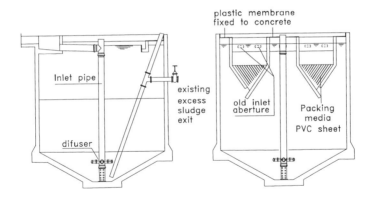

Figure 1. Imhoff tank modifications.

3. EFFLUENT CHARACTERISATION

An ultrasonic flow meter, two temperature sensors (for air and wastewater) and one pH probe connected to a Logger apparatus were used to characterize waste water. Two automatic samplers provided averaged samples proportional to wastewater flow rate and time-programmed samples of the Imhoff tank effluent.

Figure 2 depicts the averaged values of sewage flowrate, pH and temperature, obtained from continuous measurements during a daily cycle. The wastewater flow deviates from typical shape. In fact, minimum flow rate does not occur during early morning, owing to the discharge into the sewer of ground water withdrawn during the low-cost electric energy period at night. The average sewage temperature was about 18°C, varying in a small range, which is favourable to the anaerobic process. The pH values were in the normal range.

Table 1 gives the averaged values of all the physico-chemical parameters analysed during weekly monitoring. The effluent has a medium organic content, substantial suspended solids and good nutrients concentration.

The physico-chemical composition corresponds to a medium-strong sewage, as defined by Tchobanoglous (1991). The apparent averaged efficiency of organic matter removal (56-66%, in terms of COD) was higher than expected, possibly owing to the influence, in composite samples, of the early morning underground water flow, with no organic content (Figure 3).

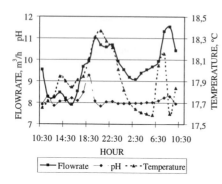

Figure 2. Hourly sewage variations of flow rate, pH and temperature.

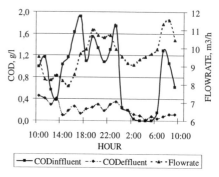

Figure 3. Hourly average variations of flow rate, influent, and effluent COD.

Table 1
Averaged parameters of sewage and Imhoff tank effluent

Parameters	Sewage	Effluent
pH	7.6	7.7
COD (mg.l^{-1})	873	256
BOD$_5$ (mg.l^{-1})	233	138
TS (mg.Kg^{-1})	998	722
VS (mg.Kg^{-1})	483	281
TSS (mg.l^{-1})	379	101
VSS (mg.l^{-1})	299	77
N$_{total}$ (mg.l^{-1})	29	20
NH$_4^+$ (mg.l^{-1})	14	18
P (mg.l^{-1})	11	7

4. CONTINUOUS ANAEROBIC TREATMENT EXPERIMENTS

A continuous laboratory experimental programme was carried out using an Anaerobic Hybrid Filter (ID 12.5 cm, 96 cm height, 13.5 litres total volume), and an UASB reactor (16.5 l total volume, 116 cm height), to simulate the potential improvements provided by the Imhoff modification. The reactors were inoculated with sludge from the Imhoff tank of Nafarros (COD = 68.89 g/l; pH = 6.97; ST = 153.5 g/Kg; SSV = 34.9 g/l). The experiments ran at ambient temperature to simulate real conditions. The applied organic load ranged from 0.1 to 7.1 Kg COD.m^{-3}.day^{-1}, corresponding to hydraulic retention time from 8 days to 8 h.

Table 2 presents averaged effluents concentrations from the anaerobic reactors. Efficiency of organic matter removal, in terms of COD and BOD, improved significantly, compared with the full scale. Anaerobic Filter performance was slightly better than of the UASB reactor.

Table 2
Averaged parameters analysed during the continuous anaerobic experiments

Parameters	Sewage	Effluent		Efficiency %	
		Filter	UASB	Filter	UASB
PH	7.42	7.59	7.74		
COD /mg.l^{-1}	854.2	167.6	171.5	80.4	79.9
COD$_{sol}$ /mg.l^{-1}	335.6	97.40	81.34	70.97	75.76
BOD$_5$ /mg.l^{-1}	328	100	106	69.5	67.7
TSS /mg.l^{-1}	408	50	67	87.87	83.58
VSS /mg.l^{-1}	267	39	49	85.22	81.65
HAc /mg.l^{-1}	41	38	56		
N$_{total}$ /mg.l^{-1}	35.11	31.33	31.40		
NH$_4^+$ /mg.l^{-1}	16.17	21.91	20.51		
Phosphorous /mg.l^{-1}	16.05	11.93	11.57		
Alkalinity /mg.l^{-1}	357	381.5	423		

Specific biogas production was in the range 0.2-0.3 m^3/kg CQOr, according to the organic load and lower than the theoretic value (0.35 m^3/Kg CQOr [Tchobanoglous, 1991]), due to the influence of dissolved methane in the effluent. The Anaerobic Filter reactor produced more gas than the UASB reactor. The averaged methane content of biogas from both reactors was about 84.9% (UASB) and 81.7% (Filter). Excluding the values obtained during an overload small period, VFA content in the effluents was always completely consumed and the pH slightly alkaline.

Owing to operational problem, only the first results of the modified full-scale reactor are available now. They strictly agree with the values from laboratory reactors.

5. BIOLOGICAL ACTIVITY TESTS

Sludge methanogenic activity of the Imhoff and laboratory reactors were compared to verify the changes in bacteria population and metabolism. The tests were performed in 250 ml flasks at 35°C, according to Soto et al. (1993), using five solutions of different concentrations of volatile fatty acids (0.5 g/l, 1.0 g/l, 2.0 g/l, 4.0 g/l of acetic acid and a mixture of 2.0 g/l acetic acid + 0.5 g/l propionic acid + 0.5 g/l n-butyric acid) fed to a fixed amount of sludge.

The methanogenic kinetics parameters were determined with Monod and Andrews equations, using the activities determined for each VFA concentration. The obtained results, presented in Table 3, indicate greater sludge activity in both laboratory reactors compared with Imhoff tank sludge. In the range of concentrations studied they do not show any VFA inhibition. Bacterial flocculation increased the sludge SSV concentration in the laboratory reactors and is the main agent responsible for the improvements. Nevertheless, the Hybrid Filter sludge gave slightly better anaerobic performance than UASB sludge. It combined a higher maximum methanogenic activity with an inferior saturation constant.

Table 3
Kinetic parameters for methanogenesis

Kinetic parameters	UASB	Filter	Imhoff
Maximum activity (gCOD$_{CH4}$.g^{-1}VSS.d^{-1})	9.337	9.634	5.555
Inhibition constant (l.g^{-1})	---	---	0.470
Saturation constant (g.l^{-1})	5.922	5.625	2.923

6. CONCLUSIONS

The modifications of the Imhoff tank of Sintra-Nafarros into anaerobic filter and UASB reactor, can substantially improve Imhoff tank performance, obtain high removal efficiencies (79.9% in UASB and 80.4% in Hybrid Filter) and produce good quality biogas (84.9% in UASB and 81.7% Hybrid Filter). The digesters also react well to overload situations. The Hybrid Filter reactor had a slightly better removal efficiency.

REFERENCES

Coulter, J.B., Soneda, S., and Ettinger, M.B. (1957). Anaerobic contact process for sewage disposal, Sewage and Industrial Wastes, 29, No. 4.

McCarty, P.L. (1981). One hundred years of anaerobic treatment, in Anaerobic Digestion, Elsevier Biomedical Press, pp. 3–22.

Soto, M., Mendez, R., and Lema, J.M. (1993). Methanogenic and non-methanogenic activity tests. Theoretical basis and experimental setup, Water Res., 27, pp. 1361–1376.

Tchobanoglous, G. (1991). Wastewater Engineering - Treatment, Disposal, Reuse, ed. by Metcalf & Eddy, Inc., 3rd Edition, McGraw-Hill Publishing.

BIOGAS PRODUCTION FROM WASTES IN PORTUGAL: PRESENT SITUATION AND PERSPECTIVES

Santino Di Berardino

Departamento de Energias Renováveis, INETI, Az. dos Lameiros, 1699 LISBOA CODEX

This work summarises Portuguese use of anaerobic digestion for waste treatment and biogas production. The process is usually applied to treat the wastes arising from medium-large and large agricultural factories, because it is difficult to spread the technology to smaller units or to implement centralised waste treatment plants. Only about 60 plants are actually in operation.

The lack of interest in developing more anaerobic digestion plants can be attributed to Portuguese environmental regulations that still permit producers to discharge into water-courses and thus promote the use of ponds as treatment systems. The existing regulations are difficult to remove owing to the opposition of producers' associations.

In addition, the concept of a centralised anaerobic system has not been accepted in Portugal and its effectiveness needs to be proved. Such technology cannot be transferred directly from other Europeans countries; it must be adapted to local constraints. In Portugal those constraints are great dilution of the wastes, unfavourable environmental regulations, lack of markets for treated solid wastes and the owners' mentality.

However, the one anaerobic centralised plant in Portugal gives some favourable information about future applications of the system. Favourable environmental and energy regulations can rapidly accelerate dissemination of the process.

1. INTRODUCTION

The anaerobic digestion is a low-energy treatment process, which converts organic matter into biogas and generates a stabilised residue that can be used as fertiliser. The biogas can be used to produce electric or thermal energy having an interesting economic value. The widespread application of anaerobic digestion in wastewater treatment is strategically important. This system can significantly contribute to sustainable growth: recycling fertilising sludge in agriculture reduces the input and flux of new matter in the nutrient cycle; biogas is a renewable energy source that reduces consumption of fossil fuels. Controlled production and use of biogas does not let methane escape to atmosphere (as in the case with residues not stabilised) thus helping to lower concentrations of one of the most active greenhouse gases.

Anaerobic digestion and gas utilisation systems have some technological complexity that makes them difficult to operate; there is a minimum size below which the system is not economically attractive and feasible. This size factor promotes use of a centralised plant that treats concentrated residues generated in many small- to medium-sized agricultural plants and transported to the central plant. The concept of centralised waste plant was developed in the 1980s, and it proved feasible in several examples in some European countries.

2. PORTUGUESE FARM RESIDUES

The process relies on the availability of organic matter. Portuguese agricultural wastes and effluent from food and drink industries produce substantial organic matter; farm wastes are the main source. According to 1989 data from a national association, livestock in Portugal (Table 1) produced greater than 3400 t/day of organic matter (Table 2).

Table 1
Number of animals existing in Portugal

Animal	Number (units)
Bovine	1 401 340
Sheep	2 921 113
Caprine	719 755
Swine	2 423 957
Horses	36 246
Donkey and mules	114 700
Turkeys	1 168 243
Chicken	28 320 020

Table 2
Organic matter from the main farm wastes

Residue	Organic Matter (t/d)
Pig manure	750
Cattle manure	2 300
Chicken manure	400
TOTAL	3 450

The energetic potential of these residues is considerable, theoretically as much as about 2 000 MWh/d by the application of anaerobic digestion. This number increases again if the process will be applied to other available organic effluents. In addition, process energy is saved by replacing aerobic processes.

Most Portuguese pig farms are very small and are family owned (Figure 1).

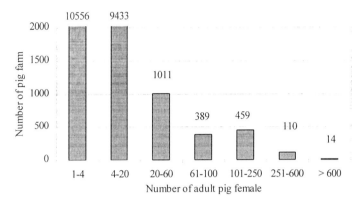

Figure 1. Distribution of pig farms by size.

The majority of the large units (from 400–600+ female reproducer pigs) are already equipped with an anaerobic treatment plant. The treatment problem lies with small- to medium-sized units, which are difficult to control and which dispose of their wastes locally. These units could feed waste to a centralised treatment plant.

3. POSSIBILITY OF APPLICATION OF CENTRALISED ANAEROBIC DIGESTION SYSTEM IN PORTUGAL

3.1. Benefits of a centralised plant

A centralised plant can obtain several advantages from wastes and overcome the problems related to the use of inorganic fertilisers, as follows:
- production of sludge fertiliser from anaerobic digestion, and storage in collective tanks;
- distribution of a controlled-quality sludge, the odours and nuisance pathogen content of which are reduced by the temperature of digestion and during storage;
- economy and flexibility of storage of the effluent due to large scale operation;
- production of large quantity of electric energy or combustible from biogas.

Anaerobic digestion can play an important role in the management of wastes and reuse of fertiliser in agriculture. In the future, renewable energy production will be more attractive, contributing to reduce existing barriers. However, this concept is still difficult to implement in Portugal owing to existing constraints that must be overcome.

3.2. Political constraints

Most of the E.C. environmental regulations have been transferred to Portugal. However it is difficult to apply the new regulations, because many local industries are not prepared for them and cannot comply in a short time. An existing law allows the disposal of pig manure effluent in watercourses even with high concentrations, which can be obtained with a pond treatment system. As a consequence, ponds proliferate in the country, frequently not cleaned, producing effluent with quite high concentrations. This legislation strays from the interests and objectives of environmental recovery, causes massive investment in treatment systems that will not be adequate in future, and delays the pollution control and the privatization of financial opportunities that is actually favoured by E.C. programs. Most farm producers are uninterested in environmentally favourable solutions, preferring to maintain the existing situation.

The ministry of the environment in 1991 transferred to Portugese regulations the Directive 86/278/EC, related with the soil protection through the use of domestic sludge in agriculture, in order to encourage the correct use in agriculture.

In 1997 the ministry transferred the European Directive 91/676/EC, related to the protection of natural water resources against pollution caused by nitrates used in agriculture and in farm industries.

The Portuguese Environmental Ministry also signed a protocol with the pig producers association in order to develop a high-quality environmental program, solve pollution problems and define a plan to adapt new environmental regulations. The plan, elaborated and proposed by the National Laboratory of Civil Engineering (LNEC)[1] according to the European laws, was favourable to land application systems. Unfortunately it was not approved by the swine producers association, which has delayed the solution of this pollution problem. Many plants are in regions lacking enough land for final disposal, so they have to ship waste or cease the activities.

This situation demonstrates the need for a great change in the farmer's mentality in order to solve environmental problems, which are actually in stand-by, and to gradually adapt customary practices to that they promote environmental standards.

In Portugal at present, the implementation of centralised anaerobic digestion plants will require more stringent environmental laws that oblige any producer to be responsible for the waste produced, and discourage any tendency to continue with water discharge. Such requirements will be costly and are politically unpopular.

A law that limits or prevents the use of chemical fertilisers, as Denmark has, can completely change one's perspective on the usefulness of organic waste, normally designated as a strong polluting agent. Wastes then can become a marketable product that compel intense co-operation between farmers and livestock producers. In this perspective, anaerobic digestion is the technological method that provides the quality necessary to satisfy land application requirements.

3.3. Technical constraint

As determined in a monitoring campaign,[2] generally Portuguese farm wastes are quite diluted and contain less than 3 - 3.5% of total solids, which impaired mesophilic anaerobic digestion in wintertime. This dilution is related to the washing systems used in piggeries and to the climate, which increases the water demand of pig populations. The excessive dilution of waste has several disadvantages, as follows:
- The wintertime digestion temperature of the slurries is generally about 15°C, which reduces to 60% the obtainable gas production.
- The biogas obtained at lower temperature is corrosive and reduces engine life.

- The volume of holding tank and digester must be bigger.
- The excessive dilution increase the transportation costs (about 1.5 ECU /m^3) by tanker trucks.

To reduce this excessive dilution, it is necessary to reduce water consumption and modify water management at the farm. Most farms have a big tank under the pit floor that stores a large quantity of wastes. This storage is a component of the farm strategy, which discharges most of the pollution load during rainy periods. In this storage tank 30 - 40% of organic matter is degraded, which reduces the potential biogas recovery. To change this situation involves high construction costs.

So the implementation of centralised plant systems requires a strategy that can change some of these adverse aspects.

Water consumption can be reduced by use of high pressure washing systems and the control of leakage. New installations can use dry cleaning systems or other processes that demand less water. Residues can be removed more frequently to reduce residence time in the storage systems. The wastes can be transported by pumping systems, in order to reduce the costs and shorten the retention time in accumulation tanks.

3.4. Economic constraint

The benefits arising from the management and control of the wastes, in addition to environmental and economic results, must also give rise to an attractive investment.

However, economic perspectives are not encouraging. The price of electric energy produced in Portugal is evaluated as high-tension electricity, which has a low price (about 0.04 ECU/kWh). Some financial inducements such as, for instance, the payment of an extra price for the renewable produced energy, can stimulate use of biogas in energy production and encourage the implementation of renewable technology management.

The lack of a market for the treated wastes does not permit any realistic income from this residue. On the other hand, if there is not enough land, the disposal of wastes can constitute a cost. Only the promulgation of a law promoting the use of digested organic waste as fertilisers can change the present situation.

4. CASE STUDIES

The only centralised anaerobic digestion plant in Portugal is in the centre of Portugal, near Rio Maior; it started up in 1992. The anaerobic digestion plant treats manure derived from several small piggeries with a total animal population of about 10 000 swine. The local natural park association manages the plant in order to protect the natural resource of the park from the pig producers.

The digester is a plug-flow rectangular anaerobic tank, which is quite efficient, economical and easy to operate. The system works in the mesophilic temperature range. The swine manure is fed into the digester, once or twice a day, to maintain a horizontal peristaltic movement in the digester liquor. The total gas available feeds a motor-generator and produces electric energy, sold to the Portuguese Electric Company, EDP. The energy produced is about 400 000 kWh/yr. The heat waste from the motor heats the digester. Wintertime digester temperature decreases to about 17-18°C.

The digester effluent is screened on a vibrating screen and then in a pond wastewater treatment system, with five serial units. Treated effluent is discharged in the local watercourse, having still high concentration.

Waste is transported by tanker trucks that are managed by the association. The cost of transportation is about 1.6 $/ m^3, the only cost paid by the pig farmer. The screened solid waste is combined with forestry residues, composted and sold for about 2.7 $/m^3.

A financial balance sheet indicates that the income is just enough to pay the current O & M costs, but is unable to finance the repair or replacement of the engine. The association supports some investment, which is admitted as social cost. The pig farmers do not want the responsibility of the plant, which reduces its ability to invest in upgrading or enlarging the system.

In the future, treated effluent will necessarily be improved or disposed of on local land, according to the regulations of future legislation. However, the existing local resources favour water reuse in agriculture or forestry irrigation.

Anaerobic digestion efficiently decreases the organic load of waste manure. Most data indicate that the efficiency of removal of T.S. and COD is in the range 80-90 %. These good results derive from the low organic load applied to the digesters.

5. CONCLUSIONS

Anaerobic centralised plant systems will become possible only when the unfavourable laws that protect pig farmers are removed, and the basic equipment needed to reuse effluent in agriculture are put into operation.

Local interest in use of biogas has been lacking, and it will be necessary to introduce inducements or grants to stimulate production.

From the technical point of view, the piggeries must gradually modify their internal washing and cleaning systems. So, still many conditions have to be changed before centralised anaerobic systems can be widely adopted.

REFERENCES

1. Caracterização do sector da suinicultura relativamente ao estado de adequação á legislação ambiental. Report LNEC pro. 606/1/12065- Lisboa 1995.
2. Levantamento e investigação experimental dos digestores anaeróbiios existentes no País - 2ª fase. Report LNETI/DER - 0076. Lisboa 1991.

COMPUTER-AIDED DESIGN MODEL FOR ANAEROBIC-PHASED-SOLIDS DIGESTER SYSTEM

Zhang Zhiqin,[a] Zhang Ruihong,[a] and Valentino Tiangco[b]

[a]Biological and Agricultural Engineering Department, University of California, Davis, California 95616
[b]Research and Development Office, California Energy Commission, 1516 Ninth St, Sacramento, California 95814

The anaerobic-phased-solids (APS) digester system is a newly developed anaerobic digestion system for converting solid wastes, such as crop residues and food wastes, into biogas for power and heat generation. A computer-aided engineering design model has been developed to design the APS-digester system and study the heat transfer from the reactors and energy production of the system. Simulation results of a case study are presented by using the model to predict the heating energy requirement and biogas energy production for anaerobic digestion of garlic waste. The important factors, such as environmental conditions, insulation properties, and characteristics of the wastes, on net energy production are also investigated.

1. INTRODUCTION

California leads the nation in agricultural production. The amounts of fruits, nuts and vegetables in California account for 50% of the national production. Each year, more than 8 million tons (dry weight) of crop residues and 1.7 million tons (dry weight) of food processing solid wastes are produced in California (California Energy Commission, 1993). The energy production from agricultural and food wastes could have a substantial effect on local energy usage.

Anaerobic digestion is a microbial biological process that converts organic wastes into biogas and stabilized residues in the absence of oxygen. The biogas, which is mainly a mixture of methane and carbon dioxide, can be used as fuel for electricity and heat generation. Organic loading rate, temperature, pH, and nutrients are the major factors that affect the rate and stability of anaerobic digestion. The most common temperatures used for digestion are 35°C (mesophilic temperature) and 55°C (thermophilic temperature). Several digester designs process solids; they include one-phase and two-phase digestion systems (Colleran et al., 1983; Ghosh et al., 1997; Khan et al., 1983). Recently, the anaerobic-phased-solids (APS) digester system was developed at the University of California at Davis for efficiently converting solid wastes, such as rice straw and food wastes, into biogas. The objectives of this study are to develop a computer model to aid the engineering design and predict the net energy production of the APS-digester system for specific applications.

The APS-digester system consists of hydrolysis reactors and biogasification reactors. Each hydrolysis reactor is designed as a solids leaching bed reactor and each biogasification reactor is designed as a suspended growth reactor. The two reactors are coupled and liquid is recirculated between them. The feedstock can be loaded into the hydrolysis reactor in batch, semi-batch, or continuous mode depending on the type of feedstock. The soluble compounds released from the solid feedstocks are collected in the hydrolysis reactor and transferred intermittently to the biogasification reactor. The digester system can be operated at either mesophilic or thermophilic temperatures. A detailed description and laboratory evaluation of the APS-digester system have been given by Zhang and Zhang (1998).

2. MODEL DEVELOPMENT

A computer model was developed to design the APS-digester system in order to simulate the heat transfer processes involved in the digester system, and to perform an energy analysis by considering biogas energy and heating requirements. Biogas energy is calculated from the biogas yield of the waste to be digested. It is usually determined with laboratory experiments at a given temperature and retention time. The heat required to operate the digester system heats influent and makes up for heat losses through the wall, top and bottom of each reactor. The calculation of net energy is given by equation (1),

$$q_{net} = q_{gas} - q_{heat} - q_{loss}, \qquad (1)$$

where q_{net} is net energy production, q_{gas} is biogas energy, q_{heat} is heat required for heating the influent, and q_{loss} is heat loss from the reactors.

Influent heating energy is calculated using the equation (2),

$$q_{heat} = mC(T_\infty - T_i), \qquad (2)$$

where m is the influent flow rate at a desirable total solids content level (lb/day), C is the specific heat (Btu/lb/°F), T_∞ is the ambient temperature, and T_i is the digestion temperature.

Heat is lost through the wall, top, and bottom of each reactor by conduction, convection, and radiation. The major factors that affect the heat loss are digestion temperature, ambient temperature and wind speed, digester configuration, insulation properties, and total solids content of the wastes. Rate of heat loss is calculated with equation (3),

$$q_{loss} = \frac{(T_i - T_s)}{R_{cond}} = \frac{T_s - T_\infty}{R_{conv}} + \frac{T_s - T_\infty}{R_{rad}}, \qquad (3)$$

where T_i and T_s are inside and outside surface temperatures of the reactor (°F), respectively.

T_∞ is the ambient temperature (°F), and R_{cond} and R_{conv} are conduction and convection resistances, respectively (°F-hr/Btu). The R_{conv} and R_{rad} are calculated as

$R_{conv} = 1/(h_{conv}A)$ and $R_{rad} = 1/(h_{rad}A)$,

where A is surface area (ft^2) of the reactors, h_{conv} is convection surface coefficient (Btu/hr-°F-ft^2) and h_{rad} is the radiation surface coefficient (Btu/hr-°F-ft^2). The convection and radiation coefficients are determined according to ASHRAE HANDBOOK (1993) and are shown in equations (4) and (5).

$$h_{conv} = C(\frac{1}{d})^{0.2}(\frac{1}{T_{ave}})^{0.181}\Delta T^{0.266}\sqrt{1+1.277\times V}, and \qquad (4)$$

$$h_{rad} = \frac{\varepsilon \times 0.1713 \times 10^{-8} \times [(T_\infty + 459.6)^4 - (T_s + 459.6)^4]}{T_\infty - T_s}, \qquad (5)$$

where d is diameter for cylinder (in inches), which equals 24 for flat surfaces and large cylinders, ΔT is surface-to-air temperature difference (°F), T_{ave} is average temperature of air film (°F), V is wind speed (mph), C is a constant that is related to shape and heat flow conditions, and ε is surface emissivity that varies with the surface materials.

To calculate the heat loss from the reactors, the convection and radiation coefficients, and surface temperature of wall, top and bottom of each reactor were solved simultaneously by using equation (3). A trial-and-error method was used to solve the non-linear equation. The model was applied for a case study to predict the heating energy requirement and biogas energy production for anaerobic digestion of garlic waste with the APS-digester system. In order to study the effects of important factors, such as environmental conditions, insulation properties, and characteristics of the wastes, on net energy production, different locations, changes of insulation thickness, and total solids content of the wastes to be digested were also investigated. The design input values, which include characteristics of the garlic waste, digester specifications, and thermal properties of the digester and insulation materials, are listed in Table 1. By changing each independent input variable, the model generates corresponding simulation results in terms of size and dimensions of the digester and detailed energy balance calculations for the system and gives a dynamic illustration to the user. A user-friendly software using Microsoft Excel was developed for this model.

3. RESULTS AND DISCUSSIONS

To demonstrate the model, a case study was conducted by designing an APS-digester system to treat 1000 lb garlic waste (based on dry weight) per day. The biogas yield of garlic waste was obtained from a previous laboratory study. The simulation results or outputs, which include size and dimensions of the digesters, biogas energy, and heating requirement, are shown in Tables 1 and 2, and Figure 1. The effects of changes of the different location, insulation thickness, and total solids content of the wastes to be digested, on the net energy production are shown in Figures 2 to 4.

As shown in Tables 1 and 2, the net energy production from the garlic waste is 1.0×10^9 Btu per year. If the energy is converted to electricity with 30% conversion efficiency, the annual power production is estimated to be 92,018 kWh. The heating requirement, which includes influent heating and heat loss from the reactors, is about 19% of total biogas production for the case study. However, the net annual energy production could decrease by 8% if the system is located in Minnesota instead of California as shown in Figure 2 if the same insulation thickness is used. That is, thicker insulation is needed to achieve the same net energy production in a cold climate. The simulation results (Figure 3) show that the total solid content in the hydrolysis reactor is also an important factor. Net energy increases with the increase of total solid content because the amount of water to be heated in the influent is reduced. Net energy production could become negative if the total solid content of the wastes to be digested is below 2%. The effect of insulation thickness on net energy production is shown in Figure 4. Net energy production increases with the increase of thickness of the insulation material. The optimum thickness should be decided by the cost of insulation and value of energy conserved.

Table 1
Input and output values for model application

INPUT		OUTPUT	
Parameters	Assigned values	Parameters	Calculated Values
Waste characteristics		Size of the digesters	
Waste generation	1000 lb/day (DW)[a]	Hydrolysis reactor	
Total solids (TS)	29.75%	- volume	2005 ft^3
Volatile solids (VS)	85.47% of TS	- height	20 ft
Digester specification		- diameter	11 ft
VS loading	5.33 lb/ft^3	Biogasification reactor	
Biogas production	7.38 ft^3 /lb VS fed	- volume	1003 ft^3
Methane content	55.58%	- height	20 ft
Retention time	10 days	- diameter	8 ft
TS in hydrolysis Reactor	10%	Energy Calculation	
Digester temperature	95°F	Influent heating	1.4×10^8 Btu/yr
System location	Sacramento, CA	Heat loss	6.3×10^7 Btu/yr
Digester Configuration	Circular tank	Biogas energy	1.25×10^9 Btu/yr
Digester materials	Stainless steel	Net energy	1.05×10^9 Btu/yr
- conductivity	8.609 Btu/hr-ft-°F		
- thickness	0.01 ft		
Insulation materials	Fiberglass		
- conductivity	0.027 Btu/hr-ft-°F		
- thickness	0.1 ft		

a: DW stands for dry weight

Table 2
The net energy analysis of anaerobic digestion of garlic waste (case study)

Month	Heating Requirement		Biogas energy	Net Energy
	Influent heating (Million Btu)	Heat loss (Million Btu)	(Million Btu)	(Million Btu)
January	15.9	7.8	104	80.4
February	14.9	7.1	104	82.2
March	13.2	6.1	104	84.8
April	12.6	5.9	104	85.6
May	9.3	4.0	104	90.8
June	8.6	3.9	104	91.6
July	7.0	2.9	104	94.2
August	7.4	3.1	104	93.6
September	8.5	3.4	104	92.2
October	11.9	5.1	104	87.1
November	13.8	6.2	104	84.0
December	16.4	7.7	104	80.1

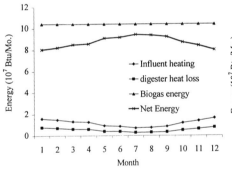

Figure 1. Heating requirement and biogas energy.

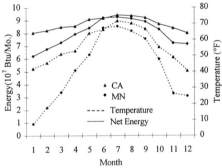

Figure 2. Effect of different location on net energy production.

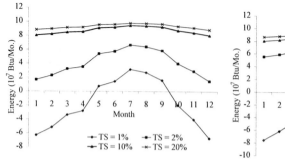

Figure 3. Influence of total solids content in hydrolysis reactor on net energy production.

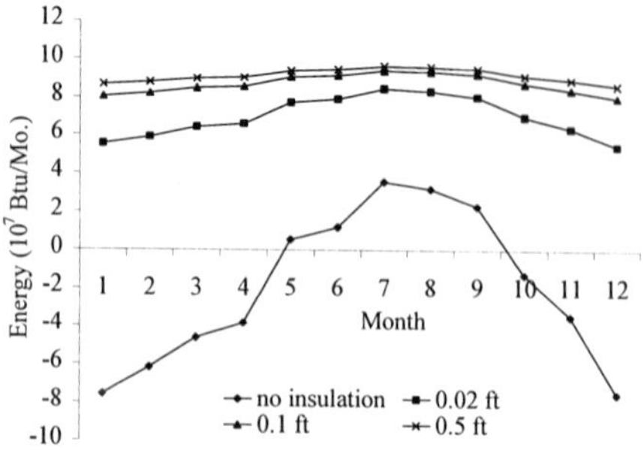

Figure 4. Changes of net energy production with different insulation thickness.

REFERENCES

ASHRAE Handbook Fundamentals (1993). American Society of Heating, Refrigerating and Air-Conditioning Engineer, Inc., 1791 Tullie Circle, N.E., Atlanta, Georgia 30329.

California Energy Commission (1993). The 1992-93 California Energy Plan.

Colleran, E., M. B Wilkie, G. Faherty, N. O'Kelly, and P.J. Renolds (1983). One- and two-stage anaerobic filter digestion of agricultural wastes. Proceedings 3rd International Symposium on Anaerobic Digestion, Boston, U.S.A., pp. 285–302.

Ghosh, S., E.R. Vieitez, T. Liu, and Y. Kato (1997). Biogasification of solid wastes by two-phase anaerobic fermentation, in Making a Business from Biomass in Energy, Environmental Chemicals, Fibers and Materials, ed. by R.P. Overend and E. Chornet, Montreal, Quebec, Canada, Elsevier.

Khan, A.W., S.S. Miller, and W.D. Murray (1983). Development of two-phased combination fermenter for the conversion of cellulose to methane, Biotechnology and Bioengineering, 25, pp. 1571–1579.

Zhang, R. and Z. Zhang (1998). Biogasification of rice straw with an anaerobic -phased solids digester system, Bioresource Technology, 68, pp. 235–245.

ACCOUNTING FOR THE FATE OF MANURE NUTRIENTS AND SOLIDS PROCESSED IN MESOPHILIC ANAEROBIC DIGESTERS

R. P. Mattocks[*] and Mark Moser[+]

[*]ENVIRONOMICS, 5700 Arlington Ave. Suite 17A, Riverdale, New York, 10471 USA
[+]RCM Inc., P.O. Box 4715, Berkeley, CA 94704 USA

Manure anaerobic digesters have had limited acceptance in the United States, because farmers fear that the system will perform differently than claimed. A mass balance methodology is proposed to reduce this sense of risk. A procedure is demonstrated to predict system outputs based on performance of other similar equipment and designs. Inputs and outputs from two digesters (1000-cow dairy digesters in Oregon and New York), in operation for over one year, were analyzed. Fresh manure is digested, producing methane from the digested manure. Digester effluent is separated into solids and liquid fractions. Solids are sold, liquid is used as a nutrient source on forage crops, and the gas is burned in an engine generator set to produce electricity and hot water. Standard laboratory tests were performed on raw manure, effluent, biogas, screened digested solids and nutrient-rich digested separated liquid. The fate of mass and nutrients entering the digester are observed in the various end products. Total mass and solids leaving the system in the various forms are found to be within ± 5% of the mass and solids entering the digester. Total nutrients leaving the system in the various forms are within ± 5% up to ± 25% of the amount entering the digester. The greatest variance is with trace elements. Nitrogen and phosphorus in the screened digested liquid was 20-30% less than the quantity entering the digester, most of the difference collected in fiber. Using this methodology, digester designers will be able to show customers the expected outputs from digestion systems. Customers will be able to make business and agronomic decisions based on real numbers drawn from other similar designs.

1. BACKGROUND

American animal production continues to concentrate into a smaller number of larger facilities, resulting in closer scrutiny of the environmental impact potential. Odor and ground and surface water quality degradation are of concern.[1] A number of technologies are available and in use to reduce environmental risks associated with manure management.

Anaerobic digestion of manure has been employed on farms in the U.S. since the early 1970s. Though many manure digesters are in use worldwide, few function in the U.S. While recognized for effectiveness in odor control, reduction of pathogenic organisms and weedseeds, installation capital costs are high compared to other manure management options. In contrast to manure digestion (i.e., methane recovery systems) few other technologies produce valuable products which offer opportunity to recover installation capital costs.[2] Dairy manure digestion systems yield methane and a fiber (bearing some resemblance to moist peat moss).

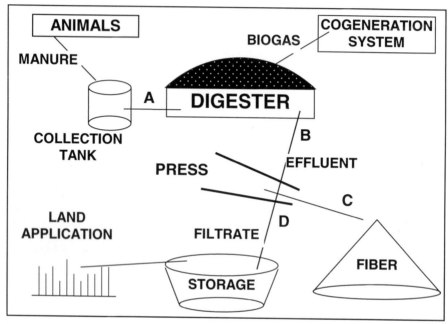

Figure 1. Conceptual diagram: Anaerobic digestion system for dairy manure.

2. SYSTEM DESCRIPTION

Figure 1 provides an overview of the anaerobic digestion systems as installed on two dairy farms. Manure is removed from animal areas, collected into a tank and pumped to the digester. Both dairies have 130' x 30' mesophilic plug flow digesters designed for manures from 1000 cows by Mark Moser, RCM digesters. Digester effluent is passed through a FAN Separator. Fiber is collected for composting and sale. Liquid passing through the FAN (filtrate) is stored in lagoons for land application.

Located in coastal Oregon, Craven Farm straddles 5 miles of prime salmon run. The initial objective of the digestion system was to reduce the risk that manure-derived pathogenic organisms might reach the river. The system had functioned at steady state for 18 months at the time of sampling.

Much of AA Dairy's land that receives manure is adjacent to the town of Candor, New York. The Amans installed a 1000-cow digester for odor control. The system was sampled after 10 months and 15 months of steady state operation.

3. OBJECTIVE

The objective is to demonstrate a methodology and report representative results.

4. METHODOLOGY

Composite samples were taken at each of the farms on four days. Four locations (Figure 1) were sampled several times during normal functioning of the system:
- manure from the collection tank as injected into the digester,
- effluent from the digester,
- fiber exiting the FAN separator,
- filtrate exiting the FAN separator.

Samples were refrigerated and shipped to Midwest Laboratories, Omaha, Nebraska for analysis the following day. Standard ASTM, AOAC and ISE laboratory procedures were employed. Manure, liquid, fiber and biogas quantities were estimated from available records.

5. RESULTS

Tables 1 and 2 report laboratory and field results. A computation is made to determine the quantity of each constituent in 1000 gallons of raw digester feed manure, the digester effluent, fiber/cu yd and filtrate. For the Aman digestion system, as an example explanation of Table 1, each 1000 gal of manure entering the digester will contain 37.6 lb of nitrogen of which 17 lb is 45% ammonia. Digester effluent will have 36.5 lb of nitrogen in a 1000 gal of which 18 lb (49%) is ammonia. Fiber from the solids separation system will have 4.7 lb of nitrogen in each cubic yard of material recovered of which 2 lb is 43% ammonia. Filtrate collected after solids separation will have 38.5 lb of nitrogen in a 1000 gal of which 23 lb is 60% ammonia. Table 2 reports similar results for the Craven Farms digestion system.

Table 3 reports a mass balance of constituents entering the system compared to the same constituents leaving the system. Total solids leaving the Aman digester in some form of product is 103% of the amount of total solids in the manure pumped into the digester.

Table 1
Aman digestion system, average values.*

	Input				Output			
	Feed average		Effluent average		Fiber average		Filtrate average	
	Analysis	lb/ 000 gal	Analysis	lb/ 000 gal	Analysis	lb/ CY	Analysis	lb/ 000 gal
Density	8.75	8750	8.9	8900	825	825	8.5	8500
Moisture	90.7%	7932	95.3%	8477	75.6%	623	95.8%	8139
T. Solids, %	9.4%	818	4.8%	423	24.4%	202	4.3%	361
VS, %TS	83.0%	679	72.5%	307	90.6%	183	63.1%	228
N, %	0.43%	37.6	0.41%	36.5	0.57%	4.7	0.45%	38.5
P_2O_5, %	0.18%	15.3	0.15%	13.4	0.32%	2.6	0.15%	13.0
K_2O, %	0.26%	22.8	0.29%	25.4	0.25%	2.1	0.30%	25.5
S, %	0.05%	4.4	0.03%	2.7	0.08%	0.6	0.03%	2.6
Mg, %	0.08%	7.0	0.07%	6.2	0.14%	1.2	0.07%	6.0
Ca, %	0.20%	17.5	0.14%	12.5	0.45%	3.7	0.14%	12.1
Na, %	0.11%	9.8	0.13%	11.6	0.09%	0.8	0.12%	10.2
Fe, ppm	64	0.56	61	0.54	149	0.12	57	0.48
Al, ppm	34	0.30	26	0.23	134	0.11	25	0.21
Mn, ppm	22	0.19	20	0.17	53	0.04	18	0.16
Cu, ppm	26	0.23	42	0.37	45	0.04	30	0.25
Zn, ppm	26	0.22	28	0.24	34	0.03	24	0.20
Ash	1.6%	139	1.3%	116.1	2.3%	18.8	1.6%	134
NH_4^+, %	0.20%	17	0.20%	18	0.24%	2	0.27%	23

*All values are "wet weight," i.e., as delivered

Similarly, total solids exiting the Craven digester in some form of product is 92%. Total N_{in} is 97% and 100% of total N_{out} for Aman and Craven, respectively.

Total solids in the Aman fiber is 33% of total solids in manure pumped into the digester. Total solids in Craven fiber is 30% of total solids in manure pumped into the digester. Aman biogas mass represented 35% of total solids in the manure pumped into the digester. Craven biogas mass represented 28% of total solids in the manure pumped into the digester. Total solids in Aman filtrate is 35% of total solids in the manure pumped into the digester. Total solids in Craven fiber is 34% of total solids in raw manure.

6. DISCUSSION

Values for the two projects show very similar trends.

Caveats: Only four composite samples were taken at each site. Additional sampling is necessary to make values statistically valid. Moreover, comparing influent and effluent samples taken on the same day fails to consider effluent of a given day corresponds to

Table 2
Craven digestion system, average values.*

	Input				Output			
	Feed average		Effluent average		Fiber average		Filtrate average	
	Analysis	lb/ 000 gal	Analysis	lb/ 000 gal	Analysis	lb/CY	Analysis	lb/ 000 gal
Density	8.6	8400	8.9	8900	910	910	8.5	8500
Moisture	87.5%	7352	93.4%	8308	77.0%	701	95.1%	8081
T. Solids, %	12.5%	1048	6.7%	592	23.0%	209	4.9%	419
V. Sol, %TS	81.6%	855	76.8%	455	86.9%	182	68.8%	288
N, %	0.46%	38.2	0.44%	39.2	0.54%	4.9	0.43%	36.3
P_2O_5, %	0.20%	16.6	0.17%	14.7	0.30%	2.7	0.16%	13.4
K_2O, %	0.31%	26.3	0.30%	26.3	0.29%	2.6	0.29%	24.9
S, %	0.05%	4.2	0.03%	2.7	0.08%	0.7	0.03%	2.6
Mg, %	0.10%	8.2	0.08%	6.9	0.15%	1.3	0.07%	6.2
Ca, %	0.18%	14.7	0.10%	9.1	0.37%	3.4	0.10%	8.1
Na, %	0.07%	5.7	0.07%	6.2	0.06%	0.5	0.07%	6.0
Fe, ppm	239	2.01	136	1.21	379	0.35	124	1.05
Al, ppm	152	1.28	97	0.86	181	0.16	91	0.77
Mn, ppm	26	0.21	15	0.13	51	0.05	14	0.11
Cu, ppm	5	0.04	6	0.06	9	0.01	6	0.05
Zn, ppm	20	0.17	21	0.18	23	0.02	20	0.17
Ash	2.2%	187	1.5%	137	3.0%	27.5	1.5%	130
NH_4^+, %	0.17%	14	0.23%	20	0.23%	2	0.23%	20

*All values are "wet weight," i.e., as delivered

manure placed in the digester more than three weeks earlier. Finally, performance of dairy digesters will vary depending on feedstuff and bedding.

6.1 Significance of results

Results from the two farms are similar, suggesting that planners may be able to use performance data to evaluate output projections provided by prospective designers.

System influent and output do not balance exactly, although they are close. Constituents are known to be settling in the digester.

Table 3 indicates several possible trends. Aman filtrate N, P, and K contains 81%, 67% and 89% of the N, P, and K in the initial manure. Craven filtrate N, P, and K contains 81%, 69% and 81% of the N, P, and K in the initial manure. More than two thirds of volatile solids exit the system through conversion to biogas.

A dairyman with scraper manure removal like those of AA Dairy and Craven Farm would have some points of comparison to claims put forth by prospective vendors when evaluating

Table 3
Mass balance: manure constituent fate in digestion system outputs.

	Manure vs product		Fiber		Biogas		Filtrate	
	Aman	Craven	Aman	Craven	Aman	Craven	Aman	Craven
Moisture	92%	108%	11%	14%	0%	0%	81%	93%
T. solids, %	103%	92%	33%	30%	35%	28%	35%	34%
Vol sol %TS	105%	95%	36%	32%	42%	35%	27%	29%
N, %	97%	100%	17%	19%	0%	0%	81%	81%
P_2O_5, %	90%	93%	23%	24%	0%	0%	67%	69%
K_2O, %	101%	95%	12%	15%	0%	0%	89%	81%
S, %	66%	77%	20%	25%	0%	0%	46%	52%
Mg, %	90%	88%	23%	24%	0%	0%	67%	64%
Ca, %	83%	81%	28%	35%	0%	0%	55%	47%
Na, %	92%	104%	10%	14%	0%	0%	82%	89%
Fe, ppm	98%	70%	30%	26%	0%	0%	68%	44%
Al, ppm	106%	71%	50%	19%	0%	0%	56%	51%
Mn, ppm	95%	78%	31%	32%	0%	0%	64%	46%
Cu, ppm	109%	131%	22%	29%	0%	0%	87%	102%
Zn, ppm	88%	104%	17%	19%	0%	0%	71%	85%
Ash	94%	81%	18%	22%	0%	0%	76%	59%
NH_4^+, %	120%	152%	16%	20%	0%	0%	104%	132%

digestion system designs. Estimates could be made of gas production, fiber production and the quantity and quality of the filtrate after effluent screening. Financial projections could incorporate these product volumes. Agronomic application planning could reflect the nutrient levels expected in the filtrate. Allowances would have to be made for the differences between the two farms and those of the potential installers.

7. ACKNOWLEDGEMENTS

Assistance provided by AA Dairy and Craven Farms is appreciated. Financial assistance for laboratory expenses from the Tioga County Soil and Water District is acknowledged.

REFERENCES

1. U.S. Department of Agriculture (1998). U.S. Environmental Protection Agency Unified National Strategy for Animal Feeding Operations, September 11, 1998.

2. Moser, M.A., R.P. Mattocks, S. Gettier, and K. Roos (1998). Keeping the neighbors happy—reducing odor while making biogas. Animal Production Systems and the Environment, Ames, Iowa.

743

AgSTAR CHARTER FARM PROGRAM: EXPERIENCE WITH FIVE FLOATING LAGOON COVERS

K.F. Roos,[a] M.A. Moser,[b] and A.G. Martin[c]

[a]United States Environmental Protection Agency, 401 M St., SW (6202-J), Washington, DC 20001, United States
[b]Resource Conservation Management, Incorporated, P.O. Box 4715, Berkeley, California 94704, United States
[c]ICF, Incorporated, 14724 Ventura Blvd., Suite 1100, Sherman Oaks, California 91403, United States

The AgSTAR Program is a voluntary program jointly administered by U.S. EPA (Environmental Protection Agency) and the USDA-NRCS (Natural Resources Conservation Service). The AgSTAR Program encourages the use of methane recovery technologies at confined animal feeding operations (CAFO) that manage manure as liquids or slurries. The Charter Farm Program promotes a broader understanding of the benefits, costs, and applications of biogas technology by developing commercial-scale demonstration systems at livestock facilities. Ten AgSTAR Charter Farms (operational since 1996) now use a variety of anaerobic digester systems.

This paper summarizes the experience of five of these farms that installed floating covers on lagoons. Two designs of floating cover are compared—bank-to-bank and modular—by discussing gas transfer and rainfall management; cover design, material warranty, and fabrication warranty; suitable materials such as HDPE and polypropylene; performance criteria; and prices based on materials, warranties, and installation. Costs ranged from $0.61/ft^2 to $5.81/ ft^2.

Long term monitoring data (Cheng et al., 1998) indicates > 90% reduction in chemical oxygen demand (COD), 65% reduction in total nitrogen, and 85% reduction in total phosphorus. This environmental performance is achieved under USDA-NRCS Interim Standard No. 360—Covered Anaerobic Lagoon.

1. INTRODUCTION

The AgSTAR Program is a voluntary program jointly administered by U.S. EPA (Environmental Protection Agency) and the USDA-NRCS (Natural Resources Conservation Service). It encourages the use of methane recovery technologies at confined animal feeding operations (CAFO) that manage manure as liquids or slurries. The AgSTAR Charter Farm

Program promotes a broader understanding of the benefits, costs, and applications of biogas technologies by developing commercial scale demonstration systems at livestock facilities in various regions in the U.S.

All farms in the AgSTAR program are Charter Farm candidates. Typically in January farms are invited to apply for Charter Farm status for the upcoming construction year by submitting a Charter Farm Application and a Charter Farm Agreement. Farms are selected on the basis of farm location, type of biogas system, applicability to other producers in the region, net methane reduction, and reasons for installing a biogas system (e.g., economic, environmental, energy). The Agreement specifies the roles of the AgSTAR technical team and farm owners in the Charter Farm project.

Since 1996, about 10 AgSTAR Charter Farms have been completed that use a variety of anaerobic digester systems: covered anaerobic lagoons, plug flow, and complete mix. This paper discusses issues that need to be considered when a covered anaerobic lagoon waste management system is considered for farm scale application, and it is based on the experience of five farms currently operating covered anaerobic lagoons.

2. COVERED ANAEROBIC LAGOONS

Covered anaerobic lagoons installed to control odor and produce energy typically consist of two lagoons operating in series. The first lagoon treats primary waste and has a dedicated volume with a fixed operating depth to biologically stabilize waste. Biogas is a by-product of this anaerobic process that destroys volatile solids and reduces chemical oxygen demand (COD). A second lagoon stores waste during non-cropping periods. Long-term monitoring data (Cheng et al., 1998) indicates > 90% reduction in chemical oxygen demand (COD), 65% reduction in total nitrogen, and 85% reduction in total phosphorus from this lagoon configuration, where lagoon sizing guidelines established under USDA-NRCS Interim Standard No. 360—Covered Anaerobic Lagoon were made.

A lagoon cover collects lagoon-generated biogas and transmits to a dedicated gas take-off point for transmission to a device that uses gas—e.g., flares, boilers, absorption coolers, and engine generators. All AgSTAR Charter Farms are equipped with low-cost auto-sparking flares that combust greenhouse gases and maintain consistent odor control. Other gas-use equipment depends on the needs of the farm. The cover must continuously move biogas, under low vacuum, to the gas takeoff point, and it must manage precipitation without interfering with gas collection. The standard of care in fabrication, assembly and installation must be very high to limit air intrusion because the lagoon cover is operated under vacuum.

2.1. Lagoon cover design options

Two basic lagoon cover designs—bank-to-bank and modular—both collect biogas and reduce odor.

Bank-to-bank covers completely span the lagoon surface with a fabricated floating cover. The edges of the cover are buried in trenches around the lagoon bank. Burying the edges creates a completely anaerobic environment, allows all gases to be captured, and sheds rainfall. To ensure completely anaerobic conditions and effective biogas collection, proper cover flotation and rainwater pumping mechanisms must be incorporated into the cover design.

Modular covers use smaller sections, or modules, to cover typically 50-90% of a lagoon's surface. These covers can be secured either with bank trenching or tether ropes. Flotation may or may not be needed to remove accumulated precipitation. This design allows for the cover modules to be fabricated off-site and assembled on-site, and it also allows for a lagoon to be covered in stages, thus reducing any one-time capital outlay.

2.2. Considerations in selecting a lagoon cover supplier

A satisfactory lagoon cover must meet high standards of fabrication, materials and design. To establish a basis for selecting a lagoon cover type and supplier, Charter Farm participants asked interested suppliers to comment on the following five topics:

- Experience in providing lagoon covers at commercial livestock farms for purposes of gas collection and transfer to some gas-use device.
- A conceptual drawing showing design type and configuration lagoon specifications provided by the farmer.
- Material specification and fabrication methods (e.g., seam technique and seam testing) as related to accepted practices and standards of the industry.
- Method of warranty coverage or bonding cover performance as related to material, fabrication, installation, and design of the cover. Acceptable cover performance required
    a) < 10% air intrusion (2% oxygen) at a negative 2" water column pressure for a specified period; and
    b) continuous and unrestricted gas movement at a 2" water column pressure as per conditions specified under the warranty.
- Total cost of supplying and installing the cover.

Cover suppliers were not asked to provide gas production estimates nor be responsible for gas production levels as these are separate aspects in the overall design of a covered lagoon waste management system.

3. SUMMARY OF SUPPLIER RESPONSES

3.1. Experience

Eight suppliers listed experience with floating covers. Most suppliers had experience with more than one cover and one had installed more than 25 during nine years. Floating covers had been installed at rendering plants and at food processing and other waste lagoons, as well as at livestock lagoons. However, some lagoon covers did not collect and transmit gas.

3.2. Services and materials

Suppliers offered various services and materials. Several suppliers install the lagoon materials and provide turnkey services; they have traveling crews with equipment. Some suppliers asked the farm to perform relatively inexpensive tasks such as digging a trench or supplying a couple of man-days of farm laborers to reduce costs. Materials used were high density polyethylene (HDPE), polypropylene, XR-5, and X-210. Each material possesses different ranges of tensile strength, UV resistance and other properties. UV-resistant materials and those that float have been more successful.

3.3. Warranties and performance bonds

Warranties were offered for varying lengths of time depending on the supplier and installer. Warranties are generally limited to a percent of the original cost, reduced by product age. Materials are warranted by their manufacturer for 2-10 years, though some accompanying literature cited 20-year warranties. Fabrication was generally warranted 1 to 2 years. Only one supplier warranted cover design, for one year based on the criteria established for cover performance.

Some suppliers also provided performance bonds for the cover at an additional cost. Some Charter Farm participants have added the lagoon cover into the farm's overall damage insurance policy. This approach does not insure the cover in case of non-performance.

3.4. Costs

Total cost of a lagoon cover accounted for installation, labor and shipping. Costs ranged from $0.61/ft^2 to $5.81/ ft^2. Table 1 summarizes the warranty types and costs for five floating lagoon covers. In general, the owners selected the lower cost cover; supplier experience was a secondary concern.

4. SUMMARY OF EXPERIENCES AT FIVE FARMS

4.1. Apex Pork

APEX Pork is an 8,900-head swine finishing operation in Rio, Illinois. Manure is collected in under-barn pull plug pits. To reduce odors, a heated and mixed covered lagoon digester was installed in June 1998. A modular, reinforced X-210 cover collected the biogas from the 120 ft × 160 ft × 14 ft digester; the biogas was then combusted in a hot water boiler to provide the digester's heat. Once digester start-up was completed, biological activity in the heated mixed digester leveled off to produce approximately 36,000 ft^3/day (Moser et al., 1998). However, the cover was damaged during a windstorm, compromising its gas collection capabilities. A replacement bank-to-bank 40 mil HDPE cover, selected on the basis of price and the experience of the installer, was installed over the existing cover. The replacement installation includes a rainfall pumpoff and has performed within the required tolerances.

Table 1
Summary of materials bids

Project	Material	Unit Cost ($/ft^2)	WARRANTY Material	Fabrication	Design
*Barham	40mil HDPE	.37	2 years	2 years	2 years
*Boland	20 mil HDPE, Reinforced	.46	1 year	1 year	N/A
Barham	40 mil HDPE	.52	5 years	N/A	1 year
*Piney Woods	40 mil MDPE	.61	2 years	2 years	N/A
Barham	40 mil co-extruded HDPE	.78	No RFP	No RFP	No RFP
Barham	36 mil Polypro	.81	No RFP	No RFP	No RFP
Barham	40 mil HDPE	.87	No RFP	No RFP	No RFP
Barham	36 mil Polypropylene	.95	No RFP	No RFP	No RFP
Barham	XR-5	1.04	No RFP	No RFP	No RFP
*Apex	40 mil HDPE	1.10	2 years	2 years	2 years
Barham	30 mil Polypropylene	1.10	No RFP	No RFP	No RFP
Piney Woods	20 mil Polypropylene	1.14	N/A	N/A	N/A
Barham	60 mil HDPE	1.22	No RFP	No RFP	No RFP
Piney Woods	40 mil co-extruded HDPE	1.29	2 years	2 years	N/A
Piney Woods	36 mil Polypropylene	1.29	N/A	N/A	N/A
Barham	45 mil Polypropylene	1.30	No RFP	No RFP	No RFP
Piney Woods	36 mil Polypropylene	1.35	2 years	2 years	N/A
*Cal Poly	36 mil Polypropylene	1.75	Manufacturer	1 year	N/A
Cal Poly	Dbl. 40 mil HDPE w insul.	1.82	N/A	N/A	N/A
Apex	40 mil HDPE	2.02			
Piney Woods	36 mil Polypropylene	2.20	Manufacturer	2 years	N/A
Apex	36 mil Polypropylene	2.26			
Piney Woods	40 mil HDPE	2.28	Manufacturer	Negotiable	N/A
Piney Woods	36 mil Polypropylene	2.42	Manufacturer	N/A	N/A
Cal Poly	40 mil HDPE	2.64	Manufacturer	1 year	1 year
Piney Woods	36 mil Polypropylene	2.78	10 years	2 years	N/A
Apex	Dbl. 40 mil HDPE w insul.	3.15	N/A		
Cal Poly	36 mil Polypropylene	3.50	Manufacturer	2 years	N/A
Apex	45 mil Polypropylene	4.23			
Piney Woods	Dbl. 40 mil HDPE w insul.	5.13	N/A	N/A	N/A
Piney Woods	XR-5	5.81	N/A	N/A	N/A

* Chosen bid

4.2. Barham Farm

Barham Farm, located in Zebulon, North Carolina, is a 4,000-sow farrow-to-wean pig farm that uses a pit recharge manure management system in its barns. The pits empty into an earthen covered anaerobic treatment lagoon, designed by USDA-NRCS, and constructed in July 1996. The farm's owner was looking to offset his farm's energy requirements by using biogas generated in the lagoon to generate hot water and electricity. A modular (4-section), 20 mil X-210 cover was installed on the 300 ft × 300 ft lagoon. A 400,000 Btu boiler and 120 kW engine generator were installed in December 1996 to utilize the gas, and digester operation began in January 1997 (Moser et al., 1998). Several weeks later fabrication and material defects in the cover rendered the cover unable to collect undiluted biogas using a vacuum pump. The manufacturer replaced the cover under warranty, but similar problems emerged. The manufacturer subsequently refunded the farm's cover expenditure. The owner selected a bank-to-bank 40 mil HDPE cover based on the price and experience of the installer. In early 1998, the cover was installed with rainfall pump-off and has been delivering air-free biogas to the engine and boiler system since.

4.3. Boland Farm

Boland Farm is a 2,400-head pig nursery located in Williamsburg, Iowa; it uses pull plug pits in the nursery barns that empty into an on-site earthen storage pond measuring 140 ft × 160 ft. Odor was a problem. In May 1998, a bank-to-bank odor control X-210 cover was installed over the storage basin (Moser et al., 1998). The owner selected the cover based on price and installed the cover with local labor in about 16 hours. The biogas collected is combusted by a flare.

4.4. Cal Poly Dairy

A project on the campus of California Polytechnic State University, San Luis Obispo, curtailed odors generated by a single-cell lagoon and demonstrated methane recovery and utilization technologies. In July 1998, a two-stage anaerobic treatment lagoon system was constructed to treat and store manure flushed from the school's 300 milk-cow drylot dairy. A partially bank-buried polypropylene biogas collection cover and a waste gas flare were installed on the primary anaerobic lagoon (260 ft × 260 ft) in December 1998. The cover was selected based on local presence and experience of the installer; price was secondary. Early in 1999, as part of the project, a 40 kW micro turbine and generator will be considered to combust the biogas and generate electricity.

4.5. Piney Woods School Farm

Piney Woods School Farm is a 12-sow farrow-to-finish swine operation on the campus of Piney Woods School (Piney Woods, Mississippi). The farm flushes manure from its farrowing and finishing barns once a day into a 115 ft × 125 ft two-stage anaerobic lagoon system. The school installed a buried bank-to-bank HDPE cover on the farm's primary anaerobic lagoon in August 1998, selecting the cover based on price. It was the first installation of a floating cover by this lagoon-lining company and has worked well. A low-volume flare combusts the collected gas. The project was jointly funded by the Mississippi

Department of Economic and Community Development (DECD) and the Tennessee Valley Authority (TVA) to demonstrate the applicability of methane recovery technologies to swine farms in the Southeast.

5. SUMMARY

Five AgSTAR Charter Farms installed floating covers on lagoons in 1998. Odor control was the goal of the Apex Pork, Boland Farm and Cal Poly Dairy; Barham Farm wanted to recover methane for environmental and energy benefits; and Piney Woods wanted to demonstrate methane recovery technologies to local farms. Two designs of floating cover were available to the five farms, bank-to-bank and modular. Although most farms chose bank-to-bank covers, both designs effectively recover methane and reduce odor. In both cases, however, gas transfer and rainfall management determine the success of the cover.

Design and material and fabrication warranty and cost should also be considered. Materials suitable for biogas covers include HDPE, Polypropylene and materials such as XR-5 and X-210. The majority of suppliers include a manufacturers warranty on the cover material for 10-20 years. Warranties on workmanship were typically one to two years. However, only one cover supplier provided a warranty on air intrusion standards. Costs for covers ranged from $0.61/ft^2 to $5.81/ft^2.

REFERENCES

1. Cheng, J., K.F. Roos, and L.M. Saele (1999). An evaluation of covered anaerobic lagoon system for swine waste treatment and energy recovery, presented at North Carolina Waste Management Symposium, North Carolina State University, January 1999.
2. Moser, M.A., Dr. S.W. Gettier, R.P. Mattocks, and K.F. Roos (1998). Benefits, costs and operating experience at seven new agricultural anaerobic digesters, Proceedings, BioEnergy '98, Expanding Bioenergy Partnerships, Madison, Wisconsin, October 4-8, 1998, pp. 623–630.

METHANE FROM MANURE: AN ENERGY-SAVING SOLUTION FOR IOWA PORK PRODUCERS

Lee Vannoy

Iowa Department of Natural Resources, Wallace State Office Building, Des Moines, IA 50319

The Iowa Department of Natural Resources (Department) is supporting and developing projects designed to demonstrate on-farm methane recovery and conversion processes for odor control, pollution prevention, and energy production. If handled properly, livestock waste represents a valuable resource and a source of farm revenue instead of an environmental hazard. Each year Iowa's hog population produces about 23 million tons of raw manure. This amount of waste possesses the potential to generate the amount of energy consumed by more than 57,000 homes annually.

Two hog producers in Iowa are installing methane recovery demonstration projects with assistance from the Department's Methane Energy Recovery Program. One is installing an anaerobic sequencing batch reactor at an existing 2,800-head finishing facility. The other is a new 5,000-sow farrow-to-wean facility being constructed by a farmer-owned cooperative that will use a complete mix digester. The purpose of each system is to control odor, recover methane, convert the methane to heat or electricity for on-site use and for sale, and to mitigate environmental impacts of their operations. These projects are a few key steps in Iowa's progress toward a sustainable energy future.

1. OVERVIEW

Recent newspaper headlines depict the growing controversy surrounding the pork industry. One states that the "New fear from hog lots: Odor may spread illness,"[1] while a response to the editor indicated that the article was "filled with anecdotal scare-mongering trivia and included not a scintilla of scientific evidence."[2] Daily news coverage of the topic illustrates the variability in facts regarding the economic benefits associated with large-scale livestock production and the environmental and social concerns with odor, soil and groundwater contamination, and methane emissions.

Iowa is the number one pork-producing state. Historically, the vast majority of hogs produced in the state were raised on relatively small family farms that produced several hundred to a few thousand hogs per year. However, because economies of scale offer a lower per-unit cost of production, large-scale confinement operations producing between 3,000–10,000 hogs or more each year are becoming increasingly prevalent. Large hog operations tax traditional methods of manure and odor containment, causing public concern about the threat that large hog production facilities may pose to the environment, quality of life, and value of nearby property. Large-scale facilities must manage and properly dispose of enormous amounts of raw manure while protecting the environment and respecting the interests of nearby neighbors. This is an immense responsibility considering Iowa's 15 million hogs produce about 23 million tons of raw manure annually.[3]

In order to put this issue in perspective, consider that the waste production of an average hog is equivalent to 2.8 people.[♣] Therefore, the treatment of the waste from Iowa's 15 million hogs would require a municipal wastewater treatment plant with a capacity equivalent to a city of 42 million people which is more than a dozen times the state's human population. Also, because hog manure is not diluted by water in the same manner as municipal waste, it has a significantly higher pollution strength.[♦,5] Considering that confinement facilities traditionally have stored raw manure in sewage lagoons and the amount of waste generated, concern that a portion of this could contaminate soil and groundwater seem reasonable. However, there are significant changes occurring in the pork industry in Iowa that should alleviate some of this concern. In 1994, 93% of manure storage construction permits were for earthen structures. In 1998, that figure was reduced to 23%[6] with the change most likely due to pending legislation, recent statutory changes, and public pressure on the pork industry.

1.1. Energy production potential

The recovery potential, if all of the 23 million tons of manure were collected, would be at least 6.8 billion cubic feet of methane annually. If converted to pipeline quality or used directly on site, this methane would have a cash value of about $40.7 million or enough to heat about 66,300 Iowa homes. While other options are available for biogas utilization, electricity is the most versatile and valuable energy product from biogas.[7] If converted to electricity, 23 million tons of manure possesses the potential to generate at least 491 million kilowatt-hours with additional farm revenue potential of $43.3 million not including the significant amount of waste heat that could be recovered from engine-generators and used on site. This is the amount of electricity used by more than 57,000 Iowa homes. Carbon dioxide emissions avoided from displaced coal would amount to as much as 614,000 tons annually.[8]

2. METHANE ENERGY RECOVERY PROGRAM

Because of the importance of this issue in Iowa, the Department and other organizations are working to develop methane energy recovery demonstrations in the state. The

---

♣Value derived from [4].
♦Pollution strength measured by biochemical oxygen demand.

Department's Methane Energy Recovery Program Advisory Committee[▼] assists with program development, proposal review, and project selection. The purpose of the program is to demonstrate cost-effective methods of livestock manure digestion, methane recovery and energy conversion, and to promote the value-added benefits of methane recovery to the agricultural industry and to electric utilities.

2.1. Crawford farm demonstration

One project, which has several public and private sector partners[♠] is a full-scale, proof-of-concept demonstration at the Steve and Audrey Crawford Farm located near Nevada, Iowa. This farm has a 2,800-hog finishing operation that will use a process called the anaerobic sequencing batch reactor (ASBR).

Among the several technical advantages of the ASBR is its ability to achieve higher microbial populations than is possible with other anaerobic treatment processes. This enables the ASBR to attain a higher production of biogas with a significantly shorter process retention time.

Although the Crawford Farm Demonstration project is not yet fully operational, some important lessons have already been learned. The original project proposal included methane energy conversion into electricity by an engine-generator sized to use all anticipated (but not metered) methane produced. There are obvious pitfalls with this approach. Any excess methane being flared would be detrimental to the economic performance of the project. Conversely, if methane production falls below expectations, there would not be enough to fuel an engine on a full time basis. Without sufficient biogas storage capacity, the resulting engine short-cycling would likely increase operation and maintenance expenses and shorten its useful life. In each case, a better approach would be to meter gas flow prior to the sizing and purchase of conversion equipment. Unfortunately, this is difficult to do where demonstration funding sources emphasize short time frames to spend grant funds and install projects, leaving little or no time to monitor or collect data.

Another challenge of engine-generator sizing concerns the amount of electricity used on site. Many livestock confinement facilities with anaerobic digesters will be capable of generating far more energy from recovered methane than can be used on site. The situation is analogous to wind turbine siting and sizing. In the case of wind turbine economic feasibility two critical issues involve the local electric utility energy sales agreement and the size of the

---

[▼]Advisory committee members include individuals from the Iowa Utility Association, Iowa Association of Electric Cooperatives, Iowa Farm Bureau Federation, Iowa Department of Agriculture and Land Stewardship, Iowa Utilities Board, Iowa Energy Center, Iowa Association of Municipal Utilities, Iowa Department of Economic Development, Iowa Department of Natural Resources Energy Bureau, Iowa State University Center for Advanced Technology Development, National Pork Producers, Iowa Pork Producers, Iowa State University Ag/Biosystems Engineering, Iowa Environmental Council, Iowa State University Center for Coal and the Environment, U.S.D.A. Natural Resources Conservation Service, Conservation Districts of Iowa, and Iowa Department of Natural Resources Air Quality Bureau.

[♠] Participants include the Center for Advanced Technology Development at Iowa State University, FOX Engineering Associates, ENTEK Biosystems, L.C. (formerly AmeriScan Technologies Corporation), the Iowa Department of Economic Development, the United States Department of Agriculture (USDA) Natural Resources Conservation Service, and the Iowa Department of Natural Resources.

turbine. Rarely is it feasible to size a wind turbine to take advantage of all wind resources at a given site, no matter how favorable site conditions may be. The same seems true of sizing engine-generators. In Iowa, the utilities are required to pay alternative energy production (AEP) facilities for electricity but only at the rate of the utility's avoided cost which is about $0.015 per kilowatt-hour. However, under 'net metering'[*] the utilities are required to allow each AEP to offset energy purchased, which may be worth $0.06 to $0.08 per kilowatt-hour or more.

Generalizations about the possibility of a favorable AEP-utility agreement are impossible to make owing to a pending challenge to net metering[9] and since the pending deregulation of electric utilities has made the future of net metering uncertain. However, certain municipal utilities and electric cooperatives with limited capacity may be interested in such projects to increase generation capacity and improve customer service with support of end-of-line voltage in rural areas. Therefore, either the site must have a favorable agreement with the local utility for purchase of excess energy, or else the size of generation equipment must be matched to the needs of the farm site with no excess to be sold. Remaining methane could be processed for resale or simply flared.

The Crawford Farm team decided to switch from an engine-generator to a biogas-fueled hot water boiler. A combination of more stringent engine-generator manufacturer warranty requirements for hydrogen sulfide, a recent change in the hydrogen sulfide scrubber supplier, and operator maintenance requirements beyond Crawford's ability to perform resulted in this change.

Another challenge concerns the goal that demonstration projects result in installation of more systems. System marketing is difficult unless proven economically feasible because anaerobic digestion and methane recovery is expensive to purchase and operate. One great motivator for these systems is the odor reduction of stored and land-applied effluent. If an appropriate value could be assigned to odor reduction associated with anaerobic digestion then the economic feasibility of these systems would be improved. Unfortunately, there are no known studies that quantify the economic value of odor control.

2.2. Swine USA project

The newest project in Iowa is the result of a U.S. Department of Energy (DOE) award of $100,000 to the Methane Energy Recovery Program to implement a methane recovery and energy conversion project.[10] The Iowa methane recovery project grant was one of 17 national projects selected and is part of a DOE program to develop renewable energy applications in remote and rural areas.[11] The energy system will be installed on a newly constructed hog confinement facility. In making the award announcement Iowa Senator Tom Harkin stated that the program was "another good example of Iowa finding creative ways to produce energy from untapped resources."[12] The intended benefits of this methane energy recovery project include the following:

---

[*] A qualifying alternate energy production (AEP) facility or small hydro facility is allowed to operate in parallel, with a single meter monitoring only the net amount of electricity sold or purchased during any billing period. This process is commonly referred to as 'net billing' or 'net metering.'

- electricity production to offset operating costs,
- a potential revenue stream,
- odor reduction,
- an environmentally friendly fertilizer product.

A contract was awarded to Swine USA to demonstrate an effective methane energy recovery and conversion system at their site.[+] Swine USA is a pork production company being managed by Livestock USA, LC. Livestock USA, LC is owned by Crestland Cooperative—a farmer-owned cooperative committed to environmentally friendly pork production.

The project site is a new 5,000 sow farrow-to-wean facility near Creston, Iowa. The facility will incorporate a complete mix anaerobic digester and produce methane to be used in an engine-generator. Environmental protection is to be provided by concrete basins with one year of treated storage capacity. It will have an estimated population of over 12,000 pigs with a total weight of nearly 2.2 million pounds. Parameters for the design of the anaerobic digester are as shown in Table 1.

Table 1
Digester design parameters

Characteristic	Parameter
Manure production	78,500 pounds per day
Dilution water	17,950 gallons per day
Total digester inflow	27,370 gallons per day
Manure volatile solids	6,517 pounds per day

The complete mix digester required to handle the above manure inflow will be a 657,000 gallon concrete tank sized for a retention time of about 20 days The digester is anticipated to yield an average daily methane flow of 27,400 cubic feet with an energy content of 27.4 million Btus. The engine-generator to be installed is expected to have an average output of 82 kilowatts.[13]

The energy generated would offset part of the estimated facility electrical demand with remaining energy available for sale to the local utility. This scenario will provide the best financial return assuming that a net metering agreement can be negotiated with the local utility. Assuming energy cost offset at the facility of $0.10 per kilowatt-hour and maintenance costs of $0.015 per kilowatt-hour yields a net benefit of $0.085 per kilowatt-hour. A summary of the costs and benefits of the methane conversion to electricity, based on an anticipated amount of engine downtime for maintenance, etc., is as shown in Table 2.

[+]Project partners include Swine USA, Livestock USA, Crestland Cooperative of Creston, IA, Agricultural Engineering Associates of Uniontown, Kansas, Mark Moser and Richard Mattocks of Resource Conservation Management, Inc. of Berkeley, CA under contract to ICF, Inc., U.S. EPA AgStar Program, The Iowa Pork Council, The National Pork Producers Council, Iowa Department of Economic Development, Iowa State University Extension, Iowa Department of Agriculture and Land Stewardship, U.S.D.A. NRCS, the Iowa Department of Natural Resources, and the U.S. Department of Energy.

All manure and wastewater generated will be used to fertilize surrounding crop and grass land by liquid injection. No manure will be sold. Annual soil testing will ensure application within utilization limits of crops. Application will be timed to avoid leaching of nutrients beyond the reach of crop roots. Application rates will be determined through nutrient testing of the stored digester effluent.

Table 2
Benefit/Cost Summary

Item	Annual Benefit/Cost
Electric energy	$61,327
Maintenance	($9,199)
Engine heat recovery propane offset	$12,879
Net benefit	$65,007

3. SUMMARY

By demonstrating the use of abundant agricultural waste to produce methane, the Department hopes to encourage private and public sector investments to develop this indigenous energy resource that promises to offset the use of fossil fuel, reduce the potential for air, soil, and water pollution, control odor, and benefit the rural economy.

REFERENCES

1. Des Moines Sunday Register, October 25, 1998, p 1.
2. Des Moines Sunday Register, November 8, 1998, p. 5AA.
3. Iowa Department of Natural Resources Energy Bureau.
4. Website for Hog Watch at www.hogwatch.org/resourcecenter/counter.html.
5. Website for the University of Minnesota Biosystems and Agricultural Engineering at www.bae.umn.edu.
6. Iowa Department of Natural Resources Environmental Protection Division.
7. Leggett, J., et al, Anaerobic Digestion: Biogas Production and Odor Reduction from Manure, Agricultural and Biological Engineering, Pennsylvania State University
8. Iowa Department of Natural Resources Energy Bureau.
9. Snyder, T. (1998-99). MidAmerican Petitions FERC to Prevent Net Billing, Energy Matters: Making Renewable Energy a Reality, The Newsletter of the Iowa Renewable Energy Association, Winter 1998-99, 8, p 2.
10. Energy Department Awards States $1.4 Million for Renewable Energy Projects, DOE News, August 12, 1998.
11. $100,000 Grant for Methane Project, Energy News newsletter, Iowa Department of Natural Resources. September 1998, p 4.

12. Harkin Announces $100,000 Grant for Alternative Energy Project, press release, August 7, 1998, www.senate.gov/~harkin/releases/98/980807.html.
13. Project assessment completed by Resource Conservation Management, Inc., for SwineUSA and AgStar.

ANAEROBIC DIGESTION OF BREWERY WASTEWATER FOR POLLUTION CONTROL AND ENERGY

D. W. Williams,[a] D. Schleef,[b] and A. Schuler,[a]

[a]BioResource and Agricultural Engineering Department, California Polytechnic State University, San Luis Obispo, California 93407
[b]SLO Brewing Company, 1400 Ramada Drive, Paso Robles, California 93446

This paper describes the operation of a pilot-scale anaerobic filter to treat brewery wastewater containing approximately 6000 mg/l of COD. The 2600-liter pilot digester was loaded at up to 2.79 kg of COD per cubic meter of digester per day and an HRT of 2.2 days. Biogas containing 75% methane was produced at the rate of 0.39 cubic meters per kg of COD per day, and the COD was reduced by 87%. The digester was maintained at 36°C. and the effluent pH was above 6.8 as long as the influent pH was adjusted to above 6.0. A full-scale system that would treat 40,000 liters of wastewater per day would have a volume of 80 cubic meters and produce 90 cubic meters of biogas per day while reducing COD by 80 to 90%. Additional aeration will be required to meet the 400 mg/l COD discharge requirement.

1. INTRODUCTION

The purpose of this study was to test the anaerobic filter digestion process developed at Cal Poly to treat the wastewater from the SLO Brewery Company in Paso Robles, California. This relatively new brewery started production in 1998 with an initial beer production of 10,000 liters and wastewater production of 27,000 liters per day; average COD was 5500 mg/liter. These figures are for the plant running at approximately 50% of capacity; therefore the design parameters for the full-scale wastewater plant were based on 2.8 liters of effluent per liter of beer. At full capacity, when the brewery is producing 20,000 liters of beer per day, these figures were estimated to be 55,000 liters of wastewater.

1.1. Previous work on anaerobic filters
The basic studies of Young and McCarty (1968) were largely responsible for early development of the anaerobic filter process. In the 17 years between that report and the summary of Colleran, et al. (1983), the process was developed and commercialized to treat food processing wastewater, wheat starch wastewater, sugar refinery wastewater, and guar processing wastewater.
Since 1983, large numbers of variously configured proprietary attached growth anaerobic filter systems have been commercially installed at food processing plants, cheese plants,

pharmaceutical plants, water plants, landfills and breweries. However, literature reporting on the performance of these installations has been limited due to the commercial nature of the projects. The project manager is familiar with many of the designers and has visited some of the plants. Several design firms do most of the design work: Biotim (Belgium), Biothane (Netherlands, US), ADI, Ltd. (Canada, US), and Aqua-Tec (US). However, these firms work exclusively for large corporations because they do not see any potential profits from working on small-scale systems. There is a large digester installation at Anheuser-Busch in Van Nuys, California, which is treating the wastewater from 900,000 gallons (3.4 million liters) of beer production per day.

1.2. Cal Poly experience

The work at Cal Poly has included several pilot scale anaerobic filter studies (Williams, et al. 1989 and Williams, et al. 1992) involving the treatment of vegetable wastes, swine manure and domestic organic wastewater. The pilot scale digester used for these studies was a 2600-liter fiberglass tank with a special plastic honeycomb-type media designed for trickling filters. The studies with tomato wastewater showed that up to 99% reduction in COD was possible with an organic loading rate of 0.85 kg of COD per cubic meter of digester per day, an HRT of 24 days and an average temperature of only 23°C. This digester operated in an upflow mode with pre-heated influent and recirculation through internal plumbing. The incoming tomato wastewater had a COD or 20,000 mg/liter and total solids content of over 2%.

2. OBJECTIVES

Biological stabilization of brewery wastewater is a problem affecting brewery management. The requirement for discharge into the Paso Robles sewage treatment plant includes a BOD level of no more than 250 mg per liter. The ratio of BOD to COD for brewery wastewater is approximately 0.6, so the equivalent COD requirement would be 400 mg/liter. This project therefore investigated the use of the Cal Poly anaerobic filter system to organically stabilize the brewery wastewater and to produce usable methane. The objectives were as follows:

- modify the pilot-scale attached film digester at Cal Poly, San Luis Obispo, for high rate anaerobic digestion to brewery wastewater,
- perform digester start-up and operate it for 10 weeks while gradually increasing the loading rate,
- collect and analyze data from the digester operation and chemical testing,
- evaluate the effectiveness of the pilot-scale operation in terms of gas production and reduction on COD,
- recommend the optimum parameters for a full-scale digester design and estimate the capital and operation costs.

3. PILOT-SCALE TEST

This section describes the test plan carried out by Cal Poly personnel in completing the tasks necessary to evaluate the anaerobic filter digestion system in terms of its effectiveness in treating brewery wastewater from the SLO Brewing facility in Paso Robles, California. The equipment used for this project was a pilot-sized digester, consisting of a 2,600 liter fiberglass digester tank filled with plastic media, a heat exchanger—electric water heater combination, pumps for circulation of the digester contents and hot water, piping and valves, a gas meter, and a thermostat for temperature regulation.

3.1. Methodology

It was determined, after brief literature search regarding mesophilic digestion of brewery wastewater, that the maximum loading rate at 35°C was 10 kg of $COD/m^3/day$, and COD reduction of 85%. In this study reported by Beyen et al. (1988), the HRT was less than one day with wastewater having a COD of 1500 to 3000 mg/liter. It was also determined that the nitrogen content of the beer waste was sufficient so that no additional nutrient was required.

The start-up procedure was to first move the skid-mounted pilot unit to the SLO Brewing site at the end of July, 1998. The pilot-scale digester was then seeded with approximately 1100 liters of anaerobic digester sludge from the Paso Robles sewage treatment plant. This combined with 1400 liters of water and the heat exchanger set to warm the contents to 36°C, the circulation pump started, and this condition was maintained for one week of start-up while monitoring pH and gas production. Then on August 7, influent additions began, and during the next five weeks, an average of 112 liters of influent was added per day. The summary of the digester results is presented in Table 1.

Table 1
Summary data from SLO brewing pilot digester study

Pump Setting #	HRT, Days	Influent, L/d	Influent COD mg/l	Influent pH	Effluent COD mg/l	Effluent pH	Biogas, m^3/d	% Methane	Biogas, m^3/kg COD/day	Organic Loading, kg $COD/m^3/d$
1	23	112	4757	7.41	896	7.18	0.40	76.58	0.76	0.20
2	4.6	562	5859	7.44	855	7.17	1.42	75.40	0.43	1.27
3	2.2	1200	6050	6.31	800	7.10	2.83	74.40	0.39	2.79

The pilot scale digester was designed to require as little maintenance as possible. There were a few daily services that needed to be performed on the digester to keep it running. It should be noted that the following tasks were performed every other day until the influent flow rate was increased. After the increase the tasks were performed daily. These daily tasks included 1) recording level in influent tanks; 2) filling equalization influent tanks; 3) if necessary, adjusting pH to above 6.0 using sodium hydroxide; 4) drawing samples of influent and effluent for COD tests; 5) checking and recording digester temperature; 6) testing CO_2 content in biogas and recording gas meter reading; and 7) recording filled level of influent tank. In addition to these daily tasks there was a heating system to check and periodically maintain. These combined tasks only required an average of an hour per day (or every other day early in the testing), and were performed by one person.

3.2. Results

The initial loading rate, which corresponded to the setting number "1" on the influent pump. This rate was maintained for approximately five weeks, during which time the average biogas production was 0.4 cubic meter per day, and the COD reduction was from 4757 to 896 mg/liter, or 81%. The loading rate was then increased to setting number "2" on the influent pump, and as shown in Table 1, the HRT was dropped from 23 to under 5 days, for an average of 562 liters of influent per day. This corresponds to a loading rate of 1.27 kg COD/m^3/day, and gas production of 1.42 cubic meters of biogas containing over 75% methane.

Loading rate number 2 was maintained for five weeks, during which time the COD was reduced on average from almost 5900 mg/liter to under 900 mg/liter, or 85%. This shows that increased loading rates can result in even higher performance by the digester, possibly due to higher bacterial populations in the digester over time. Loading rate number 3 was then introduced, in which 1200 liters of influent was added daily, which corresponds to just over a two-day HRT. This rate was maintained for approximately one week, during which time the COD reduction was from over 6000 mg/liter to 800 mg/liter, or 87%. Gas production increased to over 2.8 cubic meters per day with the same methane percentage as earlier tests, 75%. During these increases in loading rates, the digester maintained stable conditions as far as pH was concerned. As shown in Table 1, the average effluent pH was at or above 7.1 for the entire test. It was, however, necessary to make sure the influent pH was always adjusted to above 6.0 for digester stability to be maintained. Because the digester effluent, approximately 800 to 900 mg/liter of COD, was above the discharge limit, 400 mg/liter, it will be necessary to have some additional aerobic treatment to bring the effluent to below the sewer plant required BOD level of 250 mg/liter(equivalent to 400 mg/liter COD).

4. FULL-SCALE DIGESTER PARAMETERS

Based upon the pilot-scale testing, the following parameters for a full-scale anaerobic treatment system for the SLO Brewery operation in Paso Robles were recommended. Assuming that the brewery produces up to 20,000 liters of beer per day, the resulting wastewater will be up to 55,000 liters per day, or approximately 2.8 liters of wastewater per liter of beer. On an annual basis this would amount to 9 million liters of beer with 25 million liters of wastewater, based on five-days-per-week operation. When averaging the wastewater for a seven-day week, the average daily wastewater to be treated would be just less than 40,000 liters per day. The average COD of this wastewater was assumed to be 6,000 mg per liter, with the pH adjusted to over 6.0 before loading into digester.

Figure 1 shows a schematic diagram of the proposed anaerobic filter digester system to treat this wastewater so that the resulting digester effluent has a COD of below 1000 mg/l and a pH of 7.0. The estimated biogas production, based upon 0.39 cubic meters per kg of COD, will be approximately 90 cubic meters per day, from a digester with a total volume of

80 cubic meters having a hydraulic retention time (HRT) of two days. The biogas, containing approximately 75% methane, will fuel 200 MJ/hour hot water boiler that produces approximately 2500 MJ per day, which is 25% more than the energy required to preheat the influent from 26 to 36 degrees C. As shown in Figure 1, the digester volume is divided into two systems that treat the wastewater in a parallel mode. There are two 40,000-liter equalization tanks, and two 40,000-liter anaerobic filter digester tanks. The effluent then flows to one aerobic treatment system to take the COD from 1000 mg/l to less than 400 mg/l.

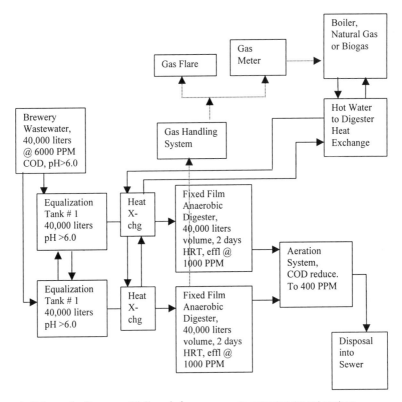

Figure 1. Schematic diagram of full-scale brewery wastewater treatment system.

5. CONCLUSIONS

Based upon the pilot-scale anaerobic filter study that was done on wastewater from a small brewery, the following conclusions were made:

- Typical brewery wastewater containing approximately 6000 mg/liter COD was successfully treated using a fixed film anaerobic digester process at 36 degrees C.
- At an HRT of two days and a biological loading rate of 2.79 kg COD per cubic meter per day, the COD was reduced to under 800 mg/liter, or an 87% reduction.
- Gas production was 2.8 cubic meters per day with 75 percent methane, for a specific biogas production of 0.39 cubic meters per kg COD per day.
- Stable conditions with respect to pH were maintained as long as the influent pH was adjusted to above 6.0; the effluent pH was at or above 7.1 for the entire test.
- A full-scale treatment system for the brewery producing 40,000 liters per day of wastewater would be an 80 cubic meter anaerobic filter tank, would produce 90 cubic meters of biogas per day, and would reduce COD by 80 to 90%. Additional aeration will be required to meet the 400 mg/l COD discharge requirement.

REFERENCES

Beyen, G., R. Kruitwagen, and L. Vriens (1988). Experience with a pilot-scale two-stage anaerobic unitank system on brewery wastewater at lower temperatures, in Poster Papers, Fifth International Symposium on Anaerobic Digestion, ed. by A. Tilche, Monduzzi Editore S.p.A., via Ferrarese 119/2-40128 Bologna, Italy.

Colleran, E. et al. (1983). One and two-stage anaerobic filter digestion of agricultural wastes, in Third International Symposium on Anaerobic Digestion—Proceedings. Published by Third International Symposium on Anaerobic Digestion, Cambridge, Massachusetts.

Williams, D.W., J. Montecalvo, and R.W. Babcock (1989). Pilot-scale digestion of cull tomatoes using an up-flow anaerobic digester, presented at the Georgia Tech 1989 Food Processing Waste Conference, Atlanta, Georgia, November.

Williams, D.W., R. Kull, and S Schwartzkopf (1992). Anaerobic treatment of organic wastes from controlled ecological life support systems, SAE Technical Paper No. 921272, presented at the 22nd International Conference on Environmental Systems. Seattle, Washington.

Young, J.C. and P.L. McCarty (1968). The anaerobic filter for waste treatment, Technical Report No. 87, Department of Civil Engineering, Stanford University, Stanford, California.

BIOGAS SLURRY UTILIZATION IN GHANA

Christine Asser

Ministry of Mines and Energy, Accra, Ghana

The biogas Technology Programme developed for villages remote from the national grid relies solely on cow-dung and human waste as raw materials for electricity generation. The technology helps control deforestation and desertification and reduces the long hours women spend looking for fuelwood for cooking and the health risk that smoke from fuelwood poses to women and children. Rural electricity generation also supplies pipe-borne water, community toilets and organic fertilizer for farmers.

The slurry, which is the by-product of the biogas, is an odourless, pathogen free organic fertilizer that is high in nitrogen, phosphorous and potassium. It is useful for crop production, fish farming and mushroom cultivation. Farmers now realize that energy can be extracted from cowdung, human waste, agricultural residua and other biomass resources and the biomass can still maintain its function as good organic manure.

1. INTRODUCTION

Because hydropower is the main source of electricity in Ghana, electricity must be rationed during drought, as happened in 1983-1984 and 1997-1998. Thus, diversification of the sources of power is sought. The energy economy of Ghana is predominantly woodfuel based. Firewood and charcoal consumption account for about 69% of the total energy consumption. Electricity and petroleum products account for 10% and 21% respectively. Agriculture is responsible for 8% of the annual petroleum consumption.[1]

Biogas technology was developed in Ghana in 1987 to generate electricity for villages 30 km or more from the national grid. The main purpose for the dissemination of the biogas technology is to improve sanitation and agriculture. The technology uses agricultural waste materials and ends with the by-product slurry as an organic fertilizer to increase crop, mushroom and fish yields.

2. BIOGAS TECHNOLOGY IN GHANA

The anaerobic fermentation of organic material waste such as human and animal excreta, agricultural waste, industrial waste, sludge, domestic sewage and garbage to emit methane gas is termed "biogas fermentation." This degradation of waste under anaerobic conditions by microbes grouped as nonmethane-producing bacteria from the fermentative (hydrolytic) and hydrogen-producing acetogenic bacteria group to the methane producing bacteria group completes a whole life-saving circle of waste treatment which would have been hazardous if left under aerobic fermentation.

$$\text{Complex Organic Matter} \xrightarrow{\text{Hydrolysis Enzymes}} \text{Simple organic compounds} \xrightarrow{\text{Acid bacteria}} \text{(VFA) Organic acids} + CO_2 \xrightarrow{\text{Acetogenic bacteria}} \begin{array}{c} NO_2 \\ CO_2 \\ \text{Acetate} \end{array} \xrightarrow{\text{Methanogens}} CH_4 + CO_2$$

In Ghana, as in other underdeveloped countries, biogas technology was introduced in the 1980s. The first biogas digester was constructed in the Shai Hills in the Greater Accra region in 1985 by the National Energy Board and the Institute of Industrial Research. The project did not succeed due to technical problems of being the first of its kind in the country, lack of supervision, and lack of financial and community support.

In 1987 the same group, with the material and technical support of Chinese and Indian Governments, established the Appolonia Biogas demonstration center now know as the Appolonia Integrated Rural Energy and Environmental project.

2.1. Target area profile

The project began with a pre-feasibility study in the Greater Accra Region to ensure easy monitoring. The Appolonia village was selected as one of the villages far from the national grid with the availability of raw material (cow dung) as feedstock for the digesters. Appolonia, a typical Ghanaian village of 900 people, relied solely on biomass for cooking and kerosene for lighting.

The village depended on rain water as the cleanest source of drinking water, and two unclean ponds. The village, which geographically lies within the coastal savanna grassland, has a large cattle population and free-range pigs and other domestic pets and animals that competed with the human population for the same drinking water source. There were no sanitation facilities.

The villagers, who are peasant farmers, also cut firewood, mainly nim trees to sell in nearby towns. The rapid rate of tree felling for sale, charcoal and fire wood was gradually leading to deforestation in the short term and desertification in the long term. In addition, the village was engaged in sand winning. It was hoped that rural biogas electrification would halt these activities.

The Ghana Government electrification programme determined that villages 30 km away from the national grid could not be connected with electricity from the hydropower plant a

Akosombo, the only source of electricity generation at the time of the study. To meet the challenges of rural electrification, biogas technology was the best option for Appolonia, due to the availability of raw material locally and from the nearby towns of Ashaman and Tema, whose human waste could be collected.

2.2. Community involvement and participation

In order to ensure the sustainability of the biogas project in the Appolonia village, the community was required to provide communal labour support. The community is economically sound in terms of trade, employment, animal rearing and crop production. A pre-feasibility study ascertained that the electricity generated will be easily affordable by all households and the community at large.

The project was sponsored by the Chinese, Indian and Ghana Governments. The Indian and Chinese Governments supported the project with technical staff, two generator sets of total capacity of 12.5KW/h, laboratory instruments, biogas gas cooking stoves and biogas lanterns. The Ghana government supported it with technical staff from the National Energy Board and the Industrial Research Institute and ¢54 million cedis (about US$180,000) for the construction of ten 50 cu.m digesters, 19 domestic biogas digesters, two biogas latrines, a project and community building, a one acre experimental farm plot and one kraal.

The community communal labour support consisted women carrying water from ponds, and sandstone and mortar for the construction of the digesters, and the men digging and carrying blocks.

2.3. What went wrong

The scientific aspect of the technology was ignored. The microbiological activity of the waste before and after fermentation was studied little or not at all.

The main objective for the Ghana Appolonia biogas project was first for generation of energy to supply electricity to the village which is far from the national grid. All other benefits such as gas for cooking, sanitation and agriculture were afterthoughts. As was also true in other countries, Ghana did not factor into the project the manufacturing of biogas accessories such as lamps and stoves that are subject to failure. When the first appliances supplied to the users, probably free, break down and replacement parts are not available, then the technology begins to suffer from neglect and abuse by users. The few people who are trained to construct, educate, and disseminate information to the public could cover only a small target area owing to lack of logistics and governmental and individual support, which led to lack of interest.

3. SLURRY AS AN ORGANIC FERTILIZER

Slurry is the by-product of the anaerobic fermentation of waste materials that generate biogas for cooking and lighting. The slurry is an organic fertilizer high in nitrogen, phosphorous and potassium (NPK) content. It has been established that after the biogas fermentation there is little loss of nutrients and low mineralization and the rapid available nitrogen is very high. In aerobic conditions, nitrogen in organic matter can be bound by mineralization into ammonium or oxidized by rainwater into nitric acid and nitrates. However when organic waste material is fermented in a biogas digester, it is in a hermetic anaerobic condition, which makes decomposition relatively slow and favours the maintenance of ammonia.[2]

Slurry is the most environmentally sound fertilizer at the farmer's disposal in Ghana. It has no smell, does not attract flies, does not pollute the atmosphere during its application, increases yield, does not pose health hazards to the user and animals around. Most harmful bacteria and parasites in the organic feed are mostly or completely killed by the biogas production process. Both liquid and solid slurry has been applied. Large quantities are needed to get an adequate volume of fertilizer for a particular crop. It is difficult to transport liquid slurry. But, drying is a major problem during the rainy season. When completely dried it forms lumps which take longer time to dissolve completely into the soil for plant use.

3.1. Sources of agricultural wastes in Ghana

The agricultural waste material are derive from three main sources: animal manure, agricultural crop wastes and agricultural-based industrial wastes. The biogas technology in Ghana solely relies on animal manure.

Animal Waste	Agricultural Waste	Industrial Waste
Cattle	Rice straw	Logs and lumber
Poultry	Maize	Sawdust
	Other vegetables	Other forestry
		Brewery

All the above can be used to generate methane gas through anaerobic reaction. The question is quantity of waste produced at each level of the groupings. A thorough study should be conducted in the country to ascertain the total organic waste emission available for processing, to confirm the organic solid based source for biogas generation for villages far from the grid.

The soil and slurry samples were tested by the Department of Soil Science of the University of Ghana, Legon. The samples were tested to ascertain the initial nutrient content of the experimental plots and the slurry and the effect of the slurry on the soil and the crop. Though liquid cow dung slurry was used in the experiment all other types of slurry available at the project site were also tested, because other farmers use other slurry available to them. Table 1 shows that the pH values for the soil is acid. The concentration of exchangeable bases. Organic carbon, total nitrogen, available phosphorous and total phosphorous is very low. The carbon nitrogen ratio values show that nutrients will be released readily for crop uptake. Generally the soil requires inorganic fertilizer or manure application to support crop growth because of the inherent poor fertility status.

Table 1
Analytical data for soil and slurry

Sample	O.C.	TN	C/N	AP	TP	Exchangeable bases				Total	pH
	----(g/kg)---			------ mg/kg------		Ca	Mg	K	Na		
						------------cmol (+) ------------					
Soil depth 0-15cm	17.7	1.8	9.8	18.1	160	1.68	1.68	0.58	0.15	4.09	6.9
15-30 cm	11.8	1.2	9.8	6.0	120	1.74	1.65	0.46	0.22	4.07	5.9
Human waste + cowdung	31.1	16.0	1.9		8,150	14.15	17.0	5.52	4.52	41.20	6.5
Cowdung slurry (liquid)	130.5	16.6	7.9		4,350	16.01	17.41	7.71	3.65	44.84	7.0
Cowdung (dry)	50.7	20.0	2.5		7,250	16.68	19.09	6.68	4.26	46.71	7.3
Cowdung (fresh)	243.4	15.1	16.0		3,750	6.68	11.08	11.30	2.26	31.29	8.0

The slurry samples have a better nutrient composition than the soil. The human waste and cow dung slurry both in liquid and solid states have near neutral pH values (7.0 to 7.3). The liquid cow dung slurry and the dry cow dung slurry appear to be of better quality and would improve the fertility of the soil and ensure greater crop yield. In the experiment the dry cow dung slurry was used.

4. MATERIALS AND METHODS

The study on the effects of biogas on sweet pepper production was conducted in March 1998 at Appolonia near the biogas project site. It was the third experiment on sweet pepper with slurry in the liquid state. The experiment was conducted here because of the nearness of the slurry source. Ploughing was done by power tiller which is the main transport for conveying cow dung from the near-by kraals for charging the digesters. The sweet pepper nurseries were made in wooden cases and were intensively cared for. The duration of the nursery was 10 weeks. Both the transplanting and nursery plants were laid in rows. The experimental design was randomized complete block design (R.C.B.D) factorial. The experimental plots were 2 m × 1.4 m, alleys between plots were 40 cm and alleys between replications were 40 cm. The total field size was 98 m². There were seven treatments (T) in kilogrammes of the following order: T = 0-Control, T = 10, T = 20, T = 30, T = 40, T = 50 and T = 60 and five replications (R) in the field layout. The land was ploughed seven days before transplanting. During this period the field was pegged to demarcate all the treatments for each trial. There were 16 plants on each plot. The transplanting was done early in the morning on the same day for all the trials. Slurry was the only organic fertilizer used in its dried state and the complete dose for each treatment was applied the same day as sweet pepper was transplanted. There was no pest control in this experiment, weed was controlled by weeding with a hoe. All other factors like temperature and moisture were constant.

5. INTERPRETATION OF DATA

Table 2
Fruit count and weight—replication

Treatment (kg)	Total no. of fruits	Average no. of fruits	Total weight (kg)	Average weight (kg)
0	73	14.7	3.73	0.746
10	350	70.0	17.380	3.476
20	530	106.0	23.720	4.744
30	663	132.6	29.051	5.81
40	766	153.2	32.305	6.461
50	830	166.0	39.160	7.832
60	673	134.6	31.950	6.39
TOTAL	3885	777.1	177.296	35.459

The minimum total number of fruits, 73, and the minimum weight, 3.730 kg, was observed with the 0 kg (T = 0) treatment which was the control. However, increased doses of slurry increased both the total number of fruits and total weight. With excessive increases of slurry doses the total number of fruits and total weight dropped, as was shown by the 60 kgs (T = 60) treatment. As this Table 2 indicates the yield of the T = 50 treatment produced the greatest quantity of fruits, 830, as well as the highest total weight, 39.160 kg. The difference in the number of fruits and total weight between the control and the T = 50 treatment was 757 and 35.430 kg respectively.

Table 3 indicates the total yield of all the treatments from the total weight of all the seven treatments of the five replications. The table also shows the total and average weights of all the harvest and the replications as shown by the last two columns. The highest total yield was achieved by the third replication and the lowest by the first replication. The highest average yield of 7.832 kg was recorded by T = 50.

Because the computed F value is larger than the tabular F value at the 1% level of significance, the treatment difference is judged to be highly significant.[3]

Table 3
Yield (kg)

Treatment	Replication					Total	Average
	1	2	3	4	5		
0	1.050	1.150	0.180	0.550	0.800	3.730	0.746
10	1.050	6.560	3.420	4.720	1.630	17.380	3.476
20	1.055	3.695	7.070	6.590	5.310	23.720	4.744
30	3.001	5.600	8.970	6.630	4.850	29.051	5.810
40	3.075	7.220	7.020	8.830	6.160	32.305	6.461
50	5.550	5.000	13.310	9.140	6.160	39.160	7.832
60	6.830	8.610	8.180	3.680	4.650	31.950	6.39
Total	21.611	37.835	48.15	40.14	29.56	177.296	35.459
P	x1	x2	x3	x4	x5	X	A

Table 4
Analysis of variance

Source Of Variation	Degree of freedom	Sum of square	Mean square	Computed F	Tabular F 5%	1%
Treatment	6	166.0	27.7	7.5**	2.51	3.67
Replication	4	59.31	14.8			
Error	24	89.51	3.7			
Total	34	314.82	9.3			

Coefficient of variance (CV) = 5.4%
** - Significant at 1% level

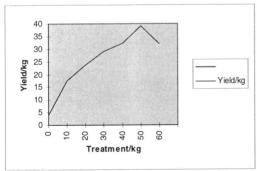

Figure 1. Regression between sweet pepper yield and slurry rate.

6. CONCLUSION

Farmers know that organic waste can be applied raw as an organic fertilizer. The experiment has created additional awareness that organic waste can also be processed by anaerobic reaction to generate gas for cooking and electricity for lighting and still maintain its quality and function as a soil conditioner and a good fertilizer. The slurry from the biogas technology is an environmentally friendly fertilizer that does not pose health hazards to animals or man. It can be applied the same day as sowing or transplanting is being done. The life of the villagers is enhanced because waste is easily disposed of to improve sanitation, and electricity is available from biogas generation. Increased slurry application shows increase in crop yield till an optimum is attained.

REFERENCE

1. Ministry of Mines and Energy (1996-2000). Energy Sector Development Programme.
2. Chengdu Biogas Research Institute of the Ministry of Agriculture (1997). P. R. C. The Biogas Technology in China.
3. Gomez, K.A. and A.A Gomez (1984). Statistical Procedures for Agricultural research.
4. Shepley, Genoa and Anne Eldridge (eds) (1986). Gardening—Expert Advice from National Garden Association.
5. Mosgaard, Christine and Birgitte Ahring (1993). Perspectives for Using Large and Medium Scale Biogas Digestors.

FEASIBILITY OF USING RICE STRAW FOR BIOGAS ENERGY PRODUCTION IN CALIFORNIA

Ruihong Zhang[a] and Jane Turnbull[b]

[a]Biological and Agricultural Engineering Department, University of California at Davis, Davis, CA 95616
[b]Peninsula Energy Partners, Los Altos, California 94022

Rice straw is abundantly produced in the Sacramento Valley of California. Anaerobic digestion offers an effective approach to convert rice straw into biogas energy. A high-rate, space-efficient anaerobic digester system, the anaerobic phased solids digester (APS-Digester), was developed recently at the University of California at Davis to convert solids wastes including rice straw into biogas and a stable soil-amendment product. Laboratory research has determined that 4.8 cu. ft. biogas can be produced from each pound of dry straw when the anaerobic digestion is carried out at 35°C with a 24-day retention time. The biogas contains about 50% methane. Pretreatment with mechanical means, chemicals and heat will increase the biogas yield. A pilot-scale APS-Digester system is currently under development and will be used to determine the optimum design and operating parameters for anaerobic digestion of rice straw and other types of feedstock. A preliminary study of the economic feasibility of energy production from rice straw showed that commercial APS-Digester systems using rice straw as feedstock to generate electric power at a cost of about $32 per ton would need to be at least 250 to 500 kW in capacity in order to be economically viable. The electricity and heat produced would be used on-site at industrial facilities, such as rice mills, food processing and packaging plants and large-scale refrigeration units, which have a consistent need for power year-round.

1. INTRODUCTION

Large quantities of rice straw are produced in California as a by-product of rice production. In the Sacramento Valley alone, 1.5 million tons of rice straws are produced each year. Open-field burning, the traditional straw disposal method, has caused widespread environmental concerns. The Rice Straw Burning Reduction Act of 1991 in California mandates rice growers to phase-down burning of rice straw; no more than 25% of the planted acreage or 125,000 acres in the Sacramento Valley, whichever is less, may be burned after the year 2000. At present, about 0.6% of the rice straw produced is disposed of or utilized off-farm and about 40% is incorporated into the soil. Rice growers are under extreme pressure to find alternative environmentally friendly methods for straw disposal and utilization. If no other practical straw disposal alternatives are developed to replace burning by the year 2000, an estimated 72.9% of the rice straw must be incorporated into the soil.

However, available research findings and experience suggest that this practice could potentially reduce crop yields, increase foliar disease, and degrade soil conditions. Other cost competitive technologies for rice straw disposal and utilization must be developed. Energy production has been identified by the state legislators and rice growers as one of the four off-farm alternatives. The other three alternatives are construction and manufacturing uses, environmental mitigation, and livestock feed. The energy content of rice straw is about 6,533 kJ/kg; the annual energy stored in rice straw of the Sacramento Valley is 8.9×0^{12} kJ. Thus, it is realistic to consider rice straw as an important renewable resource for energy generation.

Anaerobic digestion offers a promising alternative approach for production of a gaseous fuel. The digestion process uses a consortium of natural bacteria to convert a large portion of rice straw into biogas, a mixture of methane and carbon dioxide. If captured, the biogas can be used to fuel an appropriately sized fuel cell, gas turbine, or engine-generator to generate both electric power and heat to meet a local energy demand. The digested solids are expected to be marketable co-products as fertilizer and soil amendment. Recent research at the University of California at Davis (UC Davis) has led to development of a high-rate, space-efficient anaerobic digestion system, called an anaerobic phased solids digester (APS-Digester), for digesting solids wastes such as rice straw. The APS-Digester System has been extensively evaluated in the laboratory for rice straw conversion and proven to be a reliable system that can produce a medium Btu biogas (500-600 Btu/cu.ft.) from a range of biomass feedstocks, including rice and wheat straws, vegetable wastes, and paper sludge. A pilot-scale APS-Digester system is currently under development at UC Davis. This paper presents some of the laboratory test results on biogasification of rice straw with the APS-Digester system and feasibility analysis of energy generation from rice straw in California.

2. DESCRIPTION OF ANAEROBIC PHASED SOLIDS DIGESTION SYSTEM

The APS-Digester system was developed for digesting bulk solids that are not easily transportable with pumps. The APS-Digester system consists of one or more hydrolysis reactors coupled in a closed-loop configuration with a biogasification reactor (Figure 1). The hydrolysis reactor(s) processes the solid feedstock in batches or semi-batches, which reduces system labor requirements. The biogasification reactor, on the other hand, maintains an active bacterial culture and produces biogas continuously. Liquid circulates intermittently between the reactors to inoculate and hydrolyze solid feedstocks. The hydrolyzed substances pass into the biogasification reactor where the biogas is produced by methanogenic bacteria. The biogasification reactor is operated as an anaerobic sequencing batch reactor (ASBR) which allows retention of biomass (bacterial solids) through gravity settling. The working principles of ASBR are described by Zhang et al. (1997). Biogas is collected from all the reactors to be used as a fuel. The residual solids after digestion are removed from the hydrolysis reactor prior to loading a new batch of feedstock. The anaerobic digestion system includes hydrolysis and biogasification reactors, heat exchangers, feedstock loader, solid-liquid separator, and some means for short-term storage of biogas. The innovative design features allow the APS-Digester system to have greater effectiveness than existing anaerobic digestion technologies (Zhang and Zhang, 1998). Since the majority of anaerobic bacteria in the APS-Digester system are housed in a separate reactor from the feedstock solids, loading and unloading of the solids do not disrupt the anaerobic environment of the bacteria present in the biogasification reactor. Therefore, an active bacterial population is maintained in the biogasification reactor and the food to microorganisms ratio (F/M) can be more effectively managed, resulting in more efficient bacterial activities.

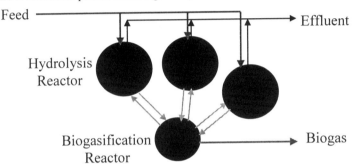

Figure 1. Anaerobic phased solids digester system.

The APS-Digester System has been extensively evaluated in the laboratory with different types of feedstock including rice and wheat straw, food wastes (garlic and onion), and paper sludge and has proved to be an easy-to-operate, reliable system. The APS-Digester System has great potential to provide an effective and reliable alternative to conventional solid waste disposal methods with the following advantages:
- Can provide an environmentally sound and sustainable means of utilizing organic wastes.
- Is able to treat a variety of organic feedstock.
- Can produce renewable bioenergy.
- Can be designed to different capacity requirements to adapt to different production scales.

A pilot-scale engineering model system is under development at UC Davis and is expected to be operating by the summer of 1999. The pilot APS-Digester system will be used to evaluate the digestion properties of various feedstock including rice straw, determine the material handling requirements for commercial installations, and develop optimum design and operational parameters for the system. The pilot system is designed with a capacity of 1,000 pounds (dry weight) per day. The approximate layout of the pilot-scale system is shown in Figure 2.

Figure 2. Layout of a pilot-scale APS-Digester system.

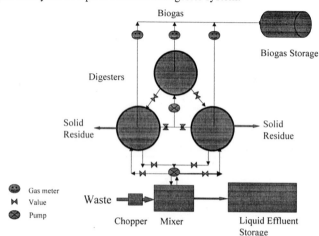

The system will consist of two hydrolysis reactors and one biogasification reactor, various equipment for material handling and gas collection and storage, gas meters, computer controls and facilities for effluent storage and disposal. The operation of the system, such as digester heating, liquid recirculation and biogas collection and storage, will be automatically controlled through a central computer station. Each of the three reactors will be a 500-gal stainless steel tank with appropriate insulation. The tank will be heated to maintain a constant temperature (35°C or 55°C). The sampling ports for solids, liquid, and biogas will be provided for sample collection and analysis.

3. TECHNICAL FEASIBLITY OF RICE STRAW DIGESTION

Rice straw is a ligno-cellulosic material, and its digestion rate largely depends on the degradation rate of polysaccharides, such as cellulose and hemicellulose. The hydrolysis of polysaccharides, which is carried out by enzymatic reactions, is the rate-controlling step for anaerobic digestion. This characteristics makes control of the digestion process relatively easy in terms of maintaining the pH within a desirable range (6.8-7.6). Laboratory experiments were performed to evaluate the digestibility of rice straw with the APS-Digester system under different operating conditions. The effects of different pretreatment methods—physical (mechanical), thermal and chemical treatment on the digestion rate of rice straw were examined at a mesophilic temperature of 35°C. The physical pretreatment include two types of size reduction methods (chopping and grinding) and two sizes (10 mm and 25 mm). Thermal and chemical pretreatment was performed by adding ammonia to the straw at 2% dry weight and heating the straw in a pressure cooker for two hours at three different temperatures (60°C, 90°C and 110°C). A combination of pretreatment conditions was tested along with no pretreatment. The detailed laboratory results have been reported by Zhang and Zhang (1998). The major findings are summarized as follows.

The average moisture content of rice straw tested was about 8% and the volatile solids (VS) was 79.5% of total solids (TS). The rice straw contained 0.46% nitrogen and 0.09% phosphorus. The carbon to nitrogen ratio (C/N) was 75.7. Therefore nitrogen needs to be supplemented in order to enhance the straw degradation. Ammonia was used in this study as the nitrogen source. The biogas yield of untreated whole rice straw with a 24-day retention time was 4.8 ft^3/lbTS or 4.4 ft^3/lb straw, assuming 8% moisture in the straw. All pretreatment methods tested improved the digestibility and biogas yield of the rice straw. A combination of grinding (10-mm length), heating (110°C), and ammonia treatment (2%) resulted in the highest biogas yield, 6.0 ft^3/lbTS, which is 20% higher than the yield of untreated whole straw. Whether to recommend the pretreatment depends on the outcome of costs and benefits analysis for specific situations. Pretreatment temperature has a significant effect on the digestibility of straw. Higher temperature resulted in higher solids reduction and biogas yield. The temperature effect was more notable when the pretreatment temperature increased from 60°C to 90°C. Higher pretreatment temperatures resulted in more acid production and lower pH levels in the hydrolysis reactor, but the biogasification reactor provided the chemical and biological buffering capacities to allow stable operation.

Generally speaking, smaller straw particles digested better; but the physical pretreatment (size reduction) had more benefit when combined with thermal pretreatment. During a 24-day digestion period, about 75-80% of the biogas was produced in the first 14 days. This implies that if the digester retention time was designed to be 14 days instead of 24 days, the digester size could be reduced by 42% with a sacrifice of 21-25% biogas production. The methane content of the biogas was about 50% for all the test conditions. Based on the laboratory data, it is estimated that about 450 kWh electrical power and 1.716×10^6 Btu heat can be produced from each ton of rice straw (with 8% moisture content) if a conversion efficiency of 35% from biogas energy to electricity and a heat recovery efficiency of 60% from an engine-generator are assumed.

4. ECONOMIC FEASIBILITY OF RICE STRAW DIGESTION

The initial markets that are targeted for the APS-digester system are located in the rice-growing counties of the Sacramento Valley of California. The biogas produced by the digester will be used to produce electricity and heat on site for rice mills, food processing and packaging plants and large-scale refrigeration units that have a consistent need for power year-round. The cost of electric energy produced by an engine generator using the biogas produced from the straw will be greater than the current avoided cost of electricity in California. The APS-Digester systems are considered to be economically competitive only when they are strategically sited in a distributed mode. The individual systems would be located as self-generation power facilities, and the power produced would offset retail power presently being purchased.

The developers of the APS-Digester system have already identified about a dozen locations in the rice-producing region of California where the plant managers have stated that they believe commercial systems would be attractive if the capital and operating and maintenance costs of APS-Digester System are optimized. In the construction of the commercial units, the capital cost of an engine-generator will be a significant factor. As there are certain economies of scale associated with the purchase of engine-generators, the developers expect that commercial systems using rice straw at a cost of about $32 per ton will need to be at least 250 to 500 kW in capacity to be economically viable. Economic viability is defined as producing electric energy at a cost less than the current retail rate. The current agricultural rate is between 10¢ and 11¢ per kWh; however, it is expected that it could decrease as much as 2.5¢ per kWh after the year 2002. The plant manager for one of the smaller mills, which has a base load of about 250 kW, has noted that he would be willing to pay as much as his current retail rate to regain the flexibility in plant operations which he gave up when that facility switched to curtailed delivery during system peaks. A 250-kW system is expected to be marginally competitive. In terms of potential sites for APS-Digester systems at rice mill sites, the developers have identified seven locations where the current management has expressed interest and, in most instances, enthusiasm. The capacity range is from 250 kW to 2.5 MW, with a total power demand of 7 MW. The annual rice straw utilization would be about 112,000 tons. In addition, there are food packagers and refrigeration plants with significant needs for electricity that could also advantageously use cost-competitive electrical energy produced on site.

REFERENCES

1. Zhang, R.H., Y. Yin, W. Sung, and R.R. Dague (1997). Anaerobic treatment of swine waste by the anaerobic sequencing batch reactor, Transactions of the ASAE, 40, pp. 761–767.
2. Zhang, Z. and R.H. Zhang (1998). Evaluation of different anaerobic digestion systems for biogasification of rice straw, Paper presented at 1998 ASAE Annual International Meeting. July 11-16, Orlando, Florida ASAE Paper No. 98-4142.
3. Zhang, R.H. and Z. Zhang (1998). Biogasification of rice straw with an anaerobic-phased solids digester system, Bioresource Technology, 68, pp. 235–245.

Biomass Transformation into Value-Added Chemicals, Liquid Fuels, and Heat and Power

Biofuels: Thermochemical Conversion and Biodiesel

UNIFIED APPROACH TO NEXT GENERATION BIOFUEL—FISCHER-TROPSCH FUEL BLENDS

G.J. Suppes and M.L. Burkhart

Department of Chemical & Petroleum Engineering, The University of Kansas, 4006 Learned, Lawrence, Kansas 66045-2223.

Fischer-Tropsch conversion of biomass gasification products to liquid hydrocarbon fuel typically includes Fischer-Tropsch synthesis followed by refining (hydrocracking and distillation) of the synthetic oil into mostly diesel, kerosene and some naphtha (a feedstock for gasoline production). Refining is assumed necessary, possibly overlooking the exceptional fuel qualities of synthetic oil for more direct utilization as a compression-ignition fuel. This paper evaluates the performance of fuels composed mostly of Fischer-Tropsch synthetic oil. Outstanding performance was achieved using mixtures of Fischer-Tropsch liquids with ethanol.

1. INTRODUCTION

To design a strategy for future fuels, it makes sense to focus on fuel properties that are suitable for the heavy-duty diesel cycle engine and use plentiful feedstocks to produce the fuel with these properties.[1] Recent trends by Ford and Chrysler on development and intended use of diesel engines in pickup trucks, SUVs, and even automobiles reemphasize the increasing importance of both compression-ignition engines and strategies to develop alternative compression-ignition fuel infrastructure. Even the National Research Council states that the compression-ignition engine is the likeliest option to meet the needs of the high-mileage Partnership for a New Generation of Vehicles (PNGV) car.[2]

If new alternative compression-ignition fuels were economically viable and could be produced from a variety of resources—coal, natural gas, biomass, and municipal solid waste—the fuel could gain the unified support of the natural gas industry, the coal industry, and biomass groups. This fuel would improve national security, lower greenhouse gas emissions, improve U.S. balance of trade, and add sustainable industries for rural America.

Fuel formulations based on Fischer-Tropsch liquids (FTL) provide the best opportunity to meet these goals. For researchers and corporations involved in research, development, and commercialization of biomass, two issues are foremost: 1) the economic viability of FTL formulations (both for potential production from biomass and for decisions related to investments in alternative technologies), and 2) the potential for biomass products to be used with FTL formulations to provide improved performance.

1.1. Economic viability

Economic feasibility studies by multiple groups indicate that Fischer-Tropsch formulations can be produced at prices comparable with the six-year average of diesel fuel prices (Table 1). They indicate the production costs for FTL, the primary component of Fischer-Tropsch fuel formulations. FTL could be produced at about $0.55 per gallon (5-year projected, large plant economics). This compares to the six-year average price for bulk diesel is $0.50 per gallon.[3] It is likely that the economics will improve when the FTL formulation fuel market is established and competition increases.

Table 1
Estimated value of FTL considering added-value[3]

Source	Technology (5 year projection, large plant economics)	Location	Projected Price $/bbl / $/gal	Projected Less U.S. Tax[a] $/bbl / $/gal
Syntroleum[4]	No air separation	Remote NG	14-21 0.33-0.50	14-21[c] 0.33-0.50
ADL[5]	Air Separation	Remote NG	18-20 0.43-0.48	18-20[c] 0.45-0.48
Bechtel[6]	Air separation with Electrical power	Remote NG (Alaskan)	24-28[a] 0.57-0.66	19-23 0.46-0.54
US-DOE/ Mitretek[7]	Coproduction/ Cofeed, Coal and Natural Gas (NG)	US Coal, Pipeline NG	33-35[b] 0.79-0.83	27-28 0.64-0.68

[a]Includes a 19% discount, estimated corporate federal income tax.[6]
[b]Published $26/bbl per barrel of equivalent crude oil plus $9.00 give $/bbl Fischer-Tropsch oil.
[c]Syntroleum and ADM estimates based on Middle or Far East production.

1.2. Desirable fuel properties

FTL refers to liquid mixtures produced by Fischer-Tropsch synthesis. These mixtures are the liquid portion of Fischer-Tropsch reactor effluent and may include portions of the waxy reactor effluent that have been hydrocracked to reduce molecular weight. Traditionally, these liquids have been treated as crude intermediates that were further reacted and refined to produce naphtha, kerosene, and diesel distillates.

Direct use of FTL as a vehicular fuel avoids the cost of refining. The FTL may rely on blend stocks to improve fuel properties such as pour point temperatures and engine emissions, and additives may be necessary to improve lubricity. Both the reaction engineering and design of the FTL process and the blendstock–additive approach will change as advanced catalysts synthesize higher value FTL that are cleaner burning and less reliant on blendstocks or additives.

Blendstocks improve liquid properties at the expense of volatility and cetane number. Volatility is a design degree-of-freedom in fuel design. Cetane numbers can be maintained at levels higher than those of standard diesel. Neat FTL fuels typically have cetane numbers near 70.

Figure 1 illustrates the conventional three-step approach of the Fischer-Tropsch process including the conventional refining to naphtha, kerosene, and diesel. Direct use of FTL without distillation would eliminate capital and operating expenses leading to maximum values and economic viabilities for FTL. Agee estimates the cost of Fischer-Tropsch diesel to be at about $1.50 at the pump in the five years.[8] In the longer term, at large-scale production, and as FTL rather than Fischer-Tropsch diesel, even lower pump costs of about $1.00 per gallon are possible. As compression-ignition fuels, FTL would be about 30% more efficient than gasoline (about $0.77 per equivalent gasoline performance, based on higher fuel economies of compression-ignition engines over spark-ignition engines).

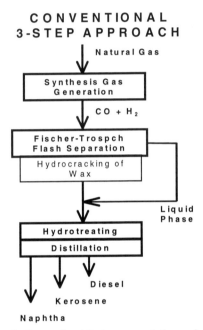

Figure 1. Conventional 3-step approach for producing vehicular fuels with Fischer-Tropsch Synthesis.

2. RESULTS AND DISCUSSION

2.1. Preliminary engine tests

FTL typically have a wider boiling-point range than diesel and hence do not fall within the conventional definition of diesel. Engine tests determined that the FTL formulations successfully powered a 453T, 2-cycle Detroit Diesel compression-ignition engine. FTL mixtures not only did not reduce thermal efficiency, ones containing ≤ 20% ethanol appeared to increase efficiency by 2–3%.

Other researchers have noted that ethanol increased thermal efficiency; it typically peaks at ethanol concentrations of 5–20%[8] in diesel. Further research is needed to evaluate the extent to which thermal efficiency is increased; even minor increases affect the economics of using ethanol at 10–20% application rates.

The engine tests[10] provided by the following information:
- FTL formulations consistently reduced NO_x emission by 10% as compared to US-1 diesel.
- FTL formulations with blend stocks that do not contain aromatics reduced particulate matter by 10–67%, depending on the blend stock (at a maximum of 33% blendstock in FTL).
- FTL formulations increased hydrocarbon formulations by 50–150 ppm. The higher the blend stock volatility the higher the hydrocarbon emissions in the exhaust.

Fortunately, the hydrocarbon emissions were below those typically exhibited by spark-ignition engines. Even a 200 ppm increase in hydrocarbon emissions may be largely irrelevant (if catalytic converters are used).

Table 2 summarizes the particulate matter (PM) emissions. The performances of the ethanol and ether formulations are nothing less than outstanding, especially as much of the remaining particulate matter may originate in the engine oil and not the fuel.

Table 2
PM as percent of US 1D PM

	LOAD	
	50%	80%
US 2D	151%	88%
US 1D	100%	100%
FTL	95%	82%
FTL with 25% gasoline	120%	77%
FTL with 20% ethanol	60%	51%
FTL with 10% ethanol and 10% diethyl ether	72%	31%
FTL with 16.% ethanol And 16.5% diethyl ether	70%	49%

These engine tests and performance data indicate that FTL formulations are not only viable fuels, but that certain formulations may substantially reduce emissions. Particulate matter and NO_x are the primary problem emissions from compression-ignition engines. The solution provided by certain of these formulations is welcome and needed. Both EPACT and CAA markets provide niche markets for these fuels.

2.2. Ethanol formulations and a unified approach

The ethanol and ethanol-diethyl ether formulations greatly reduced particulate matter emissions. Diethyl ether has an advantage as compared with ethanol in that it will not partition into a water-rich phase when the fuel is contacted with a small amount of water; however, it is more costly and more volatile.

It has been argued that ethanol in diesel does not provide distinct performance advantages, and, in fact, adds cost and can reduce performance. Two factors make applications with FLT stand out over applications with traditional diesel:

- Ethanol does not greatly reduce FTL cetane. The effect of ethanol on the cetane of FTL and US 1D is show in Figure 2. Even 10% ethanol takes conventional US 1D cetane numbers below acceptable values. Since the cetane of FTL is typically > 70, even 25% ethanol does not take the cetane below acceptable levels.
- Ethane FTL formulations do not show inconsistent phase behavior. The composition of FTL is considerably more consistent than diesel since FTL is produced in controlled environments while the crude oil originates from many sources and is of inconsistent quality. Even in the absence of water, ethanol has miscibility problems with some diesel—similar problems do not exist with FTL-ethanol formulations. FTL-ethanol-water phase behavior is similar to gasoline-ethanol-water phase behavior and should provide a similar, acceptable phase behavior for routine applications.

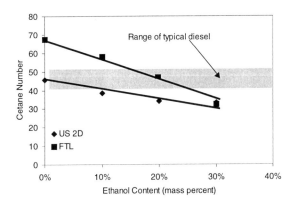

Figure 2. Impact of ethanol on the cetane of US 1D and FTL.

Even though the use of ethanol in conventional diesel can lead to cetane and phase behavior problems, reduced particulate-matter emissions is leading to its increased use.[11] In applications with FTL, the cetane problems do not exist and possible problems with phase behavior are substantially subdued.

FTL formulations provide a definite opportunity for ethanol to be used as a value-added component in a fuel. Probable advantages include reduced generation of particulate matter, reduced NO_x emissions, and increased thermal efficiency. Biomass has the potential of being used to produce both ethanol and FTL (Figure 3).

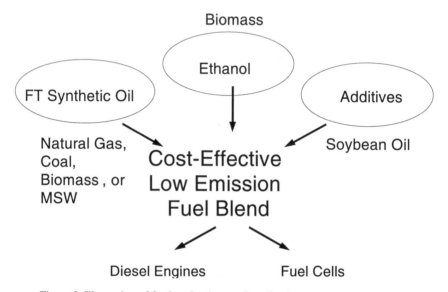

Figure 3. Illustration of feedstock mixes and applications.

Industries associated with natural gas, coal, ethanol, cellusosic feedstocks, and vegetable oils could all benefit from FTL fuels. Together, they could provide a fuel alternative that is based on production from diverse resources and would provide stable supplies and prices. Each of these industrial sectors has a unique advantage.

The fuel formulations illustrated by Figure 3 have the following advantages:
- Ease of handling (liquid)
- Compatible with existing pipelines/fuel distribution
- Low incremental vehicular cost
- Low volatility (fire safety)
- Renewable

- Long-term energy security and versatility
- High efficiency (diesel engine)
- Indigenous production potential
- Cost competitive
- Low emissions

No other alternative fuel can meet or exceed these advantages. FTL formulations offer the opportunity for a unified approach that is compatible with the interests of all the major players in the alternative fuels sector. By working together they could make a much greater impact. The fuel formulations would be both sustainable and cost-effective.

3. CONCLUSIONS

Recent breakthroughs in the production of FTL and the direct utilization of FTL as vehicular fuels have caused several major corporations to consider the use of Fischer-Tropsch fuels as an alternative to petroleum-derived fuels. Several corporations estimate that FTL can be produced for about $0.55 per gallon—a price competitive with petroleum-based diesel. Furthermore, FTL formulations are cleaner-burning than petroleum-based diesel. These formulations provide an unprecedented opportunity for natural gas, coal, and biomass industries to collaborate and achieve long-sought goals of improved national security, low greenhouse gas emissions, improved U.S. balance of trade, and additional sustainable industries for rural America.

REFERENCES

1. Eberhardt, J.J. (1997). Future Fuels for Heavy-Duty Trucks, Alternative Fuels in Trucking, NREL/BR-540-23440, October.
2. Peckham, J. (1997). Diesel likeliest option for high-mileage car says NRC. Hart's Fuels and Lubes Show Special, 1997 SAE Fuels and Lubes Meeting.
3. Suppes, G.J., J. Terry, T.M. Burkhart, and M.P. Cupps (1998). Blends of Fischer-Tropsch crude—A lower cost route to diesel fuel, Fifteenth Annual International Pittsburgh Coal Conference, Pittsburgh, Penn., September.
4. Agee, K.L., M.A. Agee, F.Y. Willingham, and E.L. Trepper (1996). Economical utilization of natural gas to produce synthetic petroleum liquids. GPA Convention Paper, March.
5. ADL (1998). Gas-to-liquids processing hits its stride, Oil & Gas Journal., June 15, pp. 34–35.
6. Choi, G.N., S.J. Kramer, S.S. Tam, J. Fox, N.L. Carr, and G.R. Wilson. Design/economics of a once-through natural gas Fischer-Tropsch plant with power co-production., Coal Liquefaction ans Solid Fuels '97 FETC conference, Pittsburgh, Pennsylvania.
7. Gray, D. and G.A. Tomlinson (1997). A novel configuration for coproducing Fischer-Tropsch fuels and electric power from coal and natural gas, Coal Liquefaction and Solid Fuels '97 FETC conference, Pittsburgh, Pennsylvania.
8. Agee, K.L. (1998). News release, http://www.media.chrysler.dom/wwwpr98/28aa.htm.

9. Heisey, J.B. and S.S. Lestz. Aqueous alcohol fumigation of a single-cylinder DI CI engine, SAE Paper 811208.
10. Suppes, G.J., C.J. Lula, M.L. Burkhart, and J.D. Swearingen (1999). Performance of Light Fischer-Tropsch Oil (LFTO) in modified off-highway diesel engine test cycle, SAE Paper in review, SAE Spring 1999 Fuels & Lubricants Meeting.
11. Marek, N. (1998). The oxydeisel fuel project. Bioenergy'98. U.S. Department of Energy, Great Lakes Regional Biomass Energy Program.

THE HTU® PROCESS FOR BIOMASS LIQUEFACTION: R&D STRATEGY AND POTENTIAL BUSINESS DEVELOPMENT

J.E. Naber,[a] F. Goudriaan,[a] S. van der Wal,[b] J.A. Zeevalkink[c] and B. van de Beld[d]

[a]Biofuel B.V., Rendorppark 30, 1963 AM Heemskerk, The Netherlands
[b]Stork Engineers & Contractors B.V., Postbox 58026, 1040 HA Amsterdam, The Netherlands
[c]TNO Environmental Sciences, Energy Research and Process Innovation (TNO-MEP), Postbox 342, 7300 AH Apeldoorn, The Netherlands
[d]Biomass Technology Group (BTG) B.V., Postbox 217, 7500 AE Enschede, The Netherlands

The HTU® process offers excellent opportunities for conversion of biomass to a transportable form of energy with a heating value approaching that of fossil fuels. It converts biomass in liquid water at temperatures of about 300°C to a 'biocrude' that resembles the atmospheric distillate of crude oil. On the basis of research carried out in 1981–1988 in the Shell Laboratory in Amsterdam, technical and economic feasibility studies were completed in 1995–1997. They concluded that the process is potentially well placed in comparison with the main alternatives for biomass conversion. Main strengths are the (prospective) economics, processability of a wide variety of (wet) feedstocks, product quality and flexibility and the potential for upscaling and rapid rate of commercial development.

A detailed R&D and business plan has been worked out and a consortium has been formed that started a 6 M$ R&D Project, supported by the Dutch Government, on November 1, 1997. The programme will run till mid-2000. Its purpose is the development of data for a reliable design of the first commercial applications. Preparatory project studies are underway to pursue opportunities in The Netherlands for a 10,000 tonne (dry basis)/a commercial demonstration, starting operation in 2001 or 2002.

1. INTRODUCTION

In the previous Biomass Conference of the Americas,[1] the HTU® process for the direct hydrothermal liquefaction of biomass was described in some detail. After discontinuation of the research at the Shell Laboratory in Amsterdam in 1988, a technical and economic feasibility study was carried out by Stork Comprimo (now Stork Engineers & Contractors) and Shell Research in 1995 to investigate prospects for the future of the HTU® technology. The study concluded that for future large-scale energy purposes (2020+) the HTU® process promises to be able to compete with fossil fuels, with clear advantages compared with alternatives for biomass conversion.

Subsequently Biofuel BV was asked to work out a development plan for finalizing the R&D required for the first commercial implementation and a business plan for the period up to 2020. The development plan comprised a critical (re-)evaluation of the technology with an update of the economics, an analysis of the strengths and weaknesses and an identification of the major R&D topics that have to be covered before the first commercial units can be designed with sufficient reliability.

The results of the study led to the formation of a consortium comprising Shell Nederland, Stork Engineers & Contractors, TNO-MEP, BTG and Biofuel. The consortium started a 6 M$ R&D programme on November 1, 1997, with support from the Dutch government in the E.E.T. Programme (Energy, Ecology, Technology). The main purpose of the R&D that will run till mid-2000 is validation of the HTU® process on a 20 kg(db)biomass/hr pilot plant scale, and the development of the necessary design data for the first commercial applications.

2. EVALUATION OF THE HTU® TECHNOLOGY

As described previously,[1-3] hydrothermal liquefaction converts biomass in liquid water at temperatures of around 300°C to a 'biocrude' that resembles the residue from atmospheric distillation of crude oil.

Based on the experimental data from the research in 1981-1988 and the results of the 1995 feasibility study, Biofuel BV critically re-evaluated the HTU® technology, targeting improved operability and robustness of the process. The main elements of the evaluation were as follows:
- *Introduction of the feed*. The challenge is to replace the digestion of biomass at 200-250°C as feed preparation by a less complicated and less expensive step for pressurizing the biomass to the required levels of 150-200 bar.
- *Heating of the reactants*. The relatively short reaction time of 5-15 min requires innovative methods for heating the biomass to 300-350°C.
- *Treating of the effluents*. The off-gas contains mainly CO_2, some volatile organics and hardly any impurities that require further treatment other than combustion in the main furnace. The effluent water contains dissolved organics and the minerals from the biomass feedstock. It can be treated to required residual levels with available conventional technology and attractive energy recovery.

The evaluation also included an elaborate analysis of strengths and weaknesses of the technology in comparison with alternatives for biomass conversion. Main strengths are the (prospective) economics as shown below (Section 4), feed flexibility, product quality and flexibility and commercialisation strategy. HTU® is clearly omnivorous with respect to feedstocks due to the processing in liquid water. There is no need for drying or costly reduction of particle size and the process is relatively insensitive to the physical characteristics of the biomass feedstocks. With respect to the product a great variety of applications can be envisaged, including light and heavy liquid fuels, solids and upgrading by hydrodeoxygenation to transportation fuels that are directly compatible with existing fuels. The feed flexibility and the characteristics of the technology enable a commercialisation strategy starting with a demonstration unit of modest size processing low-cost organic waste or agricultural residues. After the initial learning period, possibly comprising repeat modular units, the process can then be quickly and reliably scaled up to biomass throughputs of 100-500 kt(dry basis)/a.

Main weaknesses of the process are the rather demanding operating conditions and the status of the development. With the on-going research project and a quick follow-up with commercial demonstration it is expected, however, that the apparent backlog compared with alternative conversion technologies can soon be recovered.

3. R&D PROGRAMME

The go/no-go items for the viability of the process were analysed. After the critical evaluation and adaptations this now mainly includes the system for pressurizing the feed. A number of critical items were identified for which an economically less attractive fall-back is available. These include the oil–water separation, the biocrude product properties and applications, and the water effluent treatment. Additionally, although numerous items do not present technical uncertainties, data are still needed for a reliable design, e.g., phase equilibria and physical properties under reaction and separation conditions.

A sizable R&D Programme was started to speed up the development of the HTU® process and to enable implementation of the process on a commercial scale in 2001. About 25 man-years will be spent in a programme that ranges from fundamental research to commercialisation activities. Main elements in the programme are as follows:

1) Batch experiments in autoclaves of 10 ml to 2 L size to define optimal process conditions, study phase equilibria and provide support to the various projects of the R&D programme, such as reactor engineering, materials research, pilot plant design, feed/product research, process modelling and effluent treatment.
2) Design, construction and operation of a continuous pilot plant with a biomass throughput of 20 kg of biomass (dry basis)/h. The pilot plant will be started up in July 1999. At the end of the project test runs of over 1000 hours are scheduled.
3) Pressurization of biomass to 150-200 bar pressure.
4) Product research and applications. A detailed analysis of the product will be generated and a number of applications have been selected that will be validated.
5) Reactor engineering. A detailed design of the reactor section of a commercial plant will be developed.
6) Effluent water treatment. The effluent contains dissolved organics and minerals from the feedstock. Treating options will be defined that enable compliance with environmental regulations.

7) Process design and economics of a commercial HTU® process. This includes heat integration, equipment design and operability studies. A regular update of the overall economics will also be generated, based on results of the R&D.

4. COMMERCIALISATION

After a year of R&D in the on-going project the adapted concept for a simplified and more robust process has been validated. The previous cost estimate for a commercial unit of 130,000 ton wood/yr (dry basis[1]) has subsequently been updated by Stork E&C. Compared with the previous estimates capex is reduced by 25% and the biocrude production cost, excluding feedstock, is reduced from 4.7 $/GJ to 3.4 $/GJ (1$ = 1.9 NFL).

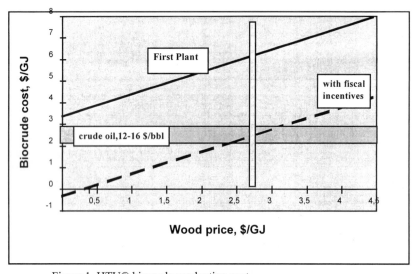

Figure 1. HTU® biocrude production cost.

Figure 1 shows the biocrude production cost as a function of the biomass feed cost for a first-of-a-kind unit with the throughput given above. The figure also shows the cost of a unit when applying the present tax and fiscal incentives that exist in The Netherlands.

These incentives translate to a total benefit of about 3.7 $/GJ. Good prospects exist for 'commercial' application with biomass feedstock costs below 2.5 $/GJ, even at today's energy prices in the range of 12-16 $/eq.oil bbl. For the future it can be expected that the learning curve upon successive commercial introductions of the HTU® process, including modular construction and upscaling, will further reduce costs and widen the scope for various biomass feedstocks.

Considering the long-term availability of fossil fuels at relatively low prices, especially natural gas, it can be expected that the major driving forces for substantial introduction of renewable energy are the environmental aspects of e.g. the emission of greenhouse gases, notably CO_2.

Figure 2 shows the costs of Figure 1 in terms of costs per ton of CO_2 avoided, when coal is replaced in power stations. Costs for a first-of-a-kind plant translate to around 40 $/ton of CO_2 avoided at about 1.5-2.5 $/GJ feedstock cost.

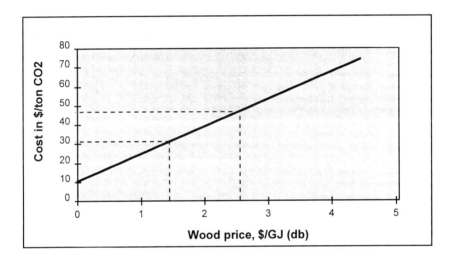

Figure 2. Cost of CO_2 avoided by HTU® processing.

It is interesting to compare these costs with the projections for various CO_2-reduction options as worked out for The Netherlands and shown in Table 1.[4]

Against the background of a total CO_2 emission in The Netherlands of around 200 Mt/a tons/a the above data suggest that biomass conversion should be able to play a significant role in achieving the 2020 targets for renewable energy in The Netherlands.

HTU® is optimally placed to start the commercial development with industrial organic waste agricultural residues. Target feedstocks have costs below about 1.5 $/GJ. In The Netherlands these comprise about 500 kt/a (dry basis) of feedstocks (such as industrial sludges, (wet) residues from horticulture, and verge grass) that generally pay a premium at the gate for disposal. In addition, about 1000 kt/a (db) of agricultural residues from sugar

Table 1
CO_2-reduction options in The Netherlands

	Cost $/tonne CO_2	Total potential M tonnes CO_2/a
Underground storage of CO_2 from H_2 manuf.	10-20	4
Fuel shift from coal to natural gas	20-40	5
H_2 from nat. gas for mixing in pipeline gas	50	5
CO_2 removal from coal-burning	50-150	30
Wind	25-100	30
Photo-voltaics (PV)	1300	30

beets and potatoes can be made available at an average alternative value of about 1.7 $/GJ. The conversion of these feedstocks is a true niche for HTU®. Project preparation studies have therefore been started for a modular commercial demonstration unit of about 10,000 tonne biomass (db)/a, to be started up in 2001 or 2002. After the initial learning period this commercial demonstration can then be followed up by a number of units with a throughput of 100-300 kt(db)/a from 2004 onwards. In this niche market HTU® can therefore contribute at least 15-20 PJ/a to the Dutch 2020 target of 75PJ/a from additional biomass conversion.

From, say, 2004 onwards further opportunities can be exploited in Europe for wastes such as olive oil wastes and possibly in synergy with bio-diesel and bio-ethanol manufacture.

With the HTU® technology well up on the learning curve there are then ample opportunities worldwide for processing of relatively low-cost residues, such as coir dust, residues from the paper and pulp industry (0.5 $/GJ in the U.S.[5]) and bagasse. Bagasse availability worldwide is about 8 EJ/a,[6] at opportunity prices of less than 1 $/GJ.[7]

At some stage in the commercial development, especially after sufficient cost reduction has been achieved, HTU® can then also be applied to and compete for sustainably grown energy crops, e.g., in large-scale energy plantations that may become realistic from 2010 onwards.

REFERENCES

1. Naber, J.E., F. Goudriaan, and A.S. Louter (1997). Proceedings of the 3rd Biomass Conference of the Americas, Montreal, August 24-29, 1997, pp. 1651–1659.
2. Goudriaan, F. and D.G.R. Peferoen (1990). Chem. Eng. Sci. 45, pp. 2729–2734.
3. Goudriaan, F., P.F.A. van Grinsven, and J.E. Naber (1994). DGMK Tagungsbericht 9401, pp. 149–162; ISBN 3-928164-70-8.
4. Symposium The Hague (November 25th 1997). "CO_2-opslag: panacee voor het klimaat probleem?"
5. Badin, J., and J. Kirschner (1998). Renewable Energy World, 1, pp. 41–45.
6. Hall, D.O., F. Rosillo-Calle, R.H. Williams, and J. Woods (1993). Renewable Energy, Sources for Fuels and Electricity, pp. 593–653; ISBN 1-85383-155-7.

7. Coelho, S., Proceedings of the 3rd Biomass Conference of the Americas, Montreal, August 24-29, 1997, p. 1631.

TRANSESTERIFICATION OF RAPESEED OILS IN SUPERCRITICAL METHANOL TO BIODIESEL FUELS

Shiro Saka and Kusdiana Dadan

Department of Socio-Environmental Science, Graduate School of Energy Science, Kyoto University, Yoshida Honmachi, Sakyo-ku, Kyoto, Japan 606-8501

The transesterification reaction of rapeseed oil was investigated. Methyl esters of this oil, known as biodiesel fuel, can be produced satisfactorily in supercritical methanol without using any catalyst. Experiments in the batch reactor at 380°C and 20 MPa with a molar ratio of 1 to 42 of oil to methanol led us to conclude that 120 s treatment is an appropriate reaction time to yield nearly as much methyl esters as the theoretical value. This oil is also clearly similar to those prepared by common methylesterification methods that use an alkaline catalyst in its production. The merit of this new process is that it requires just a much shorter time and that purification of the product is much simpler.

1. INTRODUCTION

As energy demands increase and fossil fuels are limited, research is directed towards alternative renewable fuels. Biomass has been found to produce low-molecular-weight organic liquids that can be used or proposed for vehicles. A potential diesel oil substitute is biodiesel, consisting of methyl esters of fatty acids and produced by the transesterification reaction of triglycerides derived from vegetable oils, with alcohol and with the help of a catalyst.[1]

Conventional methods for biodiesel production use a basic or acid catalyst.[2,3] By these methods, the catalyst and saponified products must, however, be removed before using the oil for technical and environmental aspects. Otherwise it can corrode engines and pollute the environment. Furthermore, this method needs a longer time reaction and a sophisticated purification procedure. With acid catalyst, reaction times of 18-19 h have been reported; basic catalyst reactions are somewhat faster depending on temperature and pressure. Biodiesel oil now is being used as alternative fuel in some countries in Europe, the USA, and Japan.

Supercritical fluid, of considerable interest for use as reaction medium because of its unique properties, has been given much attention in recent years. Saka and Ueno[4] employed supercritical water to convert various celluloses to glucose and its derivatives. The physicochemical properties of the fluid (such as density, dielectric constant and diffusivity) undergo a drastic change with pressure and temperature near the critical point of fluid.[5] The fluid can be expected to act as an acid catalyst in supercritical state where a very high reaction rate can be realized without any catalyst. Its reaction with organic compounds may yield products different from those of pyrolysis.

Based on that background, therefore, in this work we have conducted a basic study employing supercritical methanol to investigate the possibility of converting triglycerides of rapeseed oil to methyl esters as biodiesel fuels.

2. MATERIALS AND METHODS

The sample used for this work is rapeseed oil purchased from Nacalai Tesque. Rapeseed oil is the seed oil of varieties of *Brassica napus* and *B. campestris*. The air-dried seed contains about 40% oil and its crude oil consists up to 2.0% free fatty acids. The fatty acids of this oil have a chain length ranging from C_{12} to C_{18}. Therefore in all calculations the molecular weight of the oil was assumed to be 806.

The supercritical biomass conversion system employed in this work is the same system used by Saka and Ueno.[4] The system is designed to monitor both temperature and pressure in real time and in a range up to 50 MPa and 550°C of pressure and temperature, respectively. A schematic diagram of the batch type reaction system is shown in Figure 1; and the 5-ml reaction vessel is made of stainless steel (SUS 316).

The experimental procedure was as follows: A given amount of rapeseed oil and methanol with molar ratio of 1 to 42 were charged into the reaction vessel. The reaction vessel was then quickly heated up to 380°C by immersing it in a tin bath, which had been preheated at 400°C and kept for a set time interval (for supercritical treatment 10–120 s), and then it was moved into water bath to stop the reaction. The liquid was then allowed to settle for about 30 min, during which time it separated into two portions, the upper and the lower. Each portion was heated at temperature of 90°C for 20 min to remove the unreacted methanol. Each portion was weighed and its composition was analyzed by high performance liquid chromatography (Shimadzu, LC -10AT). This instrument consists of a column (STR ODS-II, 25 cm x ID 4.6 mm, Shinwa Chem. Ind. Co.) and a refractive index detector (Shimadzu, RID-10A) operated at 40°C with 1.0 ml/min flow rate of methanol as a carrier solvent.

Some methylated compounds and oils prepared by common methylesterification method with methanol and a catalyst of sodium hydroxide as well as commercial biodiesel oils were also analyzed as a standard and comparison. Commercial biodiesel oils used are E-Oil and Bio Super 3000, from Lon Ford Development Ltd., Japan and VOGEL & NOOT Technology, Biodiesel International, Austria. The former was produced from waste vegetable oil, whereas the latter is from virgin rapeseed oil.

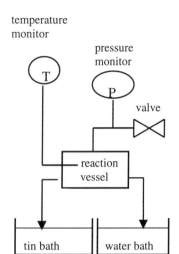
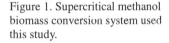

Figure 1. Supercritical methanol biomass conversion system used this study.

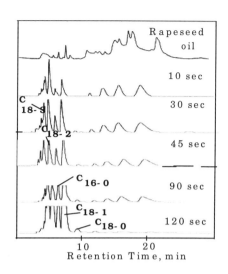

Figure 2. Chromatograms of rapeseed oil treated at various reaction times in supercritical methanol.

3. RESULTS AND DISCUSSION

Assuming that the transesterification reaction of rapeseed oil in the supercritical methanol proceeds under the same reaction mechanism as that using liquid methanol, the reaction process is as follows:[6]

$$\begin{array}{lll} CH_2CO_2R^1 & CH_2OH & R^1\,CO_2CH_3 \\ | & | & \\ CHCO_2\,R^2 + 3CH_3OH \rightleftharpoons & CHOH\; + & R^2\,CO_2CH_3 \\ | & | & \\ CH_2CO_2R^3 & CH_2OH & R^3\,CO_2CH_3 \end{array}$$

where R^1, R^2 and R^3 are alkyl groups with chain lengths ranging mainly from C_{11} to C_{17}. In this reaction, however, more methanol was used to shift the reaction equilibrium to the right side and produce more methyl esters as the proposed product.

The pattern of HPLC chromatograms in Figure 2 at various reaction times in supercritical methanol is the same for all reaction times, but gradual changes can be seen to methyl esters. The left side of those chromatograms contains methylated compounds, which are methyl palmitate (C_{16-0}), methyl stearate (C_{18-0}), methyl oleate (C_{18-1}), methyl lynoleate (C_{18-2}) and methyl lynolenate (C_{18-3}), in which subscript numbers are that of carbons in alkyl chain and that of double bond, respectively. The other side is believed to contain monoglyceride, diglyceride, and unreacted-triglyceride. Therefore, there is a high correlation between the methylated compounds and reaction time. From this work, it was concluded that a 120-s treatment is an appropriate reaction time to get nearly the same amount as the theoretical value.

Except for methyl lynoleate, the concentration of methylated compounds in percent volume increases in line with reaction time as in Figure 3; methyl oleate (C_{18-1}) shares the biggest portion followed by methyl lynoleate (C_{18-2}), methyl lynolenate (C_{18-3}), methyl palmitate (C_{16-0}) and the last is methyl stearate (C_{18-0}). From Figure 3, we also can see that the transesterification reaction actually began around 10 s treatment and by 120 s, more than 95% volume of methyl esters had been formed. It is really significant compared with other previous results. Diasakov[6] observed that methyl ester content has surpassed 85% weight after 10 h reaction time at 235°C, and Marinkovic[2] needed 3 h to yield the same result by using acid catalyst.

In Figure 4, we can see the HPLC chromatograms of 120 s treatment compared with that of other biodiesel oils. These chromatograms are clearly similar to those prepared by common methylesterification methods and commercial oils. The oils of the common method (and commercial ones also) consist of five main methyl esters: methyl oleate (C_{18-1}) makes

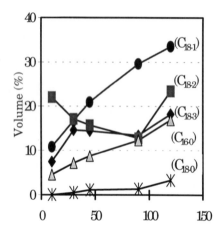

Figure 3. Volumes of various methyl esters from rapeseed oil as treated in supercritical methanol.

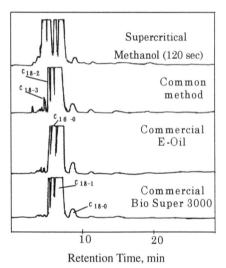

Figure 4. Comparison of chromatograms in various biodiesel oils.

up the biggest portion, followed by methyl palmitate (C_{16-0}), methyl lynoleate (C_{18-2}), methyl steorate (C_{18-0}) and methyl lynolenate (C_{18-3}) in a very small amount. The proportion of methyl esters contained in these oils is somewhat different from that of supercritically treated-oil; in the latter, methyl lynoleate (C_{18-2}) and methyl lynolenate (C_{18-3}) are higher compared with those prepared by the common method with a catalyst of sodium hydroxide. The reason for this is not known at this moment. However, our preliminary experiment indicated that free fatty acids, which are converted in the common method to be the saponified products by the alkaline catalyst, can be converted to methyl esterified compounds through the dehydration reaction in supercritical methanol. Therefore, this free fatty acid reaction may be a reason for this result.

This experiment, therefore, demonstrated that supercritical methanol can offer a promising method to convert vegetable oil by transesterification reaction and perhaps by dehydration reaction of free fatty acids to methyl esterified compounds for biodiesel fuels with higher yield in a very short time.

4. CONCLUSION

The merit of this method is that this new process just requires only a very short reaction time. In addition, because of non-catalytic process, the purification of products after the transesterification reaction is much simpler, compared with the common method in which all the catalyst and saponified products must be removed for biodiesel fuels. As the result, the reaction needs much less energy and can save energy in the manufacturing process. Therefore, this new method offers an alternative way to convert vegetable oils to methyl esters by a simpler, safer, shorter production process. Further investigations are needed in order to develop a technically applicable biodiesel oil.

REFERENCES

1. Klass, D.L. (1998). Biomass for Renewable Energy, Fuels, and Chemicals, Academic Press, New York, p. 333.
2. Marinkovic, S.S. and A. Tomasevic (1998). Fuel, 77, p. 1297.
3. Dandik, L. and H.A. Aksoy (1998). Fuel Processing Technology 57, p. 81.
4. Saka, S. and T. Ueno (1998). in Biomass for Energy and Industry, 10th European Conference and Technology Exhibition, Wurzburg, Germany, 8-11 June, p. 1815.
5. Kabyemala, B.M., T. Adschiri, R. Malaluan, and K. Arai (1997). Ind. Eng. Chem. Res., 36, p. 2205.
6. Morrison, R.T. and R.N. Boyd (1992). Organic Chemistry, Sixth edition, Prentice-Hall International, Inc., New Jersey, p. 1297.
7. Diasakov, M., A. Louloudi, and N. Papayannakos (1998). Fuel, 77, p. 1297.

PRODUCTION OF METHANOL AND HYDROGEN FROM BIOMASS VIA ADVANCED CONVERSION CONCEPTS*

Andre Faaij,[a] Eric Larson,[b] Tom Kreutz,[b] and Carlo Hamelinck[a]

[a]Department of Science Technology and Society, Faculty of Chemistry, Utrecht University, Padualaan 14 3584 CH Utrecht, The Netherlands, Tel. 31-30-2537643/00 Fax 31-30-2537601. E-mail: A.Faaij@chem.uu.nl.
[b]Center for Energy and Environmental Studies, Princeton University, P.O. Box CN 5263, Princeton, New Jersey 08544-5263

1. INTRODUCTION

Methanol and hydrogen produced from lignocellulosic biomass have been identified as promising sustainable fuels, especially when utilized in fuel cell vehicles. However, when cultivated biomass is used as a feedstock those fuels are hardly competitive with their production from natural gas and are currently more expensive than gasoline and diesel, assuming more or less proven conversion technology is applied.

This work focuses on identifying conversion concepts that could lead to higher overall efficiencies and lower costs. Improved performance may be obtained by applying improved or new (non-commercial) technologies, economies of scale and combined fuel and power production.

2. APPROACH

1. Technology assessment and selection of various concepts. The review includes technologies that are not applied commercially (examples are catalytic autothermal reforming, new shift-reaction catalysts, Liquid Phase methanol production, HT gas cleaning, high temperature gas separation techniques, and improved O_2 production).
2. Consulting of manufacturers and experts for performance and cost data of various components.

---

*The work is funded through the National Research Program on Global Air Pollution and Climate Change (NOP/MLKII) of The Netherlands, the Foundation of Technical Sciences (STW) of The Netherlands and NOVEM (Dutch Organization for Energy and Environment) who is responsible for the new research program for introduction of sustainable liquid and gaseous energy carriers in the Dutch energy system.

3. Creation of ASPEN-plus models to evaluate performance and carry out sensitivity analyses.
4. Detailed cost analyses starting from component costs (including scale factors and capacity ranges).
5. Chain analyses for calculating costs of energy services delivered, energy balance and GHG emissions.

3. RESULTS AND CONCLUSIONS

The work is ongoing. A limited set of promising conversion concepts has been identifies for further analyses. Technology reviews and preliminary performance and cost estimates suggest that investment costs could be reduced considerably combines with substantial improvements of the overall efficiency compared to concepts that are based on commercially available conversion technology. This still excludes further beneficial scale effects. Provided sufficient biomass could be supplied, costs could be reduced further by considering plants with a capacity of 1000-2000 MWth input.

TECHNICAL PERFORMANCE OF VEGETABLE OIL METHYL ESTERS WITH A HIGH IODINE NUMBER

Heinrich Prankl, Manfred Wörgetter, and Josef Rathbauer

BLT Wieselburg - Federal Institute of Agricultural Engineering, A-3250 Wieselburg, Austria

The Federal Institute of Agricultural Engineering in Austria has been gaining more experience about the technical performance of biodiesel with a high iodine number. Long-term bench tests evaluated rape seed oil methyl ester, sunflower oil methyl ester and camelina oil methyl ester with an iodine number of 107 to 150. The oil viscosity was observed and the engine parts were inspected after each run.

To demonstrate the suitability of a methyl ester with a high iodine number, a fleet of nine vehicles and one stationary engine was tested for one to three engine oil drain intervals. Camelina oil methyl ester, with a content of 37% linolenic acid (C18:3), was used. No unusual deposits were observed after dismantling the engines.

1. INTRODUCTION

Methyl ester (biodiesel) produced from several raw material sources has reached a considerable market position in the last years. The use of methyl esters in diesel engines might cause an engine oil dilution by the fuel.[1] A high content of unsaturated acids in the esters (which is expressed by the iodine number) increases the risk of polymerization in the engine oil.[4-6]

The engine oil dilution decreases viscosity. Several publications have reported a sudden increase after some time; it is supposed that the oil breakdown is related to fuel dilution followed by oxidation and polymerization of unsaturated fuel constituents.

The influence of the iodine number of biodiesel on the engine performance was investigated previously;[2,3] bench investigations were carried out on a single-cylinder-engine. Biodiesel test fuels were used with iodine numbers of 100 to 180 for long-term tests of more than 250 h. The results showed no clear difference between the test fuels with an iodine number of 100 to 140. The need for further investigations prompted a project to demonstrate whether biodiesel with an iodine number of 150 can be used as fuel for diesel engines.

2. OBJECTIVES

The aim of the project was to gain more experience about the technical performance of biodiesel with a high iodine number. The overall objectives of the project are as follows:
- analysis of the physical and chemical properties of rape seed oil methyl ester (RME), sunflower oil methyl ester (SME) and camelina oil methyl ester (CME);
- investigation of several methyl esters with a varying fatty acid distribution in long-term tests on the test bench;
- doing a fleet test with 10 vehicles using a fatty acid methyl ester with a high content of unsaturated acids over a period of 2 engine oil change intervals.

3. MATERIALS AND METHODS

3.1. Test fuels
The following methyl esters were used:
- Rapeseed oil methyl ester (RME), Iodine number 107
- Sunflower oil methyl ester (SME), Iodine number 132 and
- Camelina oil methyl ester (CME), Iodine number 150

Table 1
Analyses of the test fuels in comparison to the Austrian and the German standard for biodiesel

		RME	SME	CME	ON C1191	DIN 51606
Density 15°C	[g/cm³]	0.881	0.882	0.887	0.85-0.89	0.875-0.90
Flash point	[°C]	>150	>151	>152	>100	>110
Viscosity 40 °C	[mm2/s]	4.4	4.2	4.2	3.5-5.0	3.5-5.0
CCR (100 %)	[%mass]	0.02	0.02	0.03	0.05	0.05
Sulfated Ash	[%mass]	<0.01	<0.01	<0.01	0.02	0.03
Cetane number	[-]	60	56	52	>49	>49
Methanol content	[%mass]	0.08	0.01	0.09	0.2	0.3
Mono-glycerol	[%mass	0.18	0.41	0.20	-	0.8
Di-glycerol	[%mass]	0.12	0.10	0.05	-	0.4
Tri-glycerol	[%mass]	0.08	0.05	0.03	-	0.4
Free glycerol	[%mass]	0.01	0.01	0.003	0.02	0.02
Total glycerol	[%mass]	0.08	0.13	0.06	0.24	0.25
Iodine number	[%mass]	107	132	150	120	115
C18:3 content	[%mass]	11	0.7	37	15	-

The test fuels were produced in a 700 l pilot plant at the BLT. The high iodine number of SME is caused by a high content of linoleic acid (66% C18:2) and, in case of CME, by a high content of linolenic acid (37% C18:3).

3.2. Test engine and engine oil

A direct-injection single cylinder engine, type 1D41 Z of the Hatz company, Germany, was used. It is an air-cooled 4-stroke diesel engine with a crankshaft and cylinder head of light metal. Bore × stroke is 85 × 65 mm, the capacity is 337 cm^3, the performance is 5.6 kW at 3000 rev/min.

Engine oil: Type: OMV Truck M plus
              Specification: MIL-L-2104 E, API SG/CE, SAE 15W-40
              Viscosity 100°C: 14.0 mm^2/s

In order to have hard conditions for the tests the engine oil was pre-mixed with each test fuel. Therefore the initial viscosity was 10.5 mm^2/s (instead of 14 as for the pure engine oil).

3.3. Test cycles on the test bench

The test period comprised 256 h with each test fuel. A test cycle with variable load was chosen corresponding to ISO 8178-4:1995. To guarantee the same conditions for each test, the piston, the cylinder and the injector were changed before the start, and the temperature was controlled during the run. In the evaluation the engine parts were checked and oil samples were analyzed.

4. RESULTS

4.1. Long-term tests on the test bench

Table 2
Evaluation of engine parts

Engine part	RME	SME	CME
1st piston ring	free movable	free moveable	free moveable
1st ring groove	deposits on 80% * partly detached	deposits on 100% * partly detached	deposits on 100% * partly detached
Area 1st/2nd ring	deposits on 5% *	deposits on 10% *	deposits on 30% *
2nd piston ring	free movable	free movable	free movable
2nd ring groove	deposits on 25% *	deposits on 40% *	deposits on 50% *
Area 2nd/3rd ring	clean	clean	clean
3rd piston ring	free movable, clean	free movable, clean	free movable, clean
Injector, injector holes	clean	clean	clean

* deposits as % of the area

Different test fuels produced large differences in engine parts only on the deposits at the bottom of the 2nd ring groove and in the area between the first and second piston ring (Table 2).

Figure 1 shows the development of the viscosity. The increase of the viscosity of CME is higher than with SME and RME. A correlation of the iodine number with the average increase of the viscosity is shown in Figure 2.

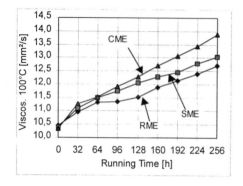

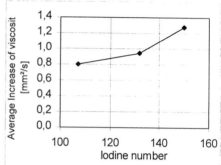

Figure 1. Viscosity vs. running time.   Figure 2. Increase of viscosity vs. iodine number.

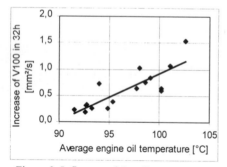

Figure 3. Influence of the oil temperature.

Oil samples were taken every 32 h in several tests with the same fuel (SME). The average temperature was determined. Figure 3 shows a clear correlation.

4.2. Fleet test

In 1998 a fleet was tested—nine vehicles (five tractors, four passenger cars) and one stationary engine for irrigation. The engines were operated during one to three engine oil drain intervals (maximum 750 h, 30.000 km) with camelina oil methyl ester (CME).

At the beginning and at the end of the test period the engines were examined carefully. The engine performance was determined on the test bench. The cylinder head was dismantled before and after the test period. Oil samples were taken every 50 h and 3000 km respectively. On some engines the oil temperature was recorded.

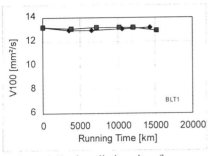
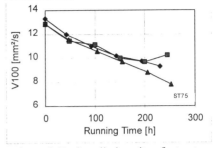

Figure 4. Engine oil viscosity of a passenger car.

Figure 5. Engine oil viscosity of a tractor.

In some engine types, oil viscosity decreased (Figure 5) because engine oil was diluted by unburned fuel.

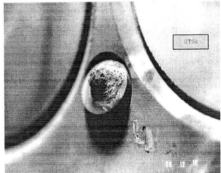

Figure 6. Combustion chamber after 760 h with CME.

Figure 7. Nozzle after 418 h with CME.

The results of the engine inspection agreed with those of former projects.[1] No unusual deposits were found on the cylinder liner, in the combustion chamber, injector or valves.

5. SUMMARY AND CONCLUSIONS

Long-term tests were carried out on the test bench with rape seed oil methyl ester, sunflower oil methyl ester and camelina oil methyl ester with an iodine number of 107 to 150. The oil viscosity was observed and the engine parts were inspected after each run. Increase in engine oil viscosity increased with iodine number. The results differ from that of a former project where nearly no differences could be found up to an iodine number of 160.[2,3]

A fleet of nine vehicles and one stationary engine was tested for one to three engine oil drain intervals. Camelina oil methyl ester was used with an iodine number of 150 and a content of 37% linolenic acid (C18:3). The engines were dismantled before and after the tests. No unusual deposits were found in the cylinder liner, the combustion chamber, the injector and the valves.

The experiments have shown that it is possible to operate an engine with a methyl ester containing more than 30% of unsaturated acids. Further investigations are necessary especially of the interaction between methyl ester fuel and engine oil.

6. ACKNOWLEDGEMENTS

The current investigations are part of an international project funded by the European Commission in the frame of the ALTENER program (XVII/4.1030/Z/96-013) 7].

Special acknowledgements are given to the European Commission; Federal Ministry of Agriculture and Forestry; Bundesversuchswirtschaften, Austria; Ölmühle Bruck, Austria; OMV, Austria; Hatz, Germany; John Deere, Austria/Germany/France; and Steyr, Austria.

REFERENCES

1. Wörgetter, M., et al. (1991). Pilotprojekt Biodiesel. Research Reports of the BLT Wieselburg. Heft 25 und 26, A-3250 Wieselburg.
2. Prankl, H. and M. Wörgetter (1995). Influence of the Iodine Number of Biodiesel to the Engine Performance. International Conference on Standardization and Analysis of Biodiesel, 6-7 Nov 1995, Vienna, ISBN 3 9014 5701 1 (1995).
3. Prankl, H. and M. Wörgetter (1996). Influence of the Iodine Number of Biodiesel to the Engine Performance. ASAE 'Liquid Fuels and Industrial Products from Renewable Resources', 15-17 Sep 1996, Nashville, Tennessee, ISBN 0-929355-79-2.
4. Blackbourn, J.H., et al. (1983). Performance of Lubricating Oils in Vegetable Oil Ester-Fuelled Diesel Engines. SAE Technical Papers 831355.
5. Siekmann, R.W., et al. (1982). The Influence of Lubricant Contamination by Methylesters of Plant Oils on Oxidation Stability an Life. Proceedings of the International Conference on Plant and Vegetable Oils as Fuels, ASAE.
6. Korus, R.A. and T.L. Mousetis (1984). Polymerization of safflower and rapeseed oils. JAOCS, 61, no. 3.
7. Prankl, H. (1997). Technical Performance of Vegetable Oil Methyl Esters with a High Iodine Number. Interim Report. BLT Wieselburg.

LIGNIN CONVERSION TO HIGH-OCTANE FUEL ADDITIVES

Joseph Shabtai,[a] Wlodzimierz Zmierczak,[a] and Subha Kadangode,[a] Esteban Chornet,[b] and David K. Johnson [b]

[a] Department of Chemical and Fuels Engineering, University of Utah, Salt Lake City, Utah 84112
[b] National Renewable Energy Laboratory, Golden, Colorado 80401

Continuing previous studies on the conversion of lignin to reformulated gasoline compositions, new lignin upgrading processes were developed that allow preferential production of specific high-octane fuel additives of two distinct types: (1) C_7-C_{10} alkylbenzenes; and (2) aryl methyl ethers, where aryl mostly = phenyl, 2-methylphenyl, 4-methylphenyl, and dimethylphenyl. Process (1) comprises base-catalyzed depolymerization (BCD) and simultaneous partial (~ 50%) deoxygenation of lignin at 270 - 290°C, in the presence of supercritical methanol as reaction medium, followed by exhaustive hydrodeoxygenation and attendant mild hydrocracking of the BCD product with sulfided catalysts to yield C_7-C_{10} alkylbenzenes as main products. Process (2) involves mild BCD at 250 - 270°C with preservation of the lignin oxygen, followed by selective C-C hydrocracking with solid superacid catalysts. This method preferentially yields a mixture of alkylated phenols, which upon acid-catalyzed etherification with methanol are converted into corresponding aryl methyl ethers (see above) possessing blending octane numbers in the range of 142-166. In a recent extension of this work, a greatly advantageous procedure for performing the BCD stage of processes (1) and (2) in water as reaction medium was developed.

1. INTRODUCTION

It was previously reported that lignin is susceptible to high-yield depolymerization and upgrading leading to reformulated gasoline compositions as final products (Shabtai, Zmierczak, and Chornet, 1997). Two processes were developed: (1) a two-stage process comprising base-catalyzed depolymerization (BCD) of the lignin feed in supercritical methanol as reaction medium, followed by deoxygenative hydroprocessing (HPR) to yield a reformulated hydrocarbon gasoline that consists of a mixture of C_5-C_{11} branched and multibranched paraffins, C_6-C_{11} mono-, di-, tri-, and polyalkylated naphthenes, and C_7-C_{11} alkylbenzenes (Shabtai, Zmierczak, and Chornet, 1998a); and (2) another two-stage process comprising mild BCD, followed by non-deoxygenative hydrotreatment and mild hydrocracking (HT) that yields a reformulated, partially oxygenated gasoline, consisting of a mixture of (substituted) phenyl methyl ethers and cycloalkyl methyl ethers, C_7-C_{10} alkylbenzenes, C_5-C_{10} branched and multibranched paraffins, and polyalkylated cycloalkanes (Shabtai, Zmierczak, and Chornet, 1998b).

Alternative procedures based on lignin as feed have been now developed, specifically oriented toward production of high-octane fuel additives of two types: (a) C_7-C_{10} alkylbenzenes; and (b) aryl methyl ethers, where aryl mostly = phenyl, methylphenyl, or dimethylphenyl.

C_7-C_{10} alkylbenzenes, in permissible concentrations of up to 25 wt%, are considered as essential components of current gasolines. Therefore, it was an objective of the present work to demonstrate that lignin could provide an abundant renewable source of such compounds as gasoline additives. Aromatic ethers such as aryl methyl ethers, owing to their extraordinarily high octane numbers, could likewise be of considerable potential value as lignin-derived fuel additives.

2. EXPERIMENTAL

2.1. Materials
Three different lignin samples were used in the study: (1) a Kraft Indulin AT sample, pretreated by washing with aqueous KOH and water; (2) an organosolve sample, supplied by REPAP Technologies, Inc.; and (3) a sample obtained as by-product in the NREL ethanol process (Wyman, 1996). Relatively small differences in overall chemical reactivity were observed for these samples.

2.2. Catalysts
The preferred base catalyst-solvent systems used in BCD runs were methanolic solutions of NaOH with concentrations in the range of 5.0 to 7.5 wt%.

Two hydrodeoxygenation catalysts possessing high C-O hydrogenolysis selectivity and low ring hydrogenation activity (Shabtai, Nag, and Massoth, 1987), i.e., CoMo/Al_2O_3 and RuMo/Al_2O_3, were used. Mild hydrocracking catalysts used for conversion of residual oligomeric components in the production of C_7-C_{10} alkylbenzenes included CoMo/SiO_2-Al_2O_3 and solid superacids. The preparation of the latter, e.g., sulfated oxides, was detailed elsewhere (Zmierczak, Xiao, and Shabtai, 1994; Shabtai, Xiao, and Zmierczak, 1997).

Selective mild hydrocracking catalysts, viz., catalysts causing selective C-C hydrogenolysis in oligomeric components of BCD products, while preserving O-containing groups, included various solid superacids, in particular Pt/SO_4^{2-}/ZrO_2, Pt/WO_4^{2-}/ZrO_2, and Pt/SO_4^{2-}/TiO_2 (Shabtai, Zmierczak, and Chornet, 1998b).

Etherification catalysts used included previously described solid superacids (see above), and some other reported sulfated oxide systems, e.g., SO_4^{2-}/MnO_x/Al_2O_3, SO_4^{2-}/MoO_x/Al_2O_3 and SO_4^{2-}/WO_x/Al_2O_3 (Mossman, 1986; 1987).

2.3. Reactors

Both 300 cc autoclaves and 50 cc Microclaves (Autoclave Engineers) were used in the BCD runs. Most of the hydroprocessing runs were performed in 50 cc Microclaves.

2.4. Experimental procedures

The base-catalyzed depolymerization (BCD), the hydroprocessing (HPR), and selective hydrocracking (HT) procedures used was detailed in a preliminary communication (Shabtai, Zmierczak, Chornet, and Johnson, 1999). A summary of these procedures is provided in Figures 1 and 2 (see next section). The etherification (ETR) procedure was modified to include continuous drying, *viz.*, removal of water produced during the reaction, and recyclization of unreacted phenolic feed and methanol.

3. RESULTS AND DISCUSSION

Figure 1 outlines the scheme of the two-stage (BCD-HPR) procedure for preferential conversion of lignin to C_7-C_{10} alkylbenzenes. Stage I of the procedure comprises BCD treatment of the wet lignin feed (permissible water/lignin weight ratios in the approximate range of 0.1 to 1.5), using an alcoholic solution of NaOH or KOH in methanol or ethanol as depolymerizing agent. The BCD temperature range is between 270-290°C and preferably around 270°C. At this temperature the methanol or ethanol medium is under supercritical condition, which is an essential requirement for effective hydrolysis or alcoholysis of the etheric linkages in the lignin structural network. The preferred range for methanol/lignin or ethanol/lignin weight ratios in the feed solution is from 2:1 to 5:1. In this range of weight ratios, and by proper reduction in the reaction time (1-5 min), the total number of alkyl substituents in the depolymerized product components can be regulated not to exceed 1 to 3 substituents per depolymerized molecule. These 1 to 3 substituents (mostly methyl groups in the presence of methanol as medium) include residual alkyl groups originally present in the monomeric lignin units, and some methyl groups inserted in these units during the BCD. Under selected processing conditions, the BCD reaction is characterized by a very high lignin conversion rate which is reflected in a high-yield (> 95 wt%) performance, both in autoclave and flow reactors. The preferred reaction time is 5 to 10 min in autoclave reactors, and 1 to 5 min in a continuous flow reactor. At such short times, and particularly for low methanol/lignin ratios, the extent of ring alkylation in the depolymerized products can be easily controlled to a desirable low level. As indicated in Figure 1, the depolymerized lignin product obtained by BCD consists of a mixture of alkylated phenols and alkoxyphenols, accompanied by smaller amounts of hydrocarbons. Description of a typical BCD run was provided elsewhere (Shabtai, Zmierczak, Chornet, and Johnson, 1999).

In Stage II of the procedure (Figure 1) the depolymerized lignin product is subjected to hydroprocessing (HPR) predominantly yielding C_7-C_{10} alkylbenzenes. HPR comprises two sequential steps: exhaustive hydrodeoxygenation (HDO) with a catalyst possessing low ring hydrogenation activity, such as $CoMo/Al_2O_3$ or $RuMo/Al_2O_3$ (Shabtai, Nag, and Massoth, 1987), followed by mild hydrocracking (HCR) with sulfided $CoMo/SiO_2$-Al_2O_3, or preferably with a solid superacid catalyst. The supplemental HCR step results in the effective conversion of some

Lignin
(wet)

↓

Stage I — Base-catalyzed depolymerization (BCD)

Catalysts: NaOH, KOH or other base
Medium: MeOH or EtOH (plus water)
Temperature, °C: 270 - 290

Depolymerized lignin
(mixture of phenols, alkoxyphenols and hydrocarbons)

↓

Stage II — Hydroprocessing (HPR), comprizing (a) exhaustive HDO; and (b) mild hydrocracking, HCR

Catalysts: (a) $CoMo/Al_2O_3$
(b) $CoMo/SiO_2$-Al_2O_3
or solid superacid
Temperature, °C: 350-385
No solvent

C_7 - C_{10} alkylbenzenes (approx. 75-90 wt%);
C_5 - C_{10} branched paraffins plus C_6 - C_{10} alkylated naphthenes (approx. 10-25 wt%)

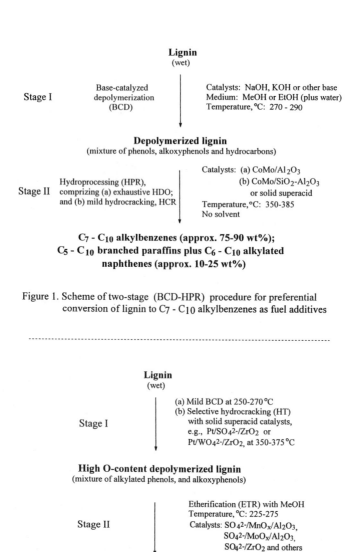

Figure 1. Scheme of two-stage (BCD-HPR) procedure for preferential conversion of lignin to C_7 - C_{10} alkylbenzenes as fuel additives

---

Lignin
(wet)

↓

Stage I

(a) Mild BCD at 250-270 °C
(b) Selective hydrocracking (HT) with solid superacid catalysts, e.g., $Pt/SO_4^{2-}/ZrO_2$ or $Pt/WO_4^{2-}/ZrO_2$, at 350-375 °C

High O-content depolymerized lignin
(mixture of alkylated phenols, and alkoxyphenols)

↓

Stage II

Etherification (ETR) with MeOH
Temperature, °C: 225-275
Catalysts: $SO_4^{2-}/MnO_x/Al_2O_3$,
$SO_4^{2-}/MoO_x/Al_2O_3$,
SO_4^{2-}/ZrO_2 and others

Aromatic ethers
(mixture of aryl methyl ethers, where aryl mainly = phenyl, 2-methylphenyl, 4-methylphenyl and dimethylphenyl)

Figure 2. Scheme of two-stage (BCD-ETR) procedure for preferential conversion of lignin to aromatic ethers as fuel additives

residual oligomeric hydrodeoxygenated products into alkylbenzenes. Work is continuing on the combination of the HDO and HCR steps into a single hydroprocessing operation. The final HPR product consists predominantly of C_7-C_{10} alkylbenzenes, accompanied by smaller amounts of C_5-C_{10} branched paraffins and C_6-C_{10} alkylated naphthenes.

Figure 2 provides the scheme of the two-stage (BCD-ETR) procedure for preferential conversion of lignin to aromatic ethers. Stage I of the procedure comprises mild BCD at 250-265°C, using low MeOH/lignin ratios (3:1 to 5:1) and short reaction time (5 to 10 min.). Under such conditions, the BCD product retains most (> 90 wt%) of the original oxygen content of the lignin feed, which is important for the high-yield production of the final etheric products. On the other hand, the BCD product under the above mild conditions contains significant concentrations (about 25-40 wt%) of incompletely depolymerized components. This content necessitates a supplemental depolymerizing treatment of the BCD product by selective C-C hydrocracking with preservation of the O-containing groups, viz., with minimal concurrent C-O hydrocracking. Such a selective treatment (HT) is achieved by the use of Pt-modified solid superacids, Pt/SO_4^{2-}/ZrO_2 or Pt/WO_4^{2-}/ZrO_2, at 350-365°C under moderate H_2 pressure—about 1500 psig. Under such conditions, essentially no oxygen is lost and the content of oligomeric components in the phenolic product is reduced to < 10 wt%. In Stage II of the procedure the distilled monomeric phenols are subjected to etherification with methanol, at 225-275°C, in the presence of a solid superacid catalyst, such as SO_4^{2-}/ZrO_2, WO_4^{2-}/ZrO_2, SO_4^{2-}/MnO_x/Al_2O_3 or SO_4^{2-}/WO_x/Al_2O_3. In order to displace the equibrium between phenols and aryl methyl ethers in the direction of the desired etheric products, water-absorbing agents, including anhydrous. $MgSO_4$ and specially designed pillared clays, are being used to remove water formed during the etherification reaction, prior to recyclization of unreacted phenols and methanol.

Figure 3 provides an example of GC/MS analysis of a vacuum distilled BCD-HT product from Kraft (Indulin AT) lignin. The BCD conditions used in this early run were: temperature, 270°C; MeOH/lignin wt ratio, 7.5; reaction time, 30 min. The BCD product obtained was used as feed for the subsequent HT step, which was performed under the following conditions: temperature, 350°C; catalyst, Pt/SO_4^{2-}/ZrO_2; feed/catalyst wt ratio, 5:1, H_2 pressure, 1500 psig; reaction time, 2 h. Under these conditions the final BCD-HT product contained about 73.5 wt% of C_1-C_3 alkylated phenols and methoxyphenols, where C_1-C_3 indicates the total number of carbons in alkyl substituents, viz., 1, 2 or 3 alkyl (predominantly methyl) groups per molecule. In recent optimization of the BCD-HT procedure, the BCD step was performed at 265°C, using a lower MeOH/lignin ratio, (5:1) and a shorter reaction time (5 min.). Further, the HT step was performed at a higher H_2 pressure, 1800 psig. Under these modified conditions a highly desirable BCD-HT product, containing 79.1 wt% of C_1-C_2 alkylated phenols and 2-methoxyphenols (where C_1-C_2 indicates mostly methyl and dimethyl substitution), was obtained. The product contained 12.9 wt% of hydrocarbons and only ~ 8 wt% of higher phenols. Etherification of the vacuum distilled product with methanol produced (yield, ~ 90 wt%) a mixture of the corresponding aryl methyl ethers, which are characterized by blending octane numbers in the approximate range of 142–166.

Figure 4 provides an example of GC/MS analysis of a BCD-HPR product from NREL lignin. The selectivity for production of C_7-C_{10} alkylbenzenes in this early run was only moderate due to the use of an extended reaction time (1 h) and a high MeOH/lignin wt ratio (7.5:1) in the BCD

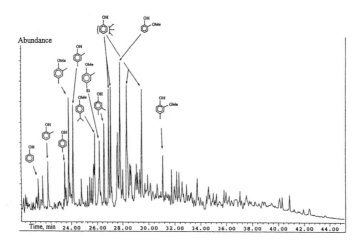

Figure 3. Example of GC/MS analysis of vacuum distilled BCD-HT product from Kraft lignin

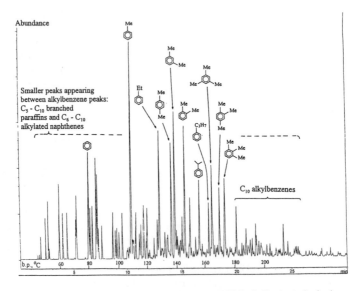

Figure 4. GC/MS analysis of BCD-HPR product from NREL lignin (lignin obtained as by-product in the NREL ethanol process)

step. The selectivity of the BCD–HPR procedure, the extent of ring alkylation and, in particular, the extent of undesirable ring hydrogenation of the C_7-C_{10} alkylbenzene products to corresponding alkylated naphthenes, are being sharply decreased in current work. Decreases are due to shorter reaction times and lower MeOH/lignin ratios in the BCD step, and to a combination of shorter reaction times and catalysts of lower ring hydrogenation activity (RuMo/SiO_2-Al_2O_3 or a solid superacid) in the HCR step. Under such modified conditions the BCD–HPR product predominantly consists of C_7-C_9 alkylbenzenes (approximately 80-85 wt%) accompanied by desirable C_6-C_{10} multibranched paraffins, 12-19 wt%, and < 4 wt% of alkylated naphthenes.

In a recent extension of the present work, the BCD reaction of Repap and NREL lignins were investigated in water (instead of methanol) as reaction medium. It was found that high lignin conversions into ether-soluble products (in the approximate range of 76-83 wt%) can be achieved under the following BCD conditions: temperature, 300-330 °C; reaction time (autoclave), 5-15 min; autogeneous pressure, 1700-1850 psig. The GC/MS of the predominant ether-soluble BCD product (after supplemental HT) obtained from Repap lignin, at 330 °C and a reaction time of 5 min, using a 7.1% aqueous solution of NaOH as reaction medium, shows that the main components of the product are phenol, methylphenols, and dimethylphenols, accompanied by smaller amounts of higher phenols, e.g., trimethylphenols and ethylmethylphenols. The extent of alkyl substitution in the phenolic products is lower than that previously observed for BCD products obtained with methanol as reaction medium. As previously found, methanol could act under certain BCD conditions as ring alkylation agent. On the other hand, in the presence of water as hydrolyzing agent and reaction medium, ring alkylation is excluded (except for a limited extent of transalkylation). HPR of the product, using conditions indicated in Figure 1, yielded a final product containing 89.4 wt% of C_7-C_{10} alkylbenzenes. GC/MS analysis of an ether/water-soluble fraction of the BCD product, obtained by continuous liquid/liquid extraction of the aqueous BCD layer with ether, showed that it mainly consists of 1,2-dihydroxybenzene (catechol) accompanied by smaller amounts of methyl-1,2-dihydroxybenzene (methylcatechol) and some lower-boiling components such as phenols and low carboxylic acids. The formation of catechol as a major BCD product from lignin in the presence of water as medium, can be rationalized in terms of hydrolysis of an aryloxy (ArO-) and an adjacent alkoxy (e.g., MeO-) group, and attendant or subsequent dealkylation reactions of C_3 ring substituents.

The results obtained show that BCD of lignin can be advantageously performed in an aqueous medium to yield a mixture of phenols having a low degree of alkyl substitution. Such phenols are effectively hydroprocessed to yield C_7-C_{10} alkylbenzenes, or etherified with methanol to produce aryl methyl ethers, two types of potential fuel additives.

4. ACKNOWLEDGMENTS

The authors wish to thank the U.S. Department of Energy, Biofuels Systems Division, for financial support through Sandia Corp. (AU-8776).

REFERENCES

Mossman, A.B., U.S. Patent 4,611,084 (1986). (1987) U.S. Patent 4,675,455.
Shabtai, J., N.K. Nag, and F.E. Massoth (1987). J. Catal., 104, pp. 413–423.
Shabtai, J., X. Xiao, and W. Zmierczak (1997). Energy Fuels, 11, pp. 76–87.
Shabtai, J., W. Zmierczak, and E. Chornet (1997). Proc. 3rd Biomass Confer. of the Americas, Montreal, Elsevier, 2, pp. 1037–1040.
Shabtai, J., W. Zmierczak, and E. Chornet (1998a). U.S. Patent Appl. No. 09/136,336.
Shabtai, J., W. Zmierczak, and E. Chornet (1998b). U.S. Provisional Patent Appl. No. 60/097,701.
Shabtai, J., W. Zmierczak, E. Chornet, and D.K. Johnson (1999). Am. Chem. Soc. Div. Fuel Chem. Prelim. Commun., 44.
Wyman, C.E. (ed.) (1996). Handbook on Bioethanol Production and Utilization, Taylor & Francis.
Zmierczak, W., X. Xiao, and J. Shabtai (1994). Energy Fuels, 8, pp. 113–116.

PERFORMANCE ADVANTAGES OF CETANE IMPROVERS PRODUCED FROM SOYBEAN OIL

G.J. Suppes, T.T. Tshung, M.H. Mason, and J.A. Heppert

Department of Chemical and Petroleum Engineering, The University of Kansas, 4006 Learned, Lawrence, Kansas 66045-2223

Synthesized nitrate derivatives of soybean oil provide a valuable additive that increases the cetane of diesel fuel. This paper presents fundamental data on how the components of soybean oil contribute differently to product performance. Performance is comparable to commercial cetane improvers. These products substantially reduce nitrogen contents as compared to 2-ethylhexyl nitrate and incorporate relatively low cost renewable feedstocks in their production.

1. INTRODUCTION

Cetane improvers (CI) are a commonly used diesel fuel additive that decrease the ignition delay time between start of fuel injection and ignition; the delay time is quantified on a cetane number scale defined by reference fuels specified in the ASTM D-613 standard. A shorter ignition delay means faster startup in cold weather, reduced NO_x emissions, and smoother engine operation. A diesel fuel's cetane number is one of five properties that can be used to justify a "premium" rating for that fuel[1]—the analogous octane rating of gasoline is used to determine quality and price of gasoline.

The predominant commercial CI is 2-ethylhexyl nitrate (EHN). Table 1 illustrates the structure of EHN as well as two alternative CI. *Ditertiary*-butyl peroxide (DTBP) is used as an alternative to EHN when the nitrogen content of EHN is undesirable in the fuel; however, DTBP is both more expensive than EHN and must be transported in a diluted form (e.g., 32% DTBP in diesel).[1] Methyl oleate dinitrate (MODN) is the dinitrate of the oleic acid methyl ester.

MODN is produced by the transesterification of oleic acid followed by the epoxidation of the carbon-carbon pi bond and then nitration with nitric acid. Although a single compound is illustrated by Table 1, the product of this synthesis process is a mixture of this product, mononitrates of similar structure, and a small quantity (< 10%) of by-products and precursors. MODN is the best of a series of products produced from soybean oil derivatives and was chosen to represent this class of CI.

Table 1
Example CI compounds

EHN: 2-ethylhexyl nitrate structure (CH₃CH₂-branched hexyl with ONO₂)

DTBP: di-tert-butyl peroxide structure ((CH₃)₃C-O-O-C(CH₃)₃ shown with CH groups)

MODN: long-chain dinitrate with terminal acetate ester structure bearing two ONO₂ groups

This paper compares the performance of the three products indicated by Table 1 and discusses the advantages and disadvantages of each. Several measures of performance can be used to characterize the performance of a CI, including
- blending cetane number (BCN), which is the increase in cetane number divided by mass fraction of CI in the fuel;
- percent ignition improvement (the PII or decrease in ignition delay time) relative to EHN;
- the ratio of either BCN or PII to the nitrogen content. This can be normalized with respect to this ratio for EHN;
- the ratio BCN or percent ignition improvement relative to the cost per mass of CI.

In practice, any of the above criteria (or others) could be used to identify the best CI for an application. To complicate the characterization even further, the BCN and PII are both functions of CI application rates. For purposes of this work, the PII and its respective ratios are used—this circumvents the need to constantly run cetane number reference standards on the constant volume combustor.

2. EXPERIMENTAL PROCEDURE

2.1. Chemical synthesis

While direct nitration of carbon-carbon double bonds with N_2O_5 leads to the formation of both nitrate and nitro groups, use of an epoxy intermediate leads to a product dominated by formation of nitrates. More specifically, the double bonds are converted to dinitrates. Nitration was performed with a mixture of acetic anhydride and nitric acid. This method is different that the dinitrogen pentoxide method is described elsewhere.[2]

2.2. Epoxidation of methyl oleate

11.0 g of methyl oleate was weighed into a 250-ml three-neck flask, equipped with a thermocouple and a magnetic stirrer. Next, 2.0 g of 88% formic acid was added with stirring. For a period of 30 min, 3.9 g of 50% hydrogen peroxide was added using a dripper at room temperature. The reaction temperature was allowed to fluctuate the first 2 h. Temperature was maintained at 35°C for the next 2 h. Total reaction time was 4 hs. Then the reaction was quenched with ice water, neutralized with dilute $NaHCO_3$, extracted into diethyl ether and washed with water. ^1H NMR analysis shows 85% conversion in the double bond region.

2.3. Hydrolysis of methyl oleate epoxide

In a 250-ml round bottom flask, 1 volume of methyl oleate epoxide was reacted with 2 volumes of 1 N HCl and 2 drops of H_2SO_4/10 ml of total volume. This mixture was stirred heated to 80°C for 3 hs, then heated to 90°C for another 1 h. The reaction was quenched with ice water, neutralized with dilute $NaHCO_3$, extracted into diethyl ether and washed with water. ^1H NMR analysis shows 100% conversion in the epoxide bond region.

2.4. Nitration of methyl oleate diol

In a 250-ml round bottom flask, 3.0 g of methyl oleate diol was mixed with an equivalent volume of CH_2Cl_2. Nitrogen gas was used to purge air from the flask. The 2.8 g 90% HNO_3 to 1.4 g 99% acetic anhydride was mixed in a graduated cylinder at below 5°C. The acid mixture was dripped onto the methyl oleate diol-CH_2Cl_2 mixture slowly as to maintain the reaction temperature below 5°C. The reactants were stirred all the time. The temperature was maintained below 5°C in the first hour, about 10°C in the second hour, about 25°C in the third hour and at 30°C in the fourth hour. Then the reaction was quenched with ice water, neutralized with dilute $NaHCO_3$, extracted into diethyl ether and washed with water. ^1HNMR and IR analysis showed 100% conversion in the hydroxyl bond region.

Isolated yield of crude product was 40%. FT-IR spectra showed a small peak at 3510 cm^{-1}, with large peaks at 2933, 2864, 1736, 1650**, 1553*, 1467, 1273**, 854**. GC-MS data confirmed the presence of nonanal, and 9-oxo-nonanoic acid, methyl ester as minor products. Both were expected from oxidative cleavage of the double bond in methyl oleate. The ^1H NMR appeared to be soy bean oil with a small set of multiplets growing in at about 2.8 ppm.

Suppes et al.[2] described the performance of MODN at about 60% to 70% of the performance of EHN. Since this initial data, several improvements have been made on the synthesis, including the following:

EHN and DTBP were both purchased from Aldrich for comparative studies.

2.5. Testing

The product of nitration was evaluated as a cetane improver in a constant volume combustor that directly measured the ignition delay times. Application rates of the cetane improvers at 0.5% in hexanes provided a good balance between typical application rates (about 0.1%) and a good signal-to-noise ratio for the method. The ignition delay time was plotted as a function of temperature to observe trends. About 10 ignition delay times were measured and averaged at 750, 800, and 850 K. The higher temperature is considered the most relevant since it is closer to the actual engine conditions. Further details on the experimental procedure can be found elsewhere.

3. RESULTS

Two batches of MODN were produced with similar chemical synthesis methods. They were both clear yellowish liquids. Batch one was produced on July 31, 1998, and tested within five days. Batch two was produced on September 15, 1998 and tested within 30 days. Most of double bonds in the methyl oleate of each batch was utilized and converted to dinitrates. The first batch, CY073198, has about 65% NO_3 and 35% NO_2 bonds. The second batch, CY091598, has about 95% NO_3 and 5% NO_2 bonds. Figures 1 and 2 show the ignition delay time trends.

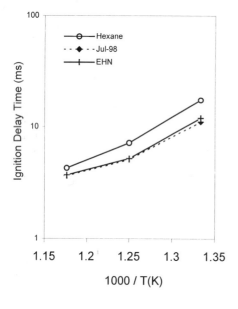
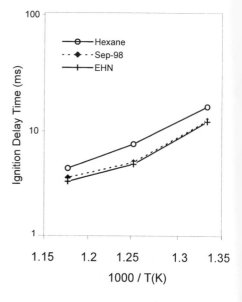

Figure 1. Performance of MODN synthesized in July of 1998 (CY073198).

Figure 2. Performance of MODN synthesized in September of 1998 (CY091598).

The first batch of MODN performed at about 110% of EHN while the second batch performed at about 85% of EHN. Elemental analysis of the products showed the MODN to contain about 49% of the nitrogen of EHN on a mass basis. Based on these results, the ratio [PII to nitrogen content]$_{MODN}$: [PII to nitrogen content]$_{EHN}$ is 2.0. This product provided about twice the performance increase for a given mass of nitrogen added to the fuel.

4. DISCUSSION

Both MODN products had ignition delay times that are very competitive to the EHN and changes in performance are believed to be primarily due to as-yet-to-be-determined variations in synthesis or storage. CY091598 was expected to perform better than CY073198 because it had a higher percentage of NO_3 bonds; however, this was not realized. The NO_3 content of the compound has been observed to correlate with good performance while the NO_2 content has little or adverse impact.

The first possible factor contributing to this observation is storage length. The ability to reduce ignition delay time may deteriorate with time. The second product was stored for a much longer than was the first product. The second possible factor is the solubility of nitrated products in straight hexane. Solubility of nitrated products tends to decrease with increasing NO_3 species. Decrease in solubility reduced the effectiveness of nitrated product in reducing ignition delay time.

The synthesis of a nitrate product from oleic acid with a performance similar to EHN and with a PII to nitrogen ratio twice that of EHN is a substantial advance in cetane improvers. For markets with regulations controlling the maximum nitrogen content of diesel, EHN is typically used until the maximum nitrogen levels are reached. For additional increases in cetane, DTBP is added. The advantages of a product with a higher PII to nitrogen content for these applications are obvious. MODN produced by the methods described herein is the best fatty-acid-based product tested to date (Table 2). In general, the nitrates of the methyl esters performed better than the products obtained by direct nitration of vegetable oils.

Furthermore, the product obtained from the methyl ester of oleic acid is better than that produced by the nitration of biodiesel (methyl esteres) produced from soybean oil. The deterioration of performance of biodiesel nitrates is attributed to by-products and solubility related to the greater degree of unsaturation (more carbon-carbon π bonds) of biodiesel. If these by-products and solubility issues can be controlled, a performance similar to MODN should be realized.

Possible advantages of a cetane improver produced from soybean with a performance similar to MODN include
- use of renewable feedstocks,
- improved biodegradability,
- reduced nitrogen content as compared to EHN,
- substantially improved PII to nitrogen content ratios, and
- development of sustainable rural economies related to these products.

Table 2
Comparison of performances of different products reported in percent of performance of EHN[2]

	750 K %	800 K %	850 K %
Tested in Hexanes			
0.5% Dinitrate	59	64	56
1% Oleic acid	2	8	4
1% Biodiesel	2	14	6
1% Dinitrate	73	56	50
1% Soybean oil nitro-nitrate	< 63	< 67	< 57
Tested in 90% hexanes + 10% ethanol			
	750 K %	800 K %	850 K %
1% OAEGN	91	48	47
0.5% OAEGN + 0.5% DEGDN	59	66	55
0.8% OAEGN + 0.2% DEGDN	78	66	66
1% OAEGN	51	78	82
0.8% OAEGN + 0.2% DEGDN	97	126	103
1% DEGDN	72	84	88
1% Castoroil nitrate	48	75	69
1% Methyl oleate nitro-nitrate	38	60	60

5. CONCLUSIONS

A nitrate product (MODN) of oleic acid has been produced with a performance comparable to the current commercial product, EHN, and with a performance-to-nitrogen-content about twice that of EHN. Synthesis methods based on production of a similar product from soybean oil could offer several competitive advantages over EHN and is currently under development.

6. ACKNOWLEDGEMENTS

This work was supported by the Kansas Soybean Commission, the Kansas Value Added Center, and the NRI Competitive Grants Program/USDA award number 97-35504-4244. Funding from The University of Kansas was also provided through the New Faculty Award,

Department of Chemical and Petroleum Engineering, and the Energy Research Center. Mike Hiskey of Los Alamos National Laboratory provided extremely helpful information about nitration chemistry.

REFERENCES

1. Peckham, J. (1998). Peroxide CI could help some diesel refiners, Growing Your Business with Premium Diesel, special report from Diesel Fuel News, Hart Publications Inc.
2. Suppes, G.J., M. Mason, Y.T. Tshung, R. Aggarwal, and J.A. Heppert (1998). Performance advantages of CI produced from soybean oil, Bioenergy'98, Madison, Wisconsin, October.

FLUIDIZED BED CATALYTIC STEAM REFORMING OF PYROLYSIS OIL FOR PRODUCTION OF HYDROGEN

S. Czernik, R. French, C. Feik, and E. Chornet*

National Renewable Energy Laboratory, 1617 Cole Boulevard, Golden, CO 80401
*Also affiliated with Université de Sherbrooke, Sherbrooke, Québec, J1K 2R1 Canada

Biomass can be an attractive alternative to fossil feedstocks for the production of hydrogen because of essentially zero net CO_2 impact. The concept proposed in this work combines fast pyrolysis of biomass and catalytic steam reforming of the pyrolysis oil or its fractions. This two-step approach has several advantages over the traditional gasification/water-gas shift technology. The most important is the potential for recovery of higher value co-products from bio-oil that could significantly improve the economics of the entire process. Steam reforming using commercial nickel-based catalysts can efficiently convert volatile oil components to hydrogen and carbon oxides. However, the non-volatile compounds such as sugars and lignin-derived oligomers tend to decompose thermally and to form carbonaceous deposits on the catalyst surface and in the reactor freeboard. To reduce this undesirable effect we employed a fluidized bed reformer configuration with fine mist feed injection to the catalyst bed. The hydrogen yields obtained from carbohydrate-derived bio-oil fraction exceeded 80% of that possible by stoichiometric conversion. Though 90% of the feed carbon was converted to CO_2 and CO, carbonaceous deposits were formed on the catalyst surface, which resulted in the gradual loss of its activity. The catalyst was easily regenerated by steam or carbon dioxide gasification of the deposits.

1. INTRODUCTION

At present, hydrogen is produced almost entirely from fossil fuels such as natural gas, naphtha, and inexpensive coal. In such a case, the same amount of CO_2 as that formed from combustion of those fuels is released during hydrogen production. Renewable biomass is an attractive alternative to fossil feedstocks because of essentially zero net CO_2 impact. Unfortunately, hydrogen content in biomass is only 6-6.5%, compared to almost 25% in natural gas. For this reason, on a cost basis, producing hydrogen by a direct conversion process such as the biomass gasification/water-gas shift cannot compete with the well-developed technology for steam reforming of natural gas. However, an integrated process, in which biomass is partly used to produce more valuable materials or chemicals with only residual fractions utilised for generation of hydrogen, can be economically viable. The proposed method, which was described earlier,[1] combines two stages: fast pyrolysis of biomass to generate bio-oil and catalytic steam reforming of the bio-oil to hydrogen and carbon dioxide. This concept has several advantages over the traditional gasification/water-gas shift technology.

First, bio-oil is much easier to transport than solid biomass, and therefore pyrolysis and reforming can be carried out at different locations to improve the economics. For instance, a series of small size pyrolysis units could be constructed at sites where low-cost feedstock is available. The oil could then be transported to a central reforming plant at a site with hydrogen storage and distribution infrastructure.

A second advantage is the potential production and recovery of higher value added co-products from bio-oil that could significantly affect the economics of the entire process. In this concept, the lignin-derived fraction would be separated from bio-oil and used as a phenol substitute in phenol-formaldehyde adhesives, while the carbohydrate-derived fraction would be catalytically steam reformed to produce hydrogen. Assuming that the phenolic fraction could be sold for $0.44/kg (approximately half of the price of phenol), the estimated cost of hydrogen from this conceptual process would be $7.7/GJ,[2] which is at the low end of the current selling prices.

In previous years we demonstrated, initially through micro-scale tests and then in bench-scale fixed-bed reactor experiments,[3] that bio-oil model compounds as well as a carbohydrate-derived fraction can be efficiently converted to hydrogen. Using commercial nickel catalysts, the hydrogen yields approached or exceeded 90% of those possible by stoichiometric conversion. The carbohydrate-derived bio-oil fraction contains a substantial amount of non-volatile compounds (sugars and oligomers) which tend to decompose thermally and carbonize before contacting the steam reforming catalyst. We managed to reduce these undesired reactions by injecting the oil fraction into the reactor as a fine mist. However, even with the large excess of steam used, the carbonaceous deposits on the catalyst and in the reactor freeboard limited the reforming time to 3-4 h. The limitations of the fixed-bed reactor were even more obvious for processing whole bio-oil. The hydrogen yield was only 41% of that stoichiometrically possible and the reforming duration was less than 45 min. For this reason we decided to employ a fluidized bed reactor configuration that should overcome at least some limitations of the fixed-bed unit. Even if carbonization of the oil cannot be avoided, still the bulk of the fluidizing catalyst would be in contact with the oil droplets fed to the reactor. This contact should greatly increase the reforming efficiency and extend the catalyst time-on-stream. Catalyst regeneration can be done by steam or carbon dioxide gasification of carbonaceous residues in a second fluidized bed reactor providing additional amounts of hydrogen.

2. EXPERIMENTAL

2.1. Materials

The bio-oil was generated from poplar wood using the NREL fast pyrolysis vortex reactor system.[4] The oil contained 46.8% carbon, 7.4% hydrogen, 45.8% oxygen and a water content of 19%. It was separated into aqueous (carbohydrate-derived) and organic (lignin-derived) fractions by adding water to the oil at a weight ratio of 2:1. The aqueous fraction (55% of the whole oil) contained 22.9% organics ($CH_{1.34}O_{0.81}$) and 77.1% water.

U91, a commercial nickel-based catalyst used for steam reforming of natural gas, was obtained from United Catalysts and ground to the particle size of 300-500 µ.

2.2. Fluidized bed reformer

The bench-scale fluidized bed reactor is shown in Figure 1.

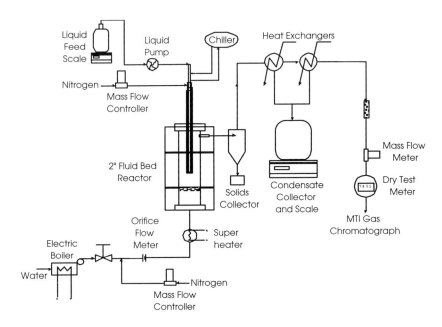

Figure 1. Schematic of the 2" fluidized bed reactor system.

A two-inch-diameter inconel reactor supplied with a porous metal distribution plate was placed inside a three-zone electric furnace. The reactor contained 150-200 g of commercial nickel-based catalyst ground to the particle size of 300-500 µ. The catalyst was fluidized using superheated steam, which is also a reactant in the reforming process. Steam was generated in a boiler and superheated to 750°C before entering the reactor at a flow rate of 2-4 g/min. Liquids were fed at a rate of 4-5 g/min using a diaphragm pump. A specially designed injection nozzle supplied with a cooling jacket was used to spray liquids into the catalyst bed. The temperature in the injector was controlled by coolant flow and maintained below the feed boiling point to prevent evaporation of volatile and deposition of nonvolatile components.

The product passed through a cyclone that captured fine catalyst particles and, possibly, char generated in the reactor. Then two heat exchangers removed excess steam. The condensate was collected in a vessel whose weight was continuously monitored. The outlet gas flow rate was measured by a mass flow meter and by a dry test meter. The gas composition was analyzed every 5 min by an MTI gas chromatograph. The analysis provided concentrations of hydrogen, carbon monoxide, carbon dioxide, methane, ethylene, and nitrogen in the outlet gas stream as a function of time of the test. The temperatures in the system as well as the flows were recorded and controlled by the G2/OPTO data acquisition and control system. Total and elemental balances were calculated as well as the yield of hydrogen generated from the feed.

3. RESULTS

Steam reforming experiments in the fluidized bed reactor were carried out at 800°C and 850°C. The steam to carbon ratio was held at 7-9 while methane-equivalent gas hourly space velocity $G_{C1}HSV$ was in the range of 1200-1500 h^{-1}. During the experiment at 800°C a slow decrease in the concentration of hydrogen and carbon dioxide and an increase of carbon monoxide and methane in the gas generated by steam reforming of the carbohydrate-derived oil fraction was observed. These changes resulted from a gradual loss of the catalyst activity, probably due to coke deposits. As a consequence of that, the yield of hydrogen produced from the oil fraction decreased from the initial value of 95% of stoichiometric (3.24 g of hydrogen from 100 g of feed) to 77% after 12 h on stream. If the reformer was followed by a water-gas shift reactor the hydrogen yields would increase to 99% at the beginning and to 84% after 12 h on stream.

The catalyst used in this experiment was regenerated by carbon dioxide and steam treatment for 2 h at 850°C and then reused in the next test, which was carried out at 850°C. We expected that at a higher temperature less char and coke would be formed or their gasification by steam would be more efficient than that achieved under the previous conditions. After regeneration the catalyst recovered its initial activity. During 8 h of reforming the bio-oil carbohydrate-derived fraction, the composition of the product gas remained constant (Figure 2). This stability indicates that no catalysts deactivated during the run. The yield of hydrogen produced from the bio-oil fraction was approximately 90% of that possible by stoichiometric conversion. It would be greater than 95% if carbon monoxide underwent the complete shift reaction with steam. Only small amounts of feed were collected as char in the cyclone and condensers, and little or no coke was deposited on the catalyst.

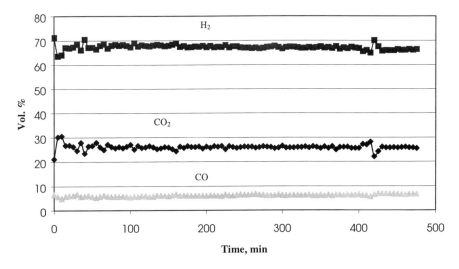

Figure 2. Reforming gas composition.

4. CONCLUSIONS

We successfully demonstrated that hydrogen could be efficiently produced by catalytic steam reforming of a carbohydrate-derived bio-oil fraction using a commercial nickel-based catalyst in a fluidized bed reactor. Greater steam excess than that used for natural gas reforming was necessary to minimize the formation of char and coke (or to gasify these carbonaceous solids) resulting from thermal decomposition of complex carbohydrate-derived compounds.

At 850°C with a steam to carbon ratio of 9, the hydrogen yield was 90% of that possible by stoichiometric conversion during 8 h of the catalyst on-stream time. This yield could be 5-7% greater if a secondary water-gas shift reactor followed the reformer.

Coke deposits were efficiently removed from the catalyst by steam and carbon dioxide gasification, which restored the initial catalytic activity.

REFERENCES

1. Wang, D., S. Czernik, D. Montané, M. Mann, and E. Chornet (1997). I&EC Research, 36, p. 1507.
2. Mann, M.K., P.L. Spath, and K. Kadam (1996). In Proceedings of the 1996 U.S. DOE Hydrogen Program Review, Miami, FL, May 1–2, 1996, NREL/CP-430-21968; pp. 249–272.
3. Wang, D., S. Czernik, and E. Chornet (1998). Energy&Fuels, 12, p. 19.
4. Scahill, J.W., J. Diebold, C.J. Feik (1997). In Developments in Thermochemical Biomass Conversion, ed. by A.V. Bridgwater and D.G.B. Boocock, Blackie Academic & Professional, London, pp. 253–266.

This project was sponsored by the Energy Technologies Advancement Program (ETAP) of the California Energy Commission. A partnership between Arcadis Geraghty & Miller and the College of Engineering - Center for Environmental Research and Technology (CE-CERT) conducted design, construction and operation of this work with the support of the Hynol Corporation, Martech International, T.R. Miles Consulting Design Engineers. The pilot plant was built at the CE-CERT facility in Riverside, California.

The reactor system was designed, constructed and assembled by the project team during this stage of the project. At the time of writing this report, the HPR system installation was nearly complete and preliminary shakedown of individual process components was underway. Following the completion of all troubleshooting and shakedown procedures, actual testing of the HPR gasifier will occur. The demonstration tests address the ETAP objectives as follows:

- Alkali removal techniques. Alkali metals, primarily potassium and sodium, will be removed by using proprietary ARCADIS Geraghty & Miller mixtures of alkali getters such as kaolinite, bauxite, and sand.
- Control of high molecular weight tars. The hydropyrolysis process will produce a number of tar compounds depending on the feedstock. Tar in the HPR effluent gas will be removed at the surface of the hot gas filter media. The cake formed on the filter surface will provide an ideal surface for the adsorption of tar compounds.
- Hot gas filtration. Filtration of the hot effluent gases leaving the HPR will be achieved through a ceramic hot gas filter. The hot gas filter system, provided by Pall Filtration Systems, is designed as an integral part of the HPR system and its performance will be evaluated during the testing phase.
- Biomass feed. The type of biomass used is an important factor in determining the overall process efficiency and feasibility. Initially clean wood sawdust (high quality fuel) will be used as the biomass feed. Typical low quality biomass waste such as grass waste fines will be also be used in the test matrix. Biomass feeding will be achieved by augering the feed from a hopper into the HPR vessel.

3. HOT GAS FILTRATION

The Energy Commission was very much interested in this project owing to the opportunity to test filtration equipment for cleaning hot flue gas. The hot gases leaving the HPR will be filtered using high-temperature filter systems. Pall Filtration Systems, Inc., provided the high-temperature filtration system for this project. High temperature filters are typically made from ceramic or sintered metal. Both materials have inherent advantages and disadvantages. For this project, Pall the use of ceramic filters with a pulse-jet backflow purging system was recommended. The key issues regarding hot gas filtration that will be addressed during this project will include corrosion of the filter media due to the presence of alkali metals such as sodium and potassium; fouling of the filter media by the particulates;

potential glass-coating of the filter media by the fluidized-bed sand; and overall lifecycle-type performance issues.

The first stage of the Hynol project included the demonstration of the HPR gasifier. Arcadis Geraghty & Miller has stated that the second and third stages of the Hynol project will involve the construction and testing of the Steam Pyrolysis Reactor (SPR) and the Methanol Synthesis Reactor (MSR).

A GREEN APPROACH FOR THE PRODUCTION OF BIO-CETANE ENHANCER FOR DIESEL FUELS

Al Wong[a] and Ed Hogan[b]

[a]Arbokem Inc., P.O. Box 95014, Vancouver V6P 6V4, Canada
[b]Natural Resources Canada, 580 Booth Street, Ottawa, Ontario K1A OE4, Canada

Biomass-oil based transportation fuel provides a practical means to reduce emission of conventional pollutants and to lower the emission of CO_2. The bio-fuel could be made by the trans-esterification of vegetable oils and fats or by the catalytic hydrotreating of vegetable or tree oils and fats.

Catalytic hydrotreatment has been tested on a pilot scale to convert many types of biomass oil into a 60-90 Cetane Number middle distillate. This bio-cetane product can be used neat as a diesel fuel or as a blending agent for ordinary diesel fuel.

Laboratory emission testing of a transit bus has indicated that significantly lower emissions of particulates, carbon monoxide, and hydrocarbons can be achieved. A 10-month on-road test of six postal delivery vans has shown that the engine fuel economy was greatly improved by a blend of petrodiesel and the bio-cetane product.

1. INTRODUCTION

The two main methods of producing biomass-based diesel fuel are trans-esterification of biomass oil with methanol, and catalytic hydrotreating of biomass oil.[1]

The notable drawbacks of the ester approach are that methanol is made most economically from fossil fuel, and that the resulting methyl ester fuel does not chemically resemble conventional diesel fuel.[1]

In contrast, the catalytic hydrotreatment approach[2] yields several fuels: 3-5% burner gas, 2-5% light distillate (b.p. < 160 deg. C), 55-65% middle distillate (b.p. 160-325 deg. C) and 20-30% residue (b.p. > 325 deg. C) and about 10% water. The middle distillate has an alkane profile that matches that of conventional diesel fuel, without the presence of undesirable branched or cyclic hydrocarbons.[3,4] The "diesel fraction" can thus be used neat or as a specific cetane enhancer.

2. RAW MATERIAL

The feedstock in both cases could be obtained from vegetable oils, marine fish oils, animal fats or tree oil. The free-market economics of feedstock is determined by the prevailing relative values of raw biomass oil and fat and of crude petroleum oil, at equi-energy content. The economic competitiveness of fuel made from biomass oil is worsened when edible virgin vegetable oil is used.[5]

The practicable feedstock for hydrotreatment to produce bio-cetane includes tall oil (a by-product of wood pulp manufacture) and used vegetable cooking oils and fats.[3,4]

It has been found recently that the hydrogen required for the hydrotreatment process can be derived entirely in situ from a portion of the non-"diesel fraction."[6] Obtaining hydrogen from fossil fuel would be avoided. In this fashion, the hydrotreatment process moves closer to an energy self-sufficient bio-fuel.

3. BIO-CETANE PERFORMANCE

3.1. Fuel economy

Field tests of a blend of 60% low-sulphur diesel fuel and 40% tall oil-based bio-cetane enhancer were excellent during a 12-month vehicle trial in Vancouver. As shown in Table 1, a fuel economy of 10% was achieved.[7] This finding concurs with the CO_2 reduction observed in a separate emission analysis from earlier engine dynamometer testing.[7]

Table 1
Comparative fuel economy of six postal delivery vans[7] in litres per 100 km

Vehicle #	84058A	84070A	84095A	84072B	84085B	84090B
Pre-test*	23.1	21.3	21.4	21.2	21.5	20.5
		Petrodiesel "A"			Cetane/Petrodiesel "A"	
July	9.25	3.84	6.18	6.10	7.40	4.96
August	32.44	23.71	25.99	36.10	30.05	29.76
September	21.68	14.09	29.08	16.75	23.82	26.44
October	21.67	36.81	17.09	24.63	20.57	18.83
November	22.91	22.41	30.42	30.62	24.23	30.75
December	33.08	28.14	17.16	16.96	19.78	9.37
January	18.34	21.89	19.04	17.36	20.51	17.94
February	22.95	26.64	20.21	22.66	21.52	27.19
March	0.00	10.54	20.60	10.38	13.81	6.73
April	0.00	18.71	16.34	12.29	7.51	23.48
Average	22.79	20.68	20.21	19.39	18.92	19.55
% diff. From pre-test period	-1.3	-2.9	-5.6	-8.6	-12.0	-4.7

* 12-month average prior to road test, using 0.05%-sulphur petrodiesel

3.2. Emission of conventional pollutants

Laboratory emission testing suggested that significant reductions in the emissions of particulates, carbon monoxide, and hydrocarbons could be achieved with the use of this specific fuel blend on transit buses and light-duty trucks.[7] Table 2 illustrates some example results from a series of chassis dynamometer tests on an Ottawa transit bus (DDC 6V92 engine).

Table 2
Chassis dynamometer testing using a blend of 60% petrodiesel and 40% bio-cetane[7]

	Test Cycle		
	NY Comp	NY Bus	CBD
Total hydrocarbons	-2.5%	-7.5%	-6.5%
Carbon monoxide	-8.5%	-16.0%	6.0%
Nitrogen oxides	-1.5%	-4.0%	1.0%
Particulates (PM-10)	-20%	4.0%	-20%

NY Comp = New York Composite driving cycle to simulate non-freeway and freeway driving conditions. Typical average speed is 14.12 km/hour. Typical fuel consumption is 81 litres per 100 km.

NY Bus = New York Bus cycle to simulate non-freeway driving. Typical average speed is 6.26 km/hour. Typical fuel consumption is 178 litres per 100 km.

CBD = Central Business District to simulate a series of "stop-go-stop" driving. Typical average speed is 19.92 km/hour. Typical fuel consumption is 84 litres per 100 km.

4. CARBON DIOXIDE EMISSION

Biomass-oil based fuel offers a practical means of reducing emission of CO_2 and the real-time recycling of CO_2 produced by the operation of conventional motor vehicles. As given in Figure 1, a life cycle analysis showed the bio-cetane diesel fuel would emit about 150 grams CO_2 equivalent per km,[7] the lowest value among the conventional fuels compared.

Electric vehicles require substantial new supplies of electricity that could not be obtained in practice from additional installations of hydroelectric generating or nuclear power facilities. New industrial electricity could of course be generated from the classical burning of fossil fuel; fossil CO_2 is produced in this instance.

A review of the alternative approach of using natural gas or fuel cell technology to power transit buses suggests that substantial emission of fossil CO_2 could still occur. Table 3 compares CO_2 emission potentials for selected means of fuelling transit buses.

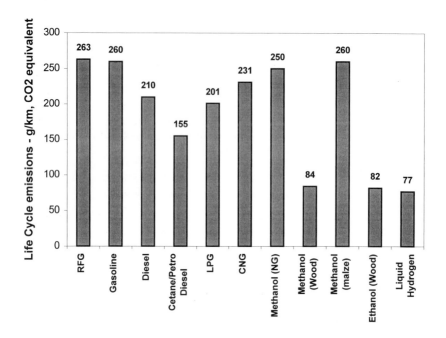

Figure 1. Comparative life-cycle greenhouse gas emissions.[7]

Table 3
Qualitative analysis of CO_2 Emission of selected fuelling approaches for the operation of a transit bus

Net CO_2 Emission	Transit Bus Operation		Practical Source of Electricity	Net CO_2 Emission
Zero	Fuel Cell A	Hydrogen		
			Hydroelectric	Zero
			Nuclear	Zero
			Fossil fuel	Yes
			Fossil fuel	Yes
Yes	Fuel Cell B	Methanol	Fossil fuel	Yes
Yes	Natural gas	Natural gas	Gas wells	Zero
Recycled (Zero)	Biomass oil	Methanol	Fossil fuel	Yes
	(Methyl ester)	Vegetable oil	Organic cropping	Recycled (Zero)
Recycled (Zero)	Biomass oil	Vegetable oil	Organic cropping	Recycled (Zero)
	(Bio-cetane)	Hydrogen	In situ	Zero

5. CONCLUDING REMARKS

Catalytic hydrotreating of biomass oil offers a practical means to deliver a transportation fuel which would meet the requirements of low emission of conventional pollutants and lower net emission of fossil carbon dioxide.

In order to achieve this goal, biomass oil feedstock could be obtained from organically grown crops in which the energy required to produce the synthetic nitrogen fertilizer would be zero. The hydrogen required to hydrotreat the biomass oil could be made in situ from one of the hydrotreated fractions. Biomass oil affords true "real time" recycling of CO_2-produced from the operation of transport vehicles.

6. ACKNOWLEDGMENT

The participation of the members of the Cetane Testing Consortium (Arbokem Inc., BC Chemicals Ltd., BC Transit, Canada Post Corp., Environment Canada, Natural Resources Canada, OC Transpo, and Petro-Canada) in the research project is gratefully acknowledged. [AK17709]

REFERENCES

1. Stumborg, M. et al. (1996). Hydroprocessed vegetable oils for diesel fuel improvement, Bioresource Technology, 56, p. 13.
2. Liu, D. et al. (1997). Production of high quality cetane enhancer from depitched tall oil, Petroleum Sci. Tech, 54, p. 12.
3. Feng, Y. et al. (1993). Chemical composition of tall oil-based cetane enhancer for diesel fuels, Proc. First Biomass Conference of the Americas, Burlington, Vermont, USA, August.
4. Wong, A. et al. (1995). Conversion of vegetable oils and animal fats into paraffinic cetane enhancer for diesel fuels, Proc. 2nd Biomass Conference of the Americas, Portland, Oregon, USA, August.
5. Wong, A. et al. (1994). Technical and economic aspects of manufacturing cetane-enhanced diesel fuel from canola oil, Proc. SCDC Bio-Oil Symposium, Saskatoon, Canada, March.
6. Bangala, D. et al. (1998). Catalytic steam reforming of residual bio-naphthas from tall oil hydrotreating process, paper presented at the AIChE Spring National Meeting, New Orleans, Louisiana, USA.
7. Wong, A. and E. Hogan (1998). Bio-based cetane enhancer for diesel fuels, Proc. Bioenergy '98 Conference, Madison, Wisconsin, USA, October.

BIOMASS CONVERSION TO FISCHER-TROPSCH LIQUIDS: PRELIMINARY ENERGY BALANCES

Eric D. Larson and Haiming Jin

Center for Energy and Environmental Studies, Princeton University, Princeton, New Jersey, USA, 08544

We present preliminary energy balances, based on a simplified calculation approach, for the production of Fischer-Tropsch (FT) liquids from biomass, coal, and natural gas. These balances are compared with balances of coal and natural gas feedstocks in process configurations that maximize F-T liquids. With biomass feedstocks, we present results for two process configurations using an atmospheric-pressure indirectly heated gasifier. In one case, production of F-T liquids is maximized. In the other case, F-T liquids are produced via one-pass of the synthesis gas through the F-T reactor, with unconverted syngas used to co-produce electricity in a gas turbine combined cycle. Additionally, we compare the energy efficiencies of "once-through" process configurations for five different biomass gasifier designs: indirectly-heated (two low-pressure designs), oxygen-blown (high-pressure), and air-blown (high-pressure and low-pressure).

1. INTRODUCTION

In Fischer-Tropsch conversion, hydrocarbons are synthesized from carbon monoxide (CO) and hydrogen (H_2) over iron or cobalt catalysts. The CO and H_2 feed gas is produced from a carbon-containing feedstock, for example by gasification of biomass or coal.

Fischer-Tropsch conversion was first used commercially in the 1930s, when Germany started producing F-T liquids from coal to fuel vehicles. A coal-to-fuels program has been operating in South Africa since the early 1950s. There is renewed interest globally today in F-T synthesis to produce liquids from remote natural gas that has little or no value because of its distance from markets (Fouda, 1998). Of particular interest is the production of middle distillate with an unusually high cetane number and with little or no sulfur or aromatics. Such a fuel can be blended with conventional diesel fuels to meet increasingly strict vehicle fuel specifications designed to reduce tailpipe emissions. Recent commercial efforts in "gas-to-liquids" (GTL) technology development include those at Exxon, Rentech, Sasol, Shell, and Syntroleum (Knott, 1997; Parkinson, 1997; Rentech, 1999). Because a large percentage of the world's gas fields are relatively small (Anonymous, 1998), one segment of the industry developing F-T synthesis processes is focussed on smaller-scale facilities (Knott, 1997; Tijm et al., 1997; Rentech, 1999). Technological developments in this sector of the industry might be especially relevant to F-T synthesis using biomass as a feedstock.

We present energy balances using a simplified calculation approach for the production of F-T liquids from biomass, coal, and natural gas. We compare our results to other published results in the case of coal and natural gas feedstocks.

2. FISCHER-TROPSCH SYNTHESIS

2.1. F-T conversion of natural gas

Picture the process of producing F-T liquids by considering conversion of natural gas (Figure 1). The first step involves producing a synthesis gas (syngas, consisting largely of CO and H_2) by reforming, e.g., steam-reforming, partial oxidation reforming, or autothermal reforming. Different reformer processes will give different syngas compositions. The molecules of greatest interest in the resulting syngas are CO and H_2. If needed, the H_2:CO ratio can be adjusted downstream of the reformer using the water-gas shift reaction. A shift reactor may not be required with natural gas, as the H_2:CO ratio will generally be high enough for direct feeding to the F-T synthesis reactor. (Carbon formation may deactivate the catalyst if the H_2:CO ratio is too low.)

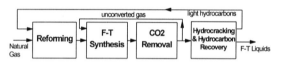

Figure 1. Simplified diagram for Fischer-Tropsch liquids production from natural gas.

In the F-T synthesis reactor, CO and H_2 combine exothermically in the presence of a catalyst under moderate pressure and temperature (20 to 35 bar and 180 to 350°C, depending on the design) to produce a mixture of straight-chain hydrocarbons, paraffins (C_nH_{2n+2}) and olefins (C_nH_{2n}) that range from methane to high molecular weight waxes. The reactor is cooled to maintain the desired reaction temperature. The output from the synthesis reactor includes a mix of hydrocarbons, CO_2, H_2O, inert species in the feed gas, and other minor compounds. Carbon dioxide is removed (e.g., by chemical absorption) to prevent it building up in recycle loops, to reduce downstream equipment sizes, and to avoid possible solidification problems during cryogenic hydrocarbon recovery. After CO_2 removal, unconverted gas can be recovered and recycled to the F-T reactor to increase overall conversion.

The relative proportion of different hydrocarbons produced by the F-T reactions is determined primarily by feed gas composition, catalyst type and loading, reaction temperature, pressure, and residence time, the net result of which can be captured in a characteristic "alpha" number for a given reactor design and operation. The α characterization derives from a single-parameter (α) model that accurately predicts empirical carbon number distributions (Anderson, 1984). The model accounts for the fact that longer hydrocarbons are formed by the linear addition of -CH_2- segments to shorter hydrocarbons in the synthesis process. The carbon number distribution predicted by this model is called the

Schulz-Flory or Anderson-Flory-Schulz distribution[1] (Eilers et al., 1990). Commercial F-T synthesis reactors are characterized by $0.65 < \alpha < 0.95$. Lower α will give a lower average molecular weight synthesis product compared to higher-α synthesis.

In GTL plants operating today, the F-T synthesis reactor is typically operated at high-end α values. The resulting waxy product is easily and with high selectivity formed into desired lighter products by subsequent hydrocracking, which involves the breaking up of the large hydrocarbon molecules into desired final products in a hydrogen-rich environment. Hydrocracking of large straight-chain hydrocarbons can be done under much less severe temperature conditions (350–400°C for cracking to C_5-C_{18} range) than is required for hydrocracking of aromatic molecules in conventional petrochemical refining. The lighter hydrocarbons leaving the hydrocracker can be recycled to the reformer for further conversion (Figure 1).

An important recent technological development in commercial F-T conversion is "liquid-phase" synthesis. In a liquid phase reactor, the feed gas is bubbled through a heavy oil (e.g., the waxy fraction of F-T liquids) in which catalyst particles are suspended. The vigorous mixing, the intimate gas-catalyst contact, and the uniform temperature distribution enable conversion of feed gas to F-T liquids in a single pass of about 80%, as measured by fraction of CO converted (Bechtel Group, 1990). This compares to less than 40% conversion with traditional fixed-bed F-T reactors, such as those used in the Shell Malaysia plant and in South Africa. Considerable recycling is required with fixed-bed reactors to achieve high overall yield. The higher gas throughput capacity per unit volume with liquid-phase synthesis reduces capital costs compared to a fixed-bed reactor, and catalyst consumption per unit of product is reduced dramatically (Jager, 1997). Liquid-phase reactors are now commercially available for F-T synthesis (Jager, 1997), and are being developed for synthesis of methanol and dimethylether (Tijm et al., 1997).

2.2. F-T conversion of coal

The main difference between a process that produces F-T liquids from coal rather than natural gas is in the syngas production step. The reforming step is replaced by a pressurized oxygen-blown gasifier when coal is used. The resulting syngas (after gas cooling and cleaning) consists almost entirely of CO and H_2. Depending on the gasifier design, the H_2/CO ratio in the syngas can be too low for F-T synthesis. The ratio is adjusted using the shift reaction, either in a shift reactor upstream of the F-T synthesis step, or by direct injection of steam into the F-T reactor, wherein the shift reaction occurs along with the F-T synthesis reactions.

2.3. F-T conversion of biomass

The process for converting biomass into F-T liquids (Figure 2) is similar in many respects to that for coal conversion. However, some methane and other light hydrocarbons are found in the product gas from most biomass gasifiers, so a hydrocarbon reforming step is needed after gasification to maximize conversion to F-T liquids. Biomass is more reactive than coal,

---

[1] The distribution of carbon number species (C_n) is given by $\log(C_n) = \log[(1-\alpha)/\alpha] + n\log(\alpha)$.

so lower temperatures can be used in the gasification step. This provides the possibility of using indirectly heated gasifiers, which produce a gas undiluted by inert nitrogen without the use of costly oxygen. (Indirectly-heated gasifiers under development include the Brightstar Synfuels Company [BSC] design [Menville, 1998], the Battelle Columbus Laboratory [BCL] design [Anson et al., 1999], the Thermochem design [MTCI, 1990], and the DMT design [Chughtai and Kubiak, 1998].) Air-blown gasification can also be used, though this requires larger downstream vessel sizes to handle the nitrogen-diluted syngas. Oxygen-blown gasification avoids nitrogen dilution, but the reduced vessel and piping costs must be evaluated against the added costs for oxygen supply.

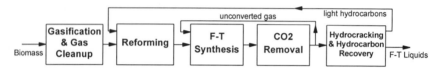

Figure 2. Simplified diagram for Fischer-Tropsch liquids production from biomass.

A simpler process design involves "once-through" F-T synthesis, wherein the reforming step is eliminated, and the syngas passes only once through the F-T synthesis reactor. Unconverted gas, rather than being recycled for further conversion, is used to fire a gas turbine to generate electricity as a co-product (Figure 3). The elimination of recycle loops compared with the "full recycle" configuration reduces investment costs for the synthesis and refining steps. Also, CO_2 removal is not needed if a high-α synthesis step is used in combination with hydrocracking. In such a configuration, only liquid-phase material will go to the hydrocracker and all gases will be directed to the combined cycle.

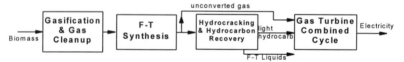

Figure 3. Simplified diagram of "once-through" co-production of electricity and F-T liquids from biomass.

3. PRELIMINARY ENERGY BALANCES

3.1. Calculation method

We estimate mass and energy balances for each major piece of process equipment as follows. For coal and biomass gasifiers, we adapt from the literature the mass and energy balances for different gasifier designs, as detailed in the notes to subsequent Tables. Given the preliminary nature of the calculations in this paper, we have not sought to verify quoted gasifier performance.

With one exception, we assume all reformers are steam reformers, with a fraction of the feed gas to the reformer diverted to a burner to provide heat to drive the endothermic reactions. The fraction of feed gas diverted varies with hydrocarbon content of the gas being reformed, and is based on work by Katofsky (1993). For the conversion of natural gas, we assume air-blown partial oxidation reforming, with performance based on A.D. Little (1994). Reforming converts all hydrocarbons to CO and H_2.

The F-T synthesis step is modeled assuming that a single pass of synthesis gas through the reactor results in 80% mass conversion of the (CO + H_2) in the feed gas. Hydrocarbons constitute between 24% and 44% of the mass of converted (CO + H_2), depending on the starting H_2:CO ratio. The balance of the converted (CO + H_2) forms H_2O and/or CO_2. The carbon-number distribution of the hydrocarbon products is given by a Schulz-Flory distribution.

The F-T synthesis step assumes an α value between 0.92 and 0.95, and the product distribution from the hydrocracker is based on published empirical results. The hydrogen requirement for the hydrocracker is estimated based on discussions with industry experts. Hydrogen for the hydrocracker is assumed to be recovered from mixed gases using pressure swing adsorption (PSA).

When a gas turbine is included in the overall process, the efficiency of converting fuel gases to electricity is assumed to be 50% on a higher heating value basis, representing a modern gas turbine/steam turbine combined cycle.

Process steam demands are assumed to be met by waste heat recovery, e.g., from reformer furnace flue gases, F-T synthesis cooling flows, product streams, etc. Process electricity demands are estimated for the main compressors (assuming single-stage adiabatic efficiency of 80%, maximum single-stage pressure ratio of 3:1, and multiple stages with intercooling for larger overall pressure ratios); for oxygen production (assuming 480 kWh/tonne O_2); for PSA (based on Katofsy [1993]), and for the balance of demands (assuming the same kWh per unit of feedstock energy as given by Williams et al. [1995] for methanol production from the same feedstock).

3.2. Energy balances for coal and natural gas

Table 1 compares our estimates of overall process energy efficiency for F-T liquids production from fossil fuels with results published by others based on detailed process modeling. In all cases the process designs incorporate recycle loops to increase F-T liquids production. The overall efficiency for converting natural gas is higher than for converting coal, as expected. Our calculated efficiency with natural gas is higher than reported by Bechtel (Table 1), but is slightly lower than indicated in a study by Gray and Tomlinson (1997). Our coal conversion efficiency agrees reasonably well with Bechtel's result (Table 1), but it is somewhat lower than indicated by Gray and Tomlinson. In any case, both of our calculated efficiencies are within one or two percentage points of other published results, providing some confidence in our simplified calculation approach.

Table 1
Energy balances for F-T liquids production from coal and from natural gas

Feedstock ⇨	Coal		Natural Gas	
	Bechtel[a]	Calcs.[b]	Bechtel[c]	Calcs.[d]
	Gasifier performance		Reformer performance	
mass % H_2	2.33		-	3.20
N_2	1.0		-	65.4
CO	90.5		-	24.2
CO_2	5.88		-	2.70
CH_4	0.013		-	0
H_2O	0.30		-	4.60
kg_{gas}/kg_{feed}	1.69		-	6.74
GJ_{gas}/GJ_{feed}	0.80		-	0.847
	Process energy demands (GJ/GJ_{feed})			
Electr. (GJ_e/GJ_{feed})	0.009[e]	0.0634[f]	not avail.	0.054[g]
Butane (GJ/GJ_{feed})	0.0296	-	0.00356	-
	Products (GJ/GJ_{feed})[h]			
Electr. (GJ_e/GJ_{feed})	0	0.00151	not avail.	0.0860
LPG (C_3-C_4)	0.016	0	0.0156	0
Naphtha (C_5-C_9)	0.274	0.608	0.209	0.553
Kerosene (C_{10}-C_{12})				
Diesel (C_{13}-C_{18})	0.298		0.337	
	Fraction of feedstock HHV converted to			
Net electricity	- 0.009	- 0.0619	0.00473	0.032
Net hydrocarbons	0.559	0.608	0.558	0.553
Overall Eff.	0.550	0.546	0.563	0.585

[a] Based on Choi, et al. (1993) using a Shell gasifier with Illinois #6 coal. See also Bechtel (1991-1994).
[b] Our calculations, with gasifier performance as for Bechtel and α=0.92 for F-T synthesis.
[c] Based on Choi, et al. (1996). Reforming is primarily by partial oxidation, but the reformate is supplemented by reformate from a separate, smaller steam reformer to increase the sythesis feed-gas H_2:CO ratio.
[d] Our calculations, assuming air-blown partial oxidation reformate composition from A.D. Little (1994), and α=0.92 for synthesis.
[e] Bechtel study assumed use of steam-driven air separation unit, so process electricity requirement is relatively low.
[f] Process electricity demands (GJ_e/GJ_{feed}) include 0.00437 for hydrogen separation using PSA, 0.055 for cryogenic oxygen production, and 0.004 for other uses.
[g] Process electricity demands (GJ_e/GJ_{feed}) are 0.0457 for compressors, 0.0057 for H_2 separation with PSA, 0.003 for other.
[h] Assumed higher heating values (MJ/kg): C_3-C_4, 50.0; C_5-C_9, 48.50; C_{10}-C_{12}, 47.94; C_{13}-C_{18}, 47.66; C_{19+}, 47.50; C_5-C_{12}, 48.36.

3.3. Energy balances for biomass

Table 2 shows results of our calculations for biomass conversion using an indirectly heated gasifier design. Results in the first column are for a "full recycle" process configuration, which maximizes F-T liquids production but generates no exportable electricity demand. The overall efficiency of producing F-T liquids in this case is about 49%.

Table 2
Calculated energy balances for F-T liquids from biomass with indirectly-heated gasifier, F-T reactor $\alpha=0.95$, and with hydrocracker

Process Design	Full Recycle	Once-through
Gasifier performance[a]		
mass % H_2	3.58	
CO	35.8	
CH_4	10.2	
CO_2	39.3	
C_2H_2	1.33	
H_2O	5.52	
N_2	4.29	
kg_{gas}/kg_{feed}	0.932	
GJ_{gas}/GJ_{feed}	0.727	
Process electricity demand		
GJ_e/GJ_{feed}	0.0373[b]	0.0251[c]
Products (GJ/GJ_{feed})		
Electr. (GJ_e/GJ_{feed})	0.0373	0.197
LPG (C_3-C_4)	0	0
Naphtha (C_5-C_9)	0.125	0.067
Kerosene (C_{10}-C_{12})	0.246	0.132
Diesel (C_{13}-C_{18})	0.123	0.0658
Waxes (C_{19+})	0	0
Fraction of feedstock HHV converted to		
Net electricity	0	0.172
Net hydrocarbons	0.494	0.265
Overall HHV eff.	0.494	0.437
Effective eff.[d]	---	0.521

[a] Gasifier performance from Menville (1998), with input biomass moisture content of 40%.
[b] Process electricity demands (GJ_e/GJ_{feed}) are 0.0220 for compressors, 0.0118 for PSA-H_2 separation, and 0.0036 for other uses.
[c] Process electricity demands (GJ_e/GJ_{feed}) are 0.0194 for compressors, 0.0021 for PSA-H_2 separation, and 0.0036 for other uses.
[d] Effective efficiency equals higher heating value of net hydrocarbons divided by HHV of biomass charged to hydrocarbon production. The latter is the total biomass input less the biomass that would be required for a stand-alone BIG/GTCC to generate the same amount of net electricity as generated at the F-T facility. Assuming a stand-alone generating efficiency of 35% (HHV), the Effective efficiency = NH/[1 - (NE/0.35)], where NH = net hydrocarbon fraction and NE = net electricity fraction.

This is a few points lower than our calculated result for coal (Table 1), owing primarily to the lower cold-gas efficiency for biomass gasification (73% versus 80% for coal) and the greater energy demand for reforming hydrocarbons in the biomass-gasifier product gas before F-T synthesis.

The second set of results in Table 2 is for a "once-through" configuration. The "once-through" biomass conversion case has an overall efficiency (counting both electricity and F-T liquids as products) about six percentage points lower than the "full recycle" case because a significant amount of syngas is converted to electricity rather than to F-T liquids, and the efficiency of converting syngas to electricity is lower than that for conversion to F-T liquids. However, the effective efficiency of producing F-T liquids in the "once-through" configuration is about the same (actually slightly higher) than the efficiency of the "full recycle" case.[2] The high effective efficiency achieved with the simpler "once-through" process configuration suggests improved economics of F-T liquids production compared to production from the "full recycle" configuration.

For different biomass gasifier designs, Table 3 shows calculated efficiencies for "once-through" process configurations that co-produce F-T liquids and electricity. Results are shown for two indirectly heated, atmospheric pressure gasifiers; one pressurized, oxygen-blown gasifier; one pressurized air-blown gasifier, and one atmospheric-pressure air-blown design.

The total overall efficiency ranges from 40% to 51% among the five cases. Differences in efficiencies between cases can be explained in terms of either differences in efficiencies of the gasification step or differences in the ratio of C_xH_y:$(CO + H_2)$ in the gasifier product gas. Higher gasification efficiency gives higher overall efficiency. A higher C_xH_y:$(CO + H_2)$ ratio means that a larger fraction of the energy in the product gas is converted to electricity and less is converted to F-T liquids. Since syngas conversion to electricity is less efficient than conversion to liquids, overall efficiency is lower for larger ratios of C_xH_y:$(CO + H_2)$. Pressurized oxygen-blown gasification gives the highest overall efficiency, followed by indirectly heated gasifiers and then air-blown gasifiers. More detailed performance analysis, together with cost assessment are needed to determine which of the different process configurations would be most cost competitive in a given application.

---

[2]The effective efficiency is the higher heating value (HHV) of F-T liquids divided by the HHV of the biomass charged to F-T liquids production. The biomass charged to F-T liquids is the total biomass input less the amount of biomass that would be required with a stand-alone gasifier/combined cycle to generate the same amount of electricity as that exported from the co-producing facility. Assuming a stand-alone generating efficiency of 35% (HHV), and using the results in Table 2, the effective efficiency of producing F-T liquids with the once-through process is $0.265/[1.0 - (0.172/0.35)] = 0.521$.

Table 3
Calculated energy balances for "once-through" co-production of F-T liquids and electricity from biomass for different gasifier designs. Alpha = 0.95 for the synthesis reaction, and hydrocracking is used to produce final products

Gasifier Design ⇨	Indirect (BSC)[a]	Indirect (BCL)[b]	Oxygen (IGT)[c]	Air, low P (TP S)[b]	Air, high P (Bio flow)[b]
	Gasifier efficiency (HHV gas output/HHV biomass input)				
GJ_{gas}/GJ_{feed}	0.727	0.707	0.761	0.700	0.700
	Fraction of biomass HHV converted to				
Net electricity	0.172	0.183	0.038	0.082	0.142
Net hydrocarbons	0.265	0.231	0.473	0.318	0.264
Overall HHV eff.	0.437	0.414	0.511	0.400	0.406
Effective eff.[d]	0.521	0.484	0.530	0.415	0.445

[a]Gasifier performance for the Brightstar Synfuels Company design, as reported by Menville (1998) with 40% feed biomass moisture.
[b]Gasifier performance as reported by Consonni and Larson (1996) with 15% feed biomass moisture content. BCL = Battelle Columbus Laboratory design; TPS = TPS Studsvik design; Bioflow = Foster Wheeler's Bioflow design.
[c]Gasifier performance as reported by Katofsky (1993) with 10% feed biomass moisture. IGT = Institute of Gas Technology gasifier design.
[d]See Table 2, note d.

4. CONCLUSIONS

A resurgence of interest in Fischer-Tropsch conversion technology is being driven by the goal of converting remote natural gas resources into marketable liquid products such as high-cetane number, low-aromatic, no-sulfur diesel blending stock for reducing diesel-engine vehicle tailpipe emissions. Because remote gas fields are typically small, much of the F-T technology development effort is aimed at making smaller scale facilities cost competitive. An important recent technology development in this regard is liquid-phase synthesis, which achieves much higher throughput per unit volume than synthesis using traditional fixed-bed reactors. Processes for converting biomass to F-T liquids can take advantage of such technological developments.

We have presented preliminary energy balances for the conversion of biomass to F-T liquids in "full-recycle" process configurations that maximize F-T liquids production and in "once-through" configurations that co-produce electricity and F-T liquids. A significant result is that the effective efficiency of producing F-T liquids using a "once-through" design is about the same as the efficiency of producing F-T liquids in a "full recycle" configuration. The simpler process configuration of the former should provide for better economics.

5. ACKNOWLEDGEMENTS

The authors thank the W. Alton Jones Foundation and the Geraldine R. Dodge Foundation for financial support for the preparation of this paper.

REFERENCES

Anderson, R.B. (1984). The Fishcer-Tropsch Synthesis, Academic Press, Inc., Orlando, Florida.

Anonymous (1998). Gas-to-LiquidsTechnology Breakthroughs: Is Overhaul of Energy Equation at Hand?, International Petroleum Encyclopedia, Penn Well Press.

Anson, D., M. Paisley, and M. Ratcliff (1999). Conditioning and Detailed Analysis of Biomass Derived Fuel Gas, presented at American Society of Mechanical Engineers' Turbo Expo '99, Indianapolis, Indiana, June.

Bechtel Group, Inc. (1990). Slurry Reactor Design Studies. Slurry vs. Fixed-Bed Reactors for Fischer-Tropsch and Methanol: Final Report, US Dept. of Energy Project No. DE-AC22-89PC89867, Pittsburgh Energy Technology Center, Pittsburgh.

Bechtel (1991, 1992, 1993, 1994). Baseline Design/Economics for Advanced Fischer-Tropsch Technology, quarterly reports under contract DE-AC22-91PC90027 to the Pittsburgh Energy Technology Center, US Department of Energy, Pittsburgh.

Choi, G.N., S.J. Kramer, S.S. Tam, R. Srivastava, and G. Stiegel (1996). Natural Gas Based Fischer-Tropsch to Liquid Fuels: Economics, presented at the Society of Petroleum Engineers 66th Western Regional Meeting, May 22-24.

Choi, G.N., S.S. Tam, J.M. Fox, S.J. Kramer, and J.J. Marano (1993). Baseline Design/Economics for Advanced Fischer-Tropsch Technology, paper presented at the 1993 USDOE Coal Liquefaction & Solid Fuels Contractors' Review Conference, Pittsburgh.

Chughtai, M.Y. and H. Kubiak (1998). Hydrogen from Biomass, in Proceedings of the 10th European Conference on Biomass for Energy and Industry, C.A.R.M.E.N., Rimpar, Germany, pp. 284–286.

Consonni, S. and E.D. Larson (1996). Gas Turbine Combined Cycles, Part A: Technologies and Performance Modeling, and Part B: Performance Calculations and Economic Assessment. ASME J. Engng. for Gas Turb. & Power, 118, pp. 507–525.

Eilers, J., S.A. Posthuma, and S.T. Sie (1990). The Shell Middle Distillate Synthesis Process (SMDS), Catalysis Letters, 7, pp. 253–270.

Fouda, S.A. (1998). Liquid Fuels from Natural Gas, Scientific American, 278(3), March, 92-95.

Gray, D. and G. Tomlinson (1997). Fischer-Tropsch Fuels from Coal and Natural Gas: Carbon Emissions Implications, Mitretek Systems, McLean, Virginia, August.

Jager, B. (1997). SASOL's Advanced Commercial Fischer-Tropsch Processes, presented at the Spring Meeting of the American Institute of Chemical Engineers, Houston, Texas, 10-13 March.

Katofsky, R.E. (1993). The Production of Fluid Fuels from Biomass, MSE Thesis, Mechanical and Aerospace Engineering Department, Princeton University, Princeton, New Jersey.

Knott, D. (1997). Gas-to-Liquids Projects Gaining Momentum as Process List Grows, Oil & Gas Journal, June 23.

Little, A.D. (1994). Multi-Fuel Reformers for Fuel Cells Used in Transportation, Appendix to Phase I Final Report for contract DE-AC02-92-CE50343, US Department of Energy, Office of Transportation Technologies.

Marano, J.J., S. Rogers, G.N. Choi, and S.J. Kramer (1994). Product Valuation of Fischer-Tropsch Derived Fuels, presented at the American Chemical Society Meeting, Washington, DC, 21-26 August.

Menville, R.L. (1998). Profiting from Biomass Fuel and Power Projects Using the Brightstar Synfuels Co. Gasifier, presented at the 17th World Energy Congress, Houston, Texas, 13-17 September.

MTCI (Manufacturing and Technology Conversion International) (1990). Testing of an Advanced Thermochemical Conversion Reactor System, Battelle Pacific Northwest Laboratory, Richland, Washington.

Parkinson, G. (1997). Fischer-Tropsch Comes Back, Chemical Engineering, April.

Rentech (1999). http://www.gastoliquids.com.

Tijm, P.J.A., W.R. Brown, E.C. Heydorn, and R.B. Moore (1997). Advances in Liquid Phase Technology, presented by Air Products and Chemicals, Inc. at the American Chemical Society Meeting, San Francisco, April.

Williams, R.H., E.D. Larson, R.E. Katofsky, and J. Chen (1995). Methanol and Hydrogen from Biomass for Transportation, with Comparisons to Methanol and Hydrogen from Natural Gas and Coal, Rpt. 292, Center for Energy and Environmental Studies, Princeton University, Princeton, New Jersey.

A PRELIMINARY ASSESSMENT OF BIOMASS CONVERSION TO FISCHER-TROPSCH COOKING FUELS FOR RURAL CHINA

Eric D. Larson and Haiming Jin

Center for Energy and Environmental Studies, Princeton University, Princeton, New Jersey 08544, USA

A variety of liquid hydrocarbons can be produced by Fischer-Tropsch synthesis from biomass. We present energy balance results for two configurations for co-producing domestic cooking fuels (synthetic LPG or kerosene) and electricity from gasified biomass. We make a preliminary estimate of the costs of co-producing electricity and LPG from corn stalks in rural Jilin Province, China. Corn stalks are burned for cooking in rural Jilin today, contributing to health problems owing to indoor air pollution.

1. INTRODUCTION

More than 2 billion people worldwide cook by direct combustion of biomass (WHO, 1997), primarily in rural areas of developing countries. The resulting indoor air pollution accounts for nearly 60% of all human exposure to particulate air pollution (Smith, 1993), contributing to health damages, especially to women and children. Fluid cooking fuels are far cleaner than solid biomass. They are also far more efficient (Figure 1), even considering biomass-to-fuels conversion losses. Thus, converting biomass to fluid fuels can reduce the harmful effects of direct biomass use, while meeting the energy demands of more people.

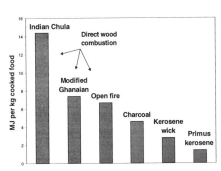

Figure 1. Energy requirements for cooking with different fuels (Dutt and Ravindranath, 1993).

In this paper, we consider the idea of producing liquid hydrocarbons from biomass by Fischer-Tropsch (F-T) synthesis for use in cooking. F-T synthesis produces hydrocarbons from CO and H_2 in carbon-rich feedstocks, including biomass. Until recently, F-T synthesis has been used to produce only vehicle fuels from coal in Germany in the 1930s and 1940s and in South Africa from 1950 to the present. Now, the oil and gas industry is interested in converting remote natural gas resources into marketable liquid products, especially high-cetane number, low-aromatic, no-sulfur diesel blending stock for reducing diesel-engine–vehicle tailpipe emissions. Larson and Jin (1999) review some fundamentals of F-T synthesis and recent developments in synthesis technology, and present comparative energy balances for F-T liquids production from natural gas, coal, and biomass. Using the approach described by Larson and Jin (1999) for calculating energy balances, we examine the production of synthetic liquefied petroleum gas (LPG--a mixture of propane and butane) or kerosene from biomass for use in cooking. We present a preliminary assessment of the cost of producing these fuels from corn stalks in rural Jilin Province, in northeast China, where direct combustion of stalks is widely used today to meet household cooking needs.

2. ENERGY BALANCES FOR F-T COOKING FUELS FROM BIOMASS

Two clean cooking fuels that can be produced from biomass by F-T synthesis are synthetic LPG (C_3-C_4 hydrocarbons) and kerosene (C_{10}-C_{12} hydrocarbons). LPG can be burned very cleanly and efficiently. Kerosene is less clean, but F-T kerosenes burn more cleanly than petroleum-derived kerosenes because of the largely paraffinic nature of F-T liquids. For example, F-T kerosenes produced at the Shell "gas-to-liquids" F-T facility in Malaysia are characterized by a "smoke point" (the height to which a flame can be adjusted in a standard burning apparatus before smoking starts) greater than 50 mm and zero sulfur content (Tijm et al., 1995). For comparison, British standards specify a minimum smoke point of 35mm and a maximum sulfur content of 0.04% by mass for kerosene used in domestic free-standing burners (without a flue). For burners connected to a flue the corresponding figures are 20 mm and 0.2% sulfur (Francis and Peters, 1980).

Figure 2 is a simplified process flow diagram for "once-through" co-production of F-T liquids and electricity. In this configuration, gasified biomass is passed once through the F-T synthesis reactor. Any gas that is not converted to liquids in the reactor is sent to a gas turbine combined cycle to generate electricity. The "once-through" configuration eliminates additional reaction steps and recycle loops that would be needed to maximize the production of liquids. Larson and Jin (1999) show that the effective efficiency of "once-through" production of F-T liquids from biomass[1] is about the same as the efficiency of producing liquids in a facility designed to maximize liquids production. They argue that the much simpler process configuration in the "once-through" design

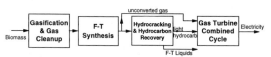

Figure 2. Simplified diagram of "once-through" co-production of electricity and F-T liquids from biomass.

---

[1] The effective efficiency is the energy contained in the F-T liquids divided by the difference in energy content of the biomass feed to the "once-through" facility and the biomass that would be needed in a stand-alone gasifier/combined cycle power plant to generate the same amount of electricity as in the "once-through" facility.

Table 1
Calculated energy balances for producing F-T cooking fuels from biomass, assuming "once-through" design with indirectly-heated gasifier

Case[a]	A	B
Gasifier performance		
mass % H_2		3.38
CO		35.8
CH_4		10.2
CO_2		39.3
C_2H_2		1.33
H_2O		5.52
N_2		4.29
kg_{gas}/kg_{feed}		0.932
GJ_{gas}/GJ_{feed}		0.727
Process electricity demand		
GJ_e/GJ_{feed}	0.025	0.029
Products (GJ/GJ_{feed})		
Electr. (GJ_e/GJ_{feed})	0.197	0.184
LPG (C_3-C_4)	0	0.287
Naphtha (C_5-C_9)	0.067	0
Kerosene (C_{10}-C_{12})	0.132	0
Diesel (C_{13}-C_{18})	0.066	0
Waxes (C_{19+})	0	0
Fraction of Feedstock HHV converted to		
Net electricity	0.172	0.155
Net hydrocarbons	0.265	0.287
Overall HHV eff.	0.437	0.442
Incremental eff.[b]	0.521	0.515

a. The process designs vary in the operation of the hydrocracker. In case A, the output of the hydrocracker is as reported by Tijm, et al. (1995). In case B, the hydrocracker maximizes C_3 and C_4 hydrocarbon output. The higher hydrogen requirement of the hydrocracker in case B is taken into account in our energy balance.

b. Incremental efficiency equals higher heating value of net hydrocarbons divided by HHV of biomass charged to hydrocarbon production. The latter is the total biomass input less the biomass that would be required for a stand-alone BIG/GTCC to generate the same amount of net electricity as generated at the F-T facility. Assuming a stand-alone generating efficiency of 35% (HHV), Incremental efficiency = NH/[1 - (NE/0.35)], where NH = net hydrocarbon fraction and NE = net electricity fraction.

should provide for better economics of liquids production.

For the analysis here, we consider two "once-through" design configurations. For the gasification step in both cases we have adopted the gasifier performance indicated by Menville (1998) for the indirectly heated biomass gasifier design of the Brightstar Synfuels Company (BSC). The BSC design is based on steam reforming of biomass. A 16 t/day capacity commercial demonstration BSC gasifier has been operating near Baton Rouge, Louisiana since 1996. For the synthesis step, we assume α = 0.95 (see Larson and Jin [1999] for discussion of α characterization of the synthesis reactor). With this assumption, the predominant hydrocarbon products of the synthesis step are high molecular weight waxes. The waxes are hydrocracked to form the final liquid products.

The two cases (Table 1) differ in the operation of the hydrocracker. In Case A, the hydrocracker output is similar to that at Shell's gas-to-liquids facility in Malaysia when operating in its "kerosene mode" (Tijm et al., 1995): about half of the hydrocracker output is kerosene, and one-quarter each are naphtha-like and diesel-like hydrocarbons. For this case, the fractions of input biomass converted to cooking fuel (kerosene) and to electricity are 13% and 17%, respectively. A substantial amount of naphtha-like and diesel-like hydrocarbons are also produced. In Case B, hydrocracking is assumed to be carried to a further extent such that all of the input wax is converted to C_3 and C_4 hydrocarbons (synthetic LPG). This configuration converts about 29% of the input biomass into cooking fuel (LPG) and 16% to electricity.[2]

3. PRELIMINARY COST ASSESSMENT FOR F-T COOKING FUELS

3.1. Context

To illustrate the potential economics of producing cooking fuels from biomass via F-T synthesis, we present a preliminary cost estimate for co-producing synthetic LPG and electricity from corn stalks in the province of Jilin in northeastern China. Jilin, with only 2% of China's population, grows 14% of China's corn. Some 35 million tonnes of corn stalks (~ 460 PJ) are generated annually with the corn harvest, about half of which are used for soil conditioning and fertilization, for livestock fodder, and for industrial feedstock (Cao, 1998). In addition, a large number of rural households also burn stalks for domestic cooking and heating, contributing to poor indoor air quality in many homes. As farmer incomes rise, coal briquettes are preferred to stalk. Coal briquettes pollute about as much as stalks, but they are more convenient to purchase as needed, rather than collecting and storing stalks. The open field burning of excess crop residues is now creating a new and serious air pollution problem.

Converting stalks to a clean cooking fuel such as LPG would help alleviate both indoor and outdoor air pollution problems. LPG is already a familiar cooking fuel in many Chinese households. Conventional LPG is widely used in urban areas, and it is estimated that (as of 1994) some 30 million rural households (16% of all rural households) also use LPG for at least some of their cooking needs (Wang, 1997).

3.2. Cost estimate

Although intensive corn production is practiced in Jilin province, the quantity of corn stalks that can be concentrated at a conversion facility is limited by transportation logistics and costs. Thus, a facility for producing LPG from corn stalks could not be large by the standards of today's gas-to-liquids industry. Based on the energy balance for Case "B" in Table 1, a corn-stalk conversion facility having LPG and electricity co-production capacities of 1500 GJ/day (or 250 barrels per day crude oil equivalent [bpdcoe]) and 9.4 MW$_e$, respectively, would require some 400 tonnes/day (5219 GJ/d) of corn stalks. Cao (1998) indicates that this amount of stalks is available within a radius of about 11 km in the corn-belt of Jilin Province.

Assuming that process technology is commercially mature, we estimate the total installed capital cost for such a facility would be about $33 million,[3] or $132,000 per bpdcoe of F-T

---

[2] In reality the ratio of LPG to electricity would be somewhat less than we calculate because hydrocracking would produce some C_1 and C_2 hydrocarbons that would be converted to electricity. In the absence of empirical data for a hydrocracker operating to maximize LPG output, we have assumed that all of the wax feed is cracked to LPG.

[3] All costs in this paper are expressed in 1998 US$. The GNP deflator has been used to convert to 1998$ costs originally given in other-year dollars.

liquids (Table 2). For comparison, the cost for a 10,000 bpd gas-to-liquids conversion facility based on technology like that of the Shell Malaysia plant (not once-through) might reach $30,000/bpd if widely implemented commercially (Tijm et al., 1995). Companies like Syntroleum, that focus on smaller-scale gas-to-liquid plants, are projecting 2500 bpd facilities costing under $30,000/bpd (Knott, 1997). In a recent study, Bechtel and Amoco estimated the capital cost for a once-through 8815 bpd GTL plant if built today to be about $48,000/bpd (Choi et al., 1997). As part of the same study, a 50,556 bpdcoe plant using coal as the feedstock and maximizing F-T liquids production (not once-through) was estimated to have a capital cost of about $64,000/bpd (Bechtel, 1998).

Other than capital, key factors in our cost analysis are labor rates, feedstock costs, and electricity sales price. Details of the cost assumptions are provided in the notes to Table 2. The average cost for operating labor for the facility is based on an assumed compensation to employees that is an estimated two times the compensation provided to young advanced-degree engineers employed in Beijing today. The cost of delivered corn stalks ($0.54/GJ) is based on a detailed cost-supply curve for the Jilin Province corn belt presented by Cao (1998). An electricity sale price of 5 ¢/kWh is shown in Table 2. For comparison, the retail price for grid-supplied electricity paid by rural consumers in Jilin province is typically about 10 ¢/kWh today (Qiang, 1998).

The calculated net cost of producing F-T LPG (including revenue from electricity sales) is $5.7/GJ, which is about 25% lower than the retail price typically paid in rural China today for conventional LPG (Table 2). Fig. 2 shows the sensitivity of the LPG cost to the assumed biomass feedstock price and the assumed electricity sales price.[4]

4. POTENTIAL IMPACT IN RURAL CHINA

Some order-of-magnitude calculations help to put in perspective the potential impact of producing F-T cooking fuels from biomass in Jilin Province and in China as a whole.

In rural Jilin Province, the current cooking fuel demand (if gas fuel were to be used) is estimated to be about 10 GJ/year per four-person household, or a total of some 36 PJ/yr for all 3.56 million rural Jilin households (Qiang, 1998). Rural electricity demand is presently some 1.2 kWh/day per household, or a total of some 1,560 GWh/yr for all rural households (Qiang, 1998). Assuming 230 PJ/year of stalks (half of the total generated) are converted to F-T LPG and electricity with efficiencies shown in Table 1 (Case B), the total production of LPG and electricity would be 66 PJ/yr and 9,900 GWh/year, or nearly twice the current rural demand for cooking fuel and six times the demand for electricity. Meeting all of current cooking fuel demand would require 73 facilities with production capacity as in Table 2.

---

[4] Our cost calculations take no account of any inherently higher value of F-T LPG over conventional LPG, as is the case with some other F-T fractions. For example, Marano et al. (1994) and Tijm et al. (1995) estimate that F-T middle distillates, because of their high cetane number and zero sulfur content, can command a premium of ~$1.2/GJ over conventional diesel.

Table 2
Cost estimate (1998$) for biomass conversion facility co-producing synthetic LPG (250 bpd crude oil equivalent capacity) and electricity

Plant Performance and Cost Assumptions[a]		Levelized LPG cost from corn stalks in rural Jilin Province context	
LPG capacity, GJ/d	1500	Assumptions	
Electric capacity, MW_e	9.4	Biomass cost, $/GJ[f]	0.54
Biomass input capacity, GJ/d	5219	Electricity sales, ¢/kWh[g]	5.0
Capacity factor	90%	Labor rate ($/yr)[h]	4200
Annual quantities		Yearly capital charge rate, %	15
LPG production (TJ/yr)	493	Levelized cost of LPG, $/GJ	
Electricity output (GWh/yr)	73.8	Capital	10.00
Biomass consumed (TJ/yr)	1715	Biomass feedstock	1.9
Installed capital cost (10^6 $)[b]		Operation & maintenance	1.3
Syngas production	12.55	Electricity revenue	-7.5
Syngas conversion/refining	12.73	Net cost of LPG production, $/GJ	5.7
Gas turb. combined cycle	7.70		
TOTAL	32.99		
Maintenance & ins. (10^3 $/yr)[c]	330		
Catalysts & chem. (10^3 $/yr)[d]	127	Retail price for LPG in rural China today:	
Operating labor (no. of employees)[e]	45	$7.7/GJ[i]	

a. The energy performance of this facility is based on the last column of Table 4.
b. Capital cost is estimated for a biomass integrated-gasifier/gas turbine combined cycle plant with an oversized syngas production area plus a syngas conversion/refining area. Installed BIG/GTCC cost is $1645/$kW_e$ (based on scaling Elliott and Booth (1993) estimate of $1500/$kW_e$ by US GNP deflator). Half the cost of a stand-alone BIG/GT power plant is assumed to be for syngas production and half for the combined cycle. Thus, the combined cycle in the table (9.363 MW_e capacity) costs 9363 kW_e x 1645/2 = $7.7 million. The syngas production area costs 9363 x $(5219/2311)^{0.6}$ x 1645/2 = $12.55 million, where 5219 GJ/day is the rated biomass consumption of the facility shown in this table and 2311 GJ/d is biomass consumption of a 9.4 MW_e stand-alone BIG/GT power plant with assumed efficiency of 35% (HHV). The cost of syngas conversion/refining is scaled using 0.6 exponent from a detailed cost estimate by Choi et al. (1997) for this area of a natural gas-based once-through F-T synthesis process.
c. Annual maintenance and insurance cost is assumed to be 1% of initial capital cost, as indicated by Bechtel (1994) for a coal-based F-T synthesis process.
d. Catalysts and chemicals are assumed to cost $0.26/GJ, based on Bechtel (1994).
e. Number of operating employees estimated based on detailed study by Bechtel (1994) that estimated 1088 employees would be needed to operate a large (300,000 GJ/day) coal-based F-T synthesis plant. Scaling from the Bechtel estimate by production capacity gives number of employees = 1088 x $(1500/300000)^{0.6}$ = 45.
f. The cost of air-dried corn stalks delivered to a conversion facility in the corn-growing region of Jilin is: Yuan RMB/tonne = 43.02 + (1.163 x r), where r is the radius (km) of delivery Cao (1998). Assuming exchange rate of 8 Yuan RMB/$, the cost of delivered stalks in US$/tonne = 5.4 + (0.145 x r). Assuming 13 GJ/t delivered stalks, stalk cost in $/GJ = 0.415 + (0.011 x r). Supplying a facility with capacity indicated in this table would require a delivery radius of about 11 km in the corn belt of Jilin (Cao, 1998).
g. Typical price for grid electricity in rural Jilin villages is 10 ¢/kWh (Qiang, 1998).
h. Assumed annual salary of $3000, plus 40% benefits. This is over twice the compensation level found today in Beijing for young engineers holding advanced university degrees.
i. Wang (1997) gives LPG price in rural China in 1996 of 3000 YuanRMB/tonne, a higher heating value of 48 GJ/tLPG, and exchange rate of 8.3 YuanRMB/US$. Converting to US$/GJ and correcting to 1998$ using the US GNP deflator gives a price of $7.7/GJ.

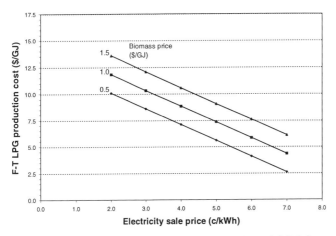

Figure 3. Estimated production cost of Fischer-Tropsch LPG from biomass in a "once-through" facility as a function of sale price for co-produced electricity. See Table 2 for details.

For China as a whole, Li, et al. (1998) project, based on detailed assessments, that some 376 million tonnes of agricultural residues (about 4900 PJ/yr) will be available for energy production in 2010 (from a total residue generation of 726 million tonnes). The available residues would be sufficient to produce some 1400 PJ/year of cooking fuels by once-through F-T synthesis, or enough to meet the cooking fuel demand (at current Jilin rates of use) of some 560 million people (about 40% of China's projected 2010 population). Electricity would be co-produced at an average rate of 24 GW_e (about 2.5 times the production rate projected for the Three Gorges hydroelectric facility).

5. CONCLUSIONS

A resurgence of interest in Fischer-Tropsch conversion technology is being driven by the goal of converting remote natural gas resources into marketable liquid products such as high-cetane number, low-aromatic, no-sulfur diesel blending stock for reducing diesel-engine vehicle tailpipe emissions. We have presented a preliminary analysis of the novel concept of converting biomass to F-T hydrocarbons suitable for use in cooking. We have examined in particular the idea of producing LPG from corn stalks in rural Jilin Province, China, as a cooking fuel for rural villages. Based on our energy balances, the supply of corn stalks in Jilin is sufficient to meet all of current rural cooking fuel demand twice over, with the co-produced electricity equivalent to six times the current rural electricity demand. Our preliminary cost analysis for small-scale co-production of synthetic LPG and electricity in Jilin is encouraging, and more detailed analysis is warranted.

6. ACKNOWLEDGEMENTS

The authors thank the W. Alton Jones Foundation and the Geraldine R. Dodge Foundation for financial support for the preparation of this paper.

REFERENCES

Bechtel (1998). Baseline Design/Economics for Advanced Fischer-Tropsch Technology, Final Reports under contract No. DE-AC22-91PC90027 to the Federal Energy Technology Center, US Department of Energy, Pittsburgh, Pennsylvania.

Bechtel (1994). Baseline Design/Economics for Advanced Fischer-Tropsch Technology, Quarterly reports under contract No. DE-AC22-91PC90027 to the Pittsburgh Energy Technology Center, US Department of Energy, Pittsburgh, Pennsylvania.

Bechtel (1998). Baseline Design/Economics for Advanced Fischer-Tropsch Technology, Final Reports under contract No. DE-AC22-91PC90027 to the Federal Energy Technology Center, US Department of Energy, Pittsburgh, Pennsylvania.

Cao, J. (1998). Evaluation on Biomass Energy Resources and Utilization Technology Prospect in Jilin Province, Proceedings of Workshop on Small-Scale Power Generation from Biomass, Working Group on Energy Strategies and Technologies, China Council for International Cooperation on Environment and Development, Changchun, China, pp. 16-25.

Choi, G.N., S.J. Kramer, S.S. Tam, J.M. Fox, N.L. Carr, and G.R. Wilson (1997). Design/Economics of a Once-Through Natural Gas Fischer-Tropsch Plant with Power Co-Production, paper presented at the 1997 USDOE Coal Liquefaction & Solid Fuels Contractors' Review Conference, Pittsburgh, Pennsylvania, 3-4 Sept.

Dutt, G.S. and N.H. Ravindranath (1993). "Bioenergy: Direct Applications in Cooking," *Renewable Energy Sources for Fuels and Electricity*, Johansson, Kelly, Reddy, and Williams (eds), Island Press, Washington, DC, pp. 653–697.

Elliot, P. and R. Booth (1993). Brazilian Biomass Power Demonstration Project, Special Project Brief, Shell International Petroleum Company, Shell Centre, London.

Francis, W. and M.C. Peters (1980). Kerosines--Properties and Specifications, in Fuels and Fuel Technology, A Summarized Manual, 2nd edition, Pergamon Press, Oxford, UK.

Knott, D. (1997). Gas-to-Liquids Projects Gaining Momentum as Process List Grows, Oil & Gas Journal, June 23.

Larson, E.D. and H. Jin (1999), Biomass Conversion to Fischer-Tropsch Liquids: Preliminary Energy Balances, this volume.

Li, J., J. Bai, and R. Overend (eds) (1998). Assessment of Biomass Resource Availability in China, China Environmental Science Press, Beijing.

Marano, J.J., S. Rogers, G.N. Choi, and S.J. Kramer (1994). Product Valuation of Fischer-Tropsch Derived Fuels, presented at the American Chemical Society Meeting, Washington, DC, 21-26 August.

Menville, R.L. (1998). Profiting from Biomass Fuel and Power Projects Using the Brightstar Synfuels Co. Gasifier, presented at the 17th World Energy Congress, Houston, Texas, 13-17 September.

Qiang, J. (1998). Vice President, Jilin Province Energy Resources Institute, Changchou, Jilin Province, China, personal communication, January.

Smith, K.R. (1993). Fuel Combustion: Air Pollution Exposures and Health in Developing Countries, Annual Review of Energy and the Environment, 18, pp. 529–566.

Tijm, P.J.A., J.M. Marriott, H. Hasenack, M.M.G. Senden, and T. van Herwijnen (1995). The Markets for Shell Middle Distillate Synthesis Products, presented at Alternate Energy '95, Vancouver, Canada, May 2-4.

Wang, X. (1997). Comparison of Constraints on Coal and Biomass Fuels Development in China's Energy Future, Ph.D. dissertation, Energy Resources Group, University of California, Berkeley.

World Health Organization (1997). Health and Environment for Sustainable Development, WHO, Geneva.

Biomass Transformation into Value-Added Chemicals, Liquid Fuels, and Heat and Power

Biofuels: Ethanol and Biotechnology

PRODUCTION OF LOW COST SUGARS FROM BIOMASS: PROGRESS, OPPORTUNITIES, AND CHALLENGES

Charles E. Wyman

Thayer School of Engineering, Dartmouth College, Hanover, New Hampshire 03755
and BC International, 990 Washington Street, Suite 104, Dedham, Massachusetts 02026

Lignocellulosic biomass (e.g., agricultural and forestry residues, waste paper, and woody and herbaceous crops) is an abundant low-cost resource that can be used to produce commodity fuels, chemicals, and materials while providing substantial environmental, economic, and strategic advantages. For biologically based conversion technologies, a key need is to produce low-cost sugars from the cellulose and hemicellulose fractions of biomass that together make up about two thirds to three quarters of the feedstock. Technologies have substantially improved that break down recalcitrant biomass into sugars and realize high product yields from the five sugars released. As a result, several firms are now pursuing commercial processes to make ethanol from biomass. Advanced pretreatment and hydrolysis process configurations may be able to further reduce biomass processing costs so that neat fuels are not competitive. Understanding the mechanisms involved in cellulose and hemicellulose hydrolysis will support such advancements and accelerate scale-up of biological processing technologies at lower costs. Development of coproducts to fully utilize the substrate will also improve process economics.

1. INTRODUCTION

Biological conversion of lignocellulosic biomass sources such as agricultural and forestry residues, portions of municipal solid waste, and herbaceous and woody crops into fuels, chemicals, and materials could provide unparalleled environmental, economic, and strategic benefits.[1-3] For example, because limited fossil fuels are needed to grow, harvest, and convert biomass for biological processes, carbon dioxide is recycled. As a result, biological processing of lignocellulosic materials can have a particularly powerful role in reducing greenhouse gas emissions.[4] Production of fuels such as bioethanol in this way is predicted to reduce such emissions by 90% or more, while biological manufacture of chemicals could actually sequester carbon. Furthermore, biomass is the only route to sustainable production of organic fuels and chemicals.

Sugar is a versatile substrate for biological production of a variety of products. Unfortunately, the cost of soluble sugars from plants such as sugar cane or sweet sorghum is too high in most of the world and growing them requires intensive agriculture.[1] Starch in grains such as corn can be readily hydrolyzed to glucose, a sugar that is easily fermented into a number of commercial products such as ethanol and lysine as well as some new chemicals under development such as 1,3-propanediol.[5] However, vital coproduct markets are limited, restricting the volume of corn that can be economically employed.[6]

On the other hand, lignocellulosic biomass is abundant and inexpensive and could support a major new industry that produced fuels, chemicals, and materials.[1-7] Yet, the recalcitrance of biomass leads to high processing costs, and better processing technologies able to produce cheap sugars are needed to realize the vast potential benefits offered by this resource. This paper reviews progress in reducing the cost of sugars from biomass and identifies new opportunities.

2. BIOMASS COSTS

Lignocellulosic biomass is a complex material.[1] About 40–50% is typically cellulose, a crystalline polymer made up of long chains of glucose covalently joined by beta linkages. Hemicellulose typically represents about 25–30% of biomass and is an amorphous polymer comprised of five sugars: xylose, arabinose, galactose, glucose, and mannose. The relative proportion of these five sugars varies from one biomass source to another, with particularly large differences between softwoods and either hardwoods or herbaceous plants. Another 10–20% of biomass is lignin, a phenyl propene glue-like material that bonds the cellulose and hemicellulose together. The remaining fraction is ash and various soluble substances termed extractives.

Biomass priced at about $40/dry ton is comparable in cost to many fossil resources.[2] However, if only the carbohydrate fraction is converted to products, the substrate cost increases proportionately on a mass basis and even more so on an energy basis because of lignin's relatively high energy content. As ethanol yields drop from the maximum theoretically possible, the revenue from ethanol declines per quantity of feedstock processed (Figure 1, compare the lower solid line for ethanol sold for use as a blend and the lower dashed line for neat fuel applications). On the other hand, if coproducts are derived from lignin and other materials, the revenues grow accordingly (Figure 1, compare the upper solid and dashed lines for selling prices of the portion not fermented). Thus, the more fully we utilize all fractions of biomass, the better the economic prospects become.

3. PROGRESS IN REDUCING BIOMASS PROCESSING COSTS

Biomass typically has been used as a feedstock to produce fermentation ethanol, also known as bioethanol. Very little attention has been given to using this resource to produce chemicals and materials, and most work has evaluated only glucose from corn wet milling plants, even in light of its price and volume limitations.[6] Thus, much of the rest of this paper will be based on studies for bioethanol, with the realization that similar implications would apply to use of biomass for biological manufacture of chemicals, materials, and other fuels.

Costs of various options for bioethanol production have been estimated.[8,9] These studies and reductions in government funding led us to concentrate on enzymatically based routes to convert cellulose into glucose, and the models were further used to track research progress (Figure 2).

The enzymatic conversion process begins with size reduction and dilute acid pretreatment to open up the biomass structure and expose the cellulose fraction to hydrolysis by enzymes. A small portion of the pretreated substrate is then used to support production of cellulase enzymes that are combined with the bulk of the pretreated material to form glucose. Yeast are added to the same vessel to ferment glucose to ethanol as soon as the sugars are released, eliminating the need for a secondary fermentation system and reducing inhibition of enzymes by the sugars they form. The fermentation broth is then distilled to separate pure ethanol from solids that can be used as boiler fuel to meet the energy needs for the process.[9]

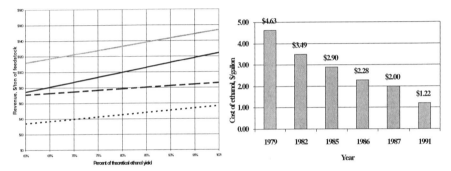

Figure 1. Effect of yield on revenues.   Figure 2. Bioethanol cost reductions.

Cost reductions (Figure 2) reflect two primary classes of advancements: reductions in the cost of breaking down carbohydrate polymers into sugars, and development of technology to ferment all five sugars in biomass to ethanol with high yields.[9] The former resulted from improved pretreatments that recover more hemicellulose sugars and increase the digestibility of cellulose from glucose; improved cellulase enzyme, and streamlined conversion processes with better rates, yields, and concentrations of ethanol production. The latter first came to fruition with the genetic engineering of various bacteria to ferment all five sugars to ethanol with high yield.[10]

4. COMMERCIALIZATION OF BIOETHANOL

Several projects are now in progress to commercialize bioethanol.[5] At the time of this writing, BC International is working with a team of engineers and contractors to retrofit a former molasses and then grain ethanol plant in Jennings, Louisiana, to make about 20 million gallons annually from sugar cane bagasse and other agricultural wastes abundant in that area. The process is built around the technology described in patent 5,000,000 and subsequent related patents to which BC International has exclusive rights. Although the details of the process are proprietary, it relies on dilute acid hydrolysis of hemicellulose to fermentable sugars followed by breakdown of the cellulose fraction to glucose for conversion to ethanol.

5. OPPORTUNITIES FOR FURTHER COST REDUCTIONS

The reductions in cost of bioethanol production are impressive, particularly in light of the limited research and development budgets available to fund such a daunting task, and bioethanol has the potential to compete in the blending market. However, the cost of bioethanol is still too high to be viable as a pure fuel produced in volume in an open market. Similarly, biomass sugars would not currently offer any economic advantages for manufacture of chemicals and materials compared to glucose from a corn wet mill.

Current costs of enzymatic bioethanol manufacture have been estimated.[12] Feedstock is the single most costly item, and although low-cost waste streams could be used initially to some advantage, there is little hope that feedstock costs can be dropped enough to make a significant impact for large-scale application of biomass conversion. Rather, feedstock improvements can assure a large feedstock supply, support construction of larger plants that gain economies of scale, and reduce environmental impacts. Pretreatment is the most costly processing step, and cellulose conversion is the next most costly step. These two steps alone represent about two thirds of the cost of biomass processing to ethanol, and cost reductions in these areas will pay major dividends.

6. COST REDUCTION STRATEGIES

Cost analyses show no fundamental barriers to reducing the cost of bioethanol production to compete as a pure fuel, provided product yields are high.[11] Sensitivity studies also demonstrate that high yields are important to realizing low costs.[9] However, high yields are not sufficient to meet competitive costs of production; advanced processing configurations are also needed.[12] Given their high costs, research must focus on new pretreatment and cellulose hydrolysis configurations that can reduce costs substantially.

Hemicellulose hydrolysis would benefit from less use of chemicals for hydrolysis and associated neutralization and conditioning, lower cost materials of construction, gains in the recovery of hemicellulose sugars, and improved digestibility of cellulose. Hydrothermal processing of biomass could realize such benefits with significant cost reductions.[12] The cost of enzymatic cellulose hydrolysis would drop if the rates of cellulose breakdown were improved, cellulose hydrolysis and fermentation were combined with hemicellulose fermentation, cellulase is produced in the same vessel as the fermentations, and cellulose digestibility is improved in pretreatment.[12] Although enzymatic routes have been favored for cellulose hydrolysis because glucose yields for conventional dilute acid plug flow and batch techniques are limited to about 70% and probably less, new process configurations could achieve high yields by using dilute acid cellulose hydrolysis processes.[13]

One of the difficulties in developing lower-cost pretreatment and cellulose hydrolysis technologies is that mechanisms are unclear, clouding the design of equipment, and few can afford to pursue advanced chemical hydrolysis technologies on a trial and error basis. Clarification of the mechanisms that govern sugar release and degradation is needed. For instance, although sugar yields are less than about 65% from steam gun hydrolysis of biomass without acid addition or about 90% with acid addition, percolation reactors can achieve yields of nearly 100%.[13,14] In addition, about half of the lignin is removed from biomass in the latter systems while very little lignin is solubilized in the steam explosion or plug flow reactors. Analogous behavior has been witnessed for cellulose hydrolysis, but many cellulose hydrolysis models are applicable to only a limited range and produce contradictory results.

7. CONCLUSIONS AND RECOMMENDATIONS

Remarkable progress has been made in technology for production of low cost sugars that can support biological conversion of biomass into a wide range of fuels, chemicals, and materials. However, innovative configurations for breakdown of the hemicellulose and cellulose fractions to sugars are essential if we are to make a full range of products on a large scale. For hemicellulose hydrolysis, the focus needs to be on development of technologies that eliminate the use of chemicals, reduce the need for exotic materials of construction, recover hemicellulose sugars at high yields, and produce highly digestible cellulose. Development of biological processing approaches that combine cellulase production with fermentation of sugars from cellulose and hemicellulose is a very promising path, but it may also be possible to substantially improve chemical hydrolysis of cellulose to realize high yields. A key need is to develop a better understanding of biomass hydrolysis that will foster development of advanced technologies and support rapid scale up to commercial plants.

REFERENCES

1. Wyman, C.E., ed. (1996). Handbook on Bioethanol: Production and Utilization. Applied Energy Technology Series, Washington, DC: Taylor & Francis, 424 pp.
2. Lynd, L.R., J.H. Cushman, R.J. Nichols, and C.E. Wyman (1991). Fuel ethanol from cellulosic biomass, Science, 251, pp. 1318–23.
3. Lynd, L.R. (1996). Overview and evaluation of fuel ethanol from cellulosic biomass: technology, economics, the environment, and policy, Annual Reviews of Energy and Environment, 21, pp. 403–65.
4. Tyson, K.S. (1993). Fuel Cycle Evaluations of Biomass-Ethanol and Reformulated Gasoline, Volume I, NREL/TP-463-4950, DE94000227, Golden, Colo: National Renewable Energy Laboratory, November.

5. McCoy, M. (1998). Biomass ethanol inches forward, Chemical and Engineering News, December 7, pp. 29–32.
6. U.S. Department of Agriculture (1989). Ethanol's role in clean air, USDA Backgrounder Series, Washington, DC: US Department of Agriculture.
7. Wyman, C.E. and B.J. Goodman (1993). Biotechnology for production of fuels, chemicals, and materials, Applied Biochemistry and Biotechnology, 39/40, pp. 41–59.
8. Wright, J.D. (1988). Ethanol from biomass by enzymatic hydrolysis, Chemical Engineering Progress, pp. 62–74, August.
9. Hinman, N.D., D.J. Schell, C.J. Riley, P.W. Bergeron, and P.J. Walter (1992). Preliminary estimate of the cost of ethanol production for SSF technology, Applied Biochemistry and Biotechnology, 34/35, pp. 639–49.
10. Ingram, L.O., T. Conway, D.P. Clark, G.W. Sewell, and J.F. Preston (1987). Genetic engineering of ethanol production in *Escherichia coli*, Applied Environmental Microbiology, 53, pp. 2420–5.
11. Wyman, C.E. (1995). Economic fundamentals of ethanol production from lignocellulosic biomass, In Enzymatic Degradation of Insoluble Carbohydrates, ACS Symposium Series 618, ed. by J.N. Saddler, M.H. Penner, Washington, DC: American Chemical Society, pp. 272–290.
12. Lynd, L.R., R,T. Elander, and C.E. Wyman (1996). Likely features and costs of mature biomass ethanol technology. Applied Biochemistry and Biotechnology, 57/58, pp. 741–61.
13. Torget, R.W., K.L. Kidam, T.-A. Hsu, G.P. Philippidis, and C.E. Wyman (1998). Prehydrolysis of Lignocellulose. US Patent No. 5,705,369. January 6.
14. Torget, R., C. Hatzis, T.K. Hayward, and T.-A. Hsu, Philippidis, G.P. (1996). Optimization of reverse-flow, two-temperature, dilute-acid pretreatment to enhance biomass conversion to ethanol, Applied Biochemistry and Biotechnology, 57/58, pp. 85–101.

STATUS OF BIOMASS CONVERSION TO ETHANOL AND OPPORTUNITIES FOR FUTURE COST IMPROVEMENTS

David Glassner

National Renewable Energy Laboratory, 1617 Cole Boulevard, Golden, Colorado 80401

Arkenol, BC International, Iogen/PetroCanada, and Masada Resources Group all have commercial development projects in North America. Process technologies to be utilized include concentrated acid, dilute acid and enzymatic cellulose hydrolysis. Although biomass conversion to ethanol is poised for commercialization, large cost reduction opportunities remain. Improvements in the feedstock composition and structure, pretreatment of biomass, cellulase enzyme, ethanol fermentation, lignin use and supporting infrastructure can all lower the cost of ethanol production from biomass. Process performance goals have been established for future process technology based on the technical tools available and process economic sensitivities. The improved process technology lowers the cost of ethanol production by 66 cents per gallon based on our biomass-to-ethanol model. This paper reviews technology performance goals that have been established for the process technology of the future, the technical tools to be applied to achieve the goals and the cost reduction resulting from the technology improvements.

COLLINS PINE/BCI BIOMASS TO ETHANOL PROJECT

M. A. Yancey,[a] N. D. Hinman,[b] J. J. Sheehan[c] and V. M. Tiangco[d]

[a]National Renewable Energy Laboratory, 1617 Cole Boulevard, Golden, Colorado, USA, 80401-3393
[b]BC International, 6860 South Yosemite Ct., Suite 200, Englewood, Colorado, USA, 80112,
[c]Plumas Corporation, P.O. Box 3880, Quincy, California 95971
[d]Research and Development Office, MS-43, California Energy Commission, 1516 Ninth St., Sacramento, California, USA, 95814

California has abundant biomass resources and a growing transportation fuels market. These two facts have helped to create an opportunity for biomass to ethanol projects within the state. One such project under development is the Collins Pine/BCI Project. Collins Pine Company and BC International (BCI) have teamed up to develop a forest biomass to ethanol facility to be co-located with Collins Pine's 12 MW, biomass-fueled electric generator in Chester, California. The Collins Pine Company (headquartered in Portland, Oregon) is an environmentally progressive lumber company that has owned and operated timberlands near Chester, California since the turn of the century. Collins manages 100,000 acres of timberland in the immediate area of the project. BCI (Dedham, Massachusetts) holds an exclusive license to a new, patented biotechnological process to convert lignocellulosic materials into ethanol and other specialty chemicals with significant cost savings and environmental benefits. The project has received a California Energy Commission PIER program award to continue the developmental work done in the Quincy Library Group's *Northeastern California Ethanol Manufacturing Feasibility Study* (November 1997). This paper provides (1) a brief overview of the biomass and transportation fuels market in California; (2) the current status of the Collins Pine/BCI biomass ethanol project; and (3) future prospects and hurdles for the project to overcome.

1. INTRODUCTION

The term 'biomass' refers to structural and non-structural carbohydrates and other compounds produced through photosynthesis consisting of plant materials and agricultural, industrial, and municipal wastes, and residues derived therefrom (Jenkins, 1997). Biomass has historically supplied food, feed, fiber, energy, and structural materials needs for humans. Biomass is the oldest known source of renewable energy that humans have been using since the discovery of fire and it has relatively high energy content. The energy content of dry biomass ranges from 7,000 Btu/lb for straws to 9,000 Btu/lb for wood.

Biomass is an attractive energy source for a number of reasons. First, it is a renewable energy source as long as we manage vegetation appropriately. Second, biomass is also more evenly distributed over the earth's surface than are finite energy sources, and it may be exploited using technologies that are less capital-intensive. Third, it provides the opportunity for local, regional, and national energy self-sufficiency across the globe. And fourth, energy derived from biomass does not have many of the negative environmental impacts associated with non-renewable energy sources.

The potential for biomass to supply much larger amounts of useful energy with reduced environmental impacts compared to fossil fuels has stimulated substantial research and development of systems to grow, harvest, handle, process, and convert biomass to heat, electricity, liquid and gaseous fuels, and other chemicals and products. The key to accessing the energy content in biomass is converting the raw material (feedstock) into a usable form, which can be accomplished through three principal routes: (1) thermochemical; (2) biochemical; and (3) physicochemical. In practice, combinations of two or more of these routes may be used in the generation of final product or products.

2. CALIFORNIA'S BIOMASS RESOURCES

California has abundant lignocellulosic biomass, over 35 million bone dry tons (BDT) per year (Tiangco et al., 1993) from agricultural and forestry residues, urban wood wastes, yard wastes, food processing wastes, chaparral and lumber mill waste.[1] An additional 18 million BDT annually can be produced from forest thinnings from both public and private land. California biomass is a growing waste disposal problem. If it is not solved, California's agriculture, forest and urban waste disposal will continue to be an environmental problem and disposal costs will continue to increase. This increased cost will be passed on to consumers. Currently, about 5 million BDT are being utilized by 29 biomass power plants producing 590 megawatts of capacity. Industry representatives indicate the cost of electricity for these facilities is over $0.06/kilowatt-hour (kWh). The future of the biomass power industry is uncertain owing to a myriad of barriers (Tiangco et al., 1994) and uncertainty of electricity deregulation.

The three major growing waste disposal problems in California are agricultural residues, forest slash and thinnings, and mixed solid wastes.

2.1. Agricultural residues

Historically, agricultural residues in California such as rice straw and orchard prunings have been disposed of by open field burning. About 9.5 million BDT of lignocellulosic agricultural residues are produced in the state annually. Because of the air pollutants released to the atmosphere through open-field burning, federal, state and local air quality agencies have been tightening the regulations on open field burning. Rice straw is the first agricultural

---

[1] Over 11 million BDT of livestock manure were deducted from the total biomass resource potential of 47 million BDT (Tiangco et al. 1993). An estimate of forest thinnings was not included in the study.

waste whose reduction in open-field burning is mandated (Rice Straw Burning Reduction Act of 1991). This Act mandates a 75%-100% reduction of open-field burning of rice straw by the year 2000. Rice straw production in California ranges from 1.0 to 1.5 million BDT/yr. Even though agriculture is California's number one industry, its population is 94+% urban and is increasing.

2.2. Forest slash and thinnings

Another related concern is the growing volume of dead and diseased trees and forest slash and thinnings that increases fuel loading and worsens forest health. If harvested and collected, forest slash and thinnings can be up to 18 million BDT/yr. Wildfire hazards are critical on many forested lands, foothill, mountain home sites and other business properties in California. Thinning out trees for timber stand improvement and removing shrubs and other flammable vegetation will help ensure the survivability of the remaining trees. Excessive amounts of fuel have built up in forests and woodlands of the state since fire is no longer allowed to perform its ecological role. When devastating fires break out, many of the native trees, homes, businesses and other properties can be destroyed in a few moments. Approximately $1 billion in costs/losses are incurred from wildfires annually. The fuel loads on California's forested lands and wooded and brush-covered parcels must be properly managed to survive the next wildfire. These forest health problems and the damage caused by catastrophic wildfires have reached a level of urgency that calls for new solutions.

2.3. Mixed solid waste

Mixed solid waste (including yard waste, construction waste and paper products) is increasingly a problem in the state. According to California Integrated Waste Management Board, Californians create nearly 2,900 pounds of household garbage and other industrial waste each and every second; a total of 45 million tons a year. Until recently, the only places to dispose of that trash were local landfills. Constraints on landfill capacity led to the passage of legislation in 1989, the California Integrated Waste Management Act. This Act mandates stringent goals for diverting solid waste from landfills into reuse, recycling or transformation to energy products. Municipalities operating solid waste disposal facilities were required to divert 25% of the waste stream to these uses beginning in 1995 and must attain 50% diversion rate by the year 2000. A subsequent law (Assembly Bill 688) set forth additional conditions for calculating the credits for diverting waste materials and limited the degree to which biomass transformation (either combustion or fuel production) could be counted as satisfying the diversion requirements.

3. CALIFORNIA'S TRANSPORTATION FUELS MARKET

California is the third largest consumer of gasoline in the world. It is surpassed only by the rest of the United States and the former Soviet Union. Californians use about 14 billion gallons of gasoline a year and another one billion gallons of diesel fuel. A cleaner-burning fuel called California reformulated gasoline, or Cal-RFG, became the only gasoline sold in the state starting June 1997. Most Cal-RFG contains an oxygenate called methyl tertiary butyl-ether (MTBE). On March 25, 1999, Governor Gray Davis issued Executive Order D-5-99 directing that MTBE be phased out of gasoline sold in California, with the oxygenate's complete removal by December 31, 2002.

With the phase-out of MTBE in the California gasoline pool, the prospects for ethanol use in California are, perhaps, more attractive than ever. With California consuming 14 billion gallons of gasoline each year, the potential market for ethanol may be significant. Even if Congress lifts the oxygenate requirement for reformulated gasolines, it is expected that some refineries will blend ethanol into their gasoline, depending on such factors as the season, location and the grade of gasoline marketed. The market for ethanol as a replacement oxygenate could be as much as 770 million gallons annually.

The use of ethanol in the western states in recent years has been mixed. Ethanol use in western states increased from 1992 to 1995, but declined after 1995, due to the introduction of reformulated gasolines in various metropolitan areas. Ethanol's effect of raising volatility at low blending levels has traditionally been less attractive to refiners compared with MTBE. Currently, California is consuming very little ethanol. Tosco is the sole refiner now blending ethanol with their gasoline—and only in a limited demonstration area in the San Francisco Bay Area (some 50+ stations). The potential of ethanol from lignocellulosic biomass is promising (Tiangco and Johannis, 1997; Hinman et al., 1998).

The future role of ethanol as an oxygenate remains dependent on resolving key economic, regulatory and institutional barriers. Pending federal and state legislation will, in part, determine the extent of ethanol use. The advancement of biomass conversion technologies and market forces will also affect the role of ethanol in the state. By joining ethanol production facilities with existing biomass power plants more economic and sustaining business ventures result. Although much is still unclear, ethanol, electricity and other products from forest and other residues look very promising, and ethanol may play a major role as a transportation fuel component as well as a means of effectively managing California's waste disposal and forest health concerns.

4. THE COLLINS PINE/BCI BIOMASS ETHANOL PROJECT

The California biomass power industry must find ways to become more competitive by reducing its electrical buss bar production costs if the industry is to survive the transition to a deregulated electricity market. The cost of electricity produced from biomass ranges from $0.064 to $0.113/kWh[2]. This compares with the market short run avoided cost of approximately $0.03/kWh, which will eventually be the basis for purchasing the output of biomass power plants.

Co-location of biomass ethanol plants with biomass power plants in California has the potential to reduce the production costs for both facilities. Co-location will create new sources of revenue for the biomass power plant through sales of electricity and steam to the ethanol facility and reduce fuel costs through sharing the feedstocks with the ethanol facility.

---

[2] "Electric Utility Restructuring and the California Biomass Energy Industry," Future Resources Associates, Inc., Berkeley, California, 1997.

That is, the ethanol facility can use part or all of the power plant's fuel, separate the sugars for ethanol and other by-product production, and return fuel in the form of lignin to fuel the biomass power plant, thus reducing its total raw material wood fuel needs. This co-location concept could define a new, more economical alternative that will make California's biomass power plants competitive by the time the fully deregulated market arrives in the year 2002.

The Collins Pine/BCI Biomass Ethanol Project will demonstrate the benefits of co-location of biomass ethanol production with biomass power production at the Collins Pine biomass power facility in Chester, California. Integrated ethanol and power production is expected to lower the power plant's buss bar electricity cost by up to $0.015/kWh. A 20 million gallon per year ethanol facility is contemplated for this project. The technical objectives for the project are to

1. define the technical interfaces between the ethanol facility and the biomass power plant. These include the technical specifications for steam, electricity, lignin and sharing biomass feedstock;
2. define the range of feedstock and lignin mixes that minimize the biomass power plant's electrical generation costs;
3. define the range of steam and power mixes that optimize the power plant's profitability;
4. determine the optimum economic cogeneration configurations between the ethanol facility and the biomass power plant.

Achieving these objectives will provide essential technical information for the development and expansion of cogeneration opportunities with other biomass power facilities in California and throughout the country.

Activities to be pursued with funding from the California Energy Commission's Public Interest Energy Research Program include the following:

- Develop a Feedstock Supply Plan for the project. The geographic areas to be covered by the Plan include the counties of Lassen and Plumas and nearby areas. The Feedstock Supply Plan will determine feedstock costs; identify the available infrastructure to collect, process, store and transport the required feedstock; assess competition for feedstock; and determine the emission offset credits generated.
- Conduct lignin fuel tests. A by-product of the ethanol process will be a lignin-rich solid residue. The cellulose and hemicellulose in the wood will be converted to ethanol and various by-products. What remains will be primarily lignin and a small amount of ash. The lignin appears to be an excellent boiler fuel, but the fuel characteristics of the lignin residue must be determined to provide design data and to prevent negative impacts on the biomass boiler operations.
- Develop the ethanol process design and evaluate ethanol–biomass power integration issues. BCI will develop a preliminary process design for the ethanol facility that will be used to determine the optimum economic cogeneration configurations between the ethanol facility and the biomass power plant. The range of steam and power requirements to be supplied from the Collins biomass power plant will be determined along with the range of feedstock/lignin mixes that optimize the power plant's electrical generation costs.

- Evaluate environmental impact issues. Meetings will be held with the Quincy Library Group (QLG) and local, state and federal authorities to learn what, if any, serious environmental impact issues could arise from the proposed project. The response will be used to decide on the key environmental mitigations required for the Chester site.
- Determine socioeconomic effects of the proposed project. The *Quincy Library Group Study* predicted a variety of local and state impacts that would result from construction and operation of a modest sized ethanol manufacturing facility at Chester and five other sites in Northern California. The social and economic framework was also reviewed. The QLG socioeconomic report will be updated based on the preliminary process design for the project.

An 18-month project duration is anticipated with a start date of May 1999.

REFERENCES

Hinman, N., V. Tiangco, M. Johannis, and M. Yancey (1998). Biomass Power and Ethanol Production: A Cogeneration Opportunity in California. For publication in the Proceedings of the Symposium on Biotechnology for Fuels and Chemicals, Applied Biochemistry and Biotechnology.

Jenkins, B.M. (1997). Introduction to Energy Systems Lecture Notes, University of California at Davis, Davis, California 95616.

Tiangco, V. and P.S. Sethi. (1993). Biomass resources in California, in the Proceedings of the First Biomass Conference of the Americas: Energy, Environment, Agriculture, and Industry. August 30-Sep 2. Vol. 1, pp. 94–99.

Tiangco, V., P. Sethi, Y. Lee, D. Yomogida, B. Huffaker, and J. Emery (1994). Technical, economic and environmental issues for sustainable generation of biomass power in California, in the Proceedings of the Sixth National Bioenergy Conference, BIOENERGY'94. Reno/Sparks, Nevada. October 2-6, pp. 539–546.

Tiangco, V.M. and M. Johannis (1997). Prospects of biomass to ethanol development in California. Making a Business from Biomass in Energy, Environment, Chemicals, Fibers and Materials, in Proceedings of the Third Biomass Conference of the Americas, Montreal, Quebec, Canada, August 24-29, p.1183 and see Addendum.

TECHNOLOGICAL AND ECOLOGICAL ASPECTS OF ETHANOL PRODUCTION FROM WOOD

Yu.I. Kholkin,[a] V.L. Makarov, V.V. Viglazov, V.A. Elkin, and H.D. Mettee[b]

[a]St. Petersburg Forest Technical Academy, Institutsky Per., 5, 194021, St. Petersburg, Russia
[b]Youngstown State University, One University Place, Youngstown, Ohio 44555

The industrial production of ethanol from wood in the former USSR started in 1935; the technology was based on results of investigations by the Leningrad Forest Technical Academy under the leadership of V.I. Sharkov. At the end of the Soviet period in the former USSR there were 46 hydrolysis plants in operation, including 18 ethanol-producing plants which produced 150-200 million liters of industrial ethanol per year. Currently the most efficient are producing ethanol, as compared with those producing fodder yeast, furfural, food xylitol, and other products. The main problems in real ethanol production are the following: (1) comparatively low yields and concentration of monosaccharides in the hydrolysates; (2) large amount of impurities in hydrolysates; (3) incomplete assimilation of substrate compounds during fermentation; (4) severe contamination of the environment by wastewater. This report presents new methods of hydrolysate and wastewater purification and a recycling scheme for water used in ethanol production.

1. HYDROLYSIS OF WOOD

Industrial plants use as raw materials softwood sawdust and chips (additional carbohydrate raw: sugar-beet molasses, grain, etc.). Diluted 0.5-0.7% sulfuric acid is used for percolation hydrolysis of wood at temperature 185-195°C in the batch reactors with volume 18-80 m^3. The yield of monosaccharides is 40-45% (as reducing substances) from dry wood (the theoretical yield is 65-70%). To increase the yield of monosaccharides, in 1950-1970s a Kansky-type glucose plant was constructed in Siberia (hydrolysis of wood by 41% HCl) and a pilot plant near Riga (hydrolysis by 75% H_2SO_4). These processes had technical and economical difficulties and only the percolation process is now used on an industrial scale.

2. SUBSTRATE PREPARATION

The hydrolysate after hydrolysis of coniferous wood has temperature ~ 180°C, pH ~ 1 and contents 3-4% monosaccharides (3/4 hexoses and 1/4 pentoses), ~ 0.5% sulfuric acid, 0.5-0.7% organic acids, and also furans, phenols, terpenes, ligno-huminic substances (LHS), and other impurities. The technological scheme of hydrolysate preparation as a rule includes cooling by self-evaporation, inversion, neutralization, addition of nitrogen, ammonium and calcium salts, settling of suspended matter, additional cooling, aeration and additional settling.[1] After these steps the substrate has a temperature of 32-35°C, pH 3.5-4.5 and a permissible concentration of toxic impurities. But the LHS content is still at a high level.

LHS are generated during wood hydrolysis as a result of condensation reactions between lignin soluble fractions and furans. LHS are colloids and suspended compounds and cause the black-brown color of hydrolysates. LHS are sorbed by microorganisms on their surfaces and are involved with assimilation-dissimilation processes. LHS are oxidation-resistant compounds and they are not oxidized during substrate fermentation and biological purification of wastewaters, which in turn leads to environmental pollution.

We suggested a new flocculation method for purifying hydrolysates from LHS and other colloidal and suspended matter. Some cationic polymer flocculants provide good decolorization and purification of hydrolysates. Cationic polyelectrolytes allow particles of LHS that have negative charge to coalesce, and as result these particles are converted to suspended matter and precipitated as sludge.

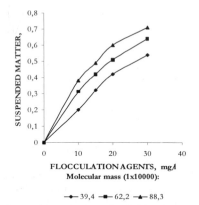

Figure 1. Conversion of colloid LHS to suspended matter by flocculant.

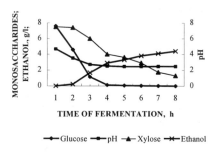

Figure 2. Conversion of xylose-glucose mixtures to ethanol.

Figure 1 shows conversion of colloid LHS to suspended matter by poly(diallyl-dimethylammonium chloride) $(C_8H_{16}NCl)_n$ of various molecular mass. These compounds are quaternary ammonium bases with a high positive electrokinetic potential. The efficiency of purification depends upon the molecular mass of flocculant, its quantity, pH, temperature and other factors. Using 10-20 g of flocculant per 1 ton (t) of hydrolysate provides 60-80% decolorization and purification. This method[2] was successfully implemented at three industrial hydrolysis plants.

3. FERMENTATION AND RECTIFICATION

Carbohydrate substrates are fermented by technical strains of *Schizosaccharomyces* or *Saccharomyces* during 5-7 h at a temperature of 32-34°C and pH 4.0-4.5. These yeasts can assimilate only hexoses, and the pentoses in waste cultural liquid are used to culture fodder yeast (as rule *Candida scottii* strains). Bioconversion of pentoses to ethanol is a very difficult problem. There are a lot of publication in this field,[3-5] but this problem still persists. We also tried to use an adapted strain of *Pachisolen tannophylus* for xylose-glucose mixtures periodic fermentation (Figure 2), but the yield of ethanol was still not sufficient. Continued fermentation increased the ethanol yields, but additional investigations are needed for practical realization of this process.

Hydrolysis plants produce high quality ethanol using a five-column rectification scheme. For example, "Extra"-grade ethanol can contain not more than 0.03% methanol, followed by impurities in mg/l: aldehydes ≤ 2, acids ≤ 10, fusel oil ≤ 3, esters ≤ 25. The commercial products are used in medicine, pharmaceuticals, perfumery, cosmetic, electronic, chemistry, and many other purposes (but not for food drink production). Fuel ethanol is not produced from wood, yet.

4. ENVIRONMENTAL PROTECTION

In hydrolysis processes 12-15 t of water are used per 1 t dry wood: the total quantity of wastewater consists of 20-50 t/t dry wood. The wastewater contains significant impurities and the cost of purification is very high. As a rule, the degree of purification is not enough and enterprises have to pay fees and penalties for environment pollution.

We suggested the recycling scheme (Figure 3) for reducing its water treatment expenses in ethanol producing. This scheme provides for the re-use of wastewater after its mechanical and biological purification. But this water has impurities as well, and their concentration will rise during multiple re-circulation in the partly closed system.

For additional purification of biologically treated waters, we compare some methods with optimal conditions for every method (Table 1). Good results are achieved by ultrafiltration using highly selective membranes and adsorption on active carbon from lignin. But from a technical point of view, the flocculation method of purification and decolorization is more convenient and effective.

Local biological anaerobic-aerobic fermentation of purified wastewater provides a degree of purification of 80-95% (as COD) and decolorization of 80-90%. But this method can be used for more effective preliminary purification of waste cultural liquid.[6] The choice of any real recycling scheme depends mostly upon the capacity of the main equipment for wastewater purification.

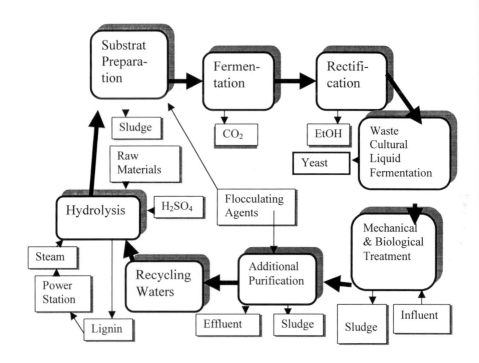

Figure 3. Recycling use of water in ethanol production.

Table 1
Additional purification of biologically treated effluents

Method of purification	Degree of Purification (%)			
	COD	BOD_5	LHS	Decolorization
Flocculation	40-50	36-45	57-59	45-55
Adsorption	45-50	40-42	44-46	60-70
Flocculation - Adsorption	87-89	84-86	89-91	90-92
Ultrafiltration	55-60	50-52	65-70	70-80
Flocculation - Ultrafiltration	61-63	51-53	79-81	86-88

REFERENCES

1. Kholkin, Yu.I. (1989). Technology of Hydrolysis Production, Lesnaja Prom., Moscow, 496 pp.
2. Viglazov, V.V., V.B. Kind, Yu.I. Kholkin, and S.V. Avdashkevich (1997). Russian Patent No. 2 077 594.
3. Hinman, N.D., J. D. Wright, W. Hoagland, and C.E. Wyman (1988). Appl. Biochem. Biotechnol., 20–21, p. 391.
4. Prior, D.A., S.G. Kilian, and J.C. Du Preer (1989). Proc. Biochem., 24, p. 21.
5. Zhang, M., C. Eddy, K. Deanda, M. Finkelstein, and S. Picataggio (1995). Science, 267, p. 240.
6. Amrani, Yu, I. Kholkin, and V.V. Makarov (1996). Sci. des Aliments, 16, p. 643.

SIMULTANEOUS SACCHARIFICATION AND CO-FERMENTATION OF PERACETIC ACID PRETREATED SUGAR CANE BAGASSE

Lincoln C. Teixeira,[a] James C. Linden,[b] and Herbert A. Schroeder[b]

[a]Department of Biotechnology and Chemical Technology, Fundação Centro Tecnológico de Minas Gerais - CETEC, Av. José Cândido da Silveira 2000, CEP 31170-000, Belo Horizonte, MG, Brazil
[b]Department of Chemical and Bioresource Engineering, Colorado State University, Fort Collins, Colorado, 80523, USA

Previous work in our laboratory[1,2] has demonstrated that peracetic acid improves the enzymatic digestibility of lignocellulosic materials. From the same studies, use of dilute alkali solutions as a pre-pretreatment prior to peracetic acid lignin oxidation increases sugar conversion yields in a synergistic, not additive, manner. Deacetylation of xylan is conducted easily by use of dilute alkali solutions at mild conditions. In this paper, the effectiveness of peracetic acid pretreatment of sugar cane bagasse combined with an alkaline pre-pretreatment is evaluated through simultaneous saccharification and co-fermentation (SSCF) procedures. A practical 92% of theoretical ethanol yield using recombinant *Zymomonas mobilis* CP4/pZB5 is achieved using 6% NaOH/15% peracetic acid pretreated substrate. No sugar accumulation is observed during SSCF; the recombinant microorganism exhibits greater glucose utilization rates than those of xylose. Acetate levels at the end of the co-fermentations are less than 0.2% (w/v). Based on demonstrated reduction of acetyl groups of the biomass, alkaline pre-pretreatments help to reduce peracetic acid requirements. The influence of deacetylation is more pronounced in combined pretreatments using lower peracetic acid loadings. Stereochemical impediments of the acetyl groups in hemicellulose on the activity of specific enzymes may be involved.

1. INTRODUCTION

Sugar cane bagasse, a residue generated by the sucrose and ethanol industry, constitutes the most important lignocellulosic material to be considered in Brazil as an alternative for the production of ethanol fuel. At present, most bagasse is burned, and because of its moisture content, it has a low fuel value. Bagasse is available at the sugar mill site at no additional cost because harvesting and transportation costs are borne by sugar production.

Conventional methods for making biomass more accessible to enzymes target lignin or hemicellulose removal. Steam explosion and dilute acids constitute the most studied pretreatment technologies. Both methods use high temperatures and pressure reactors, and fractionation procedures drive up costs. A silo type system designed to use peracetic acid is a new alternative for pretreating biomass. This method has no need for expensive reactors, energy input, or fractionation procedures.

Peracetic acid, a powerful oxidizing agent, is very selective towards the lignin structure. It cleaves aromatic nuclei in lignin, thus generating dicarboxylic acids and their lactones.[3] Peracetic acid is produced by reaction of peroxides and acetic acid. Large scale production of peracetic acid, 10,000 metric ton/year, began in 1996 in Finland.[4,5] Conveniently prepared from ethanol by aerobic fermentation, and recycled from peracetic acid consumed in the pretreatment, acetic acid is available and inexpensive, especially in countries that produce abundant ethanol. Because of the worldwide expansion of hydrogen peroxide production[6] and because it has been recognized as an environmentally friendly chemical,[7] the cost of that acid will probably decrease.

Lignocellulosic materials pretreated with peracetic acid have most of the original xylan retained without significant deacetylation. The acetyl groups cause stereochemical impediments to enzymatic degradation as reported in the literature.[8-11] This paper demonstrates the value of deacetylation in reducing pretreatment chemical costs, the efficiency of enzymatic hydrolysis, and the consequent yields in SSCF.

2. MATERIALS AND METHODS

Sugar cane bagasse was provided by Cajun Sugar Co-op, Inc., New Iberia, Louisiana, milled and screened to 20-80 mesh. The pretreatment procedures are described in previous work.[1] The enzymatic hydrolysis tests as well as the enzyme activity assays have been developed from standard procedures used at the National Renewable Energy Laboratory and are performed as described in previous work.[1]

The inoculum preparation and the procedure for performing the SSCF tests have been developed from standard procedures used at the National Renewable Energy Laboratory and are described in previous work.[2] The recombinant *Zymomonas mobilis* CP4/pZB5 supplied by the National Renewable Energy Laboratory is used in this study.[12] This organism uses both glucose and xylose as fermentation substrates, which is important because peracetic acid pretreatment does not degrade hemicellulose.

3. RESULTS AND DISCUSSION

3.1. Biomass chemical composition

The chemical composition of the sugar cane bagasse used in this study is shown in Table 1. The high extractive content is due to residual sucrose left during processing of the sugar cane. Total lignin (Klason and soluble) and other results, excluding extractive content, are calculated based on extractive free biomass.

Table 1
Chemical composition of sugar cane bagasse (dry wt. basis, %)

Cellulose	Pentosans	Lignin*	Holocellulose	Ash	Extractives
39.6	29.7	24.7	74.5	4.1	14.3

3.2. Pretreatment product composition and yield

The pretreatment yields are shown in Table 2. The average pretreatment yield of approximately 80% is greater than the 60% yield obtained by steaming exploded sugar cane bagasse.[13] The carbohydrate content increases with the extent of delignification. Klason lignin content decreases with the increase of peracetic acid concentration; soluble lignin increases with the loading concentrations except at the 60% peracetic acid loading. At this loading, the soluble lignin content is probably reduced because the oxidative fragmentation of the phenyl ring structures decreases the absorption at 205 nm.

Table 2
Partial composition and yield of pretreated sugar cane bagasse. The xylan content is reported as a linear polymer not considering any side groups

Pretreatment condition	Glucan (%)	Xylan (%)	Klason lignin (%)	Soluble lignin (%)	Pretreatment overall yield (%)
0% peracetic acid (raw bagasse)	39.9	23.4	22.0	2.1	88.2
6% peracetic acid	42.0	23.9	20.5	2.6	85.8
9% peracetic acid	43.3	24.7	18.6	3.4	84.6
15% peracetic acid	44.2	25.2	14.1	4.3	81.6
21% peracetic acid	49.1	26.5	10.8	5.1	78.4
30% peracetic acid	53.4	27.4	6.3	5.1	74.6
60% peracetic acid	55.9	27.7	4.3	3.3	68.6
6% NaOH - 15% peracetic acid	50.1	26.6	11.4	4.1	74.6
6% NaOH - 9% peracetic acid	47.9	25.5	16.5	3.5	78.0
6% NaOH - 6% peracetic acid	46.9	25.0	17.2	3.0	78.0
3% NaOH - 15% peracetic acid	47.7	25.6	15.6	4.9	77.8
3% NaOH - 9% peracetic acid	45.4	24.9	12.9	3.6	80.8

3.3 Simultaneous saccharification and co-fermentation

Raw bagasse gives a poor bioconversion to ethanol. The use of 15% peracetic acid based on oven-dry weight of biomass does not give good ethanol conversion yields when compared with the substrate treated with same amount of acid and pre-washed with alkali. The ethanol yields of sugar cane bagasse samples from different pretreatment conditions using the recently developed recombinant Z. mobilis CP4/pZB5 are given in Figure 1. Ethanol yields of greater than 90% in 10 days of fermentation are obtained from three of the 12 pretreated biomass samples. The 6% NaOH/15% peracetic acid, 21% peracetic acid, and 60% peracetic acid pretreated bagasse samples give ethanol yields of 91.9, 91.4, and 90.7% respectively.

These yields are only slightly lower than the 95.0% yield, for the same recombinant microorganism at 30°C using a glucose- and xylose-containing medium.[12] The differences in

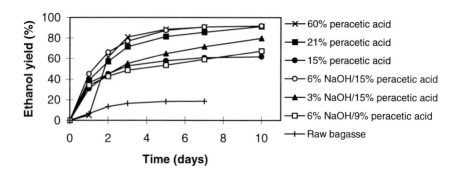

Figure 1. Ethanol yield from treated and untreated sugar cane bagasse.

yield may be due to using a concentrated medium containing monosaccharides, a form of substrate that is ready to be consumed by the recombinant bacteria. Temperature effects should also be considered. *Z. mobilis* grows better at 30°C than at 37°C, the temperature used for the SSCF. This strain can also grows satisfactorily at 37°C.[14,15] Saddler et al.[15] have reported satisfactory ethanol yields of 85.4% from steam-exploded poplar with excellent glucose consumption in six-day simultaneous saccharification and fermentation at 37°C, using *Z. mobilis* Z2 (ATCC 29191).

The SSCF kinetic curves for the fermentations of 6% NaOH/15% peracetic acid treated and of untreated bagasse are given in Figure 2. The glucose and xylose levels are very low during the co-fermentation of the treated substrate, which suggests rapid assimilation of these sugars by *Z. mobilis*. Also, the low acetyl content of the 6% NaOH/15% peracetic acid pretreated material (Table 3) contribute to a low level of acetate during the co-fermentation and consequently lower growth inhibition. In addition, the acetic acid by-product of peracetic acid reaction has been removed by efficiently washing the pretreated biomass to pH 5. Because the inhibitory metabolite is found at concentrations of only 1.21 g/L, ethanol conversion by the *Z. mobilis* using this practical substrate is quite efficient. As expected, untreated materials show poor ethanol yields as a consequence of unsatisfactory enzymatic hydrolysis.

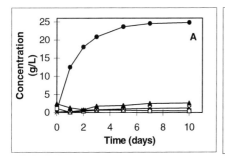 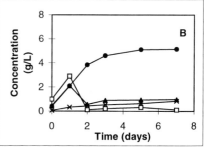

Figure 2. SSCF kinetics for sugar cane bagasse. (A) 6% NaOH/15% peracetic acid pretreated biomass. (B) Untreated biomass. Legend: ● ethanol, glucose, ▲ xylose, × acetate.

Table 3
Acetyl content and retention after pretreatment of bagasse. Values for the retained acetyl have been corrected to raw biomass considering variations in the amount of xylan in the treated substrate

Pretreatment condition	Acetate content (%)	Retained acetyl (%)
0% peracetic acid (raw bagasse)	5.92	100.0
6% NaOH/15% peracetic acid	2.95	43.8
6% NaOH/9% peracetic acid	3.07	47.6
6% NaOH/6% peracetic acid	3.20	50.8
3% NaOH/15% peracetic acid	4.51	69.6
3% NaOH/9% peracetic acid	4.76	75.6

3.4. Synergistic effect of alkali washing on enzymatic hydrolysis

Alkaline washing is very helpful prior to oxidative treatment by reducing requirements for peracetic acid. Synergism is verified by observing enzymatic hydrolysis extent as a function of acetyl group removal from xylan. Higher acid concentrations (15%) have less effect on enzymatic hydrolysis than lower concentrations, as verified by the slopes of the bioconversion curves (Figure 3). When lower amounts of peracetic acid are used the slope increases considerably indicating more influence of the alkaline washing. Extrapolation of the curves from 6% peracetic acid treated samples suggests 100% sugar conversion if acetyl groups are removed completely from xylan before the oxidative step.

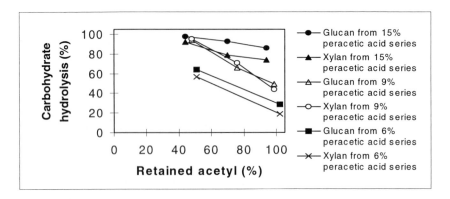

Figure 3. Carbohydrate hydrolysis as a function of retained acetyl groups in pretreated bagasse.

4. CONCLUSIONS

Peracetic acid and alkaline–peracetic acid pretreatments give higher overall solid yields than conventional pretreatments because of the mild conditions characteristic of the process. The minimum loading of peracetic acid for obtaining a satisfactory ethanol yield is 21% based on oven-dry weight of biomass.

Pre-pretreatment using a diluted (6%) sodium hydroxide prior to peracetic acid pretreatment results in substantially better glucan and xylan conversion to simple sugars using the 9 and 15% peracetic acid treatments. Consequently, smaller amounts of peracetic are needed.

Preliminary results indicate that reduction of acetyl groups in combination with slight delignification are responsible for satisfactory sugar and ethanol yields. Extrapolation of sugar cane bagasse data suggests that if the acetyl groups were totally removed from biomass by an efficient alkaline pre-pretreatment, 6% of peracetic acid could be sufficient to obtain similar results found using higher acid concentrations of 9 and 15% (oven-dry weight). The technical viability of sugar cane bagasse pretreated with 6% NaOH and 15% peracetic acid is verified through SSFC using recombinant *Z. mobilis* CP4/pZB5, with an average theoretical ethanol yield of 91.4%.

REFERENCES

1. Teixeira, L.C., J.C. Linden, and H.A. Schroeder (1999). Optimizing peracetic acid pretreatment conditions for improved simultaneous saccharification and co-fermentation (SSFC) of sugar cane bagasse to ethanol fuel, Renewable Energy, 16, pp. 1070–1073.
2. Teixeira, L.C., J.C. Linden, and H.A. Schroeder (1999). Alkaline and peracetic acid pretreatments of biomass for ethanol production, Applied Biochemistry and Biotechnology, 77–79, pp. 1–16.
3. Lai, Y.Z. and K.V. Sarkanen (1968). Delignification by peracetic acid. II. Comparative study on softwood and hardwood lignins, TAPPI, 51, pp. 449–453.
4. Anonymous (1995). Kemira peracetic unit a "first", Chemical Marketing Report, Dec. 25, p. 9.
5. Anonymous (1996). Kemira to build peracetic acid plant, Chemical Week, Jan. 3/10, p. 22.
6. Szmant, H.H. (1989). Organic Building Blocks of the Chemical Industry, Wiley-Interscience, John Wiley & Sons, New York, pp. 236–239.
7. Wilson, S. (1994). Peroxygen technology in the chemical industry, Chemistry & Industry, 7, 255–258.
8. Kong, F., C.R. Engler, and E.J. Soltes (1992). Effects of cell-wall acetate, xylan backbone, and lignin on enzymatic hydrolysis of aspen wood, Applied Biochemistry and Biotechnology, 34/35, pp. 23–35.
9. Kong, F. (1990). Effect of acetate and other cell wall components on enzymatic hydrolysis of aspen wood. M.S. thesis, Texas A&M, College Station, TX.
10. Mitchell, D.J. (1989). Acetyl xylans: The effect of acetylation on the enzymatic digestion of biomass. M.S. thesis, Colorado State University, Fort Collins, CO.
11. Grohmann, K., D.J. Mitchell, M.E. Himmel, B.E. Dale, and H.A. Schroeder (1989). The role of ester groups in resistance of plant cell wall polysaccharides to enzymatic hydrolysis. Applied Biochemistry and Biotechnology, 20/21, pp. 45–61.
12. Zhang, M., C. Eddy, K. Deanda, M. Finkelstein, and S. Picataggio (1995). Metabolic engineering of a pentose metabolism pathway in ethanogenic Zymomonas mobilis, Science, 267, pp. 240–243.
13. Dekker, R.F.H., and A.F.A. Wallis (1983). Enzymic saccharification of sugar cane bagasse pretreated by autohydrolysis-steam explosion. Biotechnology and Bioengineering, 25, pp. 3027–3048.
14. Grohmann, K. (1993). Simultaneous saccharification and fermentation of cellulosic substrates to ethanol, in Bioconversion of Forest and Agricultural Plant Residues, ed. by J.N. Saddler, Biotechnology in Agriculture Series No. 9, Wallingford, Oxon, UK, pp. 183–209.
15. Saddler, J.N., C. Hogan, M.K.-H. Chan, and G. Louis-Seize (1982). Ethanol fermentation of enzymatically hydrolyzed pretreated wood fractions using Trichoderma cellulases, *Zymomonas mobilis*, and *Saccharomyces cerevisiae*. Canadian Journal of Microbiology, 28, pp. 1311–1319.

TRANSGENIC FUNGAL-BASED CONVERSION OF WASTE STARCH TO INDUSTRIAL ENZYMES

J. Gao, B.S. Hooker, R.S. Skeen, and D.B. Anderson

Bioprocessing Group, Pacific Northwest National Laboratory, Battelle Boulevard, P.O. Box 999, MS K2-10, Richland, Washington, 99352, USA, Phone: (509) 375-6639, Fax: (509) 372-4660, E-mail: jw.gao@pnl.gov.

The production of a bacterial enzyme, beta-glucuronidase (GUS), was investigated using a genetically modified starch-degrading *Saccharomyces* strain in suspension cultures of various waste starch sources. A shuttle plasmid expression vector was constructed using a yeast episomal plasmid. The glucuronidase (*gus*) gene was placed under the control of an inducible promoter, GAL1, and terminated by a transcription terminator, T_{cyc1}. Different sources of starches including corn and waste potato starch were used for yeast biomass accumulation and glucuronidase expression studies. In addition, a thermostable bacterial cellulase, *Acidothermus cellulolyticus* E1 endoglucanase was cloned into the plasmid expression vector and expressed in the starch-degrading *Saccharomyces* strain.

1. INTRODUCTION

Vast quantities of biomass are produced in the U.S. by the agricultural and food processing industries, with only a small portion directly utilized for food, fiber, chemicals, or energy. A large fraction of the remaining biomass provides only marginal economic benefit and much of it must be managed as waste, incurring expenses to the producer or processor. For example, U.S. fruit and vegetable processors currently produce approximately 7.4 million dry metric tons of low-value byproduct (much of it sold at or near cost as livestock feed), and treat more than 300 million cubic meters of high-BOD waste waters each year. Both of these byproduct streams are usually rich in starch, which could serve as an inexpensive yet robust feedstock for biotechnological processes. Such wastes, rich in carbohydrates and other nutritional factors, can be used as a growth medium by bacteria and fungi for useful industrial enzyme production.

Natural yeast strains have been identified that can use starch as a primary growth substrate via complete or partial enzymatic hydrolysis.[1,2] These yeast strains include *Saccharomycopsis fibuligera*, *Schwanniomyces castellii*, and *Saccharomyces diastaticus*.[1,2,3] A fusion yeast cell strain of *Saccharomyces diastaticus* and *Saccharomyces cerevisiae* has also been demonstrated to degrade 60% of starch present in culture media within two days.[4] In addition, other natural *Saccharomyces* species can ferment starch and dextrin to ethanol.[5] For the past two decades, the starch-biotransforming yeast technology was mostly used for producing single-cell protein as a cattle feed supplement.[2,3] Others use this technology to improve ethanol production from starch and higher sugars.[6,7] In the present study, a starch-degrading strain was genetically modified for the expression of a bacterial enzyme, beta-glucuronidase, and a thermostable enzyme, endoglucanase, to demonstrate that this system may serve as means to produce inexpensive transgenic protein products.

2. MATERIALS AND METHODS

2.1. Bacterial strains, yeast strains, cloning techniques, and transformation

Escherichia coli strain Top 10' (Invitrogen Inc., Carlsbad, Calif.) was used as a host for routine cloning experiments. The yeast strain used in the study was a *Saccharomyces* strain provided by Dr. James R. Mattoon of University of Colorado, Colorado Springs, Colorado. Plasmid DNA purification, plasmid construction and transformation, and polymerase chain reaction (PCR) were conducted under standard molecular cloning techniques.[8] An EasySelect Expression Kit (Invitrogen) was used for preparing competent yeast cells, which were subsequently used for expression vector transformation. Upon transformation, the transformed yeast cells were plated onto a selective YPD agar plate containing 1.0% glucose, 0.5% yeast extract, 1.0% peptone, and 200 mg/l antibiotic Zeocin. After a three-day incubation period at 30°C, transformed yeast colonies were obtained on the selective agar plate.

2.2. Culture medium

The culture medium used for yeast biomass accumulation contains waste potato starch (Lamb-Weston, Richland, Washington) or corn starch (Sigma, St. Louis, Missouri) as the primary carbon source, supplemented with 1.0% peptone and 0.5% yeast extract as the nitrogen source. The expression medium contains 2% galactose, 1.0% peptone, and 0.5% yeast extract. The medium was adjusted to pH 6.0 by 6 N HCl and autoclaved at 250°F for 20 minutes. All suspension cultures were grown aerobically at 30°C in an orbital shaker shaking at 200 rpm.

2.3. Protein extraction

Intracellular protein of transformed yeast biomass was extracted using the glass-bead disintegrating method. Briefly in this method, one volume of acid-washed 500 µm glass beads (Sigma) was added to one volume cell sample in an extraction buffer containing 50 mM pH 7.0 sodium phosphate, 1 mM EDTA, 1 mM phenylmethylsulfonyl fluoride (PMSF), 10 mM beta-mercaptoethanol, and 0.1% triton X-100. Cells were disrupted by vortexing vigorously for 30 seconds for five times. Samples were kept on ice during vortexing intervals. After cell disruption, the sample was centrifuged at 4°C at 20,000×g for 5 minutes. The supernatant was saved for both protein and enzyme activity assays. The extracted protein samples were assayed for protein concentration using the Bio-Rad protein assay (Bio-Rad Laboratories, Hercules, California).

2.4. Glucuronidase activity analysis

Glucuronidase activity was assayed using an enzymatic reaction in which a substrate 4-methylumbelliferyl-beta-D-glucuronide (MUG) can be hydrolyzed by glucuronidase to a fluorescent compound, 4-methylumbelliferone (MU). One unit of glucuronidase activity is defined as the amount of glucuronidase that produces one pmole MU from MUG per minute at 37°C. The fluorescence of produced MU was assayed in a DyNA QUANT 200 fluorometer (Pharmacia Biotech, Piscataway, New Jersey). The specific activity of glucuronidase is calculated as the units of glucuronidase per milligram of total protein in the sample.

2.5. Endoglucanase activity analysis

E1 endoglucanase activity was assayed using an enzymatic reaction in which the substrate 4-methylumbelliferyl-beta-D-cellobioside (MUC) can be hydrolyzed by E1 endoglucanase to a fluorescent compound, 4-methylumbelliferone (MU). One unit of glucuronidase activity is defined as the amount of E1 endoglucanase that produces one pmole MU from MUC per minute at 55°C. The specific activity of E1 endoglucanase is calculated as the units of E1 endoglucanase per milligram of total protein in the sample.

3. RESULTS AND DISCUSSION

3.1. Expression vector construction and transformation

Plasmid vectors were constructed to effectively transform the starch-degrading *Saccharomyces* strain and select the transformants after transformation. A plasmid vector pYES2 (5,857 bp) was obtained from Invitrogen. The plasmid was modified to form plasmid pGA2026 (4,961 bp) as shown in Figure 1A by replacing the ampicillin resistance and URA3 genes in the pYES2 vector at Nhe I and BspH I restriction enzyme sites with the antibiotic Zeocin resistance gene (1,186 bp) of the plasmid vector pGAPZα-A, also obtained from Invitrogen. The plasmid vector pGA2026 enables the selection of transformed *Saccharomyces* strains without using uracil-deficient selection medium. The plasmid pGA2026 contains an expression cassette for foreign gene expression under the control of an inducible galactokinase promoter (GAL1), a T7 RNA promoter and a T_{cyc1} terminator. In addition, the plasmid also contains a 2 µm DNA fragment for plasmid replication in *Saccharomyces* strains, a ColE1 origin for plasmid replication during gene manipulation in *E. coli* strains, a f1 phage origin, and the antibiotic Zeocin resistance gene for both yeast and *E. coli* selection during gene manipulation after transformation. A bacterial *gus* gene (1,923 bp)[9] was subsequently cloned into pGA2026 at Hind III and Sac I restriction enzyme sites to form pGA2028 (6,884 bp) as shown in Figure 1B.

The starch-degrading yeast strain *Saccharomyces* was also used to express a thermostable cellulolytic enzyme gene, endoglucanase from *Acidothermus cellulolyticus*.[10] The beta-1,4-

endoglucanase (E1) precursor gene was obtained from Steven R. Thomas of the National Renewable Energy Laboratory in Golden, Colorado. The mature endoglucanase gene (1,562 bp) was cloned out by PCR from the beta-1,4-endoglucanase precursor gene and adapted with a Hind III restriction enzyme site and an initiation codon ATG at the 5' end and a Sac I restriction enzyme site at the 3' end using the following primers: 5' end forward primer-AGG CCT AAG CTT ATG GCG GGC GGC GGC TAT TGG CAC ACG; 3' end reverse primer-GTC GAC GAG CTC TTA ACT TGC TGC GCA GGC GAC TGT CGG. The PCRed mature E1 gene was digested with Hind III and Sac I restriction enzymes and cloned into the plasmid vector pGA2026 at Hind III and Sac I restriction enzyme sites to form pGA2035 (6,518 bp) as shown in Figure 1C.

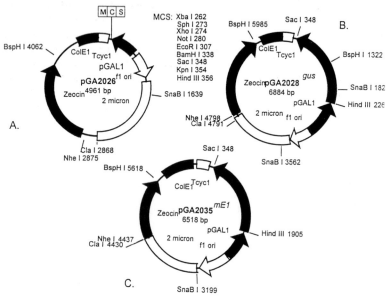

Figure 1. Vectors for starch-degrading *Saccharomyces* transformation and expression of beta-glucuronidase and E1 endoglucanase. A: plasmid vector pGA2026; B: plasmid vector pGA2028; C: plasmid vector pGA2035. MCS: multiple cloning sites.

3.2. Beta-glucuronidase (GUS) expression

In order to test glucuronidase gene expression after pGA2028 vector transformation, ten transformed yeast colonies were streak-purified on fresh selective YPD agar plates and single colonies were used in batch cultures for glucuronidase expression. The yeast colonies were first grown aerobically in 2-ml YPD medium for 16 hours. The propagated biomass was transferred into 2-ml production medium containing promoter activity inducer, galactose for GUS expression. After a 5-hour inducing period, yeast biomass was harvested and intracellular protein was extracted using the glass-bead disintegrating method. Table 1 illustrates the results of GUS specific activities of ten different transformed clones. These data demonstrate that glucuronidase can be highly expressed in the transgenic yeast host under the control of GAL1 promoter. The highest specific activity obtained in the culture was 10,057 units per mg of extracted intracellular protein. There was no extracellular GUS activity found since there was no secretion transit peptide sequence to the *gus* gene. Current studies are focused on glucuronidase expression under a constitutive promoter.

3.3. Batch expression using potato and corn starch

Starch medium was used to cultivate transformed yeast clones for GUS expression. Corn and waste potato starches were used. The culture medium is a sugar-free medium and contains 1.0% corn or potato starch supplemented with 1.0% peptone and 0.5% yeast extract. After a two-day growth period in the starch medium, the biomass was collected and GUS was induced in a production medium primarily containing 2% galactose as inducer. Cells were harvested periodically and intracellular protein was extracted in the extraction buffer. Figure 2 shows the results of GUS expression during different inducing periods. After a 4-hour induction period, the glucuronidase activity reached 13,396 units/mg of extracted intracellular protein. The GUS activities leveled off thereafter, indicating stable expression of GUS in the cultures.

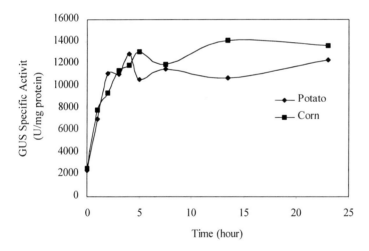

Figure 2. Time course of beta-glucuronidase (GUS) specific activity in transgenic starch degrading *Saccharomyces* under galactose induction condition after biomass propagation in a 250 ml Erlenmeyer flask containing 50 ml of potato or corn starch medium. Each data point represents the average of two duplicate cultures.

3.4. Endoglucanase expression

The starch-degrading yeast strain *Saccharomyces* was used for pGA2035 vector transformation and E1 endoglucanase expression. Upon transformation, the transformed yeast cells were plated onto selective YPD agar media containing glucose, yeast extract, peptone, and antibiotic Zeocin. After a three-day incubation period at 30°C, transformed yeast colonies were obtained on the selective culture media.

Twelve transformed yeast colonies were streak-purified on fresh selective YPD agar plates and single colonies were used in batch cultures for E1 endoglucanase expression. The yeast colonies were first grown aerobically in 2-ml YPD medium for 16 hours. The propagated biomass was transferred into 2-ml production medium containing the promoter activity inducer, galactose for E1 endoglucanase expression. After a 5-hour inducing period, yeast biomass was harvested and intracellular protein was extracted in the extraction buffer using the glass-bead disintegrating method. Table 2 shows the results of E1 endoglucanase specific activities of twelve different transformed clones. The highest specific activity obtained in the culture was 1,724 units per mg of extracted intracellular protein. However, there was no secreted E1 endoglucanase detected in the culture medium even though a native leader sequence was used in the E1 expression construct. This is probably due to the strong binding activity of the E1 cellulose-binding domain to the cell wall cellulose, inhibiting E1 secretion. Preliminary results showed that the removal of the cellulose-binding domain greatly enhanced E1 enzyme secretion.

Table 1
Transgenic glucuronidase activity in transformed starch degrading *Saccharomyces*

Clone No.	Glucuronidase Specific Activity (unit/mg)
Control[a]	16
1	3.469
2	10.057
3	4.061
4	7.013
5	5.309
6	7.786
7	6.035
8	4.439
9	4.000
10	8.346

[a]No genetic transformation

Table 2
Transgenic thermostable E1 endoglucanase activity in transformed starch degrading *Saccharomyces*

Clone No.	E1 endoglucanase Specific Activity (unit/mg)
Control[a]	7
1	1.724
2	961
3	1.241
4	1.383
5	1.294
6	1.273
7	1.257
8	1.111
9	1.501
10	1.258
11	508
12	1.476

[a]No genetic transformation

4. ACKNOWLEDGMENTS

The authors thank Dr. James R. Mattoon of University of Colorado, Colorado Springs, Colorado, for providing the *Saccharomyces* strain. The authors also thank Dr. Steven R. Thomas of National Renewable Energy Laboratory, Golden, Colorado for providing the gene of a thermostable cellulolytic enzyme, endoglucanase from *Acidothermus cellulolyticus*. This work was supported by the U.S. Department of Energy, Office of Science and Technology, under Contract DE-AC06-76RLO 1830. Pacific Northwest National Laboratory is operated by Battelle for the United States Department of Energy under contract DE-AC06-76RLO 1830.

5. REFERENCES

1. Sills, A.M. and G.G. Stewart. (1982) J. Inst. Brew., 88, p. 313.
2. Hongpattarakere, T., A.H. Kittikun (1995). J. Microbiol. Biotechnol., 11, p. 607.
3. Lemmel, S.A., R.C. Heimsch, and R.A. Korus (1980). Appl. Environ. Microbiol., 39, p. 387.
4. Kim, K., C.S. Park, and J.R. Mattoon (1988). Appl. Environ. Microbiol., 54, p. 966.
5. Laluce, C., M.C. Bertolini, J.R. Ernandes, A.V. Martini, A. Martini (1988). Appl. Environ. Microbiol., 54, p. 2447.
6. Pirselova, K., D. Smogrovicova, S. Balaz (1993). World J. Microbiol. Biotechnol., 9, p. 338.
7. Ryu, Y.W., S.H. Ko, S.Y. Byun, C. Kim (1994). Biotechnol. Lett., 16, p. 107.
8. Sambrook, J., E.F. Fritsch, and T. Maniatis (eds) (1989). Molecular Cloning: A Laboratory Manual, 2nd ed, Cold Spring Harbor Laboratory Press, Cold Spring Harbor, New York.
9. Jefferson, R.A., T.A. Kavanagh, and M. W. Bevan (1987). EMBO J., 6, p. 3901.
10. Laymon, R.A., M.E. Himmel, and S.R. Thomas (1995). *GeneBank* Database, NID 988299.

SIMULATION OF LOW-POWER AGITATION SYSTEMS FOR LARGE-SCALE BIOMASS CONVERSION REACTORS

Sonja P. Svihla, Charles K. Svihla, and Thomas R. Hanley

Chemical Engineering Department, University of Louisville, Room 427 Lutz Hall (Academic Building), Louisville, Kentucky 40292, U.S.A.

This paper presents the results of CFD simulations (FLUENT Version 4.5) conducted to identify the optimal design for an agitation system for a large-scale solids suspension problem typical of biomass conversion processes. The effect of impeller location, impeller speed, spacing between impellers, and the use of a draft tube upon the ability of the system to provide off-bottom suspension of particles within the power limits set by the process economics are investigated and critically assessed.

1. INTRODUCTION

The economics of many biomass conversion processes dictate operation on a large scale with relatively low permissible levels of power input by mechanical agitation. In many cases there is a lack of reliable scale-up data owing to the large gap between the size of pilot-scale and process-scale installations. One tool which can help ensure that the critical function of the agitation system is fulfilled for the full-scale system is computational fluid dynamics (CFD). In this approach, the flow is simulated for a proposed configuration and the results (velocities, flow patterns, etc.) are examined to determine if the design is acceptable or if it has shortcomings which might be addressed by changing one or more aspects of the design. In this research, the computational fluid dynamics software FLUENT (Version 4.5) was used to perform the simulations, while the FLUENT preprocessor MixSim (Version 1.5) was used to set up the geometry and details of the problem.

1.1. Modeling the impeller and the tank geometry

The primary difficulty of simulating flow in mixing tanks has been the complex geometry and associated boundary conditions that must be applied at the rotating impeller and stationary baffles. When no baffles are present, a simple approach such as the rotating frame method can be used. This method uses a reference frame that rotates along with the impeller, thus immobilizing the impeller which results in simple boundary conditions (zero velocities at the impeller blade and the impeller angular velocity at the wall) (FLUENT, 1995).

If wall baffles are present, the rotating frame approach cannot be used and a more complicated method, such as the sliding mesh approach, is required. In this technique, the impeller geometry is modeled rigorously using two different mesh regions to model the rotation. A rotating mesh is used to describe the impeller and its immediate surroundings, while a stationary frame is used for the baffles and tank walls. The two meshes slide past each other at a slip plane. This situation allows simple boundary conditions (zero velocities) to be applied for the impeller and baffles. Since the motion of the impeller(s) past the baffles is periodic, sliding mesh simulations are inherently time-dependent (FLUENT, 1995).

Another approach used to model a mixing tank with impeller and baffles is the velocity data method where the complicated geometry of the impeller is replaced by a disk plane (for an axial flow impeller) at which the velocities and turbulence quantities are fixed based on experimental data which were obtained previously for a particular impeller and are available in the MixSim library. The data are scaleable to the size and speed of the impeller used in the vessel. This is a time-averaged steady-state approach (FLUENT, 1995).

The results presented here were obtained using the velocity data approach. The simulations were performed for a 90° sector of the tank using the symmetry imposed by the four longitudinal baffles. The computational domain was divided into cells numbering 49 x 32 x 74 in the theta (i), radial (j), and axial (k) directions, respectively.

2. SIMULATIONS OF THE SMALL-SCALE AND THE LARGE-SCALE VESSELS

The general design for the small-scale tank are a diameter, T, of 0.738 m, a liquid volume, V, of 100 gallons (0.364 m^3); a liquid height, H, of 0.885 m (H/T = 1.2); four vertical baffles set at 90° intervals; two high efficiency hydrofoil impellers (Lightnin A310) with an impeller diameter to tank diameter ratio (D/T) of 0.4; and an ASME dish bottom. Simulations were performed for this configuration for impeller speeds of 300 rpm (Case01) and 144 rpm (Case02). The speed used in the Case02 simulation was selected based on the maximum power input per unit volume set for the large-scale simulation.

The large-scale tank is designed as follows: T = 7.378 m, V = 100,000 gallons (363.713 m^3), H = 8.854 m, four baffles, two Lightnin A310 impellers, and an ASME dish bottom. The axial locations of the impellers, the use of a draft tube, and the spacing between the two impellers were varied in the simulations. The impeller speed of 6.69 rpm used in the simulations was based on a maximum power input of 0.25 horsepower per 1000 gallons. The power input for each impeller, P (W) was calculated as (Oldshue, 1983):

$$P = P_o \rho N^3 D^5 \tag{1}$$

where P_o is a power number for the impeller, ρ is the fluid density (kg/m^3), N is the impeller speed (rps), and D is the impeller diameter (m). The power number for a Lightnin A310 impeller is 0.3 (FLUENT, 1996).

The configuration variables for all seven cases are summarized in Table 1.

Table 1
Design Specifications of the small-scale tank (Case01 and Case02) and of the large-scale tank (Case03-Case07)

	Speed (rpm)	Impellers Diameter (m)		Axial Locations (m)		Draft Tube Bottom Height (m)
		Imp. #1	Imp. #2	Lower	Upper	
Case01	300	0.2951	0.2951	0.3	0.7	-
Case02	144	0.2951	0.2951	0.3	0.7	-
Case03	6.69	2.9513	2.9513	2.95	7.08	-
Case04	6.69	2.3253	2.3253	2.95	7.08	2.0
Case05	6.69	2.9513	2.9513	1.5	7.08	-
Case06	6.69	2.9513	2.9513	1.5	6.0	-
Case07	6.69	2.3253	2.3253	1.5	6.0	1.3

3. DISCUSSION AND CONCLUSIONS

The ability of the agitation system to provide an off-bottom suspension of solids (i.e., to prevent settling) is of particular interest in the design of a reactor for biomass conversion. Therefore, the flow field near the center bottom of the tank is an important region to observe. A typical flow pattern created by two downward-pumping A310 impellers consists of a main circulation loop throughout the tank (downward flow at the center and upward flow along the sides) with a small nearly stagnant recirculation region near the center bottom of the tank. In every simulation performed here, a recirculation region was found to exist below the lower impeller near the center bottom of the tank. Various strategies were employed to either minimize this region or to create a more uniform axial velocities at the bottom center of the vessel. Some of the velocity values from the simulations are tabulated in Table 2. In Table 2, the plane I = 25 is a vertical plane located between two baffles, while the plane K = 73 is a horizontal slice adjacent to the bottom of the tank. Figures 1 and 2 show examples of the velocity fields at these planes for the Case03 simulation. Most of the observations made in this section derive from examination of similar views of the flow pattern and velocity vectors in the tank which unfortunately cannot all be presented in this brief paper.

A comparison of the results for the two small-scale simulations (Case01 at 300 rpm and Case02 at 144 rpm) reveals that an increase in the impeller speed does not by itself substantially affect the size of the recirculation region below the lower impeller. For both cases, calculation of massless particle trajectories demonstrated that particles originally located near the bottom center of the tank were found to circulate around the tank bottom while other particles originally located outward of this recirculation region traveled to the upper part of the tank following the main circulation loop. Operation at a higher impeller speed could increase the total circulation of solid particles throughout the tank but could not eliminate the recirculating region.

Table 2
Magnitudes of velocity vectors and axial velocities at the plane I = 25 and axial velocities at the bottom of the tank (K = 73)

		At the Plane I = 25		At the Plane K = 73
		Velocity Vector Magnitudes (m/s)	Axial Velocities (m/s)	Axial Velocities (m/s)
Case01	min.	1.704E-02	-9.409E-01	-3.555E-01
	max.	1.416E-00	5.839E-02	5.839E-02
Case02	min.	1.386E-02	-4.501E-01	-1.689E-01
	max.	6.798E-01	6.540E-01	3.181E-02
Case03	min.	1.070E-05	-2.028E-01	-8.613E-02
	max.	3.173E-01	3.063E-01	1.728E-02
Case04	min.	9.677E-04	-1.230E-01	-6.461E-02
	max.	2.481E-01	2.394E-01	1.357E-02
Case05	min.	1.641E-03	-2.608E-01	-1.334E-01
	max.	3.173E-01	3.063E-01	2.285E-02
Case06	min.	2.173E-03	-2.660E-01	-1.333E-01
	max.	3.173E-01	3.063E-01	2.285E-02
Case07	min.	8.472E-04	-1.654E-01	-8.460E-02
	max.	2.481E-01	2.394E-01	2.365E-02

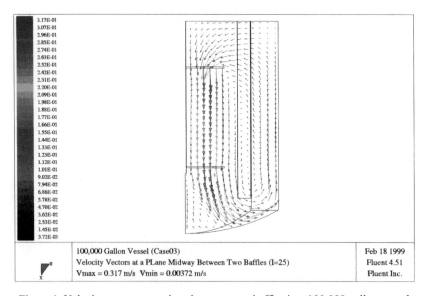

Figure 1. Velocity vectors at a plane between two baffles in a 100,000 gallon vessel.

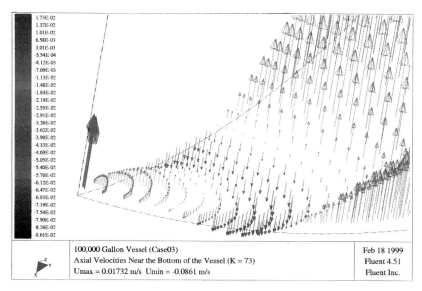

Figure 2. Axial velocities near the bottom of the tank in a 100,000 gallon vessel.

Because increasing the impeller speed cannot eliminate the recirculation region below the lower impeller, some other stratagem must be employed to limit its extent. A comparison of the simulation results for Case03 and Case04 illustrates the effect of adding a central draft tube (Case04), while holding the axial positions and impeller speed constant. When Case04 is compared to Case03, a decrease of about 21% of the maximum velocity and an insignificant change in the recirculating region near the center bottom of the tank were observed. This was not unexpected since the same impeller speed was used for Case04 with a smaller impeller diameter. This result does not imply that a draft tube would not be useful in the agitated bioreactor since other changes in the configuration can still be made. A draft tube is used to create a more uniform and stronger downflow to the center bottom of the tank so that solid particles can be swept outward and lifted up to follow the main circulation loop.

The effect of moving the lower impeller closer to the tank bottom while holding all other configuration variables constant can be elucidated by comparing the results for the Case05 and Case03 simulations. In Case05, the ratio of the off-bottom spacing of the lower impeller, Z_1, to the liquid height, H, is 0.17 while in Case03, the value of that ratio is 0.33. Moving the lower impeller to the Case05 position increased the overall minimum velocity magnitudes by 152%. An increase of 54.9% and 32.2% in the upflow (negative velocities) and the downflow (positive velocities) of the axial velocities, respectively, at the tank bottom are also observed which represents an improvement in the characteristic flow pattern in the recirculating region.

The effect of a decrease in the spacing between the two impellers can be seen by a comparison of the results for the Case05 and Case06 simulations. While an increase of 32.4% in the overall minimum velocity magnitude is found when the spacing between the two impeller is decreased by about 20% (in Case06, the upper impeller is moved downward while the position of the lower impeller is fixed), no significant change was observed in the axial velocities at the tank bottom. This result was as expected since the fluid flow pattern near the center bottom of the tank is mostly affected by the lower impeller.

A comparison of the results for Case07 and Case06 illustrates that the addition of a draft tube even with a reduction in the impeller diameter results in a slight increase of 3.5% in the maximum downward velocity near the tank center bottom. Thus, when the lower impeller is located close enough to the bottom, a draft tube can improve the flow pattern near the tank bottom by providing stronger downward flow.

Several conclusions can be drawn from the simulation results discussed herein. Lowering the lower impeller to an off-bottom spacing to liquid height ratio of 0.17 from an initial value of 0.33 provided a better velocity profile (increasing the axial velocity magnitude by 32.2 to 54.9%) to suspend solids particles off the tank bottom. When the axial location of the lower impeller is optimized, a draft tube can further improve the flow pattern near the center bottom of the tank and can ensure a good top-to bottom mixing in a tank with a high ratio of liquid height to tank diameter. For the simulations presented in this paper, the recirculating region near the center bottom of the tank could not be eliminated completely, although strategies to minimize its extent could be identified.

4. FUTURE WORK

The simulations presented in this paper were all performed using the velocity data approach for modeling the A310 impellers. In some cases (i.e., for a viscous liquid phase or for an impeller position close to a tank wall), the conditions for which the experimental data used to generate the fixed impeller velocity boundary conditions were obtained may no longer match those of the simulation. In that case, the impeller geometry must be modeled explicitly using the sliding mesh or multiple reference frame approach. This requires that the curved surface of the A310 impeller be modeled in detail since the geometry information for impellers of this type are not included in the MixSim library. To address this concern, the authors are currently working to develop computational models which include explicit representation of the geometry of impellers with curved blade surfaces.

REFERENCES

Fluent Inc. (1995). Fluent User's Guide, Fluent Inc., Lebanon, New Hampshire.
Fluent Inc. (1996). MixSim User's Guide, Fluent Inc., Lebanon, New Hampshire.
Oldshue, J.Y. (1983). Fluid Mixing Technology, McGraw-Hill, New York.

PRODUCTION OF ETHANOL, PROTEIN CONCENTRATE AND TECHNICAL FIBERS FROM CLOVER/GRASS

Stefan Grass, Graeme Hansen, Marc Sieber and Peter H. Müller

2B AG, Biomass and Bioenergy, Neugutstr. 66, 8600 Dübendorf, Switzerland
Phone: +41-1-820 19 62, Fax: +41-1-820 19 50

2B AG has developed a technology to produce ethanol, protein concentrate and technical fibers from cheap and plentiful fibrous material (e.g., grass). The technology comprises steam pre-treatment of the raw material, simultaneous saccharification and fermentation, production of cellulase enzymes and end product recovery. It is economically interesting and fully environmentally sustainable. In spring 1998, *2B* AG has taken into operation a complete demonstration and production plant in the eastern part of Switzerland. This plant uses equipment on an industrial scale.

The business of *2B* AG is to license the technology (patent pending) and to sell certain key components and engineering to customers. The *2B* AG technology offers the following benefits: (1) It is economically interesting for farmers, plant operators, and investors. (2) It supports rural structures and development. (3) The technology is clean and all its products are environmentally beneficial.

1. VISION

2B AG's vision is to economically produce ethanol and valuable by products from cheap and plentiful biomass with a technology that is fully environmentally sustainable.
The *2B* AG technology meets several future trends:
- To produce renewable energy from locally available and low cost raw material.
- To help reducing agricultural overproduction by offering a new economic use for grassland or land used currently for crop production. This results in new jobs and rural development.
- To recycle waste from the food and beverage industries, and wastes such as spent grains, reed, bagasse, producing valuable by-products instead of disposing of such waste at considerable cost.

2B AG is a young biotech company that has developed this process during the last four years. The business of *2B AG* is to license this technology and to sell certain key components and engineering to customers. *2B AG* also conducts feasibility studies for potential plant sites and new raw materials.

2. TECHNOLOGY AND PRODUCTS

The *2B AG* technology applies steam treatment to the raw material and subsequent enzymatic saccharification of the cellulosic fibers. The monomeric hexose-sugars are simultaneously converted to ethanol using yeast in a SSF process (simultaneous saccharification and fermentation). This process allows a fractionation of the raw material giving ethanol, protein concentrate and hemicellulose fibers as valuable products.

The enzymes required for the saccharification step are produced by *2B AG* using grass and other biomass feedstock. The enzymes so produced are specially suited for saccharification of a variety of oligomeric and polymeric sugars contained in grass.

Engineering solutions on the industrial level include an innovative, continuous, energy efficient pretreatment unit. Figure 1 exhibits the key process steps. The patent on the *2B AG* technology has been applied internationally.

One tonne of clover/grass dry matter yields the following products: 180-220 L ethanol, 150-250 kg protein concentrate and 200-250 kg fibers (all values on a dry basis). All plant components are converted to products and no solid waste stream leaves the process. An independent study shows an excellent energy output to input ratio of 4 to 1.

Ethanol has markets in the chemical or pharmaceutical industry, as well as the transportation sector. Protein concentrate can replace soy meal for feeding pig and poultry. The fibers are well suited for use in the insulation market. All named products have been tested by official institutes and meet or exceed the required quality criteria of these markets.

3. ECONOMICS

The economics of the process primarily depend on the following:
- The capacity of the plant and year round utilization.
- The availability of suitable infrastructure (building, energy supply, logistics).
- The market value of the end products in the respective country.
- The cost of raw material.

The *2B AG* process allows a return on investment well over 10% (without government investment help or other subsidies). This figure implies a plant processing capacity of 5–10,000 t dry matter per year and market prices prevalent in the EU.

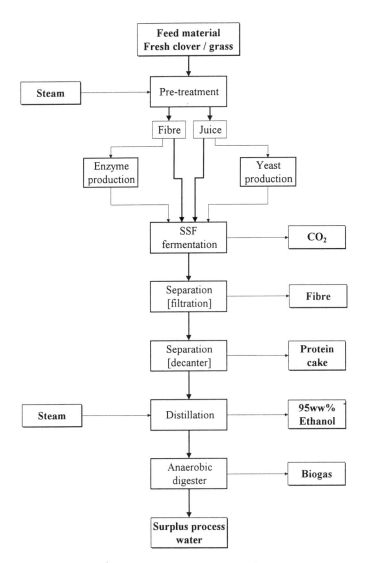

Figure 1. Process chart.

4. STATUS OF TECHNOLOGY DEVELOPMENT AND COMMERCIALIZATION

Over the last four years, *2B* AG has developed and demonstrated the process, as well as product yields and qualities. Customers for the products were successfully sought and raw material availability and cost were determined. In spring 1998, *2B* AG took into operation a demonstration and industrial pilot plant in the eastern part of Switzerland. This plant is capable of processing various qualities of grass and clover–grass (fresh, wilted or silaged) as well as other biomass feed stocks. It also incorporates an enzyme production unit for production of cellulases. All plant components are at full industrial scale and the processing capacity is 2 tons dry matter per day (10 tons of wet grass).

At the end of 1999 the first fully commercial industrial plant with an annual capacity of 5,000 tons dry matter will be set up in Switzerland.

BIBLIOGRAPHY

Bioworld (1998). grünes Erdöl, 4/98.
Die Grüne (1993). Ethanol und Protein aus Gras, June 4.
Llandfreund (1998). Wir brauchen 3,000t TS Gras, Dec. 8.
Neue Zürcher Zeitung (1997). Alkohol-Treibstoff aus Gras, June 18.
Stefan Grass (1998). Production of ethanol from grass. Proceedings, 1998 Biomass Congress Würzburg, pp. 631–633.

BARK-RICH RESIDUAL BIOMASS AS FEEDSTOCK FOR ETHANOL AND CO-PRODUCTS

J.M. Garro,[a] P. Jollez,[a] D. Cameron,[b] R. Benson,[b] W. Cruickshank,[c] G.B.B. Lê,[d] Q. Nguyen[e] and E. Chornet[e,f]

[a]Kemestrie Inc.
[b]Tembec Inc.
[c]CANMET Energy Technology Centre (CETC), Natural Ressources Canada
[d]Gouvernement du Québec, Secteur de l'énergie, Ministère des Ressources naturelles
[e]National Renewable Energy Laboratory, NREL
[f]Chemical Engineering Department, Université de Sherbrooke.

Bark-rich residues represent a readily available biomass adjacent to either sawmills, pulp and paper mills and, in general, forest operations. Such residues are currently being used for their energy value by direct combustion to generate electricity and steam. We have investigated the possible use of these residues as feedstock for ethanol, heat and power, and chemical co-products.

1. INTRODUCTION

Coniferous bark-rich residues obtained from mill operations represent low-cost feedstocks for producing ethanol by fermentation. They have a significant hexose content (from hemicelluloses and cellulose) and a substantial amount of extractives that require adaptation of micro-organisms to ferment the sugar-rich liquours. We have investigated the fractionation of the biomass in order to separate the extractives (and examine their potential for added value products), then solubilize and partially depolymerize the hemicelluloses, delignify (or not) the residual lignocellulosic material, and depolymerize the cellulose to hexose sugars. The liquors containing the hemicellulose- and cellulose-derived sugars will be subjected to fermentation.

The technological approach used is known as FIRST (Feedstock Impregnation and Rapid Steam Treatment). It is based on a diffusion model for feedstock impregnation which (a) leads to saturation of the raw material with the impregnating liquid thus removing any occluded oxygen (from air); (b) minimizes the amount of liquid wetting the feedstock and thus reduces energy (steam) costs, and (c) adds the necessary hydrolytic agents (acid or bases). The rapid steam treatment requires the addition of steam to bring the impregnated feedstock to the temperatures needed to induce destructuring, disaggregation and depolymerization.

1.1. Extractives removal

Extractives removal was carried out in two steps: initial impregnation with a solvent followed by the extraction itself. Since ethanol is available in situ from subsequent steps of the process, we restricted the solvents to water, ethanol and their mixtures (50% and 95%).

For each solvent combination, impregnation and no impregnation were compared, and catalysis in the presence or not of added inorganic acid (sulfuric) and base (caustic) was also studied.

The impregnation was carried out as follows. The liquid / solid wt ratio = 10 ; catalyst = acid or base at 3 wt% in liquid; temperature = ambient (25°C); pressure = 30 psig and time = 10 min. Extraction was performed at the reflux temperature of each solvent combination and the conditions used were liquid / solid wt ratio = 10 ; pressure = barometric and time = 1 hour. Different analyses were performed over the dry extract in order to identify polyphenols, fatty acids, sterols and essential oils.

1.2. Solubilization of hemicelluloses

The solubilization of hemicelluloses was achieved by a combination of aqueous impregnation and steam (explosion) treatment. The process begins with aqueous impregnation of the feedstock obtained after removal of the extractives. This impregnation saturates the fiber material with the water, which contains the catalyst, through complete capillary penetration. It also removes any occluded air in the capillaries resulting in the solubilization of residual extractives susceptible to hydrolysis as a function of severity, which is the key parameter that will determine solubilization yields. The process continues with a rapid steam treatment of the impregnated material at severities that induce the desired changes (hydrolysis) in the supramolecular structure and at the molecular level. The aqueous treatment (impregnation) parameters were the following: liquid / solid ratio = 10 ; catalyst (sulfuric acid or caustic) = 2 - 4 wt% in liquid ; temperature = ambient (25°C) ; pressure = 30 Psig and time = 10 min. The subsequent steam treatment was performed within a wide range of temperatures (160 – 190°C) and times (3 - 10 min).

1.3. Delignification of the residual lignocellulosic material

The treated biomass residual solid resulting from the aqueous / steam treatment can follow two different approaches :
(a) A second hydrolysis treatment to produce a final hydrolyzate of the lignocellulosic material followed by fermentation to convert hexoses in ethanol (the hydrolysis could be achieved by enzymatic methods or acid-impregnation followed by steam treatment);
(b) A base-catalyzed delignification to produce a black liquor principally constituted by lignin and a cellulose-rich residue (fiber + fines).

1.4. Cellulose upgrading (fiber and fines)

The black liquor can be subjected to further catalytic upgrading and caustic recovery while the cellulose-rich residue is to be separated into fiber (to be bleached and blended with non-bark fibers) and fines (for subsequent hydrolysis and fermentation to ethanol).

2. MATERIALS AND METHODS

Two different raw materials are reported in this paper. The first one consists of mixed softwood forest thinnings having as composition 70% white fir (*Pseudotsuga concolore*) and 30% ponderosa pine (*Pinus ponderosa*). The second is mixed bark (inner and outer) and trunks, 75% balsam fir (*Abies balsamea*) and 25% spruce (*Picea*). In both cases, to obtain a homogenous and stabilized feedstock the raw material was milled (Thomas-Willey - Model 4) and analyzed for moisture, density, granulometry, carbohydrates, lignin, ash and extractives contents. All analyses of carbohydrate content were determined by HPLC after a two-step acid hydrolysis: 72% H_2SO_4 (2 h at 30°C) followed by 4% H_2SO_4 (1 h at 121°C).

2.1. Extractives removal

A typical experiment (2 kg dry raw material) consists of two operations: impregnation and extraction. Both operations are carried out in the same vessel. Impregnation was conducted under nitrogen (30 Psig) for 10 minutes at ambient temperature. After impregnation, nitrogen was released and heating was started (steam in jacket) until the targeted temperature for the extraction, previously determined (lab scale of 75 gr), was obtained. No filtration or washing were made between impregnation and extraction. The liquid / solid ratio was fixed at 10 (wt/wt) and only when the entire treatment was completed were filtration and washings (of the recovered residual solids in the filter) done. The first washing was made with the same solvent combination and the second washing only with water. The filtered liquor was concentrated by vacuum evaporation followed by lyophilization to obtain the dry extract.

2.2. Solubilization of hemicelluloses

A typical experiment also consists of two process steps: impregnation and steam explosion. Impregnation of the residual solids recovered from the extractives separation (100 g dry basis) with the appropriate catalyst (acid or base 0.4 - 4.0 wt% in the liquid solvent) was carried out in a pressurized (30 Psig) vessel. After impregnation, the excess solvent is separated by pressing and the resulting cake (i.e., the impregnated material) is disaggregated (by hand) and subjected to a steam treatment. The latter is carried out in a jacketed vessel previously heated to the desired temperature. Live, saturated or slightly superheated, steam is admitted to the vessel and to the jacket at the conditions set for the experiment. The temperature is monitored by two thermocouples located within the sample once placed in the reactor vessel. After the desired reaction time has elapsed the blowdown valve is opened and the material is expelled out of the vessel and thus quenched. A receiver vessel (ice-salt cooled) captures the slurry obtained and cools it rapidly to avoid degradation. The slurry is filtered. The filtrate (up to 12 wt% hemicellulose-derived sugars content depending on the procedures used) is ready for fermentation after neutralization. The recovered solids, i.e., the "lignocellulose," are then ready for subsequent fractionation and processing. For analytical purposes, the slurry is filtered (Büchner) and the solid cake ("lignocellulose") is thoroughly washed with water. Liquors are mixed for analysis and concentrated to a 10 - 20 wt% dissolved solids.

3. RESULTS AND DISCUSION

3.1. Softwood forest thinnings (70% white fir and 30% ponderosa pine)

3.1.1. Extractives removal

The criteria used to determine the optimum conditions for extraction were (a) the yield of extractives and (b) the content of C6 carbohydrate in the residual solids. According to these criteria the optimum treatments were those which had a high rate of extractives (removal of more than 90% of extractives present in the initial raw material) and a high recovery of C6 carbohydrate in the solid residue (recovery of > 85% of C6 the carbohydrates initially present in the raw material). A set of experiments, performed with the milled raw material at laboratory scale (75 g) and reported in Table 1, determined the optimum conditions to be used for the extractives removal of a larger (2 kg) sample.

Table 1
Material balances following extraction with several solvent combinations

			Raw Material (%)	SOLVENT						
	ANALYSIS			H$_2$O			EtOH 50%			EtOH 95%
				1	2	3	4	5	6	7
				Ext.80 (%)	Ext.100 (%)	Imp. & Ext.80 (%)	Ext.80 (%)	Imp. H$^+$ & Ext. 80 (%)	Imp. & Ext. 80 (%)	Imp. OH$^-$ Ext. 80 (%)	Ext.80 (%)
S O L I D	Sugars	C5	8,4	8,5	8,6	8,5	10,5	8,8	9,5	6,6	9,6
		C6	53,7	48,8	45,1	47,9	53,4	50,8	51,5	39,7	50,8
		Total	62,1	57,2	53,7	56,5	63,9	59,6	61,0	46,4	60,4
	Lignin	Insoluble (%)	29,6	31,9	32,7	31,0	29,8	29,2	28,9	26,9	29,5
	Ash, %		0,8	0,5	0,8	0,6	0,8	0,8	0,7	0,5	0,7
	Other (by diff.)		2,5	7,5	11,0	8,0	5,5	4,9	3,3	17,4	5,1
L	Ext. material in liquid		4,5*	2,9**	1,8**	4,0**	4,1**	5,5**	6,0**	8,9**	4,3**

* EtOH extract; ** Weight difference between "initial dry weight" and "final dry weight" of the solid residue. Imp.—impregnation ; Ext. —extraction at the indicated temperatures (°C).

The optimum set of conditions was obtained by solvent impregnation followed by an extraction with ethanol 50% (treatment 5: 6.0% of extractives). According to the results, prior impregnation (i.e., pressure soaking) is important for the extraction process. Moreover, impregnation catalyzed by sodium hydroxide (treatment 6), results in sugar degradation and oxidation of tannins with a residual solid having reduced carbohydrate content (39.7%).

The optimum conditions were directly applied to the raw material (ground but not milled) at a larger scale (2 kg). This time the yield of extractives (3.5%) was determined by weighing the powder recovered after evaporation and lyophilization. This value is representative of the weight loss of volatiles during the concentration (evaporation + lyophilization) steps. The major component identified was lignin oligomers (85.0%), but proanthocyanidins (a natural antioxidant with a significant OTC market) accounted for 7.8% of the recovered extractives.

3.1.2. Aqueous / steam treatment

Acid catalysts level for the first aqueous/steam treatment was set at 0.4%, 0.5%, 1.0%, 2.0% and 4.0% (wt/wt of aqueous liquid) and temperature for the steam treatment ranged from 160°C to 190°C. Experiments at 160°C result in a "colloidal mixture" of difficult separation. Results at 190°C are presented in Tables 2 and 3.

Table 2
Yield of C5 and C6 in Hemis-rich liquor after aqueous / steam treatment (190°C, 3 min) at different acid loadings during impregnation

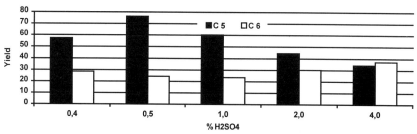

Table 3
Yield of C5 and C6 in lignocellulose after aqueous / steam treatment (190°C, 3 min) at different acid loadings during impregnation

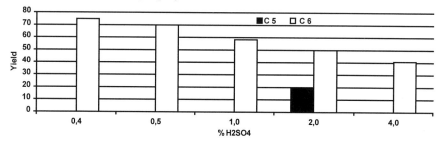

The results show that increasing the concentration of the catalyst in the aqueous treatment leads to a decrease in the recovery of C5 in the "hemicellulose-rich solution." The same tendency was observed in the potential yield of C6 in the lignocellulose, which is free of pentosans as a result of the steam treatment.

3.2. Bark (75% balsam and 25% spruce)

3.2.1. Extractives removal

As described for the previous raw material (3.1.1) a similar screening approach was used in order to determine optimum parameters for extraction. Once again optimal results were obtained under impregnation followed by extraction with ethanol 50%. At laboratory scale the yield for the material extracted calculated by weight difference was 11.9 wt% (o.d.b.). In the larger sample (2 kg) used for further treatments, the extractives recovered was 8.9% (weight of the powder obtained after evaporation and lyophilization). Lignin oligomers account for 69.0%; carbohydrates for 10.3% and proanthocyanidins for 7.5%. Fatty acids and sterols are detected only in minimal quantities.

3.2.2. Aqueous / steam treatment

Acid loading for aqueous impregnation was set at 0.5% and 1.0% and temperature for steam treatment at 180°C and 190°C. The results (Tables 4 and 5) indicate that we can obtain high solubilization of pentoses in the FIRST process using 1.0 wt% sulfuric acid in the impregnating solution and 190°C as steam treatment. Note that the sum of sugars found in the "hemicellulose-rich solution" and "wet lignocellulose" is higher than 90% at both temperatures. Little degradation of the sugars is thus observed under the conditions used.

Table 4
Yield of C5 and C6 in hemicellulose rich solution and wet lignocellulose after aqueous / steam treatment (1.0 wt% H_2SO_4 at 190°C, 3 min)

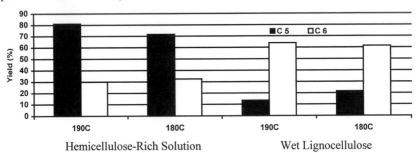

Table 5
Yield of C5 and C6 in hemicellulose rich solution and wet lignocellulose after aqueous / steam treatment (0.5 wt% H_2SO_4 at 190°C, 3 min)

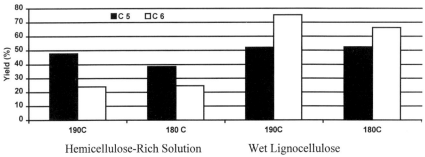

4. CONCLUSION

Favorable conditions for extractives removal are soaking and impregnating the feedstock with 50% aqueous ethanolic solution at L/S = 10 (wt/wt) ; T = 25°C ; P = 30 Psig ; t = 10 min.

These conditions gave higher yields when the raw material was appropriated ground (6 – 12 wt%) than when was not (3.5 – 9 wt%). The major product obtain is lignin (70 - 85%) as a result of the high polarity of the solvent used. Proanthocyanidins (a well-accepted natural antioxidant) accounts for 7.5% and could be successfully separated and purified.

Acid impregnation for hemicellulose solubilization gives better results than base impregnation. The latter led to degradation and reduced C6 contents. Specific conditions were determined for particular feedstocks in order to achieve the maximum solubilization of pentoses (more than 80% of the potential recovery) and hexoses (30%). The residual "lignocellulose" solids still contain 65-75% of the potential initial hexoses and 0-13% of the initial potential pentoses.

5. ACKNOWLEDGMENTS

The authors are indebted to Eva Capeck Menard, Jean Frechette, Henri Gauvin and Michel Trottier, of the GRTPC of the Department of Chemical Engineering of the Université de Sherbrooke, for their efforts and technical support in all stages of this work.

REFERENCES

1. Kazi, K.M.F., H. Gauvin, P. Jollez, and E. Chornet (1997). A diffusion mode for the impregnation of lignocellulosics, Tappi Journal, 80, pp. 209–219.
2. Montané, D., R.P. Overend, and E. Chornet (1998). Kinetic model for non-homogeneous complex systems with a time-dependent rate constant, Canadian Journal of Chemical Engineering, 76, pp. 58–68.
3. Jollez, P., E. Chornet, and R.P. Overend (1994). Steam-aqueous fractionation of sugar cane bagasse : an optimization study of process conditions at the pilot plant level, in Advances in Thermochemical Biomass Conversion, 2, pp. 1659–1669.
4. Heitz, M., E. Capek-Ménard, P.G. Koeberle, J. Gagné, E. Chornet, R.P. Overend, J.D. Taylor, and E. Yu (1991). Fractionation of populus tremuloides at the pilot plant scale: optimization of steam pretreatment conditions using the STAKE II technology, Bioresource Technology, 35, pp. 23–32.
5. Haslam, E. (1989). Plant Polyphenols. Vegetable Tannins Revisited, Cambridge University Press, Cambridge (UK).
6. Hemingway, R.W. (1989). Recent developments in the use of tannins as specialty chemicals, TAPPI Proceedings, Wood and Pulping Chemistry, pp. 377–386.

PERFORMANCE OF DIESEL-ETHANOL FUELED COMPRESSION IGNITION ENGINE

Guo Fenghua and Zhang Baozhao

Member of BERA
Agricultural Engineering Department, China National Rice Research Institute, Hangzhou, China

Performance characteristics were evaluated in a diesel engine operating on blends of ethanol and No. 2 diesel fuel or ethanol fumigation. Engine performance, energy consumption, thermal efficiency, Bosch smoke emission, combustion characteristics and noise level were compared with diesel fuel. A realistic means for use of ethanol in a unmodified diesel engine with satisfactory results can follow from this technique.

1. INTRODUCTION

Biofuel contains little sulfur and its combustion does not produce SO_2. They can be mixed with gasoline in almost any ratio and increase the oxygen content and boost the octane rating (OR = 98 ~ 108). Ethanol vehicles with spark ignition (S-I) engines were developed decades ago.[3] However, an excessively low cetane rating (CR), higher self-ignition temperature (420°C) and latent heat of vaporization (925 kJ/kg) prevented it from expanding satisfactorily in a compression ignition (C-I) engine as well as in the S-I engine.[1,2,5,7] Both stabilized ethanol blended fuel and ethanol fumigation in the intake stream were tested in a unmodified C-I engine in order to maximize use of ethanol while maintaining easy reversibility back to diesel operation.

2. EXPERIMENTS

Tests were run on a naturally aspirated, water cooled, 4-stroke, single cylinder C-I engine (Changzhou Diesel Factory, China Model S-195). Rated power was 8.82 kW/2000 rpm. Compression ratio was 20:1. Displacement of cylinder was 0.815 L. Fuel injection timing θ at 17^0 ~ 19^0CA. Mean effective pressure Pe was 6.7 MPa.

No. 2 diesel fuel, dehydrate ethanol and diesel-ethanol blend (dieschol) were adopted. Ratio of ethanol:diesel fuel by volume was expressed by E%:D%. The homogenous mixture uses rejected heat from the engine coolant to improve it, if the temperature < 15°C.

A hydraulic dynamometer, oscilloscope recorder, fuel and air flowmeter, engine speed and injection timing potentiometer, sound level meter and Bosch smoke indicator were used in the testing.

3. RESULTS AND DISCUSSION

3.1. Fuel economy

The quantity of heat Qt (kJ/h) and specific heat consumption Qe (kJ/kWh) were used as criteria. When 17%:83% dieschol was used, the Qt, Qe, effective thermal efficiency η_e and exhaust temperature Tr were nearly equal to those measured using neat diesel at full load (Figures 1, 3 and Table 1). When the mixture reached 22%:78%, however, its properties tended to be worse.

Table 1
Comparison of economic property

Fuel	Maximum power					Maximum torque				
	ge	Qe	Tr	η_e	dB(A)	ge	Qe	Tr	η_e	dB(A)
Neat diesel	247.1	10528	498	0.331	99.5	263.4	10053	482	0.328	96
12.4%:87.6%	254.1	10336	529	0.337	99.5	242.1	9836	524	0.354	96.7
23.7%:76.3%	268.0	10385	555	0.336	101					

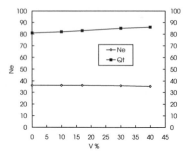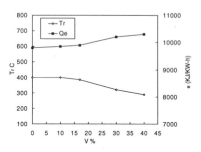

Figure 1. Parameters of S-195 C-I engine fueled with different dieschol.

3.2. Combustion character

Observation of combustion pressure on dieschol engine indicates that increasing the proportion of ethanol resulted in longer ignition delay d_i and higher peak pressure P_Z (Figure 2). This is due to the presence of less OR, higher latent heat of vaporization and lower burning velocity of dieschol.[8] P_Z increased from 7.26 MPa of neat diesel fuel to 7.34 MPa of dieschol; the rate of pressure rise $dp/d\phi$ rose from 0.529 Mpa/^{0}CA to 0.653 Mpa/^{0}CA. The $dp/d\phi$ fell to 0.478 Mpa/^{0}CA favorably to advance injection timing θ from 18.30°CA to 21.60°CA. Slight detonation occurred when the ethanol proportion went up to >30% in dieschol, while θ was unchanged.

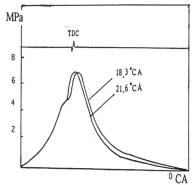

Figure 2: P-V diagram of S-195 diesel fueled with Dieschol.

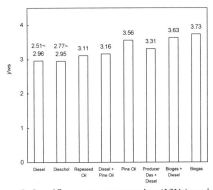

Figure 3: Specific energy consumption (J/Ws) engine of C-I engine fueled with various biomass-based fuel.

3.3. Dynamic property

If the delivery volume of each injection of dieschol was kept the same as neat diesel fuel, both brake Ne and Me would drop slightly (Table 2). This result was expected as ethanol has 6% lower density and 36.9% less heating value and lower burning velocity than those of diesel fuel.

Table 2
Comparison of dynamic property

Fuel	Power Ne (kW)	Ne%	Torque Me (NM)	Me%	Smoke Reducing %
diesel fuel	8.98		42.38		
10%:90%	8.10	-9.7	40.12	-5.3	
17%:83%	8.09	-9.9	38.28	-9.67	52
30%:70%	7.06	-21.3	33.17	-21.73	37.2
40%:60%	6.62	-26.3	31.18	-26.61	25.2

3.4. Possibility of overfueling of dieschol

It is well known that C-I engines operate with a comparatively thin mixture (equivalence ratio ϕ of 0.14 ~ 0.725) in a divided combustion chamber. Theory would suggest to make the quantity of dieschol heat fed into combustion chamber equal to that of diesel by increasing the original delivery volume of each injection. Tests showed that 9.9% power and 9.67% torque loss with 17%:83% dieschol at full load would be compensated by 8.4% dieschol overfueling. The opacity of smoke and thermal efficiency ηe were equal to or exceeded that of diesel.

3.5. The influence of inlet temperature Ts

At optimum θ, either Ts increased the trend in the dieschol engine to reduce the d_i and dp/dφ. Tests show g_e is lower by 2.6% and ηe is higher by 2.7% at Ts of 30°C than at Ts of 25°C (Figure 4). Too high Ts would decrease the charge efficiency, and thus its properties.

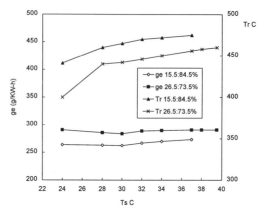

Figure 4. The influence of Ts on economic property.

3.6. Cold start ability

There is no significant problem on starting a dieschol engine at Ts > 15°C without preheating. Preheating by glow plug or warm coolant are needed to instigate ethanol fumigation at lower Ts.

3.7. Smoke emission

Smoke emissions were examined on mixtures of 17% to 40% ethanol at full load. Generally, smoke emissions decreased by 52% on 17:83% dieschol (Table 2).[1,9] Too high an ethanol content has a negative effect on the rational thermodynamic chemical reaction, such as longer d_i and higher Tz. The effect of θ, Ts, φ, injector spray pattern, combustion chamber shape... etc., on NO_x and HC emissions are complex. No carbon on injector nozzles or crankcase oil contamination were found after 500 h of testing.[6] Using blends of ethanol (15%), vegetable oil (10%) and diesel (75%) as a fuel may allow lubrication with the same engine performance.

4. CONCLUSION

Dischol of 17:83% may be used as a fuel substitute for an unmodified C-I engine. Performance of dieschol engine at full load (Qe, ηe, Pzmax, dp/d φ. Tr, dB(A) level) were nearly equal to those of diesel fuel and slightly exceeded those of some biomass based fuels (Figure 3).

The reduction of power output with increase in ethanol was expected, as ethanol has a smaller heat value than diesel fuel. It can be compensated by overfueling to equal the quantity of energy provided by diesel fuel.

Bosch smoke emission fell sharply when an appropriate concentration of dieschol was used as fuel.

Ethanol fumigation is an easy way to run a dieschol engine with acceptable performance.

The combustion characteristics of dieschol C-I engine were modified for heavy load, medium speed, better homogeneity and temperature of mixture and appropriate timing setting (to advance 3°CA ~ 5°CA beyond the manufacturer's setting).

To further improve performance, characteristics should be relevant to improving fuel ignition and durability of the engine.

5. ACKNOWLEDGMENT

Authors are grateful to Mr. Ni Zhaneng, Professor Hui Dehua in the Jiangsu Institute of Technology and Science for their friendly help in this testing.

REFERENCES

1. Fenghua, G., et al. (1988). An investigation of a C-I engine fueled with ethanol-diesel blends. Transaction of the Chinese Society of Agriculture Engineering. No.1, pp. 48–50. Beijing, China.
2. Adelman, H. Alcohols in diesel engine - A review, SAE Paper 8201544.
3. Manlu, L. and Z. Baozhao (1996). Alcohol fuel - Contender of fossil oil, Energy Engineering, China. No 1. pp. 24–27.
4. Chanceller, W. (1980). Alternate fuels for engines. The 49th Annual Agricultural Engineering Farm Machinery Conference, UC Davis, USA.
5. Fenghua, G., et al. (1996). An investigation on engine performance and operation mode with biomass based alternative fuels. International Farm Mechanization Conference. pp 96–103, Beijing, China.
6. Goetz, W., et al. From a dual fuel diesel-ethanol farm tractor to an optimum 4-stroke heavy-duty ethanol engine. ORTECH Corporation, Ontartio, Canada.
7. Baozhao, Z. (1963). A study on multifueled C-I engine. Transactions of the Chinese Society of Agriculture Machinery, 4, pp. 31–34. Nanjing, China.
8. Takeuchi, K. et al. (1982). Ignition temperature and burning velocity of alcohol, internal combustion engine, 21, 263 pp. 9–17, Japan.
9. Alperstein, M. et al. (1958). Fumigation kills smoke-improvers Diesel, SAE Transactions, Vol. 66, pp. 574–595.

STEAM PRETREATMENT CONDITIONS TO OPTIMIZE THE HEMICELLULOSE SUGAR RECOVERY AND FERMENTATION OF SOFTWOOD-DERIVED FEEDSTOCKS

Y. Cai, J. Robinson, S.M. Shevchenko, D.J. Gregg and J.N. Saddler

The University of British Columbia, 4041 - 2424 Main Mall, UBC, Vancouver, British Columbia, Canada V6T 1Z4

Despite difficulties in pretreating and fractionating softwood substrates using steam explosion, there continues to be considerable interest in using softwood feedstocks in the western portion of North America. The driving force for this interest is primarily utilization of sizeable quantities of sawmill residues or forestry trimmings. In contrast to published data on hardwoods and herbaceous materials, published data on softwood conversion to ethanol has been rather limited.

This presentation will compare the optimized (for maximum hemicellulose solubilization and monomer recovery) pretreatment conditions and fermentability of two softwood feedstocks (Douglas fir and a mixture of 70% white fir + 30% ponderosa pine). We have assessed the effect of steam explosion severity (temperature, time and concentration of SO_2) on the recovery of solids and hemicellulose sugars. The fermentability of each of the water-washed pretreated materials was determined. A further study, using these preliminary optimized conditions, investigated the effect of bark and needles on the pretreatment conditions and fermentability.

Author Index

Abatzoglou, N. 953
Abdullayev, K.M. 1185
Abrahamson, L.P. 25, 75
Adams, D. 63
Agterberg, A.E. 477
Aho, V.-J. 311, 349
Ahrendt, N. 1647
Aladjadjiyan, A.G. 249
Aldas, R.E. 1457
Aldea, M.E. 1177
Alegria, A. 715
Alig, R. 63
Alker, G.R. 133
Allen, J. 319
Alzueta, M.U. 1315
Amen-Chen, C. 521
Anderson, D.B. 895
Andries, J. 1009
Anikeev, V. 563
Anile, F. 1565
Antunes, S. 715
Arauzo, J. 947, 969, 1105, 1193
Arvelakis, S. 299
Asser, C. 765
Assink, D. 1119
Atakora, S.B. 241, 1487
Augenstein, D. 691
Auzenne, R.J. 1285
Aznar, M.P. 933, 979, 1049

Babu, R. 185
Bacher, W. 503

Bahrton, A. 549
Bajay, S.V. 495, 1557
Bakhshi, N.N. 985
Bakker, R.R. 1357, 1393, 1425
Bakker-Dhaliwal, R. 335
Ballard, B. 25
Ballester, J. 1509, 1517
Bandaranayake, W. 127
Bangala, D. 953
Barbucci, P. 1079
Basch, G. 175
Bates, R. 1357
Battista, J.R. 1309
Bauen, A. 371
Bautista, E.U. 1457
Baxter, L. 1277
Baxter, L.L. 1393
Beaton, P. 589, 1433
Beaton, P.A. 1509, 1517
Becker, D.A. 85
Beckman, D. 463
Bellan, J. 1145
Benemann, J.R. 413
Benson, R. 915
Bentzen, J.D. 1025
Bergs, M. 715
Bernheim, L.G. 69
Berni, M.D. 495, 1557
Betran, M. 1177
Bezzon, G. 527, 583, 1099
Bickelhaupt, D. 25
Bilbao, R. 947, 1177, 1315
Billa, E. 265

Bilski, R.J. 595
Black, J. 1061
Blase, M.G. 363
Bloksberg-Fireovid, R.L. 1609
Blunk, S. 1357
Blunk, S.L. 1385
Bock, B. 577
Bock, B.R. 127
Böhm, T. 273
Bonelli, P. 549
Bonelli, P.R. 1201
Bonfitto, E. 291
Boukis, I.P.H. 1253
Bouton, J.H. 147
Bowman, L. 1445
Boyd, J. 319
Boylan, D.M. 1307, 1435
Bozell, J.J. 595
Brage, C. 1017, 1033
Brammer, J.G. 281, 1119
Bransby, D.I. 325, 577, 1307, 1435, 1747
Braster, M.L. 1771
Breger, D.S. 107
Brew-Hammond, A. 241, 1487
Bridgwater, A.V. 281, 1119, 1217, 1253, 1255, 1647
Brito, A.L. 1509, 1517
Brossard, L.E. 583, 1099
Brossard Perez, L.E. 527
Brown, R.C. 961, 1329
Brown, S. 1733
Browne, M. 319
Brunner, T. 1377
Burke, H. 1075
Burkhart, M.L. 781
Bush, P.V. 1307

Cabal, H. 429
Caballero, M.A. 933, 979, 1049

Cabrita, I. 1041
Cai, Y. 929
Calvé, L. 521
Cameron, D. 915
Campos, I.A. 1685
Canella, L. 437
Capareda, S.C. 203
Carlson, W. 1357
Carpentieri, E. 247
Cassanello, M. 549
Ceamanos, J. 1177
Cerrella, E.G. 1201
Cerrella, G. 549
Chan, I.S. 1285
Chaudhry, S. 833
Cheifetz, R. 1209
Chen, C.-M. 571
Chen, G. 1017, 1033
Chornet, E. 811, 827, 915, 953
Christensen, B.H. 1739
Chum, H.L. 513
Cipollone, R. 291
Cleland, J.G. 1067
Clifton-Brown, J. 139
Cocco, D. 291
Coelho, S.T. 1685, 1777
Cooper, J.T. 1771
Corella, J. 933, 979, 1049
Correia, A.C. 343
Cortez, L.A.B. 527, 583, 1099
Costello, R. 1577
Craig, K. 1533
Cramer, J. 1739
Cruickshank, W. 915
Cukierman, A.L. 549, 1201
Cushman, J. 1639
Czernik, S. 827, 1217, 1235, 1247

da Rocha, A.O.F. 187, 233, 289

Da Rocha, B.R.P. 289
da Rocha, B.R.P. 187, 215, 233
da Silva, I.M.O. 187, 233, 289
da Silva, I.T. 187, 289
Dadan, K. 797
Dahl, J. 1377
Dahl, K. 691
Dai, L. 193
Dalai, A.K. 985
Dam-Johansen, K. 1169, 1507
Das, A. 1001
Davis, B.S. 1665
de Jong, W. 1009
de Lange, H.J. 1079
De Pillis, A.F. 1535, 1753
Dee, V. 471, 509
DeLaquil, P. 1087
Della Rocca, P.A. 1201
Demeter, C. 75
DePillis, A.F. 1573
Deutsch, S. 1001
Di Berardino, S. 715, 723
Dogru, M. 1051
Dopazo, C. 1509, 1517
Dorian, J.P. 1611
dos Santos, M.A. 1673
Downing, M. 1703, 1733, 1765
Drennan, S.A. 1285
Duffy, J. 1357

Echeverria, C. 1401
Eckels, D. 961
Eckinger, W. 1665
Edick, S. 1733
Eflin, J. 1697
El Bassam, N. 503
Eleniewski, M.A. 1343
Elkin, V.A. 881
Elliott, D.C. 595

Ellis, S. 1001
Engvall, K. 1365
Esperanza, E. 1105
Evald, A. 1481

Faaij, A. 487, 497, 803
Faaij, A.P.C. 477
Fang, J. 1547
Farao, Z. 621, 635
Farris, M.C. 1061
Feik, C. 827
Ferraro, F.M. 699
Ferreira-Dias, S. 343
Filhart, R.C. 25
Fische, F.S. 1087
Fitzpatrick, S.W. 595
Fonseca Felfli, F. 589
Foscolo, P.U. 977
Franco, C. 1041
Freeman, M.C. 1577, 1607
French, R. 827
Frye Jr, J.G. 595

Gandini, A. 613
Gao, J. 895
Garcia, L. 947
Garro, J.M. 915
Gea, G. 969, 1105
Georgali, B. 299
Ghahremani, F.G. 1495
Giaier, T.A. 1343, 1541
Gil, J. 933, 979, 1049
Gil, L. 305
Girard, P. 1161, 1269
Girouard, P. 17, 85
Glarborg, P. 1507
Glassner, D. 873
Gøbel, B. 1025
Godovikova, V.A. 161
Goldberg, P.M. 1577, 1607
Gómez, J. 1193

Gonçalves, T.A. 1679
Goor, F. 39
Goudriaan, F. 789
Graham, R.L. 63
Grass, S. 911
Green, A.E.S. 133
Gregg, D.J. 929
Grozdits, G.A. 541
Gubinsky, M. 1209
Gulyurtlu, I. 1041
Guo, F. 923

Haase, S.G. 99
Hall, D.O. 387
Hallgren, A.L. 1365
Hamelinck, C. 803
Hanley, T.R. 903
Hansen, G. 911
Hansen, U. 1241
Hartmann, H. 273
Hartsough, B.R. 1409
Hayashitani, M. 179
Heaton, R.J. 45
Hekkert, M. 487
Hektor, B. 11
Helynen, S. 1293
Henriksen, U. 1025
Heppert, J.A. 819
Heyerdahl, P.H. 1341
Hillring, B. 1527
Himmelblau, D.A. 541
Hinman, N.D. 875
Hiraki, T.T. 939
Hirata, S. 179
Hirotsu, T. 621, 635
Hodges, A.W. 91
Hogan, E. 837
Holm-Nielsen, J.B. 1579
Hoogwijk, M. 255
Hooker, B.S. 895
Hoppesteyn, P.D.J. 1009

Horowitz, G. 549
Houbak, N. 1025
Houmøller, S. 1481, 1739
Houston, A. 127
Howarth, C.R. 1051
Hughes, E. 1287
Huisman, W. 327
Humphreys, C.L. 1647
Hunter, A. 319
Hus, P.J. 1349

Ibáñez, J.C. 1315
III 1087
Ikegami, Y. 179, 621, 635
Ince, P.J. 63
Irving, J.M. 1061
Ishimura, D.M. 939, 991, 1611
Izquierdo, E. 527

James, R.A. 1607
Janes, H.W. 1555
Janse, A.M.C. 1137
Jarnefeld, J.L. 595
Jenkins, B.M. 69, 335, 1335,
   1357, 1385, 1393, 1409,
   1425
Jensen, A. 1507
Jensen, P.A. 1169
Jin, H. 843, 855
Johnson, D.A. 1665
Johnson, D.K. 811
Johnson, W. 1759
Jollez, P. 915
Jørgensen, L. 1739
Joslin, J.D. 127
Jossart, J.M. 47
Jossart, J.-M. 39
Jungmeier, G. 437

Kabeya, H. 621, 635
Kadam, K.L. 699, 1385

Kadangode, S. 811
Kajba, D. 55
Kaltschmitt, M. 371
Kaminsky, J. 1639
Kask, Ü 1691
Kaya, M.H. 1611
Kayhanian, M. 689
Kelley, S.S. 513
Kess, P.I. 629
Kholkin, Y.I. 881
Kiker, C.F. 91
Kinoshita, C.M. 939, 991, 1611
Kitagawa, R. 621, 635
Knight, C. 1285
Knopf, U.C. 601
Knudsen, N.O. 1299
Kolisis, F. 641
Komar, K. 133
Kopp, R.F. 25
Kornell, P. 1717
Koukios, E.G. 265, 299, 641
Koullas, D.P. 641
Kreutz, T. 803
Kulkarni, P. 1561, 1593
Kurano, N. 655
Kuuyuor, T. 1501
Kwant, K.W. 1629

Labrecque, M. 31
Laine, J. 407
Lamascese, N. 155
Lane, N.W. 1445
Langer, E. 1717
Langerwerf, L. 681
Langseth, D. 1703
Larson, D.L. 1099
Larson, E. 803
Larson, E.D. 247, 843, 855
Lathouwers, D. 1145
Lauer, M. 1263

Le, G.B.B. 915
Leao, A.C. 247
Lebas, E. 1371
Lechón, Y. 429
Ledent, J.F. 47
Ledent, J.-F. 39
Lee, H. 335
Lee, P.K. 663
Lee, Y. 471, 509
Lefcort, M.D. 1495
Lewandowski, I. 139
Li, J. 217
Li, Y. 533
Lim, K.O. 209
Linden, J.C. 887
Lindsey, C. 75
Ling, E. 11
Liu, X. 193
Lopes, J.P. 187, 215, 233, 289
Losavio, N. 155
Lourenço, E.V. 343
Lu, X. 521
Luengo, C.A. 589
Lundblad, T. 1703
Luo, W. 1619
Lusk, P. 707
Lusk, P.D. 1655
Lv, Z. 1619

Ma, L.Q. 133
Mäentausta, O.K. 629
Maggi, R. 1235
Maier, L. 273
Majerski, P. 1153, 1225
Makarov, V.L. 881
Malik, A.A. 1051
Mann, M.D. 1475
Mann, M.K. 379
Manning, T.O. 1555
Mansoubi, A. 691
Marandola, C. 1565

Marano, J.J. 427
Marcano, E. 667
Martin, A.G. 743
Martin, G. 1371
Martin, J. 1433
Martin, J.A. 979
Martín, J.A. 1049
Martin, V. 479
Mason, H.B. 1285
Mason, M.H. 819
Masri, M. 1561, 1679
Mastral, J.F. 1177
Masutani, S.M. 939
Mathur, M.P. 1607
Mattocks, R. 707
Mattocks, R.P. 735
Maunsbach, K. 479
Mazumbe, J.V. 343
McLaughlin, S. 649
McLaughlin, S.B. 147
Mehdi, B. 17
Meier, D. 1217, 1229, 1255
Melis, A. 413
Mercedes, S.S.P. 1777
Mettee, H.D. 881
Meuleman, B. 497
Miles, T.R. 1335
Millera, A. 1315
Mills, J. 63
Miranda, V. 187, 215, 233, 289
Miyachi, S. 655
Moens, L. 595
Monteiro, C. 187, 215, 233, 289
Moore, R. 691
Moraes, S.B. 187, 233, 289
Moreira, J.R. 1685
Morris, G. 1
Morris, K.W. 1225
Morrow, R.S. 1541
Moser, M. 735, 1695

Moser, M.A. 673, 681, 743
Mouras, S. 1161, 1269
Muiste, P. 1691
Müller, P.H. 911
Mullins, G.R. 577
Murillo, M.B. 969
Mut, R.F. 1433

Naber, J.E. 789
Nader, J. 977
Napoli, A. 1161
Nascimento, C. 305
Nesterenko, S.B. 161
Neuenschwander, G.G. 595
Neufeld, G. 1075
Neufeld, J. 1075
Neuhauser, E. 1711
Neuhauser, E.F. 75
Nguyen, Q. 915
Nicholls, D.L. 571
Nicoletti, G. 1467, 1565
Nielsen, P.S. 421, 1501, 1579
Norris, G. 673
Nousiainen, I. 311, 349
Nowak, C.A. 25, 75
Numazawa, S. 1161
Nyrönen, T. 407

Oasmaa, A. 1229, 1247
Obernberger, I. 1377, 1417
Olgun, H. 1051
Oliva, M. 1315
Olivares, E. 527, 583, 1099
Oliveira, A.C. 1685
Östman, A. 463
Otsuki, T. 621, 635
Overend, R.P. 463, 1061, 1321
Overgaard, P. 1299

Pacey, J. 691
Padilla, A. 667

Padilla, D. 667
Paisley, M.A. 1061
Pakdel, H. 521
Palmer, H. 319
Parker III, C.E. 445
Pastou, A. 641
Patel, P.S. 1533
Pavlish, J.H. 1475
Peacocke, G.V.C. 1199, 1235
Pedersen, L.T. 1507
Pepper, J.C. 1559
Peres, S. 1113
Peterson, J.M. 75
Pettry, D.E. 127
Pfaff, D. 1425
Pierce, R. 1703, 1765
Pinheiro, E.C.L. 187, 233, 289
Pinto, F. 1041
Piskorz, J. 1153, 1217, 1225
Plant, R.E. 69
Plasynski, S. 1287
Plasynski, S.I. 1577
Podesser, E. 463
Ponce, F. 1127
Ponte, M.X. 419
Porter, K. 1547
Prabhu, E. 1439
Prankl, H. 805
Prasad, R. 1601
Prine, G.M. 113
Prins, W. 1137
Proakis, G.J. 1711
Purvis, C.R. 1067
Putsche, V.L. 699

Qu, F. 193

Radlein, D. 1153
Radway, J.C. 413
Rahmani, M. 91
Ramsay, B. 1075
Randerson, P.F. 45

Rapagná, S. 977
Rathbauer, J. 805
Ray, D.E. 63
Reed, T.B. 1001, 1093
Regato, J.E. 343
Reinhardt, G.A. 393
Riddell-Black, D. 1563
Robb, M.F. 1475
Robinson, A. 1277
Robinson, J. 929
Rocha, J.D. 513
Rockwood, D.L. 113, 133
Rogers, S. 427
Rooney, T.E. 99
Roos, A. 11
Roos, K.F. 743
Rosén, C. 1017
Rosenqvist, H. 11
Roy, C. 521, 1227
Ruiz, J. 1193
Ruocco, J.J. 699
Rupp, E. 555

Sachs, K.M. 1453
Saddler, J. N. 929
Sáez, R. 429
Saka, S. 797
Salomonsen, K. 1579
Salter, E.H. 1255
Salvador, M.L. 947
Samson, R. 17
Samuel, R.W. 45
Sander, B. 1169, 1299
Sanz, M.J. 1193
Saravanan, A. 185
Sarkanen, S. 533
Sauer, I.L. 1777
Schleef, D. 759
Schmidt, D.D. 1475
Schmidt, U. 139
Schoenholtz, S. 127

Schroeder, H.A. 887
Schuck, S.M. 1595
Schuler, A. 759
Selin, P. 407
Seliverstov, G.V. 161
Sereti, V. 641
Sethi, P. 471, 509, 833
Shabtai, J. 811
Shakya, B.S. 119
Sharma, A. 1601
Sheehan, J.J. 875
Shevchenko, S.M. 929
Shishko, Y. 1209
Shoemaker, S. 1663
Shumny, V.K. 161
Sieber, M. 911
Simons, G. 471, 509, 689, 1561, 1593, 1695
Sims, R.E.H. 401
Sjöström, K. 1017, 1033
Skeen, R.S. 895
Skog, K. 63
Skrifvars, B.-J. 1365
Slater, F.M. 45
Slinksy, S.P. 63
Smeenk, J. 961
Snyder, H. 107
Solantausta, Y. 463
Southgate, D. 119
Souza, M.R. 1321
Spath, P.L. 379
Specca, D.R. 1555
Spelter, H. 63
Spinelli, R. 355
Spitzer, J. 437
Sprague, S. 1639
Srinivas, S.T. 985
Stamatis, H. 641
Stoffel, R. 1703, 1765
Strenziok, R. 1241
Stucki, K. 1357

Su, M. 193
Suarez, J.A. 589
Sunter, R. 1075
Suppes, G.J. 781, 819
Svedberg, G. 479
Svihla, C.K. 903
Svihla, S.P. 903
Sweterlitsch, J.J. 1329
Szuhaij, B.F. 663

Taliaferro, C.M. 147
Tamaro, R.F. 1401
Tantlinger, J. 1611
Tayebi, K. 175
Teixeira, F. 175
Teixeira, L.C. 887
Teodorescu, T.I. 31
Thornton, F.C. 127
Tiangco, V. 471, 509, 689, 729, 1285, 1357, 1439, 1533, 1561, 1593, 1695
Tiangco, V.M. 875, 1335, 1409, 1457
Tillman, D.A. 1287, 1309, 1349
Tolbert, V.R. 127
Tshung, T.T. 819
Tsutao, H. 621, 635
Tudan, C. 247
Tullus, H. 167
Turn, S.Q. 939, 1611
Turnbull, J. 773
Tuskan, G.A. 147
Tyler, D. 127

Ueno, Y. 655
Ugarte, De la Torre 63
Ünal, Ö 1009
Unnasch, S. 833
Uri, V. 167
Usenko, A. 1209

van de Beld, B. 789
van de Beld, L. 1119
van den Broek, R. 255, 487
van der Lans, R.P. 1507
van der Wal, S. 789
Van Dyne, D.L. 363
van Leenders, C. 1629
Vann Bush, P. 1435
Vannoy, L. 751
Varani, F.T. 699
Varela, M. 429
Varela, N. 583
Vasselli, J.J. 1711
Ventrella, D. 155
Viglazov, V.V. 881
Vilppunen, P.V. 629
Vinterbäck, J. 1527
Vivekanandan, M. 185
Vleeshouwers, L. 255
Vogel, K.P. 147
Voiles, R.E. 699
Volk, T.A. 25, 75, 1711, 1733
Vonella, A.V. 155
Vyas, S. 1725

Walbert, G.F. 1607
Walsh, M.E. 63, 85
Walt, R. 1001, 1093
Walter, A. 1127, 1321
Wang, G. 193
Wang, Y. 595
Wanner, L. 1587
Waterland, L.R. 1285
Weber, T. 1587
Weissinger, A. 1417
Wene, E. 1759
Westerhout, R.W.J. 1137
White, E.H. 25, 75
Wichert, D.B. 1535, 1573, 1753

Wickboldt, P. 1241
Wiles, D. 1075
Williams, D.W. 673, 759
Williams, R.B. 1357, 1393, 1425
Wilstrand, M.O. 455
Wiltsee, G.A. 79
Wimberly, J. 1465
Wiselogel, A.E. 99
Wiser, R. 1547
Wörgetter, M. 805, 1633
Wong, A. 605, 649, 837
Woods, J. 387
Wooley, R.J. 699
Wright, L. 1639
Wulzinger, P. 1255
Wyman, C.E. 867

Yamashita, M. 621, 635
Yan, L. 69
Yancey, M.A. 875
Yazdani, R. 689, 691
Yomogida, D. 471, 509
Yomogida, D.E. 1357, 1409
Yoshitake, M. 621, 635
Young, J. 1695
Yu, Q. 1017, 1033

Zamansky, V.M. 1335
Zan, C. 17
Zeevalkink, J.A. 789
Zemanek, G. 393
Zeng, B. 1619
Zhang, B. 923
Zhang, R. 729, 773
Zhang, Z. 729, 1619
Zhou, A. 217, 1619
Zhou, J. 939, 991
Zmierczak, W. 811
Zuman, P. 555